Mass

The **gram** is the basic unit.

1 **kilo**gram = 1000 grams	1 kg = 1000g
1 **hecto**gram = 100 grams	1 hg = 100g
1 **deka**gram = 10 grams	1 dag = 10g

1 **deci**gram $= \dfrac{1}{10}$ of a gram   1 dg = .1g

1 **centi**gram $= \dfrac{1}{100}$ of a gram   1 cg = .01g

1 **milli**gram $= \dfrac{1}{1000}$ of a gram   1 mg = .001g

Some Approximate Comparisons Between English and Metric Systems (the symbol $\approx$ means "is approximately equal to.")

Length

1 inch $\approx$ 2.5 centimeters	1 centimeter $\approx$ .4 of an inch
1 foot $\approx$ 30.5 centimeters	1 centimeter $\approx$ .03 of a foot
1 yard $\approx$ .9 of a meter	1 meter $\approx$ 1.1 yards
1 mile $\approx$ 1.6 kilometers	1 kilometer $\approx$ .6 of a mile

Volume

1 pint $\approx$ .5 of a liter	1 liter $\approx$ 2.1 pints
1 quart $\approx$ .9 of a liter	1 liter $\approx$ 1.1 quarts
1 gallon $\approx$ 3.8 liters	1 liter $\approx$ .3 of a gallon

Mass

1 ounce $\approx$ 28.35 grams	1 gram $\approx$ .04 of an ounce
1 pound $\approx$ 453.6 grams	1 gram $\approx$ .002 of a pound
1 pound $\approx$ .45 of a kilogram	1 kilogram $\approx$ 2.2 pounds

Elementary and Intermediate Algebra

A Combined Approach

The Prindle, Weber & Schmidt Series in Developmental, Business, and Technical Mathematics

Baley/Holstege, *Algebra: A First Course, Third Edition*
Cass/O'Connor, *Beginning Algebra*
Cass/O'Connor, *Beginning Algebra with Fundamentals*
Cass/O'Connor, *Fundamentals with Elements of Algebra: A Bridge to College Mathematics, Second Edition*
Ewen/Nelson, *Elementary Technical Mathematics, Fifth Edition*
Geltner/Peterson, *Geometry for College Students, Second Edition*
Johnston/Willis/Hughes, *Developmental Mathematics, Third Edition*
Johnston/Willis/Lazaris, *Essential Algebra, Sixth Edition*
Johnston/Willis/Lazaris, *Essential Arithmetic, Sixth Edition*
Johnston/Willis/Lazaris, *Intermediate Algebra, Fifth Edition*
Jordan/Palow, *Integrated Arithmetic and Algebra*
Kennedy/Green, *Prealgebra for College Students*
Lee, *Self-Paced Business Mathematics, Fourth Edition*
McCready, *Business Mathematics, Sixth Edition*
McKeague, *Basic Mathematics, Third Edition*
McKeague, *Introductory Mathematics*
McKeague, *Prealgebra, Second Edition*
Pierce/Tebeaux, *Operational Mathematics for Business, Second Edition*
Proga, *Arithmetic and Algebra, Third Edition*
Proga, *Basic Mathematics, Third Edition*
Rogers/Haney/Laird, *Fundamentals of Business Mathematics*
Szymanski, *Algebra Facts*
Szymanski, *Math Facts: Survival Guide to Basic Mathematics*
Weltman/Perez/Byrum/Polito, *Beginning Algebra: A Worktext*
Weltman/Perez/Byrum/Polito, *Intermediate Algebra: A Worktext*
Wood/Capell/Hall, *Developmental Mathematics, Fourth Edition*

The Prindle, Weber & Schmidt Series in Precalculus, Liberal Arts Mathematics, and Teacher Training

Barnett, *Analytic Trigonometry with Applications, Fifth Edition*
Bean/Sharp/Sharp, *Precalculus*
Boye/Kavanaugh/Williams, *Elementary Algebra*
Boye/Kavanaugh/Williams, *Intermediate Algebra*
Davis/Murphy/Moran, *Precalculus in Context: Functioning in the Real World*
Drooyan/Franklin, *Intermediate Algebra, Seventh Edition*
Franklin/Drooyan, *Modeling, Functions and Graphs*
Fraser, *Elementary Algebra*
Fraser, *Intermediate Algebra: An Early Functions Approach*
Gantner/Gantner, *Trigonometry*
Gobran, *Beginning Algebra, Fifth Edition*
Grady/Drooyan/Beckenbach, *College Algebra, Eighth Edition*
Hall, *Beginning Algebra*
Hall, *Intermediate Algebra*
Hall, *Algebra for College Students, Second Edition*
Hall, *College Algebra with Applications, Third Edition*
Holder, *A Primer for Calculus, Sixth Edition*
Huff/Peterson, *College Algebra Activities for the TI-81 Graphics Calculator*

Johnson/Mowry, *Mathematics: A Practical Odyssey*
Kaufmann, *Elementary Algebra for College Students, Fourth Edition*
Kaufmann, *Intermediate Algebra for College Students, Fourth Edition*
Kaufmann, *Elementary and Intermediate Algebra: A Combined Approach*
Kaufmann, *Algebra for College Students, Fourth Edition*
Kaufmann, *Algebra with Trigonometry for College Students, Third Edition*
Kaufmann, *College Algebra, Third Edition*
Kaufmann, *Trigonometry, Second Edition*
Kaufmann, *College Algebra and Trigonometry, Third Edition*
Kaufmann, *Precalculus, Second Edition*
Lavoie, *Discovering Mathematics*
Rice/Strange, *Plane Trigonometry, Sixth Edition*
Riddle, *Analytic Geometry, Fifth Edition*
Ruud/Shell, *Prelude to Calculus, Second Edition*
Sgroi/Sgroi, *Mathematics for Elementary School Teachers*
Swokowski/Cole, *Fundamentals of College Algebra, Eighth Edition*
Swokowski/Cole, *Fundamentals of Algebra and Trigonometry, Eighth Edition*
Swokowski/Cole, *Fundamentals of Trigonometry, Eighth Edition*
Swokowski/Cole, *Algebra and Trigonometry with Analytic Geometry, Eighth Edition*
Swokowski/Cole, *Precalculus: Functions and Graphs, Seventh Edition*
Weltman/Perez, *Beginning Algebra, Second Edition*
Weltman/Perez, *Intermediate Algebra, Third Edition*

The Prindle, Weber & Schmidt Series in Calculus and Upper-Division Mathematics

Althoen/Bumcrot, *Introduction to Discrete Mathematics*
Andrilli/Hecker, *Linear Algebra*
Burden/Faires, *Numerical Analysis, Fifth Edition*
Crooke/Ratcliffe, *A Guidebook to Calculus with Mathematica*
Cullen, *An Introduction to Numerical Linear Algebra*
Cullen, *Linear Algebra and Differential Equations, Second Edition*
Denton/Nasby, *Finite Mathematics, Preliminary Edition*
Dick/Patton, *Calculus*
Dick/Patton, *Single Variable Calculus*
Dick/Patton, *Technology in Calculus: A Sourcebook of Activities*
Edgar, *A First Course in Number Theory*
Eves, *In Mathematical Circles*
Eves, *Mathematical Circles Revisited*
Eves, *Mathematical Circles Squared*
Eves, *Return to Mathematical Circles*
Faires/Burden, *Numerical Methods*
Finizio/Ladas, *Introduction to Differential Equations*
Finizio/Ladas, *Ordinary Differential Equations with Modern Applications, Third Edition*
Fletcher/Hoyle/Patty, *Foundations of Discrete Mathematics*
Fletcher/Patty, *Foundations of Higher Mathematics, Second Edition*
Gilbert/Gilbert, *Elements of Modern Algebra, Third Edition*
Gordon, *Calculus and the Computer*
Hartfiel/Hobbs, *Elementary Linear Algebra*
Hill/Ellis/Lodi, *Calculus Illustrated*
Hillman/Alexanderson, *Abstract Algebra: A First Undergraduate Course, Fifth Edition*

Humi/Miller, *Boundary-Value Problems and Partial Differential Equations*
Laufer, *Discrete Mathematics and Applied Modern Algebra*
Leinbach, *Calculus Laboratories Using Derive*
Maron/Lopez, *Numerical Analysis, Third Edition*
Miech, *Calculus with Mathcad*
Mizrahi/Sullivan, *Calculus with Analytic Geometry, Third Edition*
Molluzzo/Buckley, *A First Course in Discrete Mathematics*
Nicholson, *Elementary Linear Algebra with Applications, Third Edition*
Nicholson, *Introduction to Abstract Algebra*
O'Neil, *Advanced Engineering Mathematics, Third Edition*
Pence, *Calculus Activities for Graphic Calculators*
Pence, *Calculus Activities for the TI Graphic Calculator, Second Edition*
Plybon, *An Introduction to Applied Numerical Analysis*
Powers, *Elementary Differential Equations with Boundary-Value Problems*
Powers, *Elementary Differential Equations with Linear Algebra*
Prescience Corporation, *The Student Edition of Theorist*
Riddle, *Calculus and Analytic Geometry, Fourth Edition*
Schelin/Bange, *Mathematical Analysis for Business and Economics, Second Edition*
Sentilles, *Applying Calculus in Economics and Life Science*
Swokowski/Olinick/Pence, *Calculus, Sixth Edition*
Swokowski/Olinick/Pence, *Calculus of a Single Variable, Second Edition*
Swokowski, *Calculus, Fifth Edition (Late Trigonometry Version)*
Swokowski, *Elements of Calculus with Analytic Geometry: High School Edition*
Tan, *Applied Finite Mathematics, Fourth Edition*
Tan, *Calculus for the Managerial, Life, and Social Sciences, Third Edition*
Tan, *Applied Calculus, Third Edition*
Tan, *College Mathematics, Third Edition*
Trim, *Applied Partial Differential Equations*
Venit/Bishop, *Elementary Linear Algebra, Third Edition*
Venit/Bishop, *Elementary Linear Algebra, Alternate Second Edition*
Wattenberg, *Calculus in a Real and Complex World*
Wiggins, *Problem Solver for Finite Mathematics and Calculus*
Zill, *Calculus, Third Edition*
Zill, *A First Course in Differential Equations, Fifth Edition*
Zill/Cullen, *Differential Equations with Boundary-Value Problems, Third Edition*
Zill/Cullen, *Advanced Engineering Mathematics*

The Prindle, Weber & Schmidt Series in Advanced Mathematics

Ehrlich, *Fundamental Concepts of Abstract Algebra*
Eves, *Foundations and Fundamental Concepts of Mathematics, Third Edition*
Judson, *Abstract Algebra: Theory and Applications*
Keisler, *Elementary Calculus: An Infinitesimal Approach, Second Edition*
Kirkwood, *An Introduction to Real Analysis*
Patty, *Foundations of Topology*
Ruckle, *Modern Analysis: Measure Theory and Functional Analysis with Applications*
Sieradski, *An Introduction to Topology and Homotopy*
Steinberger, *Algebra*
Strayer, *Elementary Number Theory*
Troutman/Bautista, *Linear Boundary-Value Problems*

Elementary and Intermediate Algebra

A Combined Approach

Jerome E. Kaufmann

PWS Publishing Company
Boston

PWS Publishing Company

PWS Publishing Company is a division of Wadsworth, Inc.

Portions of this book also appear in *Elementary Algebra for College Students, Fourth Edition,* and *Intermediate Algebra for College Students, Fourth Edition,* by Jerome E. Kaufmann.

Library of Congress Cataloging-in-Publication Data
Kaufmann, Jerome E.
    Elementary and intermediate algebra : a combined approach / Jerome
E. Kaufmann.
        p.   cm.
    Includes index.
    ISBN 0-534-93345-9
    1. Algebra.   I. Title.
QA152.2.K385   1993
512.9--dc20   92-28093
                                                       CIP

Acquisitions Editor: Timothy Anderson
Assistant Editor: Kelle Karshick
Text Design: Susan Graham and Helen Walden
Production and Cover Design: Helen Walden
Manufacturing Coordinator: Ellen Glisker
Typesetter: Polyglot Pte Ltd/Doyle Graphics
Printer/Binder: R.R. Donnelley and Sons
Cover Printer: Henry N. Sawyer Company
Cover Photo: Marcus Halevi

Printed in the United States of America
93 94 95 96 97—10 9 8 7 6 5 4 3 2

About the cover: Launched in 1984, the Spirit of Massachusetts is an authentic replica of the fishing schooners that sailed out of Gloucester, Boston, and Nova Scotia in the late 1800s. This two-masted, gaff-rigged schooner participates in a variety of education programs with colleges and universities, high schools and middle schools, special education classes, scouts and youth groups, and urban youth programs. Students become fully involved as sailors in the life and operation of the ship, through a rigorous program of hands-on training in the techniques and principles of traditional seamanship and navigation. For further information, contact the captain at the New England Historic Seaport, Inc., Building 1, Charlestown Navy Yard, Boston, Massachusetts 02129.

International Thomson Publishing
The trademark ITP is used under license

Contents

4   Formulas, Problem Solving, and Inequalities   129

5   Coordinate Geometry and Linear Systems   175

6   Exponents and Polynomials   239

11 Graphing Techniques 461

12 Functions 501

13 Exponential and Logarithmic Functions 558

14 Using Matrices and Determinants to Solve Linear Systems 603

15   Sequences and Series   644

Appendixes   676

Answers to Odd-Numbered Problems and All Chapter Review Problems   696

Index   747

Preface

This text presents the basic topics of both elementary and intermediate algebra. By combining these topics in one text we have been able to develop an organizational format that allows for frequent reinforcement of concepts but eliminates the need to reintroduce topics as is necessary when two separate texts are used.

The basic concepts of elementary and intermediate algebra are developed in a logical sequence, but in an easy-to-read manner, without excessive technical vocabulary and formalism. Whenever possible, the algebraic concepts are allowed to develop from their arithmetic counterparts. The following are two specific examples of this development.

1. Manipulation with simple algebraic fractions begins early (Sections 2.1 and 2.2) when reviewing operations with rational numbers.

2. Manipulation with monomials — without any of the formal vocabulary — is introduced in Section 2.4 when working with exponents.

There is a common thread throughout the book, which is *learn a skill,* then *use the skill to help solve equations and inequalities,* and then *use equations and inequalities to solve word problems.* This thread influenced some other decisions.

1. Approximately 700 word problems are scattered throughout the text. Every effort was made to start with easy ones and then to very gradually increase the level of difficulty.

2. Many problem solving suggestions are offered throughout, with special discussions in several sections. Approximately 90 examples are used to illustrate a variety of problem-solving techniques.

3. Newly acquired skills are used as soon as possible to help solve equations and inequalities, which are in turn used to solve word problems. Therefore, the concept of solving equations and inequalities is introduced early and developed throughout the text. Furthermore, systems of two linear equations in two unknowns are introduced in Chapter 5 and then used for problem-solving purposes.

As recommended by the American Mathematical Association of Two-Year Colleges, some basic geometric concepts used in subsequent courses are integrated in problem-solving settings as follows.

Section 3.2:	Complementary and supplementary angles, sum of the measures of the angles of a triangle equals 180°
Section 4.1:	Area and volume formulas
Sections 7.1, 7.2, and 7.5	More on area and volume formulas, perimeter and circumference formulas
Section 7.3:	Pythagorean Theorem
Section 10.2:	More on Pythagorean Theorem, including work with isosceles right triangles and 30°–60° right triangles.

Most of the problem sets come from my elementary and intermediate algebra books, which are now in the fourth editions. Thus, the problems have been very carefully scrutinized by numerous reviewers and users of those texts. Many of the problem sets contain a special category of problems called *Miscellaneous Problems.* These problems encompass a variety of ideas. Some of them exhibit different approaches to topics covered in the text, some bring in supplementary topics and relationships, and some of them are calculator exercises. All of them could be omitted without breaking the continuity pattern of the text; however, I feel that they do add another flexibility feature. You may wish to turn to Problem Sets 1.2, 2.1, 7.3, and 10.4 to see examples of these problems.

Chapter Review Problem Sets provide a vehicle for students to use to pull together the concepts of the chapter. *Cumulative Review Problem Sets* appear at the end of each even-numbered chapter.

Approximately one-third of the problem sets contain a few problems called *Thoughts into Words.* These problems provide the students with an opportunity to express, in written form, their thoughts about various mathematical ideas. You may wish to turn to Problem Sets 1.5, 3.2, 4.3, and 11.3 to see examples of these problems.

I tried to assign the calculator its rightful place in the study of mathematics, that is, as a tool, useful at times, unnecessary at other times. Many topics of elementary and intermediate algebra can be studied very effectively without the use of a calculator; however, there are times when a calculator is a very useful tool. For example, in our study of exponents and logarithms in Chapter 13 the calculator plays an important role. I have included exercises that simply provide practice with the use of a calculator (see Problem Sets 1.2, 2.1, and 8.2) as well as some problems that use the calculator as a teaching tool (see Problem Sets 8.4, 8.5, and 8.7). I have also added a few graphics calculator problems in Chapters 11, 12, and 13.

Supplements

The following supplements arc available to adopters:

For Instructors:

1. An **Answer Book** that contains answers for all of the problems except the Chapter Review Problem Sets, which are in the back of the book.

2. A printed **Test Bank** that contains two multiple choice and one short answer test for each chapter.

3. An EXPTest **Computerized Test Bank** for IBM PCs and compatibles that features over one thousand questions written specifically to accompany this text by Joan and Stuart Thomas of the University of Oregon. This program allows users to view and edit all tests, to add to, delete from, and modify existing questions, and to print any number of student tests.

4. A **Computerized Testing Program** for the Macintosh that features test questions written by Joan and Stuart Thomas. Questions can be stored by objectives and the user can scramble the order of questions.

For Students:

1. A **Partial Solutions Manual** (which students may purchase) that contains solutions for most of the problems numbered 1, 5, 9, 13, etc.).

2. A set of **Expert Algebra Tutor** disks for IBM PCs and compatibles by Sergei Ovchinnikov of San Francisco State University. These are tutorial software disks that define the level of tutoring needed by evaluating the user's need for further remediation or advancement in the tutoring session.

3. A set of **Algebra Lecture Series Videotapes** by John Jobe of Oklahoma State University. These videotapes can be used by students in an independent developmental math lab setting to review the topics in the textbook.

Acknowledgments

I would like to thank the following reviewers:

James Bennett
Eastfield College

John W. Coburn
Saint Louis Community College
at Florissant Valley

Joyce Huntington
Walla Walla Community College

Paul H. Meier
Columbia Basin College

Laura Moore-Mueller
Green River Community College

Nancy Nickerson
Northern Essex Community College

Barbara Jacobs
Big Bend Community College

Stephen Lane
Big Bend Community College

Marilyn McBride
Skyline College

E. James Peake
Iowa State University

Tom Reifenrath
Clark College

Harvey E. Reynolds
Golden West College

James J. Symons
DeAnza College

I am very grateful to the staff of PWS, especially Timothy Anderson and Kelle Karshick for their continuous cooperation and assistance throughout this project. I would also like to express my sincere gratitude to Helen Walden and Susan Graham. They continue to make my life as an author so much easier by carrying out the details of production in a dedicated and caring way. My thanks go out to Joan and Stuart Thomas for all of their hard work on the creation and programming of fresh questions for the computerized test banks.

Again, very special thanks are due my wife, Arlene, who spends numerous hours typing and proofreading manuscripts, answer keys, solutions manuals, and test banks.

Jerome E. Kaufmann

Chapter 1

Some Basic Concepts of Arithmetic and Algebra

Let's begin our study of algebra with a suggestion: Keep in mind that algebra is really a generalized approach to arithmetic. Many algebraic concepts are extensions of arithmetic ideas. Use your knowledge of arithmetic to help you with your study of algebra.

Algebraic Expressions

In arithmetic, we use symbols such as 4, 8, 17, and π to represent numbers. We indicate the basic operations of addition, subtraction, multiplication, and division by the symbols $+$, $-$, $\cdot$, and $\div$, respectively. Thus, we can formulate specific **numerical expressions**. For example, we can write the indicated sum of eight and four as $8 + 4$.

In algebra the concept of a **variable** provides the basis for generalizing. By using x and y to represent *any* number, we can use the expression $x + y$ to represent the indicated sum of *any two* numbers. The x and y in such an expression are called variables and the phrase $x + y$ is called an **algebraic expression**. We commonly use letters of the alphabet such as x, y, z, and w as variables. The key idea is that they represent numbers; therefore, as we review various operations and properties pertaining to numbers, we are building the foundation for our study of algebra.

Many of the notational agreements made in arithmetic are extended to algebra with a few slight modifications. The following chart summarizes these notational agreements pertaining to the four basic operations. Notice the variety of ways to write a product by including parentheses to indicate multiplication. Actually the ab form is the simplest and probably the most used form; expressions such as abc, $6x$, and $7xyz$ all indicate multiplication. Also note the various forms for indicating division; the fractional form, $\dfrac{c}{d}$, is usually used in algebra although the other forms do serve a purpose at times.

Operation	*Arithmetic*	*Algebra*	*Vocabulary*
addition	$4 + 6$	$x + y$	The *sum* of x and y
subtraction	$7 - 2$	$w - z$	The *difference* of w minus z
multiplication	$9 \cdot 8$	$a \cdot b$, $a(b)$, $(a)b$, $(a)(b)$, or ab	The *product* of a and b
division	$8 \div 2$, $\dfrac{8}{2}$, or $2\overline{)8}$	$c \div d$, $\dfrac{c}{d}$, or $d\overline{)c}$	The *quotient* of c divided by d

Now let's consider simplifying some numerical expressions involving the set of whole numbers, that is, the set $\{0, 1, 2, 3, 4, \ldots\}$. (Remember that braces are used to enclose the members or elements of a set, and the three dots denote etc.)

EXAMPLE 1   Simplify $8 + 7 - 4 + 12 - 7 + 14$.

Solution   The additions and subtractions should be performed from left to right in the order that they appear. Thus, $8 + 7 - 4 + 12 - 7 + 14$ simplifies to 30. ∎

EXAMPLE 2   Simplify $7(9 + 5)$.

Solution   The parentheses indicate the product of 7 and the quantity $9 + 5$. Perform the addition inside the parentheses first and then multiply; $7(9 + 5)$ thus simplifies to $7(14)$, which becomes 98.   ■

EXAMPLE 3   Simplify $(7 + 8) \div (4 - 1)$.

Solution   First, we perform the operations inside the parentheses. $(7 + 8) \div (4 - 1)$ thus becomes $15 \div 3$, which is 5.   ■

We frequently express a problem such as Example 3 in the form $\dfrac{7 + 8}{4 - 1}$. We don't need parentheses in this case because the fraction bar indicates that the sum of 7 and 8 is to be divided by the difference $4 - 1$. A problem may, however, contain parentheses and fraction bars, as the next example illustrates.

EXAMPLE 4   Simplify $\dfrac{(4 + 2)(7 - 1)}{9} + \dfrac{4}{7 - 3}$.

Solution   First, simplify above and below the fraction bars, and then proceed to evaluate as follows.

$$\frac{(4 + 2)(7 - 1)}{9} + \frac{4}{7 - 3} = \frac{(6)(6)}{9} + \frac{4}{4}$$

$$= \frac{36}{9} + 1 = 4 + 1 = 5 \qquad ■$$

EXAMPLE 5   Simplify $7 \cdot 9 + 5$.

Solution   If there are no parentheses to indicate otherwise, multiplication takes precedence over addition. First perform the multiplication and then do the addition: $7 \cdot 9 + 5$ therefore simplifies to $63 + 5$, which is 68.   ■

(Compare Example 2 and Example 5 and note the difference in meaning.)

EXAMPLE 6   Simplify $8 + 4 \cdot 3 - 14 \div 2$.

Solution   The multiplication and division should be done first in the order that they appear, from left to right. Thus, $8 + 4 \cdot 3 - 14 \div 2$ simplifies to $8 + 12 - 7$. We perform the addition and subtraction in the order that they appear, which simplifies $8 + 12 - 7$ to 13.   ■

EXAMPLE 7   Simplify $8 \cdot 5 \div 4 + 7 \cdot 3 - 32 \div 8 + 9 \div 3 \cdot 2$.

Solution   By performing the multiplications and divisions first in the order that they appear and then doing the additions and subtractions, our work could take on the following format.

$$8 \cdot 5 \div 4 + 7 \cdot 3 - 32 \div 8 + 9 \div 3 \cdot 2 = 10 + 21 - 4 + 6 = 33 \qquad \blacksquare$$

EXAMPLE 8   Simplify $5 + 6[2(3 + 9)]$.

Solution   We use brackets for the same purpose as parentheses. In such a problem we need to simplify *from the inside out*; perform the operations in the innermost parentheses first.

$$\begin{aligned}
5 + 6[2(3 + 9)] &= 5 + 6[2(12)] \\
&= 5 + 6[24] \\
&= 5 + 144 = 149 \qquad \blacksquare
\end{aligned}$$

Let us now summarize the ideas presented in the previous examples regarding **simplifying numerical expressions**. When simplifying a numerical expression the operations should be performed in the following order.

Order of Operations

1. Perform the operations inside the symbols of inclusion (parentheses and brackets) and above and below each fraction bar. Start with the innermost inclusion symbol.
2. Perform all multiplications and divisions in the order that they appear from left to right.
3. Perform all additions and subtractions in the order that they appear from left to right.

Use of Variables

We can use the concept of a variable to generalize from numerical expressions to algebraic expressions. Each of the following is an example of an algebraic expression.

$$3x + 2y, \qquad 5a - 2b + c, \qquad 7(w + z),$$

$$\frac{5d + 3e}{2c - d}, \qquad 2xy + 5yz, \qquad (x + y)(x - y)$$

An algebraic expression takes on a numerical value whenever each variable in the expression is replaced by a specific number. For example, if x is replaced by 9 and z by 4, the algebraic expression $x - z$ becomes the numerical expression $9 - 4$, which simplifies to 5. We say that $x - z$ **has a value** of 5 when x equals 9 and z equals 4. The value of $x - z$ when x equals 25 and z equals 12, is 13. The general algebraic expression $x - z$ has a specific value each time x and z are replaced by numbers.

Consider the following examples, which illustrate the process of finding a value of an algebraic expression. The process is often referred to as **evaluating algebraic expressions**.

EXAMPLE 9   Find the value of $3x + 2y$ when x is replaced by 5 and y by 17.

Solution   The following format is convenient for such problems.

$$3x + 2y = 3(5) + 2(17) \quad \text{when } x = 5 \text{ and } y = 17$$
$$= 15 + 34 = 49 \qquad \blacksquare$$

In the solution to Example 9 notice that parentheses were inserted to indicate multiplication as we switched from the algebraic expression to the numerical expression. We could also use the raised dot to indicate multiplication; that is, $3(5) + 2(17)$ could be written as $3 \cdot 5 + 2 \cdot 17$. Furthermore, notice that once we have a numerical expression, our previous agreements for simplifying numerical expressions are in effect.

EXAMPLE 10   Find the value of $12a - 3b$ when $a = 5$ and $b = 9$.

Solution
$$12a - 3b = 12(5) - 3(9) \quad \text{when } a = 5 \text{ and } b = 9$$
$$= 60 - 27 = 33 \qquad \blacksquare$$

EXAMPLE 11   Evaluate $4xy + 2xz - 3yz$ when $x = 8$, $y = 6$, and $z = 2$.

Solution
$$4xy + 2xz - 3yz = 4(8)(6) + 2(8)(2) - 3(6)(2) \quad \text{when } x = 8, y = 6, \text{ and } z = 2$$
$$= 192 + 32 - 36 = 188 \qquad \blacksquare$$

EXAMPLE 12   Evaluate $\dfrac{5c + d}{3c - d}$ for $c = 12$ and $d = 4$.

Solution
$$\frac{5c + d}{3c - d} = \frac{5(12) + 4}{3(12) - 4} \quad \text{for } c = 12 \text{ and } d = 4$$
$$= \frac{60 + 4}{36 - 4} = \frac{64}{32} = 2 \qquad \blacksquare$$

EXAMPLE 13   Evaluate $(2x + 5y)(3x - 2y)$ when $x = 6$ and $y = 3$.

Solution
$$(2x + 5y)(3x - 2y) = (2 \cdot 6 + 5 \cdot 3)(3 \cdot 6 - 2 \cdot 3) \quad \text{when } x = 6 \text{ and } y = 3$$
$$= (12 + 15)(18 - 6)$$
$$= (27)(12) = 324 \qquad \blacksquare$$

Problem Set 1.1

For Problems 1–34, simplify each numerical expression.

1. $9 + 14 - 7$

2. $32 - 14 + 6$

3. $7(14 - 9)$

4. $8(6 + 12)$

5. $16 + 5 \cdot 7$

6. $18 - 3(5)$

7. $4(12 + 9) - 3(8 - 4)$

8. $7(13 - 4) - 2(19 - 11)$

9. $4(7) + 6(9)$

10. $8(7) - 4(8)$

11. $6 \cdot 7 + 5 \cdot 8 - 3 \cdot 9$

12. $8(13) - 4(9) + 2(7)$

13. $(6 + 9)(8 - 4)$

14. $(15 - 6)(13 - 4)$

15. $6 + 4[3(9 - 4)]$

16. $92 - 3[2(6 - 2)]$

17. $16 \div 8 \cdot 4 + 36 \div 4 \cdot 2$

18. $7 \cdot 8 \div 4 - 72 \div 12$

19. $\dfrac{8 + 12}{4} - \dfrac{9 + 15}{8}$

20. $\dfrac{19 - 7}{6} + \dfrac{38 - 14}{3}$

21. $56 - [3(9 - 6)]$

22. $17 + 2[3(4 - 2)]$

23. $7 \cdot 4 \cdot 2 \div 8 + 14$

24. $14 \div 7 \cdot 8 - 35 \div 7 \cdot 2$

25. $32 \div 8 \cdot 2 + 24 \div 6 - 1$

26. $48 \div 12 + 7 \cdot 2 \div 2 - 1$

27. $4 \cdot 9 \div 12 + 18 \div 2 + 3$

28. $5 \cdot 8 \div 4 - 8 \div 4 \cdot 3 + 6$

29. $\dfrac{6(8 - 3)}{3} + \dfrac{12(7 - 4)}{9}$

30. $\dfrac{3(17 - 9)}{4} + \dfrac{9(16 - 7)}{3}$

31. $83 - \dfrac{4(12 - 7)}{5}$

32. $78 - \dfrac{6(21 - 9)}{4}$

33. $\dfrac{4 \cdot 6 + 5 \cdot 3}{7 + 2 \cdot 3} + \dfrac{7 \cdot 9 + 6 \cdot 5}{3 \cdot 5 + 8 \cdot 2}$

34. $\dfrac{7 \cdot 8 + 4}{5 \cdot 8 - 10} + \dfrac{9 \cdot 6 - 4}{6 \cdot 5 - 20}$

For Problems 35–54, evaluate each algebraic expression for the given values of the variables.

35. $7x + 4y$ for $x = 6$ and $y = 8$

36. $8x + 6y$ for $x = 9$ and $y = 5$

37. $16a - 9b$ for $a = 3$ and $b = 4$

38. $14a - 5b$ for $a = 7$ and $b = 9$

39. $4x + 7y + 3xy$ for $x = 4$ and $y = 9$

40. $x + 8y + 5xy$ for $x = 12$ and $y = 3$

41. $14xz + 6xy - 4yz$ for $x = 8$, $y = 5$, and $z = 7$

42. $9xy - 4xz + 3yz$ for $x = 7$, $y = 3$, and $z = 2$

43. $\dfrac{54}{n} + \dfrac{n}{3}$ for $n = 9$

44. $\dfrac{n}{4} + \dfrac{60}{n} - \dfrac{n}{6}$ for $n = 12$

45. $\dfrac{y + 16}{6} + \dfrac{50 - y}{3}$ for $y = 8$

46. $\dfrac{w + 57}{9} + \dfrac{90 - w}{7}$ for $w = 6$

47. $(x + y)(x - y)$ for $x = 8$ and $y = 3$

48. $(x + 2y)(2x - y)$ for $x = 7$ and $y = 4$

49. $(5x - 2y)(3x + 4y)$ for $x = 3$ and $y = 6$

50. $(3a + b)(7a - 2b)$ for $a = 5$ and $b = 7$

51. $6 + 3[2(x + 4)]$ for $x = 7$

52. $9 + 4[3(x + 3)]$ for $x = 6$

53. $81 - 2[5(n + 4)]$ for $n = 3$

54. $78 - 3[4(n - 2)]$ for $n = 4$

For Problems 55–60, find the value of $\dfrac{bh}{2}$ for each set of values for the variables b and h.

55. $b = 8$ and $h = 12$ **56.** $b = 6$ and $h = 14$ **57.** $b = 7$ and $h = 6$

58. $b = 9$ and $h = 4$ **59.** $b = 16$ and $h = 5$ **60.** $b = 18$ and $h = 13$

For Problems 61–66, find the value of $\dfrac{Bh}{3}$ for each set of values for the variables B and h.

61. $B = 27$ and $h = 9$ **62.** $B = 18$ and $h = 6$ **63.** $B = 25$ and $h = 12$

64. $B = 32$ and $h = 18$ **65.** $B = 36$ and $h = 7$ **66.** $B = 42$ and $h = 17$

For Problems 67–72, find the value of $\dfrac{h(b_1 + b_2)}{2}$ for each set of values for the variables h, b_1, and b_2. (Subscripts are used to indicate that b_1 and b_2 are different variables.)

67. $h = 17$, $b_1 = 14$, and $b_2 = 6$ **68.** $h = 9$, $b_1 = 12$, and $b_2 = 16$

69. $h = 8$, $b_1 = 17$, and $b_2 = 24$ **70.** $h = 12$, $b_1 = 14$, and $b_2 = 5$

71. $h = 18$, $b_1 = 6$, and $b_2 = 11$ **72.** $h = 14$, $b_1 = 9$, and $b_2 = 7$

Miscellaneous Problem

 73. You should be able to do calculations like those in Problems 1–34 *with* and *without* a calculator. Different types of calculators handle the *priority-of-operations* issue in different ways. Be sure you can do Problems 1–34 with your calculator.

1.2
Prime and Composite Numbers

We say that 6 *divides* 18 because 6 times the whole number 3 produces 18; but 6 *does not divide* 19 because there is no whole number such that 6 times the number produces 19. Likewise, 5 *divides* 35 because 5 times the whole number 7 produces 35; 5 *does not divide* 42 because there is no whole number such that 5 times the number produces 42. We can use the following general definition.

DEFINITION 1.1

> Given that a and b are whole numbers, with a not equal to zero, a *divides* b if and only if there exists a whole number k such that $a \cdot k = b$.

REMARK 1 Notice the use of variables, a, b, and k, in the statement of a *general* definition. Also note that the definition merely generalizes the concept of *divides*, which was introduced in the specific examples prior to the definition.

REMARK 2 This is a special use of the word *divides*. Occasionally, terms in mathematics are given a special meaning in the discussion of a particular topic.

The following statements further clarify Definition 1.1. Pay special attention to the italicized words for they indicate some of the various terminology used for this topic.

1. 8 *divides* 56 because $8 \cdot 7 = 56$.
2. 7 *does not divide* 38 because there is no whole number, k, such that $7 \cdot k = 38$.
3. 3 is a *factor* of 27 because $3 \cdot 9 = 27$.
4. 4 is *not a factor* of 38 because there is no whole number, k, such that $4 \cdot k = 38$.
5. 35 is a *multiple* of 5 because $5 \cdot 7 = 35$.
6. 29 is *not a multiple* of 7 because there is no whole number, k, such that $7 \cdot k = 29$.

The term "factor" is used extensively. We say that 7 and 8 are factors of 56 because $7 \cdot 8 = 56$; 4 and 14 are also factors of 56 because $4 \cdot 14 = 56$. **Factors** are sometimes called **divisors**.

Now consider two special kinds of whole numbers called **prime numbers** and **composite numbers** according to the following definition.

DEFINITION 1.2

> A **prime number** is a whole number, greater than 1, that has no factors (divisors) other than itself and 1. Whole numbers, greater than 1, that are not prime numbers are called **composite numbers**.

The prime numbers less than 50 are 2, 3, 5, 7, 11, 13, 17, 19, 23, 29, 31, 37, 41, 43, and 47. Notice that each of these has no factors other than itself and 1. An interesting point is that the set of prime numbers is an infinite set; that is, the prime numbers go on forever, and there is no *largest* prime number.

We can express every composite number as the indicated product of prime numbers. Consider the following examples.

$$4 = 2 \cdot 2, \qquad 6 = 2 \cdot 3, \qquad 8 = 2 \cdot 2 \cdot 2, \qquad 10 = 2 \cdot 5, \qquad 12 = 2 \cdot 2 \cdot 3$$

In each case we expressed a composite number as the indicated product of prime numbers. The indicated product form is sometimes called the **prime factored form** of the number.

There are various procedures to find the prime factors of a given composite number. For our purposes, the simplest technique is to factor the given composite number into any two easily recognized factors and then to continue to factor each of these until we obtain only prime factors. Consider these examples.

$$18 = 2 \cdot 9 = 2 \cdot 3 \cdot 3, \qquad\qquad 27 = 3 \cdot 9 = 3 \cdot 3 \cdot 3,$$
$$24 = 4 \cdot 6 = 2 \cdot 2 \cdot 2 \cdot 3, \qquad 150 = 10 \cdot 15 = 2 \cdot 5 \cdot 3 \cdot 5$$

It does not matter which two factors we choose first. For example, one might start by expressing 18 as $3 \cdot 6$ and then factor 6 into $2 \cdot 3$, which would produce a final result of $18 = 3 \cdot 2 \cdot 3$. Either way, 18 contains two prime factors of 3 and one prime factor of 2. The order in which we write the prime factors is not important.

Greatest Common Factor

We can use the prime factorization form of two composite numbers to conveniently find their **greatest common factor**. Consider the following example.

$$42 = 2 \cdot 3 \cdot 7,$$

$$70 = 2 \cdot 5 \cdot 7$$

Notice that 2 is a factor of both, as is 7. Therefore, 14 (the product of 2 and 7) is the greatest common factor of 42 and 70. In other words, 14 is the largest whole number that divides both 42 and 70. The following examples should further clarify the process of finding the greatest common factor of two or more numbers.

EXAMPLE 1   Find the greatest common factor of 48 and 60.

Solution

$$48 = 2 \cdot 2 \cdot 2 \cdot 2 \cdot 3,$$

$$60 = 2 \cdot 2 \cdot 3 \cdot 5$$

Since two 2s and a 3 are common to both, the greatest common factor of 48 and 60 is $2 \cdot 2 \cdot 3 = 12$. ■

EXAMPLE 2   Find the greatest common factor of 21 and 75.

Solution

$$21 = 3 \cdot 7,$$

$$75 = 3 \cdot 5 \cdot 5$$

Since only a 3 is common to both, the greatest common factor is 3. ■

EXAMPLE 3   Find the greatest common factor of 24 and 35.

Solution

$$24 = 2 \cdot 2 \cdot 2 \cdot 3,$$

$$35 = 5 \cdot 7$$

Since there are no common prime factors, the greatest common factor is 1. ■

The concept of greatest common factor can be extended to more than two numbers as the next example demonstrates.

EXAMPLE 4   Find the greatest common factor of 24, 56, and 120.

Solution
$$24 = 2 \cdot 2 \cdot 2 \cdot 3,$$
$$56 = 2 \cdot 2 \cdot 2 \cdot 7,$$
$$120 = 2 \cdot 2 \cdot 2 \cdot 3 \cdot 5$$

Since three 2s are common to the numbers, the greatest common factor of 24, 56, and 120 is $2 \cdot 2 \cdot 2 = 8$. ■

Least Common Multiple

In this section we stated earlier that 35 is a *multiple of* 5 because $5 \cdot 7 = 35$. The set of all whole numbers that are multiples of 5 consists of 0, 5, 10, 15, 20, 25, etc. In other words, 5 times each successive whole number ($5 \cdot 0 = \mathbf{0}$, $5 \cdot 1 = \mathbf{5}$, $5 \cdot 2 = \mathbf{10}$, $5 \cdot 3 = \mathbf{15}$, etc.) produces the multiples of 5. In a like manner, the set of multiples of 4 consists of 0, 4, 8, 12, 16, etc.

It is sometimes necessary to determine the smallest common *nonzero* multiple of two or more whole numbers. We use the phrase **least common multiple** to designate this nonzero number. For example, the least common multiple of 3 and 4 is 12, which means that 12 is the smallest nonzero multiple of both 3 and 4. Stated another way, 12 is the smallest nonzero whole number that is divisible by both 3 and 4. Likewise, we would say that the least common multiple of 6 and 8 is 24.

If we cannot determine the least common multiple by inspection, then the prime factorization form of composite numbers is helpful. Study the solutions to the following examples very carefully as we develop a systematic technique for finding the least common multiple of two or more numbers.

EXAMPLE 5   Find the least common multiple of 24 and 36.

Solution   Let's first express each number as a product of prime factors.
$$24 = 2 \cdot 2 \cdot 2 \cdot 3,$$
$$36 = 2 \cdot 2 \cdot 3 \cdot 3$$

Since 24 contains three 2s, the least common multiple must have three 2s. Also, since 36 contains two 3s, we need to put two 3s in the least common multiple. The least common multiple of 24 and 36 is therefore $2 \cdot 2 \cdot 2 \cdot 3 \cdot 3 = 72$. ■

If the least common multiple is not obvious by inspection, then we can proceed as follows.

Step 1. Express each number as a product of prime factors.

Step 2. The least common multiple contains each different prime factor as many times as the *most* times it appears in any one of the factorizations from step 1.

EXAMPLE 6   Find the least common multiple of 48 and 84.

Solution

$$48 = 2 \cdot 2 \cdot 2 \cdot 2 \cdot 3,$$
$$84 = 2 \cdot 2 \cdot 3 \cdot 7$$

We need four 2s in the least common multiple because of the four 2s in 48. We need one 3 because of the 3 in each of the numbers and one 7 is needed because of the 7 in 84. The least common multiple of 48 and 84 is $2 \cdot 2 \cdot 2 \cdot 2 \cdot 3 \cdot 7 = 336$.   ■

EXAMPLE 7   Find the least common multiple of 12, 18, and 28.

Solution

$$12 = 2 \cdot 2 \cdot 3,$$
$$18 = 2 \cdot 3 \cdot 3,$$
$$28 = 2 \cdot 2 \cdot 7$$

The least common multiple is $2 \cdot 2 \cdot 3 \cdot 3 \cdot 7 = 252$.   ■

EXAMPLE 8   Find the least common multiple of 8 and 9.

Solution

$$8 = 2 \cdot 2 \cdot 2,$$
$$9 = 3 \cdot 3$$

The least common multiple is $2 \cdot 2 \cdot 2 \cdot 3 \cdot 3 = 72$.   ■

Problem Set 1.2

For Problems 1–20, classify each statement as true or false.

1. 8 divides 56   **2.** 9 divides 54
3. 6 does not divide 54   **4.** 7 does not divide 42
5. 96 is a multiple of 8   **6.** 78 is a multiple of 6
7. 54 is not a multiple of 4   **8.** 64 is not a multiple of 6
9. 144 is divisible by 4   **10.** 261 is divisible by 9
11. 173 is divisible by 3   **12.** 149 is divisible by 7
13. 11 is a factor of 143   **14.** 11 is a factor of 187
15. 9 is a factor of 119   **16.** 8 is a factor of 98
17. 3 is a prime factor of 57   **18.** 7 is a prime factor of 91
19. 4 is a prime factor of 48   **20.** 6 is a prime factor of 72

For Problems 21–30, classify each number as prime or composite.

21. 53   **22.** 57   **23.** 59   **24.** 61
25. 91   **26.** 81   **27.** 89   **28.** 97
29. 111   **30.** 101

For Problems 31–42, factor each composite number into a product of prime numbers. For example, $18 = 2 \cdot 3 \cdot 3$.

31. 26	**32.** 16	**33.** 36	**34.** 80
35. 49	**36.** 92	**37.** 56	**38.** 144
39. 120	**40.** 84	**41.** 135	**42.** 98

For Problems 43–54, find the greatest common factor of the given numbers.

43. 12 and 16	**44.** 30 and 36	**45.** 56 and 64
46. 72 and 96	**47.** 63 and 81	**48.** 60 and 72
49. 84 and 96	**50.** 48 and 52	**51.** 36, 72, and 90
52. 27, 54, and 63	**53.** 48, 60, and 84	**54.** 32, 80, and 96

For Problems 55–66, find the least common multiple of the given numbers.

55. 6 and 8	**56.** 8 and 12	**57.** 12 and 16
58. 9 and 12	**59.** 28 and 35	**60.** 42 and 66
61. 49 and 56	**62.** 18 and 24	**63.** 8, 12, and 28
64. 6, 10, and 12	**65.** 9, 15, and 18	**66.** 8, 14, and 24

Miscellaneous Problems

67. The numbers 0, 2, 4, 6, 8, etc. are multiples of 2. They are also called *even* numbers. Why is 2 the only even prime number?

68. Find the smallest nonzero whole number that is divisible by 2, 3, 4, 5, 6, 7, and 8.

69. Find the smallest whole number, greater than 1, that produces a remainder of 1 when divided by 2, 3, 4, 5, or 6.

70. What is the greatest common factor of x and y if x and y are both prime numbers and x does not equal y? Explain your answer.

71. What is the greatest common factor of x and y if x and y are nonzero whole numbers and y is a multiple of x? Explain your answer.

72. What is the least common multiple of x and y if they are both prime numbers and x does not equal y? Explain your answer.

73. What is the least common multiple of x and y if the greatest common factor of x and y is 1? Explain your answer.

Problems 74–83. Familiarity with a few basic divisibility rules can be helpful for determining the prime factors of some numbers. For example, if you can quickly recognize that 51 is divisible by 3, then you can divide 51 by 3 to find another factor of 17. Since 3 and 17 are both prime numbers, we have $51 = 3 \cdot 17$. The divisibility rules for 2, 3, 5, and 9 are as follows.

> **Rule for 2** A whole number is divisible by 2 if and only if the units digit of its base-ten numeral is divisible by 2. (In other words, the units digit must be 0, 2, 4, 6, or 8.)

EXAMPLES   68 is divisible by 2 because 8 is divisible by 2.
57 is not divisible by 2 because 7 is not divisible by 2.

> **Rule for 3** A whole number is divisible by 3 if and only if the sum of the digits of its base-ten numeral is divisible by 3.

EXAMPLES 51 is divisible by 3 because $5 + 1 = 6$ and 6 is divisible by 3.
144 is divisible by 3 because $1 + 4 + 4 = 9$ and 9 is divisible by 3.
133 is not divisible by 3 because $1 + 3 + 3 = 7$ and 7 is not divisible by 3.

> **Rule for 5** A whole number is divisible by 5 if and only if the units digit of its base-ten numeral is divisible by 5. (In other words, the units digit must be 0 or 5.)

EXAMPLES 115 is divisible by 5 because 5 is divisible by 5.
172 is not divisible by 5 because 2 is not divisible by 5.

> **Rule for 9** A whole number is divisible by 9 if and only if the sum of the digits of its base-ten numeral is divisible by 9.

EXAMPLES 765 is divisible by 9 because $7 + 6 + 5 = 18$ and 18 is divisible by 9.
147 is not divisible by 9 because $1 + 4 + 7 = 12$ and 12 is not divisible by 9

Use the previous divisibility rules to help determine the prime factorization of each of the following numbers.

74. 118   **75.** 76   **76.** 201   **77.** 123
78. 85   **79.** 115   **80.** 117   **81.** 441
82. 129   **83.** 153

1.3

Integers: Addition and Subtraction

"A record temperature of 35° *below* zero was recorded on this date in 1904." "The PO stock closed *down* 3 points yesterday." "On a first down sweep around left end, Moser *lost* 7 yards." "The Widget Manufacturing Company reported *assets* of 50 million dollars and *liabilities* of 53 million dollars for 1981." Such examples simply illustrate our need for negative numbers.

The number line becomes a helpful pictorial device for our work at this time. We can associate the set of whole numbers with evenly spaced points on a line as indicated in Figure 1.1.

Figure 1.1

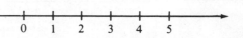

0 1 2 3 4 5

For each nonzero whole number we can associate its *negative* to the left of zero; with 1 we associate -1, with 2 we associate -2, etc. as indicated in Figure 1.2. The set of whole numbers along with $-1, -2, -3$, and so on, is called the set of **integers**.

Figure 1.2

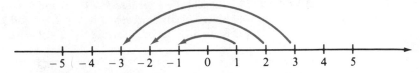

The following terminology is used with reference to the integers.

$\ldots, -3, -2, -1, 0, 1, 2, 3, \ldots$	integers
$1, 2, 3, 4, \ldots$	positive integers
$0, 1, 2, 3, 4, \ldots$	nonnegative integers
$\ldots, -3, -2, -1$	negative integers
$\ldots, -3, -2, -1, 0$	nonpositive integers

The symbol -1 can be read as *negative one*, *opposite of one*, or *additive inverse of one*. The *opposite of* and *additive inverse of* terminology is very helpful when working with variables. For example, the symbol $-x$, read as *opposite of x* or *additive inverse of x*, emphasizes an important issue. Since x can be any integer, $-x$ (the opposite of x) can be zero, positive, or negative. If x is a positive integer, then $-x$ is negative. If x is a negative integer, then $-x$ is positive. If x is zero, then $-x$ is zero. These statements could be written as follows.

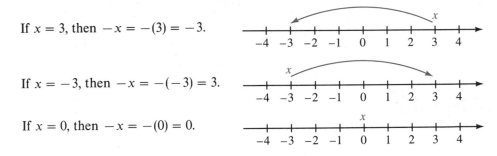

If $x = 3$, then $-x = -(3) = -3$.

If $x = -3$, then $-x = -(-3) = 3$.

If $x = 0$, then $-x = -(0) = 0$.

We should also observe that *the opposite of the opposite of any integer is the integer itself*. For example, the opposite of 3 is -3 and then the opposite of -3 is 3. Thus, we can write $-(-3) = 3$. The following general property can be stated.

PROPERTY 1.1

If a is any integer, then

$$-(-a) = a.$$

(The opposite of the opposite of any integer is the integer itself.)

Addition of Integers

The number line is also a convenient visual aid for interpreting **addition of integers**. Consider the following examples.

Problem	Number line interpretation	Sum
$3 + 2$		$3 + 2 = 5$
$3 + (-2)$		$3 + (-2) = 1$
$-3 + 2$		$-3 + 2 = -1$
$-3 + (-2)$		$-3 + (-2) = -5$

Once you acquire a feeling of movement on the number line, simply forming a mental image of this movement is sufficient. Consider the following addition problems and mentally picture the number line interpretation. Be sure that you agree with all of our answers.

$$5 + (-2) = 3, \qquad -6 + 4 = -2, \qquad -8 + 11 = 3,$$
$$-7 + (-4) = -11, \qquad -5 + 9 = 4, \qquad 9 + (-2) = 7,$$
$$14 + (-17) = -3, \qquad 0 + (-4) = -4, \qquad 6 + (-6) = 0$$

The last example illustrates a general property that should be noted: Any integer plus its opposite produces zero.

Even though all problems involving addition of integers could be done by using the number line interpretation, it is sometimes convenient to be able to give a more precise description of the addition process. For this purpose we need to briefly consider the concept of absolute value. The **absolute value** of a number is the distance between the number and zero on the number line. For example, the absolute value of 6 is 6. The absolute value of -6 is also 6. The absolute value of 0 is 0. Symbolically, absolute value is denoted with vertical bars. Thus, we write

$$|6| = 6, \qquad |-6| = 6, \qquad |0| = 0.$$

Notice that the absolute value of a positive number is the number itself, but the absolute value of a negative number is its opposite. Thus, the absolute value of any number, except 0, is positive, and the absolute value of 0 is 0.

We can describe the process of **adding integers** precisely by using the concept of absolute value as follows.

> **Two Positive Integers** The sum of two positive integers is the sum of their absolute values. (The sum of two positive integers is a positive integer.)

$$43 + 54 = |43| + |54| = 43 + 54 = 97$$

> **Two Negative Integers** The sum of two negative integers is the opposite of the sum of their absolute values. (The sum of two negative integers is a negative integer.)

$$
\begin{aligned}
(-67) + (-93) &= -(|-67| + |-93|) \\
&= -(67 + 93) \\
&= -160
\end{aligned}
$$

> **One Positive and One Negative Integer** The sum of a positive and a negative integer can be found by subtracting the smaller absolute value from the larger absolute value and giving the result the sign of the original number having the larger absolute value. If the integers have the same absolute value, then their sum is 0.

$$
\begin{aligned}
82 + (-40) &= |82| - |-40| \\
&= 82 - 40 \\
&= 42
\end{aligned}
\qquad
\begin{aligned}
74 + (-90) &= -(|-90| - |74|) \\
&= -(90 - 74) \\
&= -16
\end{aligned}
$$

$$
\begin{aligned}
(-17) + 17 &= |-17| - |17| \\
&= 17 - 17 \\
&= 0
\end{aligned}
$$

> **Zero and Another Integer** The sum of 0 and any integer is the integer itself.

$$0 + (-46) = -46$$
$$72 + 0 = 72$$

The following examples further demonstrate how to add integers using the concept of absolute value. Be sure that you agree with each of the results.

$$-18 + (-56) = -(|-18| + |-56|) \qquad -71 + (-32) = -(|-71| + |-32|)$$
$$= -(18 + 56) \qquad\qquad\qquad = -(71 + 32)$$
$$= -74 \qquad\qquad\qquad\qquad = -103$$

$$64 + (-49) = |64| - |-49| \qquad\qquad -56 + 93 = |93| - |-56|$$
$$= 64 - 49 \qquad\qquad\qquad\qquad = 93 - 56$$
$$= 15 \qquad\qquad\qquad\qquad\qquad = 37$$

$$-114 + 48 = -(|-114| - |48|) \qquad 45 + (-73) = -(|-73| - |45|)$$
$$= -(114 - 48) \qquad\qquad\qquad = -(73 - 45)$$
$$= -66 \qquad\qquad\qquad\qquad\qquad = -28$$

It is true that this *absolute value approach* does precisely describe the process of adding integers, but don't forget about the number line interpretation. Included in the next problem set are other physical models for interpreting addition of integers. Some people find these models very helpful.

Subtraction of Integers

The following examples illustrate a relationship between addition and subtraction of *whole numbers*.

$$7 - 2 = 5 \quad \text{because } 2 + 5 = 7,$$
$$9 - 6 = 3 \quad \text{because } 6 + 3 = 9,$$
$$5 - 1 = 4 \quad \text{because } 1 + 4 = 5$$

This same relationship between addition and subtraction holds for *all integers*.

$$5 - 6 = -1 \qquad \text{because } 6 + (-1) = 5,$$
$$-4 - 9 = -13 \quad \text{because } 9 + (-13) = -4,$$
$$-3 - (-7) = 4 \quad \text{because } -7 + 4 = -3,$$
$$8 - (-3) = 11 \quad \text{because } -3 + 11 = 8$$

Now consider a further observation:

$$5 - 6 = -1 \qquad \text{and} \qquad 5 + (-6) = -1,$$
$$-4 - 9 = -13 \qquad \text{and} \qquad -4 + (-9) = -13,$$
$$-3 - (-7) = 4 \qquad \text{and} \qquad -3 + 7 = 4,$$
$$8 - (-3) = 11 \qquad \text{and} \qquad 8 + 3 = 11.$$

The previous examples help us realize that we can state the subtraction of integers in terms of addition of integers. More precisely, a general description for subtraction of integers follows.

Subtraction of Integers If a and b are integers, then $a - b = a + (-b)$.

It may be helpful for you to read $a - b = a + (-b)$ as "a minus b is equal to a plus the opposite of b." Every subtraction problem can be changed to an equivalent addition problem as illustrated by the following examples.

$$6 - 13 = 6 + (-13) = -7,$$

$$9 - (-12) = 9 + 12 = 21,$$

$$-8 - 13 = -8 + (-13) = -21,$$

$$-7 - (-8) = -7 + 8 = 1$$

It should be apparent that addition of integers is a key operation. Being able to effectively add integers is a necessary skill for further algebraic work.

Evaluating Algebraic Expressions

Let's conclude this section by evaluating some algebraic expressions using negative and positive integers.

EXAMPLE 1   Evaluate each algebraic expression for the given values of the variables.

(a) $x - y$ for $x = -12$ and $y = 20$

(b) $-a + b$ for $a = -8$ and $b = -6$

(c) $-x - y$ for $x = 14$ and $y = -7$

Solution

(a) $x - y = -12 - 20$ when $x = -12$ and $y = 20$

$$= -12 + (-20)$$

$$= -32$$

(b) $-a + b = -(-8) + (-6)$ when $a = -8$ and $b = -6$

$$= 8 + (-6)$$

$$= 2$$

(c) $-x - y = -14 - (-7)$ when $x = 14$ and $y = -7$

$$= -14 + 7$$

$$= -7$$ ■

Problem Set 1.3

For Problems 1–10, use the number line interpretation to find each sum.

1. $5 + (-3)$ **2.** $7 + (-4)$ **3.** $-6 + 2$ **4.** $-9 + 4$

5. $-3 + (-4)$ **6.** $-5 + (-6)$ **7.** $8 + (-2)$ **8.** $12 + (-7)$

9. $5 + (-11)$ **10.** $4 + (-13)$

For Problems 11–30, find each sum.

11. $17 + (-9)$ **12.** $16 + (-5)$ **13.** $8 + (-19)$

14. $9 + (-14)$ **15.** $-7 + (-8)$ **16.** $-6 + (-9)$

17. $-15 + 8$ **18.** $-22 + 14$ **19.** $-13 + (-18)$

20. $-15 + (-19)$ **21.** $-27 + 8$ **22.** $-29 + 12$

23. $32 + (-23)$ **24.** $27 + (-14)$ **25.** $-25 + (-36)$

26. $-34 + (-49)$ **27.** $54 + (-72)$ **28.** $48 + (-76)$

29. $-34 + (-58)$ **30.** $-27 + (-36)$

For Problems 31–50, subtract as indicated.

31. $3 - 8$ **32.** $5 - 11$ **33.** $-4 - 9$

34. $-7 - 8$ **35.** $5 - (-7)$ **36.** $9 - (-4)$

37. $-6 - (-12)$ **38.** $-7 - (-15)$ **39.** $-11 - (-10)$

40. $-14 - (-19)$ **41.** $-18 - 27$ **42.** $-16 - 25$

43. $34 - 63$ **44.** $25 - 58$ **45.** $45 - 18$

46. $52 - 38$ **47.** $-21 - 44$ **48.** $-26 - 54$

49. $-53 - (-24)$ **50.** $-76 - (-39)$

For Problems 51–66, add or subtract as indicated.

51. $6 - 8 - 9$ **52.** $5 - 9 - 4$

53. $-4 - (-6) + 5 - 8$ **54.** $-3 - 8 + 9 - (-6)$

55. $5 + 7 - 8 - 12$ **56.** $-7 + 9 - 4 - 12$

57. $-6 - 4 - (-2) + (-5)$ **58.** $-8 - 11 - (-6) + (-4)$

59. $-6 - 5 - 9 - 8 - 7$ **60.** $-4 - 3 - 7 - 8 - 6$

61. $7 - 12 + 14 - 15 - 9$ **62.** $8 - 13 + 17 - 15 - 19$

63. $-11 - (-14) + (-17) - 18$ **64.** $-15 + 20 - 14 - 18 + 9$

65. $16 - 21 + (-15) - (-22)$ **66.** $17 - 23 - 14 - (-18)$

The horizontal format is used extensively in algebra, but occasionally the vertical format shows up. Some exposure to the vertical format is therefore needed. Find the following *sums* for Problems 67–78.

67. 5 **68.** 8 **69.** -13 **70.** -14
    $\underline{-9}$ $\underline{-13}$ $\underline{-18}$ $\underline{-28}$

71. -18 **72.** -17 **73.** -21 **74.** -15
    $\underline{9}$ $\underline{9}$ $\underline{39}$ $\underline{32}$

75. 27	**76.** 31	**77.** -53	**78.** 47
-19	-18	24	-28

For Problems 79–90, do the *subtraction* problems in vertical format.

79. 5	**80.** 8	**81.** 6	**82.** 13
12	19	-9	-7

83. -7	**84.** -6	**85.** 17	**86.** 18
-8	-5	-19	-14

87. -23	**88.** -27	**89.** -12	**90.** -13
16	15	12	-13

For Problems 91–100, evaluate each algebraic expression for the given values of the variables.

91. $x - y$ for $x = -6$ and $y = -13$ **92.** $x - y$ for $x = -7$ and $y = -9$

93. $-x + y - z$ for $x = 3, y = -4$, and $z = -6$

94. $x - y + z$ for $x = 5, y = 6$, and $z = -9$

95. $-x - y - z$ for $x = -2, y = 3$, and $z = -11$

96. $-x - y + z$ for $x = -8, y = -7$, and $z = -14$

97. $-x + y + z$ for $x = -11, y = 7$, and $z = -9$

98. $-x - y - z$ for $x = 12, y = -6$, and $z = -14$

99. $x - y - z$ for $x = -15, y = 12$, and $z = -10$

100. $x + y - z$ for $x = -18, y = 13$, and $z = 8$

Miscellaneous Problems

A game such as football can also be used to interpret addition of integers. A gain of 3 yards on one play followed by a loss of 5 yards on the next play places the ball 2 yards *behind* the initial line of scrimmage and this could be expressed as $3 + (-5) = -2$. Use this "football interpretation" to find the following sums (Problems 101–110).

101. $4 + (-7)$ **102.** $3 + (-5)$ **103.** $-4 + (-6)$ **104.** $-2 + (-5)$

105. $-5 + 2$ **106.** $-10 + 6$ **107.** $-4 + 15$ **108.** $-3 + 22$

109. $-12 + 17$ **110.** $-9 + 21$

Profits and losses pertaining to investments also provide a good physical model for interpreting addition of integers. A loss of $25 on one investment along with a profit of $60 on a second investment produces an overall profit of $35. This could be written as $-25 + 60 = 35$. Use this *profit and loss* interpretation to help find the following sums (Problems 111–120).

111. $60 + (-125)$ **112.** $50 + (-85)$ **113.** $-55 + (-45)$

114. $-120 + (-220)$ **115.** $-70 + 45$ **116.** $-125 + 45$

117. $-120 + 250$ **118.** $-75 + 165$ **119.** $145 + (-65)$

120. $275 + (-195)$

121. To find the sum of 65, 68, 71, 73, and 69 we might consider the deviation of each of these numbers from some other number, say 70. That is, consider $-5 + (-2) + 1 +$

$3 + (-1) = -4$. Thus, the sum of the original numbers is $(5 \cdot 70) + (-4) = 346$. Use this procedure to find the following sums.

(a) $84 + 78 + 82 + 75$

(b) $136 + 145 + 142 + 137 + 132$

(c) $56 + 63 + 61 + 58 + 52 + 67$

(d) $119 + 123 + 121 + 116 + 114 + 128$

 122. Do Problems 51–90 with your calculator.

1.4

Integers: Multiplication and Division

Multiplication of whole numbers may be interpreted as repeated addition. For example, $3 \cdot 4$ means the sum of three 4s; thus, $3 \cdot 4 = 4 + 4 + 4 = 12$. Consider the following examples that use the repeated addition idea to find the product of a positive integer and a negative integer.

$$3(-2) = -2 + (-2) + (-2) = -6,$$

$$2(-4) = -4 + (-4) = -8,$$

$$4(-1) = -1 + (-1) + (-1) + (-1) = -4$$

Note the use of parentheses to indicate multiplication. Sometimes both numbers are enclosed in parentheses so that we have $(3)(-2)$.

When multiplying whole numbers we realize that the order in which we multiply two factors does not change the product: $2(3) = 6$ and $3(2) = 6$. Using this idea we can now handle a negative integer times a positive integer as follows.

$$(-2)(3) = (3)(-2) = (-2) + (-2) + (-2) = -6,$$

$$(-3)(2) = (2)(-3) = (-3) + (-3) = -6,$$

$$(-4)(3) = (3)(-4) = (-4) + (-4) + (-4) = -12$$

Finally, let's consider the product of two negative integers. The following pattern helps us with the reasoning for this situation.

$$4(-3) = -12,$$

$$3(-3) = -9,$$

$$2(-3) = -6,$$

$$1(-3) = -3,$$

$$0(-3) = 0, \qquad \text{The product of zero and any integer is zero.}$$

$$(-1)(-3) = ?$$

Certainly, to continue this pattern, the product of -1 and -3 has to be 3. In general, this type of reasoning would help us to realize that the product of any two negative integers is a positive integer.

Using the concept of absolute value it is now quite easy to precisely describe the **multiplication of integers**.

> **1.** The product of two positive integers or two negative integers is the product of their absolute values.
>
> **2.** The product of a positive and a negative integer (either order) is the opposite of the product of their absolute values.
>
> **3.** The product of zero and any integer is zero.

The following examples illustrate this description of multiplication.

$$(-5)(-2) = |-5| \cdot |-2| = 5 \cdot 2 = 10,$$
$$(7)(-6) = -(|7| \cdot |-6|) = -(7 \cdot 6) = -42,$$
$$(-8)(9) = -(|-8| \cdot |9|) = -(8 \cdot 9) = -72,$$
$$(-14)(0) = 0,$$
$$(0)(-28) = 0$$

These examples show a step-by-step process for multiplying integers. In reality, however, the key issue is to remember whether the product is positive or negative. In other words, we need to remember that **the product of two positive integers or two negative integers is a positive integer; and the product of a positive integer and a negative integer (either order) is a negative integer**. Then we can avoid the step-by-step analysis and simply write the results as follows.

$$(7)(-9) = -63, \qquad (8)(7) = 56,$$
$$(-5)(-6) = 30, \qquad (-4)(12) = -48$$

Division of Integers

By looking back at our knowledge of whole numbers we can get some guidance for our work with integers. We know, for example, that $\frac{8}{2} = 4$ because $2 \cdot 4 = 8$. In other words, the quotient of two whole numbers can be found by looking at a related multiplication problem. In the following examples we have used this same link between multiplication and division to determine the quotients.

$$\frac{8}{-2} = -4 \quad \text{because } (-2)(-4) = 8,$$

$$\frac{-10}{5} = -2 \quad \text{because } (5)(-2) = -10,$$

$$\frac{-12}{-4} = 3 \quad \text{because } (-4)(3) = -12,$$

$$\frac{0}{-6} = 0 \quad \text{because } (-6)(0) = 0,$$

$$\frac{-9}{0} \quad \text{is undefined because no number times zero produces } -9,$$

$$\frac{0}{0} \quad \text{is undefined because any number times zero produces zero.}$$ **Remember that division by zero is undefined!**

---

1. The quotient of two positive or two negative integers is the quotient of their absolute values.
2. The quotient of a positive integer and a negative integer (or a negative and a positive) is the opposite of the quotient of their absolute values.
3. The quotient of zero and any nonzero integer (zero divided by any nonzero integer) is zero.

The following examples illustrate this description of division.

$$\frac{-8}{-4} = \frac{|-8|}{|-4|} = \frac{8}{4} = 2,$$

$$\frac{-14}{2} = -\left(\frac{|-14|}{|2|}\right) = -\left(\frac{14}{2}\right) = -7,$$

$$\frac{15}{-3} = -\left(\frac{|15|}{|-3|}\right) = -\left(\frac{15}{3}\right) = -5,$$

$$\frac{0}{-4} = 0$$

For practical purposes, the key idea is to remember whether the quotient is positive or negative. We need to remember that **the quotient of two positive integers or two negative integers is a positive integer; and the quotient of a positive integer and a negative integer or a negative integer and a positive integer is a negative integer**. We then can simply write the quotients as follows without showing all of the steps.

$$\frac{-18}{-6} = 3, \qquad \frac{-24}{12} = -2, \qquad \frac{36}{-9} = -4$$

REMARK Occasionally, people use the phrase "two negatives make a positive." Hopefully, they realize that the reference is to multiplication and division only; in addition the sum of two negative integers is still a negative integer. It is probably best to avoid such imprecise statements.

Simplifying Numerical Expressions

Now we can simplify numerical expressions involving any or all of the four basic operations with integers. Keep in mind the agreements on the order of operations we stated in Section 1.1.

EXAMPLE 1   Simplify $-4(-3) - 7(-8) + 3(-9)$.

Solution
$$-4(-3) - 7(-8) + 3(-9) = 12 - (-56) + (-27)$$
$$= 12 + 56 + (-27)$$
$$= 41 \qquad \blacksquare$$

EXAMPLE 2   Simplify $\dfrac{-8 - 4(5)}{-4}$.

Solution
$$\frac{-8 - 4(5)}{-4} = \frac{-8 - 20}{-4}$$
$$= \frac{-28}{-4}$$
$$= 7 \qquad \blacksquare$$

Evaluating Algebraic Expressions

Evaluating algebraic expressions will often involve the use of two or more operations with integers. We use the final examples of this section to represent such situations.

EXAMPLE 3   Find the value of $3x + 2y$ when $x = 5$ and $y = -9$.

Solution
$$3x + 2y = 3(5) + 2(-9) \quad \text{when } x = 5 \text{ and } y = -9$$
$$= 15 + (-18)$$
$$= -3 \qquad \blacksquare$$

EXAMPLE 4   Evaluate $-2a + 9b$ for $a = 4$ and $b = -3$.

Solution
$$-2a + 9b = -2(4) + 9(-3) \quad \text{when } a = 4 \text{ and } b = -3$$
$$= -8 + (-27)$$
$$= -35 \qquad \blacksquare$$

EXAMPLE 5   Find the value of $\dfrac{x - 2y}{4}$ when $x = -6$ and $y = 5$.

Solution
$$\frac{x - 2y}{4} = \frac{-6 - 2(5)}{4} \quad \text{when } x = -6 \text{ and } y = 5$$
$$= \frac{-6 - 10}{4}$$
$$= \frac{-16}{4}$$
$$= -4 \qquad \blacksquare$$

Problem Set 1.4

For Problems 1–40, find the product or quotient (multiply or divide) as indicated.

1. $5(-6)$ **2.** $7(-9)$ **3.** $\dfrac{-27}{3}$ **4.** $\dfrac{-35}{5}$

5. $\dfrac{-42}{-6}$ **6.** $\dfrac{-72}{-8}$ **7.** $(-7)(8)$ **8.** $(-6)(9)$

9. $(-5)(-12)$ **10.** $(-7)(-14)$ **11.** $\dfrac{96}{-8}$ **12.** $\dfrac{-91}{7}$

13. $14(-9)$ **14.** $17(-7)$ **15.** $(-11)(-14)$ **16.** $(-13)(-17)$

17. $\dfrac{135}{-15}$ **18.** $\dfrac{-144}{12}$ **19.** $\dfrac{-121}{-11}$ **20.** $\dfrac{-169}{-13}$

21. $(-15)(-15)$ **22.** $(-18)(-18)$ **23.** $\dfrac{112}{-8}$ **24.** $\dfrac{112}{-7}$

25. $\dfrac{0}{-8}$ **26.** $\dfrac{-8}{0}$ **27.** $\dfrac{-138}{-6}$ **28.** $\dfrac{-105}{-5}$

29. $\dfrac{76}{-4}$ **30.** $\dfrac{-114}{6}$ **31.** $(-6)(-15)$ **32.** $\dfrac{0}{-14}$

33. $(-56) \div (-4)$ **34.** $(-78) \div (-6)$ **35.** $(-19) \div 0$

36. $(-90) \div 15$ **37.** $(-72) \div 18$ **38.** $(-70) \div 5$

39. $(-36)(27)$ **40.** $(42)(-29)$

For Problems 41–60, simplify each numerical expression.

41. $3(-4) + 5(-7)$

42. $6(-3) + 5(-9)$

43. $7(-2) - 4(-8)$

44. $9(-3) - 8(-6)$

45. $(-3)(-8) + (-9)(-5)$

46. $(-7)(-6) + (-4)(-3)$

47. $5(-6) - 4(-7) + 3(2)$

48. $7(-4) - 8(-7) + 5(-8)$

49. $\dfrac{13 + (-25)}{-3}$

50. $\dfrac{15 + (-36)}{-7}$

51. $\dfrac{12 - 48}{6}$

52. $\dfrac{16 - 40}{8}$

53. $\dfrac{-7(10) + 6(-9)}{-4}$

54. $\dfrac{-6(8) + 4(-14)}{-8}$

55. $\dfrac{4(-7) - 8(-9)}{11}$

56. $\dfrac{5(-9) - 6(-7)}{3}$

57. $-2(3) - 3(-4) + 4(-5) - 6(-7)$

58. $2(-4) + 4(-5) - 7(\ \ 6) - 3(9)$

59. $-1(-6) - 4 + 6(-2) - 7(-3) - 18$

60. $-9(-2) + 16 - 4(-7) - 12 + 3(-8)$

For Problems 61–76, evaluate each algebraic expression for the given values of the variables.

61. $7x + 5y$ for $x = -5$ and $y = 9$

62. $4a + 6b$ for $a = -6$ and $b = -8$

63. $9a - 2b$ for $a = -5$ and $b = 7$

64. $8a - 3b$ for $a = -7$ and $b = 9$

65. $-6x - 7y$ for $x = -4$ and $y = -6$

66. $-5x - 12y$ for $x = -5$ and $y = -7$

67. $\dfrac{5x - 3y}{-6}$ for $x = -6$ and $y = 4$

68. $\dfrac{-7x + 4y}{-8}$ for $x = 8$ and $y = 6$

69. $3(2a - 5b)$ for $a = -1$ and $b = -5$

70. $4(3a - 7b)$ for $a = -2$ and $b = -4$

71. $-2x + 6y - xy$ for $x = 7$ and $y = -7$

72. $-3x + 7y - 2xy$ for $x = -6$ and $y = 4$

73. $-4ab - b$ for $a = 2$ and $b = -14$

74. $-5ab + b$ for $a = -1$ and $b = -13$

75. $(ab + c)(b - c)$ for $a = -2$, $b = -3$, and $c = 4$

76. $(ab - c)(a + c)$ for $a = -3$, $b = 2$, and $c = 5$

For Problems 77–82, find the value of $\dfrac{5(F - 32)}{9}$ for each of the given values for F.

77. $F = 59$

78. $F = 68$

79. $F = 14$

80. $F = -4$

81. $F = -13$

82. $F = -22$

For Problems 83–88, find the value of $\dfrac{9C}{5} + 32$ for each of the given values for C.

83. $C = 25$

84. $C = 35$

85. $C = 40$

86. $C = 0$

87. $C = -10$

88. $C = -30$

Miscellaneous Problem

89. Do Problems 41–60 with your calculator.

1.5

Use of Properties

We will begin this section by listing and briefly commenting on some of the basic properties of integers. We will then show how these properties facilitate manipulation with integers and also serve as a basis for some algebraic computation.

> **Commutative Property of Addition**
>
> If a and b are integers, then
>
> $$a + b = b + a.$$
>
> **Commutative Property of Multiplication**
>
> If a and b are integers, then
>
> $$ab = ba.$$

Addition and multiplication are said to be commutative operations. This means that the order in which you add or multiply two integers does not affect the result. For example, $3 + 5 = 5 + 3$ and $7(8) = 8(7)$. It is also important to realize that subtraction and division *are not* commutative operations; order does make a difference. For example, $8 - 7 \neq 7 - 8$ and $16 \div 4 \neq 4 \div 16$.

> **Associative Property of Addition**
>
> If a, b, and c are integers, then
>
> $$(a + b) + c = a + (b + c).$$
>
> **Associative Property of Multiplication**
>
> If a, b, and c are integers, then
>
> $$(ab)c = a(bc).$$

Addition and multiplication are associative operations. The associative properties can be thought of as grouping properties. For example $(-8 + 3) + 9 = -8 + (3 + 9)$. Changing the grouping of the numbers does not affect the final result. This is also true for multiplication as $[(-6)(5)](-4) = (-6)[(5)(-4)]$ illustrates. Subtraction and division *are not* associative operations. For example, $(8 - 4) - 7 = -3$ whereas $8 - (4 - 7) = 11$. An example showing that division is not associative is $(8 \div 4) \div 2 = 1$; whereas $8 \div (4 \div 2) = 4$.

Identity Property of Addition

If a is an integer, then

$$a + 0 = 0 + a = a.$$

We refer to zero as the identity element for addition. This simply means that the sum of any integer and zero is exactly the same integer. For example, $-197 + 0 = 0 + (-197) = -197$.

Identity Property of Multiplication

If a is an integer, then

$$a(1) = 1(a) = a.$$

We call one the identity element for multiplication. The product of any integer and one is exactly the same integer. For example, $(-573)(1) = (1)(-573) = -573$.

Additive Inverse Property

For every integer a, there exists an integer $-a$, such that

$$a + (-a) = (-a) + a = 0.$$

The integer $-a$ is called the additive inverse of a or the opposite of a. Thus, 6 and -6 are additive inverses and their sum is 0. The additive inverse of 0 is 0.

Multiplication Property of Zero

If a is an integer, then

$$(a)(0) = (0)(a) = 0.$$

The product of zero and any integer is zero. For example, $(-873)(0) = (0)(-873) = 0$.

Multiplicative Property of Negative One

If a is an integer, then

$$(a)(-1) = (-1)(a) = -a.$$

The product of any integer and -1 is the opposite of the integer. For example, $(-1)(48) = (48)(-1) = -48$.

> **Distributive Property**
> If a, b, and c are integers, then
> $$a(b + c) = ab + ac.$$

The distributive property involves both addition and multiplication. We say that **multiplication distributes over addition**. For example, $3(4 + 7) = 3(4) + 3(7)$. Since $b - c = b + (-c)$, it follows that **multiplication also distributes over subtraction**. This could be stated as $a(b - c) = ab - ac$. For example, $7(8 - 2) = 7(8) - 7(2)$.

Let's now consider some examples that use the properties to help with certain types of manipulations.

EXAMPLE 1   Find the sum $(43 + (-24)) + 24$.

Solution   In such a problem it is much more advantageous to group -24 and 24. Thus,

$$(43 + (-24)) + 24 = 43 + ((-24) + 24) \qquad \text{associative property for addition}$$
$$= 43 + 0.$$
$$= 43.$$ ∎

EXAMPLE 2   Find the product $[(-17)(25)](4)$.

Solution   In this problem it would be easier to group 25 and 4. Thus,

$$[(-17)(25)](4) = (-17)[(25)(4)] \qquad \text{associative property for multiplication}$$
$$= (-17)(100)$$
$$= -1700.$$ ∎

EXAMPLE 3   Find the sum $17 + (-24) + (-31) + 19 + (-14) + 29 + 43$.

Solution   Certainly we could add in the order that the numbers appear. However, since addition is *commutative* and *associative* we could change the order and group any convenient way. For example, we could add all of the positive integers and add all of the negative integers and then add these two results. It might be convenient to use the vertical format as follows.

17		
19	-24	
29	-31	108
43	-14	-69
108	-69	39

∎

For a problem such as Example 3 it might be advisable to first work out the problem by adding in the order that the numbers appear and then to use the rearranging and regrouping idea as a check. Don't forget the link between addition and subtraction. A problem such as $18 - 43 + 52 - 17 - 23$ can be changed to $18 + (-43) + 52 + (-17) + (-23)$.

EXAMPLE 4   Simplify $(-75)(-4 + 100)$.

Solution   For such a problem it might be convenient to apply the *distributive property* and then to simplify.

$$
\begin{aligned}
(-75)(-4 + 100) &= (-75)(-4) + (-75)(100) \\
&= 300 + (-7500) \\
&= -7200
\end{aligned}
$$

∎

EXAMPLE 5   Simplify $19(-26 + 25)$.

Solution   For this problem we are better off *not* to apply the distributive property, but simply to add the numbers inside the parentheses first and then to find the indicated product. Thus,

$$19(-26 + 25) = 19(-1) = -19.$$

∎

EXAMPLE 6   Simplify $27(104) + 27(-4)$.

Solution   Keep in mind that the *distributive property* allows us to change from the form $a(b + c)$ to $ab + ac$ or from $ab + ac$ to $a(b + c)$. In this problem we want to use the latter change. Thus,

$$
\begin{aligned}
27(104) + 27(-4) &= 27(104 + (-4)) \\
&= 27(100) = 2700.
\end{aligned}
$$

∎

Examples 4, 5, and 6 demonstrate an important issue. Sometimes the form $a(b + c)$ is most convenient, but at other times the form $ab + ac$ is better. A suggestion in regard to this issue, as well as to the use of the other properties, is *think first*, and decide whether or not the properties can be used to make the manipulations easier.

Combining Similar Terms

Algebraic expressions such as

$$3x, \quad 5y, \quad 7xy, \quad -4abc, \quad \text{and} \quad z$$

are called **terms**. A term is an indicated product and may have any number of factors.

We call the variables in a term **literal factors** and we call the numerical factor the **numerical coefficient**. Thus, in $7xy$, the x and y are literal factors and 7 is the numerical coefficient. The numerical coefficient of the term, $-4abc$, is -4. Since $z = 1(z)$, the numerical coefficient of the term, z, is 1. Terms having the same literal factors are called **like terms** or **similar terms**. Some examples of similar terms are

$3x$ and $9x$; $14abc$ and $29abc$;

$7xy$ and $-15xy$; $4z, 9z$, and $-14z$.

We can simplify algebraic expressions containing similar terms by using a form of the distributive property. Consider the following examples.

$$3x + 5x = (3 + 5)x$$
$$= 8x$$

$$-9xy + 7xy = (-9 + 7)xy$$
$$= -2xy$$

$$18abc - 27abc = (18 - 27)abc$$
$$= (18 + (-27))abc$$
$$= -9abc$$

$$4x + x = (4 + 1)x \qquad \text{Don't forget that } x = 1(x).$$
$$= 5x$$

More complicated expressions might first require some rearranging of terms by using the commutative property.

$$7x + 3y + 9x + 5y = 7x + 9x + 3y + 5y$$
$$= (7 + 9)x + (3 + 5)y$$
$$= 16x + 8y$$

$$9a - 4 - 13a + 6 = 9a + (-4) + (-13a) + 6$$
$$= 9a + (-13a) + (-4) + 6$$
$$= (9 + (-13))a + 2$$
$$= -4a + 2$$

As you become more adept at handling the various simplifying steps, you may want to do the steps mentally and thereby go directly from the given expression to the simplified form as follows.

$$19x - 14y + 12x + 16y = 31x + 2y,$$

$$17ab + 13c - 19ab - 30c = -2ab - 17c,$$

$$9x + 5 - 11x + 4 + x - 6 = -x + 3$$

Simplifying some algebraic expressions requires repeated applications of the distributive property as the next examples demonstrate.

$$5(x - 2) + 3(x + 4) = 5(x) - 5(2) + 3(x) + 3(4)$$
$$= 5x - 10 + 3x + 12$$
$$= 5x + 3x - 10 + 12$$
$$= 8x + 2$$

$$-7(y + 1) - 4(y - 3) = -7(y) - 7(1) - 4(y) - 4(-3)$$
$$= -7y - 7 - 4y + 12 \quad \text{Be careful with this sign.}$$
$$= -7y - 4y - 7 + 12$$
$$= -11y + 5$$

$$5(x + 2) - (x + 3) = 5(x + 2) - 1(x + 3) \quad \text{Remember } -a = -1a.$$
$$= 5(x) + 5(2) - 1(x) - 1(3)$$
$$= 5x + 10 - x - 3$$
$$= 5x - x + 10 - 3$$
$$= 4x + 7$$

After you are sure of each step, you can use a more simplified format.

$$5(a + 4) - 7(a - 2) = 5a + 20 - 7a + 14$$
$$= -2a + 34$$

$$9(z - 7) + 11(z + 6) = 9z - 63 + 11z + 66$$
$$= 20z + 3$$

$$-(x - 2) + (x + 6) = -x + 2 + x + 6$$
$$= 8$$

Back to Evaluating Algebraic Expressions

To simplify by combining similar terms aids in the process of evaluating some algebraic expressions. The last examples of this section illustrate this idea.

EXAMPLE 7   Evaluate $8x - 2y + 3x + 5y$ for $x = 3$ and $y = -4$.

Solution   Let's first simplify the given expression.

$$8x - 2y + 3x + 5y = 11x + 3y$$

Now, we can evaluate for $x = 3$ and $y = -4$.

$$11x + 3y = 11(3) + 3(-4)$$
$$= 33 + (-12)$$
$$= 21 \qquad \blacksquare$$

EXAMPLE 8   Evaluate $2ab + 5c - 6ab + 12c$ for $a = 2$, $b = -3$, and $c = 7$.

Solution

$$2ab + 5c - 6ab + 12c = -4ab + 17c$$
$$= -4(2)(-3) + 17(7) \quad \text{when } a = 2, \ b = -3,$$
$$\text{and } c = 7$$
$$= 24 + 119$$
$$= 143 \qquad\qquad \blacksquare$$

EXAMPLE 9   Evaluate $8(x - 4) + 7(x + 3)$ for $x = 6$.

Solution

$$8(x - 4) + 7(x + 3) = 8x - 32 + 7x + 21$$
$$= 15x - 11$$
$$= 15(6) - 11 \quad \text{when } x = 6$$
$$= 79 \qquad\qquad \blacksquare$$

Problem Set 1.5

For Problems 1–12, state the property that justifies each statement. For example, $3 + (-4) = (-4) + 3$ because of the commutative property for addition.

1. $3(7 + 8) = 3(7) + 3(8)$
2. $(-9)(17) = 17(-9)$
3. $-2 + (5 + 7) = (-2 + 5) + 7$
4. $-19 + 0 = -19$
5. $143(-7) = -7(143)$
6. $5(9 + (-4)) = 5(9) + 5(-4)$
7. $-119 + 119 = 0$
8. $-4 + (6 + 9) = (-4 + 6) + 9$
9. $-56 + 0 = -56$
10. $5 + (-12) = -12 + 5$
11. $[5(-8)]4 = 5[-8(4)]$
12. $[6(-4)]8 = 6[-4(8)]$

For Problems 13–30, simplify each numerical expression. Don't forget to take advantage of the properties if they can be used to simplify the computation.

13. $(-18 + 56) + 18$
14. $-72 + [72 + (-14)]$
15. $36 - 48 - 22 + 41$
16. $-24 + 18 + 19 - 30$
17. $(25)(-18)(-4)$
18. $(2)(-71)(50)$
19. $(4)(-16)(-9)(-25)$
20. $(-2)(18)(-12)(-5)$
21. $37(-42 - 58)$
22. $-46(-73 - 27)$
23. $59(36) + 59(64)$
24. $-49(72) - 49(28)$
25. $15(-14) + 16(-8)$
26. $-9(14) - 7(-16)$
27. $17 + (-18) - 19 - 14 + 13 - 17$
28. $-16 - 14 + 18 + 21 + 14 - 17$

29. $-21 + 22 - 23 + 27 + 21 - 19$

30. $24 - 26 - 29 + 26 + 18 + 29 - 17 - 10$

For Problems 31–62, simplify each algebraic expression by combining similar terms.

31. $9x - 14x$ **32.** $12x - 14x + x$ **33.** $4m + m - 8m$

34. $-6m - m + 17m$ **35.** $-9y + 5y - 7y$ **36.** $14y - 17y - 19y$

37. $4x - 3y - 7x + y$ **38.** $9x + 5y - 4x - 8y$

39. $-7a - 7b - 9a + 3b - 1$ **40.** $-12a + 14b - 3a - 9b$

41. $6xy - x - 13xy + 4x$ **42.** $-7xy - 2x - xy + x$

43. $5x - 4 + 7x - 2x + 9$ **44.** $8x + 9 + 14x - 3x - 14$

45. $-2xy + 12 + 8xy - 16$ **46.** $14xy - 7 - 19xy - 6$

47. $-2a + 3b - 7b - b + 5a - 9a$

48. $-9a - a + 6b - 3a - 4b - b + a$

49. $13ab + 2a - 7a - 9ab + ab - 6a$

50. $-ab - a + 4ab + 7ab - 3a - 11ab$

51. $3(x + 2) + 5(x + 6)$ **52.** $7(x + 8) + 9(x + 1)$

53. $5(x - 4) + 6(x + 8)$ **54.** $-3(x + 2) - 4(x - 10)$

55. $9(x + 4) - (x - 8)$ **56.** $-(x - 6) + 5(x - 9)$

57. $3(a - 1) - 2(a - 6) + 4(a + 5)$

58. $-4(a + 2) + 6(a + 8) - 3(a - 6)$

59. $-2(m + 3) - 3(m - 1) + 8(m + 4)$

60. $5(m - 10) + 6(m - 11) - 9(m - 12)$

61. $(y + 3) - (y - 2) - (y + 6) - 7(y - 1)$

62. $-(y - 2) - (y + 4) - (y + 7) - 2(y + 3)$

For Problems 63–80, simplify each algebraic expression and then evaluate the resulting expression for the given values for the variables.

63. $3x + 5y + 4x - 2y$ for $x = -2$ and $y = 3$

64. $5x - 7y - 9x - 3y$ for $x = -1$ and $y = -4$

65. $5(x - 2) + 8(x + 6)$ for $x = -6$

66. $4(x - 6) + 9(x + 2)$ for $x = 7$

67. $8(x + 4) - 10(x - 3)$ for $x = -5$

68. $-(n + 2) - 3(n - 6)$ for $n = 10$

69. $(x - 6) - (x + 12)$ for $x = -3$

70. $(x + 12) - (x - 14)$ for $x = -11$

71. $2(x + y) - 3(x - y)$ for $x = -2$ and $y = 7$

72. $5(x - y) - 9(x + y)$ for $x = 4$ and $y = -4$

73. $2xy + 6 + 7xy - 8$ for $x = 2$ and $y = -4$

74. $4xy - 5 - 8xy + 9$ for $x = -3$ and $y = -3$

75. $5x - 9xy + 3x + 2xy$ for $x = 12$ and $y = -1$

76. $-9x + xy - 4xy - x$ for $x = 10$ and $y = -11$

77. $(a - b) - (a + b)$ for $a = 19$ and $b = -17$

78. $(a + b) - (a - b)$ for $a = -16$ and $b = 14$
79. $-3x + 7x + 4x - 2x - x$ for $x = -13$
80. $5x - 6x + x - 7x - x$ $2x$ for $x = -15$

Thoughts into Words

81. Explain the difference between a numerical expression and an algebraic expression.
82. State in your own words the associative property for addition.
83. State in your own words the distributive property.
84. Is $2 \cdot 3 \cdot 5 \cdot 7 \cdot 11 + 7$ a prime or composite number? Defend your answer.

Miscellaneous Problems

 85. Use your calculator to evaluate each of the following algebraic expressions for the given values of the variables.

(a) $3xy - 9x + 8y$ for $x = -12$ and $y = 16$
(b) $-17x + 12xy - 14y$ for $x = 19$ and $y = -13$
(c) $15x - 17y + 21x + 32y$ for $x = -21$ and $y = -32$
(d) $-26y - 32x - 15y - 19x$ for $x = 28$ and $y = 17$
(e) $(x - y) - (x + y)$ for $x = 41$ and $y = -38$
(f) $-14(x - 16) + 43(x + 21)$ for $x = -52$
(g) $23(x + 14) - 38(x - 54)$ for $x = 46$
(h) $-62(a - 18) - 59(a - 13)$ for $a = -67$
(i) $-14(x - y) - 19(x + y)$ for $x = 64$ and $y = 68$
(j) $25(x + y) - 76(y - x)$ for $x = 29$ and $y = -44$

Chapter 1 Summary

(1.1) To **simplify a numerical expression**, the operations should be performed in the following order.

1. Perform the operations inside the symbols of inclusion (parentheses and brackets) and above and below each fraction bar. Start with the innermost inclusion symbol.

2. Perform all multiplications and divisions in the order that they appear from left to right.

3. Perform all additions and subtractions in the order that they appear from left to right.

To **evaluate an algebraic expression**, substitute the given values for the variables into the algebraic expression and simplify the resulting numerical expression.

(1.2) A **prime number** is a whole number greater than 1 that has no factors (divisors) other than itself and 1. Whole numbers greater than 1 that are not prime numbers are called **composite numbers**. Every composite number has one and only one prime factorization.

The **greatest common factor** of 6 and 8 is 2, which means that 2 is the largest whole number divisor of both 6 and 8.

The **least common multiple** of 6 and 8 is 24, which means that 24 is the smallest nonzero multiple of both 6 and 8.

(1.3) The number line is a convenient pictorial aid for interpreting **addition of integers**.

Subtraction of integers is defined in terms of addition: $a - b$ means $a + (-b)$.

(1.4) To **multiply integers** we must remember that *the product of two positives or two negatives is positive and the product of a positive and a negative (either order) is negative.*

To **divide integers** we must remember that *the quotient of two positives or two negatives is positive and the quotient of a positive and a negative (or a negative and a positive) is negative.*

(1.5) The following basic properties help with numerical manipulations and serve as a basis for algebraic computations.

Commutative Properties $a + b = b + a$
$$ab = ba$$

Associative Properties $(a + b) + c = a + (b + c)$
$$(ab)c = a(bc)$$

Identity Properties $a + 0 = 0 + a = a$
$$a(1) = 1(a) = a$$

Additive Inverse Property $a + (-a) = (-a) + a = 0$

Multiplication Property of Zero $a(0) = 0(a) = 0$

Multiplication Property of Negative One $-1(a) = a(-1) = -a$

Distributive Properties $a(b + c) = ab + ac$
$$a(b - c) = ab - ac$$

Chapter 1 Review Problem Set

In Problems 1–10, do the indicated operations.

1. $7 + (-10)$
2. $(-12) + (-13)$
3. $8 - 13$
4. $-6 - 9$
5. $-12 - (-11)$
6. $-17 - (-19)$
7. $(13)(-12)$
8. $(-14)(-18)$
9. $(-72) \div (-12)$
10. $117 \div (-9)$

For Problems 11–15, classify each of the numbers as *prime* or *composite*.

11. 73   **12.** 87   **13.** 63   **14.** 81   **15.** 91

For Problems 16–20, express each of the numbers as the product of prime factors.

16. 24   **17.** 63   **18.** 57   **19.** 64   **20.** 84

21. Find the greatest common factor of 36 and 54.

22. Find the greatest common factor of 48, 60, and 84.

23. Find the least common multiple of 18 and 20.

24. Find the least common multiple of 15, 27, and 35.

For Problems 25–38, simplify each of the numerical expressions.

25. $(19 + 56) + (-9)$

26. $43 - 62 + 12$

27. $8 + (-9) + (-16) + (-14) + 17 + 12$

28. $19 - 23 - 14 + 21 + 14 - 13$

29. $3(-4) - 6$

30. $(-5)(-4) - 8$

31. $(5)(-2) + (6)(-4)$

32. $(-6)(8) + (-7)(-3)$

33. $(-6)(3) - (-4)(-5)$

34. $(-7)(9) - (6)(5)$

35. $\dfrac{4(-7) - (3)(-2)}{-11}$

36. $\dfrac{(-4)(9) + (5)(-3)}{1 - 18}$

37. $3 - 2[4(-3 - 1)]$

38. $-6 - [3(-4 - 7)]$

For Problems 39–50, simplify each algebraic expression by combining similar terms.

39. $12x + 3x - 7x$

40. $9y + 3 - 14y - 12$

41. $8x + 5y - 13x - y$

42. $9a + 11b + 4a - 17b$

43. $3ab - 4ab - 2a$

44. $5xy - 9xy + xy - y$

45. $3(x + 6) + 7(x + 8)$

46. $5(x - 4) - 3(x - 9)$

47. $-3(x - 2) - 4(x + 6)$

48. $-2x - 3(x - 4) + 2x$

49. $2(a - 1) - a - 3(a - 2)$

50. $-(a - 1) + 3(a - 2) - 4a + 1$

For Problems 51–64, evaluate each of the algebraic expressions for the given values of the variables.

51. $5x + 8y$ for $x = -7$ and $y = -3$

52. $7x - 9y$ for $x = -3$ and $y = 4$

53. $\dfrac{-5x - 2y}{-2x - 7}$ for $x = 6$ and $y = 4$

54. $\dfrac{-3x + 4y}{3x}$ for $x = -4$ and $y = -6$

55. $-2a + \dfrac{a - b}{a - 2}$ for $a = -5$ and $b = 9$

56. $\dfrac{2a + b}{b + 6} - 3b$ for $a = 3$ and $b = -4$

57. $5a + 6b - 7a - 2b$ for $a = -1$ and $b = 5$

58. $3x + 7y - 5x + y$ for $x = -4$ and $y = 3$

59. $2xy + 6 + 5xy - 8$ for $x = -1$ and $y = 1$

60. $7(x + 6) - 9(x + 1)$ for $x = -2$

61. $-3(x - 4) - 2(x + 8)$ for $x = 7$

62. $2(x - 1) - (x + 2) + 3(x - 4)$ for $x = -4$

63. $(a - b) - (a + b) - b$ for $a = -1$ and $b = -3$

64. $2ab - 3(a - b) + b + a$ for $a = 2$ and $b = -5$

Chapter 2

The Real Numbers

The suggestion at the beginning of Chapter 1 is worth repeating. Use your knowledge of arithmetic to help your study of algebra. A thorough understanding of arithmetic concepts provides a solid basis for algebraic ideas.

38

2.1
Rational Numbers: Multiplication and Division

Any number that can be written in the form $\dfrac{a}{b}$, where a and b are integers and b is not zero, is called a **rational number**. (The form $\dfrac{a}{b}$ is called a fraction or sometimes a common fraction.) The following are examples of rational numbers.

$$\frac{1}{2}, \quad \frac{7}{9}, \quad \frac{15}{7}, \quad \frac{-3}{4}, \quad \frac{5}{-7}, \quad \frac{-11}{-13}$$

All integers are rational numbers because every integer can be expressed as the indicated quotient of two integers. Some examples are as follows.

$$6 = \frac{6}{1} = \frac{12}{2} = \frac{18}{3}, \text{ etc.,}$$

$$27 = \frac{27}{1} = \frac{54}{2} = \frac{81}{3}, \text{ etc.,}$$

$$0 = \frac{0}{1} = \frac{0}{2} = \frac{0}{3}, \text{ etc.}$$

Our work with division involving negative integers in Chapter 1 helps with the next three examples.

$$-4 = \frac{-4}{1} = \frac{-8}{2} = \frac{-12}{3}, \text{ etc.,}$$

$$-6 = \frac{6}{-1} = \frac{12}{-2} = \frac{18}{-3}, \text{ etc.,}$$

$$10 = \frac{10}{1} = \frac{-10}{-1} = \frac{-20}{-2}, \text{ etc.}$$

Observe the following general properties.

PROPERTY 2.1

$$\frac{-a}{b} = \frac{a}{-b} = -\frac{a}{b} \quad \text{and} \quad \frac{-a}{-b} = \frac{a}{b}.$$

Therefore, a rational number such as $\dfrac{-2}{3}$ can also be written as $\dfrac{2}{-3}$ or $-\dfrac{2}{3}$. (However, we seldom express rational numbers with negative denominators.)

Multiplying Rational Numbers

We define multiplication of rational numbers in common fractional form as follows.

DEFINITION 2.1

> If a, b, c, and d are integers with b and d not equal to zero, then
>
> $$\frac{a}{b} \cdot \frac{c}{d} = \frac{a \cdot c}{b \cdot d}.$$

To multiply rational numbers in common fractional form we simply *multiply numerators and multiply denominators*. Furthermore, we see from the definition that the rational numbers are commutative and associative with respect to multiplication. We are free to rearrange and regroup factors as we do with integers.

The following examples illustrate the definition for multiplying rational numbers.

$$\frac{1}{3} \cdot \frac{2}{5} = \frac{1 \cdot 2}{3 \cdot 5} = \frac{2}{15}, \qquad \frac{3}{4} \cdot \frac{5}{7} = \frac{3 \cdot 5}{4 \cdot 7} = \frac{15}{28},$$

$$\frac{-2}{3} \cdot \frac{7}{9} = \frac{-2 \cdot 7}{3 \cdot 9} = \frac{-14}{27} \quad \text{or} \quad -\frac{14}{27},$$

$$\frac{1}{5} \cdot \frac{9}{-11} = \frac{1 \cdot 9}{5(-11)} = \frac{9}{-55} \quad \text{or} \quad -\frac{9}{55},$$

$$-\frac{3}{4} \cdot \frac{7}{13} = \frac{-3}{4} \cdot \frac{7}{13} = \frac{-3 \cdot 7}{4 \cdot 13} = \frac{-21}{52} \quad \text{or} \quad -\frac{21}{52},$$

$$\frac{3}{5} \cdot \frac{5}{3} = \frac{3 \cdot 5}{5 \cdot 3} = \frac{15}{15} = 1$$

The last example is a very special case. **If the product of two numbers is 1, the numbers are said to be reciprocals of each other.**

The following property is used frequently when working with rational numbers. It is often called the **fundamental principle of fractions**.

PROPERTY 2.2

> If b and k are nonzero integers and a is any integer, then
>
> $$\frac{a \cdot k}{b \cdot k} = \frac{a}{b}.$$

This property provides the basis for what is often called *reducing fractions to lowest terms or expressing fractions in simplest or reduced form*. Let's apply the property to a few examples.

EXAMPLE 1   Reduce $\dfrac{12}{18}$ to lowest terms.

Solution $$\frac{12}{18} = \frac{2 \cdot 6}{3 \cdot 6} = \frac{2}{3}$$ ∎

EXAMPLE 2   Change $\dfrac{14}{35}$ to simplest form.

Solution $$\frac{14}{35} = \frac{2 \cdot \not{7}}{5 \cdot \not{7}} = \frac{2}{5}$$ A common factor of 7 has been divided out of both the numerator and the denominator.   ∎

EXAMPLE 3   Express $\dfrac{-24}{32}$ in reduced form.

Solution $$\frac{-24}{32} = -\frac{3 \cdot 8}{4 \cdot 8} = -\frac{3}{4} \cdot \frac{8}{8} = -\frac{3}{4} \cdot 1 = -\frac{3}{4}$$ The multiplication property of 1 is being used.   ∎

EXAMPLE 4   Reduce $-\dfrac{72}{90}$.

Solution $$-\frac{72}{90} = -\frac{\not{2} \cdot 2 \cdot 2 \cdot \not{3} \cdot \not{3}}{\not{2} \cdot \not{3} \cdot \not{3} \cdot 5} = -\frac{4}{5}$$ The prime factored forms of the numerator and denominator may be used to help recognize common factors.   ∎

Let's pause for a moment and look back over Examples 1 through 4. The approaches used in Examples 1, 2, and 3 are very similar; however, in Example 4 we prime factored both the numerator and denominator first and then looked for common factors. It is helpful to have more than one approach at your disposal.

The fractions may contain variables in the numerator or denominator (or both), but this creates no great difficulty. Our thought processes remain the same, as these next examples illustrate. Variables appearing in denominators represent **nonzero** integers.

EXAMPLE 5   Reduce $\dfrac{9x}{17x}$.

Solution $$\frac{9x}{17x} = \frac{9 \cdot \not{x}}{17 \cdot \not{x}} = \frac{9}{17}$$ ∎

EXAMPLE 6   Simplify $\dfrac{8x}{36y}$.

Solution $\dfrac{8x}{36y} = \dfrac{\cancel{2} \cdot \cancel{2} \cdot 2 \cdot x}{\cancel{2} \cdot \cancel{2} \cdot 3 \cdot 3 \cdot y} = \dfrac{2x}{9y}$ ∎

EXAMPLE 7   Express $\dfrac{-9xy}{30y}$ in reduced form.

Solution $\dfrac{-9xy}{30y} = -\dfrac{3 \cdot \cancel{3} \cdot x \cdot \cancel{y}}{2 \cdot \cancel{3} \cdot 5 \cdot \cancel{y}} = -\dfrac{3x}{10}$ ∎

EXAMPLE 8   Reduce $\dfrac{-7abc}{-9ac}$.

Solution $\dfrac{-7abc}{-9ac} = \dfrac{7\cancel{a}b\cancel{c}}{9\cancel{a}\cancel{c}} = \dfrac{7b}{9}$ ∎

We are now ready to consider multiplication problems with the agreement that *the final answer should be expressed in reduced form.* Study the following examples carefully because we have used different formats to handle such problems.

EXAMPLE 9   Multiply $\dfrac{7}{9} \cdot \dfrac{5}{14}$.

Solution $\dfrac{7}{9} \cdot \dfrac{5}{14} = \dfrac{7 \cdot 5}{9 \cdot 14} = \dfrac{\cancel{7} \cdot 5}{3 \cdot 3 \cdot 2 \cdot \cancel{7}} = \dfrac{5}{18}$ ∎

EXAMPLE 10   Find the product of $\dfrac{8}{9}$ and $\dfrac{18}{24}$.

Solution $\dfrac{\overset{1}{\cancel{8}}}{\underset{1}{\cancel{9}}} \cdot \dfrac{\overset{2}{\cancel{18}}}{\underset{3}{\cancel{24}}} = \dfrac{2}{3}$ A common factor of 8 has been divided out of 8 and 24, and a common factor of 9 has been divided out of 9 and 18.   ∎

EXAMPLE 11   Multiply $\left(-\dfrac{6}{8}\right)\left(\dfrac{14}{32}\right)$.

Solution $\left(-\dfrac{6}{8}\right)\left(\dfrac{14}{32}\right) = -\dfrac{\overset{3}{\cancel{6}} \cdot \overset{7}{\cancel{14}}}{\underset{4}{\cancel{8}} \cdot \underset{16}{\cancel{32}}} = -\dfrac{21}{64}$ ∎

EXAMPLE 12   Multiply $\left(-\dfrac{9}{4}\right)\left(-\dfrac{14}{15}\right)$.

Solution $\left(-\dfrac{9}{4}\right)\left(-\dfrac{14}{15}\right) = \dfrac{3\cdot 3\cdot 2\cdot 7}{2\cdot 2\cdot 3\cdot 5} = \dfrac{21}{10}$ We need to recognize that the product of two negative numbers is a positive number.   ■

EXAMPLE 13   Multiply $\dfrac{9x}{7y}\cdot\dfrac{14y}{45}$.

Solution $\dfrac{9x}{7y}\cdot\dfrac{14y}{45} = \dfrac{9\cdot x\cdot \overset{2}{14}\cdot y}{7\cdot y\cdot \underset{5}{45}} = \dfrac{2x}{5}$ ■

EXAMPLE 14   Multiply $\dfrac{-6c}{7ab}\cdot\dfrac{14b}{5c}$.

Solution $\dfrac{-6c}{7ab}\cdot\dfrac{14b}{5c} = -\dfrac{2\cdot 3\cdot c\cdot 2\cdot 7\cdot b}{7\cdot a\cdot b\cdot 5\cdot c} = -\dfrac{12}{5a}$ ■

Dividing Rational Numbers

The following example motivates a definition for division of rational numbers in fractional form.

$$\dfrac{\frac{3}{4}}{\frac{2}{3}} = \left(\dfrac{\frac{3}{4}}{\frac{2}{3}}\right)\left(\dfrac{\frac{3}{2}}{\frac{3}{2}}\right) = \dfrac{\left(\frac{3}{4}\right)\left(\frac{3}{2}\right)}{1} = \left(\dfrac{3}{4}\right)\left(\dfrac{3}{2}\right) = \dfrac{9}{8}$$

↑
Notice that this is a form of 1

and $\dfrac{3}{2}$ is the reciprocal of $\dfrac{2}{3}$.

In other words, $\dfrac{3}{4}$ divided by $\dfrac{2}{3}$ is equivalent to $\dfrac{3}{4}$ times $\dfrac{3}{2}$. The following definition for division should seem reasonable.

DEFINITION 2.2

If b, c, and d are nonzero integers and a is any integer, then

$$\dfrac{a}{b} \div \dfrac{c}{d} = \dfrac{a}{b}\cdot\dfrac{d}{c}.$$

Notice that to divide $\dfrac{a}{b}$ by $\dfrac{c}{d}$, we multiply $\dfrac{a}{b}$ times the reciprocal of $\dfrac{c}{d}$, which is $\dfrac{d}{c}$. The following examples demonstrate the important steps of a division problem.

$$\frac{2}{3} \div \frac{1}{2} = \frac{2}{3} \cdot \frac{2}{1} = \frac{4}{3}, \qquad \frac{5}{6} \div \frac{3}{4} = \frac{5}{6} \cdot \frac{4}{3} = \frac{5 \cdot 4}{6 \cdot 3} = \frac{5 \cdot \cancel{2} \cdot 2}{\cancel{2} \cdot 3 \cdot 3} = \frac{10}{9},$$

$$-\frac{9}{12} \div \frac{3}{6} = -\frac{\overset{3}{\cancel{9}}}{\underset{2}{\cancel{12}}} \cdot \frac{\overset{1}{\cancel{6}}}{\underset{1}{\cancel{3}}} = -\frac{3}{2},$$

$$\left(-\frac{27}{56}\right) \div \left(-\frac{33}{72}\right) = \left(-\frac{27}{56}\right)\left(-\frac{72}{33}\right) = \frac{\overset{9}{\cancel{27}} \cdot \overset{9}{\cancel{72}}}{\underset{7}{\cancel{56}} \cdot \underset{11}{\cancel{33}}} = \frac{81}{77},$$

$$\frac{\frac{6}{7}}{\frac{3}{2}} = \frac{\overset{3}{\cancel{6}}}{7} \cdot \frac{1}{\underset{1}{\cancel{2}}} \cdot \frac{3}{7}, \qquad \frac{5x}{7y} \div \frac{10}{28y} = \frac{5x}{7y} \cdot \frac{28y}{10} = \frac{5 \cdot x \cdot \overset{2}{\cancel{28}} \cdot \cancel{y}}{\cancel{7} \cdot \cancel{y} \cdot \underset{\cancel{2}}{\cancel{10}}} = 2x$$

Problem Set 2.1

For Problems 1–24, reduce each fraction to lowest terms.

1. $\dfrac{8}{12}$ **2.** $\dfrac{12}{16}$ **3.** $\dfrac{16}{24}$ **4.** $\dfrac{18}{32}$ **5.** $\dfrac{15}{9}$

6. $\dfrac{48}{36}$ **7.** $\dfrac{-8}{48}$ **8.** $\dfrac{-3}{15}$ **9.** $\dfrac{27}{-36}$ **10.** $\dfrac{9}{-51}$

11. $\dfrac{-54}{-56}$ **12.** $\dfrac{-24}{-80}$ **13.** $\dfrac{24x}{44x}$ **14.** $\dfrac{15y}{25y}$ **15.** $\dfrac{9x}{21y}$

16. $\dfrac{4y}{30x}$ **17.** $\dfrac{14xy}{35y}$ **18.** $\dfrac{55xy}{77x}$ **19.** $\dfrac{-20ab}{52bc}$ **20.** $\dfrac{-23ac}{41c}$

21. $\dfrac{-56yz}{-49xy}$ **22.** $\dfrac{-21xy}{-14ab}$ **23.** $\dfrac{65abc}{91ac}$ **24.** $\dfrac{68xyz}{85yz}$

For Problems 25–58, multiply or divide as indicated and express answers in reduced form.

25. $\dfrac{3}{4} \cdot \dfrac{5}{7}$ **26.** $\dfrac{4}{5} \cdot \dfrac{3}{11}$ **27.** $\dfrac{2}{7} \div \dfrac{3}{5}$ **28.** $\dfrac{5}{6} \div \dfrac{11}{13}$

29. $\dfrac{3}{8} \cdot \dfrac{12}{15}$ **30.** $\dfrac{4}{9} \cdot \dfrac{3}{2}$ **31.** $\dfrac{-6}{13} \cdot \dfrac{26}{9}$ **32.** $\dfrac{3}{4} \cdot \dfrac{-14}{12}$

33. $\dfrac{7}{9} \div \dfrac{5}{9}$ **34.** $\dfrac{3}{11} \div \dfrac{7}{11}$ **35.** $\dfrac{1}{4} \div \dfrac{-5}{6}$ **36.** $\dfrac{7}{8} \div \dfrac{14}{-16}$

37. $\left(-\dfrac{8}{10}\right)\left(-\dfrac{10}{32}\right)$

38. $\left(-\dfrac{6}{7}\right)\left(-\dfrac{21}{24}\right)$

39. $-9 \div \dfrac{1}{3}$

40. $-10 \div \dfrac{1}{4}$

41. $\dfrac{5x}{9y} \cdot \dfrac{7y}{3x}$

42. $\dfrac{4a}{11b} \cdot \dfrac{6b}{7a}$

43. $\dfrac{6a}{14b} \cdot \dfrac{16b}{18a}$

44. $\dfrac{5y}{8x} \cdot \dfrac{14z}{15y}$

45. $\dfrac{10x}{-9y} \cdot \dfrac{15}{20x}$

46. $\dfrac{3x}{4y} \cdot \dfrac{-8w}{9z}$

47. $ab \cdot \dfrac{2}{b}$

48. $3xy \cdot \dfrac{4}{x}$

49. $\left(-\dfrac{7x}{12y}\right)\left(-\dfrac{24y}{35x}\right)$

50. $\left(-\dfrac{10a}{15b}\right)\left(-\dfrac{45b}{65a}\right)$

51. $\dfrac{3}{x} \div \dfrac{6}{y}$

52. $\dfrac{6}{x} \div \dfrac{14}{y}$

53. $\dfrac{5x}{9y} \div \dfrac{13x}{36y}$

54. $\dfrac{3x}{5y} \div \dfrac{7x}{10y}$

55. $\dfrac{-7}{x} \div \dfrac{9}{x}$

56. $\dfrac{8}{y} \div \dfrac{28}{-y}$

57. $\dfrac{-4}{n} \div \dfrac{-18}{n}$

58. $\dfrac{-34}{n} \div \dfrac{-51}{n}$

For Problems 59–74, perform the operations as indicated and express answers in lowest terms.

59. $\dfrac{3}{4} \cdot \dfrac{8}{9} \cdot \dfrac{12}{20}$

60. $\dfrac{5}{6} \cdot \dfrac{9}{10} \cdot \dfrac{8}{7}$

61. $\left(-\dfrac{3}{8}\right)\left(\dfrac{13}{14}\right)\left(-\dfrac{12}{9}\right)$

62. $\left(-\dfrac{7}{9}\right)\left(\dfrac{5}{11}\right)\left(-\dfrac{18}{14}\right)$

63. $\left(\dfrac{3x}{4y}\right)\left(\dfrac{8}{9x}\right)\left(\dfrac{12y}{5}\right)$

64. $\left(\dfrac{2x}{3y}\right)\left(\dfrac{5y}{x}\right)\left(\dfrac{9}{4x}\right)$

65. $\left(-\dfrac{2}{3}\right)\left(\dfrac{3}{4}\right) \div \dfrac{1}{8}$

66. $\dfrac{3}{4} \cdot \dfrac{4}{5} \div \dfrac{1}{6}$

67. $\dfrac{5}{7} \div \left(-\dfrac{5}{6}\right)\left(-\dfrac{6}{7}\right)$

68. $\left(-\dfrac{3}{8}\right) \div \left(-\dfrac{4}{5}\right)\left(\dfrac{1}{2}\right)$

69. $\left(-\dfrac{6}{7}\right) \div \left(\dfrac{5}{7}\right)\left(-\dfrac{5}{6}\right)$

70. $\left(-\dfrac{4}{3}\right) \div \left(\dfrac{4}{5}\right)\left(\dfrac{3}{5}\right)$

71. $\left(\dfrac{4}{9}\right)\left(-\dfrac{9}{8}\right) \div \left(-\dfrac{3}{4}\right)$

72. $\left(-\dfrac{7}{8}\right)\left(\dfrac{4}{7}\right) \div \left(-\dfrac{3}{2}\right)$

73. $\left(\dfrac{5}{2}\right)\left(\dfrac{2}{3}\right) \div \left(-\dfrac{1}{4}\right) \div (-3)$

74. $\dfrac{1}{3} \div \left(\dfrac{3}{4}\right)\left(\dfrac{1}{2}\right) \div 2$

Miscellaneous Problems

75. The division problem $35 \div 7$ can be interpreted as "how many 7s are there in 35?" Likewise, a division problem such as $3 \div \dfrac{1}{2}$ can be interpreted as "how many one-halves in 3?" Use this "how many" interpretation to do the following division problems.

(a) $4 \div \dfrac{1}{2}$

(b) $3 \div \dfrac{1}{4}$

(c) $5 \div \dfrac{1}{8}$

(d) $6 \div \dfrac{1}{7}$

(e) $\dfrac{5}{6} \div \dfrac{1}{6}$

(f) $\dfrac{7}{8} \div \dfrac{1}{8}$

76. Estimation is important in mathematics. In each of the following, estimate whether the answer is larger than 1 or smaller than 1 by using the "how many" idea from Problem 75.

(a) $\dfrac{3}{4} \div \dfrac{1}{2}$ (b) $1 \div \dfrac{7}{8}$ (c) $\dfrac{1}{2} \div \dfrac{3}{4}$

(d) $\dfrac{8}{7} \div \dfrac{7}{8}$ (e) $\dfrac{2}{3} \div \dfrac{1}{4}$ (f) $\dfrac{3}{5} \div \dfrac{3}{4}$

77. Reduce each of the following to lowest terms. You may find it helpful to return to Problems 74–83 in Problem Set 1.2.

(a) $\dfrac{99}{117}$ (b) $\dfrac{175}{225}$ (c) $\dfrac{-111}{123}$

(d) $\dfrac{-234}{270}$ (e) $\dfrac{270}{495}$ (f) $\dfrac{324}{459}$

(g) $\dfrac{91}{143}$ (h) $\dfrac{187}{221}$

 78. If you have a calculator that handles common fractions, use it to help reduce each of the following fractions.

(a) $\dfrac{703}{798}$ (b) $\dfrac{325}{403}$ (c) $\dfrac{424}{477}$

(d) $\dfrac{141}{188}$ (e) $\dfrac{426xy}{781x}$ (f) $\dfrac{819xz}{1071xy}$

 79. Be sure that you can do Problems 25–40 and 65–74 with your calculator.

Addition and Subtraction of Rational Numbers

Suppose that it is one-fifth of a mile between your dorm and the union and two-fifths of a mile between the union and the library along a straight line as indicated in Figure 2.1. The total distance between your dorm and the library is three-fifths of a mile and we write $\dfrac{1}{5} + \dfrac{2}{5} = \dfrac{3}{5}$.

Figure 2.1

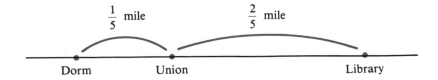

A pizza is cut into seven equal pieces and you eat two of the pieces. How much of the pizza remains? We represent the whole pizza by $\frac{7}{7}$ and then conclude that $\frac{7}{7} - \frac{2}{7} = \frac{5}{7}$ of the pizza remains.

These examples motivate the following definition for addition and subtraction of rational numbers in $\frac{a}{b}$ form.

DEFINITION 2.3

> If a, b, and c are integers and b is not zero, then
>
> $$\frac{a}{b} + \frac{c}{b} = \frac{a+c}{b}, \qquad \text{addition}$$
>
> $$\frac{a}{b} - \frac{c}{b} = \frac{a-c}{b}. \qquad \text{subtraction}$$

We say that *rational numbers **with common denominators** can be added or subtracted by adding or subtracting the numerators and placing the results over the common denominator.* Consider the following examples.

$$\frac{3}{7} + \frac{2}{7} = \frac{3+2}{7} = \frac{5}{7},$$

$$\frac{7}{8} - \frac{2}{8} = \frac{7-2}{8} = \frac{5}{8},$$

$$\frac{2}{6} + \frac{1}{6} = \frac{2+1}{6} = \frac{3}{6} = \frac{1}{2}, \qquad \text{We agree to reduce final answer.}$$

$$\frac{3}{11} - \frac{5}{11} = \frac{3-5}{11} = \frac{-2}{11} \quad \left(\text{or} -\frac{2}{11} \right),$$

$$\frac{5}{x} + \frac{7}{x} = \frac{5+7}{x} = \frac{12}{x},$$

$$\frac{9}{y} - \frac{3}{y} = \frac{9-3}{y} = \frac{6}{y}.$$

In the last two examples, the variables x and y cannot be equal to zero in order to exclude division by zero. It is always necessary to restrict denominators to nonzero values although we will not take the time or space to list such restrictions for every problem.

How do we add or subtract if the fractions do not have a common denominator? We use the fundamental principle of fractions, $\frac{a}{b} = \frac{a \cdot k}{b \cdot k}$, and obtain equivalent fractions that have a common denominator. **Equivalent fractions** are fractions that name the same number. Consider the following example, which shows the details.

EXAMPLE 1   Add $\dfrac{1}{2} + \dfrac{1}{3}$.

Solution

$$\frac{1}{2} = \frac{1 \cdot 3}{2 \cdot 3} = \frac{3}{6} \qquad \frac{1}{2} \text{ and } \frac{3}{6} \text{ are equivalent fractions naming the same number.}$$

$$\frac{1}{3} = \frac{1 \cdot 2}{3 \cdot 2} = \frac{\dfrac{2}{6}}{\dfrac{5}{6}} \qquad \frac{1}{3} \text{ and } \frac{2}{6} \text{ are equivalent fractions naming the same number.}$$

$$\frac{3}{6} + \frac{2}{6} = \frac{3+2}{6} = \frac{5}{6} \qquad \blacksquare$$

Notice that we chose 6 as our common denominator and 6 is the least common multiple of the original denominators 2 and 3. (Recall that the least common multiple is the smallest nonzero whole number divisible by the given numbers.) In general, we use the least common multiple of the denominators of the fractions to be added or subtracted as a **least common denominator** (LCD).

Recall from Section 1.2 that the least common multiple may be found either by inspection or by using the prime factorization forms of the numbers. Let's consider some examples involving these procedures.

EXAMPLE 2   Add $\dfrac{1}{4} + \dfrac{2}{5}$.

Solution   By inspection we see that the LCD is 20. Thus, both fractions can be changed to equivalent fractions having a denominator of 20.

$$\frac{1}{4} + \frac{2}{5} = \frac{1 \cdot 5}{4 \cdot 5} + \frac{2 \cdot 4}{5 \cdot 4} = \frac{5}{20} + \frac{8}{20} = \frac{13}{20}$$

use of fundamental
principle of fractions $\blacksquare$

EXAMPLE 3   Subtract $\dfrac{5}{8} - \dfrac{7}{12}$.

Solution   By inspection the LCD is 24.

$$\frac{5}{8} - \frac{7}{12} = \frac{5 \cdot 3}{8 \cdot 3} - \frac{7 \cdot 2}{12 \cdot 2} = \frac{15}{24} - \frac{14}{24} = \frac{1}{24} \qquad \blacksquare$$

If the LCD is not obvious by inspection, then we can use the technique from Chapter 1 to find the least common multiple. So we proceed as follows.

Step 1. Express each denominator as a product of prime factors.

Step 2. The LCD contains each different prime factor as many times as the *most* times it appears in any one of the factorizations from step 1.

EXAMPLE 4   Add $\dfrac{5}{18} + \dfrac{7}{24}$.

Solution   If we cannot find the LCD by inspection, then we can use the prime factorization forms.

$$\left.\begin{array}{l} 18 = 2 \cdot 3 \cdot 3 \\[4pt] 24 = 2 \cdot 2 \cdot 2 \cdot 3 \end{array}\right\} \longrightarrow \text{LCD} = 2 \cdot 2 \cdot 2 \cdot 3 \cdot 3 = 72$$

$$\frac{5}{18} + \frac{7}{24} = \frac{5 \cdot 4}{18 \cdot 4} + \frac{7 \cdot 3}{24 \cdot 3} = \frac{20}{72} + \frac{21}{72} = \frac{41}{72}$$ ∎

EXAMPLE 5   Subtract $\dfrac{3}{14} - \dfrac{8}{35}$.

Solution

$$\left.\begin{array}{l} 14 = 2 \cdot 7 \\[4pt] 35 = 5 \cdot 7 \end{array}\right\} \longrightarrow \text{LCD} = 2 \cdot 5 \cdot 7 = 70$$

$$\frac{3}{14} - \frac{8}{35} = \frac{3 \cdot 5}{14 \cdot 5} - \frac{8 \cdot 2}{35 \cdot 2} = \frac{15}{70} - \frac{16}{70} = \frac{-1}{70} \left(\text{or } -\frac{1}{70}\right)$$ ∎

EXAMPLE 6   Add $\dfrac{-5}{8} + \dfrac{3}{14}$.

Solution

$$\left.\begin{array}{l} 8 = 2 \cdot 2 \cdot 2 \\[4pt] 14 = 2 \cdot 7 \end{array}\right\} \longrightarrow \text{LCD} = 2 \cdot 2 \cdot 2 \cdot 7 = 56$$

$$\frac{-5}{8} + \frac{3}{14} = \frac{-5 \cdot 7}{8 \cdot 7} + \frac{3 \cdot 4}{14 \cdot 4} = \frac{-35}{56} + \frac{12}{56} = \frac{-23}{56} \left(\text{or } -\frac{23}{56}\right)$$ ∎

EXAMPLE 7   Add $-3 + \dfrac{2}{5}$.

Solution

$$-3 + \frac{2}{5} = \frac{-3 \cdot 5}{1 \cdot 5} + \frac{2}{5} = \frac{-15}{5} + \frac{2}{5} = \frac{-15 + 2}{5} = \frac{-13}{5} \left(\text{or } -\frac{13}{5}\right)$$ ∎

Denominators that contain variables do not complicate the situation very much, as the next examples illustrate.

EXAMPLE 8   Add $\dfrac{2}{x} + \dfrac{3}{y}$.

Solution   By inspection, the LCD is xy.

$$\frac{2}{x} + \frac{3}{y} = \frac{2 \cdot y}{x \cdot y} + \frac{3 \cdot x}{y \cdot x} = \frac{2y}{xy} + \frac{3x}{xy} = \frac{2y + 3x}{xy}$$

commutative property   ∎

EXAMPLE 9   Subtract $\dfrac{3}{8x} - \dfrac{5}{12y}$.

Solution
$$\left.\begin{array}{l} 8x = 2 \cdot 2 \cdot 2 \cdot x \\ 12y = 2 \cdot 2 \cdot 3 \cdot y \end{array}\right\} \longrightarrow \text{LCD} = 2 \cdot 2 \cdot 2 \cdot 3 \cdot x \cdot y = 24xy$$

$$\frac{3}{8x} - \frac{5}{12y} = \frac{3 \cdot 3y}{8x \cdot 3y} - \frac{5 \cdot 2x}{12y \cdot 2x} = \frac{9y}{24xy} - \frac{10x}{24xy} = \frac{9y - 10x}{24xy}$$ ∎

EXAMPLE 10   Add $\dfrac{7}{4a} + \dfrac{-5}{6bc}$.

Solution
$$\left.\begin{array}{l} 4a = 2 \cdot 2 \cdot a \\ 6bc = 2 \cdot 3 \cdot b \cdot c \end{array}\right\} \longrightarrow \text{LCD} = 2 \cdot 2 \cdot 3 \cdot a \cdot b \cdot c = 12abc$$

$$\frac{7}{4a} + \frac{-5}{6bc} = \frac{7 \cdot 3bc}{4a \cdot 3bc} + \frac{-5 \cdot 2a}{6bc \cdot 2a} = \frac{21bc}{12abc} + \frac{-10a}{12abc} = \frac{21bc - 10a}{12abc}$$ ∎

Simplifying Numerical Expressions

Let's now consider simplifying numerical expressions that contain rational numbers. As with integers, multiplications and divisions are done first and then the additions and subtractions are performed. In these next examples only the major steps are shown, so be sure that you can fill in all of the details.

EXAMPLE 11   Simplify $\dfrac{3}{4} + \dfrac{2}{3} \cdot \dfrac{3}{5} - \dfrac{1}{2} \cdot \dfrac{1}{5}$.

Solution

$$\frac{3}{4} + \frac{2}{3} \cdot \frac{3}{5} - \frac{1}{2} \cdot \frac{1}{5} = \frac{3}{4} + \frac{2}{5} - \frac{1}{10}$$

$$= \frac{15}{20} + \frac{8}{20} - \frac{2}{20}$$

$$= \frac{15 + 8 - 2}{20} = \frac{21}{20}$$ ∎

EXAMPLE 12   Simplify $\dfrac{3}{5} \div \dfrac{8}{5} + \left(-\dfrac{1}{2}\right)\left(\dfrac{1}{3}\right) + \dfrac{5}{12}$.

Solution

$$\frac{3}{5} \div \frac{8}{5} + \left(-\frac{1}{2}\right)\left(\frac{1}{3}\right) + \frac{5}{12} = \frac{3}{5} \cdot \frac{5}{8} + \left(-\frac{1}{2}\right)\left(\frac{1}{3}\right) + \frac{5}{12}$$

$$= \frac{3}{8} + \frac{-1}{6} + \frac{5}{12}$$

$$= \frac{9}{24} + \frac{-4}{24} + \frac{10}{24}$$

$$= \frac{9 + (-4) + 10}{24}$$

$$= \frac{15}{24} = \frac{5}{8} \qquad \text{Reduce!}$$ ∎

The distributive property, $a(b + c) = ab + ac$, holds true for rational numbers and (as with integers) can be used to facilitate manipulation.

EXAMPLE 13   Simplify $12\left(\dfrac{1}{3} + \dfrac{1}{4}\right)$.

Solution   For help in this situation, let's change the form by applying the distributive property.

$$12\left(\frac{1}{3} + \frac{1}{4}\right) = 12\left(\frac{1}{3}\right) + 12\left(\frac{1}{4}\right)$$

$$= 4 + 3$$

$$= 7$$ ∎

EXAMPLE 14   Simplify $\dfrac{5}{8}\left(\dfrac{1}{2} + \dfrac{1}{3}\right)$.

Solution   In this case it may be easier not to apply the distributive property but to work with the expression in its given form.

$$\frac{5}{8}\left(\frac{1}{2} + \frac{1}{3}\right) = \frac{5}{8}\left(\frac{3}{6} + \frac{2}{6}\right)$$

$$= \frac{5}{8}\left(\frac{5}{6}\right)$$

$$= \frac{25}{48}$$ ∎

Examples 13 and 14 emphasize a point made in Chapter 1. *Think first*, and decide whether or not the properties can be used to make the manipulations easier. This section concludes with an example that illustrates the combining of similar terms that have fractional coefficients.

EXAMPLE 15   Simplify $\frac{1}{2}x + \frac{2}{3}x - \frac{3}{4}x$ by combining similar terms.

Solution   The distributive property along with our knowledge of adding and subtracting rational numbers provides the basis for working this type of problem.

$$\frac{1}{2}x + \frac{2}{3}x - \frac{3}{4}x = \left(\frac{1}{2} + \frac{2}{3} - \frac{3}{4}\right)x$$

$$= \left(\frac{6}{12} + \frac{8}{12} - \frac{9}{12}\right)x$$

$$= \frac{5}{12}x$$ ∎

Problem Set 2.2

For Problems 1–64, add or subtract as indicated and express answers in lowest terms.

1. $\frac{2}{7} + \frac{3}{7}$ 2. $\frac{3}{11} + \frac{5}{11}$ 3. $\frac{7}{9} - \frac{2}{9}$ 4. $\frac{11}{13} - \frac{6}{13}$

5. $\frac{3}{4} + \frac{9}{4}$ 6. $\frac{5}{6} + \frac{7}{6}$ 7. $\frac{11}{12} - \frac{3}{12}$ 8. $\frac{13}{16} - \frac{7}{16}$

9. $\frac{1}{8} - \frac{5}{8}$ 10. $\frac{2}{9} - \frac{5}{9}$ 11. $\frac{5}{24} + \frac{11}{24}$ 12. $\frac{7}{36} + \frac{13}{36}$

13. $\frac{8}{x} + \frac{7}{x}$ 14. $\frac{17}{y} + \frac{12}{y}$ 15. $\frac{5}{3y} + \frac{1}{3y}$ 16. $\frac{3}{8x} + \frac{1}{8x}$

17. $\frac{1}{3} + \frac{1}{5}$ 18. $\frac{1}{6} + \frac{1}{8}$ 19. $\frac{15}{16} - \frac{3}{8}$ 20. $\frac{13}{12} - \frac{1}{6}$

21. $\dfrac{7}{10} + \dfrac{8}{15}$

22. $\dfrac{7}{12} + \dfrac{5}{8}$

23. $\dfrac{11}{24} + \dfrac{5}{32}$

24. $\dfrac{5}{18} + \dfrac{8}{27}$

25. $\dfrac{5}{18} - \dfrac{13}{24}$

26. $\dfrac{1}{24} - \dfrac{7}{36}$

27. $\dfrac{5}{8} - \dfrac{2}{3}$

28. $\dfrac{3}{4} - \dfrac{5}{6}$

29. $-\dfrac{2}{13} - \dfrac{7}{39}$

30. $-\dfrac{3}{11} - \dfrac{13}{33}$

31. $-\dfrac{3}{14} + \dfrac{1}{21}$

32. $-\dfrac{3}{20} + \dfrac{14}{25}$

33. $-4 - \dfrac{3}{7}$

34. $-2 - \dfrac{5}{6}$

35. $\dfrac{3}{4} - 6$

36. $\dfrac{5}{8} - 7$

37. $\dfrac{3}{x} + \dfrac{4}{y}$

38. $\dfrac{5}{x} + \dfrac{8}{y}$

39. $\dfrac{7}{a} - \dfrac{2}{b}$

40. $\dfrac{13}{a} - \dfrac{4}{b}$

41. $\dfrac{2}{x} + \dfrac{7}{2x}$

42. $\dfrac{5}{2x} + \dfrac{7}{x}$

43. $\dfrac{10}{3x} - \dfrac{2}{x}$

44. $\dfrac{13}{4x} - \dfrac{3}{x}$

45. $\dfrac{1}{x} - \dfrac{7}{5x}$

46. $\dfrac{2}{x} - \dfrac{17}{6x}$

47. $\dfrac{3}{2y} + \dfrac{5}{3y}$

48. $\dfrac{7}{3y} + \dfrac{9}{4y}$

49. $\dfrac{5}{12y} - \dfrac{3}{8y}$

50. $\dfrac{9}{4y} - \dfrac{5}{9y}$

51. $\dfrac{1}{6n} - \dfrac{7}{8n}$

52. $\dfrac{3}{10n} - \dfrac{11}{15n}$

53. $\dfrac{5}{3x} + \dfrac{7}{3y}$

54. $\dfrac{3}{2x} + \dfrac{7}{2y}$

55. $\dfrac{8}{5x} + \dfrac{3}{4y}$

56. $\dfrac{1}{5x} + \dfrac{5}{6y}$

57. $\dfrac{7}{4x} - \dfrac{5}{9y}$

58. $\dfrac{2}{7x} - \dfrac{11}{14y}$

59. $-\dfrac{3}{2x} - \dfrac{5}{4y}$

60. $-\dfrac{13}{8a} - \dfrac{11}{10b}$

61. $3 + \dfrac{2}{x}$

62. $\dfrac{5}{x} + 4$

63. $2 - \dfrac{3}{x}$

64. $-1 - \dfrac{1}{3x}$

For Problems 65–80, simplify each numerical expression expressing answers in reduced form.

65. $\dfrac{1}{4} - \dfrac{3}{8} + \dfrac{5}{12} - \dfrac{1}{24}$

66. $\dfrac{3}{4} + \dfrac{2}{3} - \dfrac{1}{6} + \dfrac{5}{12}$

67. $\dfrac{5}{6} + \dfrac{2}{3} \cdot \dfrac{3}{4} - \dfrac{1}{4} \cdot \dfrac{2}{5}$

68. $\dfrac{2}{3} + \dfrac{1}{2} \cdot \dfrac{2}{5} - \dfrac{1}{3} \cdot \dfrac{1}{5}$

69. $\dfrac{3}{4} \cdot \dfrac{6}{9} - \dfrac{5}{6} \cdot \dfrac{8}{10} + \dfrac{2}{3} \cdot \dfrac{6}{8}$

70. $\dfrac{3}{5} \cdot \dfrac{5}{7} + \dfrac{2}{3} \cdot \dfrac{3}{5} - \dfrac{1}{7} \cdot \dfrac{2}{5}$

71. $4 - \dfrac{2}{3} \cdot \dfrac{3}{5} - 6$

72. $3 + \dfrac{1}{2} \cdot \dfrac{1}{3} - 2$

73. $\dfrac{4}{5} - \dfrac{10}{12} - \dfrac{5}{6} \div \dfrac{14}{8} + \dfrac{10}{21}$

74. $\dfrac{3}{4} \div \dfrac{6}{5} + \dfrac{8}{12} \cdot \dfrac{6}{9} - \dfrac{5}{12}$

75. $24\left(\dfrac{3}{4} - \dfrac{1}{6}\right)$ Don't forget the distributive property!

76. $18\left(\dfrac{2}{3} + \dfrac{1}{9}\right)$

77. $64\left(\dfrac{3}{16} + \dfrac{5}{8} - \dfrac{1}{4} + \dfrac{1}{2}\right)$

78. $48\left(\dfrac{5}{12} - \dfrac{1}{6} + \dfrac{3}{8}\right)$

79. $\dfrac{7}{13}\left(\dfrac{2}{3} - \dfrac{1}{6}\right)$

80. $\dfrac{5}{9}\left(\dfrac{1}{2} + \dfrac{1}{4}\right)$

For Problems 81–96, simplify each algebraic expression by combining similar terms.

81. $\dfrac{1}{3}x + \dfrac{2}{5}x$

82. $\dfrac{1}{4}x + \dfrac{2}{3}x$

83. $\dfrac{1}{3}a - \dfrac{1}{8}a$

84. $\dfrac{2}{5}a - \dfrac{2}{7}a$

85. $\dfrac{1}{2}x + \dfrac{2}{3}x + \dfrac{1}{6}x$

86. $\dfrac{1}{3}x + \dfrac{2}{5}x + \dfrac{5}{6}x$

87. $\dfrac{3}{5}n - \dfrac{1}{4}n + \dfrac{3}{10}n$

88. $\dfrac{2}{5}n - \dfrac{7}{10}n + \dfrac{8}{15}n$

89. $n + \dfrac{4}{3}n - \dfrac{1}{9}n$

90. $2n - \dfrac{6}{7}n + \dfrac{5}{14}n$

91. $-n - \dfrac{7}{9}n - \dfrac{5}{12}n$

92. $-\dfrac{3}{8}n - n - \dfrac{3}{14}n$

93. $\dfrac{3}{7}x + \dfrac{1}{4}y + \dfrac{1}{2}x + \dfrac{7}{8}y$

94. $\dfrac{5}{6}x + \dfrac{3}{4}y + \dfrac{4}{9}x + \dfrac{7}{10}y$

95. $\dfrac{2}{9}x + \dfrac{5}{12}y - \dfrac{7}{15}x - \dfrac{13}{15}y$

96. $-\dfrac{9}{10}x - \dfrac{3}{14}y + \dfrac{2}{25}x + \dfrac{5}{21}y$

Thoughts into Words

97. Explain how you would do the addition problem $\dfrac{3}{8} + \dfrac{7}{18}$ without using a calculator.

98. Explain how you would do the multiplication problem $\dfrac{3}{8} \cdot \dfrac{7}{18}$ without using a calculator.

Miscellaneous Problems

99. The will of a deceased collector of antique automobiles specified that his cars be left to his three children. Half were to go to his elder son, $\dfrac{1}{3}$ to his daughter, and $\dfrac{1}{9}$ to his younger son. At the time of his death, 17 cars were in the collection. The administrator of his estate borrowed a car to make 18. Then he distributed the cars as follows.

elder son: $\dfrac{1}{2}(18) = 9$,   daughter: $\dfrac{1}{3}(18) = 6$,   younger son: $\dfrac{1}{9}(18) = 2$

This totaled 17 cars, so he then returned the borrowed car. Where is the error in this problem?

100. Be sure that you can do Problems 17–36 and 65–80 with your calculator.

101. Use your calculator to do the following addition and subtraction problems.

(a) $\dfrac{7}{9} + \dfrac{2}{15}$

(b) $\dfrac{3}{8} + \dfrac{2}{13}$

(c) $\dfrac{5}{12} + \dfrac{9}{14}$

(d) $\dfrac{9}{14} + \dfrac{7}{10}$

(e) $\dfrac{11}{13} - \dfrac{4}{11}$

(f) $\dfrac{14}{17} - \dfrac{2}{9}$

(g) $\dfrac{5}{16} - \dfrac{11}{12}$

(h) $\dfrac{2}{15} - \dfrac{7}{12}$

(i) $\dfrac{7}{8} - \dfrac{5}{12} + \dfrac{3}{14}$

(j) $\dfrac{5}{7} - \dfrac{3}{14} - \dfrac{4}{15}$

2.3

Real Numbers and Algebraic Expressions

We classify decimals—also called decimal fractions—as **terminating**, **repeating** or **nonrepeating**. Some examples of each of these classifications follow.

$$\begin{bmatrix} 0.3 \\ 0.26 \\ 0.347 \\ 0.9865 \end{bmatrix} \quad \text{terminating decimals}$$

Technically, a terminating decimal can be thought of as repeating zeros after the last digit. For example, $0.3 = 0.30 = 0.300 = 0.3000$, etc.

$$\begin{bmatrix} 0.333333\ldots \\ 0.5466666\ldots \\ 0.14141414\ldots \\ 0.237237237\ldots \end{bmatrix} \quad \text{repeating decimals}$$

A repeating decimal has a block of digits that repeats indefinitely. This repeating block of digits may contain any number of digits and may or may not begin repeating immediately after the decimal point.

$$\begin{bmatrix} 0.5918654279\ldots \\ 0.26224222722229\ldots \\ 0.145117211193111148\ldots \end{bmatrix} \quad \text{nonrepeating decimals}$$

In Section 2.1 we defined a rational number to be any number that can be written in the form $\dfrac{a}{b}$, where a and b are integers and b is not zero. **A rational number can also be defined as any number that has a terminating or repeating decimal representation.** Thus, rational numbers can be expressed in either common fractional form or decimal fraction form as the next examples illustrate.

$$\begin{bmatrix} \dfrac{3}{4} = 0.75 \\[2mm] \dfrac{1}{8} = 0.125 \\[2mm] \dfrac{5}{16} = 0.3125 \\[2mm] \dfrac{7}{25} = 0.28 \end{bmatrix} \quad \text{terminating decimals}$$

$$\left[\begin{array}{l} \dfrac{1}{3} = 0.33333\ldots \\[2mm] \dfrac{2}{3} = 0.66666\ldots \\[2mm] \dfrac{1}{6} = 0.166666\ldots \\[2mm] \dfrac{1}{12} = 0.083333\ldots \\[2mm] \dfrac{14}{99} = 0.14141414\ldots \end{array}\right] \quad \text{repeating decimals}$$

The nonrepeating decimals are called **irrational numbers** and do appear in other than decimal form. For example, $\sqrt{2}$, $\sqrt{3}$, and π are irrational numbers. A partial decimal representation for each of these is as follows.

$$\left[\begin{array}{l} \sqrt{2} = 1.414213562373\ldots \\ \sqrt{3} = 1.73205080756887\ldots \\ \pi = 3.14159265358979\ldots \end{array}\right] \quad \text{nonrepeating decimals}$$

(We will do more work with the irrationals in Chapter 9.)

The rational numbers together with the irrationals form the set of **real numbers**. The following tree diagram of the real number system is helpful for summarizing some basic ideas.

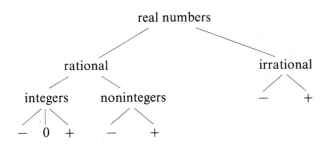

Any real number can be traced down through the diagram as follows.

5 is real, rational, an integer, and positive,

-4 is real, rational, an integer, and negative,

$\dfrac{3}{4}$ is real, rational, a noninteger, and positive,

0.23 is real, rational, a noninteger, and positive,

$-0.161616\ldots$ is real, rational, a noninteger, and negative,

$\sqrt{7}$ is real, irrational, and positive,

$-\sqrt{2}$ is real, irrational, and negative

The properties pertaining to integers we discussed in Section 1.5 are true for all real numbers and are restated here for your convenience. The multiplicative inverse property has been added to the list; a discussion follows.

Commutative Property of Addition

If a and b are real numbers, then

$$a + b = b + a.$$

Commutative Property of Multiplication

If a and b are real numbers, then

$$ab = ba.$$

Associative Property of Addition

If a, b, and c are real numbers, then

$$(a + b) + c = a + (b + c).$$

Associative Property of Multiplication

If a, b, and c are real numbers, then

$$(ab)c = a(bc).$$

Identity Property of Addition

If a is any real number, then

$$a + 0 = 0 + a = a.$$

Identity Property of Multiplication

If a is any real number, then

$$a(1) = 1(a) = a.$$

Additive Inverse Property

For every real number a, there exists a real number $-a$, such that

$$a + (-a) = (-a) + a = 0.$$

Multiplication Property of Zero

If a is any real number, then

$$a(0) = 0(a) = 0.$$

Multiplicative Property of Negative One

If a is any real number, then

$$a(-1) = -1(a) = -a.$$

Multiplicative Inverse Property

For every nonzero real number a, there exists a real number $\dfrac{1}{a}$, such that

$$a\left(\frac{1}{a}\right) = \frac{1}{a}(a) = 1.$$

Distributive Property

If a, b, and c are real numbers, then

$$a(b + c) = ab + ac.$$

The number $\dfrac{1}{a}$ is called the **multiplicative inverse** or the **reciprocal of a**. For example, the reciprocal of 2 is $\dfrac{1}{2}$ and $2\left(\dfrac{1}{2}\right) = \dfrac{1}{2}(2) = 1$. Likewise, the reciprocal of $\dfrac{1}{2}$ is $\dfrac{1}{\frac{1}{2}} = 2$. Therefore, 2 and $\dfrac{1}{2}$ are said to be reciprocals (or multiplicative inverses) of each other. Also, $\dfrac{2}{5}$ and $\dfrac{5}{2}$ are multiplicative inverses and $\left(\dfrac{2}{5}\right)\left(\dfrac{5}{2}\right) = 1$. Since division by zero is undefined, zero does not have a reciprocal.

Basic Operations with Decimals

The basic operations with decimals may be related to the corresponding operation with common fractions. For example, $0.3 + 0.4 = 0.7$ because $\dfrac{3}{10} + \dfrac{4}{10} = \dfrac{7}{10}$ and $0.37 - 0.24 = 0.13$ because $\dfrac{37}{100} - \dfrac{24}{100} = \dfrac{13}{100}$. In general, to add or subtract decimals, we add or subtract the hundredths, the tenths, the ones, the tens, and so on. To keep place values aligned, we line up the decimal points.

Addition		Subtraction	
1	1 1 1	6 16	8 11 13
2.14	5.214	7.6	9.235
3.12	3.162	4.9	6.781
5.16	7.218	2.7	2.454
10.42	8.914		
	24.508		

Examples such as the following can be used to help formulate a general rule for multiplying decimals.

Since $\dfrac{7}{10} \cdot \dfrac{3}{10} = \dfrac{21}{100}$, then $(0.7)(0.3) = 0.21$.

Since $\dfrac{9}{10} \cdot \dfrac{23}{100} = \dfrac{207}{1000}$, then $(0.9)(0.23) = 0.207$.

Since $\dfrac{11}{100} \cdot \dfrac{13}{100} = \dfrac{143}{10000}$, then $(0.11)(0.13) = 0.0143$.

In general, to multiply decimals we (1) multiply the numbers and ignore the decimal points, and then (2) insert the decimal point in the product so that the number of digits to the right of the decimal point in the product is equal to the sum of the numbers of digits to the right of the decimal point in each factor.

$$(0.7) \quad (0.3) \quad = \quad 0.21$$
$$\uparrow \qquad \uparrow \qquad\qquad \uparrow$$

one digit   one digit   two digits
to right + to right = to right

$$(0.9) \quad (0.23) \quad = \quad 0.207$$
$$\uparrow \qquad \uparrow \qquad\qquad \uparrow$$

one digit   two digits   three digits
to right + to right = to right

$$(0.11) \quad (0.13) \quad = \quad 0.0143$$
$$\uparrow \qquad \uparrow \qquad\qquad \uparrow$$

two digits   two digits   four digits
to right + to right = to right

We frequently use the vertical format when multiplying decimals.

41.2	one digit to right
0.13	two digits to right
1236	
412	
5.356	three digits to right

0.021	three digits to right
0.03	two digits to right
0.00063	five digits to right

Notice that in the last example we actually multiplied $3 \cdot 21$ and then inserted three 0s to the left so that there would be five digits to the right of the decimal point.

Once again let's look at some links between common fractions and decimals.

Since $\dfrac{6}{10} \div 2 = \dfrac{\overset{3}{\cancel{6}}}{10} \cdot \dfrac{1}{\cancel{2}} = \dfrac{3}{10}$, then $2\overline{)0.6}^{\,0.3}$;

Since $\dfrac{39}{100} \div 13 = \dfrac{\overset{3}{\cancel{39}}}{100} \cdot \dfrac{1}{\cancel{13}} = \dfrac{3}{100}$, then $13\overline{)0.39}$ with quotient 0.03;

Since $\dfrac{85}{100} \div 5 = \dfrac{\overset{17}{\cancel{85}}}{100} \cdot \dfrac{1}{\cancel{5}} = \dfrac{17}{100}$, then $5\overline{)0.85}$ with quotient 0.17.

In general, to divide a decimal by a nonzero whole number we (1) place the decimal point in the quotient directly above the decimal point in the dividend

$$\left(\text{divisor}\overline{)\overset{\text{quotient}}{\text{dividend}}}\right),$$

and then (2) divide as with whole numbers, except that in the division process, 0s are placed in the quotient immediately to the right of the decimal point in order to show the correct place value.

$$
\begin{array}{ll}
0.121 & 0.24 \\
4\overline{)0.484} & 32\overline{)7.68} \\
& 6\,4 \\
& \overline{1\,28} \\
& 1\,28 \\
\end{array}
\qquad
\begin{array}{l}
0.0\overset{\frown}{1}9 \quad \text{zero needed to show the} \\
12\overline{)0.228} \quad\;\; \text{correct place value} \\
12 \\
\overline{108} \\
108 \\
\end{array}
$$

Don't forget that *division can be checked by multiplication.* For example, since $(12)(0.019) = 0.228$ we know that our last division example is correct.

We can easily handle problems involving division by a decimal by changing to an equivalent problem that has a whole number divisor. Consider the following examples in which the original division problem has been changed to fractional form to show the reasoning involved in the procedure.

$$0.6\overline{)0.24} \to \dfrac{0.24}{0.6} = \left(\dfrac{0.24}{0.6}\right)\left(\dfrac{10}{10}\right) = \dfrac{2.4}{6} \to 6\overline{)2.4}, \quad \text{quotient } 0.4$$

$$0.12\overline{)0.156} \to \dfrac{0.156}{0.12} = \left(\dfrac{0.156}{0.12}\right)\left(\dfrac{100}{100}\right) = \dfrac{15.6}{12} \to 12\overline{)15.6}, \quad \text{quotient } 1.3$$
$$\phantom{0.12\overline{)0.156} \to \dfrac{0.156}{0.12} = } \begin{array}{r} 12 \\ \hline 36 \\ 36 \\ \hline \end{array}$$

$$1.3\overline{)0.026} \to \dfrac{0.026}{1.3} = \left(\dfrac{0.026}{1.3}\right)\left(\dfrac{10}{10}\right) = \dfrac{0.26}{13} \to 13\overline{)0.26} \quad \text{quotient } 0.02$$
$$\phantom{1.3\overline{)0.026} \to \dfrac{0.026}{1.3} = } \begin{array}{r} 26 \end{array}$$

The format commonly used with such problems is as follows.

$$
\begin{array}{r}
5.6 \\
0.21\overline{)1.17.6} \\
1\ 05 \\
\hline
12\ 6 \\
12\ 6 \\
\hline
\end{array}
$$

The arrows indicate that the divisor and dividend have been multiplied by 100, which changes the divisor to a whole number.

$$
\begin{array}{r}
0.04 \\
3.7\overline{)0.1.48} \\
1\ 48 \\
\hline
\end{array}
$$

The divisor and dividend have been multiplied by 10.

Our agreements for operating with positive and negative integers extend to all real numbers. For example, the product of two negative real numbers is a positive real number. Make sure that you agree with the following results. (You may need to do some work on scratch paper since the steps are not shown.)

$$
\begin{array}{ll}
0.24 + (-0.18) = 0.06, & (-0.4)(0.8) = -0.32, \\
-7.2 + 5.1 = -2.1, & (-0.5)(-0.13) = 0.065, \\
-0.6 + (-0.8) = -1.4, & (1.4) \div (-0.2) = -7, \\
2.4 - 6.1 = -3.7, & (-0.18) \div (0.3) = -0.6, \\
0.31 - (-0.52) = 0.83, & (-0.24) \div (-4) = 0.06 \\
(0.2)(-0.3) = -0.06, &
\end{array}
$$

Numerical and algebraic expressions may contain the decimal form as well as the fractional form of rational numbers. We continue to follow the agreement that multiplications and divisions are done *first* and then the additions and subtractions unless parentheses indicate otherwise. The following examples illustrate a variety of situations that involve both the decimal and fractional form of rational numbers.

EXAMPLE 1   Simplify $6.3 \div 7 + (4)(2.1) - (0.24) \div (-0.4)$.

Solution

$$
\begin{aligned}
6.3 \div 7 + (4)(2.1) - (0.24) \div (-0.4) &= 0.9 + 8.4 - (-0.6) \\
&= 0.9 + 8.4 + 0.6 \\
&= 9.9
\end{aligned}
$$
∎

EXAMPLE 2   Evaluate $\dfrac{3}{5}a - \dfrac{1}{7}b$ for $a = \dfrac{5}{2}$ and $b = -1$.

Solution

$$\frac{3}{5}a - \frac{1}{7}b = \frac{3}{5}\left(\frac{5}{2}\right) - \frac{1}{7}(-1) \quad \text{for } a = \frac{5}{2} \text{ and } b = -1$$

$$= \frac{3}{2} + \frac{1}{7}$$

$$= \frac{21}{14} + \frac{2}{14}$$

$$= \frac{23}{14}$$ ■

EXAMPLE 3   Evaluate $\frac{1}{2}x + \frac{2}{3}x - \frac{1}{5}x$ for $x = -\frac{3}{4}$.

Solution

First, let's *combine similar terms* by using the distributive property.

$$\frac{1}{2}x + \frac{2}{3}x - \frac{1}{5}x = \left(\frac{1}{2} + \frac{2}{3} - \frac{1}{5}\right)x$$

$$= \left(\frac{15}{30} + \frac{20}{30} - \frac{6}{30}\right)x$$

$$= \frac{29}{30}x$$

Now we can evaluate.

$$\frac{29}{30}x = \frac{29}{30}\left(-\frac{3}{4}\right) \quad \text{when } x = -\frac{3}{4}$$

$$= \frac{29}{\overset{}{\underset{10}{30}}}\left(-\frac{\overset{1}{3}}{4}\right) = -\frac{29}{40}$$ ■

EXAMPLE 4   Evaluate $2x + 3y$ for $x = 1.6$ and $y = 2.7$.

Solution

$$2x + 3y = 2(1.6) + 3(2.7) \quad \text{when } x = 1.6 \text{ and } y = 2.7$$

$$= 3.2 + 8.1$$

$$= 11.3$$ ■

EXAMPLE 5   Evaluate $0.9x + 0.7x - 0.4x + 1.3x$ for $x = 0.2$.

Solution   First, let's *combine similar terms* by using the distributive property.

$$0.9x + 0.7x - 0.4x + 1.3x = (0.9 + 0.7 - 0.4 + 1.3)x = 2.5x$$

Now we can evaluate.

$$2.5x = (2.5)(0.2) \quad \text{for } x = 0.2$$
$$= 0.5$$

Problem Set 2.3

For Problems 1–32, perform the indicated operations.

1. $0.37 + 0.25$	**2.** $7.2 + 4.9$	**3.** $2.93 - 1.48$
4. $14.36 - 5.89$	**5.** $(7.6) + (-3.8)$	**6.** $(6.2) + (-2.4)$
7. $(-4.7) + 1.4$	**8.** $(-14.1) + 9.5$	**9.** $-3.8 + 11.3$
10. $-2.5 + 14.8$	**11.** $6.6 - (-1.2)$	**12.** $18.3 - (-7.4)$
13. $-11.5 - (-10.6)$	**14.** $-14.6 - (-8.3)$	**15.** $-17.2 - (-9.4)$
16. $-21.4 - (-14.2)$	**17.** $(0.4)(2.9)$	**18.** $(0.3)(3.6)$
19. $(-0.8)(0.34)$	**20.** $(-0.7)(0.67)$	**21.** $(9)(-2.7)$
22. $(8)(-7.6)$	**23.** $(-0.7)(-64)$	**24.** $(-0.9)(-56)$
25. $(-0.12)(-0.13)$	**26.** $(-0.11)(-0.15)$	**27.** $1.56 \div 1.3$
28. $7.14 \div 2.1$	**29.** $5.92 \div (-0.8)$	**30.** $-2.94 \div 0.6$
31. $-0.266 \div (-0.7)$	**32.** $-0.126 \div (-0.9)$	

For Problems 33–46, simplify each of the numerical expressions.

33. $16.5 - 18.7 + 9.4$	**34.** $17.7 + 21.2 - 14.6$
35. $0.34 - 0.21 - 0.74 + 0.19$	**36.** $-5.2 + 6.8 - 4.7 - 3.9 + 1.3$
37. $0.76(0.2 + 0.8)$	**38.** $9.8(1.8 - 0.8)$
39. $0.6(4.1) + 0.7(3.2)$	**40.** $0.5(74) - 0.9(87)$
41. $7(0.6) + 0.9 - 3(0.4) + 0.4$	**42.** $-5(0.9) - 0.6 + 4.1(6) - 0.9$
43. $(0.96) \div (-0.8) + 6(-1.4) - 5.2$	**44.** $(-2.98) \div 0.4 - 5(-2.3) + 1.6$
45. $5(2.3) - 1.2 - 7.36 \div 0.8 + 0.2$	**46.** $0.9(12) \div 0.4 - 1.36 \div 17 + 9.2$

For Problems 47–60, simplify each algebraic expression by combining similar terms.

47. $x - 0.4x - 1.8x$	**48.** $-2x + 1.7x - 4.6x$
49. $5.4n - 0.8n - 1.6n$	**50.** $6.2n - 7.8n - 1.3n$
51. $-3t + 4.2t - 0.9t + 0.2t$	**52.** $7.4t - 3.9t - 0.6t + 4.7t$
53. $3.6x - 7.4y - 9.4x + 10.2y$	**54.** $5.7x + 9.4y - 6.2x - 4.4y$
55. $0.3(x - 4) + 0.4(x + 6) - 0.6x$	**56.** $0.7(x + 7) - 0.9(x - 2) + 0.5x$
57. $6(x - 1.1) - 5(x - 2.3) - 4(x + 1.8)$	
58. $4(x + 0.7) - 9(x + 0.2) - 3(x - 0.6)$	
59. $5(x - 0.5) + 0.3(x - 2) - 0.7(x + 7)$	
60. $-8(x - 1.2) + 6(x - 4.6) + 4(x + 1.7)$	

For Problems 61–74, evaluate each algebraic expression for the given values for the variables. Don't forget that for some problems it might be helpful to combine similar terms first and then to evaluate.

61. $x + 2y + 3z$ for $x = \dfrac{3}{4}$, $y = \dfrac{1}{3}$, and $z = -\dfrac{1}{6}$

62. $2x - y - 3z$ for $x = -\dfrac{2}{5}$, $y = -\dfrac{3}{4}$, and $z = \dfrac{1}{2}$

63. $\dfrac{3}{5}y - \dfrac{2}{3}y - \dfrac{7}{15}y$ for $y = -\dfrac{5}{2}$ **64.** $\dfrac{1}{2}x + \dfrac{2}{3}x - \dfrac{3}{4}x$ for $x = \dfrac{7}{8}$

65. $-x - 2y + 4z$ for $x = 1.7$, $y = -2.3$, and $z = 3.6$

66. $-2x + y - 5z$ for $x = -2.9$, $y = 7.4$, and $z = -6.7$

67. $5x - 7y$ for $x = -7.8$ and $y = 8.4$

68. $8x - 9y$ for $x = -4.3$ and $y = 5.2$

69. $0.7x + 0.6y$ for $x = -2$ and $y = 6$

70. $0.8x + 2.1y$ for $x = 5$ and $y = -9$

71. $1.2x + 2.3x - 1.4x - 7.6x$ for $x = -2.5$

72. $3.4x - 1.9x + 5.2x$ for $x = 0.3$

73. $-3a - 1 + 7a - 2$ for $a = 0.9$

74. $5x - 2 + 6x + 4$ for $x = -1.1$

Miscellaneous Problems

75. Without doing the actual dividing, defend the statement "$\dfrac{1}{7}$ produces a repeating decimal." [*Hint*: Think about the possible remainders when dividing by 7.]

76. Express each of the following in repeating decimal form.

(a) $\dfrac{1}{7}$ (b) $\dfrac{2}{7}$ (c) $\dfrac{4}{9}$

(d) $\dfrac{5}{6}$ (e) $\dfrac{3}{11}$ (f) $\dfrac{1}{12}$

77. (a) How can we tell that $\dfrac{5}{16}$ will produce a terminating decimal?

(b) How can we tell that $\dfrac{7}{15}$ will not produce a terminating decimal?

(c) Without using your calculator determine which of the following will produce a terminating decimal: $\dfrac{7}{8}, \dfrac{5}{12}, \dfrac{11}{16}, \dfrac{7}{24}, \dfrac{11}{75}, \dfrac{13}{32}, \dfrac{17}{40}, \dfrac{11}{30}, \dfrac{9}{20}, \dfrac{3}{64}$.

(d) Now use your calculator to check your answers for part (c).

78. Be sure that you can do Problems 1–46 with your calculator.

2.4
Exponents

We use exponents to indicate repeated multiplication. For example, we can write $5 \cdot 5 \cdot 5$ as 5^3 where the 3 indicates that 5 is to be used as a factor 3 times. The following general definition is helpful.

DEFINITION 2.4

> If n is a positive integer and b is any real number, then
> $$b^n = \underbrace{bbb \cdots b}_{n \text{ factors of } b}.$$

We refer to the b as the **base** and n as the **exponent**. The expression b^n can be read as "b to the nth **power**." We frequently associate the terms **squared** and **cubed** with exponents of 2 and 3, respectively. For example, b^2 is read as "b squared" and b^3 as "b cubed." An exponent of 1 is usually not written, so b^1 is written as b.

The following examples further clarify the concept of an exponent.

$$2^3 = 2 \cdot 2 \cdot 2 = 8, \qquad (0.6)^2 = (0.6)(0.6) = 0.36,$$

$$3^5 = 3 \cdot 3 \cdot 3 \cdot 3 \cdot 3 = 243, \qquad \left(\frac{1}{2}\right)^4 = \frac{1}{2} \cdot \frac{1}{2} \cdot \frac{1}{2} \cdot \frac{1}{2} = \frac{1}{16},$$

$$(-5)^2 = (-5)(-5) = 25, \qquad -5^2 = -(5 \cdot 5) = -25$$

We especially want to call your attention to the last two examples. Notice that $(-5)^2$ means that -5 is the base, which is to be used as a factor twice. However, -5^2 means that 5 is the base and after 5 is squared we take the opposite of that result.

Exponents provide a way of writing algebraic expressions in compact form. Sometimes we need to change from the compact form to an expanded form as these next examples demonstrate.

$$x^4 = x \cdot x \cdot x \cdot x, \qquad (2x)^3 = (2x)(2x)(2x),$$
$$2y^3 = 2 \cdot y \cdot y \cdot y, \qquad (-2x)^3 = (-2x)(-2x)(-2x),$$
$$-3x^5 = -3 \cdot x \cdot x \cdot x \cdot x \cdot x, \qquad -x^2 = -(x \cdot x),$$
$$a^2 + b^2 = a \cdot a + b \cdot b,$$

At other times we need to change from an expanded form to a more compact form using the exponent notation. The next examples illustrate this idea.

$$3 \cdot x \cdot x = 3x^2,$$
$$2 \cdot 5 \cdot x \cdot x \cdot x = 10x^3,$$
$$3 \cdot 4 \cdot x \cdot x \cdot y = 12x^2 y,$$

$$7 \cdot a \cdot a \cdot a \cdot b \cdot b = 7a^3b^2,$$
$$(2x)(3y) = 2 \cdot x \cdot 3 \cdot y = 2 \cdot 3 \cdot x \cdot y = 6xy,$$
$$(3a^2)(4a) = 3 \cdot a \cdot a \cdot 4 \cdot a = 3 \cdot 4 \cdot a \cdot a \cdot a = 12a^3,$$
$$(-2x)(3x) = -2 \cdot x \cdot 3 \cdot x = -2 \cdot 3 \cdot x \cdot x = -6x^2$$

The commutative and associative properties for multiplication have allowed us to rearrange and regroup factors in the last three examples above.

The concept of *exponent* can be used to extend our work with combining similar terms, operating with fractions, and evaluating algebraic expressions. Study the following examples very carefully; it should help you to pull together many ideas.

EXAMPLE 1   Simplify $4x^2 + 7x^2 - 2x^2$ by combining similar terms.

Solution   By applying the distributive property we obtain

$$4x^2 + 7x^2 - 2x^2 = (4 + 7 - 2)x^2$$
$$= 9x^2. \qquad \blacksquare$$

EXAMPLE 2   Simplify $-8x^3 + 9y^2 + 4x^3 - 11y^2$ by combining similar terms.

Solution   By rearranging terms and then applying the distributive property we obtain

$$-8x^3 + 9y^2 + 4x^3 - 11y^2 = -8x^3 + 4x^3 + 9y^2 - 11y^2$$
$$= (-8 + 4)x^3 + (9 - 11)y^2$$
$$= -4x^3 - 2y^2. \qquad \blacksquare$$

EXAMPLE 3   Simplify $-7x^2 + 4x + 3x^2 - 9x$.

Solution

$$-7x^2 + 4x + 3x^2 - 9x = -7x^2 + 3x^2 + 4x - 9x$$
$$= (-7 + 3)x^2 + (4 - 9)x$$
$$= -4x^2 - 5x \qquad \blacksquare$$

As soon as you feel comfortable with this process of combining similar terms, you may want to do some of the steps mentally. Then your work may appear as follows.

$$9a^2 + 6a^2 - 12a^2 = 3a^2,$$
$$6x^2 + 7y^2 - 3x^2 - 11y^2 = 3x^2 - 4y^2,$$
$$7x^2y + 5xy^2 - 9x^2y + 10xy^2 = -2x^2y + 15xy^2,$$
$$2x^3 - 5x^2 - 10x - 7x^3 + 9x^2 - 4x = -5x^3 + 4x^2 - 14x$$

The next two examples illustrate the use of exponents when reducing fractions.

EXAMPLE 4   Reduce $\dfrac{8x^2y}{12xy}$.

Solution $\dfrac{8x^2y}{12xy} = \dfrac{2 \cdot 2 \cdot 2 \cdot x \cdot x \cdot y}{2 \cdot 2 \cdot 3 \cdot x \cdot y} = \dfrac{2x}{3}$ ∎

EXAMPLE 5   Reduce $\dfrac{15a^2b^3}{25a^3b}$.

Solution $\dfrac{15a^2b^3}{25a^3b} = \dfrac{3 \cdot 5 \cdot a \cdot a \cdot b \cdot b \cdot b}{5 \cdot 5 \cdot a \cdot a \cdot a \cdot b} = \dfrac{3b^2}{5a}$ ∎

The next three examples show how exponents may be used when multiplying and dividing fractions.

EXAMPLE 6   Multiply $\left(\dfrac{4x}{6y}\right)\left(\dfrac{12y^2}{7x^2}\right)$ and express answers in reduced form.

Solution $\left(\dfrac{4x}{6y}\right)\left(\dfrac{12y^2}{7x^2}\right) = \dfrac{4 \cdot \overset{2}{12} \cdot x \cdot y \cdot y}{6 \cdot 7 \cdot y \cdot x \cdot x} = \dfrac{8y}{7x}$ ∎

EXAMPLE 7   Multiply and simplify
$\left(\dfrac{8a^3}{9b}\right)\left(\dfrac{12b^2}{16a}\right).$

Solution $\left(\dfrac{8a^3}{9b}\right)\left(\dfrac{12b^2}{16a}\right) = \dfrac{8 \cdot \overset{\overset{2}{4}}{12} \cdot a \cdot a \cdot a \cdot b \cdot b}{\underset{3}{9} \cdot \underset{2}{16} \cdot b \cdot a} = \dfrac{2a^2b}{3}$ ∎

EXAMPLE 8   Divide and express in reduced form,
$\dfrac{-2x^3}{3y^2} \div \dfrac{4}{9xy}.$

Solution $\dfrac{-2x^3}{3y^2} \div \dfrac{4}{9xy} = -\dfrac{2x^3}{3y^2} \cdot \dfrac{9xy}{4} = -\dfrac{2 \cdot \overset{3}{9} \cdot x \cdot x \cdot x \cdot x \cdot y}{3 \cdot \underset{2}{4} \cdot y \cdot y} = -\dfrac{3x^4}{2y}$ ∎

The next two examples demonstrate the use of exponents when adding and subtracting fractions.

EXAMPLE 9   Add $\dfrac{4}{x^2} + \dfrac{7}{x}$.

Solution   The LCD is x^2. Thus,

$$\frac{4}{x^2} + \frac{7}{x} = \frac{4}{x^2} + \frac{7 \cdot x}{x \cdot x} = \frac{4}{x^2} + \frac{7x}{x^2} = \frac{4 + 7x}{x^2}.$$ ∎

EXAMPLE 10   Subtract $\dfrac{3}{xy} - \dfrac{4}{y^2}$.

Solution

$$\left. \begin{array}{l} xy = x \cdot y \\[2mm] y^2 = y \cdot y \end{array} \right\} \longrightarrow \text{The LCD is } xy^2.$$

$$\frac{3}{xy} - \frac{4}{y^2} = \frac{3 \cdot y}{xy \cdot y} - \frac{4 \cdot x}{y^2 \cdot x} = \frac{3y}{xy^2} - \frac{4x}{xy^2} = \frac{3y - 4x}{xy^2}$$ ∎

Remember that exponents are used to indicate repeated multiplication. Therefore, to simplify numerical expressions containing exponents we proceed as follows.

---

1. Perform the operations inside the symbols of inclusion (parentheses and brackets) and above and below each fraction bar. Start with the innermost inclusion symbol.
2. Compute all indicated powers.
3. Perform all multiplications and divisions in the order that they appear from left to right.
4. Perform all additions and subtractions in the order that they appear from left to right.

Keep these steps in mind as we evaluate some algebraic expressions containing exponents.

EXAMPLE 11   Evaluate $3x^2 - 4y^2$ for $x = -2$ and $y = 5$.

Solution

$$3x^2 - 4y^2 - 3(-2)^2 - 4(5)^2 \quad \text{when } x = -2 \text{ and } y = 5$$
$$= 3(-2)(-2) - 4(5)(5)$$
$$= 12 - 100 = -88$$ ∎

EXAMPLE 12   Find the value of $a^2 - b^2$ when $a = \dfrac{1}{2}$ and $b = -\dfrac{1}{3}$.

Solution

$$a^2 - b^2 = \left(\frac{1}{2}\right)^2 - \left(-\frac{1}{3}\right)^2 \quad \text{when } a = \frac{1}{2} \text{ and } b = -\frac{1}{3}$$

$$= \frac{1}{4} - \frac{1}{9}$$

$$= \frac{9}{36} - \frac{4}{36}$$

$$= \frac{5}{36}$$ ■

EXAMPLE 13   Evaluate $5x^2 + 4xy$ for $x = 0.4$ and $y = -0.3$.

Solution

$$5x^2 + 4xy = 5(0.4)^2 + 4(0.4)(-0.3) \quad \text{when } x = 0.4 \text{ and } y = -0.3$$

$$= 5(0.16) + 4(-0.12)$$

$$= 0.80 + (-0.48)$$

$$= 0.32$$ ■

Problem Set 2.4

For Problems 1–20, find the value of each numerical expression. For example, $2^4 = 2 \cdot 2 \cdot 2 \cdot 2 = 16$.

1. 2^6 **2.** 2^7 **3.** 3^4 **4.** 4^3

5. $(-2)^3$ **6.** $(-2)^5$ **7.** -3^2 **8.** -3^4

9. $(-4)^2$ **10.** $(-5)^4$ **11.** $\left(\dfrac{2}{3}\right)^4$ **12.** $\left(\dfrac{3}{4}\right)^3$

13. $-\left(\dfrac{1}{2}\right)^3$ **14.** $-\left(\dfrac{3}{2}\right)^3$ **15.** $\left(-\dfrac{3}{2}\right)^2$ **16.** $\left(-\dfrac{4}{3}\right)^2$

17. $(0.3)^3$ **18.** $(0.2)^4$ **19.** $-(1.2)^2$ **20.** $-(1.1)^3$

For Problems 21–34, simplify each numerical expression.

21. $3^2 + 2^3 - 4^3$ **22.** $2^4 - 3^3 + 5^2$ **23.** $(-2)^3 - 2^4 - 3^2$

24. $(-3)^3 - 3^2 - 6^2$ **25.** $5(2)^2 - 4(2) - 1$ **26.** $7(-2)^2 - 6(-2) - 8$

27. $-2(3)^3 - 3(3)^2 + 4(3) - 6$ **28.** $5(-3)^3 - 4(-3)^2 + 6(-3) + 1$

29. $-7^2 - 6^2 + 5^2$ **30.** $-8^2 + 3^4 - 4^3$

31. $-3(-4)^2 - 2(-3)^3 + (-5)^2$ **32.** $-4(-3)^3 + 5(-2)^3 - (4)^2$

33. $\dfrac{-3(2)^4}{12} + \dfrac{5(-3)^3}{15}$ **34.** $\dfrac{4(2)^3}{16} - \dfrac{2(3)^2}{6}$

For Problems 35–46, use exponents to help express each algebraic expression in a more compact form. For example, $3 \cdot 5 \cdot x \cdot x \cdot y = 15x^2y$ and $(3x)(2x^2) = 6x^3$.

35. $9 \cdot x \cdot x$

36. $8 \cdot x \cdot x \cdot x \cdot y$

37. $3 \cdot 4 \cdot x \cdot y \cdot y$

38. $7 \cdot 2 \cdot a \cdot a \cdot b \cdot b \cdot b$

39. $-2 \cdot 9 \cdot x \cdot x \cdot x \cdot x \cdot y$

40. $-3 \cdot 4 \cdot x \cdot y \cdot z \cdot z$

41. $(5x)(3y)$

42. $(3x^2)(2y)$

43. $(6x^2)(2x^2)$

44. $(-3xy)(6xy)$

45. $(-4a^2)(-2a^3)$

46. $(-7a^3)(-3a)$

For Problems 47–58, simplify each expression by combining similar terms.

47. $3x^2 - 7x^2 - 4x^2$

48. $-2x^3 + 7x^3 - 4x^3$

49. $-12y^3 + 17y^3 - y^3$

50. $-y^3 + 8y^3 - 13y^3$

51. $7x^2 - 2y^2 - 9x^2 + 8y^2$

52. $5x^3 + 9y^3 - 8x^3 - 14y^3$

53. $\dfrac{2}{3}n^2 - \dfrac{1}{4}n^2 - \dfrac{3}{5}n^2$

54. $-\dfrac{1}{2}n^2 + \dfrac{5}{6}n^2 - \dfrac{4}{9}n^2$

55. $5x^2 - 8x - 7x^2 + 2x$

56. $-10x^2 + 4x + 4x^2 - 8x$

57. $x^2 - 2x - 4 + 6x^2 - x + 12$

58. $-3x^3 - x^2 + 7x - 2x^3 + 7x^2 - 4x$

For Problems 59–68, reduce each fraction to simplest form.

59. $\dfrac{9xy}{15x}$

60. $\dfrac{8x^2y}{14x}$

61. $\dfrac{22xy^2}{6xy^3}$

62. $\dfrac{18x^3y}{12xy^4}$

63. $\dfrac{7a^2b^3}{17a^3b}$

64. $\dfrac{9a^3b^3}{22a^4b^2}$

65. $\dfrac{-24abc^2}{32bc}$

66. $\dfrac{4a^2c^3}{-22b^2c^4}$

67. $\dfrac{-5x^4y^3}{-20x^2y}$

68. $\dfrac{-32xy^2z^4}{-48x^3y^3z}$

For Problems 69–86, perform the indicated operations and express answers in reduced form.

69. $\left(\dfrac{7x^2}{9y}\right)\left(\dfrac{12y}{21x}\right)$

70. $\left(\dfrac{3x}{8y^2}\right)\left(\dfrac{14xy}{9y}\right)$

71. $\left(\dfrac{5c}{a^2b^2}\right) \div \left(\dfrac{12c}{ab}\right)$

72. $\left(\dfrac{13ab^2}{12c}\right) \div \left(\dfrac{26b}{14c}\right)$

73. $\dfrac{6}{x} + \dfrac{5}{y^2}$

74. $\dfrac{8}{y} - \dfrac{6}{x^2}$

75. $\dfrac{5}{x^4} - \dfrac{7}{x^2}$

76. $\dfrac{9}{x} - \dfrac{11}{x^3}$

77. $\dfrac{3}{2x^3} + \dfrac{6}{x}$

78. $\dfrac{5}{3x^2} + \dfrac{6}{x}$

79. $\dfrac{-5}{4x^2} + \dfrac{7}{3x^2}$

80. $\dfrac{-8}{5x^3} + \dfrac{10}{3x^3}$

81. $\dfrac{11}{a^2} - \dfrac{14}{b^2}$

82. $\dfrac{9}{x^2} + \dfrac{8}{y^2}$

83. $\dfrac{1}{2x^3} - \dfrac{4}{3x^2}$

84. $\dfrac{2}{3x^3} - \dfrac{5}{4x}$

85. $\dfrac{3}{x} - \dfrac{4}{y} - \dfrac{5}{xy}$

86. $\dfrac{5}{x} + \dfrac{7}{y} - \dfrac{1}{xy}$

For Problems 87–100, evaluate each algebraic expression for the given values of the variables.

87. $4x^2 + 7y^2$ for $x = -2$ and $y = -3$

88. $5x^2 + 2y^3$ for $x = -4$ and $y = -1$

89. $3x^2 - y^2$ for $x = \dfrac{1}{2}$ and $y = -\dfrac{1}{3}$

90. $x^2 - 2y^2$ for $x = -\dfrac{2}{3}$ and $y = \dfrac{3}{2}$

91. $x^2 - 2xy + y^2$ for $x = -\dfrac{1}{2}$ and $y = 2$

92. $x^2 + 2xy + y^2$ for $x = -\dfrac{3}{2}$ and $y = -2$

93. $-x^2$ for $x = -8$ 94. $-x^3$ for $x = 5$

95. $-x^2 - y^2$ for $x = -3$ and $y = -4$ 96. $-x^2 + y^2$ for $x = -2$ and $y = 6$

97. $-a^2 - 3b^3$ for $a = -6$ and $b = -1$ 98. $-a^3 + 3b^2$ for $a = -3$ and $b = -5$

99. $y^2 - 3xy$ for $x = 0.4$ and $y = -0.3$

100. $x^2 + 5xy$ for $x = -0.2$ and $y = -0.6$

Miscellaneous Problems

101. Do Problems 1–34 using your calculator.

102. Use your calculator to evaluate each of the following.

    (a) 2^{10} (b) 3^7 (c) $(-2)^8$ (d) $(-2)^{11}$ (e) -4^9

    (f) -5^6 (g) $(3.14)^3$ (h) $(1.41)^4$ (i) $(1.73)^5$

2.5

Translating from English to Algebra

In order to use the tools of algebra for solving problems, we must be able to translate back and forth between the English language and the language of algebra. In this section we want to translate algebraic expressions to English phrases (word phrases) and English phrases to algebraic expressions.

Let's begin by considering the following translations from algebraic expressions to word phrases.

Algebraic expression	*Word phrase*
$x + y$	The sum of x and y
$x - y$	The difference of x minus y
$y - x$	The difference of y minus x
xy	The product of x and y
$\dfrac{x}{y}$	The quotient of x divided by y
$3x$	The product of 3 and x
$x^2 + y^2$	The sum of x squared and y squared
$2xy$	The product of 2, x, and y
$2(x + y)$	Two times the quantity x plus y
$x - 3$	Three less than x

Now let's consider the reverse process, translating from word phrases to algebraic expressions. Part of the difficulty in translating from English to algebra is the fact that different word phrases translate to the same algebraic expression. So we need to become familiar with *different ways of saying the same thing*, especially when referring to the four fundamental operations. The following examples should help to acquaint you with some of the phrases used in the basic operations.

$$\begin{bmatrix} \text{The sum of } x \text{ and } 4 \\ \text{The sum ``}x \text{ plus 4''} \end{bmatrix} \longrightarrow \quad x + 4$$

$$\begin{bmatrix} \text{The difference of 5 subtracted from } n \\ \text{The difference ``}n \text{ minus 5''} \\ \text{The difference ``}n \text{ less 5''} \\ \text{The difference ``5 less than } n\text{''} \end{bmatrix} \longrightarrow \quad n - 5$$

$$\begin{bmatrix} \text{The product of 4 and } y \\ \text{The product ``4 times } y\text{''} \end{bmatrix} \longrightarrow \quad 4y$$

$$\begin{bmatrix} \text{The quotient when } n \text{ is divided by 6} \\ \text{The quotient ``}n \text{ divided by 6''} \\ \text{The quotient of 6 divided into } n \end{bmatrix} \longrightarrow \quad \frac{n}{6}$$

Often a word phrase indicates more than one operation. Furthermore, the standard vocabulary of sum, difference, product, and quotient may be replaced by other terminology. Study the following translations very carefully.

Word phrase	*Algebraic expression*
The sum of two times x and three times y	$2x + 3y$
The sum of the squares of a and b	$a^2 + b^2$
Five times x divided by y	$\dfrac{5x}{y}$
Two more than the square of x	$x^2 + 2$
Three less than the cube of b	$b^3 - 3$
Five less than the product of x and y	$xy - 5$
Three less x	$3 - x$
Nine minus the product of x and y	$9 - xy$
Four times the sum of x and 2	$4(x + 2)$
Six times the quantity w minus 4	$6(w - 4)$

Suppose you are told that the sum of two numbers is 12 and one of the numbers is 8. What is the other number? The other number is $12 - 8$, which equals 4. Now suppose that you are told that the product of two numbers is 56 and one of the numbers is 7. What is the other number? The other number is $56 \div 7$, which equals 8. The following examples illustrate the use of these addition-subtraction and multiplication-division relationships in a more general setting.

EXAMPLE 1   The sum of two numbers is 83 and one of the numbers is x. What is the other number?

Solution   Using the addition and subtraction relationship we can represent the other number by $83 - x$.   ■

EXAMPLE 2   The difference of two numbers is 14. The smaller number is n. What is the larger number?

Solution   Since the smaller number plus the difference must equal the larger number, we can represent the larger number by $n + 14$.   ∎

EXAMPLE 3   The product of two numbers is 39 and one of the numbers is y. Represent the other number.

Solution   Using the multiplication and division relationship we can represent the other number by $\dfrac{39}{y}$.   ∎

The English statement may not contain any key words such as sum, difference, product, or quotient. Instead, the statement may describe a physical situation; from this description you need to deduce the operations involved. We make some suggestions for handling such situations in the following examples.

EXAMPLE 4   Arlene can type 70 words per minute. How many words can she type in m minutes?

Solution   In 10 minutes she would type $70(10) = 700$ words. In 50 minutes she would type $70(50) = 3500$ words. Thus, in m minutes she would type $70m$ words.   ∎

Notice the use of some specific examples ($70(10) = 700$ and $70(50) = 3500$) to help formulate the general expression. This technique of first formulating some specific examples and then generalizing can be very effective.

EXAMPLE 5   Lynn has n nickels and d dimes. Express, in cents, this amount of money.

Solution   Three nickels and 8 dimes would be $5(3) + 10(8) = 95$ cents. Thus, n nickels and d dimes would be $5n + 10d$ cents.   ∎

EXAMPLE 6   A train travels at the rate of r miles per hour. How far will it travel in 8 hours?

Solution   Suppose that a train travels at 50 miles per hour. Using the formula *distance equals rate times time*, it would travel $50 \cdot 8 = 400$ miles. Therefore, at r miles per hour, it would travel $r \cdot 8$ miles. The expression $r \cdot 8$ is usually written as $8r$.   ∎

EXAMPLE 7   The cost of a 5-pound box of candy is d dollars. How much is the cost per pound for the candy?

Solution   The price per pound is figured by dividing the total cost by the number of pounds. Therefore, the price per pound is represented by $\dfrac{d}{5}$.   ∎

The English statement to be translated to algebra may contain some geometric ideas. For example, suppose that we want to express in inches the length of a line segment that is f feet long. Since 1 foot = 12 inches, we can represent f feet by 12 times f, written as $12f$ inches.

Tables 2.1 and 2.2 list some of the basic relationships pertaining to linear measurements in the English and metric systems, respectively. (Additional listings of both systems are located in the inside back cover of this book.)

Table 2.1	*English system*
	12 inches = 1 foot
	3 feet = 36 inches = 1 yard
	5280 feet = 1760 yards = 1 mile

Table 2.2	*Metric system*
	1 kilometer = 1000 meters
	1 hectometer = 100 meters
	1 dekameter = 10 meters
	1 decimeter = 0.1 meter
	1 centimeter = 0.01 meter
	1 millimeter = 0.001 meter

EXAMPLE 8   The distance between two cities is k kilometers. Express this distance in meters.

Solution   Since 1 kilometer equals 1000 meters, we need to multiply k by 1000. Therefore, the distance in meters is represented by $1000k$.   ■

EXAMPLE 9   The length of a line segment is i inches. Express that length in yards.

Solution   To change from inches to yards, we must divide by 36. Therefore, $\dfrac{i}{36}$ represents in yards the length of the line segment.   ■

EXAMPLE 10   The width of a rectangle is w centimeters and the length is 5 centimeters less than twice the width. What is the length of the rectangle? What is the perimeter of the rectangle?

Solution   The length of the rectangle can be represented by $2w - 5$. Now we can sketch a rectangle and record the given information. The perimeter of a rectangle is the sum

of the lengths of the four sides. Therefore, the perimeter is given by $2w + 2(2w - 5)$, which can be written as $2w + 4w - 10$ and then simplified to $6w - 10$.   ■

Problem Set 2.5

For Problems 1–12, write a word phrase for each of the algebraic expressions. For example, lw can be expressed as "the product of l and w.

1. $a - b$ **2.** $x + y$ **3.** $\frac{1}{3}Bh$ **4.** $\frac{1}{2}bh$

5. $2(l + w)$ **6.** πr^2 **7.** $\frac{A}{w}$ **8.** $\frac{C}{\pi}$

9. $\frac{a + b}{2}$ **10.** $\frac{a - b}{4}$ **11.** $3y + 2$ **12.** $3(x - y)$

For Problems 13–36, translate each word phrase into an algebraic expression. For example, "the sum of x and 14" translates into $x + 14$.

13. The sum of l and w **14.** The difference of x minus y

15. The product of a and b **16.** The product of $\frac{1}{3}$, B, and h

17. The quotient of d divided by t **18.** The quotient of r divided into d

19. The product of l, w, and h **20.** The product of π and the square of r

21. The difference of x subtracted from y

22. The difference "x subtract y"

23. Two larger than the product of x and y

24. Six plus the cube of x

25. Seven minus the square of y

26. The quantity, x minus 2, cubed

27. The quantity, x minus y, divided by four

28. Eight less than x

29. Ten less x

30. Nine times the quantity, n minus four

31. Ten times the quantity, n plus two

32. The sum of four times x and five times y

33. The difference of seven subtracted from the product of x and y

34. Three times the sum of n and 2

35. Twelve less than the product of x and y

36. Twelve less the product of x and y

For Problems 37–68, answer the question with an algebraic expression.

37. The sum of two numbers is 35 and one of the numbers is n. What is the other number?

38. The sum of two numbers is 100 and one of the numbers is x. What is the other number?

39. The difference of two numbers is 45 and the smaller number is n. What is the other number?

40. The product of two numbers is 25 and one of the numbers is x. What is the other number?

41. Janet is y years old. How old will she be in 10 years?

42. Hector is y years old. How old was he 5 years ago?

43. Debra is x years old and her mother is 3 years less than twice as old as Debra. How old is Debra's mother?

44. Jack is x years old and Dudley is 1 year more than three times as old as Jack. How old is Dudley?

45. Donna has d dimes and q quarters in her bank. How much money in cents does she have?

46. Andy has c cents that is all in dimes. How many dimes does he have?

47. A car travels d miles in t hours. What is the rate of the car?

48. If g gallons of gas cost d dollars, what is the price per gallon?

49. If p pounds of candy cost d dollars, what is the price per pound?

50. Sue can type x words per minute. How many words can she type in an hour?

51. Larry's annual salary is d dollars. What is his monthly salary?

52. Nancy's monthly salary is d dollars. What is her annual salary?

53. If n represents a whole number, what represents the next larger whole number?

54. If n represents an even number, what represents the next larger even number?

55. If n represents an odd number, what represents the next larger odd number?

56. Maria is y years old and her sister is twice as old. What represents the sum of their ages?

57. Willie is y years old and his father is 2 years less than twice Willie's age. What represents the sum of their ages?

58. Harriet has p pennies, n nickels, and d dimes. How much money in cents does she have?

59. The perimeter of a rectangle is y yards and f feet. What is the perimeter expressed in inches?

60. The perimeter of a triangle is m meters and c centimeters. What is the perimeter in centimeters?

61. A rectangular plot of ground is f feet long. What is its length in yards?

62. The height of a telephone pole is f feet. What is the height in yards?

63. The width of a rectangle is w feet and its length is three times the width. What is the perimeter of the rectangle in feet?

64. The width of a rectangle is w feet and its length is 1 foot more than twice its width. What is the perimeter of the rectangle in feet?

65. The length of a rectangle is l inches and its width is 2 inches less than one-half of its length. What is the perimeter of the rectangle in inches?

66. The length of a rectangle is l inches and its width is 3 inches more than one-third of its length. What is the perimeter of the rectangle in inches?

67. The first side of a triangle is f feet long. The second side is 2 feet longer than the first side. The third side is twice as long as the second side. What is the perimeter of the triangle in inches?

68. The first side of a triangle is y yards long. The second side is 3 yards shorter than the first side. The third side is 3 times as long as the second side. What is the perimeter of the triangle in feet?

Thoughts into Words

69. How have you used the basic properties of the real numbers thus far in your study of algebra?

70. What does the phrase "translating from English to algebra" mean to you?

Chapter 2 Summary

(2.1) The property $\dfrac{a \cdot k}{b \cdot k} = \dfrac{a}{b}$ is used to express fractions in reduced form.

To **multiply** rational numbers in common fractional form we multiply numerators, multiply denominators, and express the result in reduced form.

To **divide** rational numbers in common fractional form, we multiply by the reciprocal of the divisor.

(2.2) Addition and **subtraction** of rational numbers in common fractional form are based on the following.

$$\frac{a}{b} + \frac{c}{b} = \frac{a + c}{b}, \qquad \text{addition}$$

$$\frac{a}{b} - \frac{c}{b} = \frac{a - c}{b} \qquad \text{subtraction}$$

To add or subtract fractions that do not have a common denominator, we use the fundamental principle of fractions, $\dfrac{a}{b} = \dfrac{a \cdot k}{b \cdot k}$, and obtain equivalent fractions that have a common denominator.

(2.3) To **add** or **subtract decimals**, we write the numbers in a column so that the decimal points are lined up, and then we add or subtract as we do with integers.

To **multiply decimals** we (1) multiply the numbers ignoring the decimal points, and then (2) insert the decimal point in the product so that the number of digits to the right of the decimal point in the product is equal to the sum of the numbers of digits to the right of the decimal point in each factor.

To **divide a decimal by a nonzero whole number** we (1) place the decimal point in the quotient directly above the decimal point in the dividend, and then (2) divide as with whole numbers, except that in the division process, we place zeros in the quotient immediately to the right of the decimal point (if necessary) to show the correct place value.

To **divide by a decimal** we change to an equivalent problem that has a whole number divisor.

(2.4) Expressions of the form b^n, where

$$b^n = b \cdot b \cdot b \cdots b, \qquad n \text{ factors of } b$$

are read as "b to the nth power"; b is the *base* and n is the *exponent*.

(2.5) To translate English phrases to algebraic expressions we must be familiar with the standard vocabulary of **sum**, **difference**, **product**, and **quotient** as well as other terms used to express the same ideas.

Chapter 2 Review Problem Set

For Problems 1–10, find the value of each of the following.

1. 2^6

2. $(-3)^3$

3. -4^2

4. $\left(\dfrac{3}{4}\right)^2$

5. $\left(\dfrac{1}{2} + \dfrac{2}{3}\right)^2$

6. $(0.6)^3$

7. $(0.12)^2$

8. $(0.06)^2$

9. $\left(-\dfrac{2}{3}\right)^3$

10. $\left(-\dfrac{1}{2}\right)^4$

For Problems 11–20, perform the indicated operations and express answers in reduced form.

11. $\dfrac{3}{8} + \dfrac{5}{12}$

12. $\dfrac{9}{14} - \dfrac{3}{35}$

13. $\dfrac{2}{3} + \dfrac{-3}{5}$

14. $\dfrac{7}{x} + \dfrac{9}{2y}$

15. $\dfrac{5}{xy} - \dfrac{8}{x^2}$

16. $\left(\dfrac{7y}{8x}\right)\left(\dfrac{14x}{35}\right)$

17. $\left(\dfrac{6xy}{9y^2}\right) \div \left(\dfrac{15y}{18x^2}\right)$

18. $\left(\dfrac{-3x}{12y}\right)\left(\dfrac{8y}{-7x}\right)$

19. $\left(-\dfrac{4y}{3x}\right)\left(-\dfrac{3x}{4y}\right)$

20. $\left(\dfrac{6n}{7}\right)\left(\dfrac{9n}{8}\right)$

For Problems 21–30, simplify each of the following numerical expressions.

21. $\dfrac{1}{6} + \dfrac{2}{3} \cdot \dfrac{3}{4} - \dfrac{5}{6} \div \dfrac{8}{6}$

22. $\dfrac{3}{4} \cdot \dfrac{1}{2} - \dfrac{4}{3} \cdot \dfrac{3}{2}$

23. $\dfrac{7}{9} \cdot \dfrac{3}{5} + \dfrac{7}{9} \cdot \dfrac{2}{5}$

24. $\dfrac{4}{5} \div \dfrac{1}{5} \cdot \dfrac{2}{3} - \dfrac{1}{4}$

25. $\dfrac{2}{3} \cdot \dfrac{1}{4} \div \dfrac{1}{2} + \dfrac{2}{3} \cdot \dfrac{1}{4}$

26. $0.48 + 0.72 - 0.35 - 0.18$

27. $0.81 + (0.6)(0.4) - (0.7)(0.8)$

28. $1.28 \div 0.8 - 0.81 \div 0.9 + 1.7$

29. $(0.3)^2 + (0.4)^2 - (0.6)^2$

30. $(1.76)(0.8) + (1.76)(0.2)$

For Problems 31–36, simplify each of the following algebraic expressions by combining similar terms. Express answers in reduced form when working with common fractions.

31. $\dfrac{3}{8}x^2 - \dfrac{2}{5}y^2 - \dfrac{2}{7}x^2 + \dfrac{3}{4}y^2$

32. $0.24ab + 0.73bc - 0.82ab - 0.37bc$

33. $\dfrac{1}{2}x + \dfrac{3}{4}x - \dfrac{5}{6}x + \dfrac{1}{24}x$ **34.** $1.4a - 1.9b + 0.8a + 3.6b$

35. $\dfrac{2}{5}n + \dfrac{1}{3}n - \dfrac{5}{6}n$ **36.** $n - \dfrac{3}{4}n + 2n - \dfrac{1}{5}n$

For Problems 37–42, evaluate the following algebraic expressions for the given values of the variables.

37. $\dfrac{1}{4}x - \dfrac{2}{5}y$ for $x = \dfrac{2}{3}$ and $y = -\dfrac{5}{7}$ **38.** $a^3 + b^2$ for $a = -\dfrac{1}{2}$ and $b = \dfrac{1}{3}$

39. $2x^2 - 3y^2$ for $x = 0.6$ and $y = 0.7$

40. $0.7w + 0.9z$ for $w = 0.4$ and $z = -0.7$

41. $\dfrac{3}{5}x - \dfrac{1}{3}x + \dfrac{7}{15}x - \dfrac{2}{3}x$ for $x = \dfrac{15}{17}$ **42.** $\dfrac{1}{3}n + \dfrac{2}{7}n - n$ for $n = 21$

For Problems 43–50, answer each of the following questions with an algebraic expression.

43. The sum of two numbers is 72 and one of the numbers is n. What is the other number?

44. Joan has p pennies and d dimes. How much money in cents does she have?

45. Ellen types x words in an hour. What is her typing rate per minute?

46. Harry is y years old. His brother is 3 years less than twice as old as Harry. How old is Harry's brother?

47. Larry chose a number n. Cindy chose a number 3 more than 5 times the number chosen by Larry. What number did Cindy choose?

48. The height of a file cabinet is y yards and f feet. How tall is the file cabinet in inches?

49. The length of a rectangular room is m meters. How long in centimeters is the room?

50. Corinne has n nickels, d dimes, and q quarters. How much money in cents does she have?

For Problems 51–60, translate each word phrase into an algebraic expression.

51. Five less than n

52. Five less n

53. Ten times the quantity, x minus 2

54. Ten times x minus 2

55. The difference of x minus three

56. The quotient of d divided by r

57. x squared plus nine

58. x plus nine, the quantity squared

59. The sum of the cubes of x and y

60. Four less than the product of x and y

Cumulative Review Problem Set   (Chapters 1 and 2)

For Problems 1–12, simplify each of the numerical expressions.

1. $16 - 18 - 14 + 21 - 14 + 19$

2. $7(-6) - 8(-6) + 4(-9)$

3. $6 - [3 - (10 - 12)]$

4. $-9 - 2[4 - (-10 + 6)] - 1$

5. $\dfrac{-7(-4) - 5(-6)}{-2}$

6. $\dfrac{5(-3) + (-4)(6) - 3(4)}{-3}$

7. $\dfrac{3}{4} + \dfrac{1}{3} \div \dfrac{4}{3} - \dfrac{1}{2}$

8. $\left(\dfrac{2}{3}\right)\left(-\dfrac{3}{4}\right) - \left(\dfrac{5}{6}\right)\left(\dfrac{4}{5}\right)$

9. $\left(\dfrac{1}{2} - \dfrac{2}{3}\right)^2$

10. -4^3

11. $\left(-\dfrac{1}{2}\right)^3 - \left(-\dfrac{3}{4}\right)^2$

12. $(.2)^2 - (.3)^3 + (.4)^2$

For Problems 13–20, evaluate each algebraic expression for the given values of the variables.

13. $3xy - 2x - 4y$ for $x = -6$ and $y = 7$

14. $-4x^2 y - 2xy^2 + xy$ for $x = -2$ and $y = -4$

15. $\dfrac{5x - 2y}{3x}$ for $x = \dfrac{1}{2}$ and $y = -\dfrac{1}{3}$

16. $.2x - .3y + 2xy$ for $x = .1$ and $y = .3$

17. $-7x + 4y + 6x - 9y + x - y$ for $x = -.2$ and $y = .4$

18. $\dfrac{2}{3}x - \dfrac{3}{5}y + \dfrac{3}{4}x - \dfrac{1}{2}y$ for $x = \dfrac{6}{5}$ and $y = -\dfrac{1}{4}$

19. $\dfrac{1}{5}n - \dfrac{1}{3}n + n - \dfrac{1}{6}n$ for $n = \dfrac{1}{5}$

20. $-ab + \dfrac{1}{5}a - \dfrac{2}{3}b$ for $a = -2$ and $b = \dfrac{3}{4}$

For Problems 21–24, express each of the numbers as a product of prime factors.

21. 54   **22.** 78   **23.** 91   **24.** 153

For Problems 25–28, find the greatest common factor of the given numbers.

25. 42 and 70   **26.** 63 and 81   **27.** 28, 36, and 52   **28.** 48, 66, and 78

For Problems 29–32, find the least common multiple of the given numbers.

29. 20 and 28   **30.** 40 and 100   **31.** 12, 18, and 27   **32.** 16, 20, and 80

For Problems 33–38, simplify each algebraic expression by combining similar terms.

33. $\dfrac{2}{3}x - \dfrac{1}{4}y - \dfrac{3}{4}x - \dfrac{2}{3}y$

34. $-n - \dfrac{1}{2}n + \dfrac{3}{5}n + \dfrac{5}{6}n$

35. $3.2a - 1.4b - 6.2a + 3.3b$

36. $-(n - 1) + 2(n - 2) - 3(n - 3)$

37. $-x + 4(x - 1) - 3(x + 2) - (x + 5)$

38. $2a - 5(a + 3) - 2(a - 1) - 4a$

For Problems 39–46, perform the indicated operations and express answers in reduced form.

39. $\dfrac{5}{12} - \dfrac{3}{16}$

40. $\dfrac{3}{4} - \dfrac{5}{6} - \dfrac{7}{9}$

41. $\dfrac{5}{xy} - \dfrac{2}{x} + \dfrac{3}{y}$

42. $-\dfrac{7}{x^2} + \dfrac{9}{xy}$

43. $\left(\dfrac{7x}{9y}\right)\left(\dfrac{12y}{14}\right)$

44. $\left(-\dfrac{5a}{7b^2}\right)\left(-\dfrac{8ab}{15}\right)$

45. $\left(\dfrac{6x^2y}{11}\right) \div \left(\dfrac{9y^2}{22}\right)$

46. $\left(-\dfrac{9a}{8b}\right) \div \left(\dfrac{12a}{18b}\right)$

For Problems 47–50, answer each question with an algebraic expression.

47. Hector has p pennies, n nickels, and d dimes. How much money in cents does he have?

48. Ginny chose a number n. Penny chose a number 5 less than 4 times the number chosen by Ginny. What number did Penny choose?

49. The height of a flagpole is y yards, f feet, and i inches. How tall is the flagpole in inches?

50. A rectangular room is x meters by y meters. What is its perimeter in centimeters?

Chapter 3

Equations and Problem Solving

Throughout this book we will follow a common theme: We will develop some new skills; use the skills to help solve equations and inequalities; and finally use the equations and inequalities to solve applied problems. In this chapter we want to use the skills we have developed in the first two chapters to help us solve equations and begin our work with applied problems.

3.1

Solving First-Degree Equations

The following are examples of **numerical statements**.

$$3 + 4 = 7, \qquad 5 - 2 = 3, \qquad 7 + 1 = 12$$

The first two are true statements and the third one is a false statement.

Using x as a variable, the statements

$$x + 3 = 4, \qquad 2x - 1 = 7, \qquad \text{and} \qquad x^2 = 4$$

are called **algebraic equations** in x. A number a is called a **solution** or **root** of an equation if a true numerical statement is formed when a is substituted for x. (We also say that a satisfies the equation.) For example, 1 is a solution of $x + 3 = 4$ because substituting 1 for x produces the true numerical statement $1 + 3 = 4$. The set of all solutions of an equation is called its **solution set**. Thus, the solution set of $x + 3 = 4$ is $\{1\}$. Likewise, the solution set of $2x - 1 = 7$ is $\{4\}$ and the solution set of $x^2 = 4$ is $\{-2, 2\}$. **Solving an equation** refers to the process of determining the solution set. Remember that a set which consists of no elements is called the **empty** or **null set** and is denoted by $\varnothing$. Thus, we say that the solution set of $x = x + 1$ is $\varnothing$; that is, there are no real numbers that satisfy $x = x + 1$.

In this chapter we shall consider techniques for solving **first-degree equations of one variable**. This means that the equations contain only one variable and this variable has an exponent of one. The following are examples of first degree equations of one variable.

$$3x + 4 = 7, \qquad 8w + 7 = 5w - 4,$$

$$\frac{1}{2}y + 2 = 9, \qquad 7x + 2x - 1 = 4x - 1$$

Equivalent equations are equations that have the same solution set. For example,

$$5x - 4 = 3x + 8, \qquad 2x = 12, \qquad \text{and} \qquad x = 6,$$

are all equivalent equations. This can be verified by showing that 6 is the solution for all three equations.

The general procedure for solving an equation is to continue replacing the given equation with equivalent but simpler equations until we obtain an equation of the form **variable = constant** or **constant = variable**. Thus, in the preceding example, $5x - 4 = 3x + 8$ was simplified to $2x = 12$, which was further simplified to $x = 6$, from which the solution of 6 is obvious. The exact procedure for simplifying equations becomes our next concern.

Two properties of equality play an important role in the process of solving equations. The first of these is the **addition-subtraction property of equality**, which can be stated as follows.

PROPERTY 3.1

> **Addition-Subtraction Property of Equality**
>
> For all real numbers a, b, and c,
>
>   **1.** $a = b$ if and only if $a + c = b + c$;
>
>   **2.** $a = b$ if and only if $a - c = b - c$.

Property 3.1 states that *any number can be added to or subtracted from both sides of an equation and an equivalent equation is produced.* Consider the use of this property in the next four examples.

EXAMPLE 1   Solve $x - 8 = 3$.

Solution

$$x - 8 = 3$$
$$x - 8 + 8 = 3 + 8 \qquad \text{Add 8 to both sides.}$$
$$x = 11$$

The potential solution, 11, can be *checked* by substituting it into the original equation to see if a true numerical statement is obtained.

CHECK $x - 8 = 3$

$$11 - 8 \stackrel{?}{=} 3 \qquad \text{Substitute 11 for } x.$$
$$3 = 3$$

Now we know that the solution set is $\{11\}$. ■

> **REMARK** It is true that a simple equation such as Example 1 can be solved "by inspection." That is to say, we could think "some number minus 8 produces 3." Obviously, the number is 11. However, as the equations become more complex, the technique of solving by inspection becomes ineffective. So it is necessary to develop more formal techniques for solving equations. Therefore, we will begin developing such techniques even with very simple types of equations.

EXAMPLE 2   Solve $x + 14 = -8$.

Solution

$$x + 14 = -8$$
$$x + 14 - 14 = -8 - 14 \qquad \text{Subtract 14 from both sides.}$$
$$x = -22$$

CHECK $x + 14 = -8$

$$-22 + 14 \stackrel{?}{=} -8 \qquad \text{Substitute } -22 \text{ for } x.$$
$$-8 = -8$$

The solution set is $\{-22\}$. ■

EXAMPLE 3   Solve $n - \dfrac{1}{3} = \dfrac{1}{4}$.

Solution

$$n - \frac{1}{3} = \frac{1}{4}$$

$$n - \frac{1}{3} + \frac{1}{3} = \frac{1}{4} + \frac{1}{3} \qquad \text{Add } \frac{1}{3} \text{ to both sides.}$$

$$n = \frac{3}{12} + \frac{4}{12}$$

$$n = \frac{7}{12}$$

CHECK $n - \dfrac{1}{3} = \dfrac{1}{4}$

$$\frac{7}{12} - \frac{1}{3} \overset{?}{=} \frac{1}{4} \qquad \text{Substitute } \frac{7}{12} \text{ for } n.$$

$$\frac{7}{12} - \frac{4}{12} \overset{?}{=} \frac{1}{4}$$

$$\frac{3}{12} \overset{?}{=} \frac{1}{4}$$

$$\frac{1}{4} = \frac{1}{4}$$

The solution set is $\left\{\dfrac{7}{12}\right\}$. ∎

EXAMPLE 4   Solve $0.72 = y + 0.35$.

Solution

$$0.72 = y + 0.35$$

$$0.72 - 0.35 = y + 0.35 - 0.35 \qquad \text{Subtract 0.35 from both sides.}$$

$$0.37 = y$$

CHECK $0.72 = y + 0.35$

$$0.72 \overset{?}{=} 0.37 + 0.35 \qquad \text{Substitute 0.37 for } y.$$

$$0.72 = 0.72$$

The solution set is $\{0.37\}$. ∎

Note that in Example 4 the final equation is $0.37 = y$ instead of $y = 0.37$. Technically, the **symmetry property of equality** (if $a = b$, then $b = a$) would permit us to change from $0.37 = y$ to $y = 0.37$, but such a change is not necessary to determine that the solution is 0.37.

One other comment pertaining to Property 3.1 should be made at this time. Since subtracting a number is equivalent to adding its opposite, Property 3.1 could be stated only in terms of addition. Thus, to solve an equation such as Example 4 we could add -0.35 to both sides rather than subtract 0.35 from both sides.

The other important property for solving equations is the **multiplication-division property of equality**.

PROPERTY 3.2

Multiplication-Division Property of Equality

For all real numbers, a, b, and c, where $c \neq 0$,

1. $a = b$ if and only if $ac = bc$;

2. $a = b$ if and only if $\dfrac{a}{c} = \dfrac{b}{c}$.

Property 3.2 states that *an equivalent equation is obtained whenever both sides of a given equation are multiplied or divided by the same nonzero real number.* The following examples illustrate the use of this property.

EXAMPLE 5   Solve $\dfrac{3}{4}x = 6$.

Solution

$$\frac{3}{4}x = 6$$

$$\frac{4}{3}\left(\frac{3}{4}x\right) = \frac{4}{3}(6) \qquad \text{Multiply both sides by } \frac{4}{3} \text{ since } \left(\frac{4}{3}\right)\left(\frac{3}{4}\right) = 1.$$

$$x = 8$$

$$\textbf{CHECK} \quad \frac{3}{4}x = 6$$

$$\frac{3}{4}(8) \overset{?}{=} 6 \qquad \text{Substitute 8 for } x.$$

$$6 = 6$$

The solution set is $\{8\}$.   ■

EXAMPLE 6   Solve $5x = 27$.

Solution

$$5x = 27$$

$$\frac{5x}{5} = \frac{27}{5} \qquad \text{Divide both sides by 5.}$$

$$x = \frac{27}{5} \qquad \frac{27}{5} \text{ could be expressed as } 5\frac{2}{5} \text{ or } 5.4.$$

CHECK $\qquad 5x = 27$

$$5\left(\frac{27}{5}\right) \stackrel{?}{=} 27 \qquad \text{Substitute } \frac{27}{5} \text{ for } x.$$

$$27 = 27$$

The solution set is $\left\{\dfrac{27}{5}\right\}$. ■

EXAMPLE 7   Solve $-\dfrac{2}{3}p = \dfrac{1}{2}$.

Solution

$$-\frac{2}{3}p = \frac{1}{2}$$

$$-\frac{3}{2}\left(-\frac{2}{3}p\right) = \left(-\frac{3}{2}\right)\left(\frac{1}{2}\right) \qquad \text{Multiply both sides by } -\frac{3}{2}$$
$$\text{since } \left(-\frac{3}{2}\right)\left(-\frac{2}{3}\right) = 1.$$

$$p = -\frac{3}{4}$$

CHECK $\qquad -\dfrac{2}{3}p = \dfrac{1}{2}$

$$-\frac{2}{3}\left(-\frac{3}{4}\right) \stackrel{?}{=} \frac{1}{2} \qquad \text{Substitute } -\frac{3}{4} \text{ tor } p.$$

$$\frac{1}{2} = \frac{1}{2}$$

The solution set is $\left\{-\dfrac{3}{4}\right\}$. ■

EXAMPLE 8   Solve $26 = -6x$.

Solution

$$26 = -6x$$

$$\frac{26}{-6} = \frac{-6x}{-6} \qquad \text{Divide both sides by } -6.$$

$$-\frac{26}{6} = x \qquad \frac{26}{-6} = -\frac{26}{6}$$

$$-\frac{13}{3} = x \qquad \text{Don't forget to reduce!}$$

CHECK $26 = -6x$

$$26 \stackrel{?}{=} -6\left(-\frac{13}{3}\right) \qquad \text{Substitute } -\frac{13}{3} \text{ for } x.$$

$$26 = 26$$

The solution set is $\left\{-\dfrac{13}{3}\right\}$. ■

Looking back at Examples 5–8, you will notice that we divided both sides of the equation by the coefficient of the variable whenever it was an integer; otherwise we used the multiplication part of Property 3.2. Technically, since dividing by a number is equivalent to multiplying by its reciprocal, Property 3.2 could be stated only in terms of multiplication. Thus, to solve an equation such as $5x = 27$, we could multiply both sides by $\dfrac{1}{5}$ instead of dividing both sides by 5.

Problem Set 3.1

Use the properties of equality to help solve each of the following equations.

1. $x + 9 = 17$ **2.** $x + 7 = 21$ **3.** $x + 11 = 5$

4. $x + 13 = 2$ **5.** $-7 = x + 2$ **6.** $-12 = x + 4$

7. $8 = n + 14$ **8.** $6 = n + 19$ **9.** $21 + y = 34$

10. $17 + y = 26$ **11.** $x - 17 = 31$ **12.** $x - 22 = 14$

13. $14 = x - 9$ **14.** $17 = x - 28$ **15.** $-26 = n - 19$

16. $-34 = n - 15$ **17.** $y - \dfrac{2}{3} = \dfrac{3}{4}$ **18.** $y - \dfrac{2}{5} = \dfrac{1}{6}$

19. $x + \dfrac{3}{5} = \dfrac{1}{3}$ **20.** $x + \dfrac{5}{8} = \dfrac{2}{5}$ **21.** $b + 0.19 = 0.46$

22. $b + 0.27 = 0.74$ **23.** $n - 1.7 = -5.2$ **24.** $n - 3.6 = -7.3$

25. $15 - x = 32$ **26.** $13 - x = 47$ **27.** $-14 - n - 21$

28. $-9 - n = 61$ **29.** $7x = -56$ **30.** $9x = -108$ **31.** $-6x = 102$

32. $-5x = 90$ **33.** $5x = 37$ **34.** $7x = 62$ **35.** $-18 = 6n$

36. $-52 = 13n$ **37.** $-26 = -4n$ **38.** $-56 = -6n$ **39.** $\dfrac{t}{9} = 16$

40. $\dfrac{t}{12} = 8$ **41.** $\dfrac{n}{-8} = -3$ **42.** $\dfrac{n}{-9} = -5$ **43.** $-x = 15$

44. $-x = -17$ **45.** $\dfrac{3}{4}x = 18$ **46.** $\dfrac{2}{3}x = 32$ **47.** $-\dfrac{2}{5}n = 14$

48. $-\dfrac{3}{8}n = 33$ **49.** $\dfrac{2}{3}n = \dfrac{1}{5}$ **50.** $\dfrac{3}{4}n = \dfrac{1}{8}$ **51.** $\dfrac{5}{6}n = -\dfrac{3}{4}$

52. $\dfrac{6}{7}n = -\dfrac{3}{8}$ **53.** $\dfrac{3x}{10} = \dfrac{3}{20}$ **54.** $\dfrac{5x}{12} = \dfrac{5}{36}$ **55.** $\dfrac{-y}{2} = \dfrac{1}{6}$

56. $\dfrac{-y}{4} = \dfrac{1}{9}$ **57.** $-\dfrac{4}{3}x = -\dfrac{9}{8}$ **58.** $-\dfrac{6}{5}x = -\dfrac{10}{14}$ **59.** $-\dfrac{5}{12} = \dfrac{7}{6}x$

60. $-\dfrac{7}{24} = \dfrac{3}{8}x$ **61.** $-\dfrac{5}{7}x = 1$ **62.** $-\dfrac{11}{12}x = -1$ **63.** $-4n = \dfrac{1}{3}$

64. $-6n = \dfrac{3}{4}$ **65.** $-8n = \dfrac{6}{5}$ **66.** $-12n = \dfrac{8}{3}$

3.2

Equations and Problem Solving

We often need more than one property of equality to help find the solution of an equation. Consider the following examples.

EXAMPLE 1   Solve $3x + 1 = 7$.

Solution

$$3x + 1 - 7$$

$$3x + 1 - 1 = 7 - 1 \qquad \text{Subtract 1 from both sides.}$$

$$3x = 6$$

$$\frac{3x}{3} = \frac{6}{3} \qquad \text{Divide both sides by 3.}$$

$$x = 2$$

The potential solution can be *checked* by substituting it into the original equation to see if a true numerical statement is obtained.

CHECK $3x + 1 = 7$

$$3(2) + 1 \stackrel{?}{=} 7$$

$$6 + 1 \stackrel{?}{=} 7$$

$$7 = 7$$

Now we know that the solution set is $\{2\}$.   ■

EXAMPLE 2   Solve $5x - 6 = 14$.

Solution

$$5x - 6 = 14$$

$$5x - 6 + 6 = 14 + 6 \qquad \text{Add 6 to both sides.}$$

$$5x = 20$$

$$\frac{5x}{5} = \frac{20}{5} \qquad \text{Divide both sides by 5.}$$

$$x = 4$$

CHECK $5x - 6 = 14$

$$5(4) - 6 \stackrel{?}{=} 14$$

$$20 - 6 \stackrel{?}{=} 14$$

$$14 = 14$$

The solution set is $\{4\}$.   ■

EXAMPLE 3   Solve $-3a + 4 = 22$.

Solution

$$-3a + 4 = 22$$

$$-3a + 4 - 4 = 22 - 4 \qquad \text{Subtract 4 from both sides.}$$

$$-3a = 18$$

$$\frac{-3a}{-3} = \frac{18}{-3} \qquad \text{Divide both sides by } -3.$$

$$a = -6$$

CHECK $-3a + 4 = 22$

$$-3(-6) + 4 \stackrel{?}{=} 22$$

$$18 + 4 \stackrel{?}{=} 22$$

$$22 = 22$$

The solution set is $\{-6\}$.   ■

Notice that in Examples 1, 2, and 3 we used the addition-subtraction property first and then used the multiplication-division property. In general, this sequence of steps provides the easiest format for solving such equations. Perhaps you should convince yourself of that fact by doing Example 1 again, but this time use the multiplication-division property first and then the addition-subtraction property.

EXAMPLE 4   Solve $19 = 2n + 4$.

Solution

$$19 = 2n + 4$$

$$19 - 4 = 2n + 4 - 4 \qquad \text{Subtract 4 from both sides.}$$

$$15 = 2n$$

$$\frac{15}{2} = \frac{2n}{2} \qquad \text{Divide both sides by 2.}$$

$$\frac{15}{2} = n$$

$$\textbf{CHECK} \quad 19 = 2n + 4$$

$$19 \overset{?}{=} 2\left(\frac{15}{2}\right) + 4$$

$$19 \overset{?}{=} 15 + 4$$

$$19 = 19$$

The solution set is $\left\{\dfrac{15}{2}\right\}$.   ■

Word Problems

In the last section of Chapter 2 we translated English phrases into algebraic expressions. We are now ready to extend that idea to the translation of English *sentences* into algebraic *equations*. Such translations allow us to use the concepts of algebra to solve verbal problems. Let's consider some examples.

PROBLEM 1   A certain number added to 17 yields a sum of 29. What is the number?

Solution   Let n represent the number to be found. The sentence "A certain number added to 17 yields a sum of 29" translates into the algebraic equation $17 + n = 29$. Solving this equation, we obtain

$$17 + n = 29$$

$$17 + n - 17 = 29 - 17$$

$$n = 12.$$

The solution is 12, which is the number asked for in the problem.   ■

The statement "let n represent the number to be found" is often referred to as **declaring the variable**. We need to choose a letter to use as a variable and indicate what it represents for a specific problem. This may seem like an insignificant idea, but as the problems become more complex, the process of declaring the variable becomes even more important. Furthermore, we are aware that we could solve a problem such as Problem 1 without setting up an algebraic equation. However, as problems increase in difficulty, the translation from English to an algebraic equation becomes a key issue. Therefore, even with these relatively simple problems we must concentrate on the translation process.

PROBLEM 2   Six years ago Bill was 13 years old. How old is he now?

Solution   Let y represent Bill's age now; therefore, $y - 6$ represents his age six years ago. Thus,

$$y - 6 = 13$$
$$y - 6 + 6 = 13 + 6$$
$$y = 19.$$

Bill is presently 19 years old.   ■

PROBLEM 3   Betty worked 8 hours Saturday and earned \$32. How much did she earn per hour?

Solution A   Let x represent the amount Betty earned per hour. The number of hours worked times the wage per hour yields the total earnings. Thus,

$$8x = 32$$
$$\frac{8x}{8} = \frac{32}{8}$$
$$x = 4.$$

Betty earned \$4 per hour.

Solution B   Let y represent the amount Betty earned per hour. The wage per hour equals the total wage divided by the number of hours. Thus,

$$y = \frac{32}{8}$$
$$y = 4.$$

Betty earned \$4 per hour.   ■

Sometimes more than one equation can be used to solve a problem. In Solution A the equation was set up in terms of multiplication, whereas in Solution B we were thinking in terms of division.

PROBLEM 4   If 2 is subtracted from five-times-a-certain-number the result is 28. Find the number.

Solution   Let n represent the number to be found. Translating the first sentence in the problem into an algebraic equation, we obtain

$$5n - 2 = 28.$$

Solving this equation yields

$$5n - 2 + 2 = 28 + 2$$

$$5n = 30$$

$$\frac{5n}{5} = \frac{30}{5}$$

$$n = 6.$$

The number to be found is 6.   ■

PROBLEM 5   A plumbing repair bill was $95. This included $17 for parts and an amount for 3 hours of labor. Find the hourly rate that was charged for labor.

Solution   Let l represent the hourly charge for labor; then $3l$ represents the total cost for labor. Thus, the cost for labor plus the cost for parts is the total bill of $95; so we can proceed as follows.

$$\text{cost for labor} + \text{cost for parts} = \$95$$
$$\downarrow \qquad\qquad \downarrow$$
$$3l \quad + \quad 17 \quad = 95$$

Solving this equation we obtain

$$3l = 78$$

$$\frac{3l}{3} = \frac{78}{3}$$

$$l = 26.$$

The charge for labor was $26 per hour.   ■

Problem Set 3.2

For Problems 1–40, solve each equation.

1. $2x + 5 = 13$	**2.** $3x + 4 = 19$	**3.** $5x + 2 = 32$	**4.** $7x + 3 = 24$
5. $3x - 1 = 23$	**6.** $2x - 5 = 21$	**7.** $4n - 3 = 41$	**8.** $5n - 6 = 19$
9. $6y - 1 = 16$	**10.** $4y - 3 = 14$	**11.** $2x + 3 = 22$	**12.** $3x + 1 = 21$
13. $10 = 3t - 8$	**14.** $17 = 2t + 5$	**15.** $5x + 14 = 9$	**16.** $4x + 17 = 9$
17. $18 - n = 23$	**18.** $17 - n = 29$	**19.** $-3x + 2 = 20$	**20.** $-6x + 1 = 43$

21. $7 + 4x = 29$ **22.** $9 + 6x = 23$ **23.** $16 = -2 - 9a$

24. $18 = -10 - 7a$ **25.** $-7x + 3 = -7$ **26.** $-9x + 5 = -18$

27. $17 - 2x = -19$ **28.** $18 - 3x = -24$ **29.** $-16 - 4x = 9$

30. $-14 - 6x = 7$ **31.** $-12t + 4 = 88$ **32.** $-16t + 3 = 67$

33. $14y + 15 = -33$ **34.** $12y + 13 = -15$ **35.** $32 - 16n = -8$

36. $-41 = 12n - 19$ **37.** $17x - 41 = -37$ **38.** $19y - 53 = -47$

39. $29 = -7 - 15x$ **40.** $49 = -5 - 14x$

For Problems 41–68, (a) choose a variable and indicate what it represents in the problem, (b) set up an equation that represents the situation described, and (c) solve the equation.

41. Twelve added to a certain number is 21. What is the number?

42. A certain number added to 14 is 25. Find the number.

43. Nine subtracted from a certain number is 13. Find the number.

44. A certain number subtracted from 32 is 15. What is the number?

45. Two items cost $43. If one of the items costs $25, what is the cost of the other item?

46. Eight years ago Rosa was 22 years old. Find Rosa's present age.

47. Six years from now, Nora will be 41 years old. What is her present age?

48. The Chi's bought 8 pizzas for a total of $50. What was the price per pizza?

49. Chad worked 6 hours Saturday for a total of $39. How much per hour did he earn?

50. Jill worked 8 hours Saturday at $5.50 per hour. How much did she earn?

51. If six is added to three times a certain number, the result is 24. Find the number.

52. If two is subtracted from five times a certain number, the result is 38. Find the number.

53. Nineteen is four larger than three times a certain number. Find the number.

54. If nine times a certain number is subtracted from seven, the result is 52. Find the number.

55. Forty-nine is equal to six less than five times a certain number. Find the number.

56. Seventy-one is equal to two more than three times a certain number. Find the number.

57. If one is subtracted from six times a certain number, the result is 47. Find the number.

58. Five less than four times a number equals 31. Find the number.

59. If eight times a certain number is subtracted from 27, the result is 3. Find the number.

60. Twenty is 22 less than six times a certain number. Find the number.

61. The price of a ring is $550. This price represents $50 less than twice the cost of the ring. Find the cost of the ring.

62. Todd is on a 1750 calorie-per-day diet plan. This plan permits 650 calories less than twice the number of calories permitted by Lerae's diet plan. How many calories are permitted by Lerae's plan?

63. The length of a rectangular shaped floor is 18 meters. This represents 2 meters less than five times the width of the floor. Find the width of the floor.

64. An executive is earning $45,000 per year. This represents $15,000 less than twice her salary four years ago. Find her salary four years ago.

65. The LM car dealership sold 32 cars during December of 1984. This represented 4 more than twice the number of cars they sold during December of 1983. How many cars did they sell during December of 1983?

66. A car repair bill was $119. This included $42 for parts and $22 for each hour of labor. Find the number of hours of labor.

67. An electric golf cart repair bill was $156. This included $36 for parts and $24 per hour for labor. How many hours of labor were involved?

68. Richard earned $305 last week for working 40 hours. Included in the $305 was a bonus of $25. How much was Richard's hourly pay?

Thoughts into Words

69. What is an algebraic equation?

70. Illustrate how the symmetric property of equality (if $a = b$, then $b = a$) can be used when solving equations.

71. What does the phrase "declare a variable" mean when solving a word problem?

3.3

More on Solving Equations and Problem Solving

As equations become more complex, additional steps are needed to solve them. This creates a need to carefully organize our work so as to minimize the chances for error. So let's begin this section with some suggestions for solving equations and then illustrate a *solution format* that is effective.

The process of solving first-degree equations of one variable can be summarized as follows.

Step 1. Simplify both sides of the equation as much as possible.

Step 2. Use the addition or subtraction property of equality to isolate a term containing the variable on one side and a constant on the other side of the equation.

Step 3. Use the multiplication or division property of equality to make the coefficient of the variable one.

The following examples illustrate this step-by-step process for solving equations. Study them carefully and be sure that you understand each step taken in the solving process.

EXAMPLE 1   Solve $5y - 4 + 3y = 12$.

Solution

$$5y - 4 + 3y = 12$$

$$8y - 4 = 12 \qquad \text{Combine similar terms on the left side.}$$

$$8y - 4 + 4 = 12 + 4 \qquad \text{Add 4 to both sides.}$$

$$8y = 16$$

$$\frac{8y}{8} = \frac{16}{8} \qquad \text{Divide both sides by 8.}$$

$$y = 2$$

The solution set is $\{2\}$. You can do the check alone now!   ■

EXAMPLE 2   Solve $7x - 2 = 3x + 9$.

Solution   Notice that both sides of the equation are in simplified form; thus, we can begin by applying the subtraction property of equality.

$$7x - 2 = 3x + 9$$

$$7x - 2 - 3x = 3x + 9 - 3x \qquad \text{Subtract } 3x \text{ from both sides.}$$

$$4x - 2 = 9$$

$$4x - 2 + 2 = 9 + 2 \qquad \text{Add 2 to both sides.}$$

$$4x = 11$$

$$\frac{4x}{4} = \frac{11}{4} \qquad \text{Divide both sides by 4.}$$

$$x = \frac{11}{4}$$

The solution set is $\left\{\dfrac{11}{4}\right\}$.   ■

EXAMPLE 3   Solve $5n + 12 = 9n - 16$.

Solution

$$5n + 12 = 9n - 16$$

$$5n + 12 - 9n = 9n - 16 - 9n \qquad \text{Subtract } 9n \text{ from both sides.}$$

$$-4n + 12 = -16$$

$$-4n + 12 - 12 = -16 - 12 \qquad \text{Subtract 12 from both sides.}$$

$$-4n = -28$$

$$\frac{-4n}{-4} = \frac{-28}{-4} \qquad \text{Divide both sides by } -4.$$

$$n = 7$$

The solution set is $\{7\}$.   ■

Word Problems

As we expand our skills for solving equations, we also expand our capabilities for solving verbal problems. There is no one definite procedure that will insure success at solving word problems, but the following suggestions can be helpful.

Suggestions for Solving Word Problems

1. Read the problem carefully, making certain that you understand the meanings of all the words. Be especially alert for any technical terms used in the statement of the problem.

2. Read the problem a second time (perhaps even a third time) to get an overview of the situation being described and to determine the known facts as well as what is to be found.

3. Sketch any figure, diagram, or chart that might be helpful in analyzing the problem.

4. Choose a meaningful variable to represent an unknown quantity in the problem (perhaps *t*, if time is an unknown quantity); represent any other unknowns in terms of that variable.

5. Look for a **guideline** that can be used to set up an equation. A guideline might be a formula such as "distance equals rate times time" or a statement of a relationship such as "the sum of the two numbers is 28." A guideline may also be indicated by a figure or diagram that you sketch for a particular problem.

6. Form an equation containing the variable that translates the conditions of the guideline from English to algebra.

7. Solve the equation and use the solution to determine all facts requested in the problem.

8. **Check** all answers back into the **original statement of the problem**.

If you decide not to check an answer, at least use the *reasonableness-of-answer* idea as a partial check. That is to say, ask yourself the question: Is this answer reasonable? For example, if the problem involves two investments totaling $10,000, then an answer of $12,000 for one investment is certainly *not reasonable*.

Now let's consider some problems and use these suggestions.

PROBLEM 1   Find two consecutive even numbers whose sum is 74.

Solution   To solve this problem we must know the meaning of the technical phrase "two consecutive even numbers." Two consecutive even numbers are two even numbers with one and only one whole number between them. For example, 2 and 4 are consecutive even numbers. Now we can proceed as follows. Let *n* represent the first even number; then $n + 2$ represents the next even number. Since their sum is 74, we can set up and solve the following equation.

$$n + n + 2 = 74$$

$$2n + 2 = 74$$

$$2n + 2 - 2 = 74 - 2$$

$$2n = 72$$

$$\frac{2n}{2} = \frac{72}{2}$$

$$n = 36$$

If $n = 36$, then $n + 2 = 38$; thus the numbers are 36 and 38.

> **CHECK** To check our answers for Problem 1, we must determine whether they satisfy the conditions stated in the original problem. Since 36 and 38 are two consecutive even numbers and $36 + 38 = 74$ (their sum is 74), we know that our answers are correct. ∎

Suggestion 5 in our list of problem solving suggestions was to "look for a *guideline* that can be used to set up an equation." The guideline may not be explicitly stated in the problem but may be implied by the nature of the problem. Consider the following example.

PROBLEM 2   Barry sells bicycles on a salary-plus-commission basis. He receives a monthly salary of $300 and a commission of $15 for each bicycle that he sells. How many bicycles must he sell in a month to have a total monthly salary of $750?

Solution   Let b represent the number of bicycles to be sold in a month. Then $15b$ represents his commission for those bicycles. The *guideline* "fixed salary plus commission equals total monthly salary" generates the following equation.

fixed salary + commission = total monthly salary

$$\$300 \;\; + \;\; 15b \;\; = \;\; \$750$$

Solving this equation yields

$$300 + 15b - 300 = 750 - 300$$

$$15b = 450$$

$$\frac{15b}{15} = \frac{450}{15}$$

$$b = 30.$$

He must sell 30 bicycles per month. (Does this number check?) ∎

Geometric Problems

Sometimes the guideline for setting up an equation to solve a problem is based on a geometric relationship. There are several basic geometric relationships pertaining to angle measure. Let's state some of these relationships and then consider some problems.

1. **Two** angles for which the sum of their measures is 90° (the symbol ° indicates degrees) are called **complementary angles**.
2. **Two** angles for which the sum of their measures is 180° are called **supplementary angles**.
3. The sum of the measures of the three angles of a triangle is 180°.

PROBLEM 3   One of two complementary angles is 14° larger than the other. Find the measure of each of the angles.

Solution   If we let a represent the measure of the smaller angle, then $a + 14$ represents the measure of the larger angle. Since they are complementary angles their sum is 90° and we can proceed as follows.

$$a + a + 14 = 90$$

$$2a + 14 = 90$$

$$2a + 14 - 14 = 90 - 14$$

$$2a = 76$$

$$\frac{2a}{2} = \frac{76}{2}$$

$$a = 38$$

If $a = 38$, then $a + 14 = 52$, and the angles are of measure 38° and 52°.   ■

PROBLEM 4   Find the measures of the three angles of a triangle if the second is three times the first and the third is twice the second.

Solution   If we let a represent the measure of the smallest angle, then $3a$ and $2(3a)$ represent the measures of the other two angles. Therefore, we can set up and solve the following equation.

$$a + 3a + 2(3a) = 180$$

$$a + 3a + 6a = 180$$

$$10a = 180$$

$$\frac{10a}{10} = \frac{180}{10}$$

$$a = 18$$

If $a = 18$, then $3a = 54$ and $2(3a) = 108$. So the angles have measures of 18°, 54°, and 108°.   ■

Problem Set 3.3

Solve each of the following equations.

1. $2x + 7 + 3x = 32$

2. $3x + 9 + 4x = 30$

3. $7x - 4 - 3x = -36$

4. $8x - 3 - 2x = -45$

5. $3y - 1 + 2y - 3 = 4$

6. $y + 3 + 2y - 4 = 6$

7. $5n - 2 - 8n = 31$

8. $6n - 1 - 10n = 51$

9. $-2n + 1 - 3n + n - 4 = 7$

10. $-n + 7 - 2n + 5n - 3 = -6$

11. $3x + 4 = 2x - 5$

12. $5x - 2 = 4x + 6$

13. $5x - 7 = 6x - 9$

14. $7x - 3 = 8x - 13$

15. $6x + 1 = 3x - 8$

16. $4x - 10 = x + 17$

17. $7y - 3 = 5y + 10$

18. $8y + 4 = 5y - 4$

19. $8n - 2 = 11n - 7$

20. $7n - 10 = 9n - 13$

21. $-2x - 7 = -3x + 10$

22. $-4x + 6 = -5x - 9$

23. $-3x + 5 = -5x - 8$

24. $-4x + 7 = -6x + 4$

25. $-7 - 6x = 9 - 9x$

26. $-10 - 7x = 14 - 12x$

27. $2x - 1 - x = 3x - 5$

28. $3x - 4 - 4x = 5 - 5x + 3x$

29. $5n - 4 - n = -3n - 6 + n$

30. $4x - 3 + 2x = 8x - 3 - x$

31. $-7 - 2n - 6n = 7n - 5n + 12$

32. $-3n + 6 + 5n = 7n - 8n - 9$

Solve each of the following verbal problems by setting up and solving an algebraic equation.

33. The sum of a number plus four times the number is 85. What is the number?

34. A number subtracted from three times the number yields 68. Find the number.

35. Find two consecutive odd numbers whose sum is 72.

36. Find two consecutive even numbers whose sum is 94.

37. Find three consecutive even numbers whose sum is 1ı

38. Find three consecutive odd numbers whose sum is 159.

39. Two more than three times a certain number is the same as four less than seven times the number. Find the number.

40. One more than five times a certain number is equal to eleven less than nine times the number. What is the number?

41. The sum of a number and five times the number equals eighteen less than three times the number. Find the number.

42. One of two supplementary angles is five times as large as the other. Find the measure of each angle.

43. One of two complementary angles is six less than twice the other angle. Find the measure of each angle.

44. If two angles are complementary and the difference of their measures is 62°, find the measure of each angle.

45. If two angles are supplementary and the larger angle is 20° less than three times the smaller angle, find the measure of each angle.

46. Find the measures of the three angles of a triangle if the largest is 14° less than three times the smallest and the other angle is 4° more than the smallest.

47. One of the angles of a triangle has a measure of 40°. Find the measures of the other two angles if the difference of their measures is 10°.

48. Larry sold encyclopedias one summer on a salary-plus-commission basis. He was given a flat salary of $1800 for the summer plus a commission of $75 for each set of encyclopedias that he sold. His total earnings for the summer were $2550. How many sets of encyclopedias did he sell?

49. Daniel sold some stock at $35 per share. This was $17 a share less than twice what he paid for it. What price did he pay for each share of the stock?

50. Becky sold some stock at $37 per share. This was $14 a share less than three times what she paid for it. What price did she pay for the stock?

51. Bob is paid two times his normal hourly rate for each hour worked over 40 hours in a week. Last week he earned $504 for 48 hours of work. What is his hourly wage?

52. Last week on an algebra test, the highest grade was 9 points less than three times the lowest grade. The sum of the two grades was 135. Find the lowest and highest grades on the test.

53. At a concert there were three times as many females as males. A total of 600 people attended the concert. How many males and how many females attended?

54. Suppose that a triangular lot is enclosed with 135 yards of fencing. The longest side of the lot is 5 yards more than twice the length of the shortest side. The other side is 10 yards longer than the shortest side. Find the lengths of the three sides of the lot.

55. Melton and Sanchez were opposing candidates in a school board election. Sanchez received ten more than twice the number of votes that Melton received. If there was a total of 1030 votes cast, how many did Sanchez receive?

56. The plans for an apartment complex provided for 260 apartments, with each apartment having one, two, or three bedrooms. There were to be 25 more three-bedroom than one-bedroom apartments, and the number of two-bedroom apartments was to be 15 less than three times the number of one-bedroom apartments. How many apartments of each type were in the plans?

57. A board 20 feet long is to be cut into two pieces so that one piece is 8 feet longer than the other. How long should the shorter piece be?

58. A wire 110 centimeters long was bent in the shape of a rectangle. The length of the rectangle was 10 centimeters more than twice the width. Find the dimensions of the rectangle.

59. Seventy-eight yards of fencing was purchased to enclose a rectangular garden. The length of the garden is 1 yard less than three times its width. Find the length and width of the garden.

Thoughts into Words

60. Give a step-by-step description of how you would solve the equation $3x - 2 - x = 5x + 3 + 4x$.

61. Why must potential answers to word problems be checked back into the original statement of the problem?

Miscellaneous Problems

62. Solve each of the following equations.

(a) $7x - 3 = 4x - 3$

(b) $-x - 4 + 3x = 2x - 7$

(c) $-3x + 9 - 2x = -5x + 9$

(d) $5x - 3 = 6x - 7 - x$

(e) $7x + 4 = -x + 4 + 8x$

(f) $3x - 2 - 5x = 7x - 2 - 5x$

(g) $-6x - 8 = 6x + 4$

(h) $-8x + 9 = -8x + 5$

3.4
Equations Involving Parentheses and Fractional Forms

We will use the distributive property frequently in this section as we expand our techniques for solving equations. Recall that in symbolic form the distributive property states that $a(b + c) = ab + ac$. Consider the following examples, which illustrate using this property to *remove parentheses*. Pay special attention to the last two examples that involve a negative number in front of the parentheses.

$$3(x + 2) = \quad 3 \cdot x + 3 \cdot 2 \quad = 3x + 6,$$

$$5(y - 3) = \quad 5 \cdot y - 5 \cdot 3 \quad = 5y - 15, \qquad (a(b - c) = ab - ac)$$

$$2(4x + 7) = \quad 2(4x) + 2(7) \quad = 8x + 14,$$

$$-1(n + 4) = \quad (-1)(n) + (-1)(4) \quad = -n - 4,$$

$$-6(x - 2) = \quad (-6)(x) - (-6)(2) \quad = -6x + 12$$

$\longrightarrow$ Do this step mentally!

It is often necessary to solve equations in which the variable is part of an expression enclosed in parentheses. The distributive property is used to remove the parentheses; we then proceed in the usual way. Consider the following examples. (Notice we are beginning to show *only the major steps* when solving an equation.)

EXAMPLE 1   Solve $3(x + 2) = 23$.

Solution

$$3(x + 2) = 23$$

$$3x + 6 = 23 \qquad \text{applied distributive property to left side}$$

$$3x = 17 \qquad \text{subtracted 6 from both sides}$$

$$x = \frac{17}{3} \qquad \text{divided both sides by 3}$$

The solution set is $\left\{ \frac{17}{3} \right\}$. ∎

EXAMPLE 2   Solve $4(x + 3) = 2(x - 6)$.

Solution

$$4(x + 3) = 2(x - 6)$$

$4x + 12 = 2x - 12$ applied distributive property on each side

$2x + 12 = -12$ subtracted $2x$ from both sides

$2x = -24$ subtracted 12 from both sides

$x = -12$ divided both sides by 2

The solution set is $\{-12\}$. ∎

It may be necessary to remove more than one set of parentheses and then to use the distributive property again to combine similar terms. Consider the following two examples.

EXAMPLE 3   Solve $5(w + 3) + 3(w + 1) = 14$.

Solution

$$5(w + 3) + 3(w + 1) = 14$$

$5w + 15 + 3w + 3 = 14$ applied distributive property

$8w + 18 = 14$ combined similar terms

$8w = -4$ subtracted 18 from both sides

$w = -\dfrac{4}{8}$ divided both sides by 8

$w = -\dfrac{1}{2}$ reduced

The solution set is $\left\{-\dfrac{1}{2}\right\}$. ∎

EXAMPLE 4   Solve $6(x - 7) - 2(x - 4) = 13$.

Solution

$6(x - 7) - 2(x - 4) = 13$ ←-Be careful with this sign!

$6x - 42 - 2x + 8 = 13$ distributive property

$4x - 34 = 13$ combined similar terms

$4x = 47$ added 34 to both sides

$x = \dfrac{47}{4}$ divided both sides by 4

The solution set is $\left\{\dfrac{47}{4}\right\}$. ∎

In a previous section we solved equations like $x - \dfrac{2}{3} = \dfrac{3}{4}$ by adding $\dfrac{2}{3}$ to both sides. If an equation contains several fractions, then it is usually easier to *clear the equation of all fractions* by multiplying both sides by the least common denominator of all the denominators. Perhaps several examples will clarify this idea.

EXAMPLE 5   Solve $\dfrac{1}{2}x + \dfrac{2}{3} = \dfrac{5}{6}$.

Solution

$$\frac{1}{2}x + \frac{2}{3} = \frac{5}{6}$$

$$6\left(\frac{1}{2}x + \frac{2}{3}\right) = 6\left(\frac{5}{6}\right) \qquad \text{6 is the LCD of 2,3, and 6.}$$

$$6\left(\frac{1}{2}x\right) + 6\left(\frac{2}{3}\right) = 6\left(\frac{5}{6}\right) \qquad \text{distributive property}$$

$$3x + 4 = 5 \qquad \begin{array}{l}\text{Note how the equation has been}\\ \textit{cleared of all fractions.}\end{array}$$

$$3x = 1$$

$$x = \frac{1}{3}$$

The solution set is $\left\{\dfrac{1}{3}\right\}$. ∎

EXAMPLE 6   Solve $\dfrac{5n}{6} - \dfrac{1}{4} = \dfrac{3}{8}$.

Solution

$$\frac{5n}{6} - \frac{1}{4} = \frac{3}{8} \qquad \text{Remember } \frac{5n}{6} = \frac{5}{6}n.$$

$$24\left(\frac{5n}{6} - \frac{1}{4}\right) = 24\left(\frac{3}{8}\right) \qquad \text{24 is the LCD of 6,4, and 8.}$$

$$24\left(\frac{5n}{6}\right) - 24\left(\frac{1}{4}\right) = 24\left(\frac{3}{8}\right) \qquad \text{distributive property}$$

$$20n - 6 = 9$$

$$20n = 15$$

$$n = \frac{15}{20} = \frac{3}{4}$$

The solution set is $\left\{\dfrac{3}{4}\right\}$. ∎

We use many of the ideas presented in this section to help solve the following equations. Study the solutions carefully and be sure that you can supply reasons for each step. It might be helpful to cover up the solutions and to try to solve the equations on your own.

EXAMPLE 7   Solve $\dfrac{x+3}{2} + \dfrac{x+4}{5} = \dfrac{3}{10}$.

Solution

$$\frac{x+3}{2} + \frac{x+4}{5} = \frac{3}{10}$$

$$10\left(\frac{x+3}{2} + \frac{x+4}{5}\right) = 10\left(\frac{3}{10}\right) \qquad \text{10 is the LCD of 2,5, and 10.}$$

$$10\left(\frac{x+3}{2}\right) + 10\left(\frac{x+4}{5}\right) = 10\left(\frac{3}{10}\right) \qquad \text{distributive property}$$

$$5(x+3) + 2(x+4) = 3$$

$$5x + 15 + 2x + 8 = 3$$

$$7x + 23 = 3$$

$$7x = -20$$

$$x = -\frac{20}{7}$$

The solution set is $\left\{-\dfrac{20}{7}\right\}$.   ∎

EXAMPLE 8   Solve $\dfrac{x-1}{4} - \dfrac{x-2}{6} = \dfrac{2}{3}$.

Solution

$$\frac{x-1}{4} - \frac{x-2}{6} = \frac{2}{3}$$

$$12\left(\frac{x-1}{4} - \frac{x-2}{6}\right) = 12\left(\frac{2}{3}\right) \qquad \text{12 is the LCD of 4,6, and 3.}$$

$$12\left(\frac{x-1}{4}\right) - 12\left(\frac{x-2}{6}\right) = 12\left(\frac{2}{3}\right) \qquad \text{distributive property}$$

$$3(x-1) - 2(x-2) = 8$$

$$3x - 3 - 2x + 4 = 8 \qquad \text{Be careful with this sign.}$$

$$x + 1 = 8$$

$$x = 7$$

The solution set is $\{7\}$.   ∎

Word Problems

We are now ready to solve some verbal problems using equations of the different types presented in this section. Again, it might be helpful for you to attempt to solve the problems on your own before looking at the book's approach.

PROBLEM 1   Loretta has 19 coins consisting of quarters and nickels, which amount to $2.35. How many coins of each kind does she have?

Solution   Let q represent the number of quarters. Then $19 - q$ represents the number of nickels. We can use the following guideline to help set up an equation.

$$\underset{\downarrow}{\underset{\text{in cents}}{\text{value of quarters}}} + \underset{\downarrow}{\underset{\text{in cents}}{\text{value of nickels}}} = \underset{\downarrow}{\underset{\text{in cents}}{\text{total value}}}$$

$$25q \quad + \quad 5(19 - q) \quad = \quad 235$$

Solving the equation we obtain

$$25q + 95 - 5q = 235$$

$$20q + 95 = 235$$

$$20q = 140$$

$$q = 7.$$

If $q = 7$, then $19 - q = 12$; so she has 7 quarters and 12 nickels.   ∎

PROBLEM 2   Find a number such that 4-less-than-two-thirds the number is equal to one-sixth the number.

Solution   Let n represent the number. Then $\dfrac{2}{3}n - 4$ represents 4 less than two-thirds the number, and $\dfrac{1}{6}n$ represents one-sixth the number.

$$\frac{2}{3}n - 4 = \frac{1}{6}n$$

$$6\left(\frac{2}{3}n - 4\right) = 6\left(\frac{1}{6}n\right)$$

$$4n - 24 = n$$

$$3n - 24 = 0$$

$$3n = 24$$

$$n = 8$$

The number is 8.   ∎

PROBLEM 3   Lance is paid $1\frac{1}{2}$ times his normal hourly rate for each hour worked over 40 hours in a week. Last week he worked 50 hours and earned $462. What is his normal hourly rate?

Solution   Let x represent his normal hourly rate. Then $\frac{3}{2}x$ represents $1\frac{1}{2}$ times his normal hourly rate. The following guideline can be used to help set up the equation.

regular wages for first 40 hours	+	wages for 10 hours of overtime	=	total wages
↓		↓		↓

$$40x \quad + \quad 10\left(\frac{3}{2}x\right) \quad = \quad 462$$

Solving this equation we obtain

$$40x + 15x = 462$$
$$55x = 462$$
$$x = 8.40.$$

His normal hourly rate is $8.40.   ■

PROBLEM 4   Find three consecutive whole numbers such that the sum of the first plus twice the second plus three times the third is 134.

Solution   Let n represent the first whole number. Then $n + 1$ represents the second whole number and $n + 2$ represents the third whole number.

$$n + 2(n + 1) + 3(n + 2) = 134$$
$$n + 2n + 2 + 3n + 6 = 134$$
$$6n + 8 = 134$$
$$6n = 126$$
$$n = 21$$

The numbers are 21, 22, and 23.   ■

Problem Set 3.4

Solve each of the following equations.

1. $7(x + 2) = 21$ 2. $4(x + 4) = 24$
3. $5(x - 3) = 35$ 4. $6(x - 2) = 18$
5. $-3(x + 5) = 12$ 6. $-5(x - 6) = -15$
7. $4(n - 6) = 5$ 8. $3(n + 4) = 7$

9. $6(n + 7) = 8$

10. $8(n - 3) = 12$

11. $-10 = -5(t - 8)$

12. $-16 = -4(t + 7)$

13. $5(x - 4) = 4(x + 6)$

14. $7(x + 3) = 6(x - 8)$

15. $8(x + 1) = 9(x - 2)$

16. $4(x - 7) = 5(x + 2)$

17. $8(t + 5) = 6(t - 6)$

18. $7(t - 5) = 5(t + 3)$

19. $3(2t + 1) = 4(3t - 2)$

20. $2(3t - 1) = 3(3t - 2)$

21. $-2(x - 6) = -(x - 9)$

22. $-(x + 7) = -2(x + 10)$

23. $-3(t - 4) - 2(t + 4) = 9$

24. $5(t - 4) - 3(t - 2) = 12$

25. $3(n - 10) - 5(n + 12) = -86$

26. $4(n + 9) - 7(n - 8) = 83$

27. $3(x + 1) + 4(2x - 1) = 5(2x + 3)$

28. $4(x - 1) + 5(x + 2) = 3(x - 8)$

29. $-(x + 2) + 2(x - 3) = -2(x - 7)$

30. $-2(x + 6) + 3(3x - 2) = -3(x - 4)$

31. $5(2x - 1) - (3x + 4) = 4(x + 3) - 27$

32. $3(4x + 1) - 2(2x + 1) = -2(x - 1) - 1$

33. $-(a - 1) - (3a - 2) = 6 + 2(a - 1)$

34. $3(2a - 1) - 2(5a + 1) = 4(3a + 4)$

35. $3(x - 1) + 2(x - 3) = -4(x - 2) + 10(x + 4)$

36. $-2(x - 4) - (3x - 2) = -2 + (-6x + 2)$

37. $3 - 7(x - 1) = 9 - 6(2x + 1)$

38. $8 - 5(2x + 1) = 2 - 6(x - 3)$

39. $\dfrac{3}{4}x - \dfrac{2}{3} = \dfrac{5}{6}$

40. $\dfrac{1}{2}x - \dfrac{4}{3} = -\dfrac{5}{6}$

41. $\dfrac{5}{6}x + \dfrac{1}{4} = -\dfrac{9}{4}$

42. $\dfrac{3}{8}x + \dfrac{1}{6} = -\dfrac{7}{12}$

43. $\dfrac{1}{2}x - \dfrac{3}{5} = \dfrac{3}{4}$

44. $\dfrac{1}{4}x - \dfrac{2}{5} = \dfrac{5}{6}$

45. $\dfrac{n}{3} + \dfrac{5n}{6} = \dfrac{1}{8}$

46. $\dfrac{n}{6} + \dfrac{3n}{8} = \dfrac{5}{12}$

47. $\dfrac{5y}{6} - \dfrac{3}{5} = \dfrac{2y}{3}$

48. $\dfrac{3y}{7} + \dfrac{1}{2} = \dfrac{y}{4}$

49. $\dfrac{h}{6} + \dfrac{h}{8} = 1$

50. $\dfrac{h}{4} + \dfrac{h}{3} = 1$

51. $\dfrac{x + 2}{3} + \dfrac{x + 3}{4} = \dfrac{13}{3}$

52. $\dfrac{x - 1}{4} + \dfrac{x + 2}{5} = \dfrac{39}{20}$

53. $\dfrac{x - 1}{5} - \dfrac{x + 4}{6} = -\dfrac{13}{15}$

54. $\dfrac{x + 1}{7} - \dfrac{x - 3}{5} = \dfrac{4}{5}$

55. $\dfrac{x + 8}{2} - \dfrac{x + 10}{7} = \dfrac{3}{4}$

56. $\dfrac{x + 7}{3} - \dfrac{x + 9}{6} = \dfrac{5}{9}$

57. $\dfrac{x - 2}{8} - 1 = \dfrac{x + 1}{4}$

58. $\dfrac{x - 4}{2} + 3 = \dfrac{x - 2}{4}$

59. $\dfrac{x + 1}{4} = \dfrac{x - 3}{6} + 2$

60. $\dfrac{x + 3}{5} = \dfrac{x - 6}{2} + 1$

Solve each of the following problems by setting up and solving an appropriate algebraic equation.

61. Find two consecutive whole numbers such that the smaller plus four times the larger equals 39.

62. Find two consecutive whole numbers such that the smaller subtracted from five times the larger equals 57.

63. Find three consecutive whole numbers such that twice the sum of the two smallest numbers is ten more than three times the largest number.

64. Find four consecutive whole numbers such that the sum of the first three equals the fourth number.

65. The sum of two numbers is 17. If twice the smaller is one more than the larger, find the numbers.

66. The sum of two numbers is 53. If three times the smaller is one less than the larger, find the numbers.

67. Find a number such that 20 more than one-third of the number equals three-fourths of the number.

68. The sum of three-eighths of a number and five-sixths of the same number is 29. Find the number.

69. The difference of two numbers is six. One-half of the larger number is five larger than one-third of the smaller.

70. The difference of two numbers is 16. Three-fourths of the larger number is 14 larger than one-half of the smaller. Find the numbers.

71. Suppose that a board 20 feet long is cut into two pieces. Four times the length of the shorter piece is 4 feet less than three times the length of the longer piece. Find the length of each piece.

72. Ellen is paid "time and a half" for each hour over 40 hours worked in a week. Last week she worked 44 hours and earned $299. What is her normal hourly rate?

73. Lucy has 69 coins consisting of nickels, dimes, and quarters. She has five more dimes than nickels and the number of quarters is four more than three times the number of nickels. How many coins of each kind does she have?

74. Suppose that Julian has 44 coins consisting of pennies and nickels. If the number of nickels is two more than twice the number of pennies, find the number of coins of each kind.

75. Max has a collection of 210 coins consisting of nickels, dimes, and quarters. He has twice as many dimes as nickels and ten more quarters than dimes. How many coins of each kind does he have?

76. Ginny has a collection of 425 coins consisting of pennies, nickels, and dimes. She has 50 more nickels than pennies and 25 more dimes than nickels. How many coins of each kind does she have?

77. Mary has 18 coins consisting of dimes and quarters amounting to $3.30. How many coins of each kind does she have?

78. Ike has some nickels and dimes amounting to $2.90. The number of dimes is one less than twice the number of nickels. How many coins of each kind does he have?

79. A collection of pennies, nickels, and dimes totals $30.75. There are twice as many nickels as pennies and three times as many dimes as pennies. How many coins of each kind are in the collection?

80. Tickets for a concert were priced at $3 for students and $5 for nonstudents. There were 1500 tickets sold for a total of $5000. How many student tickets were sold?

81. The supplement of an angle is 30° more than twice its complement. Find the measure of the angle.

82. The sum of the measure of an angle and three times its complement is 202°. Find the measure of the angle.

83. In triangle ABC, the measure of angle A is 2° less than one-fifth of the measure of angle C. The measure of angle B is 5° less than one-half of the measure of angle C. Find the measures of the three angles of the triangle.

84. If one-fourth of the complement of an angle plus one-fifth of the supplement of the angle equals 36°, find the measure of the angle.

85. The supplement of an angle is 10° less than three times its complement. Find the size of the angle.

86. In triangle ABC, the measure of angle C is eight times the measure of angle A and the measure of angle B is 10° more than the measure of angle C. Find the measure of each angle of the triangle.

Miscellaneous Problems

87. Solve each of the following equations.

(a) $-2(x - 1) = -2x + 2$

(b) $3(x + 4) = 3x - 4$

(c) $5(x - 1) = -5x - 5$

(d) $\dfrac{x - 3}{3} + 4 = 3$

(e) $\dfrac{x + 2}{3} + 1 = \dfrac{x - 2}{3}$

(f) $\dfrac{x - 1}{5} - 2 = \dfrac{x - 11}{5}$

(g) $4(x - 2) - 2(x + 3) = 2(x + 6)$

(h) $5(x + 3) - 3(x - 5) = 2(x + 15)$

(i) $7(x - 1) + 4(x - 2) = 15(x - 1)$

88. Find three consecutive integers such that the sum of the smallest and largest is equal to twice the middle integer.

3.5

Ratios, Proportions, and Percents

In Figure 3.1, as gear A revolves 4 times, gear B will revolve 3 times. We say that the *gear ratio* of A to B is 4 to 3 or the gear ratio of B to A is 3 to 4. Mathematically, a **ratio** *is the comparison of two numbers by division*. The gear ratio of A to B can be

Figure 3.1

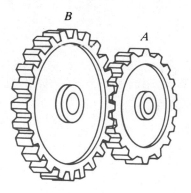

written as

$$4 \text{ to } 3 \quad \text{or} \quad 4:3 \quad \text{or} \quad \frac{4}{3}.$$

Ratios are commonly expressed as fractions in reduced form. For example, if there are 7500 females and 5000 males at a certain university, then the ratio of females to males is $\dfrac{7500}{5000} = \dfrac{3}{2}$.

A statement of equality between two ratios is called a **proportion**. For example,

$$\frac{2}{3} = \frac{8}{12}$$

is a proportion that states that the ratios $\dfrac{2}{3}$ and $\dfrac{8}{12}$ are equal. In the general proportion

$$\frac{a}{b} = \frac{c}{d}, \qquad b \neq 0 \text{ and } d \neq 0,$$

if we multiply both sides of the equation by the common denominator, bd, we obtain

$$(bd)\left(\frac{a}{b}\right) = (bd)\left(\frac{c}{d}\right)$$

$$ad = bc.$$

Let's state this as a property of proportions.

$$\frac{a}{b} = \frac{c}{d} \quad \text{if and only if } ad = bc, \quad \text{where } b \neq 0 \text{ and } d \neq 0$$

EXAMPLE 1   Solve $\dfrac{x}{20} = \dfrac{3}{4}$.

Solution

$$\dfrac{x}{20} = \dfrac{3}{4}$$

$$4x = 60 \qquad \text{Cross products are equal.}$$

$$x = 15$$

The solution set is $\{15\}$.   ■

EXAMPLE 2   Solve $\dfrac{x-3}{5} - \dfrac{x+2}{4}$.

Solution

$$\dfrac{x-3}{5} = \dfrac{x+2}{4}$$

$$4(x-3) = 5(x+2) \qquad \text{Cross products are equal.}$$

$$4x - 12 = 5x + 10 \qquad \text{distributive property}$$

$$-12 = x + 10 \qquad \text{subtracted } 4x \text{ from both sides}$$

$$-22 = x \qquad \text{subtracted 10 from both sides}$$

The solution set is $\{-22\}$.   ■

If a variable appears in one or both of the denominators, then a proper restriction should be made to avoid division by zero, as the next example illustrates.

EXAMPLE 3   Solve $\dfrac{7}{a-2} = \dfrac{4}{a+3}$.

Solution

$$\dfrac{7}{a-2} = \dfrac{4}{a+3}, \qquad a \neq 2 \text{ and } a \neq -3$$

$$7(a+3) = 4(a-2) \qquad \text{Cross products are equal.}$$

$$7a + 21 = 4a - 8 \qquad \text{distributive property}$$

$$3a + 21 = -8 \qquad \text{subtracted } 4a \text{ from both sides}$$

$$3a = -29 \qquad \text{subtracted 21 from both sides}$$

$$a = -\dfrac{29}{3} \qquad \text{divided both sides by 3}$$

The solution set is $\left\{ -\dfrac{29}{3} \right\}$.   ■

EXAMPLE 4   Solve $\frac{x}{4} + 3 = \frac{x}{5}$.

Solution   This is *not* a proportion, so let's multiply both sides by 20 to clear the equation of all fractions.

$$\frac{x}{4} + 3 = \frac{x}{5}$$

$$20\left(\frac{x}{4} + 3\right) = 20\left(\frac{x}{5}\right) \qquad \text{Multiply both sides by 20.}$$

$$20\left(\frac{x}{4}\right) + 20(3) = 20\left(\frac{x}{5}\right) \qquad \text{applied the distributive property}$$

$$5x + 60 = 4x$$

$$x + 60 = 0 \qquad \text{subtracted } 4x \text{ from both sides}$$

$$x = -60 \qquad \text{subtracted 60 from both sides}$$

The solution set is $\{-60\}$. ■

REMARK Example 4 demonstrates the importance of "thinking first before pushing the pencil." Since the equation was not in the form of a proportion, we needed to revert back to a previous technique for solving such equations.

Problem Solving Using Proportions

Some word problems can be conveniently set up and solved using the concepts of ratio and proportion. Consider the following examples.

PROBLEM 1   On a certain map, 1 inch represents 20 miles. If two cities are $6\frac{1}{2}$ inches apart on the map, find the number of miles between the cities.

Solution   Let m represent the number of miles between the two cities. Now let's set up a proportion for which one ratio compares distance in inches on the map, and the other ratio compares *corresponding* distance in miles on land.

$$\frac{1}{6\frac{1}{2}} = \frac{20}{m}$$

Solving this equation by equating the cross products produces

$$m(1) = \left(6\frac{1}{2}\right)(20)$$

$$m = \left(\frac{13}{2}\right)(20) = 130.$$

The distance between the two cities is 130 miles.   ■

PROBLEM 2   A sum of \$1750 is to be divided between two people in the ratio of 3 to 4. How much does each person receive?

Solution   Let d represent the amount of money to be received by one person. Then $1750 - d$ represents the amount for the other person. The following proportion can be set up and solved.

$$\frac{d}{1750 - d} = \frac{3}{4}$$

$$4d = 3(1750 - d)$$

$$4d = 5250 - 3d$$

$$7d = 5250$$

$$d = 750$$

If $d = 750$, then $1750 - d = 1000$; therefore, one person receives \$750 and the other person receives \$1000.   ■

Percent

The word *percent* means "per one hundred" and we use the symbol % to express it. For example, 7 percent is written as 7% and means $\frac{7}{100}$ or 0.07. In other words, percent is a special kind of ratio, namely, one in which the denominator is always 100. Proportions provide a convenient basis for changing common fractions to percents. Consider the following examples.

EXAMPLE 5   Express $\frac{7}{20}$ as a percent.

Solution   We are asking "what number compares to 100 as 7 compares to 20?" Therefore, if we let n represent that number, the following proportion can be set up and solved.

$$\frac{n}{100} = \frac{7}{20}$$

$$20n = 700$$

$$n = 35$$

Thus, $\frac{7}{20} = \frac{35}{100} = 35\%$. ∎

EXAMPLE 6   Express $\frac{5}{6}$ as a percent.

Solution $$\frac{n}{100} = \frac{5}{6}$$

$$6n = 500$$

$$n = \frac{500}{6} = \frac{250}{3} = 83\frac{1}{3}$$

Therefore, $\frac{5}{6} = 83\frac{1}{3}\%$. ∎

Some Basic Percent Problems

What is 8% of 35? Fifteen percent of what number is 24? Twenty-one is what percent of 70? These are often referred to as the three basic types of percent problems. Each of these problems can be easily solved by translating into and solving a simple algebraic equation.

PROBLEM 3   What is 8% of 35?

Solution   Let n represent the number to be found. The "is" refers to equality and "of" means multiplication. Thus, the question translates into

$$n = (8\%)(35),$$

which can be solved as follows.

$$n = (0.08)(35)$$
$$= 2.8$$

Therefore, 2.8 is 8% of 35. ∎

PROBLEM 4   Fifteen percent of what number is 24?

Solution   Let n represent the number to be found.

$$(15\%)(n) = 24$$

$$0.15n = 24$$

$$15n = 2400 \qquad \text{multiplied both sides by 100}$$

$$n = 160$$

Therefore, 15% of 160 is 24.   ■

PROBLEM 5   Twenty-one is what percent of 70?

Solution   Let r represent the percent to be found.

$$21 = r(70)$$

$$\frac{21}{70} = r$$

$$\frac{3}{10} = r \qquad \text{Reduce!}$$

$$\frac{30}{100} = r \qquad \text{changed } \frac{3}{10} \text{ to } \frac{30}{100}$$

$$30\% = r$$

Therefore, 21 is 30% of 70.   ■

PROBLEM 6   Seventy-two is what percent of 60?

Solution   Let r represent the percent to be found.

$$72 = r(60)$$

$$\frac{72}{60} = r$$

$$\frac{6}{5} = r$$

$$\frac{120}{100} = r$$

$$120\% = r$$

Therefore, 72 is 120% of 60.   ■

Again, it is helpful to get into the habit of checking answers for "reasonableness." We also suggest that you alert yourself to a potential computational error by *estimating* the answer before you actually do the problem. For example, prior to doing Problem 6 we may have estimated as follows: Since 72 is larger than 60, we know that the answer has to be greater than 100%. Furthermore, 1.5 (or 150%) times 60 equals 90. Therefore, we can estimate the answer to be somewhere between 100% and 150%. That may seem like a rather rough estimate, but many times such an estimate will detect a computational error.

Problem Set 3.5

For Problems 1–32, solve each of the equations.

1. $\dfrac{x}{6} = \dfrac{3}{2}$

2. $\dfrac{x}{9} = \dfrac{5}{3}$

3. $\dfrac{5}{12} = \dfrac{n}{24}$

4. $\dfrac{7}{8} = \dfrac{n}{16}$

5. $\dfrac{x}{3} = \dfrac{5}{2}$

6. $\dfrac{x}{7} = \dfrac{4}{3}$

7. $\dfrac{x-2}{4} = \dfrac{x+4}{3}$

8. $\dfrac{x-6}{7} = \dfrac{x+9}{8}$

9. $\dfrac{x+1}{6} = \dfrac{x+2}{4}$

10. $\dfrac{x-2}{6} = \dfrac{x-6}{8}$

11. $\dfrac{h}{2} - \dfrac{h}{3} = 1$

12. $\dfrac{h}{5} + \dfrac{h}{4} = 2$

13. $\dfrac{x+1}{3} - \dfrac{x+2}{2} = 4$

14. $\dfrac{x-2}{5} - \dfrac{x+3}{6} = -4$

15. $\dfrac{-4}{x+2} = \dfrac{-3}{x-7}$

16. $\dfrac{-9}{x+1} = \dfrac{-8}{x+5}$

17. $\dfrac{-1}{x-7} = \dfrac{5}{x-1}$

18. $\dfrac{3}{x-10} = \dfrac{-2}{x+6}$

19. $\dfrac{3}{2x-1} = \dfrac{2}{3x+2}$

20. $\dfrac{1}{4x+3} = \dfrac{2}{5x-3}$

21. $\dfrac{n+1}{n} = \dfrac{8}{7}$

22. $\dfrac{5}{6} = \dfrac{n}{n+1}$

23. $\dfrac{x-1}{2} - 1 = \dfrac{3}{4}$

24. $-2 + \dfrac{x+3}{4} = \dfrac{5}{6}$

25. $-3 - \dfrac{x+4}{5} = \dfrac{3}{2}$

26. $\dfrac{x-5}{3} + 2 = \dfrac{5}{9}$

27. $\dfrac{n}{150-n} = \dfrac{1}{2}$

28. $\dfrac{n}{200-n} = \dfrac{3}{5}$

29. $\dfrac{300-n}{n} = \dfrac{3}{2}$

30. $\dfrac{80-n}{n} = \dfrac{7}{9}$

31. $\dfrac{-1}{5x-1} = \dfrac{-2}{3x+7}$

32. $\dfrac{-3}{2x-5} = \dfrac{-4}{x-3}$

For Problems 33–44, use proportions to change each common fraction to a percent.

33. $\dfrac{11}{20}$ 34. $\dfrac{17}{20}$ 35. $\dfrac{3}{5}$ 36. $\dfrac{7}{25}$ 37. $\dfrac{1}{6}$ 38. $\dfrac{5}{7}$

39. $\dfrac{3}{8}$ 40. $\dfrac{1}{16}$ 41. $\dfrac{3}{2}$ 42. $\dfrac{5}{4}$ 43. $\dfrac{12}{5}$ 44. $\dfrac{13}{6}$

For Problems 45–56, answer the question by setting up and solving an appropriate equation.

45. What is 7% of 38?

46. What is 35% of 52?

47. 15% of what number is 6.3?

48. 55% of what number is 38.5?

49. 76 is what percent of 95?

50. 72 is what percent of 120?

51. What is 120% of 50?

52. What is 160% of 70?

53. 46 is what percent of 40?

54. 26 is what percent of 20?

55. 160% of what number is 144?

56. 220% of what number is 66?

For Problems 57–73, solve each problem using a proportion.

57. A blueprint has a scale that 1 inch represents 6 feet. Find the dimensions of a rectangular room that measures $2\frac{1}{2}$ inches by $3\frac{1}{4}$ inches on the blueprint.

58. On a certain map, 1 inch represents 15 miles. If two cities are 7 inches apart on the map, find the number of miles between the cities.

59. A car can travel 264 miles on 12 gallons of gasoline. How far will it go on 15 gallons?

60. Jesse used 10 gallons of gasoline to drive 170 miles. How much gasoline will he need to travel 238 miles?

61. If the ratio of the length of a rectangle to its width is $\frac{5}{2}$ and the width is 24 centimeters, find its length.

62. If the ratio of the width of a rectangle to its length is $\frac{4}{5}$ and the length is 45 centimeters, find the width.

63. A saltwater solution is made by dissolving 3 pounds of salt in 10 gallons of water. At this rate, how many pounds of salt are needed for 25 gallons of water?

64. A home valued at $50,000 is assessed $900 in real estate taxes. At the same rate, how much are the taxes on a home assessed at $60,000?

65. If 20 pounds of fertilizer will cover 1500 square feet of lawn, how many pounds are needed for 2500 square feet?

66. It was reported that a flu epidemic is affecting 6 out of every 10 college students in a certain part of the country. At this rate, how many students will be affected at a university of 15,000 students?

67. A poll indicated that 3 out of every 7 voters were going to vote in an upcoming election. At this rate, how many people would be expected to vote in a city of 210,000?

68. A board 28 feet long is cut into two pieces whose lengths are in the ratio of 2 to 5. Find the lengths of the two pieces.

69. The perimeter of a rectangle is 50 inches. If the ratio of its length to its width is 3 to 2, find the dimensions of the rectangle.

70. The ratio of male to female students at a university is 5 to 4. If there is a total of 6975 students, find the number of male students and the number of female students.

71. An investment of $500 earns $45 in a year. At the same rate, how much additional money must be invested to raise the earnings to $72 per year?

72. A sum of $1250 is to be divided between two people in the ratio of 2 to 3. How much does each person receive?

73. An inheritance of $180,000 is to be divided between a child and the local cancer fund in the ratio of 5 to 1. How much money will the child receive?

Miscellaneous Problems

Solve each of the following equations. Don't forget that division by zero is undefined.

74. $\dfrac{3}{x-2} = \dfrac{6}{2x-4}$

75. $\dfrac{8}{2x+1} = \dfrac{4}{x-3}$

76. $\dfrac{5}{x-3} = \dfrac{10}{x-6}$

77. $\dfrac{6}{x-1} = \dfrac{5}{x-1}$

78. Be sure that you can do Problems 45–56 with your calculator.

79. Use your calculator to help change each of the following common fractions to a percent. Express your answers to the nearest tenth of a percent. (It would be good practice to estimate an answer for each one before you use your calculator.)

(a) $\dfrac{5}{8}$ (b) $\dfrac{5}{6}$ (c) $\dfrac{3}{7}$ (d) $\dfrac{7}{12}$ (e) $\dfrac{4}{13}$ (f) $\dfrac{8}{17}$

(g) $\dfrac{5}{32}$ (h) $\dfrac{19}{48}$ (i) $\dfrac{9}{8}$ (j) $\dfrac{13}{6}$ (k) $\dfrac{17}{12}$ (l) $\dfrac{17}{3}$

3.6

More on Percents and Problem Solving

The equation $x + 0.35 = 0.72$ can be solved by subtracting 0.35 from both sides of the equation. Another technique for solving equations containing decimals is to *clear the equation of all decimals* by multiplying both sides by an appropriate power of 10. The following examples demonstrate the use of the "clear equation of all decimals" strategy in a variety of situations.

EXAMPLE 1   Solve $0.5x = 14$.

Solution

$$0.5x = 14$$

$$5x = 140 \qquad \text{multiplied both sides by 10}$$

$$x = 28 \qquad \text{divided both sides by 5}$$

The solution set is $\{28\}$.   ■

EXAMPLE 2   Solve $x + 0.07x = 0.13$.

Solution

$$x + 0.07x = 0.13$$

$$100(x + 0.07x) = 100(0.13) \qquad \text{Multiply both sides by 100.}$$

$$100(x) + 100(0.07x) = 100(0.13) \qquad \text{distributive property}$$

$$100x + 7x = 13$$

$$107x = 13$$

$$x = \frac{13}{107}$$

The solution set is $\left\{\dfrac{13}{107}\right\}$.   ■

EXAMPLE 3   Solve $0.08y + 0.09y = 3.4$.

Solution

$$0.08y + 0.09y = 3.4$$
$$8y + 9y = 340 \qquad \text{multiplied both sides by 100}$$
$$17y = 340$$
$$y = 20$$

The solution set is $\{20\}$.   ■

EXAMPLE 4   Solve $0.10t = 560 - 0.12(t + 1000)$.

Solution

$$0.10t = 560 - 0.12(t + 1000)$$
$$10t = 56{,}000 - 12(t + 1000) \qquad \text{multiplied both sides by 100}$$
$$10t = 56{,}000 - 12t - 12{,}000 \qquad \text{distributive property}$$
$$22t = 44{,}000$$
$$t = 2000$$

The solution set is $\{2000\}$.

Problems Involving Percents

Many consumer problems can be solved with an "equation approach." For example, the following is a general guideline regarding discount sales.

> original selling price − discount = discount sale price

Let's consider some examples using our algebraic techniques along with this basic guideline.

PROBLEM 1   Amy bought a dress at a 30% discount sale for $35. What was the original price of the dress?

Solution   Let p represent the original price of the dress. We can use the basic discount guideline to set up an algebraic equation.

original selling price − discount   = discount sale price
$$\downarrow \qquad\qquad \downarrow \qquad\qquad\qquad \downarrow$$
$$(100\%)(p) \quad - (30\%)(p) = \qquad \$35$$

Solving this equation we obtain

$$(100\%)(p) - (30\%)(p) = 35$$
$$(70\%)(p) = 35$$
$$0.7p = 35$$
$$7p = 350$$
$$p = 50.$$

The original price of the dress was $50.   ■

Don't forget that if an item is on sale for 30% off, then you are going to pay $100\% - 30\% = 70\%$ of the original price. So at a 30% discount sale a $50 dress can be purchased for $(70\%)(\$50) = \$35$. (Note that we have now checked our answer for Problem 1.)

PROBLEM 2   Find the cost of a $60 pair of jogging shoes that is on sale for 20% off.

Solution   Let x represent the discount sale price. Since the shoes are on sale for 20% off, we must pay 80% of the original price.

$$x = (80\%)(60)$$
$$= (0.8)(60)$$
$$= 48$$

The sale price is $48.   ■

Another equation that we can use in consumer problems is the following.

+---+
| selling price = cost + profit |
+---+

Profit (also called *markup, markon, margin*, and *margin of profit*) may be stated in different ways. It may be stated as a percent of the selling price, a percent of the cost, or simply in terms of dollars and cents. Let's consider some problems where the profit is either a percent of the selling price or a percent of the cost.

PROBLEM 3   A retailer has some shirts that cost him $20 each. He wants to sell them at a profit of 60% of the cost. What selling price should be marked on the shirts?

Solution

Let s represent the selling price. The basic relationship "selling price equals cost plus profit" can be used as a guideline.

$$\text{selling price} = \text{cost} + \text{profit (\% of cost)}$$
$$s = \$20 + (60\%)(20)$$

Solving this equation we obtain

$$s = 20 + (60\%)(20)$$
$$s = 20 + (0.6)(20)$$
$$s = 20 + 12$$
$$s = 32.$$

The selling price should be $32.   ■

PROBLEM 4

Kathrin bought a painting for $120 and later decided to resell it. She made a profit of 40% of the selling price. What did she receive for the painting?

Solution

We can use the same basic relationship as a guideline except this time the profit is a percent of the selling price. Let s represent the selling price.

$$\text{selling price} = \text{cost} + \text{profit (\% of selling price)}$$
$$s = 120 + (40\%)(s)$$

Solving this equation we obtain

$$s = 120 + (40\%)(s)$$
$$s = 120 + 0.4s$$
$$0.6s = 120$$
$$s = \frac{120}{0.6} = 200.$$

She received $200 for the painting.   ■

Certain types of investment problems can also be translated into algebraic equations. The simple interest formula $i = Prt$, where i represents the amount of interest earned by investing P dollars at a yearly rate of r percent for t years, is used in some of these problems.

PROBLEM 5

A woman invests a total of $5000. Part of it is invested at 10% and the remainder at 12%. Her total yearly interest from the two investments is $560. How much does she have invested at each rate?

Solution   Let x represent the amount invested at 12%. Then $5000 - x$ represents the amount invested at 10%. The following guideline can be used.

$$\underset{\downarrow}{\begin{array}{c}\text{interest earned from}\\\text{12\% investment}\end{array}} + \underset{\downarrow}{\begin{array}{c}\text{interest earned from}\\\text{10\% investment}\end{array}} = \underset{\downarrow}{\begin{array}{c}\text{total interest}\\\text{earned}\end{array}}$$

$$(12\%)(x) \quad + (10\%)(\$5000 - x) = \quad \$560$$

Solving this equation yields

$$(12\%)(x) + (10\%)(5000 - x) = 560$$

$$0.12x + 0.10(5000 - x) = 560$$

$$12x + 10(5000 - x) = 56{,}000$$

$$12x + 50{,}000 - 10x = 56{,}000$$

$$2x + 50{,}000 = 56{,}000$$

$$2x = 6000$$

$$x = 3000.$$

Therefore $5000 - x = 2000$.

$3000 was invested at 12% and $2000 at 10%.   ■

PROBLEM 6   An investor invests a certain amount of money at 9%. Then he finds a better deal and invests $5000 more than that amount at 12%. His yearly income from the two investments was $1650. How much did he invest at each rate?

Solution   Let x represent the amount invested at 9%. Then $x + 5000$ represents the amount invested at 12%.

$$(9\%)(x) + (12\%)(x + 5000) = 1650$$

$$0.09x + 0.12(x + 5000) = 1650$$

$$9x + 12(x + 5000) = 165{,}000$$

$$9x + 12x + 60{,}000 = 165{,}000$$

$$21x + 60{,}000 = 165{,}000$$

$$21x = 105{,}000$$

$$x = 5000$$

Therefore $x + 5000 = 10{,}000$.

He invested $5000 at 9% and $10,000 at 12%.   ■

Problem Set 3.6

For Problems 1–22, solve each of the equations.

1. $x - 0.36 = 0.75$ **2.** $x - 0.15 = 0.42$ **3.** $x + 7.6 = 14.2$

4. $x + 11.8 = 17.1$ **5.** $0.62 - y = 0.14$ **6.** $7.4 - y = 2.2$

7. $0.7t = 56$ **8.** $1.3t = 39$ **9.** $x = 3.36 - 0.12x$

10. $x = 5.3 - 0.06x$ **11.** $s = 35 + 0.3s$ **12.** $s = 40 + 0.5s$

13. $s = 42 + 0.4s$ **14.** $s = 24 + 0.6s$

15. $0.07x + 0.08(x + 600) = 78$ **16.** $0.06x + 0.09(x + 200) = 63$

17. $0.09x + 0.1(2x) = 130.5$ **18.** $0.11x + 0.12(3x) = 188$

19. $0.08x + 0.11(500 - x) = 50.5$ **20.** $0.07x + 0.09(2000 - x) = 164$

21. $0.09x = 550 - 0.11(5400 - x)$ **22.** $0.08x = 580 - 0.1(6000 - x)$

For Problems 23–48, set up an equation and solve each problem.

23. Tom bought a pair of trousers at a 30% discount sale for $35. What was the original price of the trousers?

24. Louise bought a dress for $140, which represented a 20% discount of the original price. What was the original price of the dress?

25. Find the cost of a $48 sweater that is on sale for 25% off.

26. Byron purchased a bicycle at a 10% discount sale for $121.50. What was the original price of the bicycle?

27. Suppose that Jack bought a $32 putter on sale for 35% off. How much did he pay for the putter?

28. Mindy bought a 13-inch portable color TV for 20% off of the list price. The list price was $349.95. What did she pay for the TV?

29. Pierre bought a coat for $126 that was listed for $180. What rate of discount did he receive?

30. Phoebe paid $32 for a pair of gold shoes that were listed for $40. What rate of discount did she receive?

31. A retailer has some ties that cost him $5 each. He wants to sell them at a profit of 70% of the cost. What should be the selling price of the ties?

32. A retailer has some blouses that cost her $25 each. She wants to sell them at a profit of 80% of the cost. What price should she charge for the blouses?

33. The owner of a pizza parlor wants to make a profit of 55% of the cost for each pizza sold. If it costs $3 to make a pizza, at what price should it be sold?

34. Produce in a food market usually has a high markup because of loss due to spoilage. If a head of lettuce costs a retailer $.40, at what price should it be sold to realize a profit of 130% of the cost?

35. Jewelry has a very high markup rate. If a ring costs a jeweler $400, at what price should it be sold to gain a profit of 60% of the selling price?

36. If a box of candy costs a retailer $2.50 and he wants to make a profit of 50% based on the selling price, what price should he charge for the candy?

37. If the cost for a pair of shoes for a retailer is $32 and he sells them for $44.80, what is his rate of profit based on the cost?

38. A retailer has some skirts that cost her $24. If she sells them for $31.20, find her rate of profit based on the cost.

39. Suppose that Lou invested a certain amount of money at 9% interest and $250 more than that amount at 10%. His total yearly interest was $101. How much did he invest at each rate?

40. Nina received an inheritance of $12,000 from her grandmother. She invested part of it at 12% interest and the remainder at 14%. If the total yearly interest from both investments was $1580, how much did she invest at each rate?

41. Nolan received $1200 from his parents as a graduation present. He invested part of it at 9% interest and the remainder at 12%. If the total yearly interest amounted to $129, how much did he invest at each rate?

42. Sally invested a certain sum of money at 9%, twice that sum at 10%, and three times that sum at 11%. Her total yearly interest from all three investments was $310. How much did she invest at each rate?

43. If $2000 is invested at 8% interest, how much additional money must be invested at 11% interest so that the total return for both investments averages 10%?

44. Fawn invested a certain amount of money at 10% interest and $250 more than that amount at 11%. Her total yearly interest was $153.50. How much did she invest at each rate?

45. A sum of $2300 is invested, part of it at 10% interest and the remainder at 12%. If the interest earned by the 12% investment is $100 more than the interest earned by the 10% investment, find the amount invested at each rate.

46. If $3000 is invested at 9% interest, how much additional money must be invested at 12% so that the total return for both investments averages 11%?

47. How can $5400 be invested, part of it at 8% and the remainder at 10%, so that the two investments will produce the same amount of interest?

48. A sum of $6000 is invested, part of it at 9% interest and the remainder at 11%. If the interest earned by the 9% investment is $160 less than the interest earned by the 11% investment, find the amount invested at each rate.

Thoughts into Words

49. Explain the difference between arithmetic and algebra.

50. Give a step-by-step description of how you would solve the equation

$$.07x + .09(500 - x) = 40.$$

Miscellaneous Problems

51. A retailer buys an item for $40, resells it for $50, and claims that he is only making a 20% profit. Is his claim correct?

52. A store has a special discount sale of 40% off on all items. It also advertises an additional 10% off on items bought in quantities of a dozen or more. How much will it cost to buy a dozen items of some particular kind that regularly sell for $5 per item? (Be careful, a 40% discount followed by a 10% discount is not equal to a 50% discount.)

53. Some people use the following formula for determining the selling price of an item when the profit is based on a percent of the selling price.

$$\text{selling price} = \frac{\text{cost}}{100\% - \text{percent of profit}}$$

Show how this formula is developed.

Solve each of the following equations and express the solutions in decimal form. Use your calculator whenever it seems appropriate.

54. $2.4x + 5.7 = 9.6$

55. $-3.2x - 1.6 = 5.8$

56. $0.08x + 0.09(800 - x) = 68.5$

57. $0.10x + 0.12(720 - x) = 80$

58. $7x - 0.39 = 0.03$

59. $9x - 0.37 = 0.35$

60. $0.2(t + 1.6) = 3.4$

61. $0.4(t - 3.8) = 2.2$

Chapter 3 Summary

(3.1) **Numerical equations** may be true or false. **Algebraic equations** (open sentences) contain one or more variables. **Solving an equation** refers to the process of finding the number (or numbers) that makes an algebraic equation a true statement. A **first-degree equation of one variable** is an equation containing only one variable and this variable has an exponent of one.

Properties 3.1 and 3.2 provide the basis for solving equations. Be sure that you can use these properties to solve the variety of equations presented in this chapter.

(3.2) It is often necessary to use both the addition and multiplication properties of equality to solve an equation.

Be sure to *declare your variable* as you translate English sentences into algebraic equations.

(3.3) Keep the following suggestions in mind as you solve word problems.

1. Read the problem carefully.
2. Sketch any figure, diagram, or chart that might be helpful.
3. Choose a meaningful variable.
4. Look for a *guideline*.
5. Form an equation.
6. Solve the equation.
7. Check your answers.

(3.4) The **distributive property** is used to *remove parentheses*.

If an equation contains several fractions, then it is usually advisable to *clear the equation of all fractions* by multiplying both sides by the least common denominator of all of the denominators in the equation.

(3.5) A **ratio** is the comparison of two numbers by division.

A statement of equality between two ratios is a **proportion**. In a proportion, the *cross products* are equal. That is to say,

$$\text{if } \frac{a}{b} = \frac{c}{d}, \quad \text{then } ad = bc, \quad b \neq 0 \text{ and } d \neq 0.$$

The cross product property can be used to solve equations that are in the form of a proportion.

A variety of word problems can be conveniently set up and solved using proportions.

The concept of **percent** means "per one hundred" and is therefore a special ratio, namely, one having a denominator of 100. Proportions provide a convenient basis for changing common fractions to percents.

(3.6) To solve equations containing decimals, you can clear the equation of all decimals by multiplying both sides by an appropriate power of ten.

Many consumer problems involve the concept of *percent* and can be solved with an equation approach. The basic relationships *selling price equals cost plus profit* and *original selling price minus discount equals discount sale price* are used frequently.

Chapter 3 Review Problem Set

For Problems 1–25, solve each of the equations.

1. $9x - 2 = -29$

2. $-3 = -4y + 1$

3. $7 - 4x = 10$

4. $6y - 5 = 4y + 13$

5. $4n - 3 = 7n + 9$

6. $7(y - 4) = 4(y + 3)$

7. $2(x + 1) + 5(x - 3) = 11(x - 2)$

8. $-3(x + 6) = 5x - 3$

9. $\dfrac{2}{5}n - \dfrac{1}{2}n = \dfrac{7}{10}$

10. $\dfrac{3n}{4} + \dfrac{5n}{7} = \dfrac{1}{14}$

11. $\dfrac{x - 3}{6} + \dfrac{x + 5}{8} = \dfrac{11}{12}$

12. $\dfrac{n}{2} - \dfrac{n - 1}{4} = \dfrac{3}{8}$

13. $0.5x + 0.7x = 1.7$

14. $0.07t + 0.12(t - 3) = 0.59$

15. $-2(x - 4) = -3(x + 8)$

16. $3x - 4x - 2 = 7x - 14 - 9x$

17. $5(n - 1) - 4(n + 2) = -3(n - 1) + 3n + 5$

18. $\dfrac{9}{x - 3} = \dfrac{8}{x + 4}$

19. $\dfrac{-3}{x - 1} = \dfrac{-4}{x + 2}$

20. $-(t - 3) - (2t + 1) = 3(t + 5) - 2(t + 1)$

21. $\dfrac{2x - 1}{3} = \dfrac{3x + 2}{2}$

22. $0.1x + 0.12(1700 - x) = 188$

23. $x - 0.25x = 12$

24. $0.2(x - 3) = 14$

25. $3(2t - 4) + 2(3t + 1) = -2(4t + 3) - (t - 1)$

Set up an equation and solve each of the following problems (26–43).

26. Three-fourths of a number equals 18. Find the number.

27. Nineteen is two less than three times a certain number. Find the number.

28. The difference of two numbers is 21. If 12 is the smaller number, find the other number.

29. One subtracted from nine times a certain number is the same as 15 added to seven times the number. Find the number.

30. Fifteen percent of a certain number is six. Find the number.

31. The sum of two numbers is 40. Six times the smaller number equals four times the larger. Find the numbers.

32. Eighteen is what percent of 30?

33. Find a number such that two less than two-thirds of the number is one more than one-half of the number.

34. The sum of two numbers is 96 and their ratio is 5 to 7. Find the numbers.

35. Miriam has 30 coins consisting of nickels and dimes amounting to $2.60. How many coins of each kind does she have?

36. Suppose that Russ has a bunch of nickels, dimes, and quarters amounting to $15.40. The number of dimes is one more than three times the number of nickels and the number of quarters is twice the number of dimes. How many coins of each kind does he have?

37. The supplement of an angle is 14° more than three times the complement of the angle. Find the measure of the angle.

38. The ratio of the complement of an angle to the supplement of the angle is 7 to 16. Find the measure of the angle.

39. If a car uses 18 gallons of gasoline for a 369-mile trip, at the same rate of consumption how many gallons will it use on a 615-mile trip?

40. A sum of $2100 is invested, part of it at 9% interest and the remainder at 11%. If the interest earned by the 11% investment is $51 more than the interest from the 9% investment, find the amount invested at each rate.

41. A retailer has some sweaters that cost $28 each. At what price should the sweaters be sold to obtain a profit of 30% of the selling price?

42. Robin bought a dress on sale for $39 and the original price of the dress was $60. What percent of discount did she receive?

43. Pam rented a car from a rental agency that charges $25 a day and $0.20 per mile. She kept the car for 3 days and her bill was $215. How many miles did she drive during that 3-day period?

Chapter 4

Formulas, Problem Solving, and Inequalities

In this chapter we shall expand some of the ideas from the previous chapter: Solving equations and using equations to solve word problems. We shall also consider some techniques for solving inequalities.

4.1
Formulas

To find the distance traveled in 3 hours at a rate of 50 miles per hour, we multiply the rate by the time. Thus, the distance is 50(3) = 150 miles. We usually state the rule *distance equals rate times time* as a **formula**: $d = rt$. Formulas are simply rules stated in symbolic language and usually expressed as equations. Thus, the formula $d = rt$ is an equation involving three variables, d, r, and t.

As we work with formulas it is often necessary to solve for some specific variable when numerical values for the remaining variables are known. Consider the following examples.

EXAMPLE 1   Solve $d = rt$ for r if $d = 330$ and $t = 6$.

Solution   Substituting 330 for d and 6 for t in the given formula, we obtain

$$330 = r(6).$$

Solving this equation yields

$$330 = 6r$$

$$55 = r.$$ ∎

EXAMPLE 2   Solve $C = \dfrac{5}{9}(F - 32)$ for F if $C = 10$. (This formula expresses the relationship between the Fahrenheit and Celsius temperature scales.)

Solution   Substituting 10 for C, we obtain

$$10 = \frac{5}{9}(F - 32).$$

Solving this equation produces

$$\frac{9}{5}(10) = \frac{9}{5}\left(\frac{5}{9}\right)(F - 32) \qquad \text{Multiply both sides by } \frac{9}{5}.$$

$$18 = F - 32$$

$$50 = F.$$ ∎

Sometimes it may be convenient to change a formula's form by using the properties of equality. For example, the formula, $d = rt$, can be changed as follows.

$$d = rt$$

$$\frac{d}{r} = \frac{rt}{r} \qquad \text{Divide both sides by } r.$$

$$\frac{d}{r} = t$$

We say that the formula $d = rt$ **has been solved for the variable t**. The formula can also be **solved for r** as follows.

$$d = rt$$

$$\frac{d}{t} = \frac{rt}{t} \qquad \text{Divide both sides by } t.$$

$$\frac{d}{t} = r$$

Geometric Formulas

There are several formulas in geometry that we use quite often. Let's briefly review them at this time; they will be used periodically throughout the remainder of the text. (These formulas along with some others are also listed in the inside back cover of this text.)

$$A = lw \qquad P = 2l + 2w$$

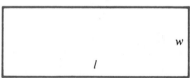

Rectangle

    A area
    P perimeter
    l length
    w width

Triangle

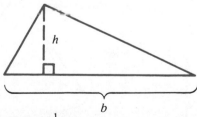

$$A = \frac{1}{2} bh$$

A area
b base
h altitude (height)

Trapezoid

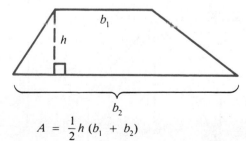

$$A = \frac{1}{2} h (b_1 + b_2)$$

A area
b_1, b_2 bases
h altitude

Parallelogram

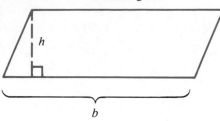

$A = bh$

A area
b base
h altitude (height)

Circle

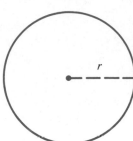

$A = \pi r^2$ $C = 2\pi r$

A area
C circumference
r radius

Sphere

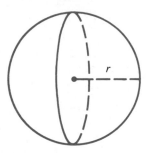

$S = 4\pi r^2$ $V = \dfrac{4}{3}\pi r^3$

S surface area
V volume
r radius

Prism

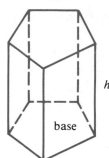

base

$V = Bh$

V volume
B area of base
h altitude (height)

Pyramid

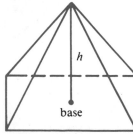

base

$V = \dfrac{1}{3}Bh$

V volume
B area of base
h altitude (height)

Right circular cylinder

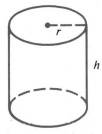

$V = \pi r^2 h$ $S = 2\pi r^2 + 2\pi rh$

V volume
S total surface area
r radius
h altitude (height)

Right circular cone

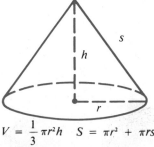

$V = \dfrac{1}{3}\pi r^2 h$ $S = \pi r^2 + \pi rs$

V volume
S total surface area
r radius
h altitude (height)
s slant height

EXAMPLE 3   Solve $C = 2\pi r$ for r.

Solution
$$C = 2\pi r$$

$$\frac{C}{2\pi} = \frac{2\pi r}{2\pi} \qquad \text{Divide both sides by } 2\pi.$$

$$\frac{C}{2\pi} = r \qquad\qquad\qquad\qquad\qquad\qquad\qquad \blacksquare$$

EXAMPLE 4   Solve $V = \dfrac{1}{3}Bh$ for h.

Solution
$$V = \frac{1}{3}Bh$$

$$3(V) = 3\left(\frac{1}{3}Bh\right) \qquad \text{Multiply both sides by 3.}$$

$$3V = Bh$$

$$\frac{3V}{B} = \frac{Bh}{B} \qquad \text{Divide both sides by } B.$$

$$\frac{3V}{B} = h \qquad\qquad\qquad\qquad\qquad\qquad \blacksquare$$

EXAMPLE 5   Solve $P = 2l + 2w$ for w.

Solution
$$P = 2l + 2w$$

$$P - 2l = 2l + 2w - 2l \qquad \text{Subtract } 2l \text{ from both sides.}$$

$$P - 2l = 2w$$

$$\frac{P - 2l}{2} = \frac{2w}{2} \qquad\qquad \text{Divide both sides by 2.}$$

$$\frac{P - 2l}{2} = w \qquad\qquad\qquad\qquad\qquad\qquad \blacksquare$$

EXAMPLE 6   Find the total surface area of a right circular cylinder having a radius of 10 inches and a height of 14 inches.

Solution   Let's sketch a right circular cylinder and record the given information.

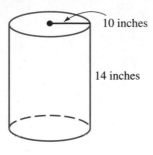

10 inches

14 inches

Substituting 10 for r and 14 for h in the formula for finding total surface area of a right circular cylinder we obtain

$$S = 2\pi r^2 + 2\pi rh$$
$$= 2\pi(10)^2 + 2\pi(10)(4)$$
$$= 200\pi + 280\pi$$
$$= 480\pi.$$

The total surface area is 480π square inches.   ■

   In Example 6 we used the figure to record the given information and it also served as a reminder of the geometric figure under consideration. Now let's consider an example where a figure is very useful in the analysis of the problem.

EXAMPLE 7   A sidewalk 3 feet wide surrounds a rectangular plot of ground that measures 75 feet by 100 feet. Find the area of the sidewalk.

Solution   Let's sketch a figure and record the given information.

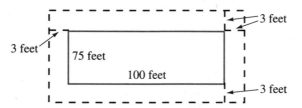

3 feet

3 feet

75 feet

100 feet

3 feet

The area of the sidewalk can be found by subtracting the area of the rectangular plot from the area of the plot plus the sidewalk (the large dashed rectangle). The width of the large rectangle is $75 + 3 + 3 = 81$ feet and its length is $100 + 3 + 3 = 106$ feet.

$$A = (81)(106) - (75)(100)$$
$$= 8585 - 7500 = 1086$$

The area of the sidewalk is 1086 square feet.   ■

In Chapter 5 you will be working with equations that contain two variables. At times you will need to solve for one variable in terms of the other variable; that is, change the form of the equation as we have been doing with formulas. The following examples illustrate once again how we can use the properties of equality for such situations.

EXAMPLE 8   Solve $3x + y = 4$ for x.

Solution

$$3x + y = 4$$

$$3x + y - y = 4 - y \qquad \text{Subtract } y \text{ from both sides.}$$

$$3x = 4 - y$$

$$\frac{3x}{3} = \frac{4 - y}{3} \qquad \text{Divide both sides by 3.}$$

$$x = \frac{4 - y}{3}$$ ∎

EXAMPLE 9   Solve $4x - 5y = 7$ for y.

Solution

$$4x - 5y = 7$$

$$4x - 5y - 4x = 7 - 4x \qquad \text{Subtract } 4x \text{ from both sides.}$$

$$-5y = 7 - 4x$$

$$\frac{-5y}{-5} = \frac{7 - 4x}{-5} \qquad \text{Divide both sides by } -5.$$

$$y = \frac{7 - 4x}{-5}\left(\frac{-1}{-1}\right) \qquad \begin{array}{l}\text{Multiply numerator and denominator}\\ \text{of fraction on the right by } -1.\end{array}$$

$$y = \frac{4x - 7}{5} \qquad \begin{array}{l}\text{We commonly do this so that the}\\ \text{denominator is positive.}\end{array}$$ ∎

EXAMPLE 10   Solve $y = mx + b$ for m

Solution

$$y = mx + b$$

$$y - b = mx + b - b \qquad \text{Subtract } b \text{ from both sides.}$$

$$y - b = mx$$

$$\frac{y - b}{x} = \frac{mx}{x} \qquad \text{Divide both sides by } x.$$

$$\frac{y - b}{x} = m$$ ∎

Problem Set 4.1

For Problems 1–10, solve for the specified variable using the given facts.

1. Solve $d = rt$ for t if $d = 336$ and $r = 48$.

2. Solve $d = rt$ for r if $d = 486$ and $t = 9$.

3. Solve $i = Prt$ for P if $i = 200$, $r = 0.08$, and $t = 5$.

4. Solve $i = Prt$ for t if $i = 540$, $P = 750$, and $r = 0.09$.

5. Solve $F = \frac{9}{5}C + 32$ for C if $F = 68$.

6. Solve $C = \frac{5}{9}(F - 32)$ for F if $C = 15$.

7. Solve $V = \frac{1}{3}Bh$ for B if $V = 112$ and $h = 7$.

8. Solve $V = \frac{1}{3}Bh$ for h if $V = 216$ and $B = 54$.

9. Solve $A = P + Prt$ for t if $A = 652$, $P = 400$, and $r = 0.07$.

10. Solve $A = P + Prt$ for P if $A = 1032$, $r = 0.06$, and $t = 12$.

For Problems 11–32, use the geometric formulas given in this section to help solve the problems.

11. Find the perimeter of a rectangle that is 14 centimeters long and 9 centimeters wide.

12. If the perimeter of a rectangle is 80 centimeters and its length is 24 centimeters, find its width.

13. If the perimeter of a rectangle is 108 inches and its length is $3\frac{1}{4}$ feet, find its width in inches.

14. How many yards of fencing would it take to enclose a rectangular plot of ground that is 69 feet long and 42 feet wide?

15. A dirt path 4 feet wide surrounds a rectangular garden that is 38 feet long and 17 feet wide. Find the area of the dirt path.

16. Find the area of a cement walk 3 feet wide that surrounds a rectangular plot of ground 86 feet long and 42 feet wide.

17. Suppose that paint costs $2.00 per liter and that one liter will cover 9 square meters of surface. We are going to paint (on one side only) 50 pieces of wood of the same size, which are rectangular in shape, and have a length of 60 centimeters and a width of 30 centimeters. What will the cost for the paint be?

18. A lawn is in the shape of a triangle with one side 130 feet long and the altitude to that side 60 feet long. Will one sack of fertilizer, which covers 4000 square feet, be enough to fertilize the lawn?

19. Find the length of an altitude of a trapezoid with bases of 8 inches and 20 inches and an area of 98 square inches.

20. A flower garden is in the shape of a trapezoid with bases of 6 yards and 10 yards. The distance between the bases is 4 yards. Find the area of the garden.

21. The diameter of a metal washer is 4 centimeters. The diameter of the hole is 2 centimeters. How many square centimeters of metal are there in 50 washers? Express the answer in terms of π.

22. Find the area of a circular plot of ground that has a radius of length 14 meters. Use $3\frac{1}{7}$ as an approximation for π.

23. Find the area of a circular region that has a diameter of 1 yard. Express the answer in terms of π.

24. Find the area of a circular region if the circumference is 12π units. Express the answer in terms of π.

25. Find the total surface area and volume of a sphere that has a radius 9 inches long. Express the answers in terms of π.

26. A circular pool is 34 feet in diameter and has a flagstone walk around it that is 3 feet wide. Find the area of the walk. Express the answer in terms of π.

27. Find the volume and total surface area of a right circular cylinder that has a radius of 8 feet and a height of 18 feet. Express answers in terms of π.

28. Find the total surface area and volume of a sphere that has a diameter 12 centimeters long. Express the answers in terms of π.

29. If the volume of a right circular cone is 324π cubic inches and a radius of the base is 9 inches long, find the height of the cone.

30. Find the volume and total surface area of a tin can if the radius of the base is 3 centimeters and the height of the can is 10 centimeters. Express answers in terms of π.

31. If the total surface area of a right circular cone is 65π square feet and a radius of the base is 5 feet long, find the slant height of the cone.

32. If the total surface area of a right circular cylinder is 104π square meters and a radius of the base is 4 meters long, find the height of the cylinder.

For Problems 33–44, solve each of the formulas for the indicated variable.

33. $V = Bh$ for h volume of a prism

34. $A = lw$ for l area of a rectangle

35. $V = \frac{1}{3}Bh$ for B volume of a pyramid

36. $A = \frac{1}{2}bh$ for h area of a triangle

37. $P = 2l + 2w$ for w perimeter of a rectangle

38. $V = \pi r^2 h$ for h volume of a cylinder

39. $V = \frac{1}{3}\pi r^2 h$ for h volume of a cone

40. $i = Prt$ for t simple interest formula

41. $F = \dfrac{9}{5}C + 32$ for C Celsius to Fahrenheit

42. $A = P + Prt$ for t simple interest formula

43. $A = 2\pi r^2 + 2\pi rh$ for h surface area of a cylinder

44. $C = \dfrac{5}{9}(F - 32)$ for F Fahrenheit to Celsius

For Problems 45–60, solve each equation for the indicated variable.

45. $3x + 7y = 9$ for x **46.** $5x + 2y = 12$ for x

47. $9x - 6y = 13$ for y **48.** $3x - 5y = 19$ for y

49. $-2x + 11y = 14$ for x **50.** $-x + 14y = 17$ for x

51. $y = -3x - 4$ for x **52.** $y = -7x + 10$ for x

53. $\dfrac{x - 2}{4} = \dfrac{y - 3}{6}$ for y **54.** $\dfrac{x + 1}{3} = \dfrac{y - 5}{2}$ for y

55. $ax - by - c = 0$ for y **56.** $ax + by = c$ for y

57. $\dfrac{x + 6}{2} = \dfrac{y + 4}{5}$ for x **58.** $\dfrac{x - 3}{6} = \dfrac{y - 4}{8}$ for x

59. $m = \dfrac{y - b}{x}$ for y **60.** $y = mx + b$ for x

Miscellaneous Problems

For each of the following problems, use 3.14 as an approximation for π. Use your calculator whenever it seems appropriate.

61. Find the area of a circular plot of ground that has a radius 16.3 meters long. Express your answer to the nearest tenth of a square meter.

62. Find the area, to the nearest tenth of a square centimeter, of the "shaded ring" in the accompanying figure.

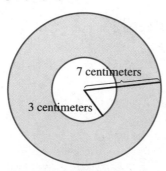

63. Find the area, to the nearest square inch, of each of the following pizzas.

       10-inch diameter _____

       12-inch diameter _____

       14-inch diameter _____

64. Find the total surface area, to the nearest square centimeter, of the accompanying tin can.

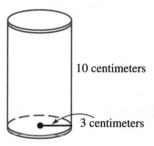

10 centimeters

3 centimeters

65. Find the total surface area, to the nearest square centimeter, of a baseball that has a radius of 4 centimeters.

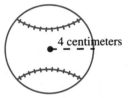

4 centimeters

66. Find the volume, to the nearest cubic inch, of a softball that has a diameter of 5 inches.

5-inch diameter

67. Find the volume, to the nearest cubic meter, of the accompanying figure.

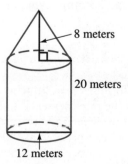

8 meters

20 meters

12 meters

4.2

Problem Solving

Let's begin this section by restating the suggestions for solving word problems we offered in Section 3.3.

Suggestions for Solving Word Problems

1. Read the problem carefully and make certain that you understand the meanings of all the words. Be especially alert for any technical terms used in the statement of the problem.

2. Read the problem a second time (perhaps even a third time) to get an overview of the situation being described and to determine the known facts as well as what is to be found.

3. Sketch any figure, diagram, or chart that might be helpful in analyzing the problem.

4. Choose a meaningful variable to represent an unknown quantity in the problem (perhaps t, if time is the unknown quantity); represent any other unknowns in terms of that variable.

5. Look for a guideline that can be used to set up an equation. A guideline might be a formula such as "selling price equals cost plus profit," or a relationship such as "interest earned from a 9% investment plus interest earned from a 10% investment equals total amount of interest earned." A guideline may also be indicated by a figure or diagram that you sketch for a particular problem.

6. Form an equation that contains the variable which translates the conditions of the guideline from English to algebra.

7. Solve the equation and use the solution to determine all facts requested in the problem.

8. **Check all answers back into the original statement of the problem.**

Again we emphasize the importance of suggestion number 5. Determining the guideline to follow when setting up the equation is a vital part of the analysis of a problem. Sometimes the guideline is a formula, such as one of the formulas presented in the previous section and accompanying problem set. Let's consider a problem of that type.

PROBLEM 1   How long will it take $500 to double itself if it is invested at 8% simple interest?

Solution   Let's use the basic simple interest formula, $i = Prt$, where i represents interest, P is the principal (money invested), r is the rate (percent), and t is the time in years. For $500 to "double itself" means that we want $500 to earn another $500 in interest. Thus, using $i = Prt$ as a guideline we can proceed as follows.

$$i = Prt$$

$$500 = 500(8\%)(t)$$

Now let's solve this equation.

$$500 = 500(.08)(t)$$

$$1 = .08t$$

$$100 = 8t$$

$$\frac{100}{8} = t$$

$$12\frac{1}{2} = t$$

It will take $12\frac{1}{2}$ years.   ■

If the problem involves a geometric formula, then a sketch of the figure is helpful for recording the given information and analyzing the problem. The next example illustrates this idea.

PROBLEM 2   The length of a football field is 40 feet more than twice its width and the perimeter of the field is 1040 feet. Find the length and width of the field.

Solution   Since the length is stated in terms of the width, we can let w represent the width and then $2w + 40$ represents the length.

$$2w + 40$$

A guideline for this problem is the perimeter formula $P = 2l + 2w$. Thus, the following equation can be set up and solved.

$$P = 2l + 2w$$

$$1040 = 2(2w + 40) + 2w$$

$$1040 = 4w + 80 + 2w$$

$$1040 = 6w + 80$$

$$960 = 6w$$

$$160 = w$$

If $w = 160$, then $2w + 40 = 2(160) + 40 = 360$. Thus, the football field is 360 feet long and 160 feet wide. ∎

Sometimes the formulas we use when we are analyzing a problem are different than those we use as a guideline for setting up the equation. For example, uniform motion problems involve the formula $d = rt$, but the main guideline for setting up an equation for such problems is usually a statement about either *times*, *rates*, or *distances*. Let's consider an example to illustrate this.

PROBLEM 3 Jerry leaves city A on a moped traveling toward city B at 18 miles per hour. At the same time, Cindy leaves city B on a moped traveling toward city A at 23 miles per hour. The distance between the two cities is 123 miles. How long will it take before Jerry and Cindy meet on their mopeds?

REMARK Before we present a solution to this problem we want to emphasize a point we made earlier. *Estimating* an answer before attempting to solve a problem may, in fact, help to determine the guideline to be used. To estimate an answer for Problem 3, suppose both Jerry and Cindy travel at 20 miles per hour. This means that they are approaching each other at 40 miles per hour; so it should take approximately 3 hours to travel 123 miles.

Solution First, let's sketch a diagram.

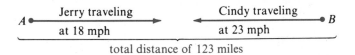

Let t represent the time that Jerry travels. Then t also represents the time that Cindy travels.

distance Jerry travels + distance Cindy travels = total distance

$$18t \quad + \quad 23t \quad = \quad 123$$

Solving this equation yields

$$18t + 23t = 123$$

$$41t = 123$$

$$t = 3.$$

They both travel for 3 hours. ∎

Some people find it helpful to use a chart to organize the known and unknown facts in a uniform motion problem. Let's illustrate this with an example.

PROBLEM 4

A car leaves a town at 60 kilometers per hour. How long will it take a second car traveling at 75 kilometers per hour to catch the first car if the second car leaves 1 hour later?

Solution

Let t represent the time of the second car. Then $t + 1$ represents the time of the first car since it travels one hour longer. We can now record the information of the problem in a chart as follows.

	Rate	*Time*	*Distance*
first car	60	$t + 1$	$60(t + 1)$
second car	75	t	$75t$

$d = rt$

Since the second car is to overtake the first car, the distances must be equal.

distance of second car equals distance of first car

$$75t \qquad = \qquad 60(t + 1)$$

Solving this equation yields

$$75t = 60(t + 1)$$
$$75t = 60t + 60$$
$$15t = 60$$
$$t = 4.$$

The second car should overtake the first car in 4 hours. (Check the answer!)   ∎

We would like to offer one bit of advice at this time. Don't become discouraged if solving word problems is giving you trouble. Problem solving is not a skill that can be developed overnight. It takes time, patience, hard work, and an open mind. Keep giving it your best shot and gradually you should become more confident in your approach to such problems. Furthermore, we realize that some (perhaps many) of these problems may not seem "practical" to you. However, keep in mind that the real goal here is to develop problem solving techniques. Finding and using a guideline, sketching a figure to record information and help in the analysis, estimating an answer before attempting to solve the problem, and using a chart to record information are the key issues.

Problem Set 4.2

For Problems 1–12, solve each of the equations. These equations are of the types you will be using in Problems 13–40.

1. $950(.12)t = 950$

2. $1200(.09)t = 1200$

3. $l + \dfrac{1}{4}l - 1 = 19$

4. $l + \dfrac{2}{3}l + 1 = 41$

5. $500(.08)t = 1000$

6. $800(.11)t = 1600$

7. $s + (2s - 1) + (3s - 4) = 37$

8. $s + (3s - 2) + (4s - 4) = 42$

9. $\dfrac{5}{2}r + \dfrac{5}{2}(r + 6) = 135$

10. $\dfrac{10}{3}r + \dfrac{10}{3}(r - 3) = 90$

11. $24\left(t - \dfrac{2}{3}\right) = 18t + 8$

12. $16t + 8\left(\dfrac{9}{2} - t\right) = 60$

Set up an equation and solve each of the following problems. Keep in mind the suggestions offered in this section.

13. How long will it take $750 to double itself if it is invested at 8% simple interest?

14. How many years will it take $1000 to double itself if it is invested at 10% simple interest?

15. How long will it take $800 to triple itself if it is invested at 10% simple interest?

16. How many years will it take $500 to earn $750 in interest if it is invested at 6% simple interest?

17. The length of a rectangle is three times its width. If the perimeter of the rectangle is 112 inches, find its length and width.

18. The width of a rectangle is one-half of its length. If the perimeter of the rectangle is 54 feet, find its length and width.

19. Suppose that the length of a rectangle is 2 centimeters less than three times its width. The perimeter of the rectangle is 92 centimeters. Find the length and width of the rectangle.

20. Suppose that the length of a certain rectangle is 1 meter more than five times its width. The perimeter of the rectangle is 98 meters. Find the length and width of the rectangle.

21. The width of a rectangle is 3 inches less than one-half of its length. If the perimeter of the rectangle is 42 inches, find the area of the rectangle.

22. The width of a rectangle is 1 foot more than one-third of its length. If the perimeter of the rectangle is 74 feet, find the area of the rectangle.

23. The perimeter of a triangle is 100 feet. The longest side is 3 feet less than twice the shortest side, and the third side is 7 feet longer than the shortest side. Find the lengths of the sides of the triangle.

24. A triangular plot of ground has a perimeter of 54 yards. The longest side is twice the shortest side and the third side is two yards longer than the shortest side. Find the lengths of the sides of the triangle.

25. The second side of a triangle is 1 centimeter more than three times the first side. The third side is 2 centimeters longer than the second side. If the perimeter is 46 centimeters, find the length of each side of the triangle.

26. The second side of a triangle is 3 meters less than twice the first side. The third side is 4 meters longer than the second side. If the perimeter is 58 meters, find the length of each side of the triangle.

27. The perimeter of an equilateral triangle is 4 centimeters more than the perimeter of a square and the length of a side of the triangle is 4 centimeters more than the length of a side of the square. Find the length of a side of the equilateral triangle. (An equilateral triangle has three sides of the same length.)

28. Suppose that a square and an equilateral triangle have the same perimeter. Each side of the equilateral triangle is 6 centimeters longer than each side of the square. Find the length of each side of the square. (An equilateral triangle has three sides of the same length).

29. Suppose that the length of a radius of a circle is the same as the length of a side of a square. If the circumference of the circle is 15.96 centimeters longer than the perimeter of the square, find the length of a radius of the circle. (Use 3.14 as an approximation for π.)

30. The circumference of a circle is 2.24 centimeters more than six times the length of a radius. Find the radius of the circle. (Use 3.14 as an approximation for π.)

31. Sandy leaves a town traveling in her car at a rate of 45 miles per hour. One hour later, Monica leaves the same town traveling the same route at a rate of 50 miles per hour. How long will it take Monica to overtake Sandy?

32. Two cars start from the same place traveling in opposite directions. One car travels 4 miles per hour faster than the other car. Find their speeds if after 5 hours they are 520 miles apart.

33. The distance between city A and city B is 325 miles. A freight train leaves city A and travels toward city B at 40 miles per hour. At the same time, a passenger train leaves city B and travels toward city A at 90 miles per hour. How long will it take the two trains to meet?

34. Kirk starts jogging at 5 miles per hour. One-half hour later Nancy starts jogging on the same route at 7 miles per hour. How long will it take Nancy to catch Kirk?

35. A car leaves a town at 40 miles per hour. Two hours later a second car leaves the town traveling the same route and overtakes the first car in 5 hours and 20 minutes. How fast was the second car traveling?

36. Two airplanes leave St. Louis at the same time and fly in opposite directions. If one travels at 500 kilometers per hour and the other at 600 kilometers per hour, how long will it take for them to be 1925 kilometers apart?

37. Two trains leave at the same time, one traveling east and the other traveling west. At the end of $9\frac{1}{2}$ hours they are 1292 miles apart. If the rate of the train traveling east is 8 miles per hour faster than the other train, find their rates.

38. Dawn starts on a 58-mile trip on her moped at 20 miles per hour. After a time the motor "kicks out" and she pedals the remainder of the trip at 12 miles per hour. The entire trip took $3\frac{1}{2}$ hours. How far had Dawn traveled when the motor on the moped quit running?

39. Jeff leaves home and rides his bicycle out into the country for 3 hours. On his return trip along the same route, it takes him three-quarters of an hour longer. If his rate on the return trip was 2 miles per hour slower than on the trip out into the country, find the total roundtrip distance.

40. In $1\frac{1}{4}$ hours more time, Rita, riding her bicycle at 12 miles per hour, rode 2 miles further than Sonya, who was riding her bicycle at 16 miles per hour. How long did each girl ride?

4.3
More about Problem Solving

Let's begin this section by considering another important but often overlooked facet of problem solving: The idea of *looking back* over your solution and considering some of the following questions.

1. Is your answer to the problem a reasonable answer? Does it agree with the estimated answer you arrived at before doing the problem?

2. Have you checked your answer by substituting it back into the conditions stated in the problem?

3. Do you now see another plan that could be used to solve the problem? Perhaps there is even another guideline that could be used.

4. Do you now see that this problem is closely related to another problem that you have previously solved?

5. Have you "tucked away for future reference" the technique used to solve this problem?

Looking back over the solution of a newly solved problem can often lay important groundwork for solving problems in the future.

Now let's consider three problems that are often referred to as mixture problems. There is no basic formula that applies for all of these problems, but the suggestion that *you think in terms of a pure substance* is often helpful in setting up a guideline. For example, a phrase such as "30% solution of acid" means that 30% of the amount of solution is acid and the remaining 70% is water.

PROBLEM 1

How many milliliters of pure acid must be added to 150 milliliters of a 30% solution of acid to obtain a 40% solution?

Solution

The key idea for solving such a problem is to recognize the following guideline.

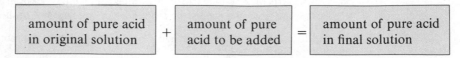

$$\begin{array}{c} \text{amount of pure acid} \\ \text{in original solution} \end{array} + \begin{array}{c} \text{amount of pure} \\ \text{acid to be added} \end{array} = \begin{array}{c} \text{amount of pure acid} \\ \text{in final solution} \end{array}$$

Let p represent the amount of pure acid to be added. Then using the guideline, we can form the following equation.

$$(30\%)(150) + p = 40\%(150 + p)$$

Now let's solve this equation to determine the amount of pure acid to be added.

$$(.30)(150) + p = .40(150 + p)$$

$$45 + p = 60 + .4p$$

$$.6p = 15$$

$$p = \frac{15}{.6} = 25$$

We must add 25 milliliters of pure acid. (Perhaps you should check this answer.)

∎

PROBLEM 2   Suppose that you have a supply of a 30% solution of alcohol and a 70% solution. How many quarts of each should be mixed to produce a 20-quart solution that is 40% alcohol?

Solution   We can use a guideline similar to the one in Problem 1.

$$\boxed{\text{pure alcohol in } 30\% \text{ solution}} + \boxed{\text{pure alcohol in } 70\% \text{ solution}} = \boxed{\text{pure alcohol in } 40\% \text{ solution}}$$

Let x represent the amount of 30% solution. Then $20 - x$ represents the amount of 70% solution. Now using the guideline, we translate to

$$(30\%)(x) + (70\%)(20 - x) = (40\%)(20).$$

Solving this equation we obtain

$$0.30x + 0.70(20 - x) = 8$$

$$30x + 70(20 - x) = 800$$

$$30x + 1400 - 70x = 800$$

$$-40x = -600$$

$$x = 15.$$

Therefore, $20 - x = 5$. We should mix 15 quarts of the 30% solution with 5 quarts of the 70% solution.

∎

PROBLEM 3   A 4-gallon radiator is full and contains a 40% solution of antifreeze. How much needs to be drained out and replaced with pure antifreeze to obtain a 70% solution?

Solution   The following guideline can be used.

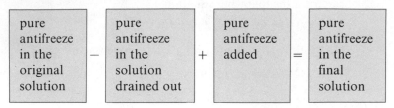

Let x represent the amount of pure antifreeze to be added. Then x also represents the amount of the 40% solution to be drained out. Thus, the guideline translates into the following equation.

$$(40\%)(4) - (40\%)(x) + x = (70\%)(4)$$

Solving this equation we obtain

$$.4(4) - .4x + x = .7(4)$$

$$1.6 + .6x = 2.8$$

$$.6x = 1.2$$

$$x = 2.$$

Therefore, we must drain out 2 gallons of the 40% solution and add 2 gallons of pure antifreeze. (Checking this answer would be a worthwhile exercise for you!)   ■

Now let's consider a problem where the process of representing the various unknown quantities in terms of one variable is the key to solving the problem.

PROBLEM 4   Jody is 6 years younger than her sister Cathy, and in 7 years Jody will be three-fourths as old as Cathy. Find their present ages.

Solution   By letting c represent Cathy's present age we can represent all of the unknown quantities as follows.

c: Cathy's present age $c + 7$:   Cathy's age in 7 years
$c - 6$:   Jody's present age $c - 6 + 7$ or $c + 1$:   Jody's age in 7 years

The statement that Jody's age in 7 years will be three-fourths of Cathy's age at that time serves as the guideline. So we can set up and solve the following equation.

$$c + 1 = \frac{3}{4}(c + 7)$$

$$4c + 4 = 3(c + 7)$$

$$4c + 4 = 3c + 21$$

$$c = 17$$

Therefore, Cathy's present age is 17 and Jody's present age is $17 - 6 = 11$.   ■

Problem Set 4.3

For Problems 1–12, solve each of the equations. You will be using these types of equations in Problems 13–37.

1. $.3x + .7(20 - x) = .4(20)$

2. $.4x + .6(50 - x) = .5(50)$

3. $.2(20) + x = .3(20 + x)$

4. $.3(32) + x = .4(32 + x)$

5. $.7(15) - x = .6(15 - x)$

6. $.8(25) - x = .7(25 - x)$

7. $.4(10) - .4x + x = .5(10)$

8. $.2(15) - .2x + x = .4(15)$

9. $20x + 12\left(4\frac{1}{2} - x\right) = 70$

10. $30x + 14\left(3\frac{1}{2} - x\right) = 97$

11. $3t = \frac{11}{2}\left(t - \frac{3}{2}\right)$

12. $5t = \frac{7}{3}\left(t + \frac{1}{2}\right)$

Set up an equation and solve each of the following problems.

13. How many milliliters of pure acid must be added to 100 milliliters of a 10% acid solution to obtain a 20% solution?

14. How many liters of pure alcohol must be added to 20 liters of a 40% solution to obtain a 60% solution?

15. How many centiliters of distilled water must be added to 10 centiliters of a 50% acid solution to obtain a 20% acid solution?

16. How many milliliters of distilled water must be added to 50 milliliters of a 40% acid solution to reduce it to a 10% acid solution?

17. We want to mix some 30% alcohol solution with some 50% alcohol solution to obtain 10 quarts of a 35% solution. How many quarts of each kind should we use?

18. We have a 20% alcohol solution and a 50% solution. How many pints must be used from each to obtain 8 pints of a 30% solution?

19. How much water needs to be removed from 20 gallons of a 30% salt solution to change it to a 40% salt solution?

20. How much water needs to be removed from 30 liters of a 20% salt solution to change it to a 50% salt solution?

21. Suppose that a 12-quart radiator contains a 20% solution of antifreeze. How much solution needs to be drained out and replaced with pure antifreeze to obtain a 40% solution of antifreeze?

22. A tank contains 50 gallons of a 40% solution of antifreeze. How much solution needs to be drained out and replaced with pure antifreeze to obtain a 50% solution?

23. How many gallons of a 15% salt solution must be mixed with 8 gallons of a 20% salt solution to obtain a 17% salt solution?

24. How many liters of a 10% salt solution must be mixed with 15 liters of a 40% salt solution to obtain a 20% salt solution?

25. Thirty ounces of a punch containing 10% grapefruit juice is added to 50 ounces of punch containing 20% grapefruit juice. Find the percent of grapefruit juice in the resulting mixture.

26. Suppose that 20 gallons of a 20% salt solution is mixed with 30 gallons of a 25% salt solution. What is the percent of salt in the resulting solution?

27. Suppose that the perimeter of a square equals the perimeter of a rectangle. The width of the rectangle is 9 inches less than twice the side of the square and the length of the rectangle is 3 inches less than twice the side of the square. Find the dimensions of the square and the rectangle.

28. The perimeter of a triangle is 40 centimeters. The longest side is 1 centimeter more than twice the shortest side. The other side is 2 centimeters shorter than the longest side. Find the lengths of the three sides.

29. Butch starts walking from point A at 2 miles per hour. One-half hour later, Dick starts walking from point A at $3\frac{1}{2}$ miles per hour and follows the same route. How long will it take Dick to catch up with Butch?

30. Karen, riding her bicycle at 15 miles per hour, rode 10 miles farther than Michelle, who was riding her bicycle at 14 miles per hour. Karen rode for 30 minutes longer than Michelle. How long did Michelle and Karen each ride their bicycles?

31. Pam is half as old as her brother Bill. Six years ago Bill was four times older than Pam. How old is each now?

32. The sum of the present ages of Tom and his father is 100 years. Ten years ago Tom's father was three times as old as Tom was at that time. Find their present ages.

33. The difference of the present ages of Abby and her mother is 21 years. In ten years Abby's mother's age will be three years less than twice Abby's age at that time. Find their present ages.

34. Kelly has a coin collection in which a penny is twice as old as a dime. In 10 years, the age of the penny will be $1\frac{1}{2}$ times the age of the dime at that time. Find the present ages of the penny and the dime.

35. The sum of the ages of our two cars is 11 years. In five years, the age of the older car will be three years more than twice the age of the newer car at that time. Find the present ages of the two cars.

36. Mitch is three-fourths as old as his sister Candy. In 8 years he will be five-sixths as old as Candy. Find their present ages.

37. Steve rode his bicycle from one town to another at 14 miles per hour. If he had increased his speed to 16 miles per hour he could have made the trip in one-half hour less time. How far apart are the towns?

Thoughts into Words

38. Give a step-by-step description of how you would solve the equation $C = \frac{5}{9}(F - 32)$ for F.

39. What similarities and what differences do you see between arithmetic and algebra?

40. What new ideas pertaining to problem solving have you acquired thus far in this course?

4.4
Inequalities

Just as we use the symbol $=$ to represent *is equal to*, we also use the symbols $<$ and $>$ to represent *is less than* and *is greater than*, respectively. The following are examples of **statements of inequality**. Notice that the first four are true statements and the last two are false.

$$6 + 4 > 7,$$
$$8 - 2 < 14,$$
$$4 \cdot 8 > 4 \cdot 6,$$
$$5 \cdot 2 < 5 \cdot 7,$$
$$5 + 8 > 19,$$
$$9 - 2 < 3$$

Algebraic inequalities contain one or more variables. The following are examples of algebraic inequalities.

$$x + 3 > 4,$$
$$2x - 1 < 6,$$
$$x^2 + 2x - 1 > 0,$$
$$2x + 3y < 7,$$
$$7ab < 9$$

An algebraic inequality such as $x + 1 > 2$ is neither true nor false as it stands; it is called an **open sentence**. Each time a number is substituted for x, the algebraic inequality $x + 1 > 2$ becomes a numerical statement that is either true or false. For example, if $x = 0$, then $x + 1 > 2$ becomes $0 + 1 > 2$, which is false. If $x = 2$, then $x + 1 > 2$ becomes $2 + 1 > 2$, which is true. **Solving an inequality** refers to the process of finding the numbers that make an algebraic inequality a true numerical statement. We say that such numbers, called the **solutions of the inequality**, satisfy the inequality. The set of all solutions of an inequality is called its **solution set**. We often state solution sets for inequalities with **set builder notation**. For example, the solution set for $x + 1 > 2$ is the set of real numbers greater than 1, expressed as $\{x \mid x > 1\}$. The set builder notation $\{x \mid x > 1\}$ is read as "the set of all x such that x is greater than 1." We sometimes graph solution sets for inequalities on a number line; the solution set for $\{x \mid x > 1\}$ is pictured below.

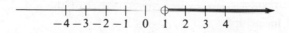

The unshaded circle around 1 indicates that 1 is not a solution; the shaded portion indicates that all numbers to the right of (greater than) 1 are solutions. We refer to the shaded portion of the number line as the *graph* of the solution set $\{x\,|\,x > 1\}$.

The following examples contain some simple algebraic inequalities, their solution sets, and the graphs of the solution sets. Look them over very carefully because they also introduce some additional symbolism.

Algebraic inequality	*Solution set*	*Graph of solution set*
$x < 2$	$\{x\,|\,x < 2\}$	
$x > -1$	$\{x\,|\,x > -1\}$	
$1 > x$	$\{x\,|\,x < 1\}$	
$3 < x$	$\{x\,|\,x > 3\}$	
$x \geq 1$ ($\geq$ is read "greater than or equal to")	$\{x\,|\,x \geq 1\}$	
$x \leq 2$ ($\leq$ is read "less than or equal to")	$\{x\,|\,x \leq 2\}$	

The general process for solving inequalities closely parallels that for solving equations. We continue to replace the given inequality with equivalent, but simpler inequalities. For example,

$$2x + 1 > 9, \tag{1}$$

$$2x > 8, \tag{2}$$

$$x > 4 \tag{3}$$

are all equivalent inequalities; that is, they have the same solutions. Thus, to solve (1) we can solve (3), which is obviously all numbers greater than 4. The exact procedure for simplifying inequalities is primarily based on two properties and they become our topics of discussion at this time. The first of these is the **addition-subtraction property of inequality**.

PROPERTY 4.1

For all real numbers a, b, and c,

1. $a > b$ if and only if $a + c > b + c$,
2. $a > b$ if and only if $a - c > b - c$.

Property 4.1 states that any number can be added to or subtracted from both sides of an inequality and an equivalent inequality is produced. The property is stated in terms of $>$, but analogous properties exist for $<$, $\geq$, and $\leq$. Consider the use of this property in the next three examples.

EXAMPLE 1   Solve $x - 3 > -1$ and graph the solutions.

Solution

$$x - 3 > -1$$
$$x - 3 + 3 > -1 + 3 \qquad \text{Add 3 to both sides.}$$
$$x > 2$$

The solution set is $\{x \mid x > 2\}$ and it can be graphed as follows.

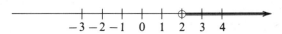

∎

EXAMPLE 2   Solve $x + 4 \leq 5$ and graph the solutions.

Solution

$$x + 4 \leq 5$$
$$x + 4 - 4 \leq 5 - 4 \qquad \text{Subtract 4 from both sides.}$$
$$x \leq 1$$

The solution set is $\{x \mid x \leq 1\}$ and it can be graphed as follows.

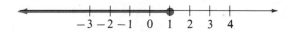

∎

EXAMPLE 3   Solve $5 > 6 + x$ and graph the solutions.

Solution

$$5 > 6 + x$$
$$5 - 6 > 6 + x - 6 \qquad \text{Subtract 6 from both sides.}$$
$$-1 > x$$

Since $-1 > x$ is equivalent to $x < -1$, the solution set is $\{x \mid x < -1\}$ and it can be graphed as follows.

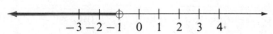

∎

Now let's look at some numerical examples to see what happens when both sides of an inequality are multiplied or divided by some number.

$$4 > 3 \quad \longrightarrow \quad 5(4) > 5(3),$$

$$-2 > -3 \quad \longrightarrow \quad 4(-2) > 4(-3),$$

$$6 > 4 \quad \longrightarrow \quad \frac{6}{2} > \frac{4}{2},$$

$$8 > -2 \quad \longrightarrow \quad \frac{8}{4} > \frac{-2}{4}$$

Notice that multiplying or dividing both sides of an inequality by a positive number produces an inequality of the same sense. This means that if the original inequality is "greater than," then the new inequality is "greater than," and if the original is "less than," then the resulting inequality is "less than."

Now note what happens when we multiply or divide both sides by a negative number.

$$3 < 5 \quad \text{but} \quad -2(3) > -2(5),$$

$$-4 < 1 \quad \text{but} \quad -5(-4) > -5(1),$$

$$4 > 2 \quad \text{but} \quad \frac{4}{-2} < \frac{2}{-2},$$

$$-3 > -6 \quad \text{but} \quad \frac{-3}{-3} < \frac{-6}{-3}$$

Multiplying or dividing both sides of an inequality *by a negative number reverses the sense of the inequality*. The following property summarizes these ideas.

PROPERTY 4.2

> **(a)** For all real numbers a, b, and c, *with $c > 0$,*
>
>   **1.** $a > b$ if and only if $ac > bc$,
>
>   **2.** $a > b$ if and only if $\dfrac{a}{c} > \dfrac{b}{c}$.
>
> **(b)** For all real numbers a, b, and c, *with $c < 0$,*
>
>   **1.** $a > b$ if and only if $ac < bc$,
>
>   **2.** $a > b$ if and only if $\dfrac{a}{c} < \dfrac{b}{c}$.

Similar properties hold if each inequality is reversed or if $>$ is replaced with $\geq$ and $<$ with $\leq$. For example, if $a \leq b$ and $c < 0$, then $ac \geq bc$ and $\dfrac{a}{c} \geq \dfrac{b}{c}$.

Observe the use of Property 4.2 in the next three examples.

EXAMPLE 4   Solve $2x > 4$.

Solution

$$2x > 4$$

$$\frac{2x}{2} > \frac{4}{2} \qquad \text{Divide both sides by 2.}$$

$$x > 2$$

The solution set is $\{x \mid x > 2\}$. ∎

EXAMPLE 5   Solve $\frac{3}{4}x \le \frac{1}{5}$.

Solution

$$\frac{3}{4}x \le \frac{1}{5}$$

$$\frac{4}{3}\left(\frac{3}{4}x\right) \le \frac{4}{3}\left(\frac{1}{5}\right) \qquad \text{Multiply both sides by } \frac{4}{3}.$$

$$x \le \frac{4}{15}$$

The solution set is $\left\{x \mid x \le \frac{4}{15}\right\}$. ∎

EXAMPLE 6   Solve $-3x > 9$.

Solution

$$-3x > 9$$

$$\frac{-3x}{-3} < \frac{9}{-3} \qquad \text{Divide both sides by } -3\text{, which reverses the sense.}$$

$$x < -3$$

The solution set is $\{x \mid x < -3\}$. ∎

As we mentioned earlier, many of the same techniques used to solve equations may be used to solve inequalities. However, we must be extremely careful when applying Property 4.2. Study the following examples and notice the similarities between solving equations and solving inequalities.

EXAMPLE 7   Solve $4x - 3 > 9$.

Solution

$$4x - 3 > 9$$

$$4x - 3 + 3 > 9 + 3 \qquad \text{Add 3 to both sides.}$$

$$4x > 12$$

$$\frac{4x}{4} > \frac{12}{4}$$ Divide both sides by 4.

$$x > 3$$

The solution set is $\{x \mid x > 3\}$.   ■

EXAMPLE 8   Solve $-3n + 5 < 11$.

Solution

$$-3n + 5 < 11$$

$$-3n + 5 - 5 < 11 - 5$$ Subtract 5 from both sides.

$$-3n < 6$$

$$\frac{-3n}{-3} > \frac{6}{-3}$$ Divide both sides by -3, which reverses the sense.

$$n > -2$$

The solution set is $\{n \mid n > -2\}$.   ■

Problem Set 4.4

For Problems 1–10, determine whether each numerical inequality is *true* or *false*.

1. $2(3) - 4(5) < 5(3) - 2(-1) + 4$

2. $5 + 6(-3) - 8(-4) > 17$

3. $\dfrac{2}{3} - \dfrac{3}{4} + \dfrac{1}{6} > \dfrac{1}{5} + \dfrac{3}{4} - \dfrac{7}{10}$

4. $\dfrac{1}{2} + \dfrac{1}{3} < \dfrac{1}{3} + \dfrac{1}{4}$

5. $\left(-\dfrac{1}{2}\right)\left(\dfrac{4}{9}\right) > \left(\dfrac{3}{5}\right)\left(-\dfrac{1}{3}\right)$

6. $\left(\dfrac{5}{6}\right)\left(\dfrac{8}{12}\right) < \left(\dfrac{3}{7}\right)\left(\dfrac{14}{15}\right)$

7. $\dfrac{3}{4} + \dfrac{2}{3} \div \dfrac{1}{5} > \dfrac{2}{3} + \dfrac{1}{2} \div \dfrac{3}{4}$

8. $1.9 - 2.6 - 3.4 < 2.5 - 1.6 - 4.2$

9. $0.16 + 0.34 > 0.23 + 0.17$

10. $(0.6)(1.4) > (0.9)(1.2)$

For Problems 11–22, state the solution set and graph it on a number line.

11. $x > -2$

12. $x > -4$

13. $x \le 3$

14. $x \le 0$

15. $2 < x$

16. $-3 \le x$

17. $-2 \ge x$

18. $1 > x$

19. $-x > 1$

20. $-x < 2$

21. $-2 < -x$

22. $-1 > -x$

For Problems 23–60, solve each of the inequalities.

23. $x + 6 < -14$

24. $x + 7 > -15$

25. $x - 4 \geq -13$

26. $x - 3 \leq -12$

27. $4x > 36$

28. $3x < 51$

29. $6x < 20$

30. $8x > 28$

31. $-5x > 40$

32. $-4x < 24$

33. $-7n \leq -56$

34. $-9n \geq -63$

35. $48 > -14n$

36. $36 < -8n$

37. $16 < 9 + n$

38. $19 > 27 + n$

39. $3x + 2 > 17$

40. $2x + 5 < 19$

41. $4x - 3 \leq 21$

42. $5x - 2 \geq 28$

43. $-2x - 1 \geq 41$

44. $-3x - 1 \leq 35$

45. $6x + 2 < 18$

46. $8x + 3 > 25$

47. $3 > 4x - 2$

48. $7 < 6x - 3$

49. $-2 < -3x + 1$

50. $-6 > -2x + 4$

51. $-38 \geq -9t - 2$

52. $36 \geq -7t + 1$

53. $5x - 4 - 3x > 24$

54. $7x - 8 - 5x < 38$

55. $4x + 2 - 6x < -1$

56. $6x + 3 - 8x > -3$

57. $-5 \geq 3t - 4 - 7t$

58. $6 \leq 4t - 7t - 10$

59. $-x - 4 - 3x > 5$

60. $-3 - x - 3x < 10$

Miscellaneous Problems

Solve each of the following inequalities.

61. $x + 3 < x - 4$

62. $x - 4 < x + 6$

63. $2x + 4 > 2x - 7$

64. $5x + 2 > 5x + 7$

65. $3x - 4 - 3x > 6$

66. $-2x + 7 + 2x > 1$

67. $-5 \leq -4x - 1 + 4x$

68. $-7 \geq 5x - 2 - 5x$

4.5

Inequalities, Compound Inequalities, and Problem Solving

Let's begin this section by solving three inequalities using the same basic steps as we did with equations. Again, be careful when applying the multiplication and division properties of inequality.

EXAMPLE 1   Solve $5x + 8 \leq 3x - 10$.

Solution

$$5x + 8 \leq 3x - 10$$

$$5x + 8 - 3x \leq 3x - 10 - 3x \qquad \text{Subtract } 3x \text{ from both sides.}$$

$$2x + 8 \leq -10$$

$$2x + 8 - 8 \leq -10 - 8 \qquad \text{Subtract 8 from both sides.}$$

$$2x \leq -18$$

$$\frac{2x}{2} \leq \frac{-18}{2} \qquad \text{Divide both sides by 2.}$$

$$x \leq -9$$

The solution set is $\{x \mid x \leq -9\}$.   ■

EXAMPLE 2   Solve $4(x + 3) + 3(x - 4) \geq 2(x - 1)$.

Solution

$$4(x + 3) + 3(x - 4) \geq 2(x - 1)$$

$$4x + 12 + 3x - 12 \geq 2x - 2 \qquad \text{distributive property}$$

$$7x \geq 2x - 2 \qquad \text{combined similar terms}$$

$$7x - 2x \geq 2x - 2 - 2x \qquad \text{Subtract } 2x \text{ from both sides.}$$

$$5x \geq -2$$

$$\frac{5x}{5} \geq \frac{-2}{5} \qquad \text{Divide both sides by 5.}$$

$$x \geq -\frac{2}{5}$$

The solution set is $\left\{ x \mid x \geq -\frac{2}{5} \right\}$.   ■

EXAMPLE 3   Solve $-\dfrac{3}{2}n + \dfrac{1}{6}n < \dfrac{3}{4}$.

Solution

$$-\frac{3}{2}n + \frac{1}{6}n < \frac{3}{4}$$

$$12\left(-\frac{3}{2}n + \frac{1}{6}n \right) < 12\left(\frac{3}{4} \right) \qquad \text{Multiply both sides by 12, which} \atop \text{is the LCD of all denominators.}$$

$$12\left(-\frac{3}{2}n \right) + 12\left(\frac{1}{6}n \right) < 12\left(\frac{3}{4} \right) \qquad \text{distributive property}$$

$$-18n + 2n < 9$$

$$-16n < 9$$

$$\frac{-16n}{-16} > \frac{9}{-16}$$ Divide both sides by -16, which reverses the inequality.

$$n > -\frac{9}{16}$$

The solution set is $\left\{ n \mid n > -\frac{9}{16} \right\}$.   ∎

Checking solutions for an inequality presents a problem. Obviously, we cannot check all of the infinitely many solutions for a particular inequality. However, by checking at least one solution, especially when the multiplication property has been used, we might catch a mistake of forgetting to reverse an inequality sign. In Example 3 we are claiming that "all numbers greater than $-\frac{9}{16}$ will satisfy the original inequality." Let's check one number, say 0.

$$-\frac{3}{2}n + \frac{1}{6}n < \frac{3}{4}$$

$$-\frac{3}{2}(0) + \frac{1}{6}(0) \overset{?}{<} \frac{3}{4}$$

$$0 < \frac{3}{4}$$

Therefore, 0 satisfies the original inequality. Had we forgotten to reverse the inequality sign when both sides were divided by -16, then our answer would have been $n < -\frac{9}{16}$ and we would have detected such an error by the check.

Compound Statements

The words *and* and *or* are used in mathematics to form **compound statements**. The following are examples of some compound numerical statements using *and*. Such statements are called **conjunctions**. We agree to call a conjunction true only if all of its component parts are true.

1. $3 + 6 = 9$ and $4 < 7$ (True)
2. $4 > 2$ and $0 > -2$ (True)
3. $8 + 7 = 15$ and $2 < 1$ (False)
4. $9 > 14$ and $13 < 18$ (False)
5. $5 > 8$ and $-6 > 0$ (False)

Compound statements using *or* are called **disjunctions**. A disjunction is true if at least one of its component parts is true. Said another way, a disjunction is false only if all of its component parts are false. Consider the following disjunctions.

1. $17 > 14$ or $18 < 25$ (True)
2. $19 > 12$ or $13 > 17$ (True)
3. $4 + 6 = 8$ or $0 < 1$ (True)
4. $7 < -4$ or $9 < 2$ (False)

Now let's consider finding solutions for some compound statements that involve algebraic inequalities. Keep in mind that the previous discussion regarding labeling conjunctions and disjunctions true or false forms the basis for our reasoning.

EXAMPLE 4   Graph the solutions for $x > 2$ and $x < 5$.

Solution   The key word is *and*, so we need to satisfy both inequalities. Thus, all numbers between 2 and 5 are solutions and we can graph them as follows.

EXAMPLE 5   Graph the solutions for $x < -1$ or $x > 3$.

Solution   The key word is *or*, so any numbers satisfying either (or both) inequalities are solutions. Thus, all numbers greater than 3 along with all numbers less than -1 are solutions and we can graph them as follows.

EXAMPLE 6   Graph the solutions for $x \geq -2$ and $x < 1$.

Solution   Since it is an *and* statement, we need to satisfy both inequalities. Thus, all numbers between -2 and 1, including -2 (but not including 1), are solutions. Notice on the graph that we shade the circle at -2, but do not shade the circle at 1.

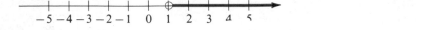

EXAMPLE 7   Graph the solutions for $x > 1$ or $x > 4$.

Solution   Since it is an *or* statement, numbers that satisfy either inequality (or both) are solutions. Thus, all numbers greater than 1 are included and the graph is as follows.

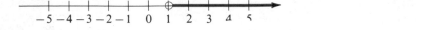

EXAMPLE 8   Graph the solutions for $x > 1$ and $x > 4$.

Solution   This is an *and* statement; so both inequalities must be satisfied. Thus, all numbers greater than 4 will work and the graph is as follows.

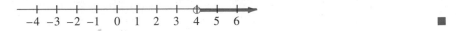

REMARK Take a good look at Examples 7 and 8. Note that in both examples the inequalities are the same but Example 7 is a disjunction and Example 8 is a conjunction.

Let's consider some word problems that translate into inequality statements. The *suggestions for solving word problems* given in Section 4.2 continue to apply here except that the situation being described in a problem will translate into an inequality instead of an equation.

PROBLEM 1   Debbie had scores of 95, 82, 93, and 84 on her first four exams of the semester. What score must she obtain on the fifth exam to have an average of 90 or better for the five exams?

Solution   Let s represent the score needed on the fifth exam. Since the average is computed by adding all five scores and dividing by 5 (the number of scores), we have the following inequality to solve.

$$\frac{95 + 82 + 93 + 84 + s}{5} \geq 90$$

Solving this inequality, we obtain

$$\frac{354 + s}{5} \geq 90 \qquad \text{simplified numerator of left side}$$

$$5\left(\frac{354 + s}{5}\right) \geq 5(90) \qquad \text{Multiply both sides by 5.}$$

$$354 + s \geq 450$$

$$354 + s - 354 \geq 450 - 354 \qquad \text{Subtract 354 from both sides.}$$

$$s \geq 96.$$

She must receive a score of 96 or better on the fifth exam.   ■

PROBLEM 2   The Cubs have won 40 baseball games and have lost 62 games. They have 60 more games to play. To win more than 50% of all of their games, how many of the 60 games remaining must they win?

Solution

Let w represent the number of games they must win out of the 60 games remaining. Since they are playing a total of $40 + 62 + 60 = 162$ games, to win more than 50% of their games they would need to win more than 81 games. Thus, we have the following inequality.

$$w + 40 > 81$$

Solving this yields

$$w > 41.$$

They need to win at least 42 of the 60 games remaining. ■

Problem Set 4.5

Solve each of the following inequalities.

1. $3x + 4 > x + 8$ **2.** $5x + 3 < 3x + 11$ **3.** $7x - 2 < 3x - 6$

4. $8x - 1 > 4x - 21$ **5.** $6x + 7 > 3x - 3$ **6.** $7x + 5 < 4x - 12$

7. $5n - 2 \le 6n + 9$ **8.** $4n - 3 \ge 5n + 6$ **9.** $2t + 9 \ge 4t - 13$

10. $6t + 14 \le 8t - 16$ **11.** $-3x - 4 < 2x + 7$ **12.** $-x - 2 > 3x - 7$

13. $-4x + 6 > -2x + 1$ **14.** $-6x + 8 < -4x + 5$ **15.** $5(x - 2) \le 30$

16. $4(x + 1) \ge 16$ **17.** $2(n + 3) > 9$ **18.** $3(n - 2) < 7$

19. $-3(y - 1) < 12$ **20.** $-2(y + 4) > 18$ **21.** $-2(x + 6) > -17$

22. $-3(x - 5) < -14$ **23.** $3(x - 2) < 2(x + 1)$ **24.** $5(x + 3) > 4(x - 2)$

25. $4(x + 3) > 6(x - 5)$ **26.** $6(x - 1) < 8(x + 5)$

27. $3(x - 4) + 2(x + 3) < 24$ **28.** $2(x + 1) + 3(x + 2) > -12$

29. $5(n + 1) - 3(n - 1) > -9$ **30.** $4(n - 5) - 2(n - 1) < 13$

31. $\dfrac{1}{2}n - \dfrac{2}{3}n \ge -7$ **32.** $\dfrac{3}{4}n + \dfrac{1}{6}n \le 1$ **33.** $\dfrac{3}{4}n - \dfrac{5}{6}n < \dfrac{3}{8}$

34. $\dfrac{2}{3}n - \dfrac{1}{2}n > \dfrac{1}{4}$ **35.** $\dfrac{3x}{5} - \dfrac{2}{3} > \dfrac{x}{10}$ **36.** $\dfrac{5x}{4} + \dfrac{3}{8} < \dfrac{7x}{12}$

37. $n \ge 3.4 + 0.15n$ **38.** $x \ge 2.1 + 0.3x$

39. $0.09t + 0.1(t + 200) > 77$ **40.** $0.07t + 0.08(t + 100) > 38$

41. $0.06x + 0.08(250 - x) \ge 19$ **42.** $0.08x + 0.09(2x) \le 130$

43. $\dfrac{x - 1}{2} + \dfrac{x + 3}{5} > \dfrac{1}{10}$ **44.** $\dfrac{x + 3}{4} + \dfrac{x - 5}{7} < \dfrac{1}{28}$ **45.** $\dfrac{x + 2}{6} - \dfrac{x + 1}{5} < -2$

46. $\dfrac{x - 6}{8} - \dfrac{x + 2}{7} > -1$ **47.** $\dfrac{n + 3}{3} + \dfrac{n - 7}{2} > 3$ **48.** $\dfrac{n - 4}{4} + \dfrac{n - 2}{3} < 4$

49. $\dfrac{x - 3}{7} - \dfrac{x - 2}{4} \le \dfrac{9}{14}$ **50.** $\dfrac{x - 1}{5} - \dfrac{x + 2}{6} \ge \dfrac{7}{15}$

Graph the solutions for each of the following compound inequalities.

51. $x > -1$ and $x < 2$ **52.** $x > 1$ and $x < 4$

53. $x < -2$ or $x > 1$

54. $x < 0$ or $x > 3$

55. $x > -2$ and $x \leq 2$

56. $x \geq -1$ and $x < 3$ 1

57. $x > -1$ and $x > 2$

58. $x < -2$ and $x < 3$ 1

59. $x > -4$ or $x > 0$

60. $x < 2$ or $x < 4$

61. $x > 3$ and $x < -1$

62. $x < -3$ and $x > 6$ 1

63. $x \leq 0$ or $x \geq 2$

64. $x \leq -2$ or $x \geq 1$

65. $x > -4$ or $x < 3$

66. $x > -1$ or $x < 2$

Solve each of the following problems by setting up and solving an appropriate inequality.

67. Five more than three times a number is greater than 26. Find the numbers.

68. Fourteen increased by twice a number is less than or equal to three times the number. Find the numbers that satisfy this relationship.

69. Suppose that the perimeter of a rectangle is to be no greater than 70 inches and the length of the rectangle must be 20 inches. Find the largest possible value for the width of the rectangle.

70. One side of a triangle is three times as long as another side. The third side is 15 centimeters long. If the perimeter of the triangle is to be no greater than 75 centimeters, find the largest lengths that the other two sides can be.

71. Sue bowled 132 and 160 in her first two games. What must she bowl in the third game to have an average of at least 150 for the three games?

72. Mike has scores of 87, 81, and 74 on his first three algebra tests. What score must he make on the fourth test to have an average of 85 or better for the four tests?

73. Lance has scores of 96, 90, and 94 on his first three algebra exams. What must he average on the last two exams to have an average of better than 92 for all five exams?

74. The Mets have won 45 baseball games and lost 55 games. To win more than 50% of all their games, how many of the 62 games remaining must they win?

75. Mona has $1000 to invest. If she invests $500 at 8% interest, at what rate must she invest the other $500 so that the two investments yield more than $100 of yearly interest?

76. The average height of the two forwards and the center of a basketball team is 6 feet and 8 inches. What must the average height of the two guards be so that the team average is at least 6 feet and 4 inches?

77. Scott shot rounds of 82, 84, 78, and 79 on the first four days of the golf tournament. What must he shoot on the fifth day to average 80 or less for the five days?

78. Lance has $500 to invest. If he invests $300 at 9% interest, at what rate must he invest the other $200 so that the two investments yield more than $47 in yearly interest?

Thoughts into Words

79. Give a step-by-step description of how you would solve the inequality $3x - 4 < 5(x - 2)$.

80. Explain the difference between a conjunction and a disjunction. Give a nonmathematical example of each one.

4.6

Equations and Inequalities Involving Absolute Value

In Chapter 1 we used the concept of absolute value to describe addition and multi-plication of integers. The absolute value of a number is the distance between the number and zero on a number line. For example, the absolute value of -6 is 6 (written as $|-6| = 6$) because there are 6 units between -6 and 0 on a number line. The concept of absolute value can be defined more formally as follows.

DEFINITION 4.1

For all real numbers a,

1. If $a \geq 0$, then $|a| = a$.
2. If $a < 0$, then $|a| = -a$.

Using Definition 4.1 we obtain

$$|4| = 4 \quad \text{by applying part 1,}$$

$$|0| = 0 \quad \text{by applying part 1,}$$

$$|-2| = -(-2) = 2 \quad \text{by applying part 2.}$$

Notice that the absolute value of a positive number is the number itself, but the absolute value of a negative number is its opposite. Thus, the absolute value of any number, except 0, is positive and the absolute value of 0 is 0.

Both the distance interpretation of absolute value and Definition 4.1 can be used to solve a variety of equations and inequalities that involve absolute value. First, let's consider some equations.

EXAMPLE 1   Solve $|x| = 2$.

Solution A   Thinking in terms of the distance interpretation for absolute value we are looking for all real numbers that are two units away from zero. Thus, the solution set is $\{-2, 2\}$.

Solution B   Let's apply Definition 4.1.

> ***Part 1*** If $x \geq 0$, then $|x| = x$; so the equation $|x| = 2$ becomes $x = 2$, which is satisfied by 2.
>
> ***Part 2*** If $x < 0$, then $|x| = -x$; so the equation $|x| = 2$ becomes $-x = 2$, which is equivalent to $x = -2$, and this is satisfied by -2. The solu-tion set is $\{-2, 2\}$. ∎

As illustrated in Example 1, we do have two basic approaches to absolute value problems. However, for our purposes in this text, the distance interpretation is usually the easier approach.

EXAMPLE 2   Solve $|x + 4| = 1$.

Solution   The number, $x + 4$, must be -1 or 1. Therefore,

$$|x + 4| = 1 \quad \text{implies}$$

$$x + 4 = -1 \quad \text{or} \quad x + 4 = 1$$

$$x = -5 \quad \text{or} \quad x = -3.$$

The solution set is $\{-5, -3\}$.   ∎

Now let's state a general property that should seem reasonable from the previous examples. This property can be verified by using Definition 4.1.

PROPERTY 4.3

> $|ax + b| = k$ is equivalent to $ax + b = -k \ or \ ax + b = k$, where k is a positive number.

Example 3 demonstrates a format that can be used for solving equations of the form $|ax + b| = k$.

EXAMPLE 3   Solve $|5x + 3| = 7$.

Solution
$$|5x + 3| = 7$$

$$5x + 3 = -7 \quad \text{or} \quad 5x + 3 = 7$$

$$5x = -10 \quad \text{or} \quad 5x = 4$$

$$x = -2 \quad \text{or} \quad x = \frac{4}{5}$$

The solution set is $\left\{ -2, \frac{4}{5} \right\}$.   ∎

"Less Than" Statements Involving Absolute Value

The distance interpretation for absolute value also provides a good basis for solving some inequalities that involve absolute value. Consider the following examples.

EXAMPLE 4   Solve $|x| < 2$ and graph the solution set.

Solution   The number, x, must be *less than two units away from zero*. Thus, $|x| < 2$ is equivalent to

$$x > -2 \quad \text{and} \quad x < 2.$$

The solution set is $\{x \mid x > -2 \text{ and } x < 2\}$ and its graph is as follows.

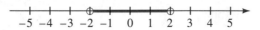

$$-5 \;\; -4 \;\; -3 \;\; -2 \;\; -1 \;\;\; 0 \;\;\; 1 \;\;\; 2 \;\;\; 3 \;\;\; 4 \;\;\; 5$$

■

EXAMPLE 5   Solve $|x + 3| < 1$ and graph the solution set.

Solution   The number, $x + 3$, must be *less than one unit away from zero*. Therefore, $|x + 3| < 1$ is equivalent to $x + 3 > -1$ and $x + 3 < 1$. Solving this conjunction yields

$$x + 3 > -1 \quad \text{and} \quad x + 3 < 1$$
$$x > -4 \quad \text{and} \quad x < -2.$$

The solution set is $\{x \mid x > -4 \text{ and } x < -2\}$ and its graph is as follows.

$$-5 \;\; -4 \;\; -3 \;\; -2 \;\; -1 \;\;\; 0 \;\;\; 1 \;\;\; 2 \;\;\; 3 \;\;\; 4 \;\;\; 5$$

■

Take another look at Examples 4 and 5. The following general property should seem reasonable. It can be verified by applying Definition 4.1.

PROPERTY 4.4

$|ax + b| < k$ *is equivalent to* $ax + b > -k$ *and* $ax + b < k$, *where k is a positive number.*

An analogous property can be stated using the relation $\leq$. Property 4.4 can be used as follows.

EXAMPLE 6   Solve $|3x - 1| \leq 8$ and graph the solution set.

Solution
$$|3x - 1| \leq 8$$
$$3x - 1 \geq -8 \quad \text{and} \quad 3x - 1 \leq 8$$
$$3x \geq -7 \quad \text{and} \quad 3x \leq 9$$
$$x \geq -\frac{7}{3} \quad \text{and} \quad x \leq 3$$

The solution set is $\left\{x \,\middle|\, x \geq -\dfrac{7}{3} \text{ and } x \leq 3\right\}$ and its graph is as follows.

"Greater Than" Statements Involving Absolute Value

Now let's return to the distance interpretation for absolute value and solve some *greater than* inequalities involving absolute value.

EXAMPLE 7   Solve $|x| > 1$ and graph the solution set.

Solution   The number, x, must be *more than one unit away from zero.* Therefore, $|x| > 1$ is equivalent to $x < -1$ or $x > 1$. The solution set is $\{x \,|\, x < -1 \text{ or } x > 1\}$ and its graph is as follows.

EXAMPLE 8   Solve $|x - 1| > 3$ and graph the solution set.

Solution   The number, $x - 1$, must be *more than three units away from zero.* Thus, $|x - 1| > 3$ is equivalent to $x - 1 < -3$ or $x - 1 > 3$. Solving this disjunction yields

$$x - 1 < -3 \qquad \text{or} \qquad x - 1 > 3$$
$$x < -2 \qquad \text{or} \qquad x > 4.$$

The solution set is $\{x \,|\, x < -2 \text{ or } x > 4\}$ and its graph is as follows.

Examples 7 and 8 lead to the following general property. This property can also be verified by applying Definition 4.1.

PROPERTY 4.5 $|ax + b| > k$ is equivalent to $ax + b < -k$ or $ax + b > k$, where k is a positive number.

An analogous property exists using the relation $\geq$. Let's consider an example using Property 4.5.

EXAMPLE 9   Solve $|4x + 1| \geq 5$ and graph the solution set.

Solution

$$|4x + 1| \geq 5$$

$$4x + 1 \leq -5 \quad \text{or} \quad 4x + 1 \geq 5$$

$$4x \leq -6 \quad \text{or} \quad 4x \geq 4$$

$$x \leq -\frac{6}{4} \quad \text{or} \quad x \geq 1$$

$$x \leq -\frac{3}{2} \quad \text{or} \quad x \geq 1$$

The solution set is $\left\{ x \,\middle|\, x \leq -\dfrac{3}{2} \text{ or } x \geq 1 \right\}$ and its graph is as follows.

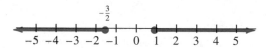

Properties 4.3, 4.4, and 4.5 provide the basis for solving a variety of equations and inequalities involving absolute value. However, if at any time you become doubtful as to what property applies, don't forget the distance interpretation. Furthermore, note that in each of the properties, k is a positive number. If k is a nonpositive number, we can determine the solution sets by inspection, as indicated by the following examples.

The solution set of $|x + 3| = 0$ is $\{-3\}$ because the number $x + 3$ has to be 0.

$|2x - 5| = -3$ has *no solutions* because the absolute value (distance) cannot be negative. (The solution set is $\varnothing$).

$|x - 7| < -4$ has *no solutions* because we cannot obtain an absolute value less than -4. (The solution set is $\varnothing$).

$|2x - 1| > -1$ is *satisfied by all real numbers* because the absolute value of $(2x - 1)$, regardless of what number is substituted for x, will always be greater than -1. (The solution set is $\{x \,|\, x \text{ is a real number}\}$.)

Problem Set 4.6

For Problems 1–26, solve and graph the solutions for each of the following.

1. $|x| = 4$ **2.** $|x| = 3$

3. $|x| < 1$ **4.** $|x| < 4$

5. $|x| \geq 2$ **6.** $|x| \geq 1$

7. $|x + 2| = 1$ **8.** $|x + 3| = 2$

9. $|x - 1| = 2$

10. $|x - 2| = 1$

11. $|x - 2| \leq 2$

12. $|x + 1| \leq 3$

13. $|x + 1| > 3$

14. $|x - 3| > 1$

15. $|2x + 1| = 3$

16. $|3x - 1| = 5$

17. $|5x - 2| = 4$

18. $|4x + 3| = 8$

19. $|2x - 3| \geq 1$

20. $|2x + 1| \geq 3$

21. $|4x + 3| < 2$

22. $|5x - 2| < 8$

23. $|3x + 6| = 0$

24. $|4x - 3| = 0$

25. $|3x - 2| > 0$

26. $|2x + 7| < 0$

For Problems 27–50, solve each of the following.

27. $|3x - 1| = 17$

28. $|4x + 3| = 27$

29. $|2x + 1| > 9$

30. $|3x - 4| > 20$

31. $|3x - 5| < 19$

32. $|5x + 3| < 14$

33. $|-3x - 1| = 17$

34. $|-4x + 7| = 26$

35. $|4x - 7| \leq 31$

36. $|5x - 2| \leq 21$

37. $|5x + 3| \geq 18$

38. $|2x - 11| \geq 4$

39. $|-x - 2| < 4$

40. $|-x - 5| < 7$

41. $|-2x + 1| > 6$

42. $|-3x + 2| > 8$

43. $\left| x - \dfrac{3}{4} \right| = \dfrac{2}{3}$

44. $\left| x + \dfrac{1}{2} \right| = \dfrac{3}{5}$

45. $\left| \dfrac{x - 3}{4} \right| < 2$

46. $\left| \dfrac{x + 2}{3} \right| < 1$

47. $\left| \dfrac{2x + 1}{2} \right| \geq 1$

48. $\left| \dfrac{3x - 1}{4} \right| \geq 3$

49. $|2x - 1| + 1 \leq 6$

50. $|4x + 3| - 2 > 5$

For Problems 51–58, solve each of the following *by inspection*.

51. $|7x| = 0$

52. $|3x - 1| = -4$

53. $|x - 6| > -4$

54. $|3x + 1| > -3$

55. $|x + 4| < -7$

56. $|5x - 2| < -2$

57. $|x + 6| \leq 0$

58. $|x + 7| > 0$

Miscellaneous Problems

A conjunction such as "$x > -2$ and $x < 4$" can be written in a more compact form $-2 < x < 4$, which is read as "-2 is less than x and x is less than 4." In other words, x is clamped between -2 and 4. The compact form is very convenient for solving conjunctions as follows.

$$-3 < 2x - 1 < 5$$

$$-2 < 2x < 6 \qquad \text{Add 1 to the left side, middle, and right side.}$$

$$-1 < x < 3 \qquad \text{Divide through by 2.}$$

Thus, the solution set can be expressed as $\{x \mid -1 < x < 3\}$.

For Problems 59–68, solve the compound inequalities using the compact form.

59. $-2 < x - 6 < 8$ **60.** $-1 < x + 3 < 9$

61. $1 \le 2x + 3 \le 11$ **62.** $-2 \le 3x - 1 \le 14$

63. $-4 < \dfrac{x-1}{3} < 2$ **64.** $2 < \dfrac{x+1}{4} < 5$

65. $|x + 4| < 3$ [*Hint:* $|x + 4| < 3$ implies $-3 < x + 4 < 3$.]

66. $|x - 6| < 5$ **67.** $|2x - 5| < 7$

68. $|3x + 2| < 14$

Chapter 4 Summary

(4.1) **Formulas** are rules stated in symbolic form. A formula such as $P = 2l + 2w$ can be solved for $l\left(l = \dfrac{P - 2w}{2}\right)$ or for $w\left(w = \dfrac{P - 2l}{2}\right)$ by applying the properties of equality.

(4.2) and (4.3) Don't forget the following suggestions for solving word problems.

1. Read the problem carefully.
2. Sketch a figure, diagram, or chart that might be helpful to organize the facts.
3. Choose a meaningful variable.
4. Look for a guideline.
5. Form an equation or inequality.
6. Solve the equation or inequality.
7. Check your answers.

(4.3) and (4.4) Properties 4.1 and 4.2 provide the basis for solving inequalities. Be sure that you can use these properties to solve the variety of inequalities presented in this chapter.

We can use many of the same techniques used to solve equations to solve inequalities *but* we must be very careful when multiplying or dividing both sides of an inequality by the same number. *Don't forget* that when miltiplying or dividing both sides of an inequality by a *negative number*, the resulting inequality *reverses sense*.

(4.5) The words **and** and **or** are used to form compound inequalities.

To solve inequalities involving **and** we must satisfy all of the conditions. Thus, the compound inequality $x > 1$ **and** $x < 3$ is satisfied by all numbers between 1 and 3.

To solve inequalities involving **or** we must satisfy one or more of the conditions. Thus, the compound inequality $x < -1$ or $x > 2$ is satisfied by (a) all numbers less than -1, or (b) all numbers greater than 2, or (c) both (a) and (b).

(4.6) We can interpret the **absolute value** of a number on the number line as the distance between that number and zero. The following properties form the basis for solving equations and inequalities involving absolute value.

$$\left.\begin{array}{l} \textbf{1.}\ |ax + b| = k \text{ is equivalent to } ax + b = -k \text{ or } ax + b = k. \\[4pt] \textbf{2.}\ |ax + b| < k \text{ is equivalent to } -k < ax + b < k. \\[4pt] \textbf{3.}\ |ax + b| > k \text{ is equivalent to } ax + b < -k \text{ or } ax + b > k. \end{array}\right\} k > 0$$

Chapter 4 Review Problem Set

1. Solve $P = 2l + 2w$ for w if $P = 50$ and $l = 19$.

2. Solve $F = \dfrac{9}{5}C + 32$ for C if $F = 77$.

3. Solve $A = P + Prt$ for t.

4. Solve $2x - 3y = 13$ for x.

5. Find the area of a trapezoid that has one base 8 inches long and the other base 14 inches long, if the altitude between the two bases is 7 inches.

6. If the area of a triangle is 27 square centimeters and the length of one side is 9 centimeters, find the length of the altitude to that side.

7. If the total surface area of a right circular cylinder is 152π square feet and a radius of a base is 4 feet long, find the height of the cylinder.

Solve each of the following inequalities.

8. $3x - 2 > 10$

9. $-2x - 5 < 3$

10. $2x - 9 \geq x + 4$

11. $3x + 1 \leq 5x - 10$

12. $6(x - 3) > 4(x + 13)$

13. $2(x + 3) + 3(x - 6) < 14$

14. $\dfrac{2n}{5} - \dfrac{n}{4} < \dfrac{3}{10}$

15. $\dfrac{n + 4}{5} + \dfrac{n - 3}{6} > \dfrac{7}{15}$

16. $s \geq 4.5 + 0.25s$

17. $0.07t + 0.09(500 - t) \geq 43$

18. $-16 < 8 + 2y - 3y$

19. $-24 > 5x - 4 - 7x$

20. $-3(n - 4) > 5(n + 2) + 3n$

21. $-4(n - 2) - (n - 1) < -4(n + 6)$

22. $\dfrac{3}{4}n - 6 \leq \dfrac{2}{3}n + 4$

23. $\dfrac{1}{2}n - \dfrac{1}{3}n - 4 \geq \dfrac{3}{5}n + 2$

24. $-12 > -4(x - 1) + 2$

25. $36 < -3(x + 2) - 1$

Graph the solutions for each of the following compound inequalities.

26. $x > -3$ and $x < 2$

27. $x < -1$ or $x > 4$

28. $x < 2$ or $x > 0$

29. $x > 1$ and $x > 0$

Solve and graph the solutions for each of the following.

30. $|3x - 5| = 7$

31. $|x - 4| < 1$

32. $|2x - 1| \geq 3$

33. $|3x - 2| \leq 4$

34. $|2x - 1| = 9$

35. $|5x - 2| \geq 6$

Set up an equation or an inequality and solve each of the following problems.

36. Suppose that the length of a certain rectangle is 5 meters more than twice the width. The perimeter of the rectangle is 46 meters. Find the length and width of the rectangle.

37. Two airplanes leave Chicago at the same time and fly in opposite directions. If one travels at 350 miles per hour and the other at 400 miles per hour, how long will it take them to be 1125 miles apart?

38. How many liters of pure alcohol must be added to 10 liters of a 70% solution to obtain a 90% solution?

39. Monica has scores of 83, 89, 78, and 86 on her first four exams. What score must she receive on the fifth exam so that her average for all five exams is 85 or better?

40. One angle of a triangle has a measure of 47°. Of the other two angles, one of them is 3° less than three times the other angle. Find the measures of the two remaining angles.

41. Connie rides out into the country on her bicycle at a rate of 10 miles per hour. An hour later Jay leaves from the same place that Connie did and rides his bicycle along the same route at 12 miles per hour. How long will it take Jay to catch Connie?

42. How many gallons of a 10% salt solution must be mixed with 12 gallons of a 15% salt solution to obtain a 12% salt solution?

43. Susan's average score for her first three psychology exams is 84. What must she get on the fourth exam so that her average for the four exams is 85 or better?

44. Suppose that 20 ounces of a punch containing 20% orange juice is added to 30 ounces of punch containing 30% orange juice. Find the percent of orange juice in the resulting mixture.

45. The present ages of Bill and Dan are 20 years and 30 years, respectively. Will there be or was there ever a time when Bill's age is (was) one-half of Dan's age?

Cumulative Review Problem Set   (Chapters 1–4)

For Problems 1–10, simplify each algebraic expression by combining similar terms.

1. $7x - 9x - 14x$

2. $-10a - 4 + 13a + a - 2$

3. $5(x - 3) + 7(x + 6)$

4. $3(x - 1) - 4(2x - 1)$

5. $-3n - 2(n - 1) + 5(3n - 2) - n$

6. $6n + 3(4n - 2) - 2(2n - 3) - 5$

7. $\dfrac{1}{2}x - \dfrac{3}{4}x + \dfrac{2}{3}x - \dfrac{1}{6}x$

8. $\dfrac{1}{3}n - \dfrac{4}{15}n + \dfrac{5}{6}n - n$

9. $0.4x + 0.7x - 0.8x + x$

10. $0.5(x - 2) + 0.4(x + 3) - 0.2x$

For Problems 11–20, evaluate each of the algebraic expressions for the given values of the variables.

11. $5x - 7y + 2xy$ for $x = -2$ and $y = 5$

12. $2ab - a + 6b$ for $a = 3$ and $b = -4$

13. $-3(x - 1) + 2(x + 6)$ for $x = -5$

14. $5(n + 3) - (n + 4) - n$ for $n = 7$

15. $\dfrac{3x - 2y}{2x - 3y}$ for $x = 3$ and $y = -6$

16. $\dfrac{3}{4}n - \dfrac{1}{3}n + \dfrac{5}{6}n$ for $n = -\dfrac{2}{3}$

17. $2a^2 - 4b^2$ for $a = 0.2$ and $b = -0.3$

18. $x^2 - 3xy - 2y^2$ for $x = \dfrac{1}{2}$ and $y = \dfrac{1}{4}$

19. $5x - 7y - 8x + 3y$ for $x = 9$ and $y = -8$

20. $\dfrac{3a - b - 4a + 3b}{a - 6b - 4b - 3a}$ for $a = -1$ and $b = 3$

For Problems 21–26, evaluate each of the expressions.

21. 3^4

22. -2^6

23. $\left(\dfrac{2}{3}\right)^3$

24. $\left(-\dfrac{1}{2}\right)^5$

25. $\left(\dfrac{1}{2} + \dfrac{1}{3}\right)^2$

26. $\left(\dfrac{3}{4} - \dfrac{7}{8}\right)^3$

For Problems 27–40, solve each of the equations.

27. $-5x + 2 = 22$

28. $3x - 4 = 7x + 4$

29. $7(n - 3) = 5(n + 7)$

30. $2(x - 1) - 3(x - 2) = 12$

31. $\dfrac{2}{5}x - \dfrac{1}{3} = \dfrac{1}{3}x + \dfrac{1}{2}$

32. $\dfrac{t - 2}{4} + \dfrac{t + 3}{3} - \dfrac{1}{6}$

33. $\dfrac{2n - 1}{5} - \dfrac{n + 2}{4} = 1$

34. $0.09x + 0.12(500 - x) = 54$

35. $|2x - 7| = 3$

36. $|3x + 5| = 7$

37. $-5(n - 1) - (n - 2) = 3(n - 1) - 2n$

38. $\dfrac{-2}{x - 1} = \dfrac{-3}{x + 4}$

39. $0.2x + 0.1(x - 4) = 0.7x - 1$

40. $-(t - 2) + (t - 4) = 2\left(t - \dfrac{1}{2}\right) - 3\left(t + \dfrac{1}{3}\right)$

For Problems 41–50, solve each of the inequalities.

41. $4x - 6 > 3x + 1$

42. $-3x - 6 < 12$

43. $-2(n - 1) \le 3(n - 2) + 1$

44. $\dfrac{2}{7}x - \dfrac{1}{4} \ge \dfrac{1}{4}x + \dfrac{1}{2}$

45. $0.08t + 0.1(300 - t) > 28$ **46.** $-4 > 5x - 2 - 3x$

47. $|3x - 1| < 5$ **48.** $|2x + 3| > 4$

49. $\dfrac{2}{3}n - 2 \geq \dfrac{1}{2}n + 1$ **50.** $-3 < -2(x - 1) - x$

For Problems 51–58, set up an equation or an inequality and solve each problem.

51. Dawn's salary this year is $32,000. This represents $2000 more than twice her salary five years ago. Find her salary five years ago.

52. One of two supplementary angles is 45° less than four times the other angle. Find the measure of each angle.

53. Jaamal has 25 coins consisting of nickels and dimes amounting to $2.10. How many coins of each kind does he have?

54. Hana bowled 144 and 176 in her first two games. What must she bowl in the third game to have an average of at least 150 for the three games?

55. A board 30 feet long is cut into two pieces whose lengths are in the ratio of 2 to 3. Find the lengths of the two pieces.

56. A retailer has some shoes that cost him $32 per pair. He wants to sell them at a profit of 20% of the selling price. What price should he charge for the shoes?

57. Two cars start from the same place traveling in opposite directions. One car travels 5 miles per hour faster than the other car. Find their speeds if after six hours they are 570 miles apart.

58. How many liters of pure alcohol must be added to 15 liters of a 20% solution to obtain a 40% solution?

Chapter 5

Coordinate Geometry and Linear Systems

On a real number line a one-to-one correspondence is established between the set of real numbers and the points on a line. To each point on a line there corresponds a real number and to each real number there corresponds a point on the line. The real number line was used in Chapter 4 to graph the solutions for inequalities of **one variable**. The idea of associating points with numbers can be extended so that a correspondence is established between pairs of real numbers and points in a plane. This will allow us to graph solutions of equations and inequalities that contain **two variables**.

175

The Rectangular Coordinate System

Consider two number lines (one vertical and one horizontal) perpendicular to each other at the point we associate with zero on both lines (Figure 5.1). We refer to these number lines as the horizontal and vertical **axes** or together as the **coordinate axes**. They partition the plane into four parts called **quadrants**. The quadrants are numbered counterclockwise from I to IV as indicated in Figure 5.1. The point of intersection of the two axes is called the **origin**.

Figure 5.1

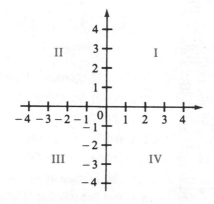

It is now possible to set up a one-to-one correspondence between ordered pairs of real numbers and the points in a plane. To each ordered pair of real numbers there corresponds a unique point in the plane and to each point there corresponds a unique ordered pair of real numbers. We have indicated a part of this correspondence in Figure 5.2. The ordered pair (3, 1) corresponds to point A and means

Figure 5.2

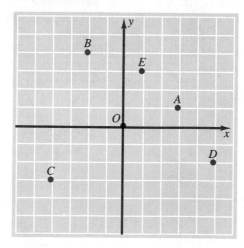

$A(3, 1)$
$B(-2, 4)$
$C(-4, -3)$
$D(5, -2)$
$O(0, 0)$
$E(1, 3)$

that the point A is located 3 units to the right of and one unit up from the origin. (The ordered pair $(0, 0)$ corresponds to the origin.) The ordered pair $(-2, 4)$ corresponds to point B and that means that point B is located 2 units to the left and 4 units up from the origin. Make sure that you agree with all of the other locations in Figure 5.2.

In general, we refer to the real numbers a and b in an ordered pair (a, b), associated with a point as the **coordinates of the point**. The first number, a, called the **abscissa**, is the directed distance of the point from the vertical axis measured parallel to the horizontal axis. The second number, b, called the **ordinate**, is the directed distance of the point from the horizontal axis measured parallel to the vertical axis (Figure 5.3(a)). Thus, in the first quadrant all points have a positive abscissa and a positive ordinate. In the second quadrant all points have a negative abscissa and a positive ordinate. We have indicated the sign situations for all four quadrants in Figure 5.3(b). This system of associating points with ordered pairs of real numbers is called the **Cartesian coordinate system** or the **rectangular coordinate system**.

Figure 5.3

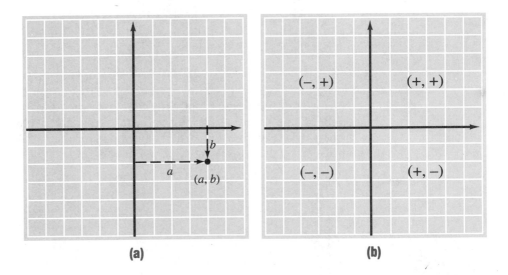

(a)   (b)

Historically, the Cartesian coordinate system provided the basis for the development of a branch of mathematics called **analytic geometry**; at the present time it is often referred to as **coordinate geometry**. With this system, René Descartes, a 17th century French mathematician was able to transform geometric problems into an algebraic setting and then use the tools of algebra to solve the problems.

Basically, there are two kinds of problems in analytic geometry.

1. Given an algebraic equation, find its geometric graph.
2. Given a set of conditions pertaining to a geometric figure, find its algebraic equation.

We will consider a few problems of type 1 at this time.

Let's begin by considering the solutions for the equation $y = x + 3$. A *solution* of an equation in two variables is an ordered pair of real numbers that satisfies the equation. When using the variables x and y, we agree that the first number of an ordered pair is a value for x and the second number is a value for y. We see that $(1, 4)$ is a solution for $y = x + 3$ because when x is replaced by 1 and y by 4, the true numerical statement $4 = 1 + 3$ is obtained. Likewise, $(-1, 2)$ is a solution for $y = x + 3$ because $2 = -1 + 3$ is a true statement. Infinitely many pairs of real numbers that satisfy $y = x + 3$ can be found by arbitrarily choosing values for x and for each value of x chosen, determining a corresponding value for y. Let's use a table to record some of the solutions for $y = x + 3$.

Choose x	Determine y from $y = x + 3$	Solution for $y = x + 3$
0	3	$(0, 3)$
1	4	$(1, 4)$
3	6	$(3, 6)$
5	8	$(5, 8)$
-1	2	$(-1, 2)$
-3	0	$(-3, 0)$
-5	-2	$(-5, -2)$

Now we can locate the point associated with each ordered pair on a rectangular coordinate system using the horizontal axis as the **x-axis** and the vertical axis as the **y-axis** (Figure 5.4(a)). The straight line in Figure 5.4(b) that contains the points is called the *graph of the equation $y = x + 3$*.

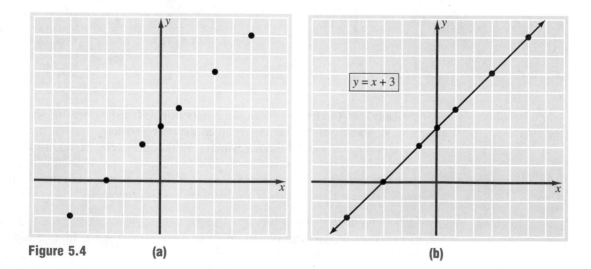

Figure 5.4　　**(a)**　　　　　　　　　　　　　　**(b)**

The following examples further illustrate the process of graphing equations.

EXAMPLE 1   Graph $y = x^2$.

Solution   First we set up a table of some of the solutions.

x	y	Solutions (x, y)
0	0	$(0, 0)$
1	1	$(1, 1)$
2	4	$(2, 4)$
3	9	$(3, 9)$
-1	1	$(-1, 1)$
-2	4	$(-2, 4)$
-3	9	$(-3, 9)$

Then we plot points associated with the solutions as in Figure 5.5(a). Finally, we connect the points with a smooth curve as in Figure 5.5(b). This curve is called a **parabola** and it will receive much more attention in a later chapter.

Figure 5.5

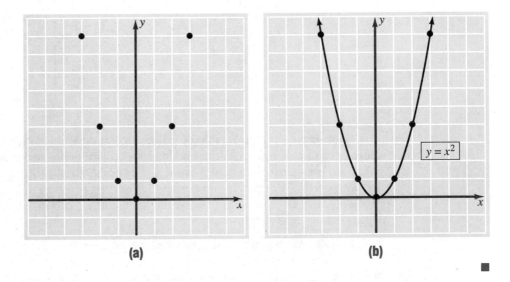

(a)   (b)

How many solutions do we need to have in a table of values? There is no definite answer to this question other than *a sufficient number so that the graph of the equation can be determined.* In other words, we need to plot points until we can tell the nature of the curve.

EXAMPLE 2   Graph $2x + 3y = 6$.

Solution   First, let's change the form of the equation to make it easier to find solutions. We can either solve for x in terms of y or for y in terms of x. With the latter, we obtain

$$2x + 3y = 6$$
$$3y = 6 - 2x$$
$$y = \frac{6 - 2x}{3}.$$

Now we can set up a table of values. Plotting these points and connecting them produces Figure 5.6.

Figure 5.6

x	y
0	2
3	0
6	−2
−3	4
−6	6

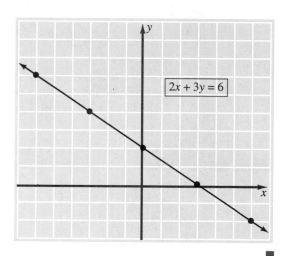

$2x + 3y = 6$

Look carefully at the table of values in Example 2. Notice that we chose values of x so that integers were obtained for y. This is not necessary but it does make things easier from a computational viewpoint. Also, plotting points associated with pairs of integers is more exact than getting involved with fractions.

To graph an equation in two variables x and y, keep in mind the following steps.

1. Solve the equation for y in terms of x or for x in terms of y, if it is not already in such a form.
2. Set up a table of ordered pairs that satisfy the equation.
3. Plot the points associated with the ordered pairs.
4. Connect the points with a smooth curve.

Let's conclude this section with two more examples that illustrate step 1.

EXAMPLE 3   Solve $4x + 9y = 12$ for y.

Solution

$$4x + 9y = 12$$
$$9y = 12 - 4x \qquad \text{subtracted } 4x \text{ from both sides}$$
$$y = \frac{12 - 4x}{9} \qquad \text{divided both sides by 9}$$

■

EXAMPLE 4   Solve $4x - 5y = 6$ for y.

Solution

$$4x - 5y = 6$$
$$-5y = 6 - 4x \qquad \text{subtracted } 4x \text{ from both sides}$$
$$y = \frac{6 - 4x}{-5} \qquad \text{divided both sides by } -5$$
$$y = \frac{4x - 6}{5} \qquad \frac{6 - 4x}{-5} \text{ can be changed to } \frac{-6 + 4x}{5} \text{ by multiplying numerator and denominator by } -1$$

■

Problem Set 5.1

For Problems 1–10, solve the given equation for the variable indicated.

1. $3x + 7y = 13$ for y
2. $5x + 9y = 17$ for y
3. $x - 3y = 9$ for x
4. $2x - 7y = 5$ for x
5. $-x + 5y = 14$ for y
6. $-2x - y = 9$ for y
7. $-3x + y = 7$ for x
8. $-x - y = 9$ for x
9. $-2x + 3y = -5$ for y
10. $3x - 4y = -7$ for y

For Problems 11–34, graph each of the equations.

11. $y = x + 1$
12. $y = x + 4$
13. $y = x - 2$
14. $y = -x - 1$
15. $y = (x - 2)^2$
16. $y = (x + 1)^2$
17. $y = x^2 - 2$
18. $y = x^2 + 1$
19. $y = \frac{1}{2}x + 3$
20. $y = \frac{1}{2}x - 2$
21. $x + 2y = 4$
22. $x + 3y = 6$
23. $2x - 5y = 10$
24. $5x - 2y = 10$
25. $y = x^3$
26. $y = x^4$
27. $y = -x^2$
28. $y = -x^3$

29. $y = x$ **30.** $y = -x$

31. $y = -3x + 2$ **32.** $3x - y = 4$

33. $y = 2x^2$ **34.** $y = -3x^2$

5.2

Linear Equations in Two Variables

The following table summarizes some of our results from graphing equations in the previous section and accompanying problem set.

Equation	Type of graph produced
$y = x + 3$	straight line
$y = x^2$	parabola
$2x + 3y = 6$	straight line
$y = -3x + 2$	straight line
$y = x^2 - 2$	parabola
$y = (x - 2)^2$	parabola
$5x - 2y = 10$	straight line
$3x - y = 4$	straight line
$y = x^3$	no name will be given at this time, but not a straight line
$y = x$	straight line
$y = \dfrac{1}{2}x + 3$	straight line

In the above table pay special attention to the equations that produced a straight line graph. They are called **linear equations in two variables**. In general, any equation of the form $Ax + By = C$, where A, B, and C are constants (A and B not both zero) and x and y are variables, is a linear equation in two variables and its graph is a straight line.

We should clarify two points about the previous description of a linear equation in two variables. First, the choice of x and y for variables is arbitrary. Any two letters could be used to represent the variables. An equation such as $3m + 2n = 7$ can be considered a linear equation in two variables. So that we are not constantly changing the labeling of the coordinate axes when graphing equations, however, it is much easier to use the same two variables in all equations. Thus, we will go along with convention and use x and y as our variables. Secondly, the statement "any equation of the form $Ax + By = C$" technically means "any equation of the form $Ax + By = C$ or *equivalent* to the form." For example, the equation $y = x + 3$, which has a straight line graph, is equivalent to $-x + y = 3$.

The knowledge that any equation of the form $Ax + By = C$ produces a straight line graph, along with the fact that two points determine a straight line, makes graphing linear equations in two variables a simple process. We merely find two solutions, plot the corresponding points, and connect the points with a straight line. It is probably wise to find a third point as a check point. Let's consider an example.

EXAMPLE 1   Graph $2x - 3y = 6$.

Solution

$$\text{Let } x = 0, \text{ then } 2(0) - 3y = 6$$
$$-3y = 6$$
$$y = -2.$$

Thus, $(0, -2)$ is a solution.

$$\text{Let } y = 0, \text{ then } 2x - 3(0) = 6$$
$$2x = 6$$
$$x = 3.$$

Thus, $(3, 0)$ is a solution.

$$\text{Let } x = -3, \text{ then } 2(-3) - 3y = 6$$
$$-6 - 3y = 6$$
$$-3y = 12$$
$$y = -4.$$

Thus, $(-3, -4)$ is a solution. Plotting the points associated with these three solutions and connecting them with a straight line produces the graph of $2x - 3y = 6$ in Figure 5.7.

Figure 5.7

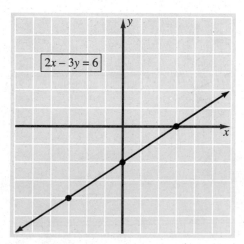

$2x - 3y = 6$

Let us briefly review our approach to Example 1. Notice that we did not begin our solution by either solving for y in terms of x or for x in terms of y. The reason for this is that we know the graph is a straight line and therefore there is no need for an extensive table of values. Thus, there is no real benefit to changing the form of the original equation. The first two solutions indicate where the line intersects the coordinate axes. The ordinate of the point $(0, -2)$ is called the **y-intercept** and the abscissa of the point $(3, 0)$ the **x-intercept** of this graph. That is, the graph of the equation $2x - 3y = 6$ has a y-intercept of -2 and an x-intercept of 3. In general, the intercepts are often easy to find. You can let $x = 0$ and solve for y to find the y-intercept and let $y = 0$ and solve for x to find the x-intercept. The third solution, $(-3, -4)$, served as a check point. If it had not been on the line determined by the two intercepts, then we would have known that an error had been committed.

EXAMPLE 2   Graph $x + 2y = 4$.

Solution   Without showing all of our work, the following table indicates the intercepts and a check point.

x	y	
0	2	intercepts
4	0	
2	1	check points

Plotting the points $(0, 2)$, $(4, 0)$, and $(2, 1)$, and connecting them with a straight line produces the graph in Figure 5.8.

Figure 5.8

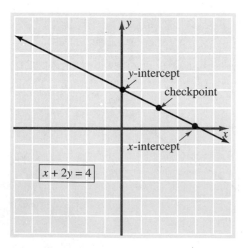

EXAMPLE 3   Graph $2x + 3y = 7$.

Solution   The intercepts and a check point are given in the following table. Finding intercepts may involve fractions, but the computation is usually rather easy. The points from the table are plotted and the graph of $2x + 3y = 7$ is shown in Figure 5.9.

Figure 5.9

x	y	
0	$\dfrac{7}{3}$	intercepts
$\dfrac{7}{2}$	0	
2	1	check point

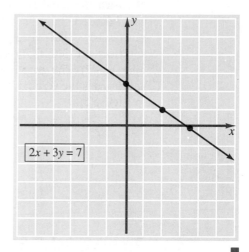

$2x + 3y = 7$

■

EXAMPLE 4   Graph $y = 2x$.

Solution   Notice that $(0, 0)$ is a solution; thus, this line intersects both axes at the origin. Since both the x-intercept and y-intercept are determined by the origin, $(0, 0)$, another point is necessary to graph the line. Then a third point should be found as a check point. These results are summarized in the following table. The graph of $y = 2x$ is shown in Figure 5.10.

Figure 5.10

x	y	
0	0	intercept
2	4	additional point
-1	-2	check point

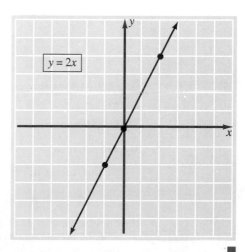

$y = 2x$

■

EXAMPLE 5   Graph $x = 3$.

Solution   Since we are considering linear equations in *two variables*, the equation
$x = 3$ is equivalent to $x + 0(y) = 3$.
Now we can see that any value of y
can be used, but the x-value must
always be 3. Therefore, some of the
solutions are $(3, 0), (3, 1), (3, 2), (3, -1)$,
and $(3, -2)$. The graph of all of the
solutions is the vertical line indicated
in Figure 5.11.

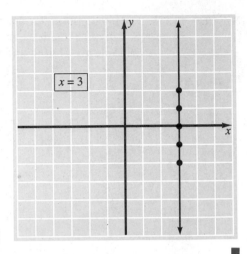

Figure 5.11

Problem Set 5.2

Use the techniques of this section to help sketch a graph for each of the following equations.

1. $x + y = 2$ **2.** $x + y = 4$ **3.** $x - y = 3$ **4.** $x - y = 1$

5. $x - y = -4$ **6.** $-x + y = 5$ **7.** $x + 2y = 2$ **8.** $x + 3y = 5$

9. $3x - y = 6$ **10.** $2x - y = -4$ **11.** $3x - 2y = 6$ **12.** $2x - 3y = 4$

13. $x - y = 0$ **14.** $x + y = 0$ **15.** $y = 3x$ **16.** $y = -2x$

17. $x = -2$ **18.** $y = 3$ **19.** $y = 0$ **20.** $x = 0$

21. $y = -2x - 1$ **22.** $y = 3x - 4$ **23.** $y = \dfrac{1}{2}x + 1$

24. $y = \dfrac{2}{3}x - 2$ **25.** $y = -\dfrac{1}{3}x - 2$ **26.** $y = -\dfrac{3}{4}x - 1$

27. $4x + 5y = -10$ **28.** $3x + 5y = -9$ **29.** $-2x + y = -4$

30. $-3x + y = -5$ **31.** $3x - 4y = 7$ **32.** $4x - 3y = 10$

33. $y + 4x = 0$ **34.** $y - 5x = 0$ **35.** $x = 2y$ **36.** $x = -3y$

37. The equation $C = \dfrac{5}{9}(F - 32)$ can be used to convert degrees in Fahrenheit to degrees in

Celsius. Let the horizontal axis represent F and the vertical axis, C, and graph the equation $C = \dfrac{5}{9}(F - 32)$.

38. Suppose that the daily profit from a small pizza stand is given by the equation $p = 2n - 4$
where n represents the number of pizzas sold in a day and p represents the number of
dollars of profit. Label the horizontal axis n and the vertical axis p and graph the equation $p = 2n - 4$ for nonnegative values of n.

39. The cost (c) of producing n plastic toys per day is given by the equation $c = 3n + 5$. Label the horizontal axis n and the vertical axis c and graph the equation for nonnegative values of n.

Thoughts into Words

40. Explain how you would graph the equation $-x + 2y = 4$.

41. Explain how you would graph the equation $y = x^3 + 1$.

42. What does it mean to say that "two points determine a line?"

Miscellaneous Problems

From our previous work with absolute value we know that $|x + y| = 2$ is equivalent to $x + y = 2$ or $x + y = -2$. Therefore, the graph of $|x + y| = 2$ consists of the two lines $x + y = 2$ and $x + y = -2$. Graph each of the following.

43. $|x + y| = 1$ **44.** $|x - y| = 2$ **45.** $|2x + y| = 4$ **46.** $|3x - y| = 6$

5.3

The Slope of a Line

In Figure 5.12 note that the line associated with $4x - y = 4$ is "steeper" than the line associated with $2x - 3y = 6$. Mathematically, the concept of **slope** is used to discuss

Figure 5.12

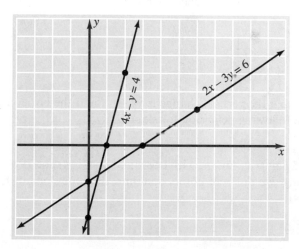

the "steepness" of lines. The slope of a line is the ratio of the vertical change compared to the horizontal change as we move from one point on a line to another point. We indicate this in Figure 5.13 using the points P_1 and P_2.

Figure 5.13

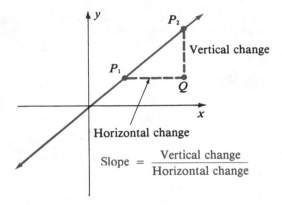

$$\text{Slope } = \frac{\text{Vertical change}}{\text{Horizontal change}}$$

A precise definition for slope can be given by considering the coordinates of the points P_1, P_2, and Q in Figure 5.14. Since P_1 and P_2 represent any two points on the line, we will assign the coordinates (x_1, y_1) to P_1 and (x_2, y_2) to P_2. The point Q is the same distance from the y-axis as P_2 and the same distance from the x-axis as P_1. Thus, we assign the coordinates (x_2, y_1) to Q (Figure 5.14). It should now be

Figure 5.14

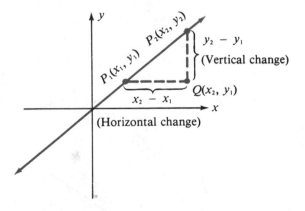

apparent that the vertical change is $y_2 - y_1$ and the horizontal change is $x_2 - x_1$. Thus, the following definition for slope is given.

DEFINITION 5.1

> If points P_1 and P_2 with coordinates (x_1, y_1) and (x_2, y_2) respectively, are any two different points on a line, then the slope of the line (denoted by m) is
>
> $$m = \frac{y_2 - y_1}{x_2 - x_1}, \qquad x_1 \neq x_2.$$

Using Definition 5.1, you can easily determine the slope of a line if the coordinates of two points on the line are known.

EXAMPLE 1   Find the slope of the line determined by each of the following pairs of points.

        (a) $(2, 1)$ and $(4, 6)$ **(b)** $(3, 2)$ and $(-4, 5)$

        (c) $(-4, -3)$ and $(-1, -3)$

Solution   **(a)** Let $(2, 1)$ be P_1 and $(4, 6)$ be P_2 (Figure 5.15).

$$m = \frac{y_2 - y_1}{x_2 - x_1} = \frac{6 - 1}{4 - 2} = \frac{5}{2}$$

Figure 5.15

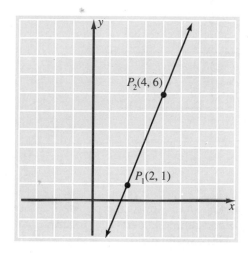

(b) Let $(3, 2)$ be P_1 and $(-4, 5)$ be P_2 (Figure 5.16).

$$m = \frac{y_2 - y_1}{x_2 - x_1} = \frac{5 - 2}{-4 - 3} = \frac{3}{-7} = -\frac{3}{7}$$

Figure 5.16

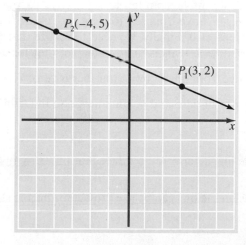

(c) Let $(-4, -3)$ be P_1 and $(-1, -3)$ be P_2 (Figure 5.17).

$$m = \frac{y_2 - y_1}{x_2 - x_1} = \frac{-3 - (-3)}{-1 - (-4)} = \frac{0}{3} = 0$$

Figure 5.17

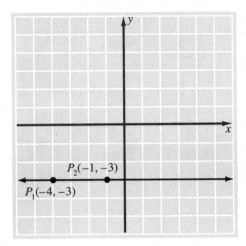

The designation of P_1 and P_2 for such problems is arbitrary and will not affect the value of the slope. For example, in part (a) of Example 1 suppose that we let $(4, 6)$ be P_1 and $(2, 1)$ be P_2. Then we obtain

$$m = \frac{y_2 - y_1}{x_2 - x_1} = \frac{1 - 6}{2 - 4} = \frac{-5}{-2} = \frac{5}{2}.$$

The various parts of Example 1 also illustrate the three basic possibilities for slope; that is, the slope of a line can be *positive*, *negative*, or *zero*. A line that has a positive slope rises as we move from left to right as in part (a). A line that has a negative slope falls as we move from left to right, as in part (b). A horizontal line, as in part (c), has a slope of 0. Finally, we need to realize that **the concept of slope is undefined for vertical lines.** This is due to the fact that for any vertical line the change in x as we move from one point to another is zero. Thus, the ratio $\dfrac{y_2 - y_1}{x_2 - x_1}$ will have a denominator of zero and be undefined. So in Definition 5.1 the restriction $x_1 \neq x_2$ is made.

EXAMPLE 2   Find the slope of the line determined by the equation $3x + 4y = 12$.

Solution   Since any two points on the line can be used to determine the slope of the line, let's find the intercepts.

If $x = 0$, then $3(0) + 4y = 12$

$$4y = 12$$

$$y = 3.$$

If $y = 0$, then $3x + 4(0) = 12$

$$3x = 12$$

$$x = 4.$$

Using $(0, 3)$ as P_1 and $(4, 0)$ as P_2, we obtain

$$m = \frac{y_2 - y_1}{x_2 - x_1} = \frac{0 - 3}{4 - 0} = \frac{-3}{4} = -\frac{3}{4}.$$ ∎

We need to emphasize one final idea pertaining to the concept of slope. The slope of a line is a **ratio** of vertical change compared to horizontal change. A slope of $\frac{3}{4}$ means that for every 3 units of vertical change there must be a corresponding 4 units of horizontal change. So starting at some point on the line, we could move to other points on the line as follows.

$\frac{3}{4} = \frac{6}{8}$:   by moving 6 units *up* and 8 units to the *right*;

$\frac{3}{4} = \frac{15}{20}$:   by moving 15 units *up* and 20 units to the *right*;

$\frac{3}{4} = \frac{\frac{3}{2}}{2}$:   by moving $1\frac{1}{2}$ units *up* and 2 units to the *right*;

$\frac{3}{4} = \frac{-3}{-4}$:   by moving 3 units *down* and 4 units to the *left*.

Likewise, a slope of $-\frac{5}{6}$ indicates that starting at some point on the line we could move to other points on the line as follows.

$-\frac{5}{6} = \frac{-5}{6}$:   by moving 5 units *down* and 6 units to the *right*;

$-\frac{5}{6} = \frac{5}{-6}$:   by moving 5 units *up* and 6 units to the *left*;

$-\frac{5}{6} = \frac{-10}{12}$:   by moving 10 units *down* and 12 units to the *right*;

$-\frac{5}{6} = \frac{15}{-18}$:   by moving 15 units *up* and 18 units to the *left*.

Problem Set 5.3

For Problems 1–20, find the slope of the line determined by each pair of points.

1. $(7, 5), (3, 2)$
2. $(9, 10), (6, 2)$
3. $(-1, 3), (-6, -4)$
4. $(-2, 5), (-7, -1)$
5. $(2, 8), (7, 2)$
6. $(3, 9), (8, 4)$
7. $(-2, 5), (1, -5)$
8. $(-3, 4), (2, -6)$
9. $(4, -1), (-4, -7)$
10. $(5, -3), (-5, -9)$
11. $(3, -4), (2, -4)$
12. $(-3, -6), (5, -6)$
13. $(-6, -1), (-2, -7)$
14. $(-8, -3), (-2, -11)$
15. $(-2, 4), (-2, -6)$
16. $(-4, -5), (-4, 9)$
17. $(-1, 10), (-9, 2)$
18. $(-2, 12), (-10, 2)$
19. $(a, b), (c, d)$
20. $(a, 0), (0, b)$

21. Find y if the line through $(7, 8)$ and $(2, y)$ has a slope of $\dfrac{4}{5}$.

22. Find y if the line through $(12, 14)$ and $(3, y)$ has a slope of $\dfrac{4}{3}$.

23. Find x if the line through $(-2, -4)$ and $(x, 2)$ has a slope of $-\dfrac{3}{2}$.

24. Find x if the line through $(6, -4)$ and $(x, 6)$ has a slope of $-\dfrac{5}{4}$.

For Problems 25–32, you are given one point on a line and the slope of the line. Find the coordinates of three other points on the line.

25. $(3, 2), m = \dfrac{2}{3}$
26. $(4, 1), m = \dfrac{5}{6}$
27. $(-2, -4), m = \dfrac{1}{2}$
28. $(-6, -2), m = \dfrac{2}{5}$
29. $(-3, 4), m = -\dfrac{3}{4}$
30. $(-2, 6), m = -\dfrac{3}{7}$
31. $(4, -5), m = -2$
32. $(6, -2), m = 4$

For Problems 33–40, sketch the line determined by each pair of points and decide whether the slope of the line is *positive*, *negative*, or *zero*.

33. $(2, 8), (7, 1)$
34. $(1, -2), (7, -8)$
35. $(-1, 3), (-6, -2)$
36. $(7, 3), (4, -6)$
37. $(-2, 4), (6, 4)$
38. $(-3, -4), (5, -4)$
39. $(-3, 5), (2, -7)$
40. $(-1, -1), (1, -9)$

For Problems 41–60, find the coordinates of two points on the given line and then use those coordinates to find the slope of the line.

41. $3x + 2y = 6$
42. $4x + 3y = 12$

43. $5x - 4y = 20$ **44.** $7x - 3y = 21$

45. $x + 5y = 6$ **46.** $2x + y = 4$

47. $2x - y = 7$ **48.** $x - 4y = -6$

49. $y = 3$ **50.** $x = 6$

51. $-2x + 5y = 9$ **52.** $-3x - 7y = 10$

53. $6x - 5y = -30$ **54.** $7x - 6y = -42$

55. $y = -3x - 1$ **56.** $y = -2x + 5$

57. $y = 4x$ **58.** $y = 6x$

59. $y = \dfrac{2}{3}x - \dfrac{1}{2}$ **60.** $y = -\dfrac{3}{4}x + \dfrac{1}{5}$

Miscellaneous Problems

61. The concept of slope is used for highway construction. The "grade" of a highway expressed as a percent means the number of feet that the highway changes in elevation for each 100 feet of horizontal change.

    (a) A certain highway has a 2% grade. How many feet does it rise in a horizontal distance of 1 mile? (1 mile = 5280 feet.)

    (b) The grade of a highway up a hill is 30%. How much change in horizontal distance is there if the vertical height of the hill is 75 feet?

62. Slope is often expressed as the ratio "rise to run" in construction of steps.

    (a) If the ratio "rise to run" is to be $\dfrac{3}{5}$ for some steps and the "rise" is 19 centimeters, find the measure of the "run" to the nearest centimeter.

    (b) If the ratio "rise to run" is to be $\dfrac{2}{3}$ for some steps and the "run" is 28 centimeters, find the "rise" to the nearest centimeter.

63. Suppose that county ordinance requires a $2\frac{1}{4}\%$ "fall" for a sewage pipe from the house to the main pipe at the street. How much vertical drop must there be for a horizontal distance of 45 feet? Express the answer to the nearest tenth of a foot.

5.4
Writing Equations of Lines

As we stated in Section 5.1, basically there are two types of problems in analytic or coordinate geometry.

    1. Given an algebraic equation, find its geometric graph.

    2. Given a set of conditions pertaining to a geometric figure, determine its algebraic equation.

Problems of type 1 were a part of the first two sections of this chapter. At this time we want to consider a few problems of type 2 that deal with straight lines. In other words, given certain facts about a line, we need to be able to write its algebraic equation.

EXAMPLE 1   Find the equation of the line that has a slope of $\frac{3}{4}$ and contains the point $(1, 2)$.

Solution   First, let's draw the line as indicated in Figure 5.18. Since the slope is $\frac{3}{4}$, a second point can be determined by moving 3 units up and 4 units to the right of the given point $(1, 2)$. (The point $(5, 5)$ merely helps to draw the line; it will not be used in analyzing the problem.) Now let's choose a point (x, y) that represents all points on the line other than the given point $(1, 2)$. The slope determined by $(1, 2)$ and (x, y) is $\frac{3}{4}$.

Thus,

$$\frac{y - 2}{x - 1} = \frac{3}{4}$$

$$3(x - 1) = 4(y - 2)$$

$$3x - 3 = 4y - 8$$

$$3x - 4y = -5.$$

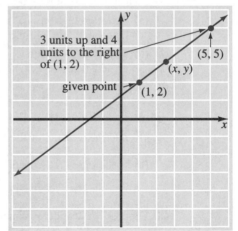

Figure 5.18

EXAMPLE 2   Find the equation of the line containing $(3, 4)$ and $(-2, 5)$.

Solution   First, let's draw the line determined by the two given points (Figure 5.19). Since we know two points, we can find the slope.

$$m = \frac{y_2 - y_1}{x_2 - x_1} = \frac{5 - 4}{-2 - 3} = \frac{1}{-5} = -\frac{1}{5}$$

Now we can use the same approach as in Example 1. Form an equation using a variable point (x, y), one of the two given points, (we chose P_1), and the slope of $-\frac{1}{5}$.

$$\frac{y-4}{x-3} = \frac{1}{-5} \qquad -\frac{1}{5} = \frac{1}{-5}$$

$$1(x-3) = -5(y-4)$$

$$x - 3 = -5y + 20$$

$$x + 5y = 23$$

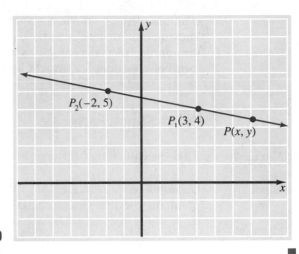

Figure 5.19

EXAMPLE 3   Find the equation of the line that has a slope of $\frac{1}{4}$ and a y-intercept of 2.

Solution   A y-intercept of 2 means that the point $(0, 2)$ is on the line. Since the slope is $\frac{1}{4}$, we can find another point by moving 1 unit up and 4 units to the right of $(0, 2)$. The line is drawn in Figure 5.20. Choosing a variable point (x, y) we can proceed as in the previous examples.

$$\frac{y-2}{x-0} = \frac{1}{4}$$

$$1(x-0) = 4(y-2)$$

$$x = 4y - 8$$

$$x - 4y = -8$$

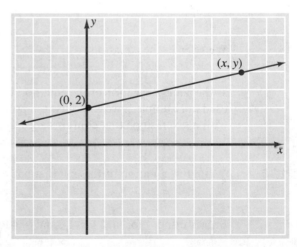

Figure 5.20

Perhaps it would be helpful for you to pause for a moment and look back over Examples 1, 2, and 3. Notice that we used the same basic approach in all three examples; that is, we chose a variable point (x, y) and used it, along with another known point, to determine the equation of the line. We should also recognize that the approach we take in these examples can be generalized to produce some special forms of equations for straight lines.

Point-Slope Form

EXAMPLE 4   Find the equation of the line that has a slope of m and contains the point (x_1, y_1).

Solution   Choosing (x, y) to represent any other point on the line (Figure 5.21) the slope of the line is given by

$$m = \frac{y - y_1}{x - x_1}$$

from which we obtain

$$y - y_1 = m(x - x_1).$$

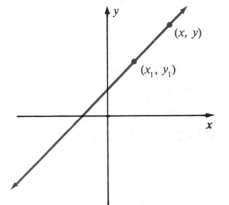

Figure 5.21

We refer to the equation $\boxed{y - y_1 = m(x - x_1)}$

as the **point-slope form** of the equation of a straight line. Instead of the approach we used in Example 1, we could use the point-slope form to write the equation of a line with a given slope that contains a given point, as the next example illustrates.

EXAMPLE 5   Write the equation of the line that has a slope of $\frac{3}{5}$ and contains the point $(2, -4)$.

Solution   Substituting $\frac{3}{5}$ for m, 2 for x_1, and -4 for y_1 in the point-slope form, we obtain

$$y - y_1 = m(x - x_1)$$
$$y - (-4) = \frac{3}{5}(x - 2)$$

$$y + 4 = \frac{3}{5}(x - 2)$$

$$5(y + 4) = 3(x - 2)$$

$$5y + 20 = 3x - 6$$

$$26 = 3x - 5y$$

∎

Slope-Intercept Form

Now consider the equation of a line that has a slope of m and a y-intercept of b (Figure 5.22).

Figure 5.22

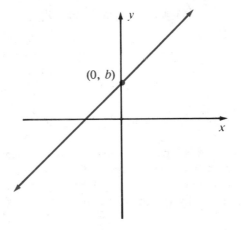

A y-intercept of b means that the line contains the point $(0, b)$; therefore, we can use the point-slope form as follows.

$$y - y_1 = m(x - x_1)$$

$$y - b = m(x - 0) \qquad y_1 = b \text{ and } x_1 = 0$$

$$y - b = mx$$

$$y = mx + b$$

The equation

$$\boxed{y = mx + b}$$

is called the **slope-intercept form** of the equation of a straight line. We use it for two primary purposes, as the next two examples illustrate.

EXAMPLE 6   Find the equation of the line that has a slope of $\frac{1}{4}$ and a y-intercept of 2.

Solution   This is a restatement of Example 3, but this time we will use the slope-intercept form of a line ($y = mx + b$) to write its equation. From the statement of the problem we know that $m = \frac{1}{4}$ and $b = 2$. Thus, substituting these values for m and b in $y = mx + b$, we obtain

$$y = mx + b$$

$$y = \frac{1}{4}x + 2$$

$$4y = x + 8$$

$$x - 4y = -8. \qquad \text{same result as in Example 3} \qquad \blacksquare$$

> **REMARK** Sometimes we leave answers in slope-intercept form. We did not do that in Example 6 because we wanted to show that it was the same result as in Example 3.

EXAMPLE 7   Find the slope of the line whose equation is $2x + 3y = 4$.

Solution   In Section 5.3 we solved this type of problem when we determined the coordinates of two points on the line and then used the slope formula. Now we have another approach for such a problem. Let's change the given equation to slope-intercept form; that is to say, let's solve the equation for y in terms of x.

$$2x + 3y = 4$$

$$3y = -2x + 4$$

$$y = -\frac{2}{3}x + \frac{4}{3}$$

Compare this result to $y = mx + b$, and you see that $m = -\frac{2}{3}$ and $b = \frac{4}{3}$. $\qquad \blacksquare$

In general, **if the equation of a nonvertical line is written in slope-intercept form, the coefficient of x is the slope of the line and the constant term is the y-intercept.** (Remember that the concept of slope is not defined for a vertical line.)

Parallel and Perpendicular Lines

We can use two important relationships between lines and their slopes to solve certain kinds of problems. It can be shown that nonvertical parallel lines have the same slope, and that two nonvertical lines are perpendicular if the product of their slopes is -1. (Details for verifying these facts are left to another course.) In other words, if two lines have slopes m_1 and m_2, respectively, then:

1. The two lines are parallel if and only if $m_1 = m_2$;

2. The two lines are perpendicular if and only if $(m_1)(m_2) = -1$.

The following examples demonstrate the use of these properties.

EXAMPLE 8   **(a)** Verify that the graphs of $2x + 3y = 7$ and $4x + 6y = 11$ are parallel lines.

(b) Verify that the graphs of $8x - 12y = 3$ and $3x + 2y = 2$ are perpendicular lines.

Solution   **(a)** Let's change each equation to slope-intercept form.

$$2x + 3y = 7 \longrightarrow 3y = -2x + 7$$

$$y = -\frac{2}{3}x + \frac{7}{3}$$

$$4x + 6y = 11 \longrightarrow 6y = -4x + 11$$

$$y = -\frac{4}{6}x + \frac{11}{6}$$

$$y = -\frac{2}{3}x + \frac{11}{6}$$

Both lines have a slope of $-\dfrac{2}{3}$ and different y-intercepts. Therefore, the two lines are parallel.

(b) Solving each equation for y in terms of x, we obtain

$$8x - 12y = 3 \longrightarrow -12y = -8x + 3$$

$$y = \frac{8}{12}x - \frac{3}{12}$$

$$y = \frac{2}{3}x - \frac{1}{4}$$

$$3x + 2y = 2 \longrightarrow 2y = -3x + 2$$

$$y = -\frac{3}{2}x + 1.$$

Because $\left(\dfrac{2}{3}\right)\left(-\dfrac{3}{2}\right) = -1$ (the product of the two slopes is -1), the lines are perpendicular.   ∎

REMARK The statement "the product of two slopes is -1" is equivalent to saying that the two slopes are negative reciprocals of each other; that is,

$$m_1 = -\frac{1}{m_2}.$$

EXAMPLE 9   Find the equation of the line that contains the point $(1, 4)$ and is parallel to the line determined by $x + 2y = 5$.

Solution   First, let's draw a figure to help in our analysis of the problem (Figure 5.23). Since the line through $(1, 4)$ is to be parallel to the line determined by $x + 2y = 5$, it must have the same slope. So let's find the slope by changing $x + 2y = 5$ to the slope-intercept form.

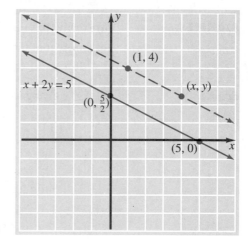

$$x + 2y = 5$$

$$2y = -x + 5$$

$$y = -\frac{1}{2}x + \frac{5}{2}$$

Figure 5.23

The slope of both lines is $-\frac{1}{2}$. Now we can choose a variable point (x, y) on the line through $(1, 4)$ and proceed as we did in earlier examples.

$$\frac{y - 4}{x - 1} = \frac{1}{-2}$$

$$1(x - 1) = -2(y - 4)$$

$$x - 1 = -2y + 8$$

$$x + 2y = 9 \qquad\qquad\blacksquare$$

EXAMPLE 10   Find the equation of the line that contains the point $(-1, -2)$ and is perpendicular to the line determined by $2x - y = 6$.

Solution   First, let's draw a figure to help in our analysis of the problem (Figure 5.24). Because the line through $(-1, -2)$ is to be perpendicular to the line determined by $2x - y = 6$, its slope must be the negative reciprocal of the slope of $2x - y = 6$. So let's find the slope of $2x - y = 6$ by changing it to the slope-intercept form.

$$2x - y = 6$$
$$-y = -2x + 6$$
$$y = 2x - 6 \qquad \text{The slope is 2.}$$

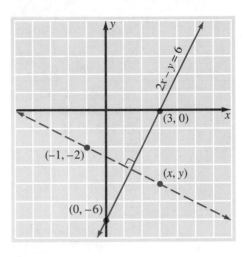

Figure 5.24

The slope of the desired line is $-\dfrac{1}{2}$ (the negative reciprocal of 2) and we can proceed as before by using a variable point (x, y).

$$\frac{y + 2}{x + 1} = \frac{1}{-2}$$
$$1(x + 1) = -2(y + 2)$$
$$x + 1 = -2y - 4$$
$$x + 2y = -5 \qquad\qquad\qquad\qquad\qquad\qquad\qquad\qquad\quad \blacksquare$$

We use two forms of equations of straight lines extensively. They are the **standard form** and the **slope-intercept form** and we describe them as follows.

Standard Form $Ax + By = C$, where B and C are integers and A is a nonnegative integer (A and B not both zero).

Slope-Intercept Form $y = mx + b$ where m is a real number representing the slope and b is a real number representing the y-intercept.

Problem Set 5.4

For Problems 1–12, find the equation of the line that contains the given point and has the given slope. Express equations in the form $Ax + By = C$, where A, B, and C are integers.

1. $(2, 3)$, $m = \dfrac{2}{3}$ **2.** $(5, 2)$, $m = \dfrac{3}{7}$ **3.** $(-3, -5)$, $m = \dfrac{1}{2}$

4. $(5, -6)$, $m = \dfrac{3}{5}$ **5.** $(-4, 8)$, $m = -\dfrac{1}{3}$ **6.** $(-2, -4)$, $m = -\dfrac{5}{6}$

7. $(3, -7)$, $m = 0$ **8.** $(-3, 9)$, $m = 0$ **9.** $(0, 0)$, $m = -\dfrac{4}{9}$

10. $(0, 0)$, $m = \dfrac{5}{11}$ **11.** $(-6, -2)$, $m = 3$ **12.** $(2, -10)$, $m = -2$

For Problems 13–22, find the equation of the line that contains the two given points. Express equations in the form $Ax + By = C$, where A, B, and C are integers.

13. $(2, 3)$ and $(7, 10)$ **14.** $(1, 4)$ and $(9, 10)$

15. $(3, -2)$ and $(-1, 4)$ **16.** $(-2, 8)$ and $(4, -2)$

17. $(-1, -2)$ and $(-6, -7)$ **18.** $(-8, -7)$ and $(-3, -1)$

19. $(0, 0)$ and $(-3, -5)$ **20.** $(5, -8)$ and $(0, 0)$

21. $(0, 4)$ and $(7, 0)$ **22.** $(-2, 0)$ and $(0, -9)$

For Problems 23–32, find the equation of the line with the given slope and y-intercept. Leave your answers in slope-intercept form.

23. $m = \dfrac{3}{5}$ and $b = 2$ **24.** $m = \dfrac{5}{9}$ and $b = 4$

25. $m = 2$ and $b = -1$ **26.** $m = 4$ and $b = -3$

27. $m = -\dfrac{1}{6}$ and $b = -4$ **28.** $m = -\dfrac{5}{7}$ and $b = -1$

29. $m = -1$ and $b = \dfrac{5}{2}$ **30.** $m = -2$ and $b = \dfrac{7}{3}$

31. $m = -\dfrac{5}{9}$ and $b = -\dfrac{1}{2}$ **32.** $m = -\dfrac{7}{12}$ and $b = -\dfrac{2}{3}$

For Problems 33–44, determine the slope and y-intercept of the line represented by the given equation.

33. $y = -2x - 5$ **34.** $y = \dfrac{2}{3}x + 4$ **35.** $3x - 5y = 15$

36. $7x + 5y = 35$ **37.** $-4x + 9y = 18$ **38.** $-6x + 7y = -14$

39. $-y = -\dfrac{3}{4}x + 4$ **40.** $5x - 2y = 0$ **41.** $-2x - 11y = 11$

42. $-y = \dfrac{2}{3}x + \dfrac{11}{2}$ **43.** $9x + 7y = 0$ **44.** $-5x - 13y = 26$

For Problems 45–50, determine whether each pair of lines are (a) parallel, (b) perpendicular, or (c) intersecting lines that are not perpendicular.

45. $5x - 2y = 6$
$2x + 5y = 9$

46. $2x - y = 9$
$6x - 3y = 5$

47. $4x - 3y = 12$
$3x - 4y = 12$

48. $9x + 2y = 18$
$2x - 9y = 13$

49. $x - 3y = 7$
$5x - 15y = 9$

50. $7x - 2y = 14$
$7x + 2y = 5$

51. Write the equation of the line that contains the point $(4, 3)$ and is parallel to the line $2x - 3y = 6$.

52. Write the equation of the line that contains the point $(-1, 3)$ and is perpendicular to the line $3x - y = 4$.

53. Write the equation of the line that contains the point $(1, -6)$ and is perpendicular to the line $3x + y = 4$.

54. Write the equation of the line that contains the point $(-2, -4)$ and is parallel to the line $5x + 2y = 6$.

55. Write the equation of the line that contains the origin and is parallel to the line $-2x - 9y = 4$.

56. Write the equation of the line that contains the origin and is perpendicular to the line $-2x + 3y = 8$.

Thoughts into Words

57. Explain the concept of slope in your own words.

58. What does "coordinate geometry" mean to you at this time?

5.5

Solving Linear Systems by Graphing

Suppose that we graph $x - 2y = 4$ and $x + 2y = 8$ on the same set of axes, as indicated in Figure 5.25. The ordered pair, $(6, 1)$, which is associated with the point of intersection of the two lines, satisfies both equations. That is to say, $(6, 1)$ is the

Figure 5.25

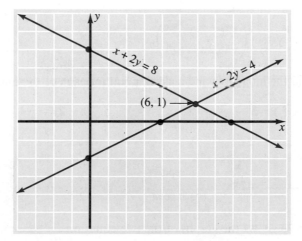

solution for $x - 2y = 4$ *and* for $x + 2y = 8$. To check this, we can substitute 6 for x and 1 for y in both equations.

$x - 2y = 4$ becomes $6 - 2(1) = 4$,

$x + 2y = 8$ becomes $6 + 2(1) = 8$

Thus, we say that $\{(6, 1)\}$ is the solution set of the system

$$\begin{pmatrix} x - 2y = 4 \\ x + 2y = 8 \end{pmatrix}.$$

Two or more linear equations in two variables considered together are called a **system of linear equations**. The following are systems of linear equations.

$$\begin{pmatrix} x - 2y = 4 \\ x + 2y = 8 \end{pmatrix}, \qquad \begin{pmatrix} 5x - 3y = 9 \\ 3x + 7y = 12 \end{pmatrix}, \qquad \begin{pmatrix} 4x - \;\; y = \;\; 5 \\ 2x + \;\; y = \;\; 9 \\ 7x - 2y = 13 \end{pmatrix}$$

To **solve a system of linear equations** means to find all of the ordered pairs that are solutions of all of the equations in the system. In this book we shall consider only systems of *two* linear equations in two variables. There are several techniques for solving systems of linear equations. We will use three of them in this chapter—a graphing method in this section, and two other methods in the following sections.

To solve a system of linear equations by **graphing**, you proceed as we did in the opening discussion of this section. We graph the equations on the same set of axes and then the ordered pairs associated with any points of intersection are the solutions to the system. Let's consider another example.

EXAMPLE 1   Solve the system $\begin{pmatrix} x + \;\; y = \;\; 5 \\ x - 2y = -4 \end{pmatrix}$.

Solution   Let's find the intercepts and a check point for each of the lines.

$x + y = 5$

x	y	
0	5	intercepts
5	0	
2	3	check point

$x - 2y = -4$

x	y	
0	2	intercepts
−4	0	
−2	1	check point

Figure 5.26 shows the graphs of the two equations.

Figure 5.26

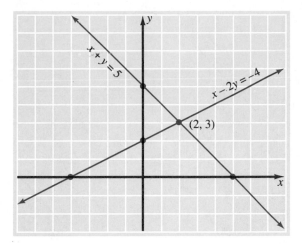

It appears that $(2, 3)$ is the solution of the system. To check it we can substitute 2 for x and 3 for y in both equations.

$$x + y \quad \text{becomes } 2 + 3 = 5, \qquad \text{a true statement}$$

$$x - 2y = -4 \quad \text{becomes } 2 - 2(3) = -4 \qquad \text{a true statement}$$

Therefore, $\{(2, 3)\}$ is the solution set. ∎

It should be evident that solving systems of equations by graphing requires accurate graphs. In fact, unless the solutions are integers it is really quite difficult to obtain exact solutions from a graph. For this reason the systems in this section have integral solutions. Furthermore, checking a solution takes on additional significance when using the graphing approach. By checking you can be absolutely sure that you are *reading* the correct solution from the graph.

Figure 5.27 shows the three possible cases for the graph of a system of two linear equations in two variables.

Case I. The graphs of the two equations are two lines intersecting in one point. There is *one solution* and we call the system a **consistent system**.

Case II. The graphs of the two equations are parallel lines. There is *no solution* and we call the system an **inconsistent system**.

Case III. The graphs of the two equations are the same line. There *are infinitely many* solutions to the system. Any pair of real numbers that satisfies one of the equations will also satisfy the other equation, and we say the equations are **dependent**.

Case I

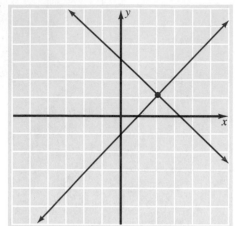

Case II

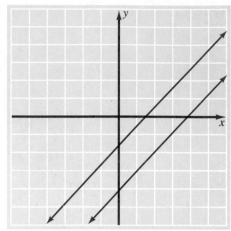

Case III

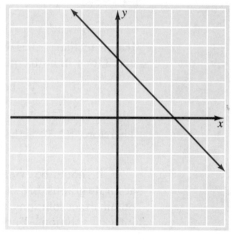

Figure 5.27

Thus, as we solve a system of two linear equations in two variables we know what to expect. The system will have no solutions, one ordered pair as a solution, or infinitely many ordered pairs as solutions. Most of the systems that we will be working with in this text will have one solution.

An example of case 1 was given in Example 1 (Figure 5.26). The next two examples illustrate the other cases.

EXAMPLE 2   Solve the system $\begin{pmatrix} 2x + 3y = 6 \\ 2x + 3y = 12 \end{pmatrix}$.

Solution

2x + 3y = 6				2x + 3y = 12	
x		*y*		*x*	*y*
0		2		0	4
3		0		6	0
−3		4		3	2

Figure 5.28 shows the graph of the system. Since the lines are parallel, there is *no solution* to the system. The solution set is ∅.

Figure 5.28

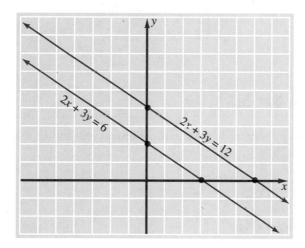

EXAMPLE 3   Solve the system $\left(\begin{array}{l} x + y - 3 \\ 2x + 2y = 6 \end{array}\right)$.

Solution

x + y = 3				2x + 2y = 6	
x		*y*		*x*	*y*
0		3		0	3
3		0		3	0
1		2		1	2

Figure 5.29 shows the graph of this system. Since the graph of both equations is the same line, there are *infinitely many solutions* to the system. Any ordered pair of real numbers that satisfies one equation will also satisfy the other equation.

Figure 5.29

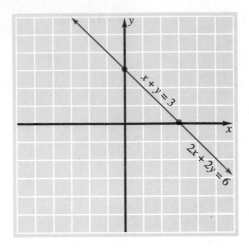

Problem Set 5.5

For Problems 1–10, decide whether the given ordered pair is a solution of the given system of equations.

1. $\begin{pmatrix} 5x + y = 9 \\ 3x - 2y = 4 \end{pmatrix}$ $(1, 4)$

2. $\begin{pmatrix} 3x - 2y = 8 \\ 2x + y = 3 \end{pmatrix}$ $(2, -1)$

3. $\begin{pmatrix} x - 3y = 17 \\ 2x + 5y = -21 \end{pmatrix}$ $(2, -5)$

4. $\begin{pmatrix} 2x + 7y = -5 \\ 4x - y = 6 \end{pmatrix}$ $(1, -1)$

5. $\begin{pmatrix} y = 2x \\ 3x - 4y = 5 \end{pmatrix}$ $(-1, -2)$

6. $\begin{pmatrix} -x + 2y = 8 \\ x - 2y = 8 \end{pmatrix}$ $(-2, 3)$

7. $\begin{pmatrix} 6x - 5y = 5 \\ 3x + 4y = -4 \end{pmatrix}$ $(0, -1)$

8. $\begin{pmatrix} y = 3x - 1 \\ 5x - 2y = -1 \end{pmatrix}$ $(3, 8)$

9. $\begin{pmatrix} -3x - y = 4 \\ -2x + 3y = -23 \end{pmatrix}$ $(4, -5)$

10. $\begin{pmatrix} 5x + 4y = -15 \\ 2x - 7y = -6 \end{pmatrix}$ $(-3, 0)$

For Problems 11–30, use the graphing method to solve each system.

11. $\begin{pmatrix} x + y = 1 \\ x - y = 3 \end{pmatrix}$

12. $\begin{pmatrix} x - y = 2 \\ x + y = -4 \end{pmatrix}$

13. $\begin{pmatrix} x + 2y = 4 \\ 2x - y = 3 \end{pmatrix}$

14. $\begin{pmatrix} 2x - y = -8 \\ x + y = 2 \end{pmatrix}$

15. $\begin{pmatrix} x + 3y = 6 \\ x + 3y = 3 \end{pmatrix}$

16. $\begin{pmatrix} y = -2x \\ y - 3x = 0 \end{pmatrix}$

17. $\begin{pmatrix} x + y = 0 \\ x - y = 0 \end{pmatrix}$

18. $\begin{pmatrix} 3x - y = 3 \\ 3x - y = -3 \end{pmatrix}$

19. $\begin{pmatrix} 3x - 2y = 5 \\ 2x + 5y = -3 \end{pmatrix}$ 20. $\begin{pmatrix} 2x + 3y = 1 \\ 4x - 3y = -7 \end{pmatrix}$

21. $\begin{pmatrix} y = 2x + 3 \\ 6x + 3y = 9 \end{pmatrix}$ 22. $\begin{pmatrix} y = 2x + 5 \\ x + 3y = -6 \end{pmatrix}$

23. $\begin{pmatrix} y = 5x - 2 \\ 4x + 3y = 13 \end{pmatrix}$ 24. $\begin{pmatrix} y = x - 2 \\ 2x - 2y = 4 \end{pmatrix}$

25. $\begin{pmatrix} y = 4 - 2x \\ y = 7 - 3x \end{pmatrix}$ 26. $\begin{pmatrix} y = 3x + 4 \\ y = 5x + 8 \end{pmatrix}$

27. $\begin{pmatrix} y = 2x \\ 3x - 2y = -2 \end{pmatrix}$ 28. $\begin{pmatrix} y = 3x \\ 4x - 3y = 5 \end{pmatrix}$

29. $\begin{pmatrix} 7x - 2y = -8 \\ x = -2 \end{pmatrix}$ 30. $\begin{pmatrix} 3x + 8y = -1 \\ y = -2 \end{pmatrix}$

5.6

The Elimination-by-Addition Method

We have used the addition property of equality (if $a = b$, then $a + c = b + c$) to help solve equations containing one variable. An extension of the addition property forms the basis for another method of solving systems of linear equations. Property 5.1 states that two equations can be added and the resulting equation will be equivalent to the original ones. Let's use this property to help solve a system of equations.

PROPERTY 5.1

For all real numbers a, b, c, and d, if $a = b$, and $c = d$, then

$$a + c = b + d.$$

EXAMPLE 1   Solve $\begin{pmatrix} x + y = 12 \\ x - y = 2 \end{pmatrix}$.

Solution

$$x + y = 12$$
$$\underline{x - y = 2}$$
$$2x = 14 \qquad \text{This is the result of adding the two equations.}$$

Solving this new equation in one variable we obtain

$$2x = 14$$
$$\boxed{x = 7.}$$

Now we can substitute the value of 7 for x in one of the original equations. Thus,

$$x + y = 12$$

$$7 + y = 12$$

$$\boxed{y = 5.}$$

To check we see that

$$x + y = 12 \quad \text{and} \quad x - y = 2,$$

$$7 + 5 = 12 \quad \text{and} \quad 7 - 5 = 2.$$

Thus, the solution set of the system is $\{(7, 5)\}$. ∎

Note in Example 1 that by adding the two original equations, a simple equation that contains only one variable was obtained. Adding equations to eliminate a variable is the key idea behind the **elimination-by-addition method** for solving systems of linear equations. The next example further illustrates this point.

EXAMPLE 2   Solve $\begin{pmatrix} 2x + 3y = -26 \\ 4x - 3y = \quad 2 \end{pmatrix}$.

Solution

$$2x + 3y = -26$$

$$\underline{4x - 3y = \quad 2}$$

$$6x \quad\quad = -24 \quad\quad \text{added the two equations}$$

$$\boxed{x = -4}$$

Substitute -4 for x in one of the two original equations.

$$2x + 3y = -26$$

$$2(-4) + 3y = -26$$

$$-8 + 3y = -26$$

$$3y = -18$$

$$\boxed{y = -6}$$

The solution set is $\{(-4, -6)\}$. ∎

It may be necessary to change the form of one, or perhaps both, of the original equations before adding them. The next example demonstrates this idea.

EXAMPLE 3   Solve $\begin{pmatrix} y = x - 21 \\ x + y = -3 \end{pmatrix}$.

Solution

$$y = x - 21 \quad \underrightarrow{\text{subtract } x \text{ from both sides}} \quad -x + y = -21$$
$$x + y = -3 \quad \underrightarrow{\text{leave alone}} \quad x + y = -3$$
$$2y = -24$$
$$\boxed{y = -12}$$

Substitute -12 for y in one of the original equations.

$$x + y = -3$$
$$x + (-12) = -3$$
$$\boxed{x = 9}$$

The solution set is $\{(9, -12)\}$. ∎

Frequently, the multiplication property of equality needs to be applied first so that adding the equations will eliminate a variable, as the next example illustrates.

EXAMPLE 4   Solve $\begin{pmatrix} 2x + 5y = 29 \\ 3x - y = 1 \end{pmatrix}$.

Solution

Notice that adding the equations as they are would not eliminate a variable. However, we should observe that multiplying the bottom equation by 5 and then adding this newly formed, but equivalent, equation to the top equation will eliminate the y's.

$$2x + 5y = 29 \quad \underrightarrow{\text{leave alone}} \quad 2x + 5y = 29$$
$$3x - y = 1 \quad \underrightarrow{\text{multiply both sides by 5}} \quad 15x - 5y = 5$$
$$17x \quad\quad = 34$$
$$\boxed{x = 2}$$

Substitute 2 for x in one of the original equations.

$$3x - y = 1$$
$$3(2) - y = 1$$
$$6 - y = 1$$
$$-y = -5$$
$$\boxed{y = 5}$$

The solution set is $\{(2, 5)\}$. ∎

Notice in these problems that after finding the value of one of the variables, we substitute this number into one of the *original* equations to find the value of the other variable. It doesn't matter which of the two original equations you use, so pick the one easiest to solve.

Sometimes the multiplication property needs to be applied to both equations. Let's look at an example of this type.

EXAMPLE 5   Solve $\begin{pmatrix} 2x + 3y = 4 \\ 9x - 2y = -13 \end{pmatrix}$.

Solution A

$2x + 3y = 4$ $\quad\underrightarrow{\text{multiply both sides by 9}}\quad$ $18x + 27y = 36$

$9x - 2y = -13$ $\quad\underrightarrow{\text{multiply both sides by } -2}\quad$ $-18x + 4y = 26$

$$31y = 62$$

$$\boxed{y = 2.}$$

Substitute 2 for y in one of the original equations.

$$2x + 3y = 4$$
$$2x + 3(2) = 4$$
$$2x + 6 = 4$$
$$2x = -2$$
$$\boxed{x = -1}$$

The solution set is $\{(-1, 2)\}$.

Solution B

$2x + 3y = 4$ $\quad\underrightarrow{\text{multiply both sides by 2}}\quad$ $4x + 6y = 8$

$9x - 2y = -13$ $\quad\underrightarrow{\text{multiply both sides by 3}}\quad$ $27x - 6y = -39$

$$31x = -31$$

$$\boxed{x = -1}$$

Substitute -1 for x in one of the original equations.

$$2x + 3y = 4$$
$$2(-1) + 3y = 4$$
$$-2 + 3y = 4$$
$$3y = 6$$
$$\boxed{y = 2}$$

The solution set is $\{(-1, 2)\}$.   ■

Look back over Solutions A and B for Example 5 carefully. Especially notice the first steps where the multiplication property of equality is applied to the two equations. In Solution A we multiplied by numbers so that adding the resulting equations eliminated the x variable. In Solution B we multiplied so that the y variable was eliminated when we added the resulting equations. Either approach will work; pick the one that involves the easiest computation.

EXAMPLE 6   Solve $\begin{pmatrix} 3x - 4y = 7 \\ 5x + 3y = 9 \end{pmatrix}$.

$3x - 4y = 7$ __multiply both sides by 3__→ $9x - 12y = 21$

$5x + 3y = 9$ __multiply both sides by 4__→ $20x + 12y = 36$

$$29x \qquad\quad = 57$$

$$x = \frac{57}{29}$$

Since substituting $\dfrac{57}{29}$ for x into one of the original equations will produce some messy calculations, let's solve for y by eliminating the x's.

$3x - 4y = 7$ __multiply both sides by -5__→ $-15x + 20y = -35$

$5x + 3y = 9$ __multiply both sides by 3__→ $15x + 9y = 27$

$$29y = -8$$

$$y = -\frac{8}{29}$$

The solution set is $\left\{ \left(\dfrac{57}{29}, -\dfrac{8}{29} \right) \right\}$.   ■

Problem Solving

Many word problems that we solved earlier in this text using one equation in one variable can also be solved using a system of two linear equations in two variables. In fact, many times you may find that it seems quite natural to use two variables. It may also seem more meaningful at times to use variables other than x and y. Let's consider some examples.

PROBLEM 1   The cost of 3 tennis balls and 2 golf balls is $7. Furthermore, the cost of 6 tennis balls and 3 golf balls is $12. Find the cost of 1 tennis ball and the cost of 1 golf ball.

Solution

Let's use t to represent the cost of one tennis ball and g the cost of one golf ball. The problem translates into the following system of equations.

$$3t + 2g = 7 \qquad \text{The cost of 3 tennis balls and 2 golf balls is \$7.}$$

$$6t + 3g = 12 \qquad \text{The cost of 6 tennis balls and 3 golf balls is \$12.}$$

Solving this system by the addition method, we obtain

$$3t + 2g = 7 \quad \underrightarrow{\text{multiply both sides by } -2} \quad -6t - 4g = -14$$

$$6t + 3g = 12 \quad \underrightarrow{\text{leave alone}} \quad \underline{6t + 3g = \quad 12}$$

$$-g = -2$$

$$\boxed{g = 2}$$

Substitute 2 for g in one of the original equations.

$$3t + 2g = 7$$

$$3t + 2(2) = 7$$

$$3t + 4 = 7$$

$$3t = 3$$

$$\boxed{t = 1}$$

The cost of a tennis ball is \$1 and the cost of a golf ball is \$2.   ■

PROBLEM 2

Niki invested \$500, part of it at 9% interest and the rest at 10%. Her total interest earned for a year was \$48. How much did she invest at each rate?

Solution

Let x represent the amount invested at 9%. Let y represent the amount invested at 10%. The problem translates into the following system.

$$x + y = 500 \qquad \text{Niki invested \$500.}$$

$$0.09x + 0.10y = 48 \qquad \text{Her total interest earned for a year was \$48.}$$

Multiplying the second equation by 100 produces $9x + 10y = 4800$. Then, we have the following equivalent system to solve.

$$\begin{pmatrix} x + y = 500 \\ 9x + 10y = 4800 \end{pmatrix}$$

Using the addition method, we can proceed as follows.

$$x + y = 500 \quad \underrightarrow{\text{multiply both sides by } -9} \quad -9x - 9y = -4500$$

$$9x + 10y = 4800 \quad \underrightarrow{\text{leave alone}} \quad \underline{9x + 10y = \quad 4800}$$

$$\boxed{y = \quad 300}$$

Substituting 300 for y in $x + y = 500$ yields

$$x + y = 500$$
$$x + 300 = 500$$
$$\boxed{x = 200.}$$

Therefore, we know that \$200 was invested at 9% interest and \$300 at 10%.   ∎

Before you tackle the word problems in this next problem set, it might be helpful to review the problem solving suggestions we offered in Section 4.2. Those suggestions continue to apply here except that now we have the flexibility of using two equations and two unknowns. Don't forget that to check a word problem you need to see if your answers satisfy the conditions stated in the problem.

Problem Set 5.6

For Problems 1–24, solve each system using the elimination-by-addition method.

1. $\begin{pmatrix} x + y = 14 \\ x - y = -2 \end{pmatrix}$

2. $\begin{pmatrix} x - y = -14 \\ x + y = 6 \end{pmatrix}$

3. $\begin{pmatrix} x + 4y = -21 \\ 3x - 4y = 1 \end{pmatrix}$

4. $\begin{pmatrix} -3x + 2y = -21 \\ 3x - 7y = 36 \end{pmatrix}$

5. $\begin{pmatrix} y = 6 - x \\ x - y = -18 \end{pmatrix}$

6. $\begin{pmatrix} x + y = -10 \\ x = y + 6 \end{pmatrix}$

7. $\begin{pmatrix} 5x + y = 23 \\ 3x - 2y = 19 \end{pmatrix}$

8. $\begin{pmatrix} 4x - 5y = -36 \\ x + 2y = 30 \end{pmatrix}$

9. $\begin{pmatrix} x + 2y = 5 \\ 3x - 2y = 6 \end{pmatrix}$

10. $\begin{pmatrix} 2x - y = 3 \\ -2x + 5y = 7 \end{pmatrix}$

11. $\begin{pmatrix} y = -x \\ 2x - y = -2 \end{pmatrix}$

12. $\begin{pmatrix} 3x = 2y \\ 8x + 20y = 19 \end{pmatrix}$

13. $\begin{pmatrix} 4x + 5y = 9 \\ 5x - 6y = -50 \end{pmatrix}$

14. $\begin{pmatrix} 2x + 7y = 46 \\ 3x - 4y = -18 \end{pmatrix}$

15. $\begin{pmatrix} 9x - 7y = 29 \\ 5x - 3y = 17 \end{pmatrix}$

16. $\begin{pmatrix} 7x + 5y = -6 \\ 4x + 3y = -4 \end{pmatrix}$

17. $\begin{pmatrix} 6x + 5y = -6 \\ 8x - 3y = 21 \end{pmatrix}$

18. $\begin{pmatrix} 3x - 8y = -23 \\ 7x + 4y = -48 \end{pmatrix}$

19. $\begin{pmatrix} 2x - 7y = -1 \\ 9x + 4y = -2 \end{pmatrix}$

20. $\begin{pmatrix} 6x - 2y = -3 \\ 5x - 9y = 1 \end{pmatrix}$

21. $\begin{pmatrix} x + \quad y = 750 \\ .07x + .08y = \quad 57.5 \end{pmatrix}$

22. $\begin{pmatrix} x + \quad y = 700 \\ .06x + .09y = \quad 54 \end{pmatrix}$

23. $\begin{pmatrix} .09x + .11y = 31 \\ y = x + 100 \end{pmatrix}$

24. $\begin{pmatrix} .08x + .1y = 56 \\ y = 2x \end{pmatrix}$

For Problems 25–40, solve each problem by setting up and solving a system of two linear equations in two variables.

25. The sum of two numbers is 30 and their difference is 12. Find the numbers.

26. The sum of two numbers is 20. If twice the smaller is subtracted from the larger, the result is 2. Find the numbers.

27. The difference of two numbers is 7. If three times the smaller is subtracted from twice the larger, the result is 6. Find the numbers.

28. The difference of two numbers is 17. If the larger is increased by three times the smaller, the result is 37. Find the numbers.

29. One number is twice another number. The sum of three times the smaller and five times the larger is 78. Find the numbers.

30. One number is three times another number. The sum of four times the smaller and seven times the larger is 175. Find the numbers.

31. Three lemons and 2 apples cost $1.05. Two lemons and 3 apples cost $1.20. Find the cost of 1 lemon and also the cost of 1 apple.

32. Betty bought 30 stamps for $6. Some of them were 17-cent stamps and the rest were 22-cent stamps. How many of each kind did she buy?

33. Larry has $1.45 in change in his pocket, consisting of dimes and quarters. If he has a total of 10 coins, how many of each kind does he have?

34. Suppose that Cindy has 90 cents in change in her purse, consisting of nickels and dimes. If she has a total of 13 coins, how many of each kind does she have in her purse?

35. Suppose that a library buys a total of 35 books, which cost $462. Some of the books cost $12 per book and the rest cost $14 per book. How many books of each price did they buy?

36. Some punch that contains 10% grapefruit juice is mixed with some punch that contains 5% grapefruit juice to produce 5 quarts of punch that is 8% grapefruit juice. How many quarts of 10% and 5% grapefruit juice must be used?

37. A 10% salt solution is to be mixed with a 15% salt solution to produce 10 gallons of a 13% salt solution. How many gallons of the 10% solution and how many gallons of the 15% solution will be needed?

38. The income from a student production was $21,000. The price of a student ticket was $3 and nonstudent tickets were $5 each. Five thousand people attended the production. How many tickets of each kind were sold?

39. Sidney invested $1300, part of it at 10% and the rest at 12%. If his total yearly interest was $146, how much did he invest at each rate?

40. Heather invested $1100, part of it at 8% and the rest at 9%. Her total interest earned for a year was $95. How much did she invest at each rate?

Miscellaneous Problems

For Problems 41–44, solve each system using the method of elimination-by-addition.

41. $\begin{pmatrix} \dfrac{1}{2}x - \dfrac{1}{3}y = -2 \\ \dfrac{3}{2}x + \dfrac{2}{3}y = 34 \end{pmatrix}$

42. $\begin{pmatrix} x - 2y = 0 \\ 3x + 5y = 0 \end{pmatrix}$

43. $\begin{pmatrix} x - 4y = 6 \\ 2x - 8y = 3 \end{pmatrix}$

44. $\begin{pmatrix} y = 1 - 2x \\ 6x + 3y = 3 \end{pmatrix}$

5.7
The Substitution Method

There is a third method of solving systems of equations that is called the **substitution method**. Like the addition method, it produces exact solutions and can be used on any system of linear equations; however, we will find that some systems lend themselves more to the substitution method than others. Let's consider a few examples to demonstrate the use of the substitution method.

EXAMPLE 1   Solve $\begin{pmatrix} y = x + 10 \\ x + y = 14 \end{pmatrix}$.

Solution   Since the first equation states that y equals $x + 10$, we can substitute $x + 10$ for y in the second equation.

$$x + y = 14 \quad \xrightarrow{\text{substitute} \quad x + 10 \text{ for } y} \quad x + (x + 10) = 14$$

Now we have an equation with one variable that can be solved in the usual way.

$$x + (x + 10) = 14$$
$$2x + 10 = 14$$
$$2x = 4$$
$$\boxed{x = 2}$$

Substituting 2 for x in one of the original equations we can find the value of y.

$$y = x + 10$$
$$y = 2 + 10$$
$$\boxed{y = 12}$$

The solution set is $\{(2, 12)\}$. ∎

EXAMPLE 2 Solve $\begin{pmatrix} 3x + 5y = -7 \\ x = 2y + 5 \end{pmatrix}$.

Solution Since the second equation states that x equals $2y + 5$ we can substitute $2y + 5$ for x in the first equation.

$$3x + 5y = -7 \quad \underrightarrow{\text{substitute } 2y + 5 \text{ for } x} \quad 3(2y + 5) + 5y = -7$$

Solving this equation we obtain

$$3(2y + 5) + 5y = -7$$
$$6y + 15 + 5y = -7$$
$$11y + 15 = -7$$
$$11y = -22$$
$$\boxed{y = -2.}$$

Substituting -2 for y in one of the two original equations produces

$$x = 2y + 5$$
$$x = 2(-2) + 5$$
$$x = -4 + 5$$
$$\boxed{x = 1.}$$

The solution set is $\{(1, -2)\}$. ∎

Note that the key idea behind the substitution method is the elimination of a variable, but the elimination is done by a substitution rather than by addition of the equations. The substitution method is especially convenient to use if at least one of the equations is of the form *y equals* or *x equals*. In Example 1, the first equation is of the form *y equals* and in Example 2 the second equation is of the form *x equals*. Let's consider another example using the substitution method.

EXAMPLE 3   Solve $\left(\begin{array}{l} 2x + 3y = -30 \\ y = \dfrac{2}{3}x - 6 \end{array} \right)$.

Solution   The second equation allows us to substitute $\dfrac{2}{3}x - 6$ for y in the first equation.

$$2x + 3y = -30 \quad \xrightarrow{\text{substitute} \frac{2}{3}x - 6 \text{ for } y} \quad 2x + 3\left(\frac{2}{3}x - 6\right) = -30$$

Solving this equation produces

$$2x + 3\left(\frac{2}{3}x - 6\right) = -30$$

$$2x + 2x - 18 = -30$$

$$4x - 18 = -30$$

$$4x = -12$$

$$\boxed{x = -3.}$$

Now we can substitute -3 for x in one of the original equations.

$$y = \frac{2}{3}x - 6$$

$$y = \frac{2}{3}(-3) - 6$$

$$y = -2 - 6$$

$$\boxed{y = -8}$$

The solution set is $\{(-3, -8)\}$.   ∎

   It may be necessary to change the form of one of the equations before a substitution can be made. The following examples clarify this point.

EXAMPLE 4   Solve $\left(\begin{array}{l} 4x - 5y = 55 \\ x + y = -2 \end{array} \right)$.

Solution   The form of the second equation can easily be changed to make it ready for the substitution method.

$$x + y = -2$$

$$y = -2 - x \qquad \text{added } -x \text{ to both sides}$$

Now, $-2 - x$ can be substituted for y in the first equation.

$$4x - 5y = 55 \quad \xrightarrow{\text{substitute} -2 - x \text{ for } y} \quad 4x - 5(-2 - x) = 55$$

Solving this equation, we obtain

$$4x - 5(-2 - x) = 55$$
$$4x + 10 + 5x = 55$$
$$9x + 10 = 55$$
$$9x = 45$$
$$\boxed{x = 5.}$$

Substituting 5 for x in one of the *original* equations produces

$$x + y = -2$$
$$5 + y = -2$$
$$\boxed{y = -7.}$$

The solution set is $\{(5, -7)\}$. ∎

In Example 4, we could have started by changing the form of the first equation to make it ready for substitution. However, this would have created a fractional form $\left(\text{either } y = \dfrac{4x - 55}{5} \text{ or } x = \dfrac{5y + 55}{4}\right)$ to substitute. We were able to avoid any messy calculations with fractions by changing the form of the second equation instead of the first. Sometimes, when using the substitution method, you cannot avoid the use of fractional forms. The next example is a case in point.

EXAMPLE 5   Solve $\begin{pmatrix} 3x + 2y = 8 \\ 2x - 3y = -38 \end{pmatrix}$.

Solution   Looking ahead we see that changing the form of either equation will produce a fractional form. Therefore, let's merely pick the first equation and solve for y.

$$3x + 2y = 8$$
$$2y = 8 - 3x \qquad \text{added } -3x \text{ to both sides}$$
$$y = \frac{8 - 3x}{2}. \qquad \text{multiplied both sides by } \frac{1}{2}$$

Now we can substitute $\dfrac{8 - 3x}{2}$ for y in the second equation and determine the value of x.

$$2x - 3y = -38$$

$$2x - 3\left(\frac{8 - 3x}{2}\right) = -38$$

$$2x - \frac{24 - 9x}{2} = -38$$

$$4x - 24 + 9x = -76 \qquad \text{multiplied both sides by 2}$$

$$13x - 24 = -76$$

$$13x = -52$$

$$\boxed{x = -4}$$

Substituting -4 for x in one of the original equations, we obtain

$$3x + 2y = 8$$

$$3(-4) + 2y = 8$$

$$-12 + 2y = 8$$

$$2y = 20$$

$$\boxed{y = 10.}$$

The solution set is $\{(-4, 10)\}$. ∎

Which Method to Use

We have now studied three methods of solving systems of linear equations—the *graphing method*, the *elimination-by-addition method*, and the *substitution method*. As we indicated earlier, the graphing method is quite restrictive and works well only if the solutions are integers or if only approximate answers are needed. Both the elimination-by-addition and the substitution methods can be used to obtain exact solutions for any system of linear equations in two variables. The method you choose may depend upon the original form of the equations. Let's consider two examples to clarify this point.

EXAMPLE 6   Solve $\begin{pmatrix} 7x - 5y = -52 \\ y = 3x - 4 \end{pmatrix}$.

Solution   Since the second equation indicates that $3x - 4$ can be substituted for y, this system lends itself to the substitution method.

$$7x - 5y = -52 \quad \xrightarrow{\text{substitute } 3x - 4 \text{ for } y} \quad 7x - 5(3x - 4) = -52$$

Solving this equation, we obtain

$$7x - 5(3x - 4) = -52$$
$$7x - 15x + 20 = -52$$
$$-8x + 20 = -52$$
$$-8x = -72$$
$$\boxed{x = 9.}$$

Substituting 9 for x in one of the original equations produces

$$y = 3x - 4$$
$$y = 3(9) - 4$$
$$y = 27 - 4$$
$$\boxed{y = 23.}$$

The solution set is $\{(9, 23)\}$.   ∎

EXAMPLE 7   Solve $\begin{pmatrix} 10x + 7y = 19 \\ 2x - 6y = -11 \end{pmatrix}$.

Solution   Because changing the form of either of the two equations in preparation for the substitution method would produce a fractional form, we are probably better off to use the addition method. Furthermore, we should notice that the coefficients of x lend themselves to the addition method.

$$
\begin{array}{lll}
10x + 7y = 19 & \xrightarrow{\text{leave alone}} & 10x + 7y = 19 \\
2x - 6y = -11 & \xrightarrow{\text{multiply by } -5} & -10x + 30y = 55 \\
& & \overline{37y = 74} \\
& & \boxed{y = 2}
\end{array}
$$

Substituting 2 for y in the first equation of the given system produces

$$10x + 7y = 19$$
$$10x + 7(2) = 19$$
$$10x + 14 = 19$$
$$10x = 5$$
$$x = \frac{5}{10} = \frac{1}{2}.$$

The solution set is $\left\{ \left(\frac{1}{2}, 2 \right) \right\}$.   ∎

In Section 5.5 we explained that you can tell by graphing the equations whether the system has no solutions, one solution, or infinitely many solutions. That is, the two lines may be parallel (no solutions), or they may intersect in one point (one solution), or they may coincide (infinitely many solutions). From a practical viewpoint, the systems that have one solution deserve most of our attention. However, we do need to be able to deal with the other situations as they arise. Let's use two examples to demonstrate what occurs when we hit a "no solution" or "infinitely many solutions" situation when we are using either the addition or substitution method.

EXAMPLE 8   Solve the system $\begin{pmatrix} y = 2x - 1 \\ 6x - 3y = 7 \end{pmatrix}$.

Solution   Since the first equation indicates that $2x - 1$ can be substituted for y, this system lends itself to the substitution method.

$$6x - 3y = 7 \quad \underrightarrow{\text{substitute } 2x - 1 \text{ for } y} \quad 6x - 3(2x - 1) = 7$$

Now let's solve this equation.

$$6x - 6x + 3 = 7$$
$$0 + 3 = 7$$
$$3 = 7$$

The false numerical statement, $3 = 7$, implies that the system has *no solutions*. Thus, the solution set is $\varnothing$. (You may want to graph the two lines to verify this conclusion!) ∎

EXAMPLE 9   Solve the system $\begin{pmatrix} 2x - 3y = 4 \\ 10x - 15y = 20 \end{pmatrix}$.

Solution   Let's use the elimination-by-addition method.

$$2x - 3y = 4 \quad \underrightarrow{\text{multiply both sides by } -5} \quad -10x + 15y = -20$$
$$10x - 15y = 20 \quad \underrightarrow{\text{leave alone}} \quad \underline{10x - 15y = 20}$$
$$0 + 0 = 0$$

The true numerical statement, $0 + 0 = 0$, implies that the system has *infinitely many solutions*. Any ordered pair that satifies one of the equations will also satisfy the other equation. ∎

Problem Solving

Let's conclude this section with three more word problems.

PROBLEM 1   The length of a rectangle is 1 centimeter less than three times the width. The perimeter of the rectangle is 94 centimeters. Find the length and width of the rectangle.

Solution   Let w represent the width of the rectangle. Let l represent the length of the rectangle.

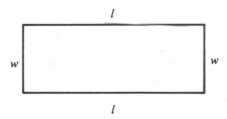

The problem translates into the following system of equations.

$l = 3w - 1,$ The length of a rectangle is 1 centimeter less than 3 times the width.

$2l + 2w = 94.$ The perimeter of the rectangle is 94 centimeters.

Multiplying both sides of the second equation by one-half produces the equivalent equation $l + w = 47$; so we have the following system to solve.

$$\begin{pmatrix} l = 3w - 1 \\ l + w = 47 \end{pmatrix}$$

The first equation indicates we can substitute $3w - 1$ for l in the second equation.

$l + w = 47$ $\xrightarrow{\text{substitute } 3w - 1 \text{ for } l}$ $3w - 1 + w = 47$

Solving this equation yields

$$3w - 1 + w = 47$$
$$4w - 1 = 47$$
$$4w = 48$$
$$\boxed{w = 12.}$$

Substituting 12 for w in one of the original equations produces

$$l = 3w - 1$$
$$l = 3(12) - 1$$
$$l = 36 - 1$$
$$\boxed{l = 35.}$$

The rectangle is 12 centimeters wide and 35 centimeters long.   ■

PROBLEM 2   The sum of the digits of a two-digit number is 7. The tens digit is 1 more than twice the units digit. Find the number.

Solution   Let t represent the tens digit. Let u represent the units digit.

$$t + u = 7, \qquad \text{The sum of the digits is 7.}$$
$$t = 2u + 1 \qquad \text{The tens digit is one more than twice the units digit.}$$

The second equation indicates that $2u + 1$ can be substituted for t in the first equation.

$$t + u = 7$$
$$(2u + 1) + u = 7$$
$$3u + 1 = 7$$
$$3u = 6$$
$$\boxed{u = 2}$$

Substituting 2 for u in one of the original equations yields

$$t = 2u + 1$$
$$t = 2(2) + 1$$
$$\boxed{t = 5.}$$

The tens digit is 5 and the units digit is 2; the number is 52.   ■

The two-variable expression $10t + u$ can be used to represent any two-digit number. The t represents the tens digit and u the units digit. For example, if $t = 4$ and $u = 7$, then $10t + u$ becomes $10(4) + 7$ or 47. Now let's consider the following problem.

PROBLEM 3   A two-digit number is 7 times its units digit. The sum of the digits is 8. Find the number.

Solution   Let t represent the tens digit. Let u represent the units digit.

$10t + u = 7u,$ A two-digit number is 7 times its units digit.

$t + u = 8$ The sum of the digits is 8.

The first equation can be simplified to $5t - 3u = 0$; thus we have the following system to solve.

$$\begin{pmatrix} 5t - 3u = 0 \\ t + \ u = 8 \end{pmatrix}$$

Using the elimination-by-addition method, we proceed as follows.

$5t - 3u = 0$ $\underrightarrow{\text{leave alone}}$ $5t - 3u = \ 0$

$t + u = 8$ $\underrightarrow{\text{multiply both sides by 3}}$ $3t + 3u = 24$

$$8t \qquad = 24$$

$$\boxed{t = 3}$$

Substituting 3 for t in $t + u = 8$ yields

$t + u = 8$

$3 + u = 8$

$\boxed{u = 5.}$

The tens digit is 3 and the units digit is 5; the number is 35.   ∎

Problem Set 5.7

For Problems 1–26, solve each system using the substitution method.

1. $\begin{pmatrix} y = 2x - 1 \\ x + y = 14 \end{pmatrix}$

2. $\begin{pmatrix} y = 3x + 4 \\ x + y = 52 \end{pmatrix}$

3. $\begin{pmatrix} x - y = -14 \\ y = -3x - 2 \end{pmatrix}$

4. $\begin{pmatrix} x - y = -23 \\ y = -2x + 5 \end{pmatrix}$

5. $\begin{pmatrix} 4x - 3y = -6 \\ y = -2x + 7 \end{pmatrix}$

6. $\begin{pmatrix} 8x - y = -8 \\ y = 4x + 5 \end{pmatrix}$

7. $\begin{pmatrix} x + \ y = 1 \\ 3x + 6y = 7 \end{pmatrix}$

8. $\begin{pmatrix} 2x - 4y = -9 \\ x + \ y = \ 3 \end{pmatrix}$

9. $\begin{pmatrix} 2x - y = 12 \\ x = \dfrac{3}{4}y \end{pmatrix}$

10. $\begin{pmatrix} 4x - 5y = 6 \\ y = \dfrac{2}{3}x \end{pmatrix}$

11. $\begin{pmatrix} y = \dfrac{3}{2}x \\ 6x - 5y = 15 \end{pmatrix}$

12. $\begin{pmatrix} x = \dfrac{3}{5}y \\ 4x - 3y = 12 \end{pmatrix}$

13. $\begin{pmatrix} 4y - 1 = x \\ 2x - 8y = 3 \end{pmatrix}$

14. $\begin{pmatrix} y = 5x - 2 \\ y = 2x + 7 \end{pmatrix}$

15. $\begin{pmatrix} 7x + 2y = -2 \\ 6x + 5y = 18 \end{pmatrix}$

16. $\begin{pmatrix} 2x - 3y = 31 \\ 3x + 5y = -20 \end{pmatrix}$

17. $\begin{pmatrix} 8x - 3y = -9 \\ x + 5y = -71 \end{pmatrix}$

18. $\begin{pmatrix} 9x - 2y = -18 \\ 4x - y = -7 \end{pmatrix}$

19. $\begin{pmatrix} 4x - 6y = 1 \\ 2x + 3y = 4 \end{pmatrix}$

20. $\begin{pmatrix} 2x + 3y = 1 \\ 4x + 6y = 2 \end{pmatrix}$

21. $\begin{pmatrix} 5x + 7y = 3 \\ 3x - 2y = 0 \end{pmatrix}$

22. $\begin{pmatrix} 7x - 3y = 4 \\ 2x + 5y = 0 \end{pmatrix}$

23. $\begin{pmatrix} .05x + .07y = 33 \\ y = x + 300 \end{pmatrix}$

24. $\begin{pmatrix} .06x + .08y = 15 \\ y = x + 100 \end{pmatrix}$

25. $\begin{pmatrix} x + y = 13 \\ .05x + .1y = 1.15 \end{pmatrix}$

26. $\begin{pmatrix} x + y = 17 \\ .1x + .25y = 3.2 \end{pmatrix}$

For Problems 27–46, solve each system using either the elimination-by-addition or the substitution method, whichever seems more appropriate to you.

27. $\begin{pmatrix} 5x - 4y = 14 \\ 7x + 3y = -32 \end{pmatrix}$

28. $\begin{pmatrix} 2x + 3y = 13 \\ 3x - 5y = -28 \end{pmatrix}$

29. $\begin{pmatrix} 2x + 9y = 6 \\ y = -x \end{pmatrix}$

30. $\begin{pmatrix} y = 3x + 2 \\ 4x - 3y = -21 \end{pmatrix}$

31. $\begin{pmatrix} x + y = 22 \\ 6x + .5y = 12 \end{pmatrix}$

32. $\begin{pmatrix} x + y = 13 \\ .4x + .5y = 6 \end{pmatrix}$

33. $\begin{pmatrix} 4x - y = 0 \\ 7x + 2y = 9 \end{pmatrix}$

34. $\begin{pmatrix} x - 2y = 0 \\ 5x = 8y + 12 \end{pmatrix}$

35. $\begin{pmatrix} 2x + y = 1 \\ 6x - 7y = -57 \end{pmatrix}$

36. $\begin{pmatrix} 7x - 9y = 11 \\ x - 3y = 3 \end{pmatrix}$

37. $\begin{pmatrix} 6x - y = -1 \\ 10x + 2y = 13 \end{pmatrix}$

38. $\begin{pmatrix} x + 4y = -5 \\ 6x - 5y = -1 \end{pmatrix}$

39. $\begin{pmatrix} 4x + 8y = 20 \\ x + 2y = 5 \end{pmatrix}$

40. $\begin{pmatrix} x = 5y - 5 \\ 2x - 10y = 2 \end{pmatrix}$

41. $\begin{pmatrix} 3x - 8y = -5 \\ x = 2y \end{pmatrix}$

42. $\begin{pmatrix} x + y = -3 \\ 5x + 6y = -22 \end{pmatrix}$

43. $\begin{pmatrix} 5y - 2x = -4 \\ 10y = 3x + 4 \end{pmatrix}$

44. $\begin{pmatrix} 3x - 7 = 2y + 15 \\ 3x = 6y + 18 \end{pmatrix}$

45. $\begin{pmatrix} x = -y - 1 \\ 6x - 5y = 4 \end{pmatrix}$

46. $\begin{pmatrix} 4x + 7y = 1 \\ y = 2x + 3 \end{pmatrix}$

For Problems 47–60, solve each problem by setting up and solving a system of two linear equations in two variables. Use either the addition or the substitution method to solve the systems.

47. Suppose that the sum of two numbers is 46 and the difference of the numbers is 22. Find the numbers.

48. The sum of two numbers is 52. The larger number is two more than four times the smaller number. Find the numbers.

49. Suppose that a certain motel rents double rooms at $28 per day and single rooms at $19 per day. If a total of 50 rooms was rented one day for $1265, how many of each kind were rented?

50. The total receipts from a concert amounted to $2600. Student tickets were sold at $4 each and nonstudent tickets at $6 each. The number of student tickets sold was five times the number of nonstudent tickets sold. How many student tickets and how many nonstudent tickets were sold?

51. The sum of the digits of a two-digit number is 9. The two-digit number is nine times its units digit. Find the number.

52. The sum of the digits of a two-digit number is 13. If the digits are reversed, the new number is nine larger than the original number. Find the number.

53. Suppose that Larry has a number of dimes and quarters totaling $12.05. The number of quarters is five more than twice the number of dimes. How many coins of each kind does he have?

54. Suppose that Sue has three times as many nickels as pennies in her collection. Together her pennies and nickels have a value of $4.80. How many pennies and how many nickels does she have?

55. The sum of the digits of a two-digit number is 12. The tens digit is three times as large as the units digit. Find the number.

56. The units digit of a two-digit number is one more than four times the tens digit. The sum of the digits is 11. Find the number.

57. Tina invested some money at 8% interest and some money at 9%. She invested $250 more at 9% than she invested at 8%. Her total yearly interest from the two investments was $48. How much did Tina invest at each rate?

58. One solution contains 40% alcohol and a second solution contains 70% alcohol. How many liters of each solution should be mixed to make 30 liters that contain 50% alcohol?

59. One solution contains 30% alcohol and a second solution contains 70% alcohol. How many liters of each solution should be mixed to make 10 liters containing 40% alcohol?

60. The length of a rectangle is 3 meters less than four times the width. The perimeter of the rectangle is 74 meters. Find the length and width of the rectangle.

Thoughts into Words

61. Give a step-by-step description of how you would solve the system $\begin{pmatrix} 2x - 3y = 5 \\ 4x + 7y = 9 \end{pmatrix}$ using the elimination-by-addition method.

62. Give a step-by-step description of how you would solve the system $\begin{pmatrix} 5x - 4y = 10 \\ 3x - y = 6 \end{pmatrix}$ using the substitution method.

Miscellaneous Problems

For Problems 63–66, use the substitution method to solve each system.

63. $\left(\begin{array}{l} y = \dfrac{2}{3}x - \dfrac{1}{2} \\[2mm] \dfrac{3}{2}x - y = -\dfrac{1}{3} \end{array} \right)$

64. $\left(\begin{array}{l} 5x - 3y = 0 \\ 4x + 7y = 0 \end{array} \right)$

65. $\left(\begin{array}{l} y = 2x - 1 \\ 6x - 3y = 3 \end{array} \right)$

66. $\left(\begin{array}{l} x = -4y + 5 \\ 2x + 8y = -1 \end{array} \right)$

5.8

Graphing Linear Inequalities

Linear inequalities in two variables are of the form $Ax + By > C$ or $Ax + By < C$, where A, B, and C are real numbers. (Combined linear equality and inequality statements are of the form $Ax + By \geq C$ or $Ax + By \leq C$.) Graphing linear inequalities is almost as easy as graphing linear equations. The following discussion leads into a simple step-by-step process. Let's consider the following equation and related inequalities.

$$x - y = 2$$

$$x - y > 2$$

$$x - y < 2$$

The graph of $x - y = 2$ is shown in Figure 5.30. The line divides the plane into two half-planes, one above the line and one below the line. In Figure 5.31(a) we have indicated coordinates for several points above the line. Note that for each point,

Figure 5.30

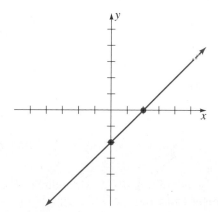

Figure 5.31

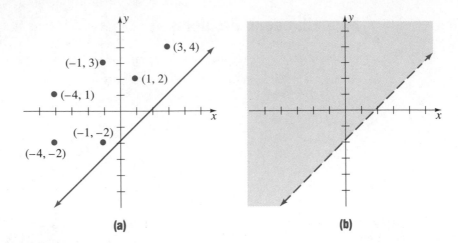

the ordered pair of real numbers satisfies the inequality $x - y < 2$. This is true for *all points* in the half-plane above the line. Therefore, the graph of $x - y < 2$ is the half-plane above the line, indicated by the shaded region in Figure 5.31(b). We use a dashed line to indicate that points on the line do not satisfy $x - y < 2$.

   In Figure 5.32(a) the coordinates of several points below the line $x - y = 2$ have been indicated. Note that for each point, the ordered pair of real numbers satisfies the inequality $x - y > 2$. This is true for *all points* in the half-plane below the line. Therefore, the graph of $x - y > 2$ is the half-plane below the line, indicated by the shaded region in Figure 5.32(b).

Figure 5.32

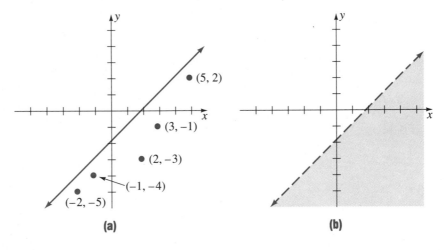

   Based on the previous discussion, we suggest the following steps for graphing linear inequalities.

   1. Graph the corresponding equality. Use a solid line if equality is included in the given statement and a dashed line if equality is not included.

2. Choose a "test point" not on the line and substitute its coordinates into the inequality statement. (The origin is a convenient point to use if it is not on the line.)

3. The graph of the given inequality is
   (a) the half-plane that contains the test point if the inequality is satisfied by the coordinates of the point, or,
   (b) the half-plane that does not contain the test point if the inequality is not satisfied by the coordinates of the point.

Let's apply these steps to some examples.

EXAMPLE 1   Graph $2x + y > 4$.

Solution   *Step 1.* Graph $2x + y = 4$ as a dashed line since equality is not included in the given statement $2x + y > 4$ (Figure 5.33(a)).

Figure 5.33

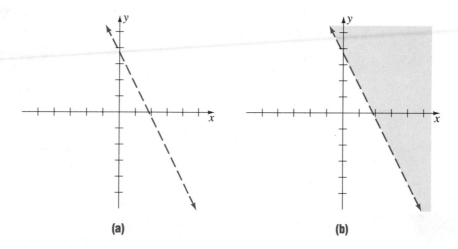

(a)   (b)

Step 2. Choose the origin as a test point and substitute its coordinates into the inequality.

$$2x + y > 4 \quad \text{becomes} \quad 2(0) + 0 > 4,$$

which is a false statement.

Step 3. Since the test point did not satisfy the given inequality, the graph is the half-plane that does not contain the test point. Thus, the graph of $2x + y > 4$ is the half-plane above the line, indicated in Figure 5.33(b). ∎

EXAMPLE 2   Graph $y \leq 2x$.

Solution   ***Step 1.*** Graph $y = 2x$ as a solid line since equality is included in the given statement (Figure 5.34(a)).

Figure 5.34

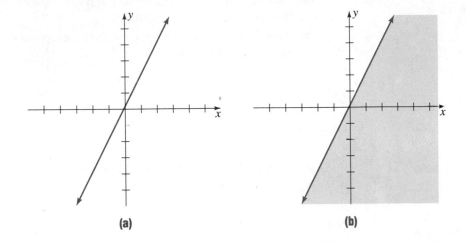

(a)   (b)

Step 2. Since the origin is on the line, we need to choose another point as a test point. Let's use $(3, 2)$.

$$y \leq 2x \quad \text{becomes } 2 \leq 2(3),$$

which is a true statement.

Step 3. Since the test point satisfies the given inequality, the graph is the half-plane that contains the test point. Thus, the graph of $y \leq 2x$ is the line, along with the half-plane below the line, indicated in Figure 5.34(b). ∎

Systems of Linear Inequalities

It is now easy to use a graphing approach to solve a system of linear inequalities. For example, the solution set of a system of linear inequalities, such as

$$\begin{pmatrix} x + y < 1 \\ x - y > 1 \end{pmatrix},$$

is the intersection of the solution sets of the individual inequalities. In Figure 5.35(a) we indicated the solution set for $x + y < 1$ and in Figure 5.35(b) we indicated the solution set for $x - y > 1$. Then, in Figure 5.35(c) we shaded the region that represents the intersection of the two shaded regions in (a) and (b); thus, it is the solution of the given system. The shaded region in Figure 5.35(c) consists of all points that are below the line $x + y = 1$ *and also* below the line $x - y = 1$.

Figure 5.35

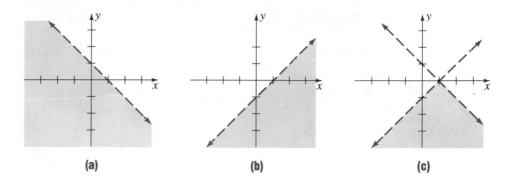

(a)   (b)   (c)

Let's solve another system of linear inequalities.

EXAMPLE 3   Solve the system

$$\begin{pmatrix} 2x + 3y \le 6 \\ x - 4y < 4 \end{pmatrix}.$$

Solution   Let's first graph the individual inequalities. The solution set for $2x + 3y \le 6$ is shown in Figure 5.36(a) and the solution set for $x - 4y < 4$ is shown in Figure 5.36(b). (Note that a solid line is used in (a) and a dashed line in (b).) Then, in Figure 5.36(c) we shaded the intersection of the graphs in (a) and (b). Thus, we represented the solution set for the given system by the shaded region in Figure 5.36(c). This region consists of all points that are on or below the line $2x + 3y = 6$ *and also* are above the line $x - 4y = 4$.

Figure 5.36

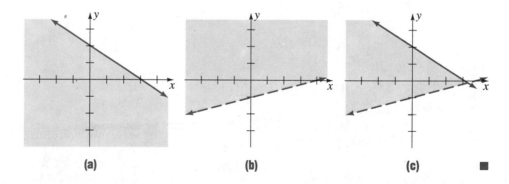

(a)   (b)   (c)   ■

REMARK Remember that the shaded region in Figure 5.36(c) represents the solution set of the given system. Parts (a) and (b) were drawn only to help determine the final shaded region. With some practice, you may be able to go directly to part (c) without actually sketching the graphs of the individual inequalities.

Problem Set 5.8

For Problems 1–20, graph each of the inequalities.

1. $x + y > 1$
2. $2x + y > 4$
3. $3x + 2y < 6$
4. $x + 3y < 3$
5. $2x - y \geq 4$
6. $x - 2y \geq 2$
7. $4x - 3y \leq 12$
8. $3x - 4y \leq 12$
9. $y > -x$
10. $y < x$
11. $2x - y \geq 0$
12. $3x - y \leq 0$
13. $-x + 2y < -2$
14. $-2x + y > -2$
15. $y \leq \dfrac{1}{2}x - 2$
16. $y \geq -\dfrac{1}{2}x + 1$
17. $y \geq -x + 4$
18. $y \leq -x - 3$
19. $3x + 4y > -12$
20. $4x + 3y > -12$

For Problems 21–30, indicate the solution set for each system of linear inequalities by shading the appropriate region.

21. $\begin{pmatrix} 2x + 3y > 6 \\ x - y < 2 \end{pmatrix}$
22. $\begin{pmatrix} x - 2y < 4 \\ 3x + y > 3 \end{pmatrix}$
23. $\begin{pmatrix} x - 3y \geq 3 \\ 3x + y \leq 3 \end{pmatrix}$
24. $\begin{pmatrix} 4x + 3y \leq 12 \\ 4x - y \geq 4 \end{pmatrix}$
25. $\begin{pmatrix} y \geq 2x \\ y < x \end{pmatrix}$
26. $\begin{pmatrix} y \leq -x \\ y > -3x \end{pmatrix}$
27. $\begin{pmatrix} y < -x + 1 \\ y > -x - 1 \end{pmatrix}$
28. $\begin{pmatrix} y > x - 2 \\ y < x + 3 \end{pmatrix}$
29. $\begin{pmatrix} y < \dfrac{1}{2}x + 2 \\ y < \dfrac{1}{2}x - 1 \end{pmatrix}$
30. $\begin{pmatrix} y > -\dfrac{1}{2}x - 2 \\ y > -\dfrac{1}{2}x + 1 \end{pmatrix}$

Miscellaneous Problems

For Problems 31–34, indicate the solution set for each system of linear inequalities by shading the appropriate region.

31. $\begin{pmatrix} y > x + 1 \\ y < x - 1 \end{pmatrix}$
32. $\begin{pmatrix} x \geq 0 \\ y \geq 0 \\ 3x + 4y \leq 12 \\ 2x + y \leq 4 \end{pmatrix}$
33. $\begin{pmatrix} x \geq 0 \\ y \geq 0 \\ 2x + y \leq 4 \\ 2x - 3y \leq 6 \end{pmatrix}$
34. $\begin{pmatrix} x \geq 0 \\ y \geq 0 \\ 3x + 5y \geq 15 \\ 5x + 3y \geq 15 \end{pmatrix}$

Chapter 5 Summary

(5.1) The **Cartesian** or **rectangular coordinate system** involves a one-to-one correspondence between ordered pairs of real numbers and the points of a plane. The system provides the basis for a study of coordinate geometry, which is a link between algebra and geometry that deals with two basic kinds of problems.

1. Given an algebraic equation find its geometric graph.
2. Given a set of conditions pertaining to a geometric figure, find its algebraic equation.

One graphing technique is to plot a sufficient number of points until the graph of the equation is determined.

(5.2) Any equation of the form $Ax + By = C$, where A, B, and C are constants (A and B not both zero) and x and y are variables, is a **linear equation in two variables** and its graph is a **straight line**.

To **graph** a linear equation we can find two solutions (the intercepts are usually easy to determine), plot the corresponding points, and connect the points with a straight line.

(5.3) If points P_1 and P_2 with coordinates (x_1, y_1) and (x_2, y_2), respectively, are any two points on a line, then the **slope** of the line (denoted by m) is given by

$$m = \frac{y_2 - y_1}{x_2 - x_1}, \qquad x_1 \neq x_2.$$

The slope of a line is a **ratio** of vertical change relative to horizontal change. The slope of a line can be negative, positive, or zero. The concept of slope is not defined for vertical lines.

(5.4) You should review Examples 1, 2, and 3 of this section to pull together a general approach to writing equations of lines given certain conditions. The equation

$$y - y_1 = m(x - x_1)$$

is called the **point-slope form** of the equation of a straight line. The equation

$$y = mx + b$$

is called the **slope-intercept form** of the equation of a straight line. If the equation of a nonvertical line is written in slope-intercept form, then the coefficient of x is the slope of the line and the constant term is the y-intercept.

(5.5) Solving a system of two linear equations by graphing produces one of the following three possibilities.

1. The graphs of the two equations are two intersecting lines, which indicates **one solution** for the system, called a **consistent system**.

2. The graphs of the two equations are two parallel lines, which indicates **no solution** for the system, called an **inconsistent system**.

3. The graphs of the two equations are the same line, which indicates **infinitely many solutions** for the system. We refer to the equations as a set of **dependent** equations.

(5.6) The **elimination-by-addition method** for solving a system of two linear equations relies on the property "if $a = b$ and $c = d$, then $a + c = b + d$". Equations are added to eliminate a variable.

(5.7) The **substitution method** for solving a system of two linear equations involves solving one of the two equations for one variable and **substituting** this expression into the other equation.

To review the problem of choosing a method for solving a particular system of two linear equations, return to Section 5.7 and study Examples 6 and 7 one more time.

(5.8) **Linear inequalities** in two variables are of the form $Ax + By > C$ or $Ax + By < C$. To **graph a linear inequality**, we suggest the following steps.

1. First, graph the corresponding equality. Use a solid line if equality is included in the original statement and a dashed line if equality is not included.

2. Choose a test point not on the line and substitute its coordinates into the inequality.

3. The graph of the original inequality is
   (a) the half-plane that contains the test point, if the inequality is satisfied by that point, or
   (b) the half-plane that does not contain the test point, if the inequality is not satisfied by the point.

The solution set of a system of linear inequalities is the intersection of the solution sets of the individual inequalities.

Chapter 5 Review Problem Set

For Problems 1–6, graph each of the equations.

1. $2x - 5y = 10$

2. $y = -\dfrac{1}{3}x + 1$

3. $y = 2x^2 + 1$

4. $y = -x^2 - 2$

5. $2x - 3y = 0$

6. $y = -x^3$

7. Find the slope of the line determined by $(3, -4)$ and $(-2, 5)$.

8. Find the slope of the line $5x - 6y = 30$.

9. Write the equation of the line that has a slope of $-\dfrac{5}{7}$ and contains the point $(2, -3)$.

10. Write the equation of the line that contains $(2, 5)$ and $(-1, -3)$.

11. Write the equation of the line that has a slope of $\dfrac{2}{9}$ and a y-intercept of -1.

12. Write the equation of the line that contains $(2, 4)$ and which is perpendicular to the x-axis.

13. Solve the system $\begin{pmatrix} 2x + y = 4 \\ x - y = 5 \end{pmatrix}$ by using the graphing method.

For Problems 14–25, solve each system by using either the addition method or the substitution method.

14. $\begin{pmatrix} 2x - y = 1 \\ 3x - 2y = -5 \end{pmatrix}$

15. $\begin{pmatrix} 2x + 5y = 7 \\ x = -3y + 1 \end{pmatrix}$

16. $\begin{pmatrix} 3x + 2y = 7 \\ 4x - 5y = 3 \end{pmatrix}$

17. $\begin{pmatrix} 9x + 2y = 140 \\ x + 5y = 135 \end{pmatrix}$

18. $\begin{pmatrix} \dfrac{1}{2}x + \dfrac{1}{4}y = -5 \\ \dfrac{2}{3}x - \dfrac{1}{2}y = 0 \end{pmatrix}$

19. $\begin{pmatrix} x + y = 1000 \\ 0.07x + 0.09y = 82 \end{pmatrix}$

20. $\begin{pmatrix} y = 5x + 2 \\ 10x - 2y = 1 \end{pmatrix}$

21. $\begin{pmatrix} 5x - 7y = 9 \\ y = 3x - 2 \end{pmatrix}$

22. $\begin{pmatrix} 10t + u = 6u \\ t + u = 12 \end{pmatrix}$

23. $\begin{pmatrix} t = 2u \\ 10t + u - 36 = 10u + t \end{pmatrix}$

24. $\begin{pmatrix} u = 2t + 1 \\ 10t + u + 10u + t = 110 \end{pmatrix}$

25. $\begin{pmatrix} y = -\dfrac{2}{3}x \\ \dfrac{1}{3}x - y = -9 \end{pmatrix}$

For Problems 26 and 27, graph each inequality.

26. $-2x + y < 4$

27. $3x + 2y \geq -6$

28. Solve, by graphing, the system of inequalities

$$\begin{pmatrix} -x + 2y > 2 \\ 3x - y > 3 \end{pmatrix}.$$

Solve each of the following problems by setting up and solving a system of two linear equations in two variables.

29. The sum of two numbers is 113. The larger number is one less than twice the smaller number. Find the numbers.

30. Last year Mark invested a certain amount of money at 9% annual interest and $50 more than that amount at 11%. He received $55.50 in interest. How much did he invest at each rate?

31. Cindy has 43 coins consisting of nickels and dimes. The total value of the coins is $3.40. How many coins of each kind does she have?

32. The length of a rectangle is 1 inch more than three times the width. If the perimeter of the rectangle is 50 inches, find the length and width.

33. The sum of the digits of a two-digit number is 13 and the units digit is one more than twice the tens digit. Find the number.

34. The sum of the digits of a two-digit number is 9. The number formed by reversing the digits is 27 larger than the original number. Find the original number.

35. Two angles are complementary and one of them is 6° less than twice the other one. Find the measure of each angle.

36. Two angles are supplementary and the larger angle is 20° less than three times the smaller angle. Find the measure of each angle.

37. Four cheeseburgers and five milkshakes cost a total of $8.35. Two milkshakes cost $.35 more than one cheeseburger. Find the cost of a cheeseburger and also find the cost of a milkshake.

38. Three cans of prune juice and two cans of tomato juice cost $3.85. On the other hand, two cans of prune juice and three cans of tomato juice cost $3.55. Find the cost per can of each.

Chapter 6

Exponents and Polynomials

We need to master certain arithmetic skills before we can use arithmetic to solve problems. A similar situation exists in algebra. There are basic algebraic skills we need in order to use algebra to solve certain kinds of problems. This chapter focuses primarily on the development of some algebraic skills that will be useful when we study a class of algebraic expressions called polynomials.

6.1

Addition and Subtraction of Polynomials

In the previous chapters, algebraic expressions such as $4x$, $5y$, $-6ab$, $7x^2$, and $-9xy^2z^3$ were called *terms*. Recall that a term is an indicated product that may contain any number of factors. The variables in a term are called *literal factors* and the numerical factor is called the *numerical coefficient* of the term. Thus, in $-6ab$, the a and b are literal factors and the numerical coefficient is -6. Terms that have the same literal factors are similar, or like, terms.

Terms that contain variables with only whole numbers as exponents are called **monomials**. The previously listed terms, $4x$, $5y$, $-6ab$, $7x^2$, and $-9xy^2z^3$ are all monomials. (We shall work with some algebraic expressions later, such as $7x^{-1}y^{-1}$ and $4a^{-2}b^{-3}$, which are not monomials.)

The **degree of a monomial** is the sum of the exponents of the literal factors.

$4xy$ is of degree 2;

$5x$ is of degree 1;

$14a^2b$ is of degree 3;

$-17xy^2z^3$ is of degree 6;

$-9y^4$ is of degree 4.

If the monomial contains only one variable, then the exponent of the variable is the degree of the monomial. Any nonzero constant term is said to be of degree zero.

A **polynomial** is a monomial or a finite sum (or difference) of monomials. The **degree of a polynomial** is the degree of the term with the highest degree in the polynomial. Some special classifications of polynomials are made according to the number of terms. We call a one-term polynomial a **monomial**, a two-term polynomial a **binomial**, and a three-term polynomial a **trinomial**. The following examples illustrate some of this terminology.

The polynomial $5x^3y^4$ is a monomial of degree 7;

The polynomial $4x^2y - 3xy$ is a binomial of degree 3;

The polynomial $5x^2 - 6x + 4$ is a trinomial of degree 2;

The polynomial $9x^4 - 7x^3 + 6x^2 + x - 2$ is given no special name, but is of degree 4.

Adding Polynomials

In the preceding chapters you have already worked many problems involving addition and subtraction of polynomials. For example, simplifying $4x^2 + 6x + 7x^2 - 2x$ to $11x^2 + 4x$ by combining similar terms can actually be considered the addition problem $(4x^2 + 6x) + (7x^2 - 2x)$. At this time we simply want to review and extend some of those ideas.

EXAMPLE 1   Add $5x^2 + 7x - 2$ and $9x^2 - 12x + 13$.

Solution   We commonly use the horizontal format for such work. Thus,

$$(5x^2 + 7x - 2) + (9x^2 - 12x + 13) = (5x^2 + 9x^2) + (7x - 12x) + (-2 + 13)$$
$$= 14x^2 - 5x + 11. \qquad \blacksquare$$

The commutative, associative, and distributive properties provide the basis for rearranging, regrouping, and combining similar terms.

EXAMPLE 2   Add $5x - 1$, $3x + 4$, and $9x - 7$.

Solution
$$(5x - 1) + (3x + 4) + (9x - 7) = (5x + 3x + 9x) + (-1 + 4 + (-7))$$
$$= 17x - 4 \qquad \blacksquare$$

EXAMPLE 3   Add $-x^2 + 2x - 1$, $2x^3 - x + 4$, and $-5x + 6$.

Solution
$$(-x^2 + 2x - 1) + (2x^3 - x + 4) + (-5x + 6)$$
$$= 2x^3 - x^2 + 2x - x - 5x - 1 + 4 + 6$$
$$= 2x^3 - x^2 - 4x + 9 \qquad \blacksquare$$

Subtracting Polynomials

Recall from Chapter 1 that $a - b = a + (-b)$. We define subtraction as *adding the opposite*. This same idea extends to polynomials in general. The opposite of a polynomial can be formed by taking the opposite of each term. For example, the opposite of $(2x^2 - 7x + 3)$ is $-2x^2 + 7x - 3$. Symbolically, we express this as

$$-(2x^2 - 7x + 3) = -2x^2 + 7x - 3.$$

Now consider the following subtraction problems.

EXAMPLE 4   Subtract $2x^2 + 9x - 3$ from $5x^2 - 7x - 1$.

Solution   Using the horizontal format, we have

$$(5x^2 - 7x - 1) - (2x^2 + 9x - 3) = (5x^2 - 7x - 1) + (-2x^2 - 9x + 3)$$
$$= (5x^2 - 2x^2) + (-7x - 9x) + (-1 + 3)$$
$$= 3x^2 - 16x + 2. \qquad \blacksquare$$

EXAMPLE 5   Subtract $-8y^2 - y + 5$ from $2y^2 + 9$.

Solution
$$(2y^2 + 9) - (-8y^2 - y + 5) = (2y^2 + 9) + (8y^2 + y - 5)$$
$$= (2y^2 + 8y^2) + (y) + (9 - 5)$$
$$= 10y^2 + y + 4 \qquad \blacksquare$$

Later, when dividing polynomials, we will need to subtract polynomials using a vertical format; therefore, let's consider two such examples.

EXAMPLE 6   Subtract $3x^2 + 5x - 2$ from $9x^2 - 7x - 1$.

Solution

$$9x^2 - 7x - 1$$
$$\underline{3x^2 + 5x - 2}$$

Notice which polynomial goes on the bottom and the alignment of similar terms in columns.

Now we can *mentally form the opposite of the bottom polynomial* and add.

$$9x^2 - \;\;7x - 1$$
$$\underline{3x^2 + \;\;5x - 2}$$
$$6x^2 - 12x + 1$$ The opposite of $3x^2 + 5x - 2$ is $-3x^2 - 5x + 2$.   ∎

EXAMPLE 7   Subtract $15y^3 + 5y^2 + 3$ from $13y^3 + 7y - 1$.

Solution

$$13y^3 \qquad\quad + 7y - 1$$ The similar terms are arranged in columns.
$$\underline{15y^3 + 5y^2 \qquad + 3}$$
$$-2y^3 - 5y^2 + 7y - 4$$ We mentally formed the opposite of the bottom polynomial and added.   ∎

We can use the distributive property along with the properties $a = 1(a)$ and $-a = -1(a)$ when adding and subtracting polynomials. The next examples illustrate this approach.

EXAMPLE 8   Perform the indicated operations.

$$(3x - 4) + (2x - 5) - (7x - 1)$$

Solution

$$(3x - 4) + (2x - 5) - (7x - 1)$$
$$= 1(3x - 4) + 1(2x - 5) - 1(7x - 1)$$
$$= 1(3x) - 1(4) + 1(2x) - 1(5) - 1(7x) - 1(-1)$$
$$= 3x - 4 + 2x - 5 - 7x + 1$$
$$= 3x + 2x - 7x - 4 - 5 + 1$$
$$= -2x - 8$$ ∎

Certainly we can do some of the steps mentally; the next example illustrates a possible format.

EXAMPLE 9   Perform the indicated operations.

$$(-y^2 + 5y - 2) - (-2y^2 + 8y + 6) + (4y^2 - 2y - 5)$$

Solution

$$(-y^2 + 5y - 2) - (-2y^2 + 8y + 6) + (4y^2 - 2y - 5)$$
$$= -y^2 + 5y - 2 + 2y^2 - 8y - 6 + 4y^2 - 2y - 5$$
$$= -y^2 + 2y^2 + 4y^2 + 5y - 8y - 2y - 2 - 6 - 5$$
$$= 5y^2 - 5y - 13$$ ∎

When we use the horizontal format, as in Examples 8 and 9, we use parentheses to indicate a quantity. In Example 8 the quantities $(3x - 4)$ and $(2x - 5)$ are to be added and from this result we are to subtract the quantity $(7x - 1)$. Brackets, [ ], are also sometimes used as grouping symbols, especially if there is a need to indicate quantities within quantities. To remove the grouping symbols, perform the indicated operations, starting with the innermost set of symbols. Let's consider two examples of this type.

EXAMPLE 10   Perform the indicated operations.

$$3x - [2x + (3x - 1)]$$

Solution   First, we need to add the quantities $2x$ and $(3x - 1)$.

$$3x - [2x + (3x - 1)] = 3x - [2x + 3x - 1]$$
$$= 3x - [5x - 1]$$

Now we need to subtract the quantity $[5x - 1]$ from $3x$.

$$3x - [5x - 1] = 3x - 5x + 1$$
$$= -2x + 1 \qquad \blacksquare$$

EXAMPLE 11   Perform the indicated operations.

$$8 - [7x - [2 + (x - 1)] + 4x]$$

Solution   Start with the innermost set of grouping symbols (the parentheses) and proceed as follows.

$$8 - [7x - [2 + (x - 1)] + 4x] = 8 - [7x - [2 + x - 1] + 4x]$$
$$= 8 - [7x - [x + 1] + 4x]$$
$$= 8 - [7x - x - 1 + 4x]$$
$$= 8 - [10x - 1]$$
$$= 8 - 10x + 1$$
$$= -10x + 9 \qquad \blacksquare$$

Problem Set 6.1

For Problems 1–8, determine the degree of each polynomial.

1. $7x^2y + 6xy$ 2. $4xy - 7x$ 3. $5x^2 - 9$

4. $8x^2y^2 - 2xy^2 - x$ 5. $5x^3 - x^2 - x + 3$ 6. $8x^4 - 2x^2 + 6$

7. $5xy$ 8. $-7x + 4$

For Problems 9–22, add the polynomials.

9. $3x + 4$ and $5x + 7$ 10. $3x - 5$ and $2x - 9$

11. $-5y - 3$ and $9y + 13$

12. $x^2 - 2x - 1$ and $-2x^2 + x + 4$

13. $-2x^2 + 7x - 9$ and $4x^2 - 9x - 14$

14. $3a^2 + 4a - 7$ and $-3a^2 - 7a + 10$

15. $5x - 2, 3x - 7,$ and $9x - 10$

16. $-x - 4, 8x + 9,$ and $-7x - 6$

17. $2x^2 - x + 4, -5x^2 - 7x - 2,$ and $9x^2 + 3x - 6$

18. $-3x^2 + 2x - 6, 6x^2 + 7x + 3,$ and $-4x^2 - 9$

19. $-4n^2 - n - 1$ and $4n^2 + 6n - 5$

20. $-5n^2 + 7n - 9$ and $-5n - 4$

21. $2x^2 - 7x - 10, -6x - 2,$ and $-9x^2 + 5$

22. $7x - 11, -x^2 - 5x + 9,$ and $-4x + 5$

For Problems 23–34, subtract the polynomials using a horizontal format.

23. $7x + 1$ from $12x + 6$ 24. $10x + 3$ from $14x + 13$

25. $5x - 2$ from $3x - 7$ 26. $7x - 2$ from $2x + 3$

27. $-x - 1$ from $-4x + 6$ 28. $-3x + 2$ from $-x - 9$

29. $x^2 - 7x + 2$ from $3x^2 + 8x - 4$

30. $2x^2 + 6x - 1$ from $8x^2 - 2x + 6$

31. $-2n^2 - 3n + 4$ from $3n^2 - n + 7$

32. $3n^2 - 7n - 9$ from $-4n^2 + 6n + 10$

33. $-4x^3 - x^2 + 6x - 1$ from $-7x^3 + x^2 + 6x - 12$

34. $-4x^2 + 6x - 2$ from $-3x^3 + 2x^2 + 7x - 1$

For Problems 35–44, subtract the polynomials using a vertical format.

35. $3x - 2$ from $12x - 4$ 36. $-4x + 6$ from $7x - 3$

37. $-5a - 6$ from $-3a + 9$ 38. $7a - 11$ from $-2a - 1$

39. $8x^2 - x + 6$ from $6x^2 - x + 11$

40. $3x^2 - 2$ from $-2x^2 + 6x - 4$

41. $-2x^3 - 6x^2 + 7x - 9$ from $4x^3 + 6x^2 + 7x - 14$

42. $4x^3 + x - 10$ from $3x^2 - 6$

43. $2x^2 - 6x - 14$ from $4x^3 - 6x^2 + 7x - 2$

44. $3x - 7$ from $7x^3 + 6x^2 - 5x - 4$

For Problems 45–64, perform the indicated operations.

45. $(5x + 3) - (7x - 2) + (3x + 6)$

46. $(3x - 4) + (9x - 1) - (14x - 7)$

47. $(-x - 1) - (-2x + 6) + (-4x - 7)$

48. $(-3x + 6) + (-x - 8) - (-7x + 10)$

49. $(x^2 - 7x - 4) + (2x^2 - 8x - 9) - (4x^2 - 2x - 1)$

50. $(3x^2 + x - 6) - (8x^2 - 9x + 1) - (7x^2 + 2x - 6)$

51. $(-x^2 - 3x + 4) + (-2x^2 - x - 2) - (-4x^2 + 7x + 10)$

52. $(-3x^2 - 2) + (7x^2 - 8) - (9x^2 - 2x - 4)$

53. $(3a - 2b) - (7a + 4b) - (6a - 3b)$

54. $(5u + 7b) + (-8a - 2b) - (5a + 6b)$

55. $(n - 6) - (2n^2 - n + 4) + (n^2 - 7)$

56. $(3n + 4) - (n^2 - 9n + 10) - (-2n + 4)$

57. $7x + [3x - (2x - 1)]$ **58.** $-6x + [-2x - (5x + 2)]$

59. $-7n - [4n - (6n - 1)]$ **60.** $9n - [3n - (5n + 4)]$

61. $(5a - 1) - [3a + (4a - 7)]$ **62.** $(-3a + 4) - [-7a + (9a - 1)]$

63. $13x - [5x - [4x - (x - 6)]]$

64. $-10x - [7x - [3x - (2x - 3)]]$

65. Subtract $5x - 3$ from the sum of $4x - 2$ and $7x + 6$.

66. Subtract $7x + 5$ from the sum of $9x - 4$ and $-3x - 2$.

67. Subtract the sum of $-2n - 5$ and $-n + 7$ from $-8n + 9$.

68. Subtract the sum of $7n - 11$ and $-4n - 3$ from $13n - 4$.

6.2

Multiplying Monomials

In Section 2.4 we used exponents and some of the basic properties of real numbers to simplify algebraic expressions into a more compact form. For example,

$$(3x)(4xy) = 3 \cdot 4 \cdot x \cdot x \cdot y = 12x^2y.$$

Actually, we were **multiplying monomials**, and it is this topic that we will pursue further at this time. Multiplying monomials is made easier by using some basic properties of exponents. These properties are the direct result of the definition of an exponent. The following examples lead into the first property.

$$x^2 \cdot x^3 = (x \cdot x)(x \cdot x \cdot x) = x^5;$$
$$a^3 \cdot a^4 = (a \cdot a \cdot a)(a \cdot a \cdot a \cdot a) = a^7;$$
$$b \cdot b^2 = (b)(b \cdot b) = b^3.$$

In general,

$$b^n \cdot b^m = \underbrace{(b \cdot b \cdot b \cdots b)}_{\substack{n \text{ factors} \\ \text{of } b}}\underbrace{(b \cdot b \cdot b \cdots b)}_{\substack{m \text{ factors} \\ \text{of } b}}$$

$$= \underbrace{b \cdot b \cdot b \cdots b}_{\substack{(n + m) \text{ factors} \\ \text{of } b}}$$

$$= b^{n+m}.$$

PROPERTY 6.1

> If b is any real number and n and m are positive integers, then
>
> $$b^n \cdot b^m = b^{n+m}.$$

Property 6.1 states that when multiplying powers with the same base, add exponents.

EXAMPLE 1   Multiply each of the following.

        (a) $x^4 \cdot x^3$ (b) $a^8 \cdot a^7$

Solution   (a) $x^4 \cdot x^3 = x^{4+3} = x^7$ (b) $a^8 \cdot a^7 = a^{8+7} = a^{15}$ ■

Another property of exponents is demonstrated by the following examples.

$$(x^2)^3 = x^2 \cdot x^2 \cdot x^2 = x^{2+2+2} = x^6;$$
$$(a^3)^2 = a^3 \cdot a^3 = a^{3+3} = a^6;$$
$$(b^3)^4 = b^3 \cdot b^3 \cdot b^3 \cdot b^3 = b^{3+3+3+3} = b^{12}.$$

In general,

$$(b^n)^m = \underbrace{b^n \cdot b^n \cdot b^n \cdots b^n}_{m \text{ factors of } b^n}$$

$$= b^{\overbrace{n+n+n+\cdots+n}^{m \text{ of these } n\text{'s}}}$$

$$= b^{mn}.$$

PROPERTY 6.2

> If b is any real number and m and n are positive integers, then
>
> $$(b^n)^m = b^{mn}.$$

Property 6.2 states that when raising a power to a power, multiply exponents.

EXAMPLE 2   Raise each of the following to the indicated power.

        (a) $(x^4)^3$ (b) $(a^5)^6$

Solution   (a) $(x^4)^3 = x^{3\cdot4} = x^{12}$ (b) $(a^5)^6 = a^{6\cdot5} = a^{30}$ ■

The third property of exponents to be used in this section raises a monomial to a power.

$$(2x)^3 = (2x)(2x)(2x) = 2 \cdot 2 \cdot 2 \cdot x \cdot x \cdot x = 2^3 \cdot x^3;$$
$$(3a^4)^2 = (3a^4)(3a^4) = 3 \cdot 3 \cdot a^4 \cdot a^4 = (3)^2(a^4)^2;$$
$$(-2xy^5)^2 = (-2xy^5)(-2xy^5) = (-2)(-2)(x)(x)(y^5)(y^5) = (-2)^2(x)^2(y^5)^2$$

In general,

$$(ab)^n = \underbrace{ab \cdot ab \cdot ab \cdots ab}_{n \text{ factors of } ab}$$

$$= \underbrace{(a \cdot a \cdot a \cdots a)}_{n \text{ factors of } a}\underbrace{(b \cdot b \cdot b \cdots b)}_{n \text{ factors of } b}$$

$$= a^n b^n.$$

PROPERTY 6.3

> If a and b are real numbers and n is a positive integer, then
>
> $$(ab)^n = a^n b^n.$$

Property 6.3 states that when raising a monomial to a power, raise each factor to that power.

EXAMPLE 3   Raise each of the following to the indicated power.

  (a) $(2x^2y^3)^4$ **(b)** $(-3ab^5)^3$

Solution

  (a) $(2x^2y^3)^4 = (2)^4(x^2)^4(y^3)^4 = 16x^8y^{12}$
  (b) $(-3ab^5)^3 = (-3)^3(a^1)^3(b^5)^3 = -27a^3b^{15}$ ∎

Consider the following examples in which we use the properties of exponents to help simplify the process of multiplying monomials.

1. $(3x^3)(5x^4) = 3 \cdot 5 \cdot x^3 \cdot x^4$
        $= 15x^7.$ $x^3 \cdot x^4 = x^{3+4} = x^7$

2. $(-4a^2b^3)(6ab^2) = -4 \cdot 6 \cdot a^2 \cdot a \cdot b^3 \cdot b^2$
        $= -24a^3b^5.$

3. $(xy)(7xy^5) = 1 \cdot 7 \cdot x \cdot x \cdot y \cdot y^5$ The numerical coefficient of xy is 1.
        $= 7x^2y^6.$

4. $\left(\frac{3}{4}x^2y^3\right)\left(\frac{1}{2}x^3y^5\right) = \frac{3}{4} \cdot \frac{1}{2} \cdot x^2 \cdot x^3 \cdot y^3 \cdot y^5$

        $= \frac{3}{8}x^5y^8.$

It is a simple process to raise a monomial to a power when using the properties of exponents. Study the following examples.

5. $(2x^3)^4 = (2)^4(x^3)^4$ by using $(ab)^n = a^n b^n$

$\qquad\qquad = (2)^4(x^{12})$ by using $(b^n)^m = b^{mn}$

$\qquad\qquad = 16x^{12}$.

6. $(-2a^4)^5 = (-2)^5(a^4)^5$

$\qquad\qquad = -32a^{20}$.

7. $\left(\dfrac{2}{5}x^2 y^3\right)^3 = \left(\dfrac{2}{5}\right)^3 (x^2)^3 (y^3)^3$

$\qquad\qquad\qquad = \dfrac{8}{125} x^6 y^9$.

8. $(0.2a^6 b^7)^2 = (0.2)^2(a^6)^2(b^7)^2$

$\qquad\qquad\quad = 0.04a^{12} b^{14}$.

Sometimes problems involve first raising monomials to a power and then multiplying the resulting monomials, as in the following examples.

9. $(3x^2)^3(2x^3)^2 = (3)^3(x^2)^3(2)^2(x^3)^2$

$\qquad\qquad\qquad = (27)(x^6)(4)(x^6)$

$\qquad\qquad\qquad = 108x^{12}$.

10. $(-x^2 y^3)^5(-2x^2 y)^2 = (-1)^5(x^2)^5(y^3)^5(-2)^2(x^2)^2(y)^2$

$\qquad\qquad\qquad\qquad = (-1)(x^{10})(y^{15})(4)(x^4)(y^2)$

$\qquad\qquad\qquad\qquad = -4x^{14} y^{17}$.

The distributive property and the properties of exponents form a basis for finding the product of a monomial and a polynomial. The following examples illustrate these ideas.

11. $(3x)(2x^2 + 6x + 1) = (3x)(2x^2) + (3x)(6x) + (3x)(1)$

$\qquad\qquad\qquad\qquad = 6x^3 + 18x^2 + 3x$.

12. $(5a^2)(a^3 - 2a^2 - 1) = (5a^2)(a^3) - (5a^2)(2a^2) - (5a^2)(1)$

$\qquad\qquad\qquad\qquad = 5a^5 - 10a^4 - 5a^2$.

13. $(-2xy)(6x^2 y - 3xy^2 - 4y^3)$

$\qquad = (-2xy)(6x^2 y) - (-2xy)(3xy^2) - (-2xy)(4y^3)$

$\qquad = -12x^3 y^2 + 6x^2 y^3 + 8xy^4$.

Once you feel comfortable with this process, you may want to perform most of the work mentally and simply write down the final result. See if you understand the following examples.

14. $3x(2x + 3) = 6x^2 + 9x.$

15. $-4x(2x^2 - 3x - 1) = -8x^3 + 12x^2 + 4x.$

16. $ab(3a^2b - 2ab^2 - b^3) = 3a^3b^2 - 2a^2b^3 - ab^4.$

Problem Set 6.2

For Problems 1–30, multiply using the properties of exponents to help with the manipulation.

1. $(5x)(9x)$

2. $(7x)(8x)$

3. $(3x^2)(7x)$

4. $(9x)(4x^3)$

5. $(-3xy)(2xy)$

6. $(6xy)(-3xy)$

7. $(-2x^2y)(-7x)$

8. $(-5xy^2)(-4y)$

9. $(4a^2b^2)(-12ab)$

10. $(-3a^3b)(13ab^2)$

11. $(-xy)(-5x^3)$

12. $(-7y^2)(-x^2y)$

13. $(8ab^2c)(13a^2c)$

14. $(9abc^3)(14bc^2)$

15. $(5x^2)(2x)(3x^3)$

16. $(4x)(2x^2)(6x^4)$

17. $(4xy)(-2x)(7y^2)$

18. $(5y^2)(-3xy)(5x^2)$

19. $(-2ab)(-ab)(-3b)$

20. $(-7ab)(-4a)(-ab)$

21. $(6cd)(-3c^2d)(-4d)$

22. $(2c^3d)(-6d^3)(-5cd)$

23. $\left(\dfrac{2}{3}xy\right)\left(\dfrac{3}{5}x^2y^4\right)$

24. $\left(-\dfrac{5}{6}x\right)\left(\dfrac{8}{3}x^2y\right)$

25. $\left(-\dfrac{7}{12}a^2b\right)\left(\dfrac{8}{21}b^4\right)$

26. $\left(-\dfrac{9}{5}a^3b^4\right)\left(-\dfrac{15}{6}ab^2\right)$

27. $(0.4x^5)(0.7x^3)$

28. $(-1.2x^4)(0.3x^2)$

29. $(-4ab)(1.6a^3b)$

30. $(-6a^2b)(-1.4a^2b^4)$

For Problems 31–46, raise each monomial to the indicated power. Use the properties of exponents to help with the manipulation.

31. $(2x^4)^2$

32. $(3x^3)^2$

33. $(-3a^2b^3)^2$

34. $(-8a^4b^5)^2$

35. $(3x^2)^3$

36. $(2x^4)^3$

37. $(-4x^4)^3$

38. $(-3x^3)^3$

39. $(9x^4y^5)^2$

40. $(8x^6y^4)^2$

41. $(2x^2y)^4$

42. $(2x^2y^3)^5$

43. $(-3a^3b^2)^4$

44. $(-2a^4b^2)^4$

45. $(-x^2y)^6$

46. $(-x^2y^3)^7$

For Problems 47–60, multiply by using the distributive property.

47. $5x(3x + 2)$

48. $7x(2x + 5)$

49. $3x^2(6x - 2)$

50. $4x^2(7x - 2)$

51. $-4x(7x^2 - 4)$

52. $-6x(9x^2 - 5)$

53. $2x(x^2 - 4x + 6)$

54. $3x(2x^2 - x + 5)$

55. $-6a(3a^2 - 5a - 7)$

56. $-8a(4a^2 - 9a - 6)$

57. $7xy(4x^2 - x + 5)$

58. $5x^2y(3x^2 + 7x - 9)$

59. $-xy(9x^2 - 2x - 6)$

60. $xy^2(6x^2 - x - 1)$

For Problems 61–70, remove parentheses by multiplying and then simplify by combining similar terms. For example,

$$3(x - y) + 2(x - 3y) = 3x - 3y + 2x - 6y$$
$$= 5x - 9y.$$

61. $5(x + 2y) + 4(2x + 3y)$

62. $3(2x + 5y) + 2(4x + y)$

63. $4(x - 3y) - 3(2x - y)$

64. $2(5x - 3y) - 5(x + 4y)$

65. $2x(x^2 - 3x - 4) + x(2x^2 + 3x - 6)$

66. $3x(2x^2 - x + 5) - 2x(x^2 + 4x + 7)$

67. $3[2x - (x - 2)] - 4(x - 2)$

68. $2[3x - (2x + 1)] - 2(3x - 4)$

69. $-4(3x + 2) - 5[2x - (3x + 4)]$

70. $-5(2x - 1) - 3[x - (4x - 3)]$

For Problems 71–80, perform the indicated operations and simplify.

71. $(3x)^2(2x^3)$

72. $(-2x)^3(4x^5)$

73. $(-3x)^3(-4x)^2$

74. $(3xy)^2(2x^2y)^4$

75. $(5x^2y)^2(xy^2)^3$

76. $(-x^2y)^3(6xy)^2$

77. $(-a^2bc^3)^3(a^3b)^2$

78. $(ab^2c^3)^4(-a^2b)^3$

79. $(-2x^2y^2)^4(-xy^3)^3$

80. $(-3xy)^3(-x^2y^3)^4$

Miscellaneous Problems

For Problems 81–90, find each of the indicated products. Assume that the variables in the exponents represent positive integers. For example,

$$(x^{2n})(x^{4n}) = x^{2n+4n} = x^{6n}.$$

81. $(x^n)(x^{3n})$

82. $(x^{2n})(x^{5n})$

83. $(x^{2n-1})(x^{3n+2})$

84. $(x^{5n+2})(x^{n-1})$

85. $(x^3)(x^{4n-5})$

86. $(x^{6n-1})(x^4)$

87. $(2x^n)(3x^{2n})$

88. $(4x^{3n})(-5x^{7n})$

89. $(-6x^{2n+4})(5x^{3n-4})$

90. $(-3x^{5n-2})(-4x^{2n+2})$

6.3
Multiplying Polynomials

In general, to go from multiplying a monomial and a polynomial to multiplying two polynomials requires the use of the distributive property. Consider the following examples.

EXAMPLE 1   Find the product of $(x + 3)$ and $(y + 4)$.

Solution
$$(x + 3)(y + 4) = x(y + 4) + 3(y + 4)$$
$$= x(y) + x(4) + 3(y) + 3(4)$$
$$= xy + 4x + 3y + 12$$ ■

Notice that each term of the first polynomial is multiplied times each term of the second polynomial.

EXAMPLE 2   Find the product of $(x - 2)$ and $(y + z + 5)$.

Solution
$$(x - 2)(y + z + 5) = x(y) + x(z) + x(5) - 2(y) - 2(z) - 2(5)$$
$$= xy + xz + 5x - 2y - 2z - 10$$ ■

Frequently, multiplying polynomials will produce similar terms that can be combined to simplify the resulting polynomial.

EXAMPLE 3   Multiply $(x + 3)(x + 2)$.

Solution
$$(x + 3)(x + 2) = x(x + 2) + 3(x + 2)$$
$$= x^2 + 2x + 3x + 6$$
$$= x^2 + 5x + 6$$ ■

EXAMPLE 4   Multiply $(x - 4)(x + 9)$.

Solution
$$(x - 4)(x + 9) = x(x + 9) - 4(x + 9)$$
$$= x^2 + 9x - 4x - 36$$
$$= x^2 + 5x - 36$$ ■

EXAMPLE 5   Multiply $(x + 4)(x^2 + 3x + 2)$.

Solution
$$(x + 4)(x^2 + 3x + 2) = x(x^2 + 3x + 2) + 4(x^2 + 3x + 2)$$
$$= x^3 + 3x^2 + 2x + 4x^2 + 12x + 8$$
$$= x^3 + 7x^2 + 14x + 8$$ ■

EXAMPLE 6   Multiply $(2x - y)(3x^2 - 2xy + 4y^2)$.

Solution

$$(2x - y)(3x^2 - 2xy + 4y^2) = 2x(3x^2 - 2xy + 4y^2) - y(3x^2 - 2xy + 4y^2)$$
$$= 6x^3 - 4x^2y + 8xy^2 - 3x^2y + 2xy^2 - 4y^3$$
$$= 6x^3 - 7x^2y + 10xy^2 - 4y^3$$ ∎

Perhaps the most frequently used type of multiplication problem is the product of two binomials. It will be a big help later if you can become proficient at multiplying binomials without showing all of the intermediate steps. This is quite easy to do by developing a three-step shortcut pattern demonstrated by the following examples.

EXAMPLE 7   Multiply $(x + 5)(x + 7)$.

Solution

$$(x + 5)(x + 7) = x^2 + 12x + 35.$$

Step 1. Multiply $x \cdot x$.
Step 2. Multiply $5 \cdot x$ and $7 \cdot x$ and combine.
Step 3. Multiply $5 \cdot 7$. ∎

EXAMPLE 8   Multiply $(x - 8)(x + 3)$.

Solution

$$(x - 8)(x + 3) = x^2 - 5x - 24.$$ ∎

EXAMPLE 9   Multiply $(3x + 2)(2x - 5)$.

Solution

$$(3x + 2)(2x - 5) = 6x^2 - 11x - 10.$$ ∎

Now see if *you* can use the pattern to find the following products.

$(x + 3)(x + 7) = $ _____

$(3x + 1)(2x + 5) = $ _____

$(x - 2)(x - 3) = $ _____

$(4x + 5)(x - 2) = $ _____

Your answers should be $x^2 + 10x + 21$, $6x^2 + 17x + 5$, $x^2 - 5x + 6$, and $4x^2 - 3x - 10$.

Keep in mind that the shortcut pattern discussed above applies only to finding the product of two binomials. For other situations, such as finding the product of a binomial and a trinomial, we would suggest showing the intermediate steps as follows.

$$\begin{aligned}(x + 3)(x^2 + 6x - 7) &= x(x^2) + x(6x) - x(7) + 3(x^2) + 3(6x) - 3(7) \\ &= x^3 + 6x^2 - 7x + 3x^2 + 18x - 21 \\ &= x^3 + 9x^2 + 11x - 21\end{aligned}$$

Perhaps we could omit the first step and shorten the form as follows.

$$\begin{aligned}(x - 4)(x^2 - 5x - 6) &= x^3 - 5x^2 - 6x - 4x^2 + 20x + 24 \\ &= x^3 - 9x^2 + 14x + 24\end{aligned}$$

Remember that you are multiplying each term of the first polynomial times each term of the second polynomial and combining similar terms.

Exponents are also used to indicate repeated multiplication of polynomials. For example, $(x + 4)(x + 4)$ can be written as $(x + 4)^2$. Thus, to square a binomial we simply write it as the product of two equal binomials and apply the shortcut pattern.

$$(x + 4)^2 = (x + 4)(x + 4) = x^2 + 8x + 16;$$
$$(x - 5)^2 = (x - 5)(x - 5) = x^2 - 10x + 25;$$
$$(2x + 3)^2 = (2x + 3)(2x + 3) = 4x^2 + 12x + 9$$

When squaring binomials, be careful not to forget the middle term. That is to say, $(x + 3)^2 \neq x^2 + 3^2$; instead, $(x + 3)^2 = (x + 3)(x + 3) = x^2 + 6x + 9$.

The following example suggests a format to use when cubing a binomial.

$$\begin{aligned}(x + 4)^3 &= (x + 4)(x + 4)(x + 4) \\ &= (x + 4)(x^2 + 8x + 16) \\ &= x(x^2 + 8x + 16) + 4(x^2 + 8x + 16) \\ &= x^3 + 8x^2 + 16x + 4x^2 + 32x + 64 \\ &= x^3 + 12x^2 + 48x + 64\end{aligned}$$

Problem Set 6.3

For Problems 1–10, find the indicated products by applying the distributive property. For example,

$$(x + 1)(y + 5) = x(y) + x(5) + 1(y) + 1(5)$$
$$= xy + 5x + y + 5.$$

1. $(x + 2)(y + 3)$

3. $(x - 4)(y + 1)$

5. $(x - 5)(y - 6)$

7. $(x + 2)(y + z + 1)$

8. $(x + 4)(y - z + 4)$

9. $(2x + 3)(3y + 1)$

10. $(3x - 2)(2y - 5)$

2. $(x + 3)(y + 6)$

4. $(x - 5)(y + 7)$

6. $(x - 7)(y - 9)$

For Problems 11–36, find the indicated products by applying the distributive property and combining similar terms. Use the following format to show your work.

$$(x + 3)(x + 8) = x(x) + x(8) + 3(x) + 3(8)$$
$$= x^2 + 8x + 3x + 24$$
$$= x^2 + 11x + 24$$

11. $(x + 3)(x + 7)$

13. $(x + 8)(x - 3)$

15. $(x - 7)(x + 1)$

17. $(n - 4)(n - 6)$

19. $(3n + 1)(n + 6)$

21. $(5x - 2)(3x + 7)$

23. $(x + 3)(x^2 + 4x + 9)$

24. $(x + 2)(x^2 + 6x + 2)$

25. $(x + 4)(x^2 - x - 6)$

26. $(x + 5)(x^2 - 2x - 7)$

27. $(x - 5)(2x^2 + 3x - 7)$

28. $(x - 4)(3x^2 + 4x - 6)$

29. $(2a - 1)(4a^2 - 5a + 9)$

30. $(3a - 2)(2a^2 - 3a - 5)$

31. $(3a + 5)(a^2 - a - 1)$

32. $(5a + 2)(a^2 + a - 3)$

33. $(x^2 + 2x + 3)(x^2 + 5x + 4)$

34. $(x^2 - 3x + 4)(x^2 + 5x - 2)$

35. $(x^2 - 6x - 7)(x^2 + 3x - 9)$

36. $(x^2 - 5x - 4)(x^2 + 7x - 8)$

12. $(x + 4)(x + 2)$

14. $(x + 9)(x - 6)$

16. $(x - 10)(x + 8)$

18. $(n - 3)(n - 7)$

20. $(4n + 3)(n + 6)$

22. $(3x - 4)(7x + 1)$

For Problems 37–80, find the indicated products by using the shortcut pattern for multiplying binomials.

37. $(x + 2)(x + 9)$

38. $(x + 3)(x + 8)$

39. $(x + 6)(x - 2)$

40. $(x + 8)(x - 6)$

41. $(x + 3)(x - 11)$

42. $(x + 4)(x - 10)$

43. $(n - 4)(n - 3)$

44. $(n - 5)(n - 9)$

45. $(n + 6)(n + 12)$

46. $(n + 8)(n + 13)$

47. $(y + 3)(y - 7)$

48. $(y + 2)(y - 12)$

49. $(y - 7)(y - 12)$

50. $(y - 4)(y - 13)$

51. $(x - 5)(x + 7)$

52. $(x - 1)(x + 9)$

53. $(x - 14)(x + 8)$

54. $(x - 15)(x + 6)$

55. $(a + 10)(a - 9)$

56. $(a + 7)(a - 6)$

57. $(2a + 1)(a + 6)$

58. $(3a + 2)(a + 4)$

59. $(5x - 2)(x + 7)$

60. $(2x - 3)(x + 8)$

61. $(3x - 7)(2x + 1)$

62. $(5x - 6)(4x + 3)$

63. $(4a + 3)(3a - 4)$

64. $(5a + 4)(4a - 5)$

65. $(6n - 5)(2n - 3)$

66. $(4n - 3)(6n - 7)$

67. $(7x - 4)(2x + 3)$

68. $(8x - 5)(3x + 7)$

69. $(5 - x)(9 - 2x)$

70. $(4 - 3x)(2 + x)$

71. $(-2x + 3)(4x - 5)$

72. $(-3x + 1)(9x - 2)$

73. $(-3x - 1)(3x - 4)$

74. $(-2x - 5)(4x + 1)$

75. $(8n + 3)(9n - 4)$

76. $(6n + 5)(9n - 7)$

77. $(3 - 2x)(9 - x)$

78. $(5 - 4x)(4 - 5x)$

79. $(-4x + 3)(-5x - 2)$

80. $(-2x + 7)(-7x - 3)$

For Problems 81–96, find the indicated products. Remember that $(a + b)^2$ means $(a + b)(a + b)$ and $(a + b)^3$ means $(a + b)(a + b)(a + b)$.

81. $(x + 7)^2$

82. $(x + 3)^2$

83. $(x - 5)^2$

84. $(x - 6)^2$

85. $(3x + 2)^2$

86. $(5x + 1)^2$

87. $(4x - 3)^2$

88. $(3x - 7)^2$

89. $(-1 - x)^2$

90. $(-2 - 3x)^2$

91. $(x + 2)^3$

92. $(x + 5)^3$

93. $(x - 3)^3$

94. $(x - 1)^3$

95. $(2x + 3)^3$

96. $(3x + 2)^3$

Thoughts into Words

97. Illustrate how the distributive property is used when multiplying polynomials.

98. How would you explain to someone why the product of x^3 and x^4 is x^7 and not x^{12}?

Miscellaneous Problems

For Problems 99–108, find the indicated products. Assume all variables that appear as exponents represent positive integers.

99. $(x^n + 5)(x^n + 7)$ 100. $(x^n - 6)(x^n - 4)$ 101. $(x^a + 8)(x^a - 4)$

102. $(x^a + 4)(x^a - 9)$ 103. $(x^{2n} + 3)(x^{2n} - 4)$ 104. $(x^{3n} - 5)(x^{3n} + 2)$

105. $(2x^a + 1)(3x^a - 4)$ 106. $(4x^a - 3)(3x^a + 5)$ 107. $(4x^n - 3)(5x^n - 6)$

108. $(x^n + 1)(6x^n - 5)$

6.4
Special Product Patterns and Dividing by Monomials

When multiplying binomials, some special patterns occur that are helpful to recognize. These patterns can be used to find products, and some of them are also used later when factoring polynomials. Let's begin with a list of the patterns and then illustrate each of them in more detail.

$$(a + b)^2 = a^2 + 2ab + b^2$$
$$(a - b)^2 = a^2 - 2ab + b^2$$
$$(a + b)(a - b) = a^2 - b^2$$
$$(a + b)^3 = a^3 + 3a^2b + 3ab^2 + b^3$$
$$(a - b)^3 = a^3 - 3a^2b + 3ab^2 - b^3$$

PATTERN

$$\begin{aligned}(a + b)^2 &= (a + b)(a + b) \\ &= a(a + b) + b(a + b) \\ &= a^2 + ab + ab + b^2 \\ &= a^2 + 2ab + b^2\end{aligned}$$

Examples

$$(x + 4)^2 = \boxed{x^2 + 2(x)(4) + 4^2} = x^2 + 8x + 16$$

$$(2x + 3y)^2 = \boxed{(2x)^2 + 2(2x)(3y) + (3y)^2} = 4x^2 + 12xy + 9y^2$$

REMARK The steps enclosed in the dashed boxes might be performed mentally.

PATTERN

$$\begin{aligned}(a - b)^2 &= (a - b)(a - b) \\ &= a(a - b) - b(a - b) \\ &= a^2 - ab - ab + b^2 \\ &= a^2 - 2ab + b^2\end{aligned}$$

Examples

$$(x - 8)^2 = \boxed{x^2 - 2(x)(8) + 8^2}$$
$$= x^2 - 16x + 64$$
$$(4a - 9b)^2 = \boxed{(4a)^2 - 2(4a)(9b) + (9b)^2}$$
$$= 16a^2 - 72ab + 81b^2$$

PATTERN

$$(a + b)(a - b) = a(a - b) + b(a - b)$$
$$= a^2 - ab + ab - b^2$$
$$= a^2 - b^2$$

Examples

$$(x + 7)(x - 7) = \boxed{x^2 - 7^2} = x^2 - 49$$
$$(3x + 2y)(3x - 2y) = \boxed{(3x)^2 - (2y)^2} = 9x^2 - 4y^2$$

PATTERN

$$(a + b)^3 = (a + b)(a + b)(a + b)$$
$$= (a + b)(a^2 + 2ab + b^2)$$
$$= a(a^2 + 2ab + b^2) + b(a^2 + 2ab + b^2)$$
$$= a^3 + 2a^2b + ab^2 + a^2b + 2ab^2 + b^3$$
$$= a^3 + 3a^2b + 3ab^2 + b^3$$

Examples

$$(x + 4)^3 = x^3 + 3x^2(4) + 3x(4)^2 + 4^3$$
$$= x^3 + 12x^2 + 48x + 64$$

$$(3x + 2)^3 = (3x)^3 + 3(3x)^2(2) + 3(3x)(2)^2 + 2^3$$
$$= 27x^3 + 54x^2 + 36x + 8$$

PATTERN

$$(a - b)^3 = (a - b)(a - b)(a - b)$$
$$= (a - b)(a^2 - 2ab + b^2)$$
$$= a(a^2 - 2ab + b^2) - b(a^2 - 2ab + b^2)$$
$$= a^3 - 2a^2b + ab^2 - a^2b + 2ab^2 - b^3$$
$$= a^3 - 3a^2b + 3ab^2 - b^3$$

Examples

$$(x - 5)^3 = x^3 - 3x^2(5) + 3x(5)^2 - 5^3$$
$$= x^3 - 15x^2 + 75x - 125$$

$$(3x - 2y)^3 = (3x)^3 - 3(3x)^2(2y) + 3(3x)(2y)^2 - (2y)^3$$
$$= 27x^3 - 54x^2y + 36xy^2 - 8y^3$$

Finally, we need to realize that if patterns are forgotten or do not apply, then we can revert back to applying the distributive property.

$$(2x - 1)(x^2 - 4x + 6) = 2x(x^2 - 4x + 6) - 1(x^2 - 4x + 6)$$
$$= 2x^3 - 8x^2 + 12x - x^2 + 4x - 6$$
$$= 2x^3 - 9x^2 + 16x - 6$$

Dividing by Monomials

To develop an effective process for dividing by a monomial we must rely on yet another property of exponents. This property is also a direct consequence of the definition of exponent and is illustrated by the following examples.

$$\frac{x^5}{x^2} = \frac{\cancel{x} \cdot \cancel{x} \cdot x \cdot x \cdot x}{\cancel{x} \cdot \cancel{x}} = x^3,$$

$$\frac{a^4}{a^3} = \frac{\cancel{a} \cdot \cancel{a} \cdot \cancel{a} \cdot a}{\cancel{a} \cdot \cancel{a} \cdot \cancel{a}} = a,$$

$$\frac{y^7}{y^3} = \frac{\cancel{y} \cdot \cancel{y} \cdot \cancel{y} \cdot y \cdot y \cdot y \cdot y}{\cancel{y} \cdot \cancel{y} \cdot \cancel{y}} = y^4,$$

$$\frac{x^4}{x^4} = \frac{\cancel{x} \cdot \cancel{x} \cdot \cancel{x} \cdot \cancel{x}}{\cancel{x} \cdot \cancel{x} \cdot \cancel{x} \cdot \cancel{x}} = 1,$$

$$\frac{y^3}{y^3} = \frac{\cancel{y} \cdot \cancel{y} \cdot \cancel{y}}{\cancel{y} \cdot \cancel{y} \cdot \cancel{y}} = 1$$

PROPERTY 6.4

If b is any nonzero real number and n and m are positive integers, then

1. $\dfrac{b^n}{b^m} = b^{n-m}$ when $n > m$;

2. $\dfrac{b^n}{b^m} = 1$ when $n = m$.

(The situation when $n < m$ will be discussed in a later section.)

Applying Property 6.4 to the previous examples yields

$$\frac{x^5}{x^2} = x^{5-2} = x^3,$$

$$\frac{a^4}{a^3} = a^{4-3} = a^1, \qquad \text{usually written as } a$$

$$\frac{y^7}{y^3} = y^{7-3} = y^4,$$

$$\frac{x^4}{x^4} = 1,$$

$$\frac{y^3}{y^3} = 1.$$

Property 6.4 along with our knowledge of dividing integers provides the basis for dividing a monomial by another monomial. Consider the following examples.

$$\frac{16x^5}{2x^3} = 8x^{5-3} = 8x^2, \qquad \frac{-81a^{12}}{-9a^4} = 9a^{12-4} = 9a^8,$$

$$\frac{-35x^9}{5x^4} = -7x^{9-4} = -7x^5, \qquad \frac{45x^4}{9x^4} = 5 \qquad \frac{x^4}{x^4} = 1,$$

$$\frac{56y^6}{-7y^2} = -8y^{6-2} = -8y^4, \qquad \frac{54x^3y^7}{-6xy^5} = -9x^{3-1}y^{7-5} = -9x^2y^2$$

Recall that $\frac{a}{c} + \frac{b}{c} = \frac{a+b}{c}$. This same property $\left(\text{except viewed as }\right.$ $\left.\frac{a+b}{c} = \frac{a}{c} + \frac{b}{c}\right)$ serves as the basis for dividing a polynomial by a monomial. Consider the following examples.

$$\frac{25x^3 + 10x^2}{5x} = \frac{25x^3}{5x} + \frac{10x^2}{5x} = 5x^2 + 2x;$$

$$\frac{-35x^8 - 28x^6}{7x^3} = \frac{-35x^8}{7x^3} - \frac{28x^6}{7x^3} = -5x^5 - 4x^3. \qquad \frac{a-b}{c} = \frac{a}{c} - \frac{b}{c}$$

To divide a polynomial by a monomial we simply divide each term of the polynomial by the monomial. Here are some additional examples.

$$\frac{12x^3y^2 - 14x^2y^5}{-2xy} = \frac{12x^3y^2}{-2xy} - \frac{14x^2y^5}{-2xy} = -6x^2y + 7xy^4;$$

$$\frac{48ab^5 + 64a^2b}{-16ab} = \frac{48ab^5}{-16ab} + \frac{64a^2b}{-16ab} = -3b^4 - 4a;$$

$$\frac{33x^6 - 24x^5 - 18x^4}{3x} = \frac{33x^6}{3x} - \frac{24x^5}{3x} - \frac{18x^4}{3x}$$
$$= 11x^5 - 8x^4 - 6x^3.$$

As with many skills, once you feel comfortable with the process, you may want to mentally perform some of the steps. Your work could take on the following format.

$$\frac{24x^4y^5 - 56x^3y^9}{8x^2y^3} = 3x^2y^2 - 7xy^6;$$

$$\frac{13a^2b - 12ab^2}{-ab} = -13a + 12b$$

Problem Set 6.4

For Problems 1–36, find the indicated products using the special patterns discussed in this section whenever they apply.

1. $(x + 9)^2$ **2.** $(x + 1)^2$ **3.** $(x - 7)^2$ **4.** $(x - 10)^2$

5. $(5x + 2)^2$ **6.** $(4x + 7)^2$ **7.** $(2x - 7)^2$ **8.** $(3x - 1)^2$

9. $(x - 2)(x^2 - 4x - 7)$ **10.** $(x + 3)(x^2 + 8x - 1)$ **11.** $(x + 9)(x - 9)$

12. $(t - 3)(t + 3)$ **13.** $(t + 11)(t - 11)$ **14.** $(5t + 2)(5t - 2)$

15. $(6t - 1)(6t + 1)$ **16.** $(x + 1)^3$ **17.** $(x + 6)^3$

18. $(n - 2)^3$ **19.** $(n - 1)^3$ **20.** $(2n + 5)^3$

21. $(3n + 4)^3$ **22.** $(2x - 1)(2x^2 + x - 1)$ **23.** $(3x + 2)(2x^2 - x - 4)$

24. $(-1 + 2x)^2$ **25.** $(-2 + 7x)^2$ **26.** $(4 - 5x)^2$

27. $(3 - 7x)^2$ **28.** $(2 + 9n)(2 - 9n)$ **29.** $(1 - 7n)(1 + 7n)$

30. $(x^2 - 2x - 6)(3x + 5)$ **31.** $(x^2 + 6x - 7)(2x - 3)$ **32.** $(4y - 3)^3$

33. $(3y - 5)^3$ **34.** $(2 - n)^3$ **35.** $(1 - n)^3$ **36.** $(x - 3y)^3$

For Problems 37–56, divide the monomials.

37. $\dfrac{x^{10}}{x^2}$ **38.** $\dfrac{x^{12}}{x^5}$ **39.** $\dfrac{4x^3}{2x}$ **40.** $\dfrac{8x^5}{4x^3}$

41. $\dfrac{-16n^6}{2n^2}$ **42.** $\dfrac{-54n^8}{6n^4}$ **43.** $\dfrac{72x^3}{-9x^3}$ **44.** $\dfrac{84x^5}{-7x^5}$

45. $\dfrac{65x^2y^3}{5xy}$ **46.** $\dfrac{70x^3y^4}{5x^2y}$ **47.** $\dfrac{-91a^4b^6}{-13a^3b^4}$ **48.** $\dfrac{-72a^5b^4}{-12ab^2}$

49. $\dfrac{18x^2y^6}{xy^2}$ **50.** $\dfrac{24x^3y^4}{x^2y^2}$ **51.** $\dfrac{32x^6y^2}{-x}$ **52.** $\dfrac{54x^5y^3}{-y^2}$

53. $\dfrac{-96x^5y^7}{12y^3}$ **54.** $\dfrac{-84x^4y^9}{14x^4}$ **55.** $\dfrac{-ab}{ab}$ **56.** $\dfrac{6ab}{-ab}$

For Problems 57–82, perform the indicated divisions of polynomials by monomials.

57. $\dfrac{8x^4 + 12x^5}{2x^2}$ **58.** $\dfrac{12x^3 + 16x^6}{4x}$ **59.** $\dfrac{9x^6 - 24x^4}{3x^3}$

60. $\dfrac{35x^8 - 45x^6}{5x^4}$ **61.** $\dfrac{-28n^5 + 36n^2}{4n^2}$ **62.** $\dfrac{-42n^6 + 54n^4}{6n^4}$

63. $\dfrac{35x^6 - 56x^5 - 84x^3}{7x^2}$ **64.** $\dfrac{27x^7 - 36x^5 - 45x^3}{3x}$

65. $\dfrac{-24n^8 + 48n^5 - 78n^3}{-6n^3}$ **66.** $\dfrac{-56n^9 + 84n^6 - 91n^2}{-7n^2}$

67. $\dfrac{-60a^7 - 96a^3}{-12a}$ **68.** $\dfrac{-65a^8 - 78a^4}{-13a^2}$ **69.** $\dfrac{27x^2y^4 - 45xy^4}{-9xy^3}$

70. $\dfrac{-40x^4y^7 + 64x^5y^8}{-8x^3y^4}$ **71.** $\dfrac{48a^2b^2 + 60a^3b^4}{-6ab}$ **72.** $\dfrac{45a^3b^4 - 63a^2b^6}{-9ab^2}$

73. $\dfrac{12a^2b^2c^2 - 52a^2b^3c^5}{-4a^2bc}$ **74.** $\dfrac{48a^3b^2c + 72a^2b^4c^5}{-12ab^2c}$ **75.** $\dfrac{9x^2y^3 - 12x^3y^4}{-xy}$

76. $\dfrac{-15x^3y + 27x^2y^4}{xy}$ **77.** $\dfrac{-42x^6 - 70x^4 + 98x^2}{14x^2}$

78. $\dfrac{-48x^8 - 80x^6 + 96x^4}{16x^4}$ **79.** $\dfrac{15a^3b - 35a^2b - 65ab^2}{-5ab}$

80. $\dfrac{-24a^4b^2 + 36a^3b - 48a^2b}{-6ab}$ **81.** $\dfrac{-xy + 5x^2y^3 - 7x^2y^6}{xy}$

82. $\dfrac{-9x^2y^3 - xy + 14xy^4}{-xy}$

Miscellaneous Problems

83. Some of the patterns discussed in this section can be used to do mental arithmetic. For example, $(18)(22)$ can be viewed as $(20 - 2)(20 + 2)$, which equals $400 - 4$ or 396. Likewise, 41^2 can be viewed as $(40 + 1)^2$, which equals $40^2 + 2(40)(1) + 1^2$ or 1681. Do the following manipulations mentally by using the appropriate pattern.

(a) $(29)(31)$ (b) $(52)(48)$ (c) $(63)(57)$ (d) 61^2

(e) 79^2 (f) 91^2 (g) 52^2 (h) 48^2

6.5
Dividing by Binomials

Perhaps the easiest way to explain the process of dividing a polynomial by a binomial is to work a few examples and describe the step-by-step procedure as we go along.

EXAMPLE 1   Divide $x^2 + 5x + 6$ by $x + 2$.

Solution

Step 1. Use the conventional long division format from arithmetic and arrange both the dividend and the divisor in descending powers of the variable.

$x + 2\overline{)x^2 + 5x + 6}$

Step 2. Find the first term of the quotient by dividing the first term of the dividend by the first term of the divisor.

$$x + 2\overline{)\overset{x}{x^2 + 5x + 6}} \qquad \frac{x^2}{x} = x$$

Step 3. Multiply the entire divisor by the term of the quotient found in step 2 and place this product so as to be subtracted from the dividend.

$$\begin{array}{r} x \\ x+2\overline{)x^2+5x+6} \\ \underline{x^2+2x} \end{array}$$

$(x(x+2) = x^2 + 2x)$

Step 4. Subtract.

Remember to add the opposite! $\longrightarrow$

$$\begin{array}{r} x \\ x+2\overline{)x^2+5x+6} \\ \underline{x^2+2x} \\ 3x+6 \end{array}$$

Step 5. Repeat the process beginning with step 2; use the polynomial that resulted from the subtraction in step 4 as a new dividend.

$$\begin{array}{r} x+3 \\ x+2\overline{)x^2+5x+6} \\ \underline{x^2+2x} \\ 3x+6 \\ \underline{3x+6} \end{array}$$

$\dfrac{3x}{x} = 3$

$3(x+2) = 3x+6$

Thus, $(x^2 + 5x + 6) \div (x + 2) = x + 3$; this can be checked by multiplying $(x + 2)$ and $(x + 3)$.

$$(x + 2)(x + 3) = x^2 + 5x + 6 \qquad \blacksquare$$

A division problem such as $(x^2 + 5x + 6) \div (x + 2)$ can also be written as $\dfrac{x^2 + 5x + 6}{x + 2}$. Using this format the final result for Example 1 could be expressed as $\dfrac{x^2 + 5x + 6}{x + 2} = x + 3$. (Technically, the restriction $x \neq -2$ should be made to avoid division by zero.)

In general, to check a division problem we can multiply the divisor times the quotient and add the remainder. This can be expressed as

dividend = (divisor)(quotient) + remainder.

Sometimes the remainder is expressed as a fractional part of the divisor. The relationship then becomes

$$\frac{\text{dividend}}{\text{divisor}} = \text{quotient} + \frac{\text{remainder}}{\text{divisor}}.$$

EXAMPLE 2   Divide $2x^2 - 3x - 20$ by $x - 4$.

Solution   ***Step 1.*** $x - 4\overline{)2x^2 - 3x - 20}$

Step 2.
$$\frac{2x}{x - 4 \overline{) 2x^2 - 3x - 20}}$$

$$\frac{2x^2}{x} = 2x$$

Step 3.
$$\frac{2x}{x - 4 \overline{) 2x^2 - 3x - 20}}$$
$$\underline{2x^2 - 8x}$$

$$2x(x - 4) = 2x^2 - 8x$$

Step 4.
$$\frac{2x}{x - 4 \overline{) 2x^2 - 3x - 20}}$$
$$\underline{2x^2 - 8x}$$
$$5x - 20$$

Step 5.
$$\frac{2x + 5}{x - 4 \overline{) 2x^2 - 3x - 20}}$$
$$\underline{2x^2 - 8x}$$

$$\frac{5x}{x} = 5$$

$$5x - 20$$
$$\underline{5x - 20}$$

$$5(x - 4) = 5x - 20$$

CHECK $(x - 4)(2x + 5) = 2x^2 - 3x - 20$

Therefore, $\dfrac{2x^2 - 3x - 20}{x - 4} = 2x + 5.$ ∎

Now let's continue to think in terms of the step-by-step division process but organize our work in the typical long division format.

EXAMPLE 3   Divide $12x^2 + x - 6$ by $3x - 2$.

Solution
$$\frac{4x + 3}{3x - 2 \overline{) 12x^2 + x - 6}}$$
$$\underline{12x^2 - 8x}$$
$$9x - 6$$
$$\underline{9x - 6}$$

CHECK $(3x - 2)(4x + 3) = 12x^2 + x - 6$

Therefore, $\dfrac{12x^2 + x - 6}{3x - 2} = 4x + 3.$ ∎

Each of the next three examples illustrates another point regarding the division process. Study them carefully; then you should be ready to work the exercises in the next problem set.

EXAMPLE 4   Perform the division $(7x^2 - 3x - 4) \div (x - 2)$.

Solution

$$
\begin{array}{r}
7x\ \ +11 \\
x-2\overline{)7x^2-\ 3x-\ 4} \\
\underline{7x^2-14x} \\
11x-\ 4 \\
\underline{11x-22} \\
18
\end{array}
$$
   $\longleftarrow$ a remainder of 18

CHECK Just as in arithmetic, we check by *adding* the remainder to the product of the divisor and quotient.

$$(x-2)(7x+11)+18 \stackrel{?}{=} 7x^2-3x-4$$

$$7x^2-3x-22+18 \stackrel{?}{=} 7x^2-3x-4$$

$$7x^2-3x-4 = 7x^2-3x-4$$

Therefore, $\dfrac{7x^2-3x-4}{x-2} = 7x+11+\dfrac{18}{x-2}.$ ■

EXAMPLE 5   Perform the division $\dfrac{x^3-8}{x-2}.$

Solution

$$
\begin{array}{r}
x^2+2x\ \ +4 \\
x-2\overline{)x^3+0x^2+0x-8} \\
\underline{x^3-2x^2} \\
2x^2+0x-8 \\
\underline{2x^2-4x} \\
4x-8 \\
\underline{4x-8}
\end{array}
$$
   $\longleftarrow$ Notice the inserting of x squared and x terms with zero coefficients.

CHECK $(x-2)(x^2+2x+4) \stackrel{?}{=} x^3-8$

$$x^3+2x^2+4x-2x^2-4x-8 \stackrel{?}{=} x^3-8$$

$$x^3-8 = x^3-8$$

Therefore, $\dfrac{x^3-8}{x-2} = x^2+2x+4.$ ■

EXAMPLE 6   Perform the division $\dfrac{x^3+5x^2-3x-4}{x^2+2x}.$

Solution

$$
\begin{array}{r}
x\ \ +3 \\
x^2+2x\overline{)x^3+5x^2-3x-4} \\
\underline{x^3+2x^2} \\
3x^2-3x-4 \\
\underline{3x^2+6x} \\
-9x-4
\end{array}
$$
   $\longleftarrow$ a remainder of $-9x-4$

The division process is stopped when the degree of the remainder is less than the degree of the divisor.

CHECK $(x^2 + 2x)(x + 3) + (-9x - 4) \overset{?}{=} x^3 + 5x^2 - 3x - 4$

$$x^3 + 3x^2 + 2x^2 + 6x - 9x - 4 \overset{?}{=} x^3 + 5x^2 - 3x - 4$$

$$x^3 + 5x^2 - 3x - 4 = x^3 + 5x^2 - 3x - 4.$$

Therefore, $\dfrac{x^3 + 5x^2 - 3x - 4}{x^2 + 2x} = x + 3 + \dfrac{-9x - 4}{x^2 + 2x}.$ ∎

REMARK If the divisor is of the form $x - k$, where the coefficient of the x-term is one, then the format of the division process described in this section can be simplified by a procedure called **synthetic division**. This procedure is outlined in Appendix A.

Problem Set 6.5

Perform the following divisions.

1. $(x^2 + 16x + 48) \div (x + 4)$

2. $(x^2 + 15x + 54) \div (x + 6)$

3. $(x^2 - 5x - 14) \div (x - 7)$

4. $(x^2 + 8x - 65) \div (x - 5)$

5. $(x^2 + 11x + 28) \div (x + 3)$

6. $(x^2 + 11x + 15) \div (x + 2)$

7. $(x^2 - 4x - 39) \div (x - 8)$

8. $(x^2 - 9x - 30) \div (x - 12)$

9. $(5n^2 - n - 4) \div (n - 1)$

10. $(7n^2 - 61n - 90) \div (n - 10)$

11. $(8y^2 + 53y - 19) \div (y + 7)$

12. $(6y^2 + 47y - 72) \div (y + 9)$

13. $(20x^2 - 31x - 7) \div (5x + 1)$

14. $(27x^2 + 21x - 20) \div (3x + 4)$

15. $(6x^2 + 25x + 8) \div (2x + 7)$

16. $(12x^2 + 28x + 27) \div (6x + 5)$

17. $(2x^3 - x^2 - 2x - 8) \div (x - 2)$

18. $(3x^3 - 7x^2 - 26x + 24) \div (x - 4)$

19. $(5n^3 + 11n^2 - 15n - 9) \div (n + 3)$

20. $(6n^3 + 29n^2 - 6n - 5) \div (n + 5)$

21. $(n^3 - 40n + 24) \div (n - 6)$

22. $(n^3 - 67n - 24) \div (n + 8)$

23. $(x^3 - 27) \div (x - 3)$

24. $(x^3 + 8) \div (x + 2)$

25. $\dfrac{27x^3 - 64}{3x - 4}$

26. $\dfrac{8x^3 + 27}{2x + 3}$

27. $\dfrac{1 + 3n^2 - 2n}{n + 2}$

28. $\dfrac{x + 5 + 12x^2}{3x - 2}$

29. $\dfrac{9t^2 + 3t + 4}{-1 + 3t}$

30. $\dfrac{4n^2 + 6n - 1}{4 + 2n}$

31. $\dfrac{6n^3 - 5n^2 - 7n + 4}{2n - 1}$ **32.** $\dfrac{21n^3 + 23n^2 - 9n - 10}{3n + 2}$

33. $\dfrac{4x^3 + 23x^2 - 30x + 32}{x + 7}$

34. $\dfrac{5x^3 - 12x^2 + 13x - 14}{x - 1}$

35. $(x^3 + 2x^2 - 3x - 1) \div (x^2 - 2x)$

36. $(x^3 - 6x^2 - 2x + 1) \div (x^2 + 3x)$

37. $(2x^3 - 4x^2 + x - 5) \div (x^2 + 4x)$

38. $(2x^3 - x^2 - 3x + 5) \div (x^2 + x)$

39. $(x^4 - 16) \div (x + 2)$ **40.** $(x^4 - 81) \div (x - 3)$

Thoughts into Words

41. Describe the process of multiplying two polynomials.

42. Describe how you would do the division problem $3x^3 + 2x - 4$ divided by $x - 4$.

Chapter 6 Summary

(6.1) Terms that contain variables with only whole numbers as exponents are called **monomials**. A **polynomial** is a monomial or a finite sum (or difference) of monomials. Polynomials of one term, two terms, and three terms are called **monomials**, **binomials**, and **trinomials**, respectively.

Addition and subtraction of polynomials are based on the distributive property and the idea of combining similar terms.

(6.2) and (6.3) The following properties of exponents serve as a basis for multiplying polynomials.

    1. $b^n \cdot b^m = b^{n+m}$.

    2. $(b^n)^m = b^{mn}$.

    3. $(ab)^n = a^n b^n$.

(6.4) The following special patterns can be helpful when multiplying binomials.

$$(a + b)^2 = a^2 + 2ab + b^2$$
$$(a - b)^2 = a^2 - 2ab + b^2$$
$$(a + b)(a - b) = a^2 - b^2$$
$$(a + b)^3 = a^3 + 3a^2b + 3ab^2 + b^3$$
$$(a - b)^3 = a^3 - 3a^2b + 3ab^2 - b^3$$

The following properties of exponents serve as a basis for dividing monomials.

1. $\dfrac{b^n}{b^m} = b^{n-m}$ when $n > m$.

2. $\dfrac{b^n}{b^m} = 1$ when $n = m$.

Dividing a polynomial by a monomial is based on the property $\dfrac{a+b}{c} = \dfrac{a}{c} + \dfrac{b}{c}$.

(6.5) To review the division of a polynomial by a binomial, return to Section 6.5 and study the examples carefully.

Chapter 6 Review Problem Set

Perform the following additions and subtractions.

1. $(5x^2 - 6x + 4) + (3x^2 - 7x - 2)$

2. $(7y^2 + 9y - 3) - (4y^2 - 2y + 6)$

3. $(2x^2 + 3x - 4) + (4x^2 - 3x - 6) - (3x^2 - 2x - 1)$

4. $(-3x^2 - 2x + 4) - (x^2 - 5x - 6) - (4x^2 + 3x - 8)$

Remove parentheses and combine similar terms.

5. $5(2x - 1) + 7(x + 3) - 2(3x + 4)$

6. $3(2x^2 - 4x - 5) - 5(3x^2 - 4x + 1)$

7. $6(y^2 - 7y - 3) - 4(y^2 + 3y - 9)$

8. $3(a - 1) - 2(3a - 4) - 5(2a + 7)$

9. $-(a + 4) + 5(-a - 2) - 7(3a - 1)$

10. $-2(3n - 1) - 4(2n + 6) + 5(3n + 4)$

11. $3(n^2 - 2n - 4) - 4(2n^2 - n - 3)$

12. $-5(-n^2 + n - 1) + 3(4n^2 - 3n - 7)$

Find the indicated products.

13. $(5x^2)(7x^4)$

14. $(-6x^3)(9x^5)$

15. $(-4xy^2)(-6x^2y^3)$

16. $(2a^3b^4)(-3ab^5)$

17. $(2a^2b^3)^3$

18. $(-3xy^2)^2$

19. $5x(7x + 3)$

20. $(-3x^2)(8x - 1)$

Find the indicated products. Be sure to simplify answers.

21. $(x + 9)(x + 8)$

22. $(3x + 7)(x + 1)$

23. $(x - 5)(x + 2)$

24. $(y - 4)(y - 9)$

25. $(2x - 1)(7x + 3)$

26. $(4a - 7)(5a + 8)$

27. $(3a - 5)^2$

28. $(x + 6)(2x^2 + 5x - 4)$

29. $(5n - 1)(6n + 5)$

30. $(3n + 4)(4n - 1)$

31. $(2n + 1)(2n - 1)$

32. $(4n - 5)(4n + 5)$

33. $(2a + 7)^2$

34. $(3a + 5)^2$

35. $(x - 2)(x^2 - x + 6)$

36. $(2x - 1)(x^2 + 4x + 7)$

37. $(a + 5)^3$

38. $(a - 6)^3$

39. $(x^2 - x - 1)(x^2 + 2x + 5)$

40. $(n^2 + 2n + 4)(n^2 - 7n - 1)$

Perform the following divisions.

41. $\dfrac{36x^4y^5}{-3xy^2}$

42. $\dfrac{-56a^5b^7}{-8a^2b^3}$

43. $\dfrac{-18x^4y^3 - 54x^6y^2}{6x^2y^2}$

44. $\dfrac{-30a^5b^{10} + 39a^4b^8}{-3ab}$

45. $\dfrac{56x^4 - 40x^3 - 32x^2}{4x^2}$

46. $(x^2 + 9x - 1) \div (x + 5)$

47. $(21x^2 - 4x - 12) \div (3x + 2)$

48. $(2x^3 - 3x^2 + 2x - 4) \div (x - 2)$

Cumulative Review Problem Set   (Chapters 1–6)

For Problems 1–6, evaluate each of the algebraic expressions for the given values of the variables.

1. $\dfrac{2x + 3y}{x - y}$ for $x = \dfrac{1}{2}$ and $y = -\dfrac{1}{3}$

2. $\dfrac{2}{5}n - \dfrac{1}{3}n - n + \dfrac{1}{2}n$ for $n = -\dfrac{3}{4}$

3. $\dfrac{3a - 2b - 4a + 7b}{-a - 3a + b - 2b}$ for $a = -1$ and $b = -\dfrac{1}{3}$

4. $-2(x - 4) + 3(2x - 1) - (3x - 2)$ for $x = -2$

5. $(x^2 + 2x - 4) - (x^2 - x - 2) + (2x^2 - 3x - 1)$ for $x = -1$

6. $2(n^2 - 3n - 1) - (n^2 + n + 4) - 3(2n - 1)$ for $n = 3$

For Problems 7–12, evaluate each of the expressions.

7. 2^7

8. -3^4

9. $\left(\dfrac{1}{4}\right)^3$

10. $\left(-\dfrac{2}{3}\right)^5$

11. $\left(\dfrac{1}{2} - \dfrac{1}{5}\right)^2$

12. $\left(\dfrac{1}{3} - \dfrac{1}{2}\right)^3$

For Problems 13–16, solve each of the equations.

13. $-3(x - 1) + 2(x + 3) = -4$

14. $\dfrac{3n + 1}{5} + \dfrac{n - 2}{3} = \dfrac{2}{15}$

15. $0.06x + 0.08(1500 - x) = 110$

16. $|-2x - 3| = 6$

For Problems 17–22, find the indicated products.

17. $(3x^2y^3)(-5xy^4)$

18. $-3xy(2x - 5y)$

19. $(5x - 2)(3x - 1)$

20. $(2x + 9)^2$

21. $(n - 7)^3$

22. $(2x - 5)(x^2 + x - 4)$

For Problems 23–27, perform the indicated divisions.

23. $\dfrac{-52x^3y^4}{13xy^2}$

24. $\dfrac{-126a^3b^5}{-9a^2b^3}$

25. $\dfrac{56xy^2 - 64x^3y - 72x^4y^4}{8xy}$

26. $(2x^3 + 2x^2 - 19x - 21) \div (x + 3)$

27. $(3x^3 + 17x^2 + 6x - 4) \div (3x - 1)$

For Problems 28–32, solve each of the inequalities.

28. $-5x + 3 > -4x + 5$

29. $\dfrac{3x}{4} - \dfrac{x}{2} \le \dfrac{5x}{6} - 1$

30. $0.08(700 - x) + 0.11x \ge 65$

31. $|3x - 2| > 10$

32. $|2x + 7| < 7$

33. Find the slope of the line determined by the points $(2, -3)$ and $(-4, 5)$.

34. Find the slope of the line determined by the equation $-x - 3y = 4$.

35. Write the equation of the line that contains the points $(1, 3)$ and $(-2, -5)$.

36. Solve each of the following systems of equations.

(a) $\begin{pmatrix} x = 2y - 12 \\ 5x + 4y = 10 \end{pmatrix}$

(b) $\begin{pmatrix} 4x - 7y = 26 \\ 3x + 2y = 5 \end{pmatrix}$

37. Graph the equation $-x - 2y = 4$.

38. Graph the equation $y = -x^4 - 2$.

39. Graph the inequality $3x - 2y < -6$.

For Problems 40–46, set up an equation or a system of equations and solve each problem.

40. The sum of four and three times a certain number is the same as the sum of the number and ten. Find the number.

41. Fifteen percent of some number is six. Find the number.

42. Lou has 18 coins consisting of dimes and quarters. If the total value of the coins is $3.30, how many coins of each denomination does he have?

43. A sum of $1500 is invested, part of it at 8% interest and the remainder at 9%. If the total interest amounts to $128, find the amount invested at each rate.

44. How many gallons of water must be added to 15 gallons of a 12% salt solution to change it to a 10% salt solution?

45. Two airplanes leave Atlanta at the same time and fly in opposite directions. If one travels at 400 miles per hour and the other at 450 miles per hour, how long will it take them to be 2975 miles apart?

46. The length of a rectangle is one meter more than twice its width. If the perimeter of the rectangle is 44 meters, find the length and width.

Chapter 7

Factoring, Solving Equations, and Problem Solving

The distributive property has played an important role in combining similar terms and multiplying polynomials. In this chapter we shall see yet another use of the distributive property as we learn how to **factor polynomials**. The ability to factor polynomials will allow us to solve other kinds of equations, which will in turn permit us to solve more kinds of word problems.

7.1
Factoring by Using the Distributive Property

In Chapter 1 we found the *greatest common factor* of two or more whole numbers by inspection or by using the prime factored form of the numbers. For example, by inspection we see that the greatest common factor of 8 and 12 is 4. This means that 4 is the largest whole number that is a factor of both 8 and 12. If it was difficult to determine the highest common factor by inspection then we used the prime factorization technique as follows.

$$42 = 2 \cdot 3 \cdot 7; \qquad 70 = 2 \cdot 5 \cdot 7$$

From this we see that $2 \cdot 7 = 14$ is the greatest common factor of 42 and 70.

It is meaningful to extend the concept of greatest common factor to monomials. Consider the following example.

EXAMPLE 1   Find the greatest common factor of $8x^2$ and $12x^3$.

Solution
$$8x^2 = 2 \cdot 2 \cdot 2 \cdot x \cdot x,$$
$$12x^3 = 2 \cdot 2 \cdot 3 \cdot x \cdot x \cdot x$$

Therefore, the greatest common factor is $2 \cdot 2 \cdot x \cdot x = 4x^2$.   ∎

By the greatest common factor of two or more monomials we mean the monomial with the largest numerical coefficient and highest power of the variables, which is a factor of the given monomials.

EXAMPLE 2   Find the greatest common factor of $16x^2y$, $24x^3y^2$, and $32xy$.

Solution
$$16x^2y = 2 \cdot 2 \cdot 2 \cdot 2 \cdot x \cdot x \cdot y,$$
$$24x^3y^2 = 2 \cdot 2 \cdot 2 \cdot 3 \cdot x \cdot x \cdot x \cdot y \cdot y,$$
$$32xy = 2 \cdot 2 \cdot 2 \cdot 2 \cdot 2 \cdot x \cdot y$$

Therefore, the greatest common factor is $2 \cdot 2 \cdot 2 \cdot x \cdot y = 8xy$.   ∎

You have used the distributive property to multiply a polynomial by a monomial. For example,

$$3x(x + 2) = 3x^2 + 6x.$$

Suppose that you start with $3x^2 + 6x$ and want to express it in factored form. Use the distributive property in the form $ab + ac = a(b + c)$.

$$3x^2 + 6x = 3x(x) + 3x(2) \qquad \text{3x is the greatest common factor of } 3x^2 \text{ and } 6x.$$
$$= 3x(x + 2) \qquad \text{Use the distributive property.}$$

The next four examples further illustrate this process of **factoring out the greatest common monomial factor**.

EXAMPLE 3   Factor $12x^3 - 8x^2$.

Solution

$$12x^3 - 8x^2 = 4x^2(3x) - 4x^2(2)$$
$$= 4x^2(3x - 2) \qquad ab - ac = a(b - c) \qquad ■$$

EXAMPLE 4   Factor $12x^2y + 18xy^2$.

Solution

$$12x^2y + 18xy^2 = 6xy(2x) + 6xy(3y)$$
$$= 6xy(2x + 3y) \qquad ■$$

EXAMPLE 5   Factor $24x^3 + 30x^4 - 42x^5$.

Solution

$$24x^3 + 30x^4 - 42x^5 = 6x^3(4) + 6x^3(5x) - 6x^3(7x^2)$$
$$= 6x^3(4 + 5x - 7x^2) \qquad ■$$

EXAMPLE 6   Factor $9x^2 + 9x$.

Solution

$$9x^2 + 9x = 9x(x) + 9x(1)$$
$$= 9x(x + 1) \qquad ■$$

We want to emphasize the point made prior to Example 3. It is important to realize that we are factoring out the *greatest* common monomial factor. An expression such as $9x^2 + 9x$ in Example 6 could be factored as $9(x^2 + x)$, $3(3x^2 + 3x)$, $3x(3x + 3)$, or even as $\frac{1}{2}(18x^2 + 18x)$, but it is the form $9x(x + 1)$ that we want to obtain. We can accomplish this by factoring out the greatest common monomial factor; we sometimes refer to this process as **factoring completely**. A polynomial with integral coefficients is in completely factored form if:

1. It is expressed as a product of polynomials with integral coefficients;
2. No polynomial, other than a monomial, within the factored form can be further factored into polynomials with integral coefficients.

Thus, $9(x^2 + x)$, $3(3x^2 + 3x)$, and $3x(3x + 3)$ are not completely factored because they violate part 2. The form $\frac{1}{2}(18x^2 + 18x)$ violates both parts 1 and 2.

Sometimes there may be a **common binomial factor** rather than a common monomial factor. For example, each of the two terms of $x(y + 2) + z(y + 2)$ has a binomial factor of $(y + 2)$. Thus, we can factor $(y + 2)$ from each term and our result is as follows.

$$x(y + 2) + z(y + 2) = (y + 2)(x + z)$$

Consider a few more examples involving a common binomial factor.

$$a(b + c) - d(b + c) = (b + c)(a - d),$$

$$x(x + 2) + 3(x + 2) = (x + 2)(x + 3),$$

$$x(x + 5) - 4(x + 5) = (x + 5)(x - 4)$$

It may be that the original polynomial exhibits no apparent common monomial or binomial factor, which is the case with

$$ab + 3a + bc + 3c.$$

However, by factoring a from the first two terms and c from the last two terms, we see that

$$ab + 3a + bc + 3c = a(b + 3) + c(b + 3).$$

Now a common binomial factor of $(b + 3)$ is obvious and we can proceed as before.

$$a(b + 3) + c(b + 3) = (b + 3)(a + c)$$

This factoring process is called **factoring by grouping**. Let's consider two more examples of factoring by grouping.

$$x^2 - x + 5x - 5 = x(x - 1) + 5(x - 1)$$ factor x from first two terms and 5 from last two terms

$$= (x - 1)(x + 5)$$ factor common binomial factor of $(x - 1)$ from both terms

$$6x^2 - 4x - 3x + 2 = 2x(3x - 2) - 1(3x - 2)$$ factor $2x$ from first two terms and -1 from last two terms

$$= (3x - 2)(2x - 1)$$ factor common binomial factor of $(3x - 2)$ from both terms

Back to Solving Equations

Suppose we are told that the product of two numbers is zero. What do we know about the numbers? Do you agree we can conclude that at least one of the numbers must be zero? The following property formalizes this idea.

PROPERTY 7.1

> For all real numbers a and b,
>
> $ab = 0$ if and only if $a = 0$ or $b = 0$.

Property 7.1 provides us with another technique for solving equations.

EXAMPLE 7   Solve $x^2 + 6x = 0$.

Solution

$$x^2 + 6x = 0$$

$$x(x + 6) = 0$$

$$x = 0 \quad \text{or} \quad x + 6 = 0 \qquad \text{Property 7.1}$$

$$x = 0 \quad \text{or} \quad x = -6$$

The solution set is $\{-6, 0\}$. (Be sure to check both of these in the original equation.)   ■

EXAMPLE 8   Solve $x^2 = 12x$.

Solution

$$x^2 = 12x$$

$$x^2 - 12x = 0 \qquad \text{added} -12x \text{ to both sides}$$

$$x(x - 12) = 0$$

$$x = 0 \quad \text{or} \quad x - 12 = 0 \qquad \text{Property 7.1}$$

$$x = 0 \quad \text{or} \quad x = 12$$

The solution set is $\{0, 12\}$.   ■

REMARK Notice in Example 8 we *did not* divide both sides of the original equation by x. This would cause us to lose the solution of 0.

EXAMPLE 9   Solve $4x^2 - 3x = 0$.

Solution

$$4x^2 - 3x = 0$$

$$x(4x - 3) = 0$$

$$x = 0 \quad \text{or} \quad 4x - 3 = 0$$

$$x = 0 \quad \text{or} \quad 4x = 3$$

$$x = 0 \quad \text{or} \quad x = \frac{3}{4}$$

The solution set is $\left\{0, \dfrac{3}{4}\right\}$.   ■

Each time that we extend our equation solving capabilities, we also gain more techniques for solving problems. Let's solve a geometric problem using the ideas from this section.

PROBLEM 1   The area of a square is numerically equal to twice its perimeter. Find the length of a side of the square.

Solution   Sketch a square and let s represent the length of each side. Then the area is represented by s^2 and the perimeter by $4s$. Thus,

$$s^2 = 2(4s)$$

$$s^2 = 8s$$

$$s^2 - 8s = 0$$

$$s(s - 8) = 0$$

$$s = 0 \quad \text{or} \quad s - 8 = 0$$

$$s = 0 \quad \text{or} \quad s = 8.$$

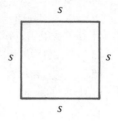

Since 0 is not a reasonable answer to the problem, the solution is 8. (Be sure to check this solution in the original statement of the problem!)   ■

Problem Set 7.1

For Problems 1–10, find the greatest common factor of the given expressions.

1. $24y$ and $30xy$
2. $32x$ and $40xy$
3. $60x^2y$ and $84xy^2$
4. $72x^3$ and $63x^2$
5. $42ab^3$ and $70a^2b^2$
6. $48a^2b^2$ and $96ab^4$
7. $6x^3, 8x,$ and $24x^2$
8. $72xy, 36x^2y,$ and $84xy^2$
9. $16a^2b^2, 40a^2b^3,$ and $56a^3b^4$
10. $70a^3b^3, 42a^2b^4,$ and $49ab^5$

For Problems 11–46, factor each polynomial completely.

11. $8x + 12y$
12. $18x + 24y$
13. $14xy - 21y$
14. $24x - 40xy$
15. $18x^2 + 45x$
16. $12x + 28x^3$
17. $12xy^2 - 30x^2y$
18. $28x^2y^2 - 49x^2y$
19. $36a^2b - 60a^3b^4$
20. $65ab^3 - 45a^2b^2$
21. $16xy^3 + 25x^2y^2$
22. $12x^2y^2 + 29x^2y$
23. $64ab - 72cd$
24. $45xy - 72zw$
25. $9a^2b^4 - 27a^2b$
26. $7a^3b^5 - 42a^2b^6$
27. $52x^4y^2 + 60x^6y$
28. $70x^5y^3 - 42x^8y^2$
29. $40x^2y^2 + 8x^2y$
30. $84x^2y^3 + 12xy^3$
31. $12x + 15xy + 21x^2$
32. $30x^2y + 40xy + 55y$
33. $2x^3 - 3x^2 + 4x$

34. $x^4 + x^3 + x^2$

35. $44y^5 - 24y^3 - 20y^2$

36. $14a - 18a^3 - 26a^5$

37. $14a^2b^3 + 35ab^2 - 49a^3b$

38. $24a^3b^2 + 36a^2b^4 - 60a^4b^3$

39. $x(y + 1) + z(y + 1)$

40. $a(c + d) + 2(c + d)$

41. $a(b - 4) - c(b - 4)$

42. $x(y - 6) - 3(y - 6)$

43. $x(x + 3) + 6(x + 3)$

44. $x(x - 7) + 9(x - 7)$

45. $2x(x + 1) - 3(x + 1)$

46. $4x(x + 8) - 5(x + 8)$

For Problems 47–60, use the process of *factoring by grouping* to factor each of the following.

47. $5x + 5y + bx + by$

48. $7x + 7y + zx + zy$

49. $bx - by - cx + cy$

50. $2x - 2y - ax + ay$

51. $ac + bc + a + b$

52. $x + y + ax + ay$

53. $x^2 + 5x + 12x + 60$

54. $x^2 + 3x + 7x + 21$

55. $x^2 - 2x - 8x + 16$

56. $x^2 - 4x - 9x + 36$

57. $2x^2 + x - 10x - 5$

58. $3x^2 + 2x - 18x - 12$

59. $6n^2 - 3n - 8n + 4$

60. $20n^2 + 8n - 15n - 6$

For Problems 61–80, solve each equation.

61. $x^2 - 8x = 0$

62. $x^2 - 12x = 0$

63. $x^2 + x = 0$

64. $x^2 + 7x = 0$

65. $n^2 = 5n$

66. $n^2 = -2n$

67. $2y^2 - 3y = 0$

68. $4y^2 - 7y = 0$

69. $7x^2 = -3x$

70. $5x^2 = -2x$

71. $3n^2 + 15n = 0$

72. $6n^2 - 24n = 0$

73. $4x^2 = 6x$

74. $12x^2 = 8x$

75. $7x - x^2 = 0$

76. $9x - x^2 = 0$

77. $13x = x^2$

78. $15x = -x^2$

79. $5x = -2x^2$

80. $7x = -5x^2$

For Problems 81–87, set up an equation and solve each problem.

81. The square of a number equals nine times that number. Find the number.

82. Suppose that four times the square of a number equals 20 times that number. What is the number?

83. The area of a square is numerically equal to five times its perimeter. Find the length of a side of the square.

84. The area of a square is 14 times as large as the area of a triangle. One side of the triangle is 7 inches long and the altitude to that side is the same length as a side of the square. Find the length of a side of the square. Also find the areas of both figures and be sure that your answer checks.

85. Suppose that the area of a circle is numerically equal to the perimeter of a square and that the length of a radius of the circle is equal to the length of a side of the square. Find the length of a side of the square. Express your answer in terms of π.

86. One side of a parallelogram, an altitude to that side, and one side of a rectangle all have the same measure. If an adjacent side of the rectangle is 20 centimeters long and the area of the rectangle is twice the area of the parallelogram, find the area of both figures.

87. The area of a rectangle is twice the area of a square. If the rectangle is 6 inches long and the width of the rectangle is the same as the length of a side of the square, find the dimensions of both the rectangle and the square.

Miscellaneous Problems

88. The total surface area of a right circular cylinder is given by the formula $A = 2\pi r^2 + 2\pi rh$, where r represents the radius of a base and h represents the height of the cylinder. For computational purposes it may be more convenient to change the form of the right side of the formula by factoring it.

$$A = 2\pi r^2 + 2\pi rh$$
$$= 2\pi r(r + h)$$

Use $A = 2\pi r(r + h)$ to find the total surface area of each of the following cylinders. Also, use $\dfrac{22}{7}$ as an approximation for π.

(a) $r = 7$ centimeters   and $h = 12$ centimeters;

(b) $r = 14$ meters   and $h = 20$ meters;

(c) $r = 3$ feet   and $h = 4$ feet;

(d) $r = 5$ yards   and $h = 9$ yards.

89. The formula $A = P + Prt$ yields the total amount of money accumulated (A) when P dollars is invested at r percent simple interest for t years. For computational purposes it may be convenient to change the right side of the formula by factoring.

$$A = P + Prt$$
$$= P(1 + rt)$$

Use $A = P(1 + rt)$ to find the total amount of money accumulated for each of the following investments.

(a) $100 at 8% for 2 years;

(b) $200 at 9% for 3 years;

(c) $500 at 10% for 5 years;

(d) $1000 at 10% for 10 years.

For Problems 90–93, solve each equation for the indicated variable.

90. $ax + bx = c$ for x

91. $b^2x^2 - cx = 0$ for x

92. $5ay^2 = by$ for y

93. $y + ay - by - c = 0$ for y

Factoring the Difference of Two Squares

In Section 6.4 we noted some special multiplication patterns. One of these patterns was

$$(a - b)(a + b) = a^2 - b^2.$$

We can view this same pattern as the factoring pattern that follows.

> **Difference of Two Squares**
>
> $$a^2 - b^2 = (a - b)(a + b)$$

To apply the pattern is a fairly simple process, as these next examples illustrate. The steps inside the box are often performed mentally.

$$x^2 - 36 = \boxed{(x)^2 - (6)^2} = (x - 6)(x + 6);$$
$$4x^2 - 25 = \boxed{(2x)^2 - (5)^2} = (2x - 5)(2x + 5);$$
$$9x^2 - 16y^2 = \boxed{(3x)^2 - (4y)^2} = (3x - 4y)(3x + 4y);$$
$$64 - y^2 = \boxed{(8)^2 - (y)^2} = (8 - y)(8 + y).$$

Since multiplication is commutative, the order of writing the factors is not important. For example, $(x - 6)(x + 6)$ can also be written as $(x + 6)(x - 6)$.

You must be careful not to assume an analogous factoring pattern for the *sum* of two squares; it does not exist. For example, $x^2 + 4 \neq (x + 2)(x + 2)$ since $(x + 2)(x + 2) = x^2 + 4x + 4$. We say that the **sum of two squares is not factorable using integers**. The phrase "using integers" is necessary because $x^2 + 4$ could be written as $\frac{1}{2}(2x^2 + 8)$, but such *factoring* is of no help. Furthermore, we do not consider $(1)(x^2 + 4)$ as factoring $x^2 + 4$.

It is possible that both the technique of *factoring out a common monomial factor* and the pattern *difference of two squares* can be applied to the same polynomial. In general, it is best to first look for a common monomial factor.

EXAMPLE 1   Factor $2x^2 - 50$.

Solution

$$\begin{aligned}
2x^2 - 50 &= 2(x^2 - 25) && \text{common factor of 2} \\
&= 2(x - 5)(x + 5). && \text{difference of squares}
\end{aligned}$$

∎

By expressing $2x^2 - 50$ as $2(x - 5)(x + 5)$ we say that it has been **factored completely**. That means the factors 2, $x - 5$, and $x + 5$ cannot be factored any further using integers.

EXAMPLE 2   Factor completely $18y^3 - 8y$.

Solution

$$18y^3 - 8y = 2y(9y^2 - 4) \qquad \text{common factor of } 2y$$
$$= 2y(3y - 2)(3y + 2) \qquad \text{difference of squares} \qquad \blacksquare$$

Sometimes it is possible to apply the difference-of-squares pattern more than once.

EXAMPLE 3   Factor completely $x^4 - 16$.

Solution

$$x^4 - 16 = (x^2 + 4)(x^2 - 4)$$
$$= (x^2 + 4)(x + 2)(x - 2) \qquad \blacksquare$$

The following examples should help you to summarize our factoring ideas thus far.

$$5x^2 + 20 = 5(x^2 + 4);$$
$$25 - y^2 = (5 - y)(5 + y);$$
$$3 - 3x^2 = 3(1 - x^2) = 3(1 + x)(1 - x);$$
$$36x^2 - 49y^2 = (6x - 7y)(6x + 7y);$$
$$a^2 + 9 \text{ is not factorable using integers;}$$
$$9x + 17y \text{ is not factorable using integers.}$$

Solving Equations

Each time that we pick up a new factoring technique we also develop more power for solving equations. Let's consider how we can use the difference-of-squares factoring pattern to help solve certain kinds of equations.

EXAMPLE 4   Solve $x^2 = 25$.

Solution

$$x^2 = 25$$
$$x^2 - 25 = 0$$
$$(x + 5)(x - 5) = 0$$

$x + 5 = 0 \qquad$ or $\qquad x - 5 = 0 \qquad$ Remember: $ab = 0$ if and only if
$\qquad\qquad\qquad\qquad\qquad\qquad\qquad\qquad\quad a = 0$ or $b = 0$.
$\qquad x = -5 \qquad$ or $\qquad\quad x = 5$

The solution set is $\{-5, 5\}$. Check these answers! $\blacksquare$

EXAMPLE 5   Solve $9x^2 = 25$.

Solution

$$9x^2 = 25$$

$$9x^2 - 25 = 0$$

$$(3x + 5)(3x - 5) = 0$$

$$3x + 5 = 0 \qquad \text{or} \qquad 3x - 5 = 0$$

$$3x = -5 \qquad \text{or} \qquad 3x = 5$$

$$x = -\frac{5}{3} \qquad \text{or} \qquad x = \frac{5}{3}$$

The solution set is $\left\{ -\dfrac{5}{3}, \dfrac{5}{3} \right\}$.   ■

EXAMPLE 6   Solve $5y^2 = 20$.

Solution

$$5y^2 = 20$$

$$\frac{5y^2}{5} = \frac{20}{5} \qquad \text{Divide both sides by 5.}$$

$$y^2 = 4$$

$$y^2 - 4 = 0$$

$$(y + 2)(y - 2) = 0$$

$$y + 2 = 0 \qquad \text{or} \qquad y - 2 = 0$$

$$y = -2 \qquad \text{or} \qquad y = 2$$

The solution set is $\{-2, 2\}$. Check it!   ■

EXAMPLE 7   Solve $x^3 - 9x = 0$.

Solution

$$x^3 - 9x = 0$$

$$x(x^2 - 9) = 0$$

$$x(x - 3)(x + 3) = 0$$

$$x = 0 \qquad \text{or} \qquad x - 3 = 0 \qquad \text{or} \qquad x + 3 = 0$$

$$x = 0 \qquad \text{or} \qquad x = 3 \qquad \text{or} \qquad x = -3$$

The solution set is $\{-3, 0, 3\}$.   ■

The more we know about solving equations, the better off we are when we solve word problems.

PROBLEM 1   The combined area of two squares is 20 square centimeters. Each side of one square is twice as long as a side of the other square. Find the lengths of the sides of each square.

Solution A   Let's sketch two squares and label the sides of the smaller square s. Then the sides of the larger square are $2s$. Since the sum of the areas of the two squares is 20 square centimeters, we can set up and solve the following equation.

$$s^2 + (2s)^2 = 20$$
$$s^2 + 4s^2 = 20$$
$$5s^2 = 20$$
$$s^2 = 4$$
$$s^2 - 4 = 0$$
$$(s + 2)(s - 2) = 0$$
$$s + 2 = 0 \qquad \text{or} \qquad s - 2 = 0$$
$$s = -2 \qquad \text{or} \qquad s = 2$$

Since s represents the length of a side of a square, the solution -2 must be disregarded. Thus, one square has sides of length 2 centimeters and the other square has sides of length $2(2) = 4$ centimeters.

Solution B   Let's sketch two squares and label the sides of the smaller square x and the sides of the larger square y.

Now we can set up a system of two equations in two variables to represent the conditions stated in the problem.

combined area of 20 square centimeters ⟶ $x^2 + y^2 = 20$

each side of larger square is twice side of smaller square ⟶ $y = 2x$

We can substitute $2x$ for y in the equation $x^2 + y^2 = 20$ and solve for x.

$$x^2 + y^2 = 20$$
$$x^2 + (2x)^2 = 20$$
$$x^2 + 4x^2 = 20$$
$$5x^2 = 20$$
$$x^2 = 4$$
$$x^2 - 4 = 0$$
$$(x + 2)(x - 2) = 0$$
$$x + 2 = 0 \qquad \text{or} \qquad x - 2 = 0$$
$$x = -2 \qquad \text{or} \qquad x = 2$$

The negative solution must be disregarded; therefore, the length of each side of the smaller square is 2 centimeters and the length of each side of the larger square is $2(2) = 4$ centimeters. ∎

Take another look at both approaches to Problem 1. In Solution A we use only one variable, s, and then represent the conditions of the problem in terms of that one variable. In Solution B we use two variables, x and y, and then express the conditions of the problem using two equations. Which approach a person uses is one of personal preference. Sometimes a problem is stated in a way that lends itself to a specific approach. In other cases, either of the two approaches might seem equally reasonable.

Another point should be made about Solution B. Note that we use the substitution method for solving the system. As illustrated, this method works very well even though one of the equations is not linear.

Problem Set 7.2

For Problems 1–12, use the difference-of-squares pattern to factor each polynomial.

1. $x^2 - 1$ **2.** $x^2 - 25$ **3.** $x^2 - 100$

4. $x^2 - 121$ **5.** $x^2 - 4y^2$ **6.** $x^2 - 36y^2$

7. $9x^2 - y^2$ **8.** $49y^2 - 64x^2$ **9.** $36a^2 - 25b^2$

10. $4a^2 - 81b^2$ **11.** $1 - 4n^2$ **12.** $4 - 9n^2$

For Problems 13–40, factor each polynomial completely. Indicate any that are not factorable using integers. Don't forget to first look for a common monomial factor.

13. $5x^2 - 20$ **14.** $7x^2 - 7$ **15.** $8x^2 + 32$

16. $12x^2 + 60$ **17.** $2x^2 - 18y^2$ **18.** $8x^2 - 32y^2$

19. $x^3 - 25x$ **20.** $2x^3 - 2x$ **21.** $x^2 + 9y^2$

22. $18x - 42y$ **23.** $45x^2 - 36xy$ **24.** $16x^2 + 25y^2$

25. $36 - 4x^2$ **26.** $75 - 3x^2$ **27.** $4a^4 + 16a^2$

28. $9a^4 + 81a^2$ **29.** $x^4 - 81$ **30.** $16 - x^4$

31. $x^4 + x^2$ **32.** $x^5 + 2x^3$ **33.** $3x^3 + 48x$

34. $6x^3 + 24x$ **35.** $5x - 20x^3$ **36.** $4x - 36x^3$

37. $4x^2 - 64$ **38.** $9x^2 - 9$ **39.** $75x^3y - 12xy^3$ **40.** $32x^3y - 18xy^3$

For Problems 41–64, solve each equation.

41. $x^2 = 9$ **42.** $x^2 = 1$ **43.** $4 = n^2$

44. $144 = n^2$ **45.** $9x^2 = 16$ **46.** $4x^2 = 9$

47. $n^2 - 121 = 0$ **48.** $n^2 - 81 = 0$ **49.** $25x^2 = 4$

50. $49x^2 = 36$ **51.** $3x^2 = 75$ **52.** $7x^2 = 28$

53. $3x^3 - 48x = 0$ **54.** $x^3 - x = 0$ **55.** $n^3 = 16n$

56. $2n^3 = 8n$ **57.** $5 - 45x^2 = 0$ **58.** $3 - 12x^2 = 0$

59. $4x^3 - 400x = 0$ **60.** $2x^3 - 98x = 0$ **61.** $64x^2 = 81$

62. $81x^2 = 25$ **63.** $36x^3 = 9x$ **64.** $64x^3 = 4x$

For Problems 65–76, set up an equation or a system of equations and solve the problem.

65. Forty-nine less than the square of a number equals zero. Find the number.

66. The cube of a number equals nine times the number. Find the number.

67. Five times the cube of a number equals 80 times the number. Find the number.

68. Ten times the square of a number equals 40. Find the number.

69. The sum of the areas of two squares is 234 square inches. Each side of the larger square is five times the length of a side of the smaller square. Find the length of a side of each square.

70. The difference of the areas of two squares is 75 square feet. Each side of the larger square is twice the length of a side of the smaller square. Find the length of a side of each square.

71. The length of a certain rectangle is $2\frac{1}{2}$ times its width and the area of that same rectangle is 160 square centimeters. Find the length and width of the rectangle.

72. The width of a certain rectangle is three-fourths of its length and the area of that same rectangle is 108 square meters. Find the length and width of the rectangle.

73. The sum of the areas of two circles is 80π square meters. Find the length of a radius of each circle if one of them is twice as long as the other.

74. The area of a triangle is 98 square feet. If one side of the triangle and the altitude to that side are of equal length, find the length.

75. The total surface area of a right circular cylinder is 100π square centimeters. If a radius of the base and the altitude of the cylinder are of the same length, find the length of a radius.

76. The total surface area of a right circular cone is 192π square feet. If the slant height of the cone is equal in length to a diameter of the base, find the length of a radius.

Thoughts into Words

77. What does the expression "not factorable using integers" mean to you?

78. Illustrate how the distributive property is used for factoring purposes.

Miscellaneous Problems

The following patterns can be used to factor the sum and difference of two cubes.

$$a^3 + b^3 = (a + b)(a^2 - ab + b^2)$$

$$a^3 - b^3 = (a - b)(a^2 + ab + b^2)$$

Consider the following examples.

$$x^3 + 8 = (x)^3 + (2)^3 = (x + 2)(x^2 - 2x + 4)$$

$$x^3 - 1 = (x)^3 - (1)^3 = (x - 1)(x^2 + x + 1)$$

Use the sum and difference-of-cubes patterns to factor each of the following.

79. $x^3 + 1$ **80.** $x^3 - 8$

81. $n^3 - 27$ **82.** $n^3 + 64$

83. $8x^3 + 27y^3$ **84.** $27a^3 - 64b^3$

85. $1 - 8x^3$ **86.** $1 + 27a^3$

87. $x^3 + 8y^3$ **88.** $8x^3 - y^3$

89. $a^3b^3 - 1$ **90.** $27x^3 - 8y^3$

91. $8 + n^3$ **92.** $125x^3 + 8y^3$

93. $27n^3 - 125$ **94.** $64 + x^3$

7.3
Factoring Trinomials of the Form $x^2 + bx + c$

Expressing a trinomial as the product of two binomials is one of the most common types of factoring used in algebra. In this section we want to consider trinomials where the coefficient of the squared term is 1; that is, trinomials of the form $x^2 + bx + c$.

Again, to develop a factoring technique we first look at some multiplication ideas. Consider the product $(x + r)(x + s)$ and use the distributive property to show how each term of the resulting trinomial is formed.

$$(x + r)(x + s) = x(x) + \underbrace{x(s) + r(x)}_{} + r(s)$$
$$\downarrow \qquad\qquad\qquad \downarrow$$
$$x^2 \quad + \quad (s + r)x \quad + \quad rs$$

Notice that the coefficient of the middle term is the *sum* of r and s and the last term is the *product* of r and s. These two relationships are used in the following examples.

EXAMPLE 1   Factor $x^2 + 7x + 12$.

Solution   We need to complete the following with two numbers whose sum is 7 and whose product is 12.

$$x^2 + 7x + 12 = (x + \text{_____})(x + \text{_____})$$

This can be done by setting up a small table as follows.

Product	Sum
$1(12) = 12$	$1 + 12 = 13$
$2(6) = 12$	$2 + 6 = 8$
$3(4) = 12$	$3 + 4 = 7$

The bottom line contains the numbers that we need. Thus,

$$x^2 + 7x + 12 = (x + 3)(x + 4).$$ ∎

EXAMPLE 2   Factor $x^2 - 11x + 24$.

Solution   We need two numbers whose product is 24 and whose sum is -11.

Product	Sum
$(-1)(-24) = 24$	$-1 + (-24) = -25$
$(-2)(-12) = 24$	$-2 + (-12) = -14$
$(-3)(-8) = 24$	$-3 + (-8) = -11$
$(-4)(-6) = 24$	$-4 + (-6) = -10$

The third line contains the numbers that we want.

$$x^2 - 11x + 24 = (x - 3)(x - 8)$$ ∎

EXAMPLE 3   Factor $x^2 + 3x - 10$.

Solution   We need two numbers whose product is -10 and whose sum is 3.

Product	Sum
$1(-10) = -10$	$1 + (-10) = -9$
$-1(10) = -10$	$-1 + 10 = 9$
$2(-5) = -10$	$2 + (-5) = -3$
$-2(5) = -10$	$-2 + 5 = 3$

The bottom line is the key line. Thus,

$$x^2 + 3x - 10 = (x + 5)(x - 2).$$ ∎

EXAMPLE 4   Factor $x^2 - 2x - 8$.

Solution   We need two numbers whose product is -8 and whose sum is -2.

Product	Sum
$1(-8) = -8$	$1 + (-8) = -7$
$-1(8) = -8$	$-1 + 8 \ = 7$
$2(-4) = -8$	$2 + (-4) = -2$
$-2(4) = -8$	$-2 + 4 \ = 2$

The third line has the desired information.

$$x^2 - 2x - 8 = (x - 4)(x + 2)$$ ∎

The tables in the four previous examples were used to illustrate one way of organizing your thoughts for such problems. We have shown complete tables. That is, for Example 4 the bottom line was included even though the desired numbers were obtained in the third line. If you use such tables, then keep in mind that as soon as the desired numbers are obtained, the table need not be completed any further. Furthermore, many times you may be able to find the numbers without using a table. The key ideas are the product and sum relationships.

EXAMPLE 5   Factor $x^2 - 13x + 12$.

Solution

Product	Sum
$(-1)(-12) = 12$	$(-1) + (-12) = -13$

We need not complete the table.

$$x^2 - 13x + 12 = (x - 1)(x - 12)$$ ∎

EXAMPLE 6   Factor $x^2 - x - 56$.

Solution   Notice that the coefficient of the middle term is -1. Therefore, we are looking for two numbers whose product is -56; since their sum is -1, the absolute value of the negative number must be one larger than the absolute value of the positive number. The numbers are -8 and 7 and we have

$$x^2 - x - 56 = (x - 8)(x + 7).$$ ∎

EXAMPLE 7   Factor $x^2 + 10x + 12$.

Solution

Product	*Sum*
$1(12) = 12$	$1 + 12 = 13$
$2(6) = 12$	$2 + 6 = 8$
$3(4) = 12$	$3 + 4 = 7$

Since the table is complete and no two factors of 12 produce a sum of 10, we conclude that

$$x^2 + 10x + 12$$

is not factorable using integers.   ∎

In a problem such as Example 7 we need to be sure that all possibilities have been tried before we conclude that the trinomial is not factorable.

Back to Solving Equations

The property "$ab = 0$ if and only if $a = 0$ or $b = 0$" continues to play an important role as we solve equations that involve the factoring ideas of this section. Consider the following examples.

EXAMPLE 8   Solve $x^2 + 8x + 15 = 0$.

Solution

$$x^2 + 8x + 15 = 0$$

$$(x + 3)(x + 5) = 0$$ Factor the left side.

$$x + 3 = 0 \quad \text{or} \quad x + 5 = 0$$ Use $ab = 0$ if and only if $a = 0$ or $b = 0$.

$$x = -3 \quad \text{or} \quad x = -5$$

The solution set is $\{-5, -3\}$.   ∎

EXAMPLE 9   Solve $x^2 + 5x - 6 = 0$.

Solution

$$x^2 + 5x - 6 = 0$$

$$(x + 6)(x - 1) = 0$$

$$x + 6 = 0 \quad \text{or} \quad x - 1 = 0$$

$$x = -6 \quad \text{or} \quad x = 1.$$

The solution set is $\{-6, 1\}$.   ∎

EXAMPLE 10   Solve $y^2 - 4y = 45$.

Solution

$$y^2 - 4y = 45$$
$$y^2 - 4y - 45 = 0$$
$$(y - 9)(y + 5) = 0$$
$$y - 9 = 0 \quad \text{or} \quad y + 5 = 0$$
$$y = 9 \quad \text{or} \quad y = -5.$$

The solution set is $\{-5, 9\}$.   ■

Don't forget that we can always check to be absolutely sure of our solutions. Let's check the solutions for Example 10.
If $y = 9$, then $y^2 - 4y = 45$ becomes

$$9^2 - 4(9) \stackrel{?}{=} 45$$
$$81 - 36 \stackrel{?}{=} 45$$
$$45 = 45.$$

If $y = -5$, then $y^2 - 4y = 45$ becomes

$$(-5)^2 - 4(-5) \stackrel{?}{=} 45$$
$$25 + 20 \stackrel{?}{=} 45$$
$$45 = 45.$$

Back to Problem Solving

The more we know about factoring and solving equations the more word problems we can solve.

PROBLEM 1   Find two consecutive integers whose product is 72.

Solution   Let n represent one integer. Then $n + 1$ represents the next integer.

$$n(n + 1) = 72 \qquad \text{The product of the two integers is 72.}$$
$$n^2 + n = 72$$
$$n^2 + n - 72 = 0$$
$$(n + 9)(n - 8) = 0$$
$$n + 9 = 0 \quad \text{or} \quad n - 8 = 0$$
$$n = -9 \quad \text{or} \quad n = 8$$

If $n = -9$, then $n + 1 = -9 + 1 = -8$.

If $n = 8$, then $n + 1 = 8 + 1 = 9$.

Thus, the consecutive integers are -9 and -8 or 8 and 9.   ∎

PROBLEM 2   A rectangular plot is 6 meters longer than it is wide. The area of the plot is 16 square meters. Find the length and width of the plot.

Solution   Let w represent the width of the plot. Then $w + 6$ represents the length.

Using the area formula ($A = lw$), we obtain

$$w(w + 6) = 16$$
$$w^2 + 6w = 16$$
$$w^2 + 6w - 16 = 0$$
$$(w + 8)(w - 2) = 0$$
$$w + 8 = 0 \quad \text{or} \quad w - 2 = 0$$
$$w = -8 \quad \text{or} \quad w = 2.$$

The solution of -8 is not possible for the width of a rectangle, so the plot is 2 meters wide and its length ($w + 6$) is 8 meters.   ∎

The Pythagorean Theorem, an important theorem pertaining to right triangles, can also serve as a guideline for solving certain types of problems. The Pythagorean Theorem states that **in any right triangle, the square of the longest side** (*called the hypotenuse*) **is equal to the sum of the squares of the other two sides** (*called legs*). Let's use this relationship to help solve a problem.

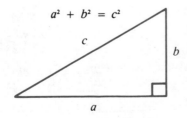

PROBLEM 3

Suppose that the lengths of the three sides of a right triangle are consecutive whole numbers. Find the lengths of the three sides.

Solution

Let s represent the length of the shortest leg. Then $s + 1$ represents the length of the other leg and $s + 2$ represents the length of the hypotenuse.

Using the Pythagorean Theorem as a guideline, we obtain the following equation.

$$\overbrace{s^2 + (s + 1)^2}^{\text{sum of squares of two legs}} = \overbrace{(s + 2)^2}^{\text{square of hypotenuse}}$$

Solving this equation yields

$$s^2 + s^2 + 2s + 1 = s^2 + 4s + 4$$

$$2s^2 + 2s + 1 = s^2 + 4s + 4$$

$$s^2 + 2s + 1 = 4s + 4$$

$$s^2 - 2s + 1 = 4$$

$$s^2 - 2s - 3 = 0$$

$$(s - 3)(s + 1) = 0$$

$$s - 3 = 0 \quad \text{or} \quad s + 1 = 0$$

$$s = 3 \quad \text{or} \quad s = -1.$$

The solution of -1 is not possible for the length of a side, so the shortest side (s) is of length 3. The other two sides ($s + 1$ and $s + 2$) have lengths of 4 and 5. ∎

Problem Set 7.3

For Problems 1–30, factor each trinomial completely. Indicate any that are not factorable using integers.

1. $x^2 + 10x + 24$

2. $x^2 + 9x + 14$

3. $x^2 + 13x + 40$

4. $x^2 + 11x + 24$

5. $x^2 - 11x + 18$

6. $x^2 - 5x + 4$

7. $n^2 - 11n + 28$

8. $n^2 - 7n + 10$

9. $n^2 + 6n - 27$

10. $n^2 + 3n - 18$

11. $n^2 - 6n - 40$

12. $n^2 - 4n - 45$

13. $t^2 + 12t + 24$

14. $t^2 + 20t + 96$

15. $x^2 - 18x + 72$

16. $x^2 - 14x + 32$

17. $x^2 + 5x - 66$

18. $x^2 + 11x - 42$

19. $y^2 - y - 72$

20. $y^2 - y - 30$

21. $x^2 + 21x + 80$

22. $x^2 + 21x + 90$

23. $x^2 + 6x - 72$

24. $x^2 - 8x - 36$

25. $x^2 - 10x - 48$

26. $x^2 - 12x - 64$

27. $x^2 + 3xy - 10y^2$

28. $x^2 - 4xy - 12y^2$

29. $a^2 - 4ab - 32b^2$

30. $a^2 + 3ab - 54b^2$

For Problems 31–50, solve each equation.

31. $x^2 + 10x + 21 = 0$

32. $x^2 + 9x + 20 = 0$

33. $x^2 - 9x + 18 = 0$

34. $x^2 - 9x + 8 = 0$

35. $x^2 - 3x - 10 = 0$

36. $x^2 - x - 12 = 0$

37. $n^2 + 5n - 36 = 0$

38. $n^2 + 3n - 18 = 0$

39. $n^2 - 6n - 40 = 0$

40. $n^2 - 8n - 48 = 0$

41. $t^2 + t - 56 = 0$

42. $t^2 + t - 72 = 0$

43. $x^2 - 16x + 28 = 0$

44. $x^2 - 18x + 45 = 0$

45. $x^2 + 11x = 12$

46. $x^2 + 8x = 20$

47. $x(x - 10) = -16$

48. $x(x - 12) = -35$

49. $-x^2 - 2x + 24 = 0$

50. $-x^2 + 6x + 16 = 0$

For Problems 51–68, set up an equation or a system of equations and solve each of the following problems.

51. Find two consecutive integers whose product is 56.

52. Find two consecutive odd whole numbers whose product is 63.

53. Find two consecutive even whole numbers whose product is 168.

54. One number is two larger than another number. The sum of their squares is 100. Find the numbers.

55. Find four consecutive integers such that the product of the two larger integers is 22 less than twice the product of the two smaller integers.

56. Find three consecutive integers such that the product of the two smaller integers is two more than ten times the largest integer.

57. One number is three smaller than another number. The square of the larger number is nine larger than ten times the smaller number. Find the numbers.

58. The area of the floor of a rectangular room is 84 square feet. The length of the room is 5 feet more than its width. Find the length and width of the room.

59. Suppose that the width of a certain rectangle is 3 inches less than its length. The area is numerically six less than twice the perimeter. Find the length and width of the rectangle.

60. The sum of the areas of a square and a rectangle is 64 square centimeters. The length of the rectangle is 4 centimeters more than a side of the square and the width of the rectangle is 2 centimeters more than a side of the square. Find the dimensions of the square and the rectangle.

61. The perimeter of a rectangle is 30 centimeters and the area is 54 square centimeters. Find the length and width of the rectangle. [*Hint*: Let w represent the width; then $15 - w$ represents the length.]

62. The perimeter of a rectangle is 44 inches and its area is 120 square inches. Find the length and width of the rectangle.

63. An apple orchard contains 84 trees. The number of trees per row is five more than the number of rows. Find the number of rows.

64. A room contains 54 chairs. The number of rows is three less than the number of chairs per row. Find the number of rows.

65. Suppose that one leg of a right triangle is 7 feet shorter than the other leg. The hypotenuse is 2 feet longer than the longer leg. Find the lengths of all three sides of the right triangle.

66. Suppose that one leg of a right triangle is 7 meters longer than the other leg. The hypotenuse is 1 meter longer than the longer leg. Find the lengths of all three sides of the right triangle.

67. Suppose that the length of one leg of a right triangle is 2 inches less than the length of the other leg. If the length of the hypotenuse is 10 inches, find the length of each leg.

68. The length of one leg of a right triangle is 3 centimeters more than the length of the other leg. The length of the hypotenuse is 15 centimeters. Find the lengths of the two legs.

Miscellaneous Problems

For Problems 69–72, factor each trinomial assuming that all variables appearing as exponents represent positive integers.

69. $x^{2a} + 10x^a + 24$ 70. $x^{2a} + 13x^a + 40$

71. $x^{2a} - 2x^a - 8$ 72. $x^{2a} + 6x^a - 27$

73. Suppose that we want to factor $n^2 + 26n + 168$ so that we can solve the equation $n^2 + 26n + 168 = 0$. We need to find two positive integers whose product is 168 and whose sum is 26. Since the constant term, 168, is rather large, let's look at it in prime factored form.

$$168 = 2 \cdot 2 \cdot 2 \cdot 3 \cdot 7$$

Now we can mentally form two numbers by using all of these factors in different combinations. Using two 2s and the 3 in one number, and the other 2 and the 7 in another number produces $2 \cdot 2 \cdot 3 = 12$ and $2 \cdot 7 = 14$. Therefore, we can solve the given equation as follows.

$$n^2 + 26n + 168 = 0$$

$$(n + 12)(n + 14) = 0$$

$$n + 12 = 0 \qquad \text{or} \qquad n + 14 = 0$$

$$n = -12 \qquad \text{or} \qquad n = -14$$

The solution set is $\{-14, -12\}$.
Solve each of the following equations.

(a) $n^2 + 30n + 216 = 0$ (b) $n^2 + 35n + 294 = 0$

(c) $n^2 - 40n + 384 = 0$ (d) $n^2 - 40n + 375 = 0$

(e) $n^2 + 6n - 432 = 0$ (f) $n^2 - 16n - 512 = 0$

Factoring Trinomials of the Form $ax^2 + bx + c$

Now let's consider factoring trinomials where the coefficient of the squared term is not 1. First, let's illustrate an informal trial and error technique that works quite well for certain types of trinomials. This technique simply relies on our knowledge of multiplication of binomials.

EXAMPLE 1 Factor $2x^2 + 7x + 3$.

Solution By looking at the first term, $2x^2$, and the positive signs of the other two terms, we know that the binomials are of the form

$$(2x + \underline{\hspace{1cm}})(x + \underline{\hspace{1cm}}).$$

Since the factors of the constant term, 3, are 1 and 3 we have only two possibilities to try.

$$(2x + 3)(x + 1) \qquad \text{or} \qquad (2x + 1)(x + 3)$$

By checking the middle term of both of these products we find the second one yields the correct middle term of $7x$. Therefore,

$$2x^2 + 7x + 3 = (2x + 1)(x + 3). \qquad \blacksquare$$

EXAMPLE 2 Factor $6x^2 - 17x + 5$.

Solution First, note that $6x^2$ can be written as $2x \cdot 3x$ or $6x \cdot x$. Secondly, since the middle term of the trinomial is negative and the last term is positive, we know that the binomials are of the form

$$(2x - \underline{\hspace{0.5cm}})(3x - \underline{\hspace{0.5cm}}) \qquad \text{or} \qquad (6x - \underline{\hspace{0.5cm}})(x - \underline{\hspace{0.5cm}}).$$

Since the factors of the constant term, 5, are 1 and 5 the following possibilities exist.

$$(2x - 5)(3x - 1), \qquad (2x - 1)(3x - 5),$$
$$(6x - 5)(x - 1), \qquad (6x - 1)(x - 5)$$

By checking the middle term for each of these products we find that the product $(2x - 5)(3x - 1)$ produces the desired term of $-17x$. Therefore,

$$6x^2 - 17x + 5 = (2x - 5)(3x - 1). \qquad \blacksquare$$

EXAMPLE 3 Factor $4x^2 - 4x - 15$.

Solution First, note that $4x^2$ can be written as $4x \cdot x$ or $2x \cdot 2x$. Secondly, the last term, -15, can be written as $(1)(-15), (-1)(15), (3)(-5),$ or $(-3)(5)$. Thus, we can generate the possibilities for the binomial factors as follows.

Using 1 and -15	**Using** -1 **and 15**
$(4x - 15)(x + 1)$	$(4x - 1)(x + 15)$
$(4x + 1)(x - 15)$	$(4x + 15)(x - 1)$
$(2x + 1)(2x - 15)$	$(2x - 1)(2x + 15)$

Using 3 and -5	**Using** -3 **and 5**
$(4x + 3)(x - 5)$	$(4x - 3)(x + 5)$
$(4x - 5)(x + 3)$	$(4x + 5)(x - 3)$
$\checkmark (2x - 5)(2x + 3)$	$(2x + 5)(2x - 3)$

By checking the middle term of each of these products we find that the product indicated with a check mark produces the desired middle term of $-4x$. Therefore,

$$4x^2 - 4x - 15 = (2x - 5)(2x + 3).$$ ∎

Let's pause for a moment and look back over Examples 1, 2, and 3. Obviously, Example 3 created the most difficulty because we had to consider so many possibilities. We have suggested one possible format for considering the possibilities but as you practice such problems you may develop a format of your own that works better for you. Regardless of the format that you use, the key idea is to organize your work so that you consider all possibilities. Let's look at another example.

EXAMPLE 4   Factor $4x^2 + 6x + 9$.

Solution   First, note that $4x^2$ can be written as $4x \cdot x$ or $2x \cdot 2x$. Secondly, since the middle term is positive and the last term is positive, we know that the binomials are of the form

$$(4x + \text{\_\_\_})(x + \text{\_\_\_}) \quad \text{or} \quad (2x + \text{\_\_\_})(2x + \text{\_\_\_}).$$

Since 9 can be written as $9 \cdot 1$ or $3 \cdot 3$, we have only the five following possibilities to try.

$$(4x + 9)(x + 1), \quad (4x + 1)(x + 9),$$

$$(4x + 3)(x + 3), \quad (2x + 1)(2x + 9),$$

$$(2x + 3)(2x + 3)$$

When we try all of these possibilities we find that none of them yields a middle term of $6x$. Therefore, $4x^2 + 6x + 9$ is *not factorable* using integers. ∎

REMARK Example 4 illustrates the importance of organizing your work so that you try *all* possibilities before you conclude that a particular trinomial is not factorable.

Another Approach

There is another more systematic technique that you may wish to use with some trinomials. It is an extension of the method we used in the previous section. Recall that at the beginning of Section 7.3 we looked at the following product.

$$(x + r)(x + s) = x(x) + x(s) + r(x) + r(s)$$
$$= x^2 + (\underline{s + r})x + \underline{rs}$$

sum of r and s product of r and s

Now let's look at the following product.

$$(px + r)(qx + s) = px(qx) + px(s) + r(qx) + r(s)$$
$$= (pq)x^2 + (ps + rq)x + rs$$

Notice that the product of the coefficient of the x^2 term, (pq), and the constant term, (rs), is $pqrs$. Likewise, the product of the two coefficients of x, (ps and rq), is also $pqrs$. Therefore, the two coefficients of x must have a sum of $ps + rq$ and a product of $pqrs$. This may seem a little confusing, but the next few examples illustrate how easy it is to apply.

EXAMPLE 5   Factor $3x^2 + 14x + 8$.

Solution

$3x^2 + 14x + 8$ sum of 14

product of $3 \cdot 8 = 24$

We need to find two integers whose sum is 14 and whose product is 24. Obviously, 2 and 12 satisfy these conditions. Therefore, we will express the middle term of the trinomial, $14x$, as $2x + 12x$ and proceed as follows.

$$3x^2 + 14x + 8 = 3x^2 + 2x + 12x + 8$$
$$= x(3x + 2) + 4(3x + 2)$$
$$= (3x + 2)(x + 4)$$ ∎

EXAMPLE 6   Factor $16x^2 - 26x + 3$.

Solution

$16x^2 - 26x + 3$ sum of -26

product of $16(3) = 48$

We need two integers whose sum is -26 and whose product is 48. The integers -2 and -24 satisfy these conditions and allow us to express the middle term, $-26x$, as $-2x - 24x$. Then, we can factor as follows.

$$16x^2 - 26x + 3 = 16x^2 - 2x - 24x + 3$$
$$= 2x(8x - 1) - 3(8x - 1)$$
$$= (8x - 1)(2x - 3) \qquad\blacksquare$$

EXAMPLE 7   Factor $6x^2 - 5x - 6$.

Solution

$$\underbrace{6x^2 - 5x - 6}_{} \qquad \text{sum of } -5$$

product of $6(-6) = -36$

We need two integers whose product is -36 and whose sum is -5. Furthermore, since the sum is negative, the absolute value of the negative number must be greater than the absolute value of the positive number. A little searching will determine that the numbers are -9 and 4. Thus, we can express the middle term of $-5x$ as $-9x + 4x$ and proceed as follows.

$$6x^2 - 5x - 6 = 6x^2 - 9x + 4x - 6$$
$$= 3x(2x - 3) + 2(2x - 3)$$
$$= (2x - 3)(3x + 2) \qquad\blacksquare$$

Now that we have shown you two possible techniques for factoring trinomials of the form $ax^2 + bx + c$, the ball is in your court. Practice may not make you perfect at factoring, but it will surely help. We are not promoting one technique over the other; that is an individual choice. Many people find the trial and error technique we presented first very useful if the number of possibilities for the factors is fairly small. However, as the list of possibilities grows, the second technique does have the advantage of being systematic. So perhaps having both techniques at your fingertips is your best bet.

Now We Can Solve More Equations

The ability to factor certain trinomials of the form $ax^2 + bx + c$ provides us with greater equation solving capabilities. Consider the following examples.

EXAMPLE 8   Solve $3x^2 + 17x + 10 = 0$.

Solution

$$3x^2 + 17x + 10 = 0$$

$$(x + 5)(3x + 2) = 0$$

Factoring $3x^2 + 17x + 10$ as $(x + 5)(3x + 2)$ may require some extra work on scratch paper.

$$x + 5 = 0 \qquad \text{or} \qquad 3x + 2 = 0$$

$ab = 0$ if and only if $a = 0$ or $b = 0$

$$x = -5 \qquad \text{or} \qquad 3x = -2$$

$$x = -5 \qquad \text{or} \qquad x = -\frac{2}{3}$$

The solution set is $\left\{ -5, -\dfrac{2}{3} \right\}$. Check it! ∎

EXAMPLE 9

Solve $24x^2 + 2x - 15 = 0$.

Solution

$$24x^2 + 2x - 15 = 0$$

$$(4x - 3)(6x + 5) = 0$$

$$4x - 3 = 0 \qquad \text{or} \qquad 6x + 5 = 0$$

$$4x = 3 \qquad \text{or} \qquad 6x = -5$$

$$x = \frac{3}{4} \qquad \text{or} \qquad x = -\frac{5}{6}$$

The solution set is $\left\{ -\dfrac{5}{6}, \dfrac{3}{4} \right\}$. ∎

Problem Set 7.4

For Problems 1–50, factor each of the trinomials completely. Indicate any that are not factorable using integers.

1. $3x^2 + 7x + 2$

2. $2x^2 + 9x + 4$

3. $6x^2 + 19x + 10$

4. $12x^2 + 19x + 4$

5. $4x^2 - 25x + 6$

6. $5x^2 - 22x + 8$

7. $12x^2 - 31x + 20$

8. $8x^2 - 30x + 7$

9. $5y^2 - 33y - 14$

10. $3y^2 - 2y - 8$

11. $2n^2 + 13n - 24$

12. $4n^2 + 17n - 15$

13. $2x^2 + x + 7$

14. $7x^2 + 19x + 10$

15. $18x^2 + 45x + 7$

16. $10x^2 + x - 5$

17. $7x^2 - 30x + 8$

18. $6x^2 - 17x + 12$

19. $8x^2 + 2x - 21$

20. $9x^2 + 15x - 14$

21. $9t^2 - 15t - 14$

22. $12t^2 - 20t - 25$

23. $12y^2 + 79y - 35$

24. $9y^2 + 52y - 12$

25. $6n^2 + 2n - 5$

26. $20n^2 - 27n + 9$

27. $14x^2 + 55x + 21$

28. $15x^2 + 34x + 15$

29. $20x^2 - 31x + 12$

30. $8t^2 - 3t - 4$

31. $16n^2 - 8n - 15$ **32.** $25n^2 - 20n - 12$ **33.** $24x^2 - 50x + 25$

34. $24x^2 - 41x + 12$ **35.** $2x^2 + 25x + 72$ **36.** $2x^2 + 23x + 56$

37. $21a^2 + a - 2$ **38.** $14a^2 + 5a - 24$ **39.** $12a^2 - 31a - 15$

40. $10a^2 - 39a - 4$ **41.** $4x^2 + 12x + 9$ **42.** $9x^2 - 12x + 4$

43. $6x^2 - 5xy + y^2$ **44.** $12x^2 + 13xy + 3y^2$ **45.** $20x^2 + 7xy - 6y^2$

46. $8x^2 - 6xy - 35y^2$ **47.** $5x^2 - 32x + 12$ **48.** $3x^2 - 35x + 50$

49. $8x^2 - 55x - 7$ **50.** $12x^2 - 67x - 30$

For Problems 51–80, solve each equation.

51. $2x^2 + 13x + 6 = 0$ **52.** $3x^2 + 16x + 5 = 0$ **53.** $12x^2 + 11x + 2 = 0$

54. $15x^2 + 56x + 20 = 0$ **55.** $3x^2 - 25x + 8 = 0$ **56.** $4x^2 - 31x + 21 = 0$

57. $15n^2 - 41n + 14 = 0$ **58.** $6n^2 - 31n + 40 = 0$ **59.** $6t^2 + 37t - 35 = 0$

60. $2t^2 + 15t - 27 = 0$ **61.** $16y^2 - 18y - 9 = 0$ **62.** $9y^2 - 15y - 14 = 0$

63. $9x^2 - 6x - 8 = 0$ **64.** $12n^2 + 28n - 5 = 0$ **65.** $10x^2 - 29x + 10 = 0$

66. $4x^2 - 16x + 15 = 0$ **67.** $6x^2 + 19x = -10$ **68.** $12x^2 + 17x = -6$

69. $16x(x + 1) = 5$ **70.** $5x(5x + 2) = 8$ **71.** $35n^2 - 34n - 21 = 0$

72. $18n^2 - 3n - 28 = 0$ **73.** $4x^2 - 45x + 50 = 0$ **74.** $7x^2 - 65x + 18 = 0$

75. $7x^2 + 46x - 21 = 0$ **76.** $2x^2 + 7x - 30 = 0$ **77.** $12x^2 - 43x - 20 = 0$

78. $14x^2 - 13x - 12 = 0$ **79.** $18x^2 + 55x - 28 = 0$ **80.** $24x^2 + 17x - 20 = 0$

Thoughts into Words

81. Discuss the role that factoring plays in solving equations.

82. Explain how you would go about factoring $24x^2 - 17x - 20$.

83. Explain how you would solve $(x - 3)(x + 4) = 0$ and also how you would solve $(x - 3)(x + 4) = 8$.

Miscellaneous Problems

84. Consider the following approach to factoring $20x^2 + 39x + 18$.

$$20x^2 + 39x + 18 \qquad \text{sum of 39}$$

$$\text{product of } 20(18) = 360$$

We need two integers whose sum is 39 and whose product is 360. To help find these integers, let's prime factor 360.

$$360 = 2 \cdot 2 \cdot 2 \cdot 3 \cdot 3 \cdot 5$$

Now by grouping these factors in various ways we find that $2 \cdot 2 \cdot 2 \cdot 3 = 24$, $3 \cdot 5 = 15$, and $24 + 15 = 39$. So the numbers are 15 and 24 and the middle term of the given trinomial, $39x$, can be expressed as $15x + 24x$. Therefore, we can complete the factoring as follows.

$$20x^2 + 39x + 18 = 20x^2 + 15x + 24x + 18$$
$$= 5x(4x + 3) + 6(4x + 3)$$
$$= (4x + 3)(5x + 6)$$

Factor each of the following trinomials.

(a) $20x^2 + 41x + 20$

(b) $24x^2 - 79x + 40$

(c) $30x^2 + 23x - 40$

(d) $36x^2 + 65x - 36$

7.5

Factoring, Solving Equations, and Problem Solving

Before we summarize our work with factoring techniques let's look at two more special factoring patterns. These patterns emerge when multiplying binomials. Consider the following examples.

$$(x + 5)^2 = (x + 5)(x + 5) = x^2 + 10x + 25;$$
$$(2x + 3)^2 = (2x + 3)(2x + 3) = 4x^2 + 12x + 9;$$
$$(4x + 7)^2 = (4x + 7)(4x + 7) = 16x^2 + 56x + 49.$$

In general, $(a + b)^2 = (a + b)(a + b) = a^2 + 2ab + b^2$.

$$(x - 6)^2 = (x - 6)(x - 6) = x^2 - 12x + 36;$$
$$(3x - 4)^2 = (3x - 4)(3x - 4) = 9x^2 - 24x + 16;$$
$$(5x - 2)^2 = (5x - 2)(5x - 2) = 25x^2 - 20x + 4.$$

In general, $(a - b)^2 = (a - b)(a - b) = a^2 - 2ab + b^2$. Thus, we have the following patterns.

Perfect Square Trinomials

$$a^2 + 2ab + b^2 = (a + b)^2,$$
$$a^2 - 2ab + b^2 = (a - b)^2$$

Trinomials of the form $a^2 + 2ab + b^2$ or $a^2 - 2ab + b^2$ are called **perfect square trinomials**. They are easy to recognize because of the nature of their terms. For example, $9x^2 + 30x + 25$ is a perfect square trinomial because:

1. The first term is a square: $(3x)^2$;
2. The last term is a square: $(5)^2$;
3. The middle term is twice the product of the quantities being squared in the first and last terms: $2(3x)(5)$.

Likewise, $25x^2 - 40xy + 16y^2$ is a perfect square trinomial because:

1. The first term is a square: $(5x)^2$;
2. The last term is a square: $(4y)^2$;
3. The middle term is twice the product of the quantities being squared in the first and last terms: $2(5x)(4y)$.

Once we know that we have a perfect square trinomial, then the factoring process follows immediately from the two basic patterns.

$$9x^2 + 30x + 25 = (3x + 5)^2;$$

$$25x^2 - 40xy + 16y^2 = (5x - 4y)^2.$$

Here are some additional examples of perfect square trinomials and their factored form.

$$x^2 - 16x + 64 = \boxed{(x)^2 - 2(x)(8) + (8)^2} = (x - 8)^2,$$
$$16x^2 - 56x + 49 = \boxed{(4x)^2 - 2(4x)(7) + (7)^2} = (4x - 7)^2,$$
$$25x^2 + 20xy + 4y^2 = \boxed{(5x)^2 + 2(5x)(2y) + (2y)^2} = (5x + 2y)^2,$$
$$1 + 6y + 9y^2 = \boxed{(1)^2 + 2(1)(3y) + (3y)^2} = (1 + 3y)^2,$$
$$4m^2 - 4mn + n^2 = \boxed{(2m)^2 - 2(2m)(n) + (n)^2} = (2m - n)^2$$

Perhaps you will want to do this step mentally after you feel comfortable with the process.

We have considered some basic factoring techniques in this chapter one at a time, but we must be able to apply them as needed in a variety of situations. So, let's first summarize the techniques and then consider some examples.

In this chapter we have discussed:

1. Factoring by using the distributive property to factor out the greatest common monomial or binomial factor;
2. Factoring by grouping;
3. Factoring by applying the difference-of-squares pattern;

4. Factoring by applying the perfect-square-trinomial pattern;

5. Factoring of trinomials of the form $x^2 + bx + c$ into the product of two binomials;

6. Factoring of trinomials of the form $ax^2 + bx + c$ into the product of two binomials.

As a general guideline, **always look for a greatest common monomial factor first**, and then proceed with the other factoring techniques.

In each of the following examples we have factored completely whenever possible. Study them carefully and notice the factoring techniques we used.

1. $2x^2 + 12x + 10 = 2(x^2 + 6x + 5) = 2(x + 1)(x + 5)$.

2. $4x^2 + 36 = 4(x^2 + 9)$.
Remember that the sum of two squares is not factorable using integers unless there is a common factor.

3. $4t^2 + 20t + 25 = (2t + 5)^2$.
If you fail to recognize a perfect trinomial square, no harm is done. Simply proceed to factor into the product of two binomials and then you will recognize that the two binomials are the same.

4. $x^2 - 3x - 8$ is not factorable using integers. This becomes obvious in the following table.

Product	*Sum*
$1(-8) = -8$	$1 + (-8) = -7$
$-1(8) = -8$	$-1 + 8 = 7$
$2(-4) = -8$	$2 + (-4) = -2$
$-2(4) = -8$	$-2 + 4 = 2$

No two factors of -8 produce a sum of -3.

5. $6y^2 - 13y - 28 = (2y - 7)(3y + 4)$. The binomial factors can be found as follows.

$$(y + \underline{\quad})(6y - \underline{\quad})$$

$$\text{or} \qquad\qquad 1 \cdot 28 \quad \text{or} \quad 28 \cdot 1$$

$$(y - \underline{\quad})(6y + \underline{\quad}) \qquad 2 \cdot 14 \quad \text{or} \quad 14 \cdot 2$$

$$\text{or} \qquad\qquad 4 \cdot 7 \quad \text{or} \quad \boxed{7 \cdot 4}$$

$$(2y - \underline{\quad})(3y + \underline{\quad})$$

$$\text{or}$$

$$(2y + \underline{\quad})(3y - \underline{\quad})$$

6. $32x^2 - 50y^2 = 2(16x^2 - 25y^2) = 2(4x + 5y)(4x \quad 5y)$

Solving Equations by Factoring

Each time that we considered a new factoring technique in this chapter we used that technique to help solve some equations. It is important to be able to recognize which technique works for a particular type of equation.

EXAMPLE 1   Solve $x^2 = 25x$.

Solution

$$x^2 = 25x$$
$$x^2 - 25x = 0$$
$$x(x - 25) = 0$$
$$x = 0 \quad \text{or} \quad x - 25 = 0$$
$$x = 0 \quad \text{or} \quad x = 25.$$

The solution set is $\{0, 25\}$. Check it!   ■

EXAMPLE 2   Solve $x^3 - 36x = 0$.

Solution

$$x^3 - 36x = 0$$
$$x(x^2 - 36) = 0$$
$$x(x + 6)(x - 6) = 0$$
$$x = 0 \quad \text{or} \quad x + 6 = 0 \quad \text{or} \quad x - 6 = 0 \qquad \text{If } abc = 0 \\ \text{then } a = 0 \text{ or} \\ b = 0 \text{ or } c = 0.$$
$$x = 0 \quad \text{or} \quad x = -6 \quad \text{or} \quad x = 6$$

The solution set is $\{-6, 0, 6\}$. Does it check?   ■

EXAMPLE 3   Solve $10x^2 - 13x - 3 = 0$.

Solution

$$10x^2 - 13x - 3 = 0$$
$$(5x + 1)(2x - 3) = 0$$
$$5x + 1 = 0 \quad \text{or} \quad 2x - 3 = 0$$
$$5x = -1 \quad \text{or} \quad 2x = 3$$
$$x = -\frac{1}{5} \quad \text{or} \quad x = \frac{3}{2}$$

The solution set is $\left\{-\frac{1}{5}, \frac{3}{2}\right\}$. Does it check?   ■

EXAMPLE 4   Solve $4x^2 - 28x + 49 = 0$.

Solution

$$4x^2 - 28x + 49 = 0$$
$$(2x - 7)^2 = 0$$
$$(2x - 7)(2x - 7) = 0$$
$$2x - 7 = 0 \quad \text{or} \quad 2x - 7 = 0$$
$$2x = 7 \quad \text{or} \quad 2x = 7$$
$$x = \frac{7}{2} \quad \text{or} \quad x = \frac{7}{2}$$

The solution set is $\left\{\frac{7}{2}\right\}$.   ∎

Pay special attention to the next example. We need to change the form of the original equation before the property "$ab = 0$ if and only if $a = 0$ or $b = 0$" can be applied. The uniqueness of this property is that we have an indicated product set equal to zero.

EXAMPLE 5   Solve $(x + 1)(x + 4) = 40$.

Solution

$$(x + 1)(x + 4) = 40$$
$$x^2 + 5x + 4 = 40$$
$$x^2 + 5x - 36 = 0$$
$$(x + 9)(x - 4) = 0$$
$$x + 9 = 0 \quad \text{or} \quad x - 4 = 0$$
$$x = -9 \quad \text{or} \quad x = 4$$

The solution set is $\{-9, 4\}$. Check it!   ∎

EXAMPLE 6   Solve $2n^2 + 16n - 40 = 0$.

Solution

$$2n^2 + 16n - 40 = 0$$
$$2(n^2 + 8n - 20) = 0$$
$$n^2 + 8n - 20 = 0 \qquad \text{multiplied both sides of equation by } \frac{1}{2}$$
$$(n + 10)(n - 2) = 0$$
$$n + 10 = 0 \quad \text{or} \quad n - 2 = 0$$
$$n = -10 \quad \text{or} \quad n = 2$$

The solution set is $\{-10, 2\}$. Does it check?   ∎

Problem Solving

The preface of this book states that a common thread throughout the book is "to learn a skill," then "to use that skill to help solve equations," and then "to use equations to help solve problems." This thread should be very apparent in this chapter. Our new factoring skills have provided us with more ways of solving equations, which in turn gives us more power to solve word problems. Let's conclude the chapter by solving a few more problems.

PROBLEM 1   Find two numbers whose product is 65 if one of the numbers is 3 more than twice the other number.

Solution   Let x represent one number and y the other number. The problem translates into the following two equations.

product of 65 $\longrightarrow$ $xy = 65$

one number is
3 more than twice $\longrightarrow$ $y = 2x + 3$
the other number

Now we can substitute $2x + 3$ for y in the equation $xy = 65$.

$$xy = 65$$
$$x(2x + 3) = 65$$
$$2x^2 + 3x = 65$$
$$2x^2 + 3x - 65 = 0$$
$$(2x + 13)(x - 5) = 0$$

$$2x + 13 = 0 \qquad \text{or} \qquad x - 5 = 0$$
$$2x = -13 \qquad \text{or} \qquad x = 5$$
$$x = -\frac{13}{2} \qquad \text{or} \qquad x = 5$$

If $x = -\dfrac{13}{2}$, then $y = 2x + 3 = 2\left(-\dfrac{13}{2}\right) + 3 = -10$. If $x = 5$, then $y = 2x + 3 =$ $2(5) + 3 = 13$. Therefore, the numbers are $-\dfrac{13}{2}$ and -10 or 5 and 13.   ∎

REMARK Note that we used two equations and two variables in the solution for Problem 1. This is certainly not necessary, but the wording of the problem does lend itself to this approach.

PROBLEM 2   The area of a triangular sheet of paper is 14 square inches. One side of the triangle is 3 inches longer than the altitude to that side. Find the length of the one side and the length of the altitude to that side.

Solution   Let h represent the altitude to the side. Then $h + 3$ represents the side of the triangle.

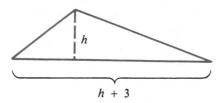

$$h + 3$$

Since the formula for finding the area of a triangle is $A = \dfrac{1}{2}bh$, we have

$$\frac{1}{2}h(h + 3) = 14$$

$$h(h + 3) = 28 \qquad \text{multiplied both sides by 2}$$

$$h^2 + 3h = 28$$

$$h^2 + 3h - 28 = 0$$

$$(h + 7)(h - 4) = 0$$

$$h + 7 = 0 \qquad \text{or} \qquad h - 4 = 0$$

$$h = -7 \qquad \text{or} \qquad h = 4$$

The solution of -7 is not reasonable. Thus, the altitude is 4 inches and the length of the side to which that altitude is drawn is 7 inches.   ∎

PROBLEM 3   A strip of uniform width is shaded along both sides and both ends of a rectangular poster 12 inches by 16 inches. How wide is the strip if one-half of the poster is shaded?

Solution   If we let x represent the width of the strip, then the following diagram depicts the situation. The area of the strip is one-half of the area of the poster; therefore, it is $\dfrac{1}{2}(12)(16) = 96$ square inches. Furthermore, we can represent the area of the strip around the poster by "the area of the poster minus the area of the unshaded portion of the poster." Thus, we can set up and solve the following equation.

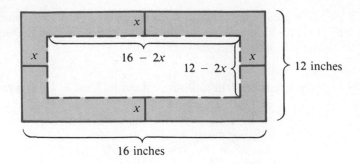

$$\text{area of poster} - \text{area of unshaded} = \text{area of strip}$$
$$\text{portion}$$

$$16(12) \quad - (16 - 2x)(12 - 2x) = \quad 96$$

$$192 - (192 - 56x + 4x^2) = 96$$

$$192 - 192 + 56x - 4x^2 = 96$$

$$-4x^2 + 56x - 96 = 0$$

$$x^2 - 14x + 24 = 0$$

$$(x - 12)(x - 2) = 0$$

$$x - 12 = 0 \quad \text{or} \quad x - 2 = 0$$

$$x = 12 \quad \text{or} \quad x = 2$$

Obviously, the strip cannot be 12 inches wide since the total width of the poster is 12 inches. Thus, we must disregard the solution of 12 and conclude that the strip is 2 inches wide. ■

Problem Set 7.5

For Problems 1–12, factor each of the perfect square trinomials.

1. $x^2 + 4x + 4$ **2.** $x^2 + 18x + 81$ **3.** $x^2 - 10x + 25$

4. $x^2 - 24x + 144$ **5.** $9n^2 + 12n + 4$ **6.** $25n^2 + 30n + 9$

7. $16a^2 - 8a + 1$ **8.** $36a^2 - 84a + 49$ **9.** $4 + 36x + 81x^2$

10. $1 - 4x + 4x^2$ **11.** $16x^2 - 24xy + 9y^2$ **12.** $64x^2 + 16xy + y^2$

For Problems 13–40, factor each polynomial completely. Indicate any that are not factorable using integers.

13. $2x^2 + 17x + 8$ **14.** $x^2 + 19x$ **15.** $2x^3 - 72x$

16. $30x^2 - x - 1$ **17.** $n^2 - 7n - 60$ **18.** $4n^3 - 100n$

19. $3a^2 - 7a - 4$ **20.** $a^2 + 7a - 30$ **21.** $8x^2 + 72$

22. $3y^3 - 36y^2 + 96y$ **23.** $9x^2 + 30x + 25$ **24.** $5x^2 - 5x - 6$

25. $15x^2 + 65x + 70$

26. $4x^2 - 20xy + 25y^2$

27. $24x^2 + 2x - 15$

28. $9x^2y - 27xy$

29. $xy + 5y - 8x - 40$

30. $xy - 3y + 9x - 27$

31. $20x^2 + 31xy - 7y^2$

32. $2x^2 - xy - 36y^2$

33. $24x^2 + 18x - 81$

34. $30x^2 + 55x - 50$

35. $12x^2 + 6x + 30$

36. $24x^2 - 8x + 32$

37. $5x^4 - 80$

38. $3x^5 - 3x$

39. $x^2 + 12xy + 36y^2$

40. $4x^2 - 28xy + 49y^2$

For Problems 41–70, solve each equation.

41. $4x^2 - 20x = 0$

42. $-3x^2 - 24x = 0$

43. $x^2 - 9x - 36 = 0$

44. $x^2 + 8x - 20 = 0$

45. $-2x^3 + 8x = 0$

46. $4x^3 - 36x = 0$

47. $6n^2 - 29n - 22 = 0$

48. $30n^2 - n - 1 = 0$

49. $(3n - 1)(4n - 3) = 0$

50. $(2n - 3)(7n + 1) = 0$

51. $(n - 2)(n + 6) = -15$

52. $(n + 3)(n - 7) = -25$

53. $2x^2 = 12x$

54. $-3x^2 = 15x$

55. $t^3 - 2t^2 - 24t = 0$

56. $2t^3 - 16t^2 - 18t = 0$

57. $12 - 40x + 25x^2 = 0$

58. $12 - 7x - 12x^2 = 0$

59. $n^2 - 28n + 192 = 0$

60. $n^2 + 33n + 270 = 0$

61. $(3n + 1)(n + 2) = 12$

62. $(2n + 5)(n + 4) = -1$

63. $x^3 = 6x^2$

64. $x^3 = -4x^2$

65. $9x^2 - 24x + 16 = 0$

66. $25x^2 + 60x + 36 = 0$

67. $x^3 + 10x^2 + 25x = 0$

68. $x^3 - 18x^2 + 81x = 0$

69. $24x^2 + 17x - 20 = 0$

70. $24x^2 + 74x - 35 = 0$

For Problems 71–88, set up an equation or a system of equations and solve each of the following problems.

71. Find two numbers whose product is 15 such that one of the numbers is seven more than four times the other number.

72. Find two numbers whose product is 12 such that one of the numbers is four less than eight times the other number.

73. Find two numbers whose product is -1. One of the numbers is three more than twice the other number.

74. The sum of the squares of three consecutive integers is 110. Find the integers.

75. One number is 1 more than twice another number. The sum of the squares of the two numbers is 97. Find the numbers.

76. One number is 1 less than three times another number. If the product of the two numbers is 102, find the numbers.

77. A room contains 54 chairs. The number of chairs per row is three less than twice the number of rows. Find the number of rows and the number of chairs per row.

78. An orchard contains 85 trees. The number of trees in each row is three less than four times the number of rows. Find the number of rows and the number of trees per row.

79. The combined area of two squares is 360 square feet. Each side of the larger square is three times as long as a side of the smaller square. How big is each square?

80. The area of a rectangular slab of sidewalk is 45 square feet. Its length is three more than four times its width. Find the length and width of the slab.

81. The length of a rectangular sheet of paper is 1 centimeter more than twice its width and the area of the rectangle is 55 square centimeters. Find the length and width of the rectangle.

82. The length of a certain rectangle is three times its width. If the length is increased by 2 inches and the width increased by 1 inch, the newly formed rectangle has an area of 70 square inches. Find the length and width of the original rectangle.

83. The area of a triangle is 51 square inches. One side of the triangle is 1 inch less than three times the length of the altitude to that side. Find the length of that side and length of the altitude to that side.

84. A square and a rectangle have equal areas. Suppose that the length of the rectangle is twice the length of a side of the square and the width of the rectangle is 4 centimeters less than the length of a side of the square. Find the dimensions of both figures.

85. A strip of uniform width is to be cut off of both sides and both ends of a sheet of paper that is 8 inches by 11 inches in order to reduce the size of the paper to an area of 40 square inches. Find the width of the strip.

86. The sum of the areas of two circles is 100π square centimeters. The length of a radius of the larger circle is 2 centimeters more than the length of a radius of the smaller circle. Find the length of a radius of each circle.

87. The sum of the areas of two circles is 180π square inches. The length of a radius of the smaller circle is 6 inches less than the length of a radius of the larger circle. Find the length of a radius of each circle.

88. A strip of uniform width is shaded along both sides and both ends of a rectangular poster that is 18 inches by 14 inches. How wide is the strip if the unshaded portion of the poster has an area of 165 square inches?

Chapter 7 Summary

(7.1) The distributive property in the form $ab + ac = a(b + c)$ provides the basis for **factoring out a greatest common monomial or binomial factor**.

Rewriting an expression such as $ab + 3a + bc + 3c$ as $a(b + 3) + c(b + 3)$, and then factoring out the common binomial factor of $b + 3$ so that $a(b + 3) + c(b + 3)$ becomes $(b + 3)(a + c)$ is called **factoring by grouping**.

The property "$ab = 0$ if and only if $a = 0$ or $b = 0$" provides us with another technique for solving equations.

(7.2) The following factoring pattern is called the **difference of two squares**.

$$(a^2 - b^2) = (a - b)(a + b)$$

(7.3) The following multiplication pattern provides a technique for factoring trinomials of the form $x^2 + bx + c$.

$$(x + r)(x + s) = x^2 + rx + sx + rs$$
$$= x^2 + (r + s)x + rs$$

sum of   product of
r and s r and s

(7.4) We presented two different techniques for factoring trinomials of the form $ax^2 + bx + c$. To review these techniques, return to Section 7.4 and study the examples.

(7.5) As a general guideline for **factoring completely**, always look for a greatest common monomial or binomial factor *first*, and then proceed with one or more of the following techniques.

1. Apply the difference-of-squares pattern.
2. Apply the perfect-square-trinomial pattern.
3. Factor a trinomial of the form $x^2 + bx + c$ into the product of two binomials.
4. Factor a trinomial of the form $ax^2 + bx + c$ into the product of two binomials.

Chapter 7 Review Problem Set

Factor completely. Indicate any that are not factorable using integers.

1. $x^2 - 9x + 14$
2. $3x^2 + 21x$
3. $9x^2 - 4$
4. $4x^2 + 8x - 5$
5. $25x^2 - 60x + 36$
6. $n^3 + 13n^2 + 40n$
7. $y^2 + 11y - 12$
8. $3xy^2 + 6x^2y$
9. $x^4 - 1$
10. $18n^2 + 9n - 5$
11. $x^2 + 7x + 24$
12. $4x^2 - 3x - 7$
13. $3n^2 + 3n - 90$
14. $x^3 - xy^2$
15. $2x^2 + 3xy - 2y^2$
16. $4n^2 - 6n - 40$
17. $5x + 5y + ax + ay$
18. $21t^2 - 5t - 4$
19. $2x^3 - 2x$
20. $3x^3 - 108x$
21. $16x^2 + 40x + 25$
22. $xy - 3x - 2y + 6$
23. $15x^2 - 7xy - 2y^2$
24. $6n^4 - 5n^3 + n^2$

Solve each of the following equations.

25. $x^2 + 4x - 12 = 0$
26. $x^2 = 11x$
27. $2x^2 + 3x - 20 = 0$
28. $9n^2 + 21n - 8 = 0$
29. $6n^2 = 24$
30. $16y^2 + 40y + 25 = 0$
31. $t^3 - t = 0$
32. $28x^2 + 71x + 18 = 0$
33. $x^2 + 3x - 28 = 0$
34. $(x - 2)(x + 2) = 21$
35. $5n^2 + 27n = 18$
36. $4n^2 + 10n = 14$
37. $2x^3 - 8x = 0$
38. $x^2 - 20x + 96 = 0$
39. $4t^2 + 17t - 15 = 0$
40. $3(x + 2) \quad x(x + 2) - 0$

41. $(2x - 5)(3x + 7) = 0$ **42.** $(x + 4)(x - 1) = 50$

43. $-7n - 2n^2 = -15$ **44.** $-23x + 6x^2 = -20$

Set up an equation or a system of equations and solve each of the following problems.

45. The larger of two numbers is one less than twice the smaller number. The difference of their squares is 33. Find the numbers.

46. The length of a rectangle is 2 centimeters less than five times the width of the rectangle. The area of the rectangle is 16 square centimeters. Find the length and width of the rectangle.

47. Suppose that the combined area of two squares is 104 square inches. Each side of the larger square is five times as long as a side of the smaller square. Find the size of each square.

48. The longer leg of a right triangle is one unit less than twice the length of the shorter leg. The hypotenuse is one unit more than twice the length of the shorter leg. Find the lengths of the three sides of the triangle.

49. The product of two numbers is 26 and one of the numbers is one larger than six times the other number. Find the numbers.

50. Find three consecutive positive odd whole numbers such that the sum of the squares of the two smaller numbers is nine more than the square of the largest number.

51. The number of books per shelf in a bookcase is one less than nine times the number of shelves. If the bookcase contains 140 books, find the number of shelves.

52. The combined area of a square and a rectangle is 225 square yards. The length of the rectangle is 8 times the width of the rectangle and the length of a side of the square is the same as the width of the rectangle. Find the dimensions of the square and the rectangle.

53. Suppose that we want to find two consecutive integers such that the sum of their squares is 613. What are they?

54. If numerically the volume of a cube equals the total surface area of the cube, find the length of an edge of the cube.

55. The combined area of two circles is 53π square meters. The length of a radius of the larger circle is 1 meter more than three times the length of a radius of the smaller circle. Find the length of a radius of each circle.

56. The product of two consecutive odd whole numbers is one less than five times their sum. Find the integers.

57. Sandy has a photograph that is 14 centimeters long and 8 centimeters wide. She wants to reduce the length and width by the same amount so that the area is decreased by 40 square centimeters. By what amount should she reduce the length and width?

58. Suppose that a strip of uniform width is plowed along both sides and both ends of a garden that is 120 feet long and 90 feet wide. How wide is the strip if the garden is one-half plowed?

Chapter 8

Exponents and Radicals

It is not uncommon in mathematics to find two separately developed concepts that are closely related to each other. In this chapter we will first develop the concepts of exponent and root individually and then show how they merge to become even more functional as a unified idea.

311

Using Integers as Exponents

Thus far in the text we have used only positive integers as exponents. In Chapter 1 the expression b^n, where b is any real number and n is a positive integer, was defined by

$$b^n = b \cdot b \cdot b \cdots b \qquad n \text{ factors of } b.$$

Then in Chapter 6 some of the parts of the following property served as a basis for manipulation with polynomials.

PROPERTY 8.1

> If m and n are positive integers and a and b are real numbers, except $b \neq 0$ whenever it appears in a denominator, then
>
> 1. $b^n \cdot b^m = b^{n+m}$;
> 2. $(b^n)^m = b^{mn}$;
> 3. $(ab)^n = a^n b^n$;
> 4. $\left(\dfrac{a}{b}\right)^n = \dfrac{a^n}{b^n}$;   Part 4 has not been previously stated.
> 5. $\dfrac{b^n}{b^m} = b^{n-m}$ when $n > m$
>
> $\dfrac{b^n}{b^m} = 1.$ when $n = m$

We are now ready to extend the concept of an exponent to include the use of zero and the negative integers as exponents.

First, let's consider the use of zero as an exponent. We want to use zero in such a way that the previously listed properties continue to hold. If "$b^n \cdot b^m = b^{n+m}$" is to hold, then $x^4 \cdot x^0 = x^{4+0} = x^4$. In other words, x^0 *acts like* 1 because $x^4 \cdot x^0 = x^4$. This line of reasoning suggests the following definition.

DEFINITION 8.1

> If b is a nonzero real number, then
> $$b^0 = 1.$$

According to Definition 8.1 the following statements are all true.

$$5^0 = 1, \qquad\qquad\qquad\qquad (-413)^0 = 1,$$

$$\left(\frac{3}{11}\right)^0 = 1, \qquad\qquad\qquad n^0 = 1, \qquad n \neq 0,$$

$$(x^3 y^4)^0 = 1, \qquad x \neq 0, y \neq 0$$

We can use a similar line of reasoning to motivate a definition for the use of negative integers as exponents. Consider the example $x^4 \cdot x^{-4}$. If "$b^n \cdot b^m = b^{n+m}$" is to hold, then $x^4 \cdot x^{-4} = x^{4+(-4)} = x^0 = 1$. Thus, x^{-4} must be the reciprocal of x^4, since their product is 1. That is to say,

$$x^{-4} = \frac{1}{x^4}.$$

This suggests the following general definition.

DEFINITION 8.2

> If n is a positive integer and b is a nonzero real number, then
>
> $$b^{-n} = \frac{1}{b^n}.$$

According to Definition 8.2 the following statements are true.

$$x^{-5} = \frac{1}{x^5}, \qquad\qquad 2^{-4} = \frac{1}{2^4} = \frac{1}{16},$$

$$10^{-2} = \frac{1}{10^2} = \frac{1}{100} \text{ or } 0.01, \qquad \frac{2}{x^{-3}} = \frac{2}{\dfrac{1}{x^3}} = (2)\left(\frac{x^3}{1}\right) = 2x^3,$$

$$\left(\frac{3}{4}\right)^{-2} = \frac{1}{\left(\dfrac{3}{4}\right)^2} = \frac{1}{\dfrac{9}{16}} = \frac{16}{9}$$

It can be verified (although it is beyond the scope of this text) that all of the parts of Property 8.1 hold for *all integers*. In fact, the following equality can replace the two separate statements for part (5).

$$\frac{b^n}{b^m} = b^{n-m} \quad \text{for all integers } n \text{ and } m$$

Let's restate Property 8.1 as it holds for all integers and include, at the right, a "name tag" for easy reference.

PROPERTY 8.2

> If m and n are integers and a and b are real numbers, except $b \neq 0$ whenever it appears in a denominator, then
>
> | | | |
> |---|---|---|
> | **1.** $b^n \cdot b^m = b^{n+m}$ | | product of two powers |
> | **2.** $(b^n)^m = b^{mn}$ | | power of a power |
> | **3.** $(ab)^n = a^n b^n$ | | power of a product |
> | **4.** $\left(\dfrac{a}{b}\right)^n = \dfrac{a^n}{b^n}$ | | power of a quotient |
> | **5.** $\dfrac{b^n}{b^m} = b^{n-m}$ | | quotient of two powers |

Having the use of all integers as exponents allows us to work with a large variety of numerical and algebraic expressions. Let's consider some examples that illustrate the use of the various parts of Property 8.2.

EXAMPLE 1   Simplify each of the following numerical expressions.

(a) $10^{-3} \cdot 10^2$ (b) $(2^{-3})^{-2}$ (c) $(2^{-1} \cdot 3^2)^{-1}$

(d) $\left(\dfrac{2^{-3}}{3^{-2}}\right)^{-1}$ (e) $\dfrac{10^{-2}}{10^{-4}}$

Solution

(a) $10^{-3} \cdot 10^2 = 10^{-3+2}$ product of two powers

$= 10^{-1}$

$= \dfrac{1}{10^1} = \dfrac{1}{10}$

(b) $(2^{-3})^{-2} = 2^{(-2)(-3)}$ power of a power

$= 2^6 = 64$

(c) $(2^{-1} \cdot 3^2)^{-1} = (2^{-1})^{-1}(3^2)^{-1}$ power of a product

$= 2^1 \cdot 3^{-2}$

$= \dfrac{2^1}{3^2} = \dfrac{2}{9}$

(d) $\left(\dfrac{2^{-3}}{3^{-2}}\right)^{-1} = \dfrac{(2^{-3})^{-1}}{(3^{-2})^{-1}}$ power of a quotient

$= \dfrac{2^3}{3^2} = \dfrac{8}{9}$

(e) $\dfrac{10^{-2}}{10^{-4}} = 10^{-2-(-4)}$ quotient of two powers

$= 10^2 = 100$ ■

EXAMPLE 2   Simplify each of the following; express final results without using zero or negative integers as exponents.

(a) $x^2 \cdot x^{-5}$ (b) $(x^{-2})^4$ (c) $(x^2 y^{-3})^{-4}$

(d) $\left(\dfrac{a^3}{b^{-5}}\right)^{-2}$ (e) $\dfrac{x^{-4}}{x^{-2}}$

Solution

(a) $x^2 \cdot x^{-5} = x^{2+(-5)}$ product of two powers

$= x^{-3}$

$= \dfrac{1}{x^3}$

(b) $(x^{-2})^4 = x^{4(-2)}$ power of a power

$\qquad\quad = x^{-8}$

$\qquad\quad = \dfrac{1}{x^8}$

(c) $(x^2 y^{-3})^{-4} = (x^2)^{-4}(y^{-3})^{-4}$ power of a product

$\qquad\qquad\quad = x^{-4(2)} y^{-4(-3)}$

$\qquad\qquad\quad = x^{-8} y^{12}$

$\qquad\qquad\quad = \dfrac{y^{12}}{x^8}$

(d) $\left(\dfrac{a^3}{b^{-5}}\right)^{-2} = \dfrac{(a^3)^{-2}}{(b^{-5})^{-2}}$ power of a quotient

$\qquad\qquad\quad = \dfrac{a^{-6}}{b^{10}}$

$\qquad\qquad\quad = \dfrac{1}{a^6 b^{10}}$

(e) $\dfrac{x^{-4}}{x^{-2}} = x^{-4-(-2)}$ quotient of two powers

$\qquad\quad = x^{-2}$

$\qquad\quad = \dfrac{1}{x^2}$ ■

EXAMPLE 3   Find the indicated products and quotients; express your results using positive integral exponents only.

$\qquad$ **(a)** $(3x^2 y^{-4})(4x^{-3} y)$ $\qquad$ **(b)** $\dfrac{12a^3 b^2}{-3a^{-1} b^5}$ $\qquad$ **(c)** $\left(\dfrac{15x^{-1} y^2}{5xy^{-4}}\right)^{-1}$

Solution   **(a)** $(3x^2 y^{-4})(4x^{-3} y) = 12x^{2+(-3)} y^{-4+1}$

$\qquad\qquad\qquad\qquad\quad = 12x^{-1} y^{-3}$

$\qquad\qquad\qquad\qquad\quad = \dfrac{12}{xy^3}$

$\qquad$ **(b)** $\dfrac{12a^3 b^2}{-3a^{-1} b^5} = -4a^{3-(-1)} b^{2-5}$

$\qquad\qquad\qquad\quad = -4a^4 b^{-3}$

$\qquad\qquad\qquad\quad = -\dfrac{4a^4}{b^3}$

(c) $\left(\dfrac{15x^{-1}y^2}{5xy^{-4}}\right)^{-1} = (3x^{-1-1}y^{2-(-4)})^{-1}$ Notice that we are first simplifying inside the parentheses.

$\qquad\qquad\qquad\qquad = (3x^{-2}y^6)^{-1}$

$\qquad\qquad\qquad\qquad = 3^{-1}x^2y^{-6}$

$\qquad\qquad\qquad\qquad = \dfrac{x^2}{3y^6}.$ ∎

The final examples of this section show the simplification of numerical and algebraic expressions that involve sums and differences. In such cases, we use Definition 8.2 to change from negative to positive exponents so that we can proceed in the usual way.

EXAMPLE 4   Simplify $2^{-3} + 3^{-1}$.

Solution $2^{-3} + 3^{-1} = \dfrac{1}{2^3} + \dfrac{1}{3^1}$

$\qquad\qquad\qquad = \dfrac{1}{8} + \dfrac{1}{3}$

$\qquad\qquad\qquad = \dfrac{3}{24} + \dfrac{8}{24}$

$\qquad\qquad\qquad = \dfrac{11}{24}$ ∎

EXAMPLE 5   Simplify $(4^{-1} - 3^{-2})^{-1}$.

Solution $(4^{-1} - 3^{-2})^{-1} = \left(\dfrac{1}{4^1} - \dfrac{1}{3^2}\right)^{-1}$ Apply $b^{-n} = \dfrac{1}{b^n}$ to 4^{-1} and to 3^{-2}.

$\qquad\qquad\qquad\qquad = \left(\dfrac{1}{4} - \dfrac{1}{9}\right)^{-1}$

$\qquad\qquad\qquad\qquad = \left(\dfrac{9}{36} - \dfrac{4}{36}\right)^{-1}$

$\qquad\qquad\qquad\qquad = \left(\dfrac{5}{36}\right)^{-1}$

$\qquad\qquad\qquad\qquad = \dfrac{1}{\left(\dfrac{5}{36}\right)^1}$ Apply $b^{-n} = \dfrac{1}{b^n}$.

$\qquad\qquad\qquad\qquad = \dfrac{1}{\dfrac{5}{36}} = \dfrac{36}{5}$ ∎

EXAMPLE 6   Express $a^{-1} + b^{-2}$ as a single fraction involving positive exponents only.

Solution

$$a^{-1} + b^{-2} = \frac{1}{a^1} + \frac{1}{b^2}$$

$$= \left(\frac{1}{a}\right)\left(\frac{b^2}{b^2}\right) + \left(\frac{1}{b^2}\right)\left(\frac{a}{a}\right) \qquad \text{Use } ab^2 \text{ as the LCD.}$$

$$= \frac{b^2}{ab^2} + \frac{a}{ab^2}$$

$$= \frac{b^2 + a}{ab^2}$$ ∎

Problem Set 8.1

Simplify each of the following numerical expressions.

1. 3^{-3}
2. 2^{-4}
3. -10^{-2}
4. 10^{-3}
5. $\frac{1}{3^{-4}}$
6. $\frac{1}{2^{-6}}$
7. $-\left(\frac{1}{3}\right)^{-3}$
8. $\left(\frac{1}{2}\right)^{-3}$
9. $\left(-\frac{1}{2}\right)^{-3}$
10. $\left(\frac{2}{7}\right)^{-2}$
11. $\left(-\frac{3}{4}\right)^{0}$
12. $\frac{1}{\left(\frac{4}{5}\right)^{-2}}$
13. $\frac{1}{\left(\frac{3}{7}\right)^{-2}}$
14. $-\left(\frac{5}{6}\right)^{0}$
15. $2^7 \cdot 2^{-3}$
16. $3^{-4} \cdot 3^6$
17. $10^{-5} \cdot 10^2$
18. $10^4 \cdot 10^{-6}$
19. $10^{-1} \cdot 10^{-2}$
20. $10^{-2} \cdot 10^{-2}$
21. $(3^{-1})^{-3}$
22. $(2^{-2})^{-4}$
23. $(5^3)^{-1}$
24. $(3^{-1})^3$
25. $(2^3 \cdot 3^{-2})^{-1}$
26. $(2^{-2} \cdot 3^{-1})^{-3}$
27. $(4^2 \cdot 5^{-1})^2$
28. $(2^{-3} \cdot 4^{-1})^{-1}$
29. $\left(\frac{2^{-1}}{5^{-2}}\right)^{-1}$
30. $\left(\frac{2^{-4}}{3^{-2}}\right)^{-2}$
31. $\left(\frac{2^{-1}}{3^{-2}}\right)^{2}$
32. $\left(\frac{3^2}{5^{-1}}\right)^{-1}$
33. $\frac{3^3}{3^{-1}}$
34. $\frac{2^{-2}}{2^3}$
35. $\frac{10^{-2}}{10^2}$
36. $\frac{10^{-2}}{10^{-5}}$
37. $2^{-2} + 3^{-2}$
38. $2^{-4} + 5^{-1}$
39. $\left(\frac{1}{3}\right)^{-1} - \left(\frac{2}{5}\right)^{-1}$
40. $\left(\frac{3}{2}\right)^{-1} - \left(\frac{1}{4}\right)^{-1}$
41. $(2^{-3} + 3^{-2})^{-1}$
42. $(5^{-1} - 2^{-3})^{-1}$

Simplify each of the following; express final results without using zero or negative integers as exponents.

43. $x^2 \cdot x^{-8}$
44. $x^{-3} \cdot x^{-4}$
45. $a^3 \cdot a^{-5} \cdot a^{-1}$
46. $b^{-2} \cdot b^3 \cdot b^{-6}$
47. $(a^{-4})^2$
48. $(b^4)^{-3}$
49. $(x^2y^{-6})^{-1}$
50. $(x^3y^{-1})^{-3}$
51. $(ab^3c^{-2})^{-4}$
52. $(a^3b^{-3}c^{-2})^{-5}$
53. $(2x^3y^{-4})^{-3}$
54. $(4x^5y^{-2})^{-2}$

55. $\left(\dfrac{x^{-1}}{y^{-4}}\right)^{-3}$ **56.** $\left(\dfrac{y^3}{x^{-4}}\right)^{-2}$ **57.** $\left(\dfrac{3a^{-2}}{2b^{-1}}\right)^{-2}$ **58.** $\left(\dfrac{2xy^2}{5a^{-1}b^{-2}}\right)^{-1}$

59. $\dfrac{x^{-6}}{x^{-4}}$ **60.** $\dfrac{a^{-2}}{a^2}$ **61.** $\dfrac{a^3b^{-2}}{a^{-2}b^{-4}}$ **62.** $\dfrac{x^{-3}y^{-4}}{x^2y^{-1}}$

Find the indicated products and quotients; express results using positive integral exponents only.

63. $(2xy^{-1})(3x^{-2}y^4)$ **64.** $(-4x^{-1}y^2)(6x^3y^{-4})$

65. $(-7a^2b^{-5})(-a^{-2}b^7)$ **66.** $(-9a^{-3}b^{-6})(-12a^{-1}b^4)$

67. $\dfrac{28x^{-2}y^{-3}}{4x^{-3}y^{-1}}$ **68.** $\dfrac{63x^2y^{-4}}{7xy^{-4}}$

69. $\dfrac{-72a^2b^{-4}}{6a^3b^{-7}}$ **70.** $\dfrac{108a^{-5}b^{-4}}{9a^{-2}b}$

71. $\left(\dfrac{35x^{-1}y^{-2}}{7x^4y^3}\right)^{-1}$ **72.** $\left(\dfrac{-48ab^2}{-6a^3b^5}\right)^{-2}$

73. $\left(\dfrac{-36a^{-1}b^{-6}}{4a^{-1}b^4}\right)^{-2}$ **74.** $\left(\dfrac{8xy^3}{-4x^4y}\right)^{-3}$

Express each of the following as a single fraction involving positive exponents only.

75. $x^{-2} + x^{-3}$ **76.** $x^{-1} + x^{-5} x$

77. $x^{-3} - y^{-1}$ **78.** $2x^{-1} - 3y^{-2}$

79. $3a^{-2} + 4b^{-1}$ **80.** $a^{-1} + a^{-1}b^{-3}$

81. $x^{-1}y^{-2} - xy^{-1}$ **82.** $x^2y^{-2} - x^{-1}y^{-3}$

83. $2x^{-1} - 3x^{-2}$ **84.** $5x^{-2}y + 6x^{-1}y^{-2}$

Thoughts into Words

85. Explain how to simplify $(2^{-1} \cdot 3^{-2})^{-1}$ and how to simplify $(2^{-1} + 3^{-2})^{-1}$.

86. Explain how you would simplify $\left(\dfrac{2}{3}\right)^{-1} + \left(\dfrac{1}{2}\right)^{-2}$.

8.2

Square Roots and Radicals

To **square a number** means to raise it to the second power, that is, to use the number as a factor twice.

$$4^2 = 4 \cdot 4 = 16, \qquad \text{read "four squared equals sixteen"}$$

$$10^2 = 10 \cdot 10 = 100,$$

$$\left(\frac{1}{2}\right)^2 = \frac{1}{2} \cdot \frac{1}{2} = \frac{1}{4},$$
$$(-3)^2 = (-3)(-3) = 9$$

A **square root of a number** is one of its two equal factors. Thus, 4 is a square root of 16 because $4 \cdot 4 = 16$. Likewise, -4 is also a square root of 16 because $(-4)(-4) = 16$. In general, a is a square root of b if $a^2 = b$. The following generalizations are a direct consequence of the previous statement.

1. Every positive real number has two square roots; one is positive and the other is negative. They are opposites of each other.

2. Negative real numbers have no real number square roots because any nonzero real number is positive when squared.

3. The square root of 0 is 0.

The symbol $\sqrt{}$, called a **radical sign**, is used to designate the nonnegative square root. The number under the radical sign is called the **radicand**. The entire expression, such as $\sqrt{16}$, is called a **radical**.

$\sqrt{16} = 4$, $\sqrt{16}$ indicates the *nonnegative* or **principal square root** of 16.

$-\sqrt{16} = -4$, $-\sqrt{16}$ indicates the negative square root of 16.

$\sqrt{0} = 0$,   Zero has only one square root. Technically, we could write $-\sqrt{0} = -0 = 0$.

$\sqrt{-4}$ is not a real number,

$-\sqrt{-4}$ is not a real number

In general, the following definition is useful.

DEFINITION 8.3

> If $a \geq 0$ and $b \geq 0$, then $\sqrt{b} = a$ if and only if $a^2 = b$; a is called the **principal square root of b**.

If a is a nonnegative integer that is the square of an integer, then $\sqrt{a}$ and $-\sqrt{a}$ are rational numbers. For example, $\sqrt{1}$, $\sqrt{4}$, and $\sqrt{25}$ are the rational numbers 1, 2, and 5, respectively. The numbers 1, 4, and 25 are called perfect squares because each one represents the square of some integer. The following chart contains the squares of the whole numbers from 1 through 20, inclusive. You should know these values

$$1^2 = 1, \qquad 5^2 = 25,$$
$$2^2 = 4, \qquad 6^2 = 36,$$
$$3^2 = 9, \qquad 7^2 = 49,$$
$$4^2 = 16, \qquad 8^2 = 64,$$

$$9^2 = 81, \qquad 15^2 = 225,$$
$$10^2 = 100, \qquad 16^2 = 256,$$
$$11^2 = 121, \qquad 17^2 = 289,$$
$$12^2 = 144, \qquad 18^2 = 324,$$
$$13^2 = 169, \qquad 19^2 = 361,$$
$$14^2 = 196, \qquad 20^2 = 400$$

so that you can immediately recognize such square roots as $\sqrt{81} = 9$, $\sqrt{144} = 12$, $\sqrt{289} = 17$, and so on from the list. Furthermore, perfect squares of multiples of 10 are easy to recognize. For example, since $30^2 = 900$, we know that $\sqrt{900} = 30$.

Knowledge of the previous listing of perfect squares also helps with square roots of some fractions. Consider the following examples.

$$\sqrt{\frac{16}{25}} = \frac{4}{5} \quad \text{because} \left(\frac{4}{5}\right)^2 = \frac{16}{25},$$

$$\sqrt{\frac{36}{49}} = \frac{6}{7} \quad \text{because} \left(\frac{6}{7}\right)^2 = \frac{36}{49},$$

$$\sqrt{.09} = .3 \quad \text{because} (.3)^2 = .09$$

If a is a positive integer that is *not* the square of an integer, then $\sqrt{a}$ and $-\sqrt{a}$ are irrational numbers. For example, $\sqrt{2}$, $-\sqrt{2}$, $\sqrt{23}$, $\sqrt{31}$, $\sqrt{52}$, and $-\sqrt{75}$ are irrational numbers. Remember that irrational numbers have nonrepeating decimal representations. For example, $\sqrt{2} = 1.414213562373\ldots$, where the decimal never repeats a block of digits.

For practical purposes, we often need to use a rational approximation of an irrational number. The calculator becomes a very useful tool for finding such approximations. Be sure that you can use your calculator to find the following approximate square roots. Each approximate square root has been rounded to the nearest thousandth. (The symbol $\approx$ means "approximately equal to.")

$$\sqrt{19} \approx 4.359, \qquad \sqrt{38} \approx 6.164, \qquad \sqrt{72} \approx 8.485, \qquad \sqrt{93} \approx 9.644$$

REMARK If you don't have your calculator available, there is a square root table in Appendix B.

Adding and Subtracting Square Roots

Recall our use of the distributive property as the basis for combining similar terms. For example,

$$3x + 2x = (3 + 2)x = 5x, \qquad 7y - 4y = (7 - 4)y = 3y,$$
$$9a^2 + 5a^2 = (9 + 5)a^2 = 14a^2.$$

In a like manner, we can often simplify expressions that contain radicals by using the distributive property as follows.

$$5\sqrt{2} + 7\sqrt{2} = (5 + 7)\sqrt{2} = 12\sqrt{2},$$
$$8\sqrt{5} - 2\sqrt{5} = (8 - 2)\sqrt{5} = 6\sqrt{5},$$
$$4\sqrt{7} + 6\sqrt{7} + 3\sqrt{11} - \sqrt{11} = (4 + 6)\sqrt{7} + (3 - 1)\sqrt{11}$$
$$= 10\sqrt{7} + 2\sqrt{11}$$

Note that **to add or subtract square roots they must have the same radicand**. Also note the form we use to indicate multiplication when a radical is involved. For example, $5 \cdot \sqrt{2}$ is written as $5\sqrt{2}$.

Now suppose that we need to evaluate $5\sqrt{2} - \sqrt{2} + 4\sqrt{2} - 2\sqrt{2}$, to the nearest tenth. We can either evaluate the expression as it stands or first simplify it by combining radicals and then evaluate that result. Let's use the latter approach. (It would probably be a good idea for you to do it both ways for checking purposes.)

$$5\sqrt{2} - \sqrt{2} + 4\sqrt{2} - 2\sqrt{2} = (5 - 1 + 4 - 2)\sqrt{2}$$
$$= 6\sqrt{2}$$
$$\approx 8.5, \quad \text{to the nearest tenth}$$

EXAMPLE 1

Find a rational approximation, to the nearest tenth, for

$$7\sqrt{3} + 9\sqrt{5} + 2\sqrt{3} - 3\sqrt{5} + 13\sqrt{3}.$$

Solution

First, let's simplify the given expression and then evaluate that result.

$$7\sqrt{3} + 9\sqrt{5} + 2\sqrt{3} - 3\sqrt{5} + 13\sqrt{3} = (7 + 2 + 13)\sqrt{3} + (9 - 3)\sqrt{5}$$
$$= 22\sqrt{3} + 6\sqrt{5}$$
$$\approx 51.5, \quad \text{to the nearest tenth} \quad \blacksquare$$

Estimation can play a meaningful role when working with radicals. For example, in the solution for Example 1 the given expression was simplified to $22\sqrt{3} + 6\sqrt{5}$. Using 2 as a whole number approximation for both $\sqrt{3}$ and $\sqrt{5}$, we can estimate an answer of $22(2) + 6(2) = 56$. This is a very rough estimate, but it does indicate that our answer of 51.5 is *reasonable*.

Furthermore, estimating radicals can often give us a better feeling of the magnitude of an expression. For example, it is easy to look at an expression such as $\sqrt{15} + \sqrt{27} - \sqrt{35}$ without any feeling as to the *size* of the number involved. However, if we use 4, 5, and 6 as whole number approximations for $\sqrt{15}$, $\sqrt{27}$, and $\sqrt{35}$, respectively, then the given expression can be approximated by $4 + 5 - 6 = 3$. Thus, a feeling of size can be associated with the expression.

Problem Set 8.2

For Problems 1–18, evaluate each radical without using a calculator or a table.

1. $\sqrt{49}$ 7
2. $\sqrt{100}$ 10
3. $-\sqrt{64}$ −8
4. $-\sqrt{36}$ −6
5. $\sqrt{121}$ 11
6. $\sqrt{144}$ 12
7. $\sqrt{3600}$ 60
8. $\sqrt{2500}$ 50
9. $-\sqrt{1600}$
10. $-\sqrt{900}$
11. $\sqrt{6400}$
12. $\sqrt{400}$
13. $\sqrt{324}$
14. $-\sqrt{361}$
15. $\sqrt{\dfrac{25}{9}}$
16. $\sqrt{\dfrac{1}{225}}$
17. $\sqrt{.16}$
18. $\sqrt{.0121}$

For Problems 19–28, use your calculator or the table in Appendix B to evaluate each radical.

19. $\sqrt{576}$
20. $\sqrt{7569}$
21. $\sqrt{2304}$
22. $\sqrt{9801}$
23. $\sqrt{784}$
24. $\sqrt{1849}$
25. $\sqrt{4225}$
26. $\sqrt{2704}$
27. $\sqrt{3364}$
28. $\sqrt{1444}$

For Problems 29–36, use your calculator or the table in Appendix B to find a rational approximation of each square root. Express your answers to the nearest hundredth.

29. $\sqrt{19}$
30. $\sqrt{34}$
31. $\sqrt{50}$
32. $\sqrt{66}$
33. $\sqrt{75}$
34. $\sqrt{90}$
35. $\sqrt{95}$
36. $\sqrt{98}$

For Problems 37–46, use your calculator or the table in Appendix B to find a whole number approximation for each of the following.

37. $\sqrt{4325}$
38. $\sqrt{7500}$
39. $\sqrt{1175}$
40. $\sqrt{1700}$
41. $\sqrt{9501}$
42. $\sqrt{8050}$
43. $\sqrt{6614}$
44. $\sqrt{5825}$
45. $\sqrt{3400}$
46. $\sqrt{2250}$

For Problems 47–58, simplify each expression by using the distributive property.

47. $7\sqrt{2} + 14\sqrt{2}$
48. $9\sqrt{3} + 4\sqrt{3}$
49. $17\sqrt{7} - 9\sqrt{7}$
50. $19\sqrt{5} - 8\sqrt{5}$
51. $6\sqrt{3} - 15\sqrt{3}$
52. $7\sqrt{6} - 21\sqrt{6}$
53. $9\sqrt{5} + 3\sqrt{5} - 6\sqrt{5}$
54. $8\sqrt{7} + 13\sqrt{7} - 9\sqrt{7}$
55. $8\sqrt{2} - 4\sqrt{3} - 9\sqrt{2} + 6\sqrt{3}$
56. $7\sqrt{5} - 9\sqrt{6} + 14\sqrt{5} - 2\sqrt{6}$
57. $6\sqrt{7} + 5\sqrt{10} - 8\sqrt{10} - 4\sqrt{7} - 11\sqrt{7} + \sqrt{10}$
58. $\sqrt{3} - \sqrt{5} + 4\sqrt{5} - 3\sqrt{3} - 9\sqrt{5} - 16\sqrt{3}$

For Problems 59–72, find a rational approximation, to the nearest one tenth, for each of the radical expressions.

59. $9\sqrt{3} + \sqrt{3}$
60. $6\sqrt{2} + 14\sqrt{2}$
61. $9\sqrt{5} - 3\sqrt{5}$
62. $18\sqrt{6} - 12\sqrt{6}$
63. $14\sqrt{2} - 15\sqrt{2}$
64. $7\sqrt{3} - 12\sqrt{3}$

65. $8\sqrt{7} - 4\sqrt{7} + 6\sqrt{7}$ **66.** $9\sqrt{5} \quad 2\sqrt{5} + \sqrt{5}$

67. $4\sqrt{3} - 2\sqrt{2}$ **68.** $3\sqrt{2} + \sqrt{3} - \sqrt{5}$

69. $9\sqrt{6} - 3\sqrt{5} + 2\sqrt{6} - 7\sqrt{5} - \sqrt{6}$

70. $8\sqrt{7} - 2\sqrt{10} + 4\sqrt{7} - 3\sqrt{10} - 7\sqrt{7} + 4\sqrt{10}$

71. $4\sqrt{11} - 5\sqrt{11} - 7\sqrt{11} + 2\sqrt{11} - 3\sqrt{11}$

72. $14\sqrt{13} - 17\sqrt{13} + 3\sqrt{13} - 4\sqrt{13} - 5\sqrt{13}$

8.3
More about Roots and Radicals

To **cube a number** means to raise it to the third power, that is, to use the number as a factor three times.

$$2^3 = 2 \cdot 2 \cdot 2 = 8, \qquad \text{read ''two cubed equals eight''}$$

$$4^3 = 4 \cdot 4 \cdot 4 = 64,$$

$$\left(\frac{2}{3}\right)^3 = \frac{2}{3} \cdot \frac{2}{3} \cdot \frac{2}{3} = \frac{8}{27},$$

$$(-2)^3 = (-2)(-2)(-2) = -8$$

A **cube root of a number** is one of its three equal factors. Thus, 2 is a cube root of 8 because $2 \cdot 2 \cdot 2 = 8$. (In fact, 2 is the only real number that is a cube root of 8.) Furthermore, -2 is a cube root of -8 because $(-2)(-2)(-2) = -8$. (In fact, -2 is the only real number that is a cube root of -8.)

In general, a is a cube root of b if $a^3 = b$. The following generalizations are a direct consequence of the previous statement.

1. Every positive real number has one positive real number cube root.

2. Every negative real number has one negative real number cube root.

3. The cube root of 0 is 0.

REMARK Technically, every nonzero real number has three cube roots, but only one of them is a real number. The other two roots are classified as complex numbers. We are restricting our work at this time to the set of real numbers.

The symbol $\sqrt[3]{}$ designates the cube root of a number. Thus, we can write

$$\sqrt[3]{8} = 2, \qquad \sqrt[3]{\frac{1}{27}} = \frac{1}{3},$$

$$\sqrt[3]{-8} = -2, \qquad \sqrt[3]{-\frac{1}{27}} = -\frac{1}{3}$$

In general, the following definition is useful.

DEFINITION 8.4

$$\sqrt[3]{b} = a \quad \text{if and only if } a^3 = b.$$

In Definition 8.4, if $b \geq 0$ then $a \geq 0$ whereas if $b < 0$ then $a < 0$. The number a is called **the principal cube root of b** or simply **the cube root of b**.

The concept of root can be extended to fourth roots, fifth roots, sixth roots, and, in general, nth roots. We can make the following generalizations.

If n is an even positive integer, then the following statements are true.

1. Every positive real number has exactly two real nth roots—one positive and one negative. For example, the real fourth roots of 16 are 2 and -2.

2. Negative real numbers do not have real nth roots. For example, there are no real fourth roots of -16.

If n is an odd positive integer greater than one, then the following statements are true.

1. Every real number has exactly one real nth root.

2. The real nth root of a positive number is positive. For example, the fifth root of 32 is 2.

3. The real nth root of a negative number is negative. For example, the fifth root of -32 is -2.

In general, the following definition is useful.

DEFINITION 8.5

$$\sqrt[n]{b} = a \quad \text{if and only if } a^n = b.$$

In Definition 8.5, if n is an even positive integer, then a and b are both nonnegative. If n is an odd positive integer greater than one, then a and b are both nonnegative or both negative. The symbol $\sqrt[n]{}$ designates the **principal nth root**. Consider the following examples.

$$\sqrt[4]{81} = 3 \quad \text{because } 3^4 = 81,$$
$$\sqrt[5]{32} = 2 \quad \text{because } 2^5 = 32,$$
$$\sqrt[5]{-32} = -2 \quad \text{because } (-2)^5 = -32$$

To complete our terminology, the n in the radical $\sqrt[n]{b}$ is called the **index** of the radical. If $n = 2$, we commonly write $\sqrt{b}$ instead of $\sqrt[2]{b}$. In the future as we use symbols such as $\sqrt[n]{b}$, $\sqrt[m]{y}$, and $\sqrt[n]{x}$, we will assume the previous agreements relative to the existence of real roots (without listing the various restrictions) unless a special restriction is necessary.

The following property is a direct consequence of Definition 8.5.

PROPERTY 8.3

> **1.** $(\sqrt[n]{b})^n = b$ n is any positive integer greater than one.
>
> **2.** $\sqrt[n]{b^n} = b$ n is any positive integer greater than one if $b \geq 0$; n is an odd positive integer greater than one if $b < 0$.

The following examples demonstrate the use of Property 8.3.

$$\sqrt{16^2} = (\sqrt{16})^2 = 4^2 = 16,$$

$$\sqrt[3]{64^3} = (\sqrt[3]{64})^3 = 4^3 = 64,$$

$$\sqrt[3]{(-8)^3} = (\sqrt[3]{-8})^3 = (-2)^3 = -8,$$

but $\sqrt{(-16)^2} \neq (\sqrt{-16})^2$ because $\sqrt{-16}$ is not a real number.

Let's use some examples to lead into the next very useful property of radicals.

$$\sqrt{4 \cdot 9} = \sqrt{36} = 6 \quad \text{and} \quad \sqrt{4} \cdot \sqrt{9} = 2 \cdot 3 = 6,$$

$$\sqrt{16 \cdot 25} - \sqrt{400} = 20 \quad \text{and} \quad \sqrt{16} \cdot \sqrt{25} = 4 \cdot 5 = 20,$$

$$\sqrt[3]{8 \cdot 27} = \sqrt[3]{216} = 6 \quad \sqrt[3]{8} \cdot \sqrt[3]{27} = 2 \cdot 3 = 6,$$

$$\sqrt[3]{(-8)(27)} = \sqrt[3]{-216} = -6 \quad \text{and} \quad \sqrt[3]{-8} \cdot \sqrt[3]{27} = (-2)(3) = -6$$

In general, we can state the following property.

PROPERTY 8.4

> $\sqrt[n]{bc} = \sqrt[n]{b}\sqrt[n]{c}$ $\sqrt[n]{b}$ and $\sqrt[n]{c}$ are real numbers

Property 8.4 states that **the nth root of a product is equal to the product of the nth roots.**

Simplest Radical Form

The definition of nth root, along with Property 8.4, provides the basis for changing radicals to simplest radical form. The concept of **simplest radical form** takes on additional meaning as we encounter more complicated expressions, but for now it simply means that the radicand is not to contain any perfect powers of the index. Let's consider some examples to clarify this idea.

EXAMPLE 1

Express each of the following in simplest radical form.

     (a) $\sqrt{8}$ **(b)** $\sqrt{45}$ **(c)** $\sqrt[3]{24}$ **(d)** $\sqrt[3]{54}$

Solution

(a) $\sqrt{8} = \sqrt{4 \cdot 2} = \sqrt{4}\sqrt{2} = 2\sqrt{2}$

            ↑
          4 is a
          perfect
          square.

(b) $\sqrt{45} = \sqrt{9 \cdot 5} = \sqrt{9}\sqrt{5} = 3\sqrt{5}$

            ↑
          9 is a
          perfect
          square.

(c) $\sqrt[3]{24} = \sqrt[3]{8 \cdot 3} = \sqrt[3]{8}\sqrt[3]{3} = 2\sqrt[3]{3}$

            ↑
          8 is a
          perfect
          cube.

(d) $\sqrt[3]{54} = \sqrt[3]{27 \cdot 2} = \sqrt[3]{27}\sqrt[3]{2} = 3\sqrt[3]{2}$

            ↑
          27 is a
          perfect
          cube. ∎

The first step in each example is to express the radicand of the given radical as the product of two factors, one of which must be a perfect nth power other than 1. Also, observe the radicands of the final radicals. In each case, the radicand *cannot* be the product of two factors; one of which must be a perfect nth power other than 1. We say that the final radicals $2\sqrt{2}$, $3\sqrt{5}$, $2\sqrt[3]{3}$, and $3\sqrt[3]{2}$ are in **simplest radical form**.

You may vary the steps somewhat in changing to simplest radical form, but the final result should be the same. Consider some different approaches to change $\sqrt{72}$ to simplest form.

$$\sqrt{72} = \sqrt{9}\sqrt{8} = 3\sqrt{8} = 3\sqrt{4}\sqrt{2} = 3 \cdot 2\sqrt{2} = 6\sqrt{2}, \quad \text{or}$$
$$\sqrt{72} = \sqrt{4}\sqrt{18} = 2\sqrt{18} = 2\sqrt{9}\sqrt{2} = 2 \cdot 3\sqrt{2} = 6\sqrt{2}, \quad \text{or}$$
$$\sqrt{72} = \sqrt{36}\sqrt{2} = 6\sqrt{2}$$

Another variation of the technique for changing radicals to simplest form is to prime factor the radicand and then to look for perfect nth powers in exponential form. The following example illustrates the use of this technique.

EXAMPLE 2

Express each of the following in simplest radical form.

    (a) $\sqrt{50}$ (b) $3\sqrt{80}$ (c) $\sqrt[3]{108}$

Solution

(a) $\sqrt{50} = \sqrt{2 \cdot 5 \cdot 5} = \sqrt{5^2}\sqrt{2} = 5\sqrt{2}$

(b) $3\sqrt{80} = 3\sqrt{2 \cdot 2 \cdot 2 \cdot 2 \cdot 5} = 3\sqrt{2^4}\sqrt{5} = 3 \cdot 2^2\sqrt{5} = 12\sqrt{5}$

(c) $\sqrt[3]{108} = \sqrt[3]{2 \cdot 2 \cdot 3 \cdot 3 \cdot 3} = \sqrt[3]{3^3}\sqrt[3]{4} = 3\sqrt[3]{4}$ ∎

More on Simplest Radical Form

Another property of nth roots is demonstrated by the following examples.

$$\sqrt{\frac{36}{9}} = \sqrt{4} = 2 \quad \text{and} \quad \frac{\sqrt{36}}{\sqrt{9}} = \frac{6}{3} = 2,$$

$$\sqrt[3]{\frac{64}{8}} = \sqrt[3]{8} = 2 \quad \text{and} \quad \frac{\sqrt[3]{64}}{\sqrt[3]{8}} = \frac{4}{2} = 2,$$

$$\sqrt[3]{\frac{-8}{64}} = \sqrt[3]{-\frac{1}{8}} = -\frac{1}{2} \quad \text{and} \quad \frac{\sqrt[3]{-8}}{\sqrt[3]{64}} = \frac{-2}{4} = -\frac{1}{2}$$

In general, we can state the following property.

PROPERTY 8.5

$$\sqrt[n]{\frac{b}{c}} = \frac{\sqrt[n]{b}}{\sqrt[n]{c}} \qquad \sqrt[n]{b} \text{ and } \sqrt[n]{c} \text{ are real numbers and } c \neq 0.$$

Property 8.5 states that **the nth root of a quotient is equal to the quotient of the nth roots**.

To evaluate radicals such as $\sqrt{\dfrac{4}{25}}$ and $\sqrt[3]{\dfrac{27}{8}}$, for which the numerator and denominator of the fractional radicand are perfect nth powers, you may use Property 8.5 or merely rely on the definition of nth root.

$$\sqrt{\frac{4}{25}} = \frac{\sqrt{4}}{\sqrt{25}} = \frac{2}{5} \qquad \text{or} \qquad \sqrt{\frac{4}{25}} = \frac{2}{5} \quad \text{because} \quad \frac{2}{5} \cdot \frac{2}{5} = \frac{4}{25},$$

$$\uparrow \qquad\qquad\qquad\qquad\qquad\qquad \uparrow$$
$$\text{Property 8.5} \qquad\qquad\qquad\qquad \text{definition of } n\text{th root}$$
$$\downarrow \qquad\qquad\qquad\qquad\qquad\qquad \downarrow$$

$$\sqrt[3]{\frac{27}{8}} = \frac{\sqrt[3]{27}}{\sqrt[3]{8}} = \frac{3}{2} \qquad \text{or} \qquad \sqrt[3]{\frac{27}{8}} = \frac{3}{2} \quad \text{because} \quad \frac{3}{2} \cdot \frac{3}{2} \cdot \frac{3}{2} = \frac{27}{8}$$

Radicals such as $\sqrt{\dfrac{28}{9}}$ and $\sqrt[3]{\dfrac{24}{27}}$ in which only the denominators of the radicand are perfect nth powers can be simplified as follows.

$$\sqrt{\frac{28}{9}} = \frac{\sqrt{28}}{\sqrt{9}} = \frac{\sqrt{28}}{3} = \frac{\sqrt{4}\sqrt{7}}{3} = \frac{2\sqrt{7}}{3},$$

$$\sqrt[3]{\frac{24}{27}} = \frac{\sqrt[3]{24}}{\sqrt[3]{27}} = \frac{\sqrt[3]{24}}{3} = \frac{\sqrt[3]{8}\sqrt[3]{3}}{3} = \frac{2\sqrt[3]{3}}{3}$$

Before considering more examples, let's summarize some ideas that pertain to the simplifying of radicals. A radical is said to be in **simplest radical form** if the following conditions are satisfied.

1. No fraction appears with a radical sign. $\sqrt{\dfrac{3}{4}}$ violates this condition.

2. No radical appears in the denominator. $\dfrac{\sqrt{2}}{\sqrt{3}}$ violates this condition.

3. No radicand when expressed in prime factored form contains a factor raised to a power equal to or greater than the index. $\sqrt{2^3 \cdot 5}$ violates this condition.

Now let's consider an example in which neither the numerator nor the denominator of the radicand is a perfect nth power.

EXAMPLE 3   Simplify $\sqrt{\dfrac{2}{3}}$.

Solution
$$\sqrt{\dfrac{2}{3}} = \dfrac{\sqrt{2}}{\sqrt{3}} = \dfrac{\sqrt{2}}{\sqrt{3}} \cdot \underset{\underset{\text{form of 1}}{\uparrow}}{\dfrac{\sqrt{3}}{\sqrt{3}}} = \dfrac{\sqrt{6}}{3}$$

■

We refer to the process we used to simplify the radical in Example 3 as **rationalizing the denominator**. Notice that the denominator becomes a rational number. The process of rationalizing the denominator can often be accomplished in more than one way, as we will see in the next example.

EXAMPLE 4   Simplify $\dfrac{\sqrt{5}}{\sqrt{8}}$.

Solution A
$$\dfrac{\sqrt{5}}{\sqrt{8}} = \dfrac{\sqrt{5}}{\sqrt{8}} \cdot \dfrac{\sqrt{8}}{\sqrt{8}} = \dfrac{\sqrt{40}}{8} = \dfrac{\sqrt{4}\sqrt{10}}{8} = \dfrac{2\sqrt{10}}{8} = \dfrac{\sqrt{10}}{4}$$

Solution B
$$\dfrac{\sqrt{5}}{\sqrt{8}} = \dfrac{\sqrt{5}}{\sqrt{8}} \cdot \dfrac{\sqrt{2}}{\sqrt{2}} = \dfrac{\sqrt{10}}{\sqrt{16}} = \dfrac{\sqrt{10}}{4}$$

Solution C
$$\dfrac{\sqrt{5}}{\sqrt{8}} = \dfrac{\sqrt{5}}{\sqrt{4}\sqrt{2}} = \dfrac{\sqrt{5}}{2\sqrt{2}} = \dfrac{\sqrt{5}}{2\sqrt{2}} \cdot \dfrac{\sqrt{2}}{\sqrt{2}} = \dfrac{\sqrt{10}}{4}$$

■

The three approaches to Example 4 again illustrate the need to think first and then push the pencil. You may find one approach easier than another.

To conclude this section, study the following examples and check the final radicals according to the three conditions previously listed for **simplest radical form**.

EXAMPLE 5   Simplify each of the following.

(a) $\dfrac{3\sqrt{2}}{5\sqrt{3}}$ (b) $\dfrac{3\sqrt{7}}{2\sqrt{18}}$ (c) $\sqrt[3]{\dfrac{5}{9}}$ (d) $\dfrac{\sqrt[3]{5}}{\sqrt[3]{16}}$

Solution

(a) $\dfrac{3\sqrt{2}}{5\sqrt{3}} = \dfrac{3\sqrt{2}}{5\sqrt{3}} \cdot \dfrac{\sqrt{3}}{\sqrt{3}} = \dfrac{3\sqrt{6}}{5\sqrt{9}} = \dfrac{3\sqrt{6}}{15} = \dfrac{\sqrt{6}}{5}$

$\uparrow$
form of 1

(b) $\dfrac{3\sqrt{7}}{2\sqrt{18}} = \dfrac{3\sqrt{7}}{2\sqrt{18}} \cdot \dfrac{\sqrt{2}}{\sqrt{2}} = \dfrac{3\sqrt{14}}{2\sqrt{36}} = \dfrac{3\sqrt{14}}{12} = \dfrac{\sqrt{14}}{4}$

$\uparrow$
form of 1

(c) $\sqrt[3]{\dfrac{5}{9}} = \dfrac{\sqrt[3]{5}}{\sqrt[3]{9}} = \dfrac{\sqrt[3]{5}}{\sqrt[3]{9}} \cdot \dfrac{\sqrt[3]{3}}{\sqrt[3]{3}} = \dfrac{\sqrt[3]{15}}{\sqrt[3]{27}} = \dfrac{\sqrt[3]{15}}{3}$

$\uparrow$
form of 1

(d) $\dfrac{\sqrt[3]{5}}{\sqrt[3]{16}} = \dfrac{\sqrt[3]{5}}{\sqrt[3]{16}} \cdot \dfrac{\sqrt[3]{4}}{\sqrt[3]{4}} = \dfrac{\sqrt[3]{20}}{\sqrt[3]{64}} = \dfrac{\sqrt[3]{20}}{4}$

$\uparrow$
form of 1   ■

Problem Set 8.3

For Problems 1–20, evaluate each of the radicals without using a calculator or a table.

1. $\sqrt[3]{64}$ 2. $\sqrt[3]{-64}$ 3. $\sqrt[3]{-125}$ 4. $\sqrt[3]{216}$

5. $-\sqrt[3]{-8}$ 6. $-\sqrt[3]{-27}$ 7. $\sqrt[3]{\dfrac{27}{64}}$ 8. $\sqrt[3]{\dfrac{8}{125}}$

9. $\sqrt[3]{8^3}$ 10. $\sqrt[3]{7^3}$ 11. $\sqrt[4]{16}$ 12. $\sqrt[4]{81}$

13. $\sqrt[4]{\dfrac{16}{81}}$ 14. $\sqrt[5]{243}$ 15. $\sqrt[5]{-243}$ 16. $\sqrt[5]{-\dfrac{1}{32}}$

17. $\sqrt[4]{16^4}$ 18. $\sqrt[3]{\dfrac{8}{64}}$ 19. $\sqrt[3]{\dfrac{64}{27}}$ 20. $\sqrt[3]{-\dfrac{1}{64}}$

For Problems 21–74, change each of the radicals to simplest radical form.

21. $\sqrt{27}$ 22. $\sqrt{48}$ 23. $\sqrt{32}$

24. $\sqrt{98}$ 25. $\sqrt{80}$ 26. $\sqrt{125}$

27. $\sqrt{160}$

28. $\sqrt{112}$

29. $4\sqrt{18}$

30. $5\sqrt{32}$

31. $-6\sqrt{20}$

32. $-4\sqrt{54}$

33. $\frac{2}{5}\sqrt{75}$

34. $\frac{1}{3}\sqrt{90}$

35. $\frac{3}{2}\sqrt{24}$

36. $\frac{3}{4}\sqrt{45}$

37. $-\frac{5}{6}\sqrt{28}$

38. $-\frac{2}{3}\sqrt{96}$

39. $\sqrt{\dfrac{19}{4}}$

40. $\sqrt{\dfrac{22}{9}}$

41. $\sqrt{\dfrac{27}{16}}$

42. $\sqrt{\dfrac{8}{25}}$

43. $\sqrt{\dfrac{75}{81}}$

44. $\sqrt{\dfrac{24}{49}}$

45. $\sqrt{\dfrac{2}{7}}$

46. $\sqrt{\dfrac{3}{8}}$

47. $\sqrt{\dfrac{2}{3}}$

48. $\sqrt{\dfrac{7}{12}}$

49. $\dfrac{\sqrt{5}}{\sqrt{12}}$

50. $\dfrac{\sqrt{3}}{\sqrt{7}}$

51. $\dfrac{\sqrt{11}}{\sqrt{24}}$

52. $\dfrac{\sqrt{5}}{\sqrt{48}}$

53. $\dfrac{\sqrt{18}}{\sqrt{27}}$

54. $\dfrac{\sqrt{10}}{\sqrt{20}}$

55. $\dfrac{\sqrt{35}}{\sqrt{7}}$

56. $\dfrac{\sqrt{42}}{\sqrt{6}}$

57. $\dfrac{2\sqrt{3}}{\sqrt{7}}$

58. $\dfrac{3\sqrt{2}}{\sqrt{6}}$

59. $-\dfrac{4\sqrt{12}}{\sqrt{5}}$

60. $\dfrac{-6\sqrt{5}}{\sqrt{18}}$

61. $\dfrac{3\sqrt{2}}{4\sqrt{3}}$

62. $\dfrac{6\sqrt{5}}{5\sqrt{12}}$

63. $\dfrac{-8\sqrt{18}}{10\sqrt{50}}$

64. $\dfrac{4\sqrt{45}}{-6\sqrt{20}}$

65. $\sqrt[3]{16}$

66. $\sqrt[3]{40}$

67. $2\sqrt[3]{81}$

68. $-3\sqrt[3]{54}$

69. $\dfrac{2}{\sqrt[3]{9}}$

70. $\dfrac{3}{\sqrt[3]{3}}$

71. $\dfrac{\sqrt[3]{27}}{\sqrt[3]{4}}$

72. $\dfrac{\sqrt[3]{8}}{\sqrt[3]{16}}$

73. $\dfrac{\sqrt[3]{6}}{\sqrt[3]{4}}$

74. $\dfrac{\sqrt[3]{4}}{\sqrt[3]{2}}$

Miscellaneous Problems

75. Use your calculator to find a rational approximation, to the nearest thousandth, for each of the following.

(a) $\sqrt{2}$

(b) $\sqrt{75}$

(c) $\sqrt{156}$

(d) $\sqrt{691}$

(e) $\sqrt{3249}$

(f) $\sqrt{45123}$

(g) $\sqrt{0.14}$

(h) $\sqrt{0.023}$

(i) $\sqrt{0.8649}$

76. Use your calculator to evaluate each of the following.

    (a) $\sqrt[3]{729}$ **(b)** $\sqrt[3]{2744}$ **(c)** $\sqrt[4]{4096}$

    (d) $\sqrt[4]{234256}$ **(e)** $\sqrt[5]{7776}$ **(f)** $\sqrt[3]{371293}$

77. See how your calculator reacts to each of the following.

    (a) $\sqrt{-4}$ **(b)** $\sqrt[3]{-8}$ **(c)** $\sqrt[4]{-16}$

8.4
Simplifying and Combining Radicals

In Section 8.2 we used the distributive property to combine radicals involving square roots. For example, $3\sqrt{7} - 5\sqrt{7} = (3 - 5)\sqrt{7} = -2\sqrt{7}$. In a like manner, radicals involving other roots can be combined as shown in the following examples.

$$7\sqrt[3]{5} - 3\sqrt[3]{5} = (7 - 3)\sqrt[3]{5} = 4\sqrt[3]{5}$$

$$3\sqrt[4]{2} + 5\sqrt[4]{2} - 2\sqrt[4]{2} = (3 + 5 - 2)\sqrt[4]{2} = 6\sqrt[4]{2}$$

Notice that *to add or subtract radicals they must have the same index and the same radicand*. Thus, we cannot simplify an expression such as $5\sqrt{2} + 7\sqrt{11}$.

Simplifying by combining radicals sometimes requires that you first express the given radicals in simplest form and then apply the distributive property. The following examples illustrate this idea.

EXAMPLE 1   Simplify $3\sqrt{8} + 2\sqrt{18} - 4\sqrt{2}$.

Solution

$$3\sqrt{8} + 2\sqrt{18} - 4\sqrt{2} = 3\sqrt{4}\sqrt{2} + 2\sqrt{9}\sqrt{2} - 4\sqrt{2}$$
$$= 6\sqrt{2} + 6\sqrt{2} - 4\sqrt{2}$$
$$= (6 + 6 - 4)\sqrt{2} = 8\sqrt{2}$$ ■

EXAMPLE 2   Simplify $\frac{1}{4}\sqrt{45} + \frac{1}{3}\sqrt{20}$.

Solution

$$\frac{1}{4}\sqrt{45} + \frac{1}{3}\sqrt{20} = \frac{1}{4}\sqrt{9}\sqrt{5} + \frac{1}{3}\sqrt{4}\sqrt{5}$$

$$= \frac{1}{4}\cdot 3 \cdot \sqrt{5} + \frac{1}{3}\cdot 2 \cdot \sqrt{5}$$

$$= \frac{3}{4}\sqrt{5} + \frac{2}{3}\sqrt{5} = \left(\frac{3}{4} + \frac{2}{3}\right)\sqrt{5}$$

$$= \left(\frac{9}{12} + \frac{8}{12}\right)\sqrt{5} = \frac{17}{12}\sqrt{5}$$ ■

EXAMPLE 3   Simplify $5\sqrt[3]{2} - 2\sqrt[3]{16} - 6\sqrt[3]{54}$.

Solution
$$
\begin{aligned}
5\sqrt[3]{2} - 2\sqrt[3]{16} - 6\sqrt[3]{54} &= 5\sqrt[3]{2} - 2\sqrt[3]{8}\sqrt[3]{2} - 6\sqrt[3]{27}\sqrt[3]{2} \\
&= 5\sqrt[3]{2} - 2 \cdot 2 \cdot \sqrt[3]{2} - 6 \cdot 3 \cdot \sqrt[3]{2} \\
&= 5\sqrt[3]{2} - 4\sqrt[3]{2} - 18\sqrt[3]{2} \\
&= (5 - 4 - 18)\sqrt[3]{2} \\
&= -17\sqrt[3]{2}
\end{aligned}
$$
∎

Radicals That Contain Variables

Before we discuss the process of simplifying *radicals that contain variables*, there is one technicality that we should call to your attention. Let's look at some examples to clarify the point. Consider the radical $\sqrt{x^2}$ as follows.

Let $x = 3$,   then $\sqrt{x^2} = \sqrt{3^2} = \sqrt{9} = 3$.

Let $x = -3$,   then $\sqrt{x^2} = \sqrt{(-3)^2} = \sqrt{9} = 3$.

Thus, if $x \geq 0$, then $\sqrt{x^2} = x$, *but if* $x < 0$, then $\sqrt{x^2} = -x$. Using the concept of absolute value we can state that *for all real numbers,* $\sqrt{x^2} = |x|$.

Now consider the radical $\sqrt{x^3}$. Since x^3 is negative when x is negative, we need to restrict x to the nonnegative reals when working with $\sqrt{x^3}$. Thus, we can write if $x \geq 0$, then $\sqrt{x^3} = \sqrt{x^2}\sqrt{x} = x\sqrt{x}$, and no absolute value sign is necessary. Finally, let's consider the radical $\sqrt[3]{x^3}$.

Let $x = 2$,   then $\sqrt[3]{x^3} = \sqrt[3]{2^3} = \sqrt[3]{8} = 2$.

Let $x = -2$,   then $\sqrt[3]{x^3} = \sqrt[3]{(-2)^3} = \sqrt[3]{-8} = -2$.

Thus, it is correct to write: $\sqrt[3]{x^3} = x$ for all real numbers, and again no absolute value sign is necessary.

The previous discussion indicates that technically every radical expression involving variables in the radicand needs to be analyzed individually as to the necessary restrictions imposed on the variables. However, to avoid considering such restrictions on a problem-to-problem basis we shall merely *assume that all variables represent positive real numbers.*

Let's consider the process of simplifying radicals that contain variables in the radicand. Study the following examples and notice that it is the same basic approach we used in Section 8.2.

EXAMPLE 4   Simplify each of the following.

    (a) $\sqrt{8x^3}$ **(b)** $\sqrt{45x^3y^7}$ **(c)** $\sqrt{180a^4b^3}$ **(d)** $\sqrt[3]{40x^4y^8}$

Solution

(a) $\sqrt{8x^3} = \sqrt{4x^2}\sqrt{2x} = 2x\sqrt{2x}$

$4x^2$ is a
perfect square.

(b) $\sqrt{45x^3y^7} = \sqrt{9x^2y^6}\sqrt{5xy} = 3xy^3\sqrt{5xy}$

$9x^2y^6$ is a
perfect square.

(c) If the numerical coefficient of the radicand is quite large, you may want to look at it in prime factored form.

$$\sqrt{180a^4b^3} = \sqrt{2 \cdot 2 \cdot 3 \cdot 3 \cdot 5 \cdot a^4 \cdot b^3}$$
$$= \sqrt{36 \cdot 5 \cdot a^4 \cdot b^3}$$
$$= \sqrt{36a^4b^2}\sqrt{5b}$$
$$= 6a^2b\sqrt{5b}$$

(d) $\sqrt[3]{40x^4y^8} = \sqrt[3]{8x^3y^6}\sqrt[3]{5xy^2} = 2xy^2\sqrt[3]{5xy^2}$ ■

$8x^3y^6$ is a
perfect cube.

Before we consider more examples, let's restate (so as to include radicands containing variables) the conditions necessary for a radical to be in *simplest radical form.*

1. A radicand contains no polynomial factor raised to a power equal to or greater than the index of the radical. $\sqrt{x^3}$ violates this condition.

2. No fraction appears within a radical sign. $\sqrt{\dfrac{2x}{3y}}$ violates this condition.

3. No radical appears in the denominator. $\dfrac{3}{\sqrt[3]{4x}}$ violates this condition.

EXAMPLE 5   Express each of the following in simplest radical form.

(a) $\sqrt{\dfrac{2x}{3y}}$ (b) $\dfrac{\sqrt{5}}{\sqrt{12a^3}}$ (c) $\dfrac{\sqrt{8x^2}}{\sqrt{27y^5}}$

(d) $\dfrac{3}{\sqrt[3]{4x}}$ (e) $\dfrac{\sqrt[3]{16x^2}}{\sqrt[3]{9y^5}}$

Solution

(a) $\sqrt{\dfrac{2x}{3y}} = \dfrac{\sqrt{2x}}{\sqrt{3y}} = \dfrac{\sqrt{2x}}{\sqrt{3y}} \cdot \dfrac{\sqrt{3y}}{\sqrt{3y}} = \dfrac{\sqrt{6xy}}{3y}$

form of 1

(b) $\dfrac{\sqrt{5}}{\sqrt{12a^3}} = \dfrac{\sqrt{5}}{\sqrt{12a^3}} \cdot \dfrac{\sqrt{3a}}{\sqrt{3a}} = \dfrac{\sqrt{15a}}{\sqrt{36a^4}} = \dfrac{\sqrt{15a}}{6a^2}$

↑

form of 1

(c) $\dfrac{\sqrt{8x^2}}{\sqrt{27y^5}} = \dfrac{\sqrt{4x^2}\sqrt{2}}{\sqrt{9y^4}\sqrt{3y}} = \dfrac{2x\sqrt{2}}{3y^2\sqrt{3y}} = \dfrac{2x\sqrt{2}}{3y^2\sqrt{3y}} \cdot \dfrac{\sqrt{3y}}{\sqrt{3y}}$

$= \dfrac{2x\sqrt{6y}}{(3y^2)(3y)} = \dfrac{2x\sqrt{6y}}{9y^3}$

(d) $\dfrac{3}{\sqrt[3]{4x}} = \dfrac{3}{\sqrt[3]{4x}} \cdot \dfrac{\sqrt[3]{2x^2}}{\sqrt[3]{2x^2}} = \dfrac{3\sqrt[3]{2x^2}}{\sqrt[3]{8x^3}} = \dfrac{3\sqrt[3]{2x^2}}{2x}$

(e) $\dfrac{\sqrt[3]{16x^2}}{\sqrt[3]{9y^5}} = \dfrac{\sqrt[3]{16x^2}}{\sqrt[3]{9y^5}} \cdot \dfrac{\sqrt[3]{3y}}{\sqrt[3]{3y}} = \dfrac{\sqrt[3]{48x^2y}}{\sqrt[3]{27y^6}} = \dfrac{\sqrt[3]{8}\sqrt[3]{6x^2y}}{3y^2} = \dfrac{2\sqrt[3]{6x^2y}}{3y^2}$ ∎

Notice that in (c) we did some simplifying first before rationalizing the denominator, whereas in (b) we proceeded immediately to rationalize the denominator. This is an individual choice and you should probably do it both ways a few times to help determine your preference.

Problem Set 8.4

Use the distributive property to help simplify each of the following. For example,

$$3\sqrt{8} - \sqrt{32} = 3\sqrt{4}\sqrt{2} - \sqrt{16}\sqrt{2}$$
$$= 3(2)\sqrt{2} - 4\sqrt{2}$$
$$= 6\sqrt{2} - 4\sqrt{2}$$
$$= (6 - 4)\sqrt{2} = 2\sqrt{2}.$$

1. $5\sqrt{18} - 2\sqrt{2}$

2. $7\sqrt{12} + 4\sqrt{3}$

3. $7\sqrt{12} + 10\sqrt{48}$

4. $6\sqrt{8} - 5\sqrt{18}$

5. $-2\sqrt{50} - 5\sqrt{32}$

6. $-2\sqrt{20} - 7\sqrt{45}$

7. $3\sqrt{20} - \sqrt{5} - 2\sqrt{45}$

8. $6\sqrt{12} + \sqrt{3} - 2\sqrt{48}$

9. $-9\sqrt{24} + 3\sqrt{54} - 12\sqrt{6}$

10. $13\sqrt{28} - 2\sqrt{63} - 7\sqrt{7}$

11. $\dfrac{3}{4}\sqrt{7} - \dfrac{2}{3}\sqrt{28}$

12. $\dfrac{3}{5}\sqrt{5} - \dfrac{1}{4}\sqrt{80}$

13. $\dfrac{3}{5}\sqrt{40} + \dfrac{5}{6}\sqrt{90}$

14. $\dfrac{3}{8}\sqrt{96} - \dfrac{2}{3}\sqrt{54}$

15. $\dfrac{3\sqrt{18}}{5} - \dfrac{5\sqrt{72}}{6} + \dfrac{3\sqrt{98}}{4}$

16. $\dfrac{-2\sqrt{20}}{3} + \dfrac{3\sqrt{45}}{4} - \dfrac{5\sqrt{80}}{6}$

17. $5\sqrt[3]{3} + 2\sqrt[3]{24} - 6\sqrt[3]{81}$ 18. $-3\sqrt[3]{2} - 2\sqrt[3]{16} + \sqrt[3]{54}$

19. $-\sqrt[3]{16} + 7\sqrt[3]{54} - 9\sqrt[3]{2}$ 20. $4\sqrt[3]{24} - 6\sqrt[3]{3} + 13\sqrt[3]{81}$

Express each of the following in simplest radical form. All variables represent positive real numbers.

21. $\sqrt{32x}$ 22. $\sqrt{50y}$ 23. $\sqrt{75x^2}$

24. $\sqrt{108y^2}$ 25. $\sqrt{20x^2y}$ 26. $\sqrt{80xy^2}$

27. $\sqrt{64x^3y^7}$ 28. $\sqrt{36x^5y^6}$ 29. $\sqrt{54a^4b^3}$

30. $\sqrt{96a^7b^8}$ 31. $\sqrt{63x^6y^8}$ 32. $\sqrt{28x^4y^{12}}$

33. $2\sqrt{40a^3}$ 34. $4\sqrt{90a^5}$ 35. $\dfrac{2}{3}\sqrt{96xy^3}$

36. $\dfrac{4}{5}\sqrt{125x^4y}$ 37. $\sqrt{\dfrac{2x}{5y}}$ 38. $\sqrt{\dfrac{3x}{2y}}$

39. $\sqrt{\dfrac{5}{12x^4}}$ 40. $\sqrt{\dfrac{7}{8x^2}}$ 41. $\dfrac{5}{\sqrt{18y}}$

42. $\dfrac{3}{\sqrt{12x}}$ 43. $\dfrac{\sqrt{7x}}{\sqrt{8y^5}}$ 44. $\dfrac{\sqrt{5y}}{\sqrt{18x^3}}$

45. $\dfrac{\sqrt{18y^3}}{\sqrt{16x}}$ 46. $\dfrac{\sqrt{2x^3}}{\sqrt{9y}}$ 47. $\dfrac{\sqrt{24a^2b^3}}{\sqrt{7ab^6}}$

48. $\dfrac{\sqrt{12a^2b}}{\sqrt{5a^3b^3}}$ 49. $\sqrt[3]{24y}$ 50. $\sqrt[3]{16x^2}$

51. $\sqrt[3]{16x^4}$ 52. $\sqrt[3]{54x^3}$ 53. $\sqrt[3]{56x^6y^8}$

54. $\sqrt[3]{81x^5y^6}$ 55. $\sqrt[3]{\dfrac{7}{9x^2}}$ 56. $\sqrt[3]{\dfrac{5}{2x}}$

57. $\dfrac{\sqrt[3]{3y}}{\sqrt[3]{16x^4}}$ 58. $\dfrac{\sqrt[3]{2y}}{\sqrt[3]{3x}}$

59. $\dfrac{\sqrt[3]{12xy}}{\sqrt[3]{3x^2y^5}}$ 60. $\dfrac{5}{\sqrt[3]{9xy^2}}$

61. $\sqrt{8x + 12y}$ [*Hint:* $\sqrt{8x + 12y} = \sqrt{4(2x + 3y)}$]

62. $\sqrt{4x + 4y}$ 63. $\sqrt{16x + 48y}$ 64. $\sqrt{27x + 18y}$

Use the distributive property to help simplify each of the following. All variables represent positive real numbers.

65. $-3\sqrt{4x} + 5\sqrt{9x} + 6\sqrt{16x}$ 66. $-2\sqrt{25x} - 4\sqrt{36x} + 7\sqrt{64x}$

67. $2\sqrt{18x} - 3\sqrt{8x} - 6\sqrt{50x}$ 68. $4\sqrt{20x} + 5\sqrt{45x} - 10\sqrt{80x}$

69. $5\sqrt{27n} - \sqrt{12n} - 6\sqrt{3n}$ 70. $4\sqrt{8n} + 3\sqrt{18n} - 2\sqrt{72n}$

71. $7\sqrt{4ab} - \sqrt{16ab} - 10\sqrt{25ab}$ 72. $4\sqrt{ab} - 9\sqrt{36ab} + 6\sqrt{49ab}$

73. $-3\sqrt{2x^3} + 4\sqrt{8x^3} - 3\sqrt{32x^3}$ 74. $2\sqrt{40x^5} - 3\sqrt{90x^5} + 5\sqrt{160x^5}$

Miscellaneous Problems

For Problems 75–80, (a) give a whole number estimate for each expression, and (b) use your calculator to find a rational approximation, to the nearest hundredth, for each expression.

75. $3\sqrt{10} + 5\sqrt{17} - 2\sqrt{48}$

76. $12\sqrt{50} - 6\sqrt{34} + 9\sqrt{62}$

77. $\dfrac{3}{4}\sqrt{18} - \dfrac{2}{3}\sqrt{80}$

78. $\dfrac{2}{5}\sqrt{26} + \dfrac{1}{2}\sqrt{37} - \dfrac{1}{3}\sqrt{83}$

79. $6\sqrt[3]{9} + 4\sqrt[3]{28} + 3\sqrt[3]{127}$

80. $5\sqrt[3]{2} - 7\sqrt[3]{65} + 8\sqrt[3]{25}$

8.5

Products and Quotients Involving Radicals

As we have seen, Property 8.4 ($\sqrt[n]{bc} = \sqrt[n]{b}\sqrt[n]{c}$) is used to express one radical as the product of two radicals and also to express the product of two radicals as one radical. In fact, we have used the property for both purposes within the framework of simplifying radicals. For example,

$$\underset{\underset{\sqrt[n]{bc}\,=\,\sqrt[n]{b}\sqrt[n]{c}}{\uparrow\qquad\uparrow}}{\frac{\sqrt{3}}{\sqrt{32}} = \frac{\sqrt{3}}{\sqrt{16}\sqrt{2}} = \frac{\sqrt{3}}{4\sqrt{2}}} = \underset{\underset{\sqrt[n]{b}\sqrt[n]{c}\,=\,\sqrt[n]{bc}}{\uparrow}}{\frac{\sqrt{3}}{4\sqrt{2}}\cdot\frac{\sqrt{2}}{\sqrt{2}}} = \underset{\underset{}{\uparrow}}{\frac{\sqrt{6}}{8}}.$$

The following examples demonstrate the use of Property 8.4 to multiply radicals and to express the product in simplest form.

EXAMPLE 1 Multiply and simplify where possible.

(a) $(2\sqrt{3})(3\sqrt{5})$ (b) $(3\sqrt{8})(5\sqrt{2})$

(c) $(7\sqrt{6})(3\sqrt{8})$ (d) $(2\sqrt[3]{6})(5\sqrt[3]{4})$

Solution

(a) $(2\sqrt{3})(3\sqrt{5}) = 2\cdot 3\cdot\sqrt{3}\cdot\sqrt{5} = 6\sqrt{15}$.

(b) $(3\sqrt{8})(5\sqrt{2}) = 3\cdot 5\cdot\sqrt{8}\cdot\sqrt{2} = 15\sqrt{16} = 15\cdot 4 = 60$.

(c) $(7\sqrt{6})(3\sqrt{8}) = 7\cdot 3\cdot\sqrt{6}\cdot\sqrt{8} = 21\sqrt{48} = 21\sqrt{16}\sqrt{3} = 21\cdot 4\cdot\sqrt{3}$
$$= 84\sqrt{3}$$

(d) $(2\sqrt[3]{6})(5\sqrt[3]{4}) = 2\cdot 5\cdot\sqrt[3]{6}\cdot\sqrt[3]{4} = 10\sqrt[3]{24}$
$$= 10\sqrt[3]{8}\sqrt[3]{3}$$
$$= 10\cdot 2\cdot\sqrt[3]{3}$$
$$= 20\sqrt[3]{3} \qquad\blacksquare$$

Recall the use of the distributive property when finding the product of a monomial and a polynomial. For example, $3x^2(2x + 7) = 3x^2(2x) + 3x^2(7) = 6x^3 + 21x^2$. In a similar manner, the distributive property, along with Property 8.4, provide the basis for finding certain special products that involve radicals. The following examples illustrate this idea.

EXAMPLE 2   Multiply and simplify where possible.

(a) $\sqrt{3}(\sqrt{6} + \sqrt{12})$ (b) $2\sqrt{2}(4\sqrt{3} - 5\sqrt{6})$

(c) $\sqrt{6x}(\sqrt{8x} + \sqrt{12xy})$ (d) $\sqrt[3]{2}(5\sqrt[3]{4} - 3\sqrt[3]{16})$

Solution

(a) $\sqrt{3}(\sqrt{6} + \sqrt{12}) = \sqrt{3}\sqrt{6} + \sqrt{3}\sqrt{12}$

$\qquad\qquad\qquad = \sqrt{18} + \sqrt{36}$

$\qquad\qquad\qquad = \sqrt{9}\sqrt{2} + 6$

$\qquad\qquad\qquad = 3\sqrt{2} + 6$

(b) $2\sqrt{2}(4\sqrt{3} - 5\sqrt{6}) = (2\sqrt{2})(4\sqrt{3}) - (2\sqrt{2})(5\sqrt{6})$

$\qquad\qquad\qquad = 8\sqrt{6} - 10\sqrt{12}$

$\qquad\qquad\qquad = 8\sqrt{6} - 10\sqrt{4}\sqrt{3}$

$\qquad\qquad\qquad = 8\sqrt{6} - 20\sqrt{3}$

(c) $\sqrt{6x}(\sqrt{8x} + \sqrt{12xy}) = (\sqrt{6x})(\sqrt{8x}) + (\sqrt{6x})(\sqrt{12xy})$

$\qquad\qquad\qquad = \sqrt{48x^2} + \sqrt{72x^2 y}$

$\qquad\qquad\qquad = \sqrt{16x^2}\sqrt{3} + \sqrt{36x^2}\sqrt{2y}$

$\qquad\qquad\qquad = 4x\sqrt{3} + 6x\sqrt{2y}$

(d) $\sqrt[3]{2}(5\sqrt[3]{4} - 3\sqrt[3]{16}) = (\sqrt[3]{2})(5\sqrt[3]{4}) - (\sqrt[3]{2})(3\sqrt[3]{16})$

$\qquad\qquad\qquad = 5\sqrt[3]{8} - 3\sqrt[3]{32}$

$\qquad\qquad\qquad = 5 \cdot 2 - 3\sqrt[3]{8}\sqrt[3]{4}$

$\qquad\qquad\qquad = 10 - 6\sqrt[3]{4}$ ∎

The distributive property also plays a central role in determining the product of two binomials. For example, $(x + 2)(x + 3) = x(x + 3) + 2(x + 3) = x^2 + 3x + 2x + 6 = x^2 + 5x + 6$. Finding the product of two binomial expressions that involve radicals can be handled in a similar fashion, as in the next examples.

EXAMPLE 3   Find the following products and simplify.

(a) $(\sqrt{3} + \sqrt{5})(\sqrt{2} + \sqrt{6})$ (b) $(2\sqrt{2} - \sqrt{7})(3\sqrt{2} + 5\sqrt{7})$

(c) $(\sqrt{8} + \sqrt{6})(\sqrt{8} - \sqrt{6})$ (d) $(\sqrt{x} + \sqrt{y})(\sqrt{x} - \sqrt{y})$

Solution

$$\textbf{(a)} \ (\sqrt{3} + \sqrt{5})(\sqrt{2} + \sqrt{6}) = \sqrt{3}(\sqrt{2} + \sqrt{6}) + \sqrt{5}(\sqrt{2} + \sqrt{6})$$
$$= \sqrt{3}\sqrt{2} + \sqrt{3}\sqrt{6} + \sqrt{5}\sqrt{2} + \sqrt{5}\sqrt{6}$$
$$= \sqrt{6} + \sqrt{18} + \sqrt{10} + \sqrt{30}$$
$$= \sqrt{6} + 3\sqrt{2} + \sqrt{10} + \sqrt{30}$$

$$\textbf{(b)} \ (2\sqrt{2} - \sqrt{7})(3\sqrt{2} + 5\sqrt{7}) = 2\sqrt{2}(3\sqrt{2} + 5\sqrt{7}) - \sqrt{7}(3\sqrt{2} + 5\sqrt{7})$$
$$= (2\sqrt{2})(3\sqrt{2}) + (2\sqrt{2})(5\sqrt{7})$$
$$- (\sqrt{7})(3\sqrt{2}) - (\sqrt{7})(5\sqrt{7})$$
$$= 12 + 10\sqrt{14} - 3\sqrt{14} - 35$$
$$= -23 + 7\sqrt{14}$$

$$\textbf{(c)} \ (\sqrt{8} + \sqrt{6})(\sqrt{8} - \sqrt{6}) = \sqrt{8}(\sqrt{8} - \sqrt{6}) + \sqrt{6}(\sqrt{8} - \sqrt{6})$$
$$= \sqrt{8}\sqrt{8} - \sqrt{8}\sqrt{6} + \sqrt{6}\sqrt{8} - \sqrt{6}\sqrt{6}$$
$$= 8 - \sqrt{48} + \sqrt{48} - 6$$
$$= 2$$

$$\textbf{(d)} \ (\sqrt{x} + \sqrt{y})(\sqrt{x} - \sqrt{y}) = \sqrt{x}(\sqrt{x} - \sqrt{y}) + \sqrt{y}(\sqrt{x} - \sqrt{y})$$
$$= \sqrt{x}\sqrt{x} - \sqrt{x}\sqrt{y} + \sqrt{y}\sqrt{x} - \sqrt{y}\sqrt{y}$$
$$= x - \sqrt{xy} + \sqrt{xy} - y$$
$$= x - y \qquad \blacksquare$$

Notice parts (**c**) and (**d**) of Example 3. They fit the special product pattern $(a + b)(a - b) = a^2 - b^2$. Furthermore, in each case the final product is in rational form. This suggests a way of rationalizing the denominator of an expression that contains a binomial denominator with radicals. Consider the following example.

EXAMPLE 4   Simplify $\dfrac{4}{\sqrt{5} + \sqrt{2}}$ by rationalizing the denominator.

Solution

$$\frac{4}{\sqrt{5} + \sqrt{2}} = \frac{4}{\sqrt{5} + \sqrt{2}} \cdot \left(\frac{\sqrt{5} - \sqrt{2}}{\sqrt{5} - \sqrt{2}}\right) \qquad \text{a form of 1}$$

$$= \frac{4(\sqrt{5} - \sqrt{2})}{(\sqrt{5} + \sqrt{2})(\sqrt{5} - \sqrt{2})} = \frac{4(\sqrt{5} - \sqrt{2})}{5 - 2}$$

$$= \frac{4(\sqrt{5} - \sqrt{2})}{3} \qquad \text{or} \qquad \frac{4\sqrt{5} - 4\sqrt{2}}{3}$$

either answer
is acceptable   ■

The next examples further illustrate the process of rationalizing and simplifying expressions that contain binomial denominators.

EXAMPLE 5   For each of the following, rationalize the denominator and simplify.

(a) $\dfrac{\sqrt{3}}{\sqrt{6}-9}$

(b) $\dfrac{7}{3\sqrt{5}+2\sqrt{3}}$

(c) $\dfrac{\sqrt{x}+2}{\sqrt{x}-3}$

(d) $\dfrac{2\sqrt{x}-3\sqrt{y}}{\sqrt{x}+\sqrt{y}}$

Solution

(a) $\dfrac{\sqrt{3}}{\sqrt{6}-9}=\dfrac{\sqrt{3}}{\sqrt{6}-9}\cdot\dfrac{\sqrt{6}+9}{\sqrt{6}+9}$

$=\dfrac{\sqrt{3}(\sqrt{6}+9)}{(\sqrt{6}-9)(\sqrt{6}+9)}$

$=\dfrac{\sqrt{18}+9\sqrt{3}}{6-81}$

$=\dfrac{3\sqrt{2}+9\sqrt{3}}{-75}$

$=\dfrac{3(\sqrt{2}+3\sqrt{3})}{(-3)(25)}$

$=-\dfrac{\sqrt{2}+3\sqrt{3}}{25}\quad$ or $\quad\dfrac{-\sqrt{2}-3\sqrt{3}}{25}$

(b) $\dfrac{7}{3\sqrt{5}+2\sqrt{3}}=\dfrac{7}{3\sqrt{5}+2\sqrt{3}}\cdot\dfrac{3\sqrt{5}-2\sqrt{3}}{3\sqrt{5}-2\sqrt{3}}$

$=\dfrac{7(3\sqrt{5}-2\sqrt{3})}{(3\sqrt{5}+2\sqrt{3})(3\sqrt{5}-2\sqrt{3})}$

$=\dfrac{7(3\sqrt{5}-2\sqrt{3})}{45-12}$

$=\dfrac{7(3\sqrt{5}-2\sqrt{3})}{33}\quad$ or $\quad\dfrac{21\sqrt{5}-14\sqrt{3}}{33}$

(c) $\dfrac{\sqrt{x}+2}{\sqrt{x}-3}=\dfrac{\sqrt{x}+2}{\sqrt{x}-3}\cdot\dfrac{\sqrt{x}+3}{\sqrt{x}+3}=\dfrac{(\sqrt{x}+2)(\sqrt{x}+3)}{(\sqrt{x}-3)(\sqrt{x}+3)}$

$=\dfrac{x+3\sqrt{x}+2\sqrt{x}+6}{x-9}$

$=\dfrac{x+5\sqrt{x}+6}{x-9}$

(d) $\dfrac{2\sqrt{x} - 3\sqrt{y}}{\sqrt{x} + \sqrt{y}} = \dfrac{2\sqrt{x} - 3\sqrt{y}}{\sqrt{x} + \sqrt{y}} \cdot \dfrac{\sqrt{x} - \sqrt{y}}{\sqrt{x} - \sqrt{y}}$

$$= \dfrac{(2\sqrt{x} - 3\sqrt{y})(\sqrt{x} - \sqrt{y})}{(\sqrt{x} + \sqrt{y})(\sqrt{x} - \sqrt{y})}$$

$$= \dfrac{2x - 2\sqrt{xy} - 3\sqrt{xy} + 3y}{x - y}$$

$$= \dfrac{2x - 5\sqrt{xy} + 3y}{x - y}$$ ∎

Problem Set 8.5

Multiply and simplify where possible.

1. $\sqrt{6}\sqrt{12}$ **2.** $\sqrt{8}\sqrt{6}$ **3.** $(3\sqrt{3})(2\sqrt{6})$

4. $(5\sqrt{2})(3\sqrt{12})$ **5.** $(4\sqrt{2})(-6\sqrt{5})$ **6.** $(-7\sqrt{3})(2\sqrt{5})$

7. $(-3\sqrt{3})(-4\sqrt{8})$ **8.** $(-5\sqrt{8})(-6\sqrt{7})$ **9.** $(5\sqrt{6})(4\sqrt{6})$

10. $(3\sqrt{7})(2\sqrt{7})$ **11.** $(2\sqrt[3]{4})(6\sqrt[3]{2})$ **12.** $(4\sqrt[3]{3})(5\sqrt[3]{9})$

13. $(4\sqrt[3]{6})(7\sqrt[3]{4})$ **14.** $(9\sqrt[3]{6})(2\sqrt[3]{9})$

Find the following products and express answers in simplest radical form. All variables represent nonnegative real numbers.

15. $\sqrt{2}(\sqrt{3} + \sqrt{5})$ **16.** $\sqrt{3}(\sqrt{7} + \sqrt{10})$

17. $3\sqrt{5}(2\sqrt{2} - \sqrt{7})$ **18.** $5\sqrt{6}(2\sqrt{5} - 3\sqrt{11})$

19. $2\sqrt{6}(3\sqrt{8} - 5\sqrt{12})$ **20.** $4\sqrt{2}(3\sqrt{12} + 7\sqrt{6})$

21. $-4\sqrt{5}(2\sqrt{5} + 4\sqrt{12})$ **22.** $-5\sqrt{3}(3\sqrt{12} - 9\sqrt{8})$

23. $3\sqrt{x}(5\sqrt{2} + \sqrt{y})$ **24.** $\sqrt{2x}(3\sqrt{y} - 7\sqrt{5})$

25. $\sqrt{xy}(5\sqrt{xy} - 6\sqrt{x})$ **26.** $4\sqrt{x}(2\sqrt{xy} + 2\sqrt{x})$

27. $\sqrt{5y}(\sqrt{8x} + \sqrt{12y^2})$ **28.** $\sqrt{2x}(\sqrt{12xy} - \sqrt{8y})$

29. $5\sqrt{3}(2\sqrt{8} - 3\sqrt{18})$ **30.** $2\sqrt{2}(3\sqrt{12} - \sqrt{27})$

31. $(\sqrt{3} + 4)(\sqrt{3} - 7)$ **32.** $(\sqrt{2} + 6)(\sqrt{2} - 2)$

33. $(\sqrt{5} - 6)(\sqrt{5} - 3)$ **34.** $(\sqrt{7} - 2)(\sqrt{7} - 8)$

35. $(3\sqrt{5} - 2\sqrt{3})(2\sqrt{7} + \sqrt{2})$ **36.** $(\sqrt{2} + \sqrt{3})(\sqrt{5} - \sqrt{7})$

37. $(2\sqrt{6} + 3\sqrt{5})(\sqrt{8} - 3\sqrt{12})$ **38.** $(5\sqrt{2} - 4\sqrt{6})(2\sqrt{8} + \sqrt{6})$

39. $(2\sqrt{6} + 5\sqrt{5})(3\sqrt{6} - \sqrt{5})$ **40.** $(7\sqrt{3} - \sqrt{7})(2\sqrt{3} + 4\sqrt{7})$

41. $(3\sqrt{2} - 5\sqrt{3})(6\sqrt{2} - 7\sqrt{3})$

42. $(\sqrt{8} \quad 3\sqrt{10})(2\sqrt{8} - 6\sqrt{10})$

43. $(\sqrt{6} + 4)(\sqrt{6} - 4)$

44. $(\sqrt{7} - 2)(\sqrt{7} + 2)$

45. $(\sqrt{2} + \sqrt{10})(\sqrt{2} - \sqrt{10})$

46. $(2\sqrt{3} + \sqrt{11})(2\sqrt{3} - \sqrt{11})$

47. $(\sqrt{2x} + \sqrt{3y})(\sqrt{2x} - \sqrt{3y})$

48. $(2\sqrt{x} - 5\sqrt{y})(2\sqrt{x} + 5\sqrt{y})$

49. $2\sqrt[3]{3}(5\sqrt[3]{4} + \sqrt[3]{6})$

50. $2\sqrt[3]{2}(3\sqrt[3]{6} - 4\sqrt[3]{5})$

51. $3\sqrt[3]{4}(2\sqrt[3]{2} - 6\sqrt[3]{4})$

52. $3\sqrt[3]{3}(4\sqrt[3]{9} + 5\sqrt[3]{7})$

For each of the following, rationalize the denominator and simplify. All variables represent positive real numbers.

53. $\dfrac{2}{\sqrt{7} + 1}$

54. $\dfrac{6}{\sqrt{5} + 2}$

55. $\dfrac{3}{\sqrt{2} - 5}$

56. $\dfrac{-4}{\sqrt{6} - 3}$

57. $\dfrac{1}{\sqrt{2} + \sqrt{7}}$

58. $\dfrac{3}{\sqrt{3} + \sqrt{10}}$

59. $\dfrac{\sqrt{2}}{\sqrt{10} - \sqrt{3}}$

60. $\dfrac{\sqrt{3}}{\sqrt{7} - \sqrt{2}}$

61. $\dfrac{\sqrt{3}}{2\sqrt{5} + 4}$

62. $\dfrac{\sqrt{7}}{3\sqrt{2} - 5}$

63. $\dfrac{6}{3\sqrt{7} - 2\sqrt{6}}$

64. $\dfrac{5}{2\sqrt{5} + 3\sqrt{7}}$

65. $\dfrac{\sqrt{6}}{3\sqrt{2} + 2\sqrt{3}}$

66. $\dfrac{3\sqrt{6}}{5\sqrt{3} - 4\sqrt{2}}$

67. $\dfrac{2}{\sqrt{x} + 4}$

68. $\dfrac{3}{\sqrt{x} + 7}$

69. $\dfrac{\sqrt{x}}{\sqrt{x} - 5}$

70. $\dfrac{\sqrt{x}}{\sqrt{x} - 1}$

71. $\dfrac{\sqrt{x} - 2}{\sqrt{x} + 6}$

72. $\dfrac{\sqrt{x} + 1}{\sqrt{x} - 10}$

73. $\dfrac{\sqrt{x}}{\sqrt{x} + 2\sqrt{y}}$

74. $\dfrac{\sqrt{y}}{2\sqrt{x} - \sqrt{y}}$

75. $\dfrac{3\sqrt{y}}{2\sqrt{x} - 3\sqrt{y}}$

76. $\dfrac{2\sqrt{x}}{3\sqrt{x} + 5\sqrt{y}}$

Miscellaneous Problems

For Problems 77–82, (a) give an estimate for each expression by using a whole number estimate for each radical, and (b) use your calculator to find a rational approximation, to the nearest hundredth, for each expression.

77. $\dfrac{5}{\sqrt{8} + \sqrt{17}}$

78. $\dfrac{2}{\sqrt{15} - \sqrt{3}}$

79. $\dfrac{\sqrt{5}}{2\sqrt{3} + 3\sqrt{10}}$

80. $\dfrac{\sqrt{10}}{4\sqrt{17} - 2\sqrt{2}}$

81. $\dfrac{3\sqrt{5}}{2 - 7\sqrt{8}}$

82. $\dfrac{4\sqrt{14}}{2\sqrt{24} - 3\sqrt{37}}$

8.6

Equations Involving Radicals

We often refer to equations that contain radicals with variables in a radicand as **radical equations**. In this section we shall discuss techniques for solving such equations that contain one or more radicals. To solve radical equations we need the following property of equality.

PROPERTY 8.6

> Let a and b be real numbers and n be a positive integer.
>
> If $a = b$, then $a^n = b^n$.

Property 8.6 states that we can *raise both sides of an equation to a positive integral power*. However, raising both sides of an equation to a positive integral power sometimes produces results that do not satisfy the original equation. Let's consider two examples to illustrate this point.

EXAMPLE 1   Solve $\sqrt{2x - 5} = 7$.

Solution

$$\sqrt{2x - 5} = 7$$
$$(\sqrt{2x - 5})^2 = 7^2 \qquad \text{Square both sides.}$$
$$2x - 5 = 49$$
$$2x = 54$$
$$x = 27$$

$$\text{CHECK} \quad \sqrt{2x - 5} = 7$$
$$\sqrt{2(27) - 5} \overset{?}{=} 7$$
$$\sqrt{49} \overset{?}{=} 7$$
$$7 = 7$$

The solution set for $\sqrt{2x - 5} = 7$ is $\{27\}$.   ∎

EXAMPLE 2   Solve $\sqrt{3a + 4} = -4$.

Solution

$$\sqrt{3a + 4} = -4$$
$$(\sqrt{3a + 4})^2 = (-4)^2 \qquad \text{Square both sides.}$$
$$3a + 4 = 16$$
$$3a = 12$$
$$a = 4$$

$$\text{CHECK} \quad \sqrt{3a + 4} = -4$$

$$\sqrt{3(4) + 4} \overset{?}{=} -4$$

$$\sqrt{16} \overset{?}{=} -4$$

$$4 \neq -4$$

Since 4 does not check, the original equation *has no real number solution*. Thus, the solution set is $\varnothing$. ∎

In general, raising both sides of an equation to a positive integral power produces an equation that has all of the solutions of the original equation; it may also have some extra solutions that will not satisfy the original equation. Such extra solutions are called **extraneous solutions**. Therefore, when using Property 8.6 you *must* check each potential solution in the original equation.

  Let's consider some examples to illustrate different situations that arise when solving radical equations.

EXAMPLE 3   Solve $\sqrt{2t - 4} = t - 2$.

Solution

$$\sqrt{2t - 4} = t - 2$$

$$(\sqrt{2t - 4})^2 = (t - 2)^2 \qquad \text{Square both sides.}$$

$$2t - 4 = t^2 - 4t + 4$$

$$0 = t^2 - 6t + 8$$

$$0 = (t - 2)(t - 4) \qquad \text{Factor the right side.}$$

$$t - 2 = 0 \quad \text{or} \quad t - 4 = 0 \qquad \text{Apply: } ab = 0 \text{ if and}$$
$$\qquad\qquad\qquad\qquad\qquad\qquad \text{only if } a = 0 \text{ or } b = 0.$$

$$t = 2 \quad \text{or} \quad t = 4$$

$$\text{CHECK} \quad \sqrt{2t - 4} = t - 2 \qquad\qquad \sqrt{2t - 4} = t - 2$$

$$\sqrt{2(2) - 4} \overset{?}{=} 2 - 2 \quad \text{or} \quad \sqrt{2(4) - 4} \overset{?}{=} 4 - 2$$

$$\sqrt{0} \overset{?}{=} 0 \qquad\qquad\qquad \sqrt{4} \overset{?}{=} 2$$

$$0 = 0 \qquad\qquad\qquad\qquad 2 = 2$$

The solution set is $\{2, 4\}$. ∎

EXAMPLE 4   Solve $\sqrt{y} + 6 = y$.

Solution

$$\sqrt{y} + 6 = y$$

$$\sqrt{y} = y - 6$$

$$(\sqrt{y})^2 = (y - 6)^2 \qquad\qquad \text{Square both sides.}$$

$$y = y^2 - 12y + 36$$

$$0 = y^2 - 13y + 36$$

$$0 = (y - 4)(y - 9) \qquad \text{Factor the right side.}$$

$$y - 4 = 0 \quad \text{or} \quad y - 9 = 0 \qquad \text{Apply: } ab = 0 \text{ if and} \\ \text{only if } a = 0 \text{ or } b = 0.$$

$$y = 4 \quad \text{or} \quad y = 9$$

CHECK

$$\sqrt{y} + 6 = y \qquad\qquad \sqrt{y} + 6 = y$$

$$\sqrt{4} + 6 \overset{?}{=} 4 \quad \text{or} \quad \sqrt{9} + 6 \overset{?}{=} 9$$

$$2 + 6 \overset{?}{=} 4 \qquad\qquad 3 + 6 \overset{?}{=} 9$$

$$8 \neq 4 \qquad\qquad\qquad 9 = 9$$

The only solution is 9; the solution set is $\{9\}$. ∎

In Example 4, note that we changed the form of the original equation $\sqrt{y} + 6 = y$ to $\sqrt{y} = y - 6$ before squaring both sides. Squaring both sides of $\sqrt{y} + 6 = y$ produces $y + 12\sqrt{y} + 36 = y^2$, a more complex equation that still contains a radical. So, again it pays to think ahead a few steps before carrying out the details. Now let's consider an example involving a cube root.

EXAMPLE 5   Solve $\sqrt[3]{n^2 - 1} = 2$.

Solution

$$\sqrt[3]{n^2 - 1} = 2$$

$$(\sqrt[3]{n^2 - 1})^3 = 2^3 \qquad \text{Cube both sides.}$$

$$n^2 - 1 = 8$$

$$n^2 - 9 = 0$$

$$(n + 3)(n - 3) = 0$$

$$n + 3 = 0 \quad \text{or} \quad n - 3 = 0$$

$$n = -3 \quad \text{or} \quad n = 3$$

CHECK

$$\sqrt[3]{n^2 - 1} = 2 \qquad\qquad \sqrt[3]{n^2 - 1} = 2$$

$$\sqrt[3]{(-3)^2 - 1} \overset{?}{=} 2 \quad \text{or} \quad \sqrt[3]{3^2 - 1} \overset{?}{=} 2$$

$$\sqrt[3]{8} \overset{?}{=} 2 \qquad\qquad \sqrt[3]{8} \overset{?}{=} 2$$

$$2 = 2 \qquad\qquad 2 = 2$$

The solution set is $\{-3, 3\}$. ∎

It may be necessary to square both sides of an equation, simplify the resulting equation, and then square both sides again. Our final example of this section illustrates this type of problem.

EXAMPLE 6   Solve $\sqrt{x + 2} = 7 - \sqrt{x + 9}$.

Solution

$$\sqrt{x + 2} = 7 - \sqrt{x + 9}$$

$$(\sqrt{x + 2})^2 = (7 - \sqrt{x + 9})^2 \qquad \text{Square both sides.}$$

$$x + 2 = 49 - 14\sqrt{x + 9} + x + 9$$

$$x + 2 = x + 58 - 14\sqrt{x + 9}$$

$$-56 = -14\sqrt{x + 9}$$

$$4 = \sqrt{x + 9}$$

$$(4)^2 = (\sqrt{x + 9})^2 \qquad \text{Square both sides.}$$

$$16 = x + 9$$

$$7 = x$$

$$\textbf{CHECK} \quad \sqrt{x + 2} = 7 - \sqrt{x + 9}$$

$$\sqrt{7 + 2} \overset{?}{=} 7 - \sqrt{7 + 9}$$

$$\sqrt{9} \overset{?}{=} 7 - \sqrt{16}$$

$$3 \overset{?}{=} 7 - 4$$

$$3 = 3$$

The solution set is $\{7\}$.   ∎

Problem Set 8.6

Solve each of the following equations. Don't forget to check each of your potential solutions.

1. $\sqrt{5x} = 10$ **2.** $\sqrt{3x} = 9$ **3.** $\sqrt{2x} + 4 = 0$ **4.** $\sqrt{4x} + 5 = 0$

5. $2\sqrt{n} = 5$ **6.** $5\sqrt{n} = 3$ **7.** $3\sqrt{n} - 2 = 0$ **8.** $2\sqrt{n} - 7 = 0$

9. $\sqrt{3y + 1} = 4$ **10.** $\sqrt{2y - 3} = 5$ **11.** $\sqrt{4y - 3} - 6 = 0$

12. $\sqrt{3y + 5} - 2 = 0$ **13.** $\sqrt{2x - 5} = -1$ **14.** $\sqrt{4x - 3} = -4$

15. $\sqrt{5x + 2} = \sqrt{6x + 1}$ **16.** $\sqrt{4x + 2} = \sqrt{3x + 4}$ **17.** $\sqrt{3x + 1} = \sqrt{7x - 5}$

18. $\sqrt{6x + 5} = \sqrt{2x + 10}$ **19.** $\sqrt{3x - 2} - \sqrt{x + 4} = 0$ **20.** $\sqrt{7x - 6} - \sqrt{5x + 2} = 0$

21. $5\sqrt{t - 1} = 6$ **22.** $4\sqrt{t + 3} = 6$ **23.** $\sqrt{x^2 + 7} = 4$

24. $\sqrt{x^2 + 3} - 2 = 0$ **25.** $\sqrt{x^2 + 13x + 37} = 1$ **26.** $\sqrt{x^2 + 5x - 20} = 2$

27. $\sqrt{x^2 - x + 1} = x + 1$ **28.** $\sqrt{n^2 - 2n - 4} = n$ **29.** $\sqrt{x^2 + 3x + 7} = x + 2$

30. $\sqrt{x^2 + 2x + 1} = x + 3$ **31.** $\sqrt{-4x + 17} = x - 3$ **32.** $\sqrt{2x - 1} = x - 2$

33. $\sqrt{n + 4} = n + 4$ **34.** $\sqrt{n + 6} = n + 6$ **35.** $\sqrt{3y} = y - 6$

36. $2\sqrt{n} = n - 3$

37. $4\sqrt{x} + 5 = x$

38. $\sqrt{-x} - 6 = x$

39. $\sqrt[3]{x-2} = 3$

40. $\sqrt[3]{x+1} = 4$

41. $\sqrt[3]{2x+3} = -3$

42. $\sqrt[3]{3x-1} = -4$

43. $\sqrt[3]{2x+5} = \sqrt[3]{4-x}$

44. $\sqrt[3]{3x-1} = \sqrt[3]{2-5x}$

45. $\sqrt{x+19} - \sqrt{x+28} = -1$

46. $\sqrt{x+4} = \sqrt{x-1} + 1$

47. $\sqrt{3x+1} + \sqrt{2x+4} = 3$

48. $\sqrt{2x-1} - \sqrt{x+3} = 1$

49. $\sqrt{n-4} + \sqrt{n+4} = 2\sqrt{n-1}$

50. $\sqrt{n-3} + \sqrt{n+5} = 2\sqrt{n}$

51. $\sqrt{t+3} - \sqrt{t-2} = \sqrt{7-t}$

52. $\sqrt{t+7} - 2\sqrt{t-8} = \sqrt{t-5}$

Thoughts into Words

53. How is the multiplication property of one used when simplifying radicals?

54. Explain the concept of "extraneous solutions."

55. Are $a = b$ and $a^2 = b^2$ equivalent statements? Explain your answer.

8.7
The Merging of Exponents and Roots

Recall that the basic properties of positive integral exponents led to a definition for the use of negative integers as exponents. In this section, the properties of integral exponents are used to form definitions for the use of rational numbers as exponents. These definitions will tie together the concepts of *exponent* and *root*.

Let's consider the following comparisons.

From our study of radicals we know that

If $(b^n)^m = b^{mn}$ is to hold when n equals a rational number of the form $\frac{1}{p}$, where p is a positive integer greater than one, then

$$(\sqrt{5})^2 = 5,$$

$$\left(5^{\frac{1}{2}}\right)^2 = 5^{2\left(\frac{1}{2}\right)} = 5^1 = 5,$$

$$(\sqrt[3]{8})^3 = 8,$$

$$\left(8^{\frac{1}{3}}\right)^3 = 8^{3\left(\frac{1}{3}\right)} = 8^1 = 8,$$

$$(\sqrt[4]{21})^4 = 21.$$

$$\left(21^{\frac{1}{4}}\right)^4 = 21^{4\left(\frac{1}{4}\right)} = 21^1 = 21.$$

It would seem reasonable to make the following definition.

DEFINITION 8.6

If b is a real number, n is a positive integer greater than one, and $\sqrt[n]{b}$ exists, then

$$b^{\frac{1}{n}} = \sqrt[n]{b}.$$

Definition 8.6 states that $b^{\frac{1}{n}}$ means *the nth root of b*. We shall assume that b and n are chosen so that $\sqrt[n]{b}$ exists. For example, $(-25)^{\frac{1}{2}}$ is not meaningful at this time because $\sqrt{-25}$ is not a real number.

Consider the following examples that demonstrate the use of Definition 8.6.

$$25^{\frac{1}{2}} = \sqrt{25} = 5, \qquad\qquad 16^{\frac{1}{4}} = \sqrt[4]{16} = 2,$$

$$8^{\frac{1}{3}} = \sqrt[3]{8} = 2, \qquad\qquad \left(\frac{36}{49}\right)^{\frac{1}{2}} = \sqrt{\frac{36}{49}} = \frac{6}{7},$$

$$(-27)^{\frac{1}{3}} = \sqrt[3]{-27} = -3$$

The following definition provides the basis for the use of *all* rational numbers as exponents.

DEFINITION 8.7

> If $\dfrac{m}{n}$ is a rational number, where n is a positive integer greater than one, and b is a real number such that $\sqrt[n]{b}$ exists, then
> $$b^{\frac{m}{n}} = \sqrt[n]{b^m} = (\sqrt[n]{b})^m.$$

In Definition 8.7, notice that the denominator of the exponent is the index of the radical; and the numerator of the exponent is either the exponent of the radicand or the exponent of the root.

Whether we use the form $\sqrt[n]{b^m}$ or $(\sqrt[n]{b})^m$ for computational purposes depends somewhat on the magnitude of the problem. Let's use both forms on two problems to illustrate this point.

$$
\begin{aligned}
8^{\frac{2}{3}} &= \sqrt[3]{8^2} \qquad \text{or} \qquad 8^{\frac{2}{3}} = (\sqrt[3]{8})^2 \\
&= \sqrt[3]{64} \qquad\qquad\qquad\quad = 2^2 \\
&= 4 \qquad\qquad\qquad\qquad\; = 4 \\
27^{\frac{2}{3}} &= \sqrt[3]{27^2} \qquad \text{or} \qquad 27^{\frac{2}{3}} = (\sqrt[3]{27})^2 \\
&= \sqrt[3]{729} \qquad\qquad\qquad = 3^2 \\
&= 9 \qquad\qquad\qquad\qquad\; = 9
\end{aligned}
$$

To compute $8^{\frac{2}{3}}$, either form seems to work about as well as the other one. However, to compute $27^{\frac{2}{3}}$, it should be obvious that $(\sqrt[3]{27})^2$ is much easier to handle than $\sqrt[3]{27^2}$.

EXAMPLE 1   Simplify each of the following numerical expressions.

(a) $25^{\frac{3}{2}}$ (b) $16^{\frac{3}{4}}$ (c) $(32)^{-\frac{2}{5}}$

(d) $(-64)^{\frac{2}{3}}$ (e) $-8^{\frac{1}{3}}$

Solution

(a) $25^{\frac{3}{2}} = (\sqrt{25})^3 = 5^3 = 125$

(b) $16^{\frac{3}{4}} = (\sqrt[4]{16})^3 = 2^3 = 8$

(c) $(32)^{-\frac{2}{5}} = \dfrac{1}{(32)^{\frac{2}{5}}} = \dfrac{1}{(\sqrt[5]{32})^2} = \dfrac{1}{2^2} = \dfrac{1}{4}$

(d) $(-64)^{\frac{2}{3}} = (\sqrt[3]{-64})^2 = (-4)^2 = 16$

(e) $-8^{\frac{1}{3}} = -\sqrt[3]{8} = -2$ ■

The basic laws of exponents stated in Property 8.2 are true for all rational exponents. Therefore, without restating Property 8.2 we shall henceforth use it for rational as well as integral exponents.

Some problems can be handled better in exponential form and others in radical form. Thus, we must be able to switch forms with a certain amount of ease. Let's consider some examples where we switch from one form to the other.

EXAMPLE 2   Write each of the following expressions in radical form.

(a) $x^{\frac{3}{4}}$ (b) $3y^{\frac{2}{5}}$

(c) $x^{\frac{1}{4}}y^{\frac{3}{4}}$ (d) $(x + y)^{\frac{2}{3}}$

Solution

(a) $x^{\frac{3}{4}} = \sqrt[4]{x^3}$ (b) $3y^{\frac{2}{5}} = 3\sqrt[5]{y^2}$

(c) $x^{\frac{1}{4}}y^{\frac{3}{4}} = (xy^3)^{\frac{1}{4}} = \sqrt[4]{xy^3}$ (d) $(x + y)^{\frac{2}{3}} = \sqrt[3]{(x + y)^2}$ ■

EXAMPLE 3   Write each of the following using positive rational exponents.

(a) $\sqrt{xy}$ (b) $\sqrt[4]{a^3b}$ (c) $4\sqrt[3]{x^2}$ (d) $\sqrt[5]{(x + y)^4}$

Solution

(a) $\sqrt{xy} = (xy)^{\frac{1}{2}} = x^{\frac{1}{2}}y^{\frac{1}{2}}$ (b) $\sqrt[4]{a^3b} = (a^3b)^{\frac{1}{4}} = a^{\frac{3}{4}}b^{\frac{1}{4}}$

(c) $4\sqrt[3]{x^2} = 4x^{\frac{2}{3}}$ (d) $\sqrt[5]{(x + y)^4} = (x + y)^{\frac{4}{5}}$ ■

The basic properties of exponents provide the basis for simplifying algebraic expressions that contain rational exponents, as these next examples illustrate.

EXAMPLE 4   Simplify each of the following. Express final results using positive exponents only.

(a) $\left(3x^{\frac{1}{2}}\right)\left(4x^{\frac{2}{3}}\right)$ (b) $\left(5a^{\frac{1}{3}}b^{\frac{1}{2}}\right)^2$ (c) $\dfrac{12y^{\frac{1}{3}}}{6y^{\frac{1}{2}}}$ (d) $\left(\dfrac{3x^{\frac{2}{5}}}{2y^{\frac{2}{3}}}\right)^4$

Solution

(a) $\left(3x^{\frac{1}{2}}\right)\left(4x^{\frac{2}{3}}\right) = 3 \cdot 4 \cdot x^{\frac{1}{2}} \cdot x^{\frac{2}{3}}$

$\qquad\qquad = 12x^{\frac{1}{2}+\frac{2}{3}}$ $b^n \cdot b^m = b^{n+m}$

$\qquad\qquad = 12x^{\frac{3}{6}+\frac{4}{6}}$

$\qquad\qquad = 12x^{\frac{7}{6}}$

(b) $\left(5a^{\frac{1}{3}}b^{\frac{1}{2}}\right)^2 = 5^2 \cdot \left(a^{\frac{1}{3}}\right)^2 \cdot \left(b^{\frac{1}{2}}\right)^2$ $(ab)^n = a^n b^n$

$\qquad\qquad = 25a^{\frac{2}{3}}b$ $(b^n)^m = b^{mn}$

(c) $\dfrac{12y^{\frac{1}{3}}}{6y^{\frac{1}{2}}} = 2y^{\frac{1}{3}-\frac{1}{2}}$ $\dfrac{b^n}{b^m} = b^{n-m}$

$\qquad\quad = 2y^{\frac{2}{6}-\frac{3}{6}}$

$\qquad\quad = 2y^{-\frac{1}{6}}$

$\qquad\quad = \dfrac{2}{y^{\frac{1}{6}}}$

(d) $\left(\dfrac{3x^{\frac{2}{5}}}{2y^{\frac{2}{3}}}\right)^4 = \dfrac{\left(3x^{\frac{2}{5}}\right)^4}{\left(2y^{\frac{2}{3}}\right)^4}$ $\left(\dfrac{a}{b}\right)^n = \dfrac{a^n}{b^n}$

$\qquad\qquad = \dfrac{3^4 \cdot \left(x^{\frac{2}{5}}\right)^4}{2^4 \cdot \left(y^{\frac{2}{3}}\right)^4}$ $(ab)^n = a^n b^n$

$\qquad\qquad = \dfrac{81x^{\frac{8}{5}}}{16y^{\frac{8}{3}}}$ $(b^n)^m = b^{mn}$ ■

The link between exponents and roots also provides a basis for multiplying and dividing some radicals even if they have a different index. The general procedure is to change from radical form to exponential form, apply the properties of exponents, and then change back to radical form. Let's consider three examples to illustrate this process.

EXAMPLE 5   Perform the indicated operations and express the answer in simplest radical form.

(a) $\sqrt{2}\,\sqrt[3]{2}$

(b) $\dfrac{\sqrt{5}}{\sqrt[3]{5}}$

(c) $\dfrac{\sqrt{4}}{\sqrt[3]{2}}$

Solution

(a) $\sqrt{2}\,\sqrt[3]{2} = 2^{\frac{1}{2}} \cdot 2^{\frac{1}{3}}$

$= 2^{\frac{1}{2}+\frac{1}{3}}$

$= 2^{\frac{3}{6}+\frac{2}{6}}$

$= 2^{\frac{5}{6}}$

$= \sqrt[6]{2^5} = \sqrt[6]{32}$

(b) $\dfrac{\sqrt{5}}{\sqrt[3]{5}} = \dfrac{5^{\frac{1}{2}}}{5^{\frac{1}{3}}}$

$= 5^{\frac{1}{2}-\frac{1}{3}}$

$= 5^{\frac{3}{6}-\frac{2}{6}}$

$= 5^{\frac{1}{6}} = \sqrt[6]{5}$

(c) $\dfrac{\sqrt{4}}{\sqrt[3]{2}} = \dfrac{4^{\frac{1}{2}}}{2^{\frac{1}{3}}}$

$= \dfrac{(2^2)^{\frac{1}{2}}}{2^{\frac{1}{3}}}$

$= \dfrac{2^1}{2^{\frac{1}{3}}}$

$= 2^{1-\frac{1}{3}}$

$= 2^{\frac{2}{3}} = \sqrt[3]{2^2} = \sqrt[3]{4}$

Problem Set 8.7

Simplify each of the following numerical expressions.

1. $81^{\frac{1}{2}}$
2. $64^{\frac{1}{2}}$
3. $27^{\frac{1}{3}}$
4. $(-32)^{\frac{1}{5}}$
5. $(-8)^{\frac{1}{3}}$
6. $\left(-\dfrac{27}{8}\right)^{\frac{1}{3}}$
7. $-25^{\frac{1}{2}}$
8. $-64^{\frac{1}{3}}$
9. $36^{-\frac{1}{2}}$
10. $81^{-\frac{1}{2}}$
11. $\left(\dfrac{1}{27}\right)^{-\frac{1}{3}}$
12. $\left(-\dfrac{8}{27}\right)^{-\frac{1}{3}}$
13. $4^{\frac{3}{2}}$
14. $64^{\frac{2}{3}}$
15. $27^{\frac{4}{3}}$
16. $4^{\frac{7}{2}}$
17. $(-1)^{\frac{7}{3}}$
18. $(-8)^{\frac{4}{3}}$

19. $-4^{\frac{5}{2}}$ **20.** $-16^{\frac{3}{2}}$ **21.** $\left(\dfrac{27}{8}\right)^{\frac{4}{3}}$

22. $\left(\dfrac{8}{125}\right)^{\frac{2}{3}}$ **23.** $\left(\dfrac{1}{8}\right)^{-\frac{2}{3}}$ **24.** $\left(-\dfrac{1}{27}\right)^{-\frac{2}{3}}$

25. $64^{-\frac{7}{6}}$ **26.** $32^{-\frac{4}{5}}$ **27.** $-25^{\frac{3}{2}}$

28. $-16^{\frac{3}{4}}$ **29.** $125^{\frac{4}{3}}$ **30.** $81^{\frac{5}{4}}$

Write each of the following in radical form. For example, $3x^{\frac{2}{3}} = 3\sqrt[3]{x^2}$.

31. $x^{\frac{4}{3}}$ **32.** $x^{\frac{2}{5}}$ **33.** $3x^{\frac{1}{2}}$

34. $5x^{\frac{1}{4}}$ **35.** $(2y)^{\frac{1}{3}}$ **36.** $(3xy)^{\frac{1}{2}}$

37. $(2x - 3y)^{\frac{1}{2}}$ **38.** $(5x + y)^{\frac{1}{3}}$ **39.** $(2a - 3b)^{\frac{2}{3}}$

40. $(5a + 7b)^{\frac{3}{5}}$ **41.** $x^{\frac{2}{3}}y^{\frac{1}{3}}$ **42.** $x^{\frac{3}{7}}y^{\frac{5}{7}}$

43. $-3x^{\frac{1}{5}}y^{\frac{2}{5}}$ **44.** $-4x^{\frac{3}{4}}y^{\frac{1}{4}}$

Write each of the following using positive rational exponents. For example,

$$\sqrt{ab} = (ab)^{\frac{1}{2}} = a^{\frac{1}{2}}b^{\frac{1}{2}}.$$

45. $\sqrt{5y}$ **46.** $\sqrt{2xy}$ **47.** $3\sqrt{y}$

48. $5\sqrt{ab}$ **49.** $\sqrt[3]{xy^2}$ **50.** $\sqrt[5]{x^2y^4}$

51. $\sqrt[4]{a^2b^3}$ **52.** $\sqrt[6]{ab^5}$ **53.** $\sqrt[5]{(2x - y)^3}$

54. $\sqrt[7]{(3x - y)^4}$ **55.** $5x\sqrt{y}$ **56.** $4y\sqrt[3]{x}$

57. $-\sqrt[3]{x + y}$ **58.** $-\sqrt[5]{(x - y)^2}$

Simplify each of the following. Express final results using positive exponents only. For example,

$$\left(2x^{\frac{1}{2}}\right)\left(3x^{\frac{1}{3}}\right) = 6x^{\frac{5}{6}}.$$

59. $\left(2x^{\frac{2}{5}}\right)\left(6x^{\frac{1}{4}}\right)$ **60.** $\left(3x^{\frac{1}{4}}\right)\left(5x^{\frac{1}{3}}\right)$ **61.** $\left(y^{\frac{3}{3}}\right)\left(y^{-\frac{1}{4}}\right)$

62. $\left(y^{\frac{3}{4}}\right)\left(y^{-\frac{1}{2}}\right)$ **63.** $\left(x^{\frac{2}{5}}\right)\left(4x^{-\frac{1}{2}}\right)$ **64.** $\left(2x^{\frac{1}{3}}\right)\left(x^{-\frac{1}{2}}\right)$

65. $\left(4x^{\frac{1}{2}}y\right)^2$ **66.** $\left(3x^{\frac{1}{4}}y^{\frac{1}{5}}\right)^3$ **67.** $(8x^6y^3)^{\frac{1}{3}}$

68. $(9x^2y^4)^{\frac{1}{2}}$ **69.** $\dfrac{24x^{\frac{3}{5}}}{6x^{\frac{1}{3}}}$ **70.** $\dfrac{18x^{\frac{1}{2}}}{9x^{\frac{1}{3}}}$

71. $\dfrac{48b^{\frac{1}{3}}}{12b^{\frac{3}{4}}}$ **72.** $\dfrac{56a^{\frac{1}{6}}}{8a^{\frac{1}{4}}}$ **73.** $\left(\dfrac{6x^{\frac{2}{5}}}{7y^{\frac{2}{3}}}\right)^2$

74. $\left(\dfrac{2x^{\frac{1}{3}}}{3y^{\frac{1}{4}}}\right)^{4}$

75. $\left(\dfrac{x^{2}}{y^{3}}\right)^{-\frac{1}{2}}$

76. $\left(\dfrac{a^{3}}{b^{-2}}\right)^{-\frac{1}{3}}$

77. $\left(\dfrac{18x^{\frac{1}{3}}}{9x^{\frac{1}{4}}}\right)^{2}$

78. $\left(\dfrac{72x^{\frac{3}{4}}}{6x^{\frac{1}{2}}}\right)^{2}$

79. $\left(\dfrac{60a^{\frac{1}{5}}}{15a^{\frac{3}{4}}}\right)^{2}$

80. $\left(\dfrac{64a^{\frac{1}{3}}}{16a^{\frac{5}{9}}}\right)^{3}$

Perform the indicated operations and express answers in simplest radical form. (See Example 5.)

81. $\sqrt[3]{3}\sqrt{3}$

82. $\sqrt{2}\sqrt[4]{2}$

83. $\sqrt[4]{6}\sqrt{6}$

84. $\sqrt[3]{5}\sqrt{5}$

85. $\dfrac{\sqrt[3]{3}}{\sqrt[4]{3}}$

86. $\dfrac{\sqrt{2}}{\sqrt[3]{2}}$

87. $\dfrac{\sqrt[3]{8}}{\sqrt[4]{4}}$

88. $\dfrac{\sqrt{9}}{\sqrt[3]{3}}$

89. $\dfrac{\sqrt[4]{27}}{\sqrt{3}}$

90. $\dfrac{\sqrt[3]{16}}{\sqrt[6]{4}}$

Miscellaneous Problems

91. If your calculator has $\boxed{y^{x}}$ and $\boxed{1/x}$ keys, then you can use them to evaluate cube roots, fourth roots, fifth roots, and so on. For example, some calculators evaluate $\sqrt[3]{4913}$ as follows:

$$4913\ \boxed{y^{x}}\ 3\ \boxed{1/x} = 17.$$

Use your calculator to evalute each of the following.

(a) $\sqrt[3]{1728}$

(b) $\sqrt[3]{5832}$

(c) $\sqrt[4]{2401}$

(d) $\sqrt[4]{65536}$

(e) $\sqrt[5]{161051}$

(f) $\sqrt[5]{6436343}$

92. Definition 8.7 stated that

$$b^{\frac{m}{n}} = \sqrt[n]{b^{m}} = (\sqrt[n]{b})^{m}.$$

Use your calculator to verify each of the following.

(a) $\sqrt[3]{27^{2}} = (\sqrt[3]{27})^{2}$

(b) $\sqrt[3]{8^{5}} = (\sqrt[3]{8})^{5}$

(c) $\sqrt[4]{16^{3}} = (\sqrt[4]{16})^{3}$

(d) $\sqrt[3]{16^{2}} = (\sqrt[3]{16})^{2}$

(e) $\sqrt[5]{9^{4}} = (\sqrt[5]{9})^{4}$

(f) $\sqrt[3]{12^{4}} = (\sqrt[3]{12})^{4}$

93. Use your calculator to evaluate each of the following.

(a) $16^{\frac{5}{2}}$

(b) $25^{\frac{7}{2}}$

(c) $16^{\frac{9}{4}}$

(d) $27^{\frac{5}{3}}$

(e) $343^{\frac{2}{3}}$

(f) $512^{\frac{4}{3}}$

94. Use your calculator to find a rational approximation, to the nearest thousandth, for each of the following.

(a) $7^{\frac{4}{3}}$ (b) $10^{\frac{4}{5}}$ (c) $12^{\frac{2}{5}}$

(d) $19^{\frac{2}{5}}$ (e) $7^{\frac{3}{4}}$ (f) $10^{\frac{5}{4}}$

8.8
Scientific Notation

Many applications of mathematics involve the use of very large and very small numbers. For example,

1. A light year—the distance that a ray of light travels in one year—is approximately 5,900,000,000,000 miles;

2. In 1982 the national debt was approximately 950,000,000,000 dollars;

3. In the metric system, a millimicron equals 0.000000001 of a meter;

4. The weight of an oxygen molecule is approximately 0.000000000000000000000053 of a gram.

Working with numbers of this type in standard form is quite cumbersome. It is much more convenient to represent very small and very large numbers in **scientific notation**, sometimes called **scientific form**. We express a number in scientific notation when we write it as a product of a number between 1 and 10 and an integral power of 10. Symbolically, a number written in scientific notation has the form

$$(N)(10)^k$$

where N is a number between 1 and 10, written in decimal form, and k is an integer. Consider the following examples that show a comparison between ordinary notation and scientific notation.

Ordinary notation	*Scientific notation*
2.14	$(2.14)(10)^0$
31.78	$(3.178)(10)^1$
412.9	$(4.129)(10)^2$
8,000,000	$(8)(10)^6$
0.14	$(1.4)(10)^{-1}$
0.0379	$(3.79)(10)^{-2}$
0.00000049	$(4.9)(10)^{-7}$

To switch from ordinary notation to scientific notation, you can use the following procedure.

Write the given number as the product of a number between 1 and 10 and a power of 10. Determine the exponent of 10 by counting the number of places that the decimal point was moved when going from the original number to the number between 1 and 10. This exponent is (a) negative if the original number is less than 1, (b) positive if the original number is greater than 10, and (c) 0 if the original number itself is between 1 and 10.

Thus, we can write

$$0.00467 = (4.67)(10)^{-3},$$

$$87,000 = (8.7)(10)^4,$$

$$3.1416 = (3.1416)(10)^0.$$

To switch from scientific notation to ordinary notation you can use the following procedure.

Move the decimal point the number of places indicated by the exponent of 10. The decimal point is moved to the right if the exponent is positive and to the left if it is negative.

Thus, we can write

$$(4.78)(10)^4 = 47,800,$$

$$(8.4)(10)^{-3} = 0.0084.$$

Scientific notation can frequently be used to simplify numerical calculations. We merely change the numbers to scientific notation and use the appropriate properties of exponents. Consider the following examples.

EXAMPLE 1   Perform the indicated operations.

(a) $(0.00024)(20,000)$ (b) $\dfrac{7,800,000}{0.0039}$

(c) $\dfrac{(0.00069)(0.0034)}{(0.0000017)(0.023)}$ (d) $\sqrt{0.000004}$

Solution

(a) $(0.00024)(20,000) = (2.4)(10)^{-4}(2)(10)^4$

$\qquad\qquad\qquad\qquad = (2.4)(2)(10)^{-4}(10)^4$

$\qquad\qquad\qquad\qquad = (4.8)(10)^0$

$\qquad\qquad\qquad\qquad = (4.8)(1)$

$\qquad\qquad\qquad\qquad = 4.8$

(b) $\dfrac{7,800,000}{0.0039} = \dfrac{(7.8)(10)^6}{(3.9)(10)^{-3}}$

$\qquad\qquad\quad = (2)(10)^9$

$\qquad\qquad\quad = 2,000,000,000$

(c) $\dfrac{(0.00069)(0.0034)}{(0.0000017)(0.023)} = \dfrac{(6.9)(10)^{-4}(3.4)(10)^{-3}}{(1.7)(10)^{-6}(2.3)(10)^{-2}}$

$$= \frac{(6\llap{$\diagdown$}.9)(3\llap{$\diagup$}.4)(10)^{-7}}{(1\llap{$\diagup$}.7)(2\llap{$\diagdown$}.3)(10)^{-8}}$$

$$= (6)(10)^{1}$$

$$= 60$$

(d) $\sqrt{0.000004} = \sqrt{(4)(10)^{-6}}$

$$= ((4)(10)^{-6})^{\frac{1}{2}}$$

$$= 4^{\frac{1}{2}}\,((10)^{-6})^{\frac{1}{2}}$$

$$= (2)(10)^{-3}$$

$$= 0.002 \qquad\qquad\qquad\blacksquare$$

Many calculators are equipped to display numbers in scientific notation. The display panel shows the number between 1 and 10 and the appropriate exponent of 10. For example, evaluating $(3,800,000)^2$ yields

$$\boxed{1.444 \quad 13}$$

Thus, $(3,800,000)^2 = (1.444)(10)^{13} = 14,440,000,000,000$.

Similarly, the answer for $(0.000168)^2$ is displayed as

$$\boxed{2.8224 \quad -08}$$

Thus, $(0.000168)^2 = (2.8224)(10)^{-8} = 0.000000028224$.

Calculators vary as to the number of digits displayed in the number between 1 and 10 when using scientific notation. For example, we used two different calculators to estimate $(6729)^6$ and obtained the following results.

$$\boxed{9.2833 \quad 22}$$

$$\boxed{9.283316768 \quad 22}.$$

Obviously, you need to know the capabilities of your calculator when working with problems in scientific notation.

Many calculators also allow the entry of a number in scientific notation. Such calculators are equipped with an enter-the-exponent key (often labeled as $\boxed{\text{EE}}$ or $\boxed{\text{E EX}}$). Thus, a number such as $(3.14)(10)^8$ might be entered as follows.

Enter	*Press*	*Display*
3.14	$\boxed{\text{EE}}$	3.14 00
8		3.14 08

Furthermore, it may be that your calculator will perform the switch from ordinary notation to scientific notation with a routine such as the following.

Enter	*Press*	*Display*	
4721	EE	4721	00
	=	4.721	03

Be sure that you know the routine that will accomplish this switch on your calculator.

It should be evident from this brief discussion that even when using a calculator, you need to have a thorough understanding of scientific notation.

Problem Set 8.8

Write each of the following in scientific notation. For example, $27,800 = (2.78)(10)^4$.

1. 89   **2.** 117   **3.** 4290

4. 812,000   **5.** 6,120,000   **6.** 72,400,000

7. 40,000,000   **8.** 500,000,000   **9.** 376.4

10. 9126.21   **11.** 0.347   **12.** 0.2165

13. 0.0214   **14.** 0.0037   **15.** 0.00005

16. 0.00000082   **17.** 0.00000000194

18. 0.00000000003

Write each of the following in ordinary notation. For example $(3.18)(10)^2 = 318$.

19. $(2.3)(10)^1$ **20.** $(1.62)(10)^2$ **21.** $(4.19)(10)^3$

22. $(7.631)(10)^4$ **23.** $(5)(10)^8$ **24.** $(7)(10)^9$

25. $(3.14)(10)^{10}$ **26.** $(2.04)(10)^{12}$

27. $(4.3)(10)^{-1}$ **28.** $(5.2)(10)^{-2}$.

29. $(9.14)(10)^{-4}$ **30.** $(8.76)(10)^{-5}$

31. $(5.123)(10)^{-8}$ **32.** $(6)(10)^{-9}$

Use scientific notation and the properties of exponents to help perform the following operations.

33. $(0.0037)(0.00002)$ **34.** $(0.00003)(0.00025)$

35. $(0.00007)(11,000)$ **36.** $(0.000004)(120,000)$

37. $\dfrac{360,000,000}{0.0012}$ **38.** $\dfrac{66,000,000,000}{0.022}$

39. $\dfrac{0.000064}{16,000}$ **40.** $\dfrac{0.00072}{0.0000024}$

41. $\dfrac{(60,000)(0.006)}{(0.0009)(400)}$ **42.** $\dfrac{(0.00063)(960,000)}{(3,200)(0.0000021)}$

43. $\dfrac{(0.0045)(60000)}{(1800)(0.00015)}$ **44.** $\dfrac{(0.00016)(300)(0.028)}{0.064}$

45. $\sqrt{9,000,000}$ **46.** $\sqrt{0.00000009}$ **47.** $\sqrt[3]{8000}$

48. $\sqrt[3]{0.001}$ **49.** $(90,000)^{\frac{3}{2}}$ **50.** $(8000)^{\frac{2}{3}}$

51. Sometimes it is more convenient to express a number as a product of a power of 10 and a number that is not between 1 and 10. For example, suppose that we want to calculate $\sqrt{640{,}000}$. We can proceed as follows.

$$\sqrt{640{,}000} = \sqrt{(64)(10)^4}$$
$$= ((64)(10)^4)^{\frac{1}{2}}$$
$$= (64)^{\frac{1}{2}}(10^4)^{\frac{1}{2}}$$
$$= (8)(10)^2$$
$$= 8(100) = 800$$

Compute each of the following.

(a) $\sqrt{49{,}000{,}000}$ (b) $\sqrt{0.0025}$ (c) $\sqrt{14400}$

(d) $\sqrt{0.000121}$ (e) $\sqrt[3]{27000}$ (f) $\sqrt[3]{0.000064}$

52. Use your calculator to evaluate each of the following. Express final answers in ordinary notation.

(a) $(27{,}000)^2$ (b) $(450{,}000)^2$ (c) $(14{,}800)^2$

(d) $(1700)^3$ (e) $(900)^4$ (f) $(60)^5$

(g) $(0.0213)^2$ (h) $(0.000213)^2$

(i) $(0.000198)^2$

(j) $(0.000009)^3$

53. Use your calculator to evaluate each of the following. Express final answers in scientific notation with the number between 1 and 10 rounded to the nearest thousandth.

(a) $(4576)^4$ (b) $(719)^{10}$ (c) $(28)^{12}$

(d) $(8619)^6$ (e) $(314)^5$ (f) $(145{,}723)^2$

54. Use your calculator to evaluate each of the following. Express final answers in ordinary notation rounded to the nearest thousandth.

(a) $(1.09)^5$ (b) $(1.08)^{10}$ (c) $(1.14)^7$

(d) $(1.12)^{20}$ (e) $(0.785)^4$ (f) $(0.492)^5$

Chapter 8 Summary

(8.1) The following properties form the basis for manipulating with exponents.

1. $b^n \cdot b^m = b^{n+m}$ product of two powers

2. $(b^n)^m = b^{mn}$ power of a power

3. $(ab)^n = a^n b^n$ power of a product

4. $\left(\dfrac{a}{b}\right)^n = \dfrac{a^n}{b^n}$ power of a quotient

5. $\dfrac{b^n}{b^m} = b^{n-m}$ quotient of two powers

(8.2–8.4) The **principal nth root of b** is designated by $\sqrt[n]{b}$, where n is the **index** and b is the **radicand**.

A radical expression is in **simplest radical form** if:

1. A radicand contains no polynomial factor raised to a power equal to or greater than the index of the radical;
2. No fraction appears within a radical sign;
3. No radical appears in the denominator.

The following properties are used to express radicals in simplest form.

$$\sqrt[n]{bc} = \sqrt[n]{b}\,\sqrt[n]{c}, \qquad \sqrt[n]{\frac{b}{c}} = \frac{\sqrt[n]{b}}{\sqrt[n]{c}}$$

Simplifying by combining radicals sometimes requires that you first express the given radicals in simplest form and then apply the distributive property.

(8.5) The distributive property and the property $\sqrt[n]{b}\,\sqrt[n]{c} = \sqrt[n]{bc}$ are used to find products of expressions that involve radicals.

The special product pattern $(a + b)(a - b) = a^2 - b^2$ suggests a procedure for **rationalizing the denominator** of an expression that contains a binomial denominator with radicals.

(8.6) Equations that contain radicals with variables in a radicand are called **radical equations**. The property, if $a = b$, then $a^n = b^n$, forms the basis for solving radical equations. Raising both sides of an equation to a positive integral power may produce **extraneous solutions**, that is, solutions that do not satisfy the original equation. Therefore, you **must check** each potential solution.

(8.7) If b is a real number, n is a positive integer greater than one, and $\sqrt[n]{b}$ exists, then

$$b^{\frac{1}{n}} = \sqrt[n]{b}.$$

Thus, $b^{\frac{1}{n}}$ means **the nth root of b**.

If $\dfrac{m}{n}$ is a rational number, when n is a positive integer greater than one, and b is a real number such that $\sqrt[n]{b}$ exists, then

$$b^{\frac{m}{n}} = \sqrt[n]{b^m} = (\sqrt[n]{b})^m.$$

Both $\sqrt[n]{b^m}$ and $(\sqrt[n]{n})^m$ can be used for computational purposes.

We need to be able to switch back and forth between **exponential** and **radical form**. The link between exponents and roots provides a basis for multiplying and dividing some radicals even if they have different indices.

(8.8) The **scientific form** of a number is expressed as

$$(N)(10)^k$$

where N is a number between 1 and 10, written in decimal form, and k is an integer. Scientific notation is often convenient to use with very small and very large numbers.

For example, .000046 can be expressed as $(4.6)(10^{-5})$ and 92,000,000 can be written as $(9.2)(10)^7$.

Scientific notation can often be used to simplify numerical calculations. For example,

$$(.000016)(30000) = (1.6)(10)^{-5}(3)(10)^4$$
$$= (4.8)(10)^{-1} = .48.$$

Chapter 8 Review Problem Set

Evaluate each of the following numerical expressions.

1. 4^{-3}

2. $\left(\dfrac{2}{3}\right)^{-2}$

3. $(3^2 \cdot 3^{-3})^{-1}$

4. $\sqrt[3]{-8}$

5. $\sqrt[4]{\dfrac{16}{81}}$

6. $4^{\frac{5}{2}}$

7. $(-1)^{-\frac{2}{3}}$

8. $\left(\dfrac{8}{27}\right)^{\frac{2}{3}}$

9. $-16^{\frac{3}{2}}$

10. $\dfrac{2^3}{2^{-2}}$

11. $(4^{-2} \cdot 4^2)^{-1}$

12. $\left(\dfrac{3^{-1}}{3^2}\right)^{-1}$

Express each of the following radicals in simplest radical form.

13. $\sqrt{54}$

14. $\sqrt{48x^3y}$

15. $\dfrac{4\sqrt{3}}{\sqrt{6}}$

16. $\sqrt{\dfrac{5}{12x^3}}$

17. $\sqrt[3]{56}$

18. $\dfrac{\sqrt[3]{2}}{\sqrt[3]{9}}$

19. $\sqrt{\dfrac{9}{5}}$

20. $\sqrt{\dfrac{3x^3}{7}}$

21. $\sqrt[3]{108x^4y^8}$

22. $\dfrac{3}{4}\sqrt{150}$

23. $\dfrac{2}{3}\sqrt{45xy^3}$

24. $\dfrac{\sqrt{8x^2}}{\sqrt{2x}}$

Multiply and simplify.

25. $(3\sqrt{8})(4\sqrt{5})$

26. $(5\sqrt[3]{2})(6\sqrt[3]{4})$

27. $3\sqrt{2}(4\sqrt{6} - 2\sqrt{7})$

28. $(\sqrt{x} + 3)(\sqrt{x} - 5)$

29. $(2\sqrt{5} - \sqrt{3})(2\sqrt{5} + \sqrt{3})$

30. $(3\sqrt{2} + \sqrt{6})(5\sqrt{2} - 3\sqrt{6})$

31. $(2\sqrt{a} + \sqrt{b})(3\sqrt{a} - 4\sqrt{b})$

32. $(4\sqrt{8} - \sqrt{2})(\sqrt{8} + 3\sqrt{2})$

Rationalize the denominator and simplify.

33. $\dfrac{4}{\sqrt{7} - 1}$

34. $\dfrac{\sqrt{3}}{\sqrt{8} + \sqrt{5}}$

35. $\dfrac{3}{2\sqrt{3} + 3\sqrt{5}}$

36. $\dfrac{3\sqrt{2}}{2\sqrt{6} - \sqrt{10}}$

Simplify each of the following and express the final results using positive exponents.

37. $(x^{-3}y^4)^{-2}$

38. $\left(\dfrac{2a^{-1}}{3b^4}\right)^{-3}$

39. $\left(4x^{\frac{1}{2}}\right)\left(5x^{\frac{1}{3}}\right)$

40. $\dfrac{42a^{\frac{3}{4}}}{6a^{\frac{1}{3}}}$

41. $\left(\dfrac{x^3}{y^4}\right)^{-\frac{1}{3}}$

42. $\left(\dfrac{6x^{-2}}{2x^4}\right)^{-2}$

Use the distributive property to help simplify each of the following.

43. $3\sqrt{45} - 2\sqrt{20} - \sqrt{80}$

44. $4\sqrt[3]{24} + 3\sqrt[3]{3} - 2\sqrt[3]{81}$

45. $3\sqrt{24} - \dfrac{2\sqrt{54}}{5} + \dfrac{\sqrt{96}}{4}$

46. $-2\sqrt{12x} + 3\sqrt{27x} - 5\sqrt{48x}$

Express each of the following as a single fraction involving positive exponents only.

47. $x^{-2} + y^{-1}$

48. $a^{-2} - 2a^{-1}b^{-1}$

Solve each of the following equations.

49. $\sqrt{7x - 3} = 4$

50. $\sqrt{2y + 1} = \sqrt{5y - 11}$

51. $\sqrt{2x} = x - 4$

52. $\sqrt{n^2 - 4n - 4} = n$

53. $\sqrt[3]{2x - 1} = 3$

54. $\sqrt{t^2 + 9t - 1} = 3$

55. $\sqrt{x^2 + 3x - 6} = x$

56. $\sqrt{x + 1} - \sqrt{2x} = -1$

Use scientific notation and the properties of exponents to help perform the following calculations.

57. $(.00002)(.0003)$

58. $(120{,}000)(300{,}000)$

59. $(.000015)(400{,}000)$

60. $\dfrac{.000045}{.0003}$

61. $\dfrac{(.00042)(.0004)}{.006}$

62. $\sqrt{.000004}$

63. $\sqrt[3]{.000000008}$

64. $(4000000)^{\frac{3}{2}}$.

Cumulative Review Problem Set   (Chapters 1–8)

For Problems 1–10, evaluate each of the numerical expressions without using a calculator or a table.

1. $\left(-\dfrac{1}{3}\right)^3$

2. $\left(\dfrac{1}{4} - \dfrac{1}{2}\right)^2$

3. $-\sqrt{\dfrac{9}{64}}$

4. $\sqrt[3]{-\dfrac{8}{27}}$

5. -2^6

6. $(-2)^6$

7. $\left(\dfrac{3}{4}\right)^{-2}$

8. $32^{-\frac{1}{5}}$

9. $-9^{\frac{3}{2}}$

10. $3^0 + 3^{-1} + 3^{-2}$

For Problems 11–16, perform the indicated operations and express answers in simplified form.

11. $(3a^2b)(-2ab)(4ab^3)$

12. $(x + 3)(2x^2 - x - 4)$

13. $(4x^3 - 17x^2 + 7x + 10) \div (4x - 5)$

14. $(3x + 1)^3$

15. $(3\sqrt{2} + 2\sqrt{5})(5\sqrt{2} - \sqrt{5})$

16. $\dfrac{3\sqrt{2}}{1 - \sqrt{3}}$

For Problems 17–32, solve each of the equations.

17. $3(x - 2) - 2(3x + 5) = 4(x - 1)$

18. $.06n + .08(n + 50) = 25$

19. $4\sqrt{x} + 5 = x$

20. $\sqrt[3]{n^2 - 1} = -1$

21. $6x^2 - 24 = 0$

22. $a^2 + 14a + 49 = 0$

23. $3n^2 + 14n - 24 = 0$

24. $-2x = 6(x - 1) + (3x - 2)$

25. $\sqrt{2x - 1} - \sqrt{x + 2} = 0$

26. $5x - 4 = \sqrt{5x - 4}$

27. $|3x - 1| = 11$

28. $(3x - 2)(4x - 1) = 0$

29. $(2x + 1)(x - 2) = 7$

30. $6x^4 - 23x^2 - 4 = 0$

31. $3n^3 + 3n = 0$

32. $n^2 - 13n - 114 = 0$

For Problems 33–38, solve each of the inequalities.

33. $6 - 2x \geq 10$

34. $4(2x - 1) < 3(x + 5)$

35. $\dfrac{n + 1}{4} + \dfrac{n - 2}{12} > \dfrac{1}{6}$

36. $|2x - 1| < 5$

37. $|3x + 2| > 11$

38. $\dfrac{1}{2}(3x - 1) - \dfrac{2}{3}(x + 4) \leq \dfrac{3}{4}(x - 1)$

39. Find the slope of the line $5x - 2y = 13$.

40. Write the equation of the line that has a slope of $\dfrac{3}{7}$ and contains the point $(1, -4)$.

41. Solve each of the following systems of equations.

(a) $\begin{pmatrix} 5x - 2y = 39 \\ 7x + \ y = 47 \end{pmatrix}$

(b) $\begin{pmatrix} 3x + 5y = \ \ 21 \\ 4x - 3y = -30 \end{pmatrix}$

For Problems 42–47, set up an equation or a system of equations and solve each problem.

42. How many liters of a 60% acid solution must be added to 14 liters of a 10% acid solution to produce a 25% acid solution?

43. A sum of $2250 is to be divided between two people in the ratio of 2 to 3. How much does each person receive?

44. The length of a picture without its border is 7 inches less than twice its width. If the border is 1 inch wide and its area is 62 square inches, what are the dimensions of the picture alone?

45. Lolita and Doug working together can paint a shed in 3 hours and 20 minutes. If Doug can paint the shed by himself in 10 hours, how long would it take Lolita to paint the shed by herself?

46. Angie bought some golf balls for $14. If each ball had cost $.25 less, she could have purchased one more ball for the same amount of money. How many golf balls did Angie buy?

47. A jogger who can run an 8-minute mile starts a half mile ahead of a jogger who can run a 6-minute mile. How long will it take the faster jogger to catch the slower jogger?

Chapter 9

Rational Expressions

Rational expressions are to algebra what rational numbers are to arithmetic. Most of the work that we will do with rational expressions in this chapter parallels the work you have previously done with arithmetic fractions. The same basic properties used to explain reducing, adding, subtracting, multiplying, and dividing arithmetic fractions will serve as a basis for our work with rational expressions. The techniques of factoring we studied in the previous chapter will also play an important role in this chapter. We will conclude the chapter by working with some fractional equations that contain rational expressions.

9.1
Simplifying Rational Expressions

A **rational expression** is the indicated quotient of two polynomials. The following are examples of rational expressions.

$$\frac{3x^2}{5}, \quad \frac{x-2}{x+3}, \quad \frac{x^2+5x-1}{x^2-9}, \quad \frac{xy^2+x^2y}{xy}, \quad \frac{a^3-3a^2-5a-1}{a^4+a^3+6}$$

Because we must avoid division by zero, no values that create a denominator of zero can be assigned to variables. Thus, the rational expression $\frac{x-2}{x+3}$ is meaningful for all values of x except for $x = -3$. Rather than making restrictions for each individual expression, we will merely assume that all denominators represent nonzero real numbers.

REMARK You may find it helpful to go back and look over the first two sections of Chapter 2. Part of our work with rational expressions in this chapter will closely parallel the work with rational numbers in those sections.

In Chapter 2 we used the properties $\frac{-a}{b} = \frac{a}{-b} = -\frac{a}{b}$ and $\frac{-a}{-b} = \frac{a}{b}$ when working with rational numbers. These same properties apply to rational expressions. For example, a rational expression such as $\frac{-2x}{3y}$ can be written as $\frac{2x}{-3y}$ or $-\frac{2x}{3y}$. Likewise, an expression such as $\frac{-4y}{-5x}$ can be written as $\frac{4y}{5x}$.

Furthermore, in Chapter 2 we used the property $\frac{ak}{bk} = \frac{a}{b}$ to reduce or simplify rational numbers in common fraction form. The following examples show some formats that can be used when reducing fractions.

$$\frac{18}{24} = \frac{3\cdot6}{4\cdot6} = \frac{3}{4}$$

$$\frac{-36}{63} = -\frac{36}{63} = -\frac{4\cdot9}{7\cdot9} = -\frac{4}{7}$$

$$\frac{75}{-90} = -\frac{75}{90} = -\frac{3\cdot5\cdot5}{2\cdot3\cdot3\cdot5} = -\frac{5}{6}$$

The property $\frac{ak}{bk} = \frac{a}{b}$ also serves as the basis for simplifying rational expressions as the next examples illustrate.

EXAMPLE 1   Simplify $\dfrac{15xy}{25y}$.

Solution
$$\frac{15xy}{25y} = \frac{3 \cdot \cancel{5} \cdot x \cdot \cancel{y}}{\cancel{5} \cdot 5 \cdot \cancel{y}} = \frac{3x}{5}$$
■

EXAMPLE 2   Simplify $\dfrac{-9}{18x^2y}$.

Solution
$$\frac{-9}{18x^2y} = -\frac{\overset{1}{\cancel{9}}}{\underset{2}{\cancel{18}}x^2y} = -\frac{1}{2x^2y}$$
A common factor of 9 was divided out of numerator and denominator.   ■

EXAMPLE 3   Simplify $\dfrac{-28a^2b^2}{-63a^2b^3}$.

Solution
$$\frac{-28a^2b^2}{-63a^2b^3} = \frac{4 \cdot \cancel{7} \cdot \cancel{a^2} \cdot \cancel{b^2}}{9 \cdot \cancel{7} \cdot \cancel{a^2} \cdot \underset{b}{\cancel{b^3}}} = \frac{4}{9b}$$
■

The factoring techniques from Chapter 7 can be used to factor numerators and/or denominators so that we can apply the property $\dfrac{a \cdot k}{b \cdot k} = \dfrac{a}{b}$. Examples 4–9 should clarify this process.

EXAMPLE 4   Simplify $\dfrac{3x + 12}{6}$.

Solution
$$\frac{3x + 12}{6} = \frac{3(x + 4)}{6} = \frac{\cancel{3}(x + 4)}{\underset{2}{\cancel{6}}}$$

$$= \frac{x + 4}{2}$$
■

EXAMPLE 5   Simplify $\dfrac{15 - 5x}{5x}$.

Solution
$$\frac{15 - 5x}{5x} = \frac{5(3 - x)}{5x} = \frac{\cancel{5}(3 - x)}{\cancel{5}x}$$

$$= \frac{3 - x}{x}$$
■

EXAMPLE 6   Simplify $\dfrac{x^2 + 4x}{x^2 - 16}$.

Solution $\dfrac{x^2 + 4x}{x^2 - 16} = \dfrac{x(\cancel{x + 4})}{(x - 4)(\cancel{x + 4})} = \dfrac{x}{x - 4}$ ∎

EXAMPLE 7   Simplify $\dfrac{4a^2 + 12a + 9}{2a + 3}$.

Solution $\dfrac{4a^2 + 12a + 9}{2a + 3} = \dfrac{(\cancel{2a + 3})(2a + 3)}{1(\cancel{2a + 3})} = \dfrac{2a + 3}{1} = 2a + 3$ ∎

EXAMPLE 8   Simplify $\dfrac{5n^2 + 6n - 8}{10n^2 - 3n - 4}$.

Solution $\dfrac{5n^2 + 6n - 8}{10n^2 - 3n - 4} = \dfrac{(\cancel{5n - 4})(n + 2)}{(\cancel{5n - 4})(2n + 1)} = \dfrac{n + 2}{2n + 1}$ ∎

EXAMPLE 9   Simplify $\dfrac{6x^3y - 6xy}{x^3 + 5x^2 + 4x}$.

Solution $\dfrac{6x^3y - 6xy}{x^3 + 5x^2 + 4x} = \dfrac{6xy(x^2 - 1)}{x(x^2 + 5x + 4)} = \dfrac{6xy(\cancel{x + 1})(x - 1)}{x(\cancel{x + 1})(x + 4)} = \dfrac{6y(x - 1)}{x + 4}$ ∎

Note that in Example 9 we left the numerator of the final fraction in factored form. This is often done if expressions other than monomials are involved. Either $\dfrac{6y(x - 1)}{x + 4}$ or $\dfrac{6xy - 6y}{x + 4}$ is an acceptable answer.

Remember that the quotient of any nonzero real number and its opposite is -1. For example, $\dfrac{6}{-6} = -1$ and $\dfrac{8}{8} = -1$. Likewise, the indicated quotient of any polynomial and its opposite is equal to -1. For example,

$\dfrac{a}{-a} = -1$ because a and $-a$ are opposites;

$\dfrac{a - b}{b - a} = -1$ because $a - b$ and $b - a$ are opposites;

$\dfrac{x^2 - 4}{4 - x^2} = -1$ because $x^2 - 4$ and $4 - x^2$ are opposites.

The final example of this section illustrates the use of this idea when simplifying rational expressions.

EXAMPLE 10 Simplify $\dfrac{6a^2 - 7a + 2}{10a - 15a^2}$.

Solution

$$\dfrac{6a^2 - 7a + 2}{10a - 15a^2} = \dfrac{(2a - 1)(3a - 2)}{5a(2 - 3a)} \qquad \dfrac{3a - 2}{2 - 3a} = -1$$

$$= (-1)\left(\dfrac{2a - 1}{5a}\right)$$

$$= -\dfrac{2a - 1}{5a} \qquad \text{or} \qquad \dfrac{1 - 2a}{5a}$$

■

Problem Set 9.1

Express each of the following rational numbers in reduced form.

1. $\dfrac{27}{36}$ 2. $\dfrac{14}{21}$ 3. $\dfrac{45}{54}$ 4. $\dfrac{-14}{42}$

5. $\dfrac{24}{-60}$ 6. $\dfrac{45}{-75}$ 7. $\dfrac{-16}{-56}$ 8. $\dfrac{-30}{-42}$

Simplify each of the following.

9. $\dfrac{12xy}{42y}$ 10. $\dfrac{21xy}{35x}$ 11. $\dfrac{18a^2}{45ab}$

12. $\dfrac{48ab}{84b^2}$ 13. $\dfrac{-14y^3}{56xy^2}$ 14. $\dfrac{-14x^2y^3}{63xy^2}$

15. $\dfrac{54c^2d}{-78cd^2}$ 16. $\dfrac{60x^3z}{-64xyz^2}$ 17. $\dfrac{-40x^3y}{-24xy^4}$

18. $\dfrac{-30x^2y^2z^2}{-35xz^3}$ 19. $\dfrac{xy}{x^2 - 2x}$ 20. $\dfrac{8}{12x - 16y}$

21. $\dfrac{x^2 + 5x}{2xy}$ 22. $\dfrac{2x^2 - 6x}{2y}$ 23. $\dfrac{3xy}{6x - 3y}$

24. $\dfrac{4x^2}{12 - 4x}$ 25. $\dfrac{12x^2 + 18x}{6x}$ 26. $\dfrac{15x^2 - 25x}{5x}$

27. $\dfrac{x^2 - 4}{x^2 + 2x}$ 28. $\dfrac{xy + y^2}{x^2 - y^2}$ 29. $\dfrac{18x + 12}{12x - 6}$

30. $\dfrac{20x + 50}{15x - 30}$ 31. $\dfrac{a^2 + 7a + 10}{a^2 - 7a - 18}$ 32. $\dfrac{a^2 + 4a - 32}{3a^2 + 26a + 16}$

33. $\dfrac{2n^2 + n - 21}{10n^2 + 33n - 7}$

34. $\dfrac{4n^2 - 15n - 4}{7n^2 - 30n + 8}$

35. $\dfrac{5x^2 + 7}{10x}$

36. $\dfrac{12x^2 + 11x - 15}{20x^2 - 23x + 6}$

37. $\dfrac{6x^2 + x - 15}{8x^2 - 10x - 3}$

38. $\dfrac{4x^2 + 8x}{x^3 + 8}$

39. $\dfrac{3x^2 - 12x}{x^3 - 64}$

40. $\dfrac{x^2 - 14x + 49}{6x^2 - 37x - 35}$

41. $\dfrac{3x^2 + 17x - 6}{9x^2 - 6x + 1}$

42. $\dfrac{9y^2 - 1}{3y^2 + 11y - 4}$

43. $\dfrac{2x^3 + 3x^2 - 14x}{x^2y + 7xy - 18y}$

44. $\dfrac{3x^3 + 12x}{9x^2 + 18x}$

45. $\dfrac{5y^2 + 22y + 8}{25y^2 - 4}$

46. $\dfrac{16x^3y + 24x^2y^2 - 16xy^3}{24x^2y + 12xy^2 - 12y^3}$

47. $\dfrac{15x^3 - 15x^2}{5x^3 + 5x}$

48. $\dfrac{5n^2 + 18n - 8}{3n^2 + 13n + 4}$

49. $\dfrac{4x^2y + 8xy^2 - 12y^3}{18x^3y - 12x^2y^2 - 6xy^3}$

50. $\dfrac{3 + x - 2x^2}{2 + x - x^2}$

51. $\dfrac{3n^2 + 14n - 24}{7n^2 + 44n + 12}$

52. $\dfrac{x^4 - 2x^2 - 15}{2x^4 + 9x^2 + 9}$

53. $\dfrac{8 + 18x - 5x^2}{10 + 31x + 15x^2}$

54. $\dfrac{6x^4 - 11x^2 + 4}{2x^4 + 17x^2 - 9}$

55. $\dfrac{27x^4 - x}{6x^3 + 10x^2 - 4x}$

56. $\dfrac{64x^4 + 27x}{12x^3 - 27x^2 - 27x}$

57. $\dfrac{-40x^3 + 24x^2 + 16x}{20x^3 + 28x^2 + 8x}$

58. $\dfrac{-6x^3 - 21x^2 + 12x}{-18x^3 - 42x^2 + 120x}$

Simplify each of the following. You will need to use factoring by grouping.

59. $\dfrac{xy + ay + bx + ab}{xy + ay + cx + ac}$

60. $\dfrac{xy + 2y + 3x + 6}{xy + 2y + 4x + 8}$

61. $\dfrac{ax - 3x + 2ay - 6y}{2ax - 6x + ay - 3y}$

62. $\dfrac{x^2 - 2x + ax - 2a}{x^2 - 2x + 3ax - 6a}$

63. $\dfrac{5x^2 + 5x + 3x + 3}{5x^2 + 3x - 30x - 18}$

64. $\dfrac{x^2 + 3x + 4x + 12}{2x^2 + 6x - x - 3}$

65. $\dfrac{2st - 30 - 12s + 5t}{3st - 6 - 18s + t}$

66. $\dfrac{nr - 6 - 3n + 2r}{nr + 10 + 2r + 5n}$

Simplify each of the following. You may want to refer to Example 8 in this section.

67. $\dfrac{5x - 7}{7 - 5x}$

68. $\dfrac{4a - 9}{9 - 4a}$

69. $\dfrac{n^2 - 49}{7 - n}$

70. $\dfrac{9 - y}{y^2 - 81}$

71. $\dfrac{2y - 2xy}{x^2y - y}$

72. $\dfrac{3x - x^2}{x^2 - 9}$

73. $\dfrac{2x^3 - 8x}{4x - x^3}$

74. $\dfrac{x^2 - (y - 1)^2}{(y - 1)^2 - x^2}$

75. $\dfrac{n^2 - 5n - 24}{40 + 3n - n^2}$

76. $\dfrac{x^2 + 2x - 24}{20 - x - x^2}$

9.2

Multiplying and Dividing Rational Expressions

Remember that to multiply rational numbers in common fraction form we simply **multiply numerators and multiply denominators** as the next examples illustrate. (The steps in the dashed boxes are usually done mentally.)

$$\frac{2}{3} \cdot \frac{4}{5} = \boxed{\frac{2 \cdot 4}{3 \cdot 5}} = \frac{8}{15},$$

$$\frac{-3}{4} \cdot \frac{5}{7} = \boxed{\frac{-3 \cdot 5}{4 \cdot 7} = \frac{-15}{28}} = -\frac{15}{28},$$

$$-\frac{5}{6} \cdot \frac{13}{3} = \boxed{\frac{-5}{6} \cdot \frac{13}{3} = \frac{-5 \cdot 13}{6 \cdot 3} = \frac{-65}{18}} \quad -\frac{65}{18}$$

We also agree, when multiplying rational numbers, to express the final product in reduced form. The following examples show some different formats to *multiply and simplify* rational numbers.

$$\frac{3}{4} \cdot \frac{4}{7} = \frac{3 \cdot \cancel{4}}{\cancel{4} \cdot 7} = \frac{3}{7};$$

$$\frac{\overset{1}{\cancel{8}}}{\underset{1}{\cancel{9}}} \cdot \frac{\overset{3}{\cancel{27}}}{\underset{4}{\cancel{32}}} = \frac{3}{4}; \qquad \text{A common factor of 9 has been divided out of 9 and 27, and a common factor of 8 has been divided out of 8 and 32.}$$

$$\left(-\frac{28}{25}\right)\left(-\frac{65}{78}\right) = \frac{\cancel{2} \cdot 2 \cdot 7 \cdot \cancel{5} \cdot \cancel{13}}{\cancel{5} \cdot 5 \cdot \cancel{2} \cdot 3 \cdot \cancel{13}} = \frac{14}{15}. \qquad \text{We should recognize that a } \textit{negative times a negative is positive.} \text{ Also, notice the use of prime factors to help recognize common factors.}$$

Multiplication of rational expressions follows the same basic pattern as multiplication of rational numbers in common fraction form. That is to say, we multiply numerators, multiply denominators, and express the final product in simplified form. The following definition formalizes this idea.

DEFINITION 9.1

For any rational expressions $\dfrac{A}{B}$ and $\dfrac{C}{D}$, where $B \neq 0$ and $D \neq 0$,

$$\frac{A}{B} \cdot \frac{C}{D} = \frac{AC}{BD}.$$

Let's consider some examples using Definition 9.1.

$$\frac{3x}{4y} \cdot \frac{8y^2}{9x} = \frac{\overset{2}{\cancel{3}} \cdot \overset{}{\cancel{8}} \cdot x \cdot \overset{y}{\cancel{y^2}}}{\underset{3}{\cancel{4} \cdot \cancel{9} \cdot \cancel{x} \cdot \cancel{y}}} = \frac{2y}{3},$$

Notice the use of the commutative property of multiplication to rearrange factors in a more convenient form for identifying common factors of the numerator and denominator.

$$\frac{-4a}{6a^2b^2} \cdot \frac{9ab}{12a^2} = -\frac{\overset{3}{\cancel{4}} \cdot \cancel{9} \cdot \cancel{a^2} \cdot \cancel{b}}{\underset{2 \quad 3 \quad a^2 \quad b}{\cancel{6} \cdot \cancel{12} \cdot \cancel{a^4} \cdot \cancel{b^2}}} = -\frac{1}{2a^2b},$$

$$\frac{12x^2y}{-18xy} \cdot \frac{-24xy^2}{56y^3} = \frac{\overset{2}{\cancel{12}} \cdot \overset{3}{\cancel{24}} \cdot \overset{x^2}{\cancel{x^3}} \cdot \cancel{y^3}}{\underset{3 \quad 7 \quad \quad y}{\cancel{18} \cdot \cancel{56} \cdot x \cdot \cancel{y^4}}} = \frac{2x^2}{7y}$$

You should recognize that the first fraction is equivalent to $-\dfrac{12x^2y}{18xy}$ and the second to $-\dfrac{24xy^2}{56y^3}$; thus, the product is positive.

If the rational expressions contain polynomials (other than monomials) that are factorable, then our work may take on the following format.

EXAMPLE 1   Multiply and simplify $\dfrac{y}{x^2 - 4} \cdot \dfrac{x+2}{y^2}$.

Solution

$$\frac{y}{x^2 - 4} \cdot \frac{x+2}{y^2} = \frac{\cancel{y}(\cancel{x+2})}{\underset{y}{\cancel{y^2}}(\cancel{x+2})(x-2)} = \frac{1}{y(x-2)} \qquad \blacksquare$$

In Example 1, notice that we combined the steps of multiplying numerators and denominators and factoring the polynomials. Also, notice that we left the final answer in factored form. Either $\dfrac{1}{y(x-2)}$ or $\dfrac{1}{xy-2y}$ would be an acceptable answer.

EXAMPLE 2   Multiply and simplify $\dfrac{x^2 - x}{x+5} \cdot \dfrac{x^2 + 5x + 4}{x^4 - x^2}$.

Solution

$$\frac{x^2 - x}{x+5} \cdot \frac{x^2 + 5x + 4}{x^4 - x^2} = \frac{\cancel{x}(\cancel{x-1})(\cancel{x+1})(x+4)}{(x+5)(\underset{x}{\cancel{x^2}})(\cancel{x-1})(\cancel{x+1})} = \frac{x+4}{x(x+5)} \qquad \blacksquare$$

EXAMPLE 3   Multiply and simplify $\dfrac{6n^2 + 7n - 5}{n^2 + 2n - 24} \cdot \dfrac{4n^2 + 21n - 18}{12n^2 + 11n - 15}$.

Solution

$$\frac{6n^2 + 7n - 5}{n^2 + 2n - 24} \cdot \frac{4n^2 + 21n - 18}{12n^2 + 11n - 15}$$

$$= \frac{(3n+5)(2n-1)(4n-3)(n+6)}{(n+6)(n-4)(3n+5)(4n-3)} = \frac{2n-1}{n-4} \qquad \blacksquare$$

Dividing Rational Expressions

Remember that to divide two rational numbers in common fraction form we invert the divisor and multiply as the next examples illustrate.

$$\frac{7}{8} \div \frac{5}{6} = \frac{7}{8} \cdot \frac{\overset{3}{\cancel{6}}}{5} = \frac{21}{20}, \qquad \frac{-5}{9} \div \frac{15}{18} = -\frac{\cancel{5}}{9} \cdot \frac{\overset{2}{\cancel{18}}}{\cancel{15}} = -\frac{2}{3},$$

$$\frac{14}{-19} \div \frac{21}{-38} = \left(-\frac{14}{19}\right) \div \left(-\frac{21}{38}\right) = \left(-\frac{\overset{2}{\cancel{14}}}{\cancel{19}}\right)\left(-\frac{\overset{2}{\cancel{38}}}{\cancel{21}}\right) = \frac{4}{3}$$

Division of rational expressions follows the same basic pattern as division of rational numbers in common fraction form and can be stated as follows.

DEFINITION 9.2

> For any rational expressions $\frac{A}{B}$ and $\frac{C}{D}$, where $B \neq 0$, $C \neq 0$, and $D \neq 0$.
>
> $$\frac{A}{B} \div \frac{C}{D} = \frac{A}{B} \cdot \frac{D}{C} = \frac{AD}{BC}.$$

Definition 9.2 states that to divide two rational expressions we **invert the divisor and multiply**. We call the expressions $\frac{C}{D}$ and $\frac{D}{C}$ **reciprocals** or **multiplicative inverses** of each other because their product is one. The following examples demonstrate the use of Definition 9.2.

EXAMPLE 4   Divide and simplify $\dfrac{16x^2y}{24xy^3} \div \dfrac{9xy}{8x^2y^2}$.

Solution

$$\frac{16x^2y}{24xy^3} \div \frac{9xy}{8x^2y^2} = \frac{16x^2y}{24xy^3} \cdot \frac{8x^2y^2}{9xy} = \frac{16 \cdot 8 \cdot x^4 \cdot y^3}{24 \cdot 9 \cdot x^2 \cdot y^4} = \frac{16x^2}{27y} \qquad \blacksquare$$

EXAMPLE 5   Divide and simplify $\dfrac{3a^2 + 12}{3a^2 - 15a} \div \dfrac{a^4 - 16}{a^2 - 3a - 10}$.

Solution

$$\frac{3a^2 + 12}{3a^2 - 15a} \div \frac{a^4 - 16}{a^2 - 3a - 10} = \frac{3a^2 + 12}{3a^2 - 15a} \cdot \frac{a^2 - 3a - 10}{a^4 - 16}$$

$$= \frac{\cancel{3}(a^2 + 4)(a - 5)(a + 2)}{\cancel{3}a(a - 5)(a^2 + 4)(a + 2)(a - 2)}$$

$$= \frac{1}{a(a - 2)} \qquad ■$$

EXAMPLE 6   Divide and simplify $\dfrac{28t^3 - 51t^2 - 27t}{49t^2 + 42t + 9} \div (4t - 9)$.

Solution

$$\frac{28t^3 - 51t^2 - 27t}{49t^2 + 42t + 9} \div \frac{4t - 9}{1} = \frac{28t^3 - 51t^2 - 27t}{49t^2 + 42t + 9} \cdot \frac{1}{4t - 9}$$

$$= \frac{t(7t + 3)(4t - 9)}{(7t + 3)(7t + 3)(4t - 9)}$$

$$= \frac{t}{7t + 3} \qquad ■$$

In a problem such as Example 6, it may be helpful to write the divisor with a denominator of 1. Thus, we write $4t - 9$ as $\dfrac{4t - 9}{1}$; then its reciprocal is obviously $\dfrac{1}{4t - 9}$.

Let's consider one final example that involves both multiplication and division.

EXAMPLE 7   Perform the indicated operations and simplify.

$$\frac{x^2 + 5x}{3x^2 - 4x - 20} \cdot \frac{x^2 y + y}{2x^2 + 11x + 5} \div \frac{xy^2}{6x^2 - 17x - 10}$$

Solution

$$\frac{x^2 + 5x}{3x^2 - 4x - 20} \cdot \frac{x^2 y + y}{2x^2 + 11x + 5} \div \frac{xy^2}{6x^2 - 17x - 10}$$

$$= \frac{x^2 + 5x}{3x^2 - 4x - 20} \cdot \frac{x^2 y + y}{2x^2 + 11x + 5} \cdot \frac{6x^2 - 17x - 10}{xy^2}$$

$$= \frac{x(x + 5)(y)(x^2 + 1)(2x + 1)(3x - 10)}{(3x - 10)(x + 2)(2x + 1)(x + 5)(x)(y^2)} = \frac{x^2 + 1}{y(x + 2)} \qquad ■$$

Problem Set 9.2

Perform the following indicated operations involving rational numbers. Express final answers in reduced form.

1. $\dfrac{7}{12} \cdot \dfrac{6}{35}$

2. $\dfrac{5}{8} \cdot \dfrac{12}{20}$

3. $\dfrac{-4}{9} \cdot \dfrac{18}{30}$

4. $\dfrac{-6}{9} \cdot \dfrac{36}{48}$

5. $\dfrac{3}{-8} \cdot \dfrac{-6}{12}$

6. $\dfrac{-12}{16} \cdot \dfrac{18}{-32}$

7. $\left(-\dfrac{5}{7}\right) \div \dfrac{6}{7}$

8. $\left(-\dfrac{5}{9}\right) \div \dfrac{10}{3}$

9. $\dfrac{-9}{5} \div \dfrac{27}{10} - \dfrac{2}{3}$

10. $\dfrac{4}{7} \div \dfrac{16}{-21}$

11. $\dfrac{4}{9} \cdot \dfrac{6}{11} \div \dfrac{4}{15}$

12. $\dfrac{2}{3} \cdot \dfrac{6}{7} \div \dfrac{8}{3}$

Perform the following indicated operations involving rational expressions. Express final answers in simplest form.

13. $\dfrac{6xy}{9y^4} \cdot \dfrac{30x^3y}{-48x}$

14. $\dfrac{-14xy^4}{18y^2} \cdot \dfrac{24x^2y^3}{35y^2}$

15. $\dfrac{5a^2b^2}{11ab} \cdot \dfrac{22a^3}{15ab^2}$

16. $\dfrac{10a^2}{5b^2} \cdot \dfrac{15b^3}{2a^4}$

17. $\dfrac{5xy}{8y^2} \cdot \dfrac{18x^2y}{15}$

18. $\dfrac{4x^2}{5y^2} \cdot \dfrac{15xy}{24x^2y^2}$

19. $\dfrac{5x^4}{12x^2y^3} \div \dfrac{9}{5xy}$

20. $\dfrac{7x^2y}{9xy^3} \div \dfrac{3x^4}{2x^2y^2}$

21. $\dfrac{9a^2c}{12bc^2} \div \dfrac{21ab}{14c^3}$

22. $\dfrac{3ab^3}{4c} \div \dfrac{21ac}{12bc^3}$

23. $\dfrac{9x^2y^3}{14x} \cdot \dfrac{21y}{15xy^2} \cdot \dfrac{10x}{12y^3}$

24. $\dfrac{5xy}{7a} \cdot \dfrac{14a^2}{15x} \cdot \dfrac{3a}{8y}$

25. $\dfrac{3x+6}{5y} \cdot \dfrac{x^2+4}{x^2+10x+16}$

26. $\dfrac{5xy}{x+6} \cdot \dfrac{x^2-36}{x^2-6x}$

27. $\dfrac{5a^2+20a}{a^3-2a^2} \cdot \dfrac{a^2-a-12}{a^2-16}$

28. $\dfrac{2a^2+6}{a^2-a} \cdot \dfrac{a^3-a^2}{8a-4}$

29. $\dfrac{3n^2+15n-18}{3n^2+10n-48} \cdot \dfrac{12n^2-17n-40}{8n^2+2n-10}$

30. $\dfrac{10n^2+21n-10}{5n^2+33n-14} \cdot \dfrac{2n^2+6n-56}{2n^2-3n-20}$

31. $\dfrac{9y^2}{x^2+12x+36} \div \dfrac{12y}{x^2+6x}$

32. $\dfrac{7xy}{x^2-4x+4} \div \dfrac{14y}{x^2-4}$

33. $\dfrac{x^2-4xy+4y^2}{7xy^2} \div \dfrac{4x^2-3xy-10y^2}{20x^2y+25xy^2}$

34. $\dfrac{x^2+5xy-6y^2}{xy^2-y^3} \cdot \dfrac{2x^2+15xy+18y^2}{xy+4y^2}$

35. $\dfrac{5-14n-3n^2}{1-2n-3n^2} \cdot \dfrac{9+7n-2n^2}{27-15n+2n^2}$

36. $\dfrac{6-n-2n^2}{12-11n+2n^2} \cdot \dfrac{24-26n+5n^2}{2+3n+n^2}$

37. $\dfrac{3x^4+2x^2-1}{3x^4+14x^2-5} \cdot \dfrac{x^4-2x^2-35}{x^4-17x^2+70}$

38. $\dfrac{2x^4+x^2-3}{2x^4+5x^2+2} \cdot \dfrac{3x^4+10x^2+8}{3x^4+x^2-4}$

39. $\dfrac{6x^2 - 35x + 25}{4x^2 - 11x - 45} \div \dfrac{18x^2 + 9x - 20}{24x^2 + 74x + 45}$

40. $\dfrac{21t^2 + 22t - 8}{5t^2 - 43t - 18} \div \dfrac{12t^2 + 7t - 12}{20t^2 - 7t - 6}$

41. $\dfrac{10t^3 + 25t}{20t + 10} \cdot \dfrac{2t^2 - t - 1}{t^5 - t}$

42. $\dfrac{t^4 - 81}{t^2 - 6t + 9} \cdot \dfrac{6t^2 - 11t - 21}{5t^2 + 8t - 21}$

43. $\dfrac{4t^2 + t - 5}{t^3 - t^2} \cdot \dfrac{t^4 + 6t^3}{16t^2 + 40t + 25}$

44. $\dfrac{9n^2 - 12n + 4}{n^2 - 4n - 32} \cdot \dfrac{n^2 + 4n}{3n^3 - 2n^2}$

45. $\dfrac{nr + 3n + 2r + 6}{nr + 3n - 3r - 9} \cdot \dfrac{n^2 - 9}{n^3 - 4n}$

46. $\dfrac{xy + xc + ay + ac}{xy - 2xc + ay - 2ac} \cdot \dfrac{2x^3 - 8x}{12x^3 + 20x^2 - 8x}$

47. $\dfrac{x^2 - x}{4y} \cdot \dfrac{10xy^2}{2x - 2} \div \dfrac{3x^2 + 3x}{15x^2y^2}$

48. $\dfrac{4xy^2}{7x} \cdot \dfrac{14x^3y}{12y} \div \dfrac{7y}{9x^3}$

49. $\dfrac{a^2 - 4ab + 4b^2}{6a^2 - 4ab} \cdot \dfrac{3a^2 + 5ab - 2b^2}{6a^2 + ab - b^2} \div \dfrac{a^2 - 4b^2}{8a + 4b}$

50. $\dfrac{2x^2 + 3x}{2x^3 - 10x^2} \cdot \dfrac{x^2 - 8x + 15}{3x^3 - 27x} \div \dfrac{14x + 21}{x^2 - 6x - 27}$

Thoughts into Words

51. How would you explain to someone the difference between a rational number and a rational expression?

52. What danger is there in using the abbreviated phrase "invert and multiply" to explain division of rational numbers?

9.3

Adding and Subtracting Rational Expressions

We can add or subtract rational numbers with a common denominator by adding or subtracting the numerators and placing the result over the common denominator. The following examples illustrate this idea.

$$\frac{2}{9} + \frac{3}{9} = \frac{2 + 3}{9} = \frac{5}{9},$$

$$\frac{7}{8} - \frac{3}{8} = \frac{7 - 3}{8} = \frac{4}{8} = \frac{1}{2}, \qquad \text{Don't forget to reduce!}$$

$$\frac{4}{6} + \frac{-5}{6} = \frac{4 + (-5)}{6} = \frac{-1}{6} = -\frac{1}{6},$$

$$\frac{7}{10} + \frac{4}{-10} - \frac{7}{10} + \frac{-4}{10} = \frac{7 + (-4)}{10} = \frac{3}{10}$$

We use this same common denominator approach when adding or subtracting rational expressions, as the following definition describes.

DEFINITION 9.3

> For any rational expressions $\dfrac{A}{B}$ and $\dfrac{C}{B}$, where $B \neq 0$.
>
> $$\frac{A}{B} + \frac{C}{B} = \frac{A+C}{B} \qquad \text{addition}$$
>
> $$\frac{A}{B} - \frac{C}{B} = \frac{A-C}{B} \qquad \text{subtraction}$$

The following examples illustrate the use of Definition 9.3.

$$\frac{3}{x} + \frac{9}{x} = \frac{3+9}{x} = \frac{12}{x},$$

$$\frac{8}{x-2} - \frac{3}{x-2} = \frac{8-3}{x-2} = \frac{5}{x-2},$$

$$\frac{9}{4y} + \frac{5}{4y} = \frac{9+5}{4y} = \frac{14}{4y} = \frac{7}{2y}, \qquad \text{Don't forget to simplify the final answer!}$$

$$\frac{n^2}{n-1} - \frac{1}{n-1} = \frac{n^2-1}{n-1} = \frac{(n+1)(n-1)}{n-1} = n+1,$$

$$\frac{6a^2}{2a+1} + \frac{13a+5}{2a+1} = \frac{6a^2+13a+5}{2a+1} = \frac{(2a+1)(3a+5)}{2a+1} = 3a+5$$

Technically, in each of the previous examples that involve rational expressions, we should restrict the variables to exclude division by zero. For example, $\dfrac{3}{x} + \dfrac{9}{x} = \dfrac{12}{x}$ is true for all real number values for x, *except* $x = 0$. Likewise, $\dfrac{8}{x-2} - \dfrac{3}{x-2} = \dfrac{5}{x-2}$ as long as x does not equal 2. Rather than taking the time and space to write down restrictions for each problem we will merely assume that such restrictions exist.

If rational numbers that do not have a common denominator are to be added or subtracted, then we apply the fundamental principle of fractions $\left(\dfrac{a}{b} = \dfrac{ak}{bk}\right)$ to obtain equivalent fractions with a common denominator. Equivalent fractions are fractions such as $\dfrac{1}{2}$ and $\dfrac{2}{4}$ that name the same number. Consider the following example.

$$\frac{1}{2} + \frac{1}{3} = \left(\frac{3}{3}\right)\left(\frac{1}{2}\right) + \left(\frac{2}{2}\right)\left(\frac{1}{3}\right)$$

$$= \frac{3}{6} + \frac{2}{6} = \frac{3+2}{6}$$

$$= \frac{5}{6}$$

Notice that we chose 6 as our common denominator and 6 is the **least common multiple** of the original denominators 2 and 3. (The least common multiple of a set of whole numbers is the smallest nonzero whole number divisible by each of the numbers.) In general, we use the least common multiple of the denominators of the fractions to be added or subtracted as a **least common denominator** (LCD).

A least common denominator may be found by inspection or by using the prime factored forms of the numbers. Let's consider an example using each of these techniques.

EXAMPLE 1   Subtract $\dfrac{5}{6} - \dfrac{3}{8}$.

Solution   By inspection we can see that the LCD is 24. Thus, both fractions can be changed to equivalent fractions, each with a denominator of 24.

$$\frac{5}{6} - \frac{3}{8} = \left(\frac{5}{6}\right)\left(\frac{4}{4}\right) - \left(\frac{3}{8}\right)\left(\frac{3}{3}\right) = \frac{20}{24} - \frac{9}{24} = \frac{11}{24}$$ ■

$$\underset{\text{form of 1}}{\uparrow} \qquad \underset{\text{form of 1}}{\uparrow}$$

In Example 1, notice that the fundamental principle of fractions, $\dfrac{a}{b} = \dfrac{a \cdot k}{b \cdot k}$, can be written as $\dfrac{a}{b} = \left(\dfrac{a}{b}\right)\left(\dfrac{k}{k}\right)$. This latter form emphasizes the fact that one is the multiplication identity element.

EXAMPLE 2   Add $\dfrac{7}{18} + \dfrac{11}{24}$.

Solution   Let's use the prime factored forms of the denominators to help find the LCD.

$$18 = 2 \cdot 3 \cdot 3, \qquad 24 = 2 \cdot 2 \cdot 2 \cdot 3$$

The LCD must contain three factors of 2 since 24 contains three 2s. The LCD must also contain two factors of 3 since 18 has two 3s. Thus, the

$$\text{LCD} = 2 \cdot 2 \cdot 2 \cdot 3 \cdot 3 = 72.$$

Now we can proceed as usual.

$$\frac{7}{18} + \frac{11}{24} = \left(\frac{7}{18}\right)\left(\frac{4}{4}\right) + \left(\frac{11}{24}\right)\left(\frac{3}{3}\right) = \frac{28}{72} + \frac{33}{72} = \frac{61}{72}$$ ∎

Adding and subtracting rational expressions with different denominators follows the same basic routine as adding or subtracting rational numbers with different denominators. Study the following examples carefully and notice the similarity to our previous work with rational numbers.

EXAMPLE 3   Add $\dfrac{x + 2}{4} + \dfrac{3x + 1}{3}$.

Solution   By inspection we see that the LCD is 12.

$$\frac{x + 2}{4} + \frac{3x + 1}{3} = \left(\frac{x + 2}{4}\right)\left(\frac{3}{3}\right) + \left(\frac{3x + 1}{3}\right)\left(\frac{4}{4}\right)$$

$$= \frac{3(x + 2)}{12} + \frac{4(3x + 1)}{12} = \frac{3x + 6 + 12x + 4}{12}$$

$$= \frac{15x + 10}{12}$$ ∎

Notice the final result in Example 3. The numerator, $15x + 10$, could be factored as $5(3x + 2)$. However, since this produces no common factors with the denominator, the fraction cannot be simplified. Thus, the final answer can be left as $\dfrac{15x + 10}{12}$, or it would also be acceptable to express it as $\dfrac{5(3x + 2)}{12}$.

EXAMPLE 4   Subtract $\dfrac{a - 2}{2} - \dfrac{a - 6}{6}$.

Solution   By inspection we see that the LCD is 6.

$$\frac{a - 2}{2} - \frac{a - 6}{6} = \left(\frac{a - 2}{2}\right)\left(\frac{3}{3}\right) - \frac{a - 6}{6}$$

$$= \frac{3(a - 2) - (a - 6)}{6}$$ Be careful with this sign as you move to the next step!

$$= \frac{3a - 6 - a + 6}{6}$$

$$= \frac{2a}{6} = \frac{a}{3}$$ Don't forget to simplify.   ∎

EXAMPLE 5   Perform the indicated operations $\dfrac{x+3}{10} + \dfrac{2x+1}{15} - \dfrac{x-2}{18}$.

Solution   If you cannot determine the LCD by inspection, then use the prime factored forms of the denominators.

$$10 = 2 \cdot 5, \qquad 15 = 3 \cdot 5, \qquad 18 = 2 \cdot 3 \cdot 3$$

The LCD must contain one factor of 2, two factors of 3, and one factor of 5. Thus, the LCD is $2 \cdot 3 \cdot 3 \cdot 5 = 90$.

$$\frac{x+3}{10} + \frac{2x+1}{15} - \frac{x-2}{18} = \left(\frac{x+3}{10}\right)\left(\frac{9}{9}\right) + \left(\frac{2x+1}{15}\right)\left(\frac{6}{6}\right) - \left(\frac{x-2}{18}\right)\left(\frac{5}{5}\right)$$

$$= \frac{9(x+3) + 6(2x+1) - 5(x-2)}{90}$$

$$= \frac{9x + 27 + 12x + 6 - 5x + 10}{90}$$

$$= \frac{16x + 43}{90} \qquad \blacksquare$$

A denominator that contains variables does not create any serious difficulties; our approach remains basically the same.

EXAMPLE 6   Add $\dfrac{3}{2x} + \dfrac{5}{3y}$.

Solution   Using an LCD of $6xy$ we can proceed as follows.

$$\frac{3}{2x} + \frac{5}{3y} = \left(\frac{3}{2x}\right)\left(\frac{3y}{3y}\right) + \left(\frac{5}{3y}\right)\left(\frac{2x}{2x}\right) = \frac{9y}{6xy} + \frac{10x}{6xy} = \frac{9y + 10x}{6xy} \qquad \blacksquare$$

EXAMPLE 7   Subtract $\dfrac{7}{12ab} - \dfrac{11}{15a^2}$.

Solution   We can prime factor the numerical coefficients of the denominators to help find the LCD.

$$\left.\begin{array}{l} 12ab = 2 \cdot 2 \cdot 3 \cdot a \cdot b \\ 15a^2 = 3 \cdot 5 \cdot a^2 \end{array}\right\} \longrightarrow \begin{array}{l} \text{LCD} = 2 \cdot 2 \cdot 3 \cdot 5 \cdot a^2 \cdot b \\ \qquad\quad = 60a^2b \end{array}$$

$$\frac{7}{12ab} - \frac{11}{15a^2} = \left(\frac{7}{12ab}\right)\left(\frac{5a}{5a}\right) - \left(\frac{11}{15a^2}\right)\left(\frac{4b}{4b}\right)$$

$$= \frac{35a}{60a^2b} - \frac{44b}{60a^2b} = \frac{35a - 44b}{60a^2b} \qquad \blacksquare$$

EXAMPLE 8   Add $\dfrac{x}{x-3} + \dfrac{4}{x}$.

Solution   By inspection the LCD is $x(x-3)$.

$$\frac{x}{x-3} + \frac{4}{x} = \left(\frac{x}{x-3}\right)\left(\frac{x}{x}\right) + \left(\frac{4}{x}\right)\left(\frac{x-3}{x-3}\right) = \frac{x^2}{x(x-3)} + \frac{4(x-3)}{x(x-3)}$$

$$= \frac{x^2 + 4x - 12}{x(x-3)} \quad \text{or} \quad \frac{(x+6)(x-2)}{x(x-3)} \qquad \blacksquare$$

EXAMPLE 9   Subtract $\dfrac{2x}{x+1} - 3$.

Solution   The LCD is $x+1$.

$$\frac{2x}{x+1} - 3 = \frac{2x}{x+1} - 3\left(\frac{x+1}{x+1}\right) = \frac{2x}{x+1} - \frac{3(x+1)}{x+1}$$

$$= \frac{2x - 3x - 3}{x+1} = \frac{-x - 3}{x+1} \qquad \blacksquare$$

Problem Set 9.3

Perform the following indicated operations involving rational numbers. Be sure to express answers in reduced form.

1. $\dfrac{1}{4} + \dfrac{5}{6}$

2. $\dfrac{3}{5} + \dfrac{1}{6}$

3. $\dfrac{7}{8} - \dfrac{3}{5}$

4. $\dfrac{7}{9} - \dfrac{1}{6}$

5. $\dfrac{6}{5} + \dfrac{1}{-4}$

6. $\dfrac{7}{8} + \dfrac{5}{-12}$

7. $\dfrac{8}{15} + \dfrac{3}{25}$

8. $\dfrac{5}{9} - \dfrac{11}{12}$

9. $\dfrac{1}{5} + \dfrac{5}{6} - \dfrac{7}{15}$

10. $\dfrac{2}{3} - \dfrac{7}{8} + \dfrac{1}{4}$

11. $\dfrac{1}{3} - \dfrac{1}{4} - \dfrac{3}{14}$

12. $\dfrac{5}{6} - \dfrac{7}{9} - \dfrac{3}{10}$

Add or subtract the following rational expressions as indicated. Be sure to express answers in simplest form.

13. $\dfrac{2x}{x-1} + \dfrac{4}{x-1}$

14. $\dfrac{3x}{2x+1} - \dfrac{5}{2x+1}$

15. $\dfrac{4a}{a+2} + \dfrac{8}{a+2}$

16. $\dfrac{6a}{a-3} - \dfrac{18}{a-3}$

17. $\dfrac{3(y-2)}{7y} + \dfrac{4(y-1)}{7y}$

18. $\dfrac{2x-1}{4x^2} + \dfrac{3(x-2)}{4x^2}$

19. $\dfrac{x-1}{2} + \dfrac{x+3}{3}$

20. $\dfrac{x-2}{4} + \dfrac{x+6}{5}$

21. $\dfrac{2a-1}{4} + \dfrac{3a+2}{6}$

22. $\dfrac{a-4}{6} + \dfrac{4a-1}{8}$

23. $\dfrac{n+2}{6} - \dfrac{n-4}{9}$

24. $\dfrac{2n+1}{9} - \dfrac{n+3}{12}$

25. $\dfrac{3x-1}{3} - \dfrac{5x+2}{5}$

26. $\dfrac{4x-3}{6} - \dfrac{8x-2}{12}$

27. $\dfrac{x-2}{5} - \dfrac{x+3}{6} + \dfrac{x+1}{15}$

28. $\dfrac{x+1}{4} + \dfrac{x-3}{6} - \dfrac{x-2}{8}$

29. $\dfrac{3}{8x} + \dfrac{7}{10x}$

30. $\dfrac{5}{6x} - \dfrac{3}{10x}$

31. $\dfrac{5}{7x} - \dfrac{11}{4y}$

32. $\dfrac{5}{12x} - \dfrac{9}{8y}$

33. $\dfrac{4}{3x} + \dfrac{5}{4y} - 1$

34. $\dfrac{7}{3x} - \dfrac{8}{7y} - 2$

35. $\dfrac{7}{10x^2} + \dfrac{11}{15x}$

36. $\dfrac{7}{12a^2} - \dfrac{5}{16a}$

37. $\dfrac{10}{7n} - \dfrac{12}{4n^2}$

38. $\dfrac{6}{8n^2} - \dfrac{3}{5n}$

39. $\dfrac{3}{n^2} - \dfrac{2}{5n} + \dfrac{4}{3}$

40. $\dfrac{1}{n^2} + \dfrac{3}{4n} - \dfrac{5}{6}$

41. $\dfrac{3}{x} - \dfrac{5}{3x^2} - \dfrac{7}{6x}$

42. $\dfrac{7}{3x^2} - \dfrac{9}{4x} - \dfrac{5}{2x}$

43. $\dfrac{6}{5t^2} - \dfrac{4}{7t^3} + \dfrac{9}{5t^3}$

44. $\dfrac{5}{7t} + \dfrac{3}{4t^2} + \dfrac{1}{14t}$

45. $\dfrac{5b}{24a^2} - \dfrac{11a}{32b}$

46. $\dfrac{9}{14x^2y} - \dfrac{4x}{7y^2}$

47. $\dfrac{7}{9xy^3} - \dfrac{4}{3x} + \dfrac{5}{2y^2}$

48. $\dfrac{7}{16a^2b} + \dfrac{3a}{20b^2}$

49. $\dfrac{2x}{x-1} + \dfrac{3}{x}$

50. $\dfrac{3x}{x-4} - \dfrac{2}{x}$

51. $\dfrac{a-2}{a} - \dfrac{3}{a+4}$

52. $\dfrac{a+1}{a} - \dfrac{2}{a+1}$

53. $\dfrac{-3}{4n+5} - \dfrac{8}{3n+5}$

54. $\dfrac{-2}{n-6} - \dfrac{6}{2n+3}$

55. $\dfrac{-1}{x+4} + \dfrac{4}{7x-1}$

56. $\dfrac{-3}{4x+3} + \dfrac{5}{2x-5}$

57. $\dfrac{7}{3x-5} - \dfrac{5}{2x+7}$

58. $\dfrac{5}{x-1} - \dfrac{3}{2x-3}$

59. $\dfrac{5}{3x-2} + \dfrac{6}{4x+5}$

60. $\dfrac{3}{2x+1} + \dfrac{2}{3x+4}$

61. $\dfrac{3x}{2x+5} + 1$

62. $2 + \dfrac{4x}{3x-1}$

63. $\dfrac{4x}{x-5} - 3$

64. $\dfrac{7x}{x+4} - 2$

65. $-1 - \dfrac{3}{2x + 1}$ ·

66. $-2 - \dfrac{5}{4x - 3}$

67. Recall that the indicated quotient of a polynomial and its opposite is -1. For example, $\dfrac{x - 2}{2 - x}$ simplifies to -1. Keep this idea in mind as you add or subtract the following rational expressions.

(a) $\dfrac{1}{x - 1} - \dfrac{x}{x - 1}$

(b) $\dfrac{3}{2x - 3} - \dfrac{2x}{2x - 3}$

(c) $\dfrac{4}{x - 4} - \dfrac{x}{x - 4} + 1$

(d) $-1 + \dfrac{2}{x - 2} - \dfrac{x}{x - 2}$

68. Consider the addition problem $\dfrac{8}{x - 2} + \dfrac{5}{2 - x}$. Notice that the denominators are opposites of each other. If the property $\dfrac{a}{-b} = -\dfrac{a}{b}$ is applied to the second fraction, we have $\dfrac{5}{2 - x} = -\dfrac{5}{x - 2}$. Thus, we proceed as follows.

$$\dfrac{8}{x - 2} + \dfrac{5}{2 - x} = \dfrac{8}{x - 2} - \dfrac{5}{x - 2} = \dfrac{8 - 5}{x - 2} = \dfrac{3}{x - 2}$$

Use this approach to do the following problems.

(a) $\dfrac{7}{x - 1} + \dfrac{2}{1 - x}$

(b) $\dfrac{5}{2x - 1} + \dfrac{8}{1 - 2x}$

(c) $\dfrac{4}{a - 3} - \dfrac{1}{3 - a}$

(d) $\dfrac{10}{a - 9} - \dfrac{5}{9 - a}$

(e) $\dfrac{x^2}{x - 1} - \dfrac{2x - 3}{1 - x}$

(f) $\dfrac{x^2}{x - 4} - \dfrac{3x - 28}{4 - x}$

9.4

More on Rational Expressions and Complex Fractions

In this section we shall expand our work with adding and subtracting rational expressions and we will also discuss the process of simplifying complex fractions. Before we begin, however, this seems like an appropriate time to offer a bit of advice regarding your study of algebra. Success in algebra depends upon having a good understanding of the concepts as well as the ability to perform the various computations. As for the computational work, you should adopt a carefully organized format that shows as many steps as *you need* in order to minimize the chances of making careless errors. Don't be eager to find shortcuts for certain computations before you have a thorough understanding of the steps involved in the process. This is especially appropriate advice at the beginning of this section.

Study the following examples very carefully. Notice the same basic procedure for each problem of: (1) finding the LCD, (2) changing each fraction to an equivalent fraction that has the LCD as its denominator, (3) adding or subtracting numerators and placing this result over the LCD, and (4) looking for possibilities to simplify the resulting fraction.

EXAMPLE 1   Add $\dfrac{8}{x^2 - 4x} + \dfrac{2}{x}$.

Solution

$$\left. \begin{array}{l} x^2 - 4x = x(x - 4) \\[2mm] x = x \end{array} \right\} \longrightarrow \text{ LCD is } x(x-4)$$

$$\frac{8}{x^2 - 4x} + \frac{2}{x} = \frac{8}{x(x - 4)} + \frac{2}{x}$$

$$= \frac{8}{x(x - 4)} + \left(\frac{2}{x}\right)\left(\frac{x - 4}{x - 4}\right)$$

$$= \frac{8}{x(x - 4)} + \frac{2(x - 4)}{x(x - 4)}$$

$$= \frac{8 + 2x - 8}{x(x - 4)}$$

$$= \frac{2x}{x(x - 4)} = \frac{2}{x - 4} \qquad \blacksquare$$

EXAMPLE 2   Subtract $\dfrac{a}{a^2 - 4} - \dfrac{3}{a + 2}$.

Solution

$$\left. \begin{array}{l} a^2 - 4 = (a + 2)(a - 2) \\[2mm] a + 2 = a + 2 \end{array} \right\} \longrightarrow \text{ LCD is } (a + 2)(a - 2)$$

$$\frac{a}{a^2 - 4} - \frac{3}{a + 2} = \frac{a}{(a + 2)(a - 2)} - \frac{3}{a + 2}$$

$$= \frac{a}{(a + 2)(a - 2)} - \left(\frac{3}{a + 2}\right)\left(\frac{a - 2}{a - 2}\right)$$

$$= \frac{a}{(a + 2)(a - 2)} - \frac{3(a - 2)}{(a + 2)(a - 2)}$$

$$= \frac{a - 3a + 6}{(a + 2)(a - 2)}$$

$$= \frac{-2a + 6}{(a + 2)(a - 2)} \qquad \text{or} \qquad \frac{-2(a - 3)}{(a + 2)(a - 2)} \qquad \blacksquare$$

EXAMPLE 3   Add $\dfrac{3n}{n^2 + 6n + 5} + \dfrac{4}{n^2 - 7n - 8}$.

Solution

$$\left.\begin{array}{r} n^2 + 6n + 5 = (n + 5)(n + 1) \\ n^2 - 7n - 8 = (n - 8)(n + 1) \end{array}\right\} \longrightarrow \text{LCD is } (n+1)(n+5)(n-8)$$

$$\dfrac{3n}{n^2 + 6n + 5} + \dfrac{4}{n^2 - 7n - 8} = \dfrac{3n}{(n + 5)(n + 1)} + \dfrac{4}{(n - 8)(n + 1)}$$

$$= \left(\dfrac{3n}{(n + 5)(n + 1)}\right)\left(\dfrac{n - 8}{n - 8}\right)$$

$$+ \left(\dfrac{4}{(n - 8)(n + 1)}\right)\left(\dfrac{n + 5}{n + 5}\right)$$

$$= \dfrac{3n(n - 8)}{(n + 5)(n + 1)(n - 8)} + \dfrac{4(n + 5)}{(n + 5)(n + 1)(n - 8)}$$

$$= \dfrac{3n^2 - 24n + 4n + 20}{(n + 5)(n + 1)(n - 8)}$$

$$= \dfrac{3n^2 - 20n + 20}{(n + 5)(n + 1)(n - 8)} \qquad\blacksquare$$

EXAMPLE 4   Perform the indicated operations.

$$\dfrac{2x^2}{x^4 - 1} + \dfrac{x}{x^2 - 1} - \dfrac{1}{x - 1}.$$

Solution

$$\left.\begin{array}{l} x^4 - 1 = (x^2 + 1)(x + 1)(x - 1) \\ x^2 - 1 = (x + 1)(x - 1) \\ x - 1 = x - 1 \end{array}\right\} \longrightarrow \text{LCD is } (x^2 + 1)(x + 1)(x - 1)$$

$$\dfrac{2x^2}{x^4 - 1} + \dfrac{x}{x^2 - 1} - \dfrac{1}{x - 1} = \dfrac{2x^2}{(x^2 + 1)(x + 1)(x - 1)} + \dfrac{x}{(x + 1)(x - 1)} - \dfrac{1}{x - 1}$$

$$= \dfrac{2x^2}{(x^2 + 1)(x + 1)(x - 1)} + \left(\dfrac{x}{(x + 1)(x - 1)}\right)\left(\dfrac{x^2 + 1}{x^2 + 1}\right)$$

$$- \left(\dfrac{1}{x - 1}\right)\left(\dfrac{(x^2 + 1)(x + 1)}{(x^2 + 1)(x + 1)}\right)$$

$$= \dfrac{2x^2}{(x^2 + 1)(x + 1)(x - 1)} + \dfrac{x(x^2 + 1)}{(x^2 + 1)(x + 1)(x - 1)}$$

$$- \dfrac{(x^2 + 1)(x + 1)}{(x^2 + 1)(x + 1)(x - 1)}$$

$$= \frac{2x^2 + x^3 + x - x^3 - x^2 - x - 1}{(x^2 + 1)(x + 1)(x - 1)}$$

$$= \frac{x^2 - 1}{(x^2 + 1)(x + 1)(x - 1)}$$

$$= \frac{(x + 1)(x - 1)}{(x^2 + 1)(x + 1)(x - 1)}$$

$$= \frac{1}{x^2 + 1} \qquad \blacksquare$$

Complex Fractions

Fractional forms that contain rational numbers or rational expressions in the numerators and/or denominators are called **complex fractions**. The following are examples of complex fractions.

$$\frac{\frac{3}{5}}{\frac{7}{8}}, \quad \frac{\frac{4}{x}}{\frac{2}{xy}}, \quad \frac{\frac{1}{2} + \frac{3}{4}}{\frac{5}{6} - \frac{3}{8}}, \quad \frac{\frac{3}{x} + \frac{2}{y}}{\frac{5}{x} - \frac{6}{y^2}}$$

It is often necessary to *simplify* a complex fraction. Let's examine some techniques for simplifying complex fractions with the four previous examples.

EXAMPLE 5   Simplify $\dfrac{\frac{3}{5}}{\frac{7}{8}}$.

Solution   This type of problem creates nothing new since it is merely a division problem. Thus,

$$\frac{\frac{3}{5}}{\frac{7}{8}} = \frac{3}{5} \div \frac{7}{8} = \frac{3}{5} \cdot \frac{8}{7} = \frac{24}{35}. \qquad \blacksquare$$

EXAMPLE 6   Simplify $\dfrac{\frac{4}{x}}{\frac{2}{xy}}$.

Solution

$$\frac{\dfrac{4}{x}}{\dfrac{2}{xy}} = \frac{4}{x} \div \frac{2}{xy} = \frac{\overset{2}{\cancel{4}}}{\cancel{x}} \cdot \frac{\cancel{x}y}{\cancel{2}} = 2y \qquad\blacksquare$$

EXAMPLE 7   Simplify $\dfrac{\dfrac{1}{2}+\dfrac{3}{4}}{\dfrac{5}{6}-\dfrac{3}{8}}$.

Let's look at two possible "attacks" for such a problem.

Solution A

$$\frac{\dfrac{1}{2}+\dfrac{3}{4}}{\dfrac{5}{6}-\dfrac{3}{8}} = \frac{\dfrac{2}{4}+\dfrac{3}{4}}{\dfrac{20}{24}-\dfrac{9}{24}} = \frac{\dfrac{5}{4}}{\dfrac{11}{24}} = \frac{5}{\cancel{4}} \cdot \frac{\overset{6}{\cancel{24}}}{11} = \frac{30}{11}$$

Solution B

The LCD of all four denominators (2, 4, 6, and 8) is 24. Multiply the entire complex fraction by a form of 1, namely, $\dfrac{24}{24}$.

$$\frac{\dfrac{1}{2}-\dfrac{3}{4}}{\dfrac{5}{6}-\dfrac{3}{8}} = \left(\frac{24}{24}\right)\frac{\dfrac{1}{2}+\dfrac{3}{4}}{\dfrac{5}{6}-\dfrac{3}{8}}$$

$$= \frac{24\left(\dfrac{1}{2}+\dfrac{3}{4}\right)}{24\left(\dfrac{5}{6}-\dfrac{3}{8}\right)}$$

$$= \frac{24\left(\dfrac{1}{2}\right)+24\left(\dfrac{3}{4}\right)}{24\left(\dfrac{5}{6}\right)-24\left(\dfrac{3}{8}\right)}$$

$$= \frac{12+18}{20-9} = \frac{30}{11} \qquad\blacksquare$$

EXAMPLE 8   Simplify $\dfrac{\dfrac{3}{x}+\dfrac{2}{y}}{\dfrac{5}{x}-\dfrac{6}{y^2}}$.

Solution A

$$\frac{\dfrac{3}{x} + \dfrac{2}{y}}{\dfrac{5}{x} - \dfrac{6}{y^2}} = \frac{\left(\dfrac{3}{x}\right)\left(\dfrac{y}{y}\right) + \left(\dfrac{2}{y}\right)\left(\dfrac{x}{x}\right)}{\left(\dfrac{5}{x}\right)\left(\dfrac{y^2}{y^2}\right) - \left(\dfrac{6}{y^2}\right)\left(\dfrac{x}{x}\right)}$$

$$= \frac{\dfrac{3y}{xy} + \dfrac{2x}{xy}}{\dfrac{5y^2}{xy^2} - \dfrac{6x}{xy^2}}$$

$$= \frac{\dfrac{3y + 2x}{xy}}{\dfrac{5y^2 - 6x}{xy^2}}$$

$$= \frac{3y + 2x}{xy} \div \frac{5y^2 - 6x}{xy^2}$$

$$= \frac{3y + 2x}{\cancel{xy}} \cdot \frac{\overset{y}{\cancel{xy^2}}}{5y^2 - 6x}$$

$$= \frac{y(3y + 2x)}{5y^2 - 6x}$$

Solution B   The LCD of all four denominators (x, y, x, and y^2) is xy^2. Multiply the entire complex fraction by a form of 1, namely, $\dfrac{xy^2}{xy^2}$.

$$\frac{\dfrac{3}{x} + \dfrac{2}{y}}{\dfrac{5}{x} - \dfrac{6}{y^2}} = \left(\frac{xy^2}{xy^2}\right)\frac{\dfrac{3}{x} + \dfrac{2}{y}}{\dfrac{5}{x} - \dfrac{6}{y^2}}$$

$$= \frac{xy^2\left(\dfrac{3}{x} + \dfrac{2}{y}\right)}{xy^2\left(\dfrac{5}{x} - \dfrac{6}{y^2}\right)}$$

$$= \frac{xy^2\left(\dfrac{3}{x}\right) + xy^2\left(\dfrac{2}{y}\right)}{xy^2\left(\dfrac{5}{x}\right) - xy^2\left(\dfrac{6}{y^2}\right)}$$

$$= \frac{3y^2 + 2xy}{5y^2 - 6x} \quad \text{or} \quad \frac{y(3y + 2x)}{5y^2 - 6x} \qquad \blacksquare$$

Certainly either approach (Solution A or Solution B) will work with problems such as Examples 7 and 8. Examine Solution B in both examples carefully. This approach works very effectively with complex fractions where the LCD of all the denominators is easy to find. (Don't be misled by the length of Solution B for Example 7. We were especially careful to show every step.) Let's conclude this section with two more examples that involve algebraic complex fractions.

EXAMPLE 9   Simplify $\dfrac{\dfrac{1}{a} + \dfrac{1}{a-1}}{\dfrac{1}{a-1} - \dfrac{1}{a}}$.

Solution   Multiply the entire complex fraction by a form of 1, namely, $\dfrac{a(a-1)}{a(a-1)}$.

$$\frac{\dfrac{1}{a} + \dfrac{1}{a-1}}{\dfrac{1}{a-1} - \dfrac{1}{a}} = \left(\frac{a(a-1)}{a(a-1)}\right) \frac{\dfrac{1}{a} + \dfrac{1}{a-1}}{\dfrac{1}{a-1} - \dfrac{1}{a}}$$

$$= \frac{a(a-1)\left(\dfrac{1}{a} + \dfrac{1}{a-1}\right)}{a(a-1)\left(\dfrac{1}{a-1} - \dfrac{1}{1}\right)}$$

$$= \frac{a(a-1)\left(\dfrac{1}{a}\right) + a(a-1)\left(\dfrac{1}{a-1}\right)}{a(a-1)\left(\dfrac{1}{a-1}\right) - a(a-1)\left(\dfrac{1}{a}\right)}$$

$$= \frac{a-1+a}{a-(a-1)} = \frac{2a-1}{a-a+1}$$

$$= \frac{2a-1}{1} \qquad \text{or} \qquad 2a-1 \qquad\blacksquare$$

EXAMPLE 10   Simplify $1 - \dfrac{n}{1 - \dfrac{1}{n}}$.

Solution   First simplify the complex fraction $\dfrac{n}{1 - \dfrac{1}{n}}$ by multiplying by $\dfrac{n}{n}$.

$$\left(\cfrac{n}{1 - \cfrac{1}{n}}\right)\left(\frac{n}{n}\right) = \frac{n^2}{n-1}$$

Now   form the subtraction.

$$\frac{}{n-1} = \left(\frac{n-1}{n-1}\right)\left(\frac{1}{1}\right) - \frac{n^2}{n-1}$$

$$= \frac{n-1}{n-1} - \frac{n^2}{n-1}$$

$$= \frac{n-1-n^2}{n-1} \quad \text{or} \quad \frac{-n^2+n-1}{n-1}$$ ■

Problem Set 9.4

Perform the indicated operations and express your answers in simplest form.

1. $\dfrac{2x}{x^2 + 4x} + \dfrac{5}{x}$

2. $\dfrac{3x}{x^2 - 6x} + \dfrac{4}{x}$

3. $\dfrac{4}{x^2 + 7x} - \dfrac{1}{x}$

4. $\dfrac{-10}{x^2 - 9x} - \dfrac{2}{x}$

5. $\dfrac{x}{x^2 - 1} + \dfrac{5}{x + 1}$

6. $\dfrac{2x}{x^2 - 16} + \dfrac{7}{x - 4}$

7. $\dfrac{6a + 4}{a^2 - 1} - \dfrac{5}{a - 1}$

8. $\dfrac{4a - 4}{a^2 - 4} - \dfrac{3}{a + 2}$

9. $\dfrac{2n}{n^2 - 25} - \dfrac{3}{4n + 20}$

10. $\dfrac{3n}{n^2 - 36} - \dfrac{2}{5n + 30}$

11. $\dfrac{5}{x} - \dfrac{5x - 30}{x^2 + 6x} + \dfrac{x}{x + 6}$

12. $\dfrac{3}{x + 1} + \dfrac{x + 5}{x^2 - 1} - \dfrac{3}{x - 1}$

13. $\dfrac{3}{x^2 + 9x + 14} + \dfrac{5}{2x^2 + 15x + 7}$

14. $\dfrac{6}{x^2 + 11x + 24} + \dfrac{4}{3x^2 + 13x + 12}$

15. $\dfrac{1}{a^2 - 3a - 10} - \dfrac{4}{a^2 + 4a - 45}$

16. $\dfrac{6}{a^2 - 3a - 54} - \dfrac{10}{a^2 + 5a - 6}$

17. $\dfrac{3a}{20a^2 - 11a - 3} + \dfrac{1}{12a^2 + 7a - 12}$

18. $\dfrac{2a}{6a^2 + 11a - 10} + \dfrac{a}{2a^2 - 3a - 20}$

19. $\dfrac{5}{x^2 + 3} - \dfrac{2}{x^2 + 4x - 21}$

20. $\dfrac{7}{x^2 + 1} - \dfrac{3}{x^2 + 7x - 60}$

21. $\dfrac{2}{y^2 + 6y - 16} - \dfrac{4}{y + 8} - \dfrac{3}{y - 2}$

22. $\dfrac{7}{y - 6} - \dfrac{10}{y + 12} + \dfrac{4}{y^2 + 6y - 72}$

23. $x - \dfrac{x^2}{x - 2} + \dfrac{3}{x^2 - 4}$

24. $x + \dfrac{5}{x^2 - 25} - \dfrac{x^2}{x + 5}$

25. $\dfrac{x + 3}{x + 10} + \dfrac{4x - 3}{x^2 + 8x - 20} + \dfrac{x - 1}{x - 2}$

26. $\dfrac{2x - 1}{x + 3} + \dfrac{x + 4}{x - 6} + \dfrac{3x - 1}{x^2 - 3x - 18}$

27. $\dfrac{n}{n - 6} + \dfrac{n + 3}{n + 8} + \dfrac{12n + 26}{n^2 + 2n - 48}$

28. $\dfrac{n - 1}{n + 4} + \dfrac{n}{n + 6} + \dfrac{2n + 18}{n^2 + 10n + 24}$

29. $\dfrac{4x - 3}{2x^2 + x - 1} - \dfrac{2x + 7}{3x^2 + x - 2} - \dfrac{3}{3x - 2}$

30. $\dfrac{2x + 5}{x^2 + 3x - 18} - \dfrac{3x - 1}{x^2 + 4x - 12} + \dfrac{5}{x - 2}$

31. $\dfrac{n}{n^2 + 1} + \dfrac{n^2 + 3n}{n^4 - 1} - \dfrac{1}{n - 1}$

32. $\dfrac{2n^2}{n^4 - 16} - \dfrac{n}{n^2 - 4} + \dfrac{1}{n + 2}$

33. $\dfrac{15x^2 - 10}{5x^2 - 7x + 2} - \dfrac{3x + 4}{x - 1} - \dfrac{2}{5x - 2}$

34. $\dfrac{32x + 9}{12x^2 + x - 6} - \dfrac{3}{4x + 3} - \dfrac{x + 5}{3x - 2}$

35. $\dfrac{t + 3}{3t - 1} + \dfrac{8t^2 + 8t + 2}{3t^2 - 7t + 2} - \dfrac{2t + 3}{t - 2}$

36. $\dfrac{t - 3}{2t + 1} + \dfrac{2t^2 + 19t - 46}{2t^2 - 9t - 5} - \dfrac{t + 4}{t - 5}$

Simplify each of the following complex fractions.

37. $\dfrac{\dfrac{1}{2} - \dfrac{1}{4}}{\dfrac{5}{8} + \dfrac{3}{4}}$

38. $\dfrac{\dfrac{3}{8} + \dfrac{3}{4}}{\dfrac{5}{8} - \dfrac{7}{12}}$

39. $\dfrac{\dfrac{3}{28} - \dfrac{5}{14}}{\dfrac{5}{7} + \dfrac{1}{4}}$

40. $\dfrac{\dfrac{5}{9}+\dfrac{7}{36}}{\dfrac{3}{18}-\dfrac{5}{12}}$

41. $\dfrac{\dfrac{5}{6y}}{\dfrac{10}{3xy}}$

42. $\dfrac{\dfrac{9}{8xy^2}}{\dfrac{5}{4x^2}}$

43. $\dfrac{\dfrac{3}{x}-\dfrac{2}{y}}{\dfrac{4}{y}-\dfrac{7}{xy}}$

44. $\dfrac{\dfrac{9}{x}+\dfrac{7}{x^2}}{\dfrac{5}{y}+\dfrac{3}{y^2}}$

45. $\dfrac{\dfrac{6}{a}-\dfrac{5}{b^2}}{\dfrac{12}{a^2}+\dfrac{2}{b}}$

46. $\dfrac{\dfrac{4}{ab}-\dfrac{3}{b^2}}{\dfrac{1}{a}+\dfrac{3}{b}}$

47. $\dfrac{\dfrac{2}{x}-3}{\dfrac{3}{y}+4}$

48. $\dfrac{1+\dfrac{3}{x}}{1-\dfrac{6}{x}}$

49. $\dfrac{3+\dfrac{2}{n+4}}{5-\dfrac{1}{n+4}}$

50. $\dfrac{4+\dfrac{6}{n-1}}{7-\dfrac{4}{n-1}}$

51. $\dfrac{5-\dfrac{2}{n-3}}{4-\dfrac{1}{n-3}}$

52. $\dfrac{\dfrac{3}{n-5}-2}{1-\dfrac{4}{n-5}}$

53. $\dfrac{\dfrac{-1}{y-2}+\dfrac{5}{x}}{\dfrac{3}{x}-\dfrac{4}{xy-2x}}$

54. $\dfrac{\dfrac{-2}{x}-\dfrac{4}{x+2}}{\dfrac{3}{x^2+2x}+\dfrac{3}{x}}$

55. $\dfrac{\dfrac{2}{x-3}-\dfrac{3}{x+3}}{\dfrac{5}{x^2-9}-\dfrac{2}{x-3}}$

56. $\dfrac{\dfrac{2}{x-y}+\dfrac{3}{x+y}}{\dfrac{5}{x+y}-\dfrac{1}{x^2-y^2}}$

57. $\dfrac{\dfrac{3a}{2-\dfrac{1}{a}}-1}{}$

58. $\dfrac{\dfrac{a}{\dfrac{1}{a}+4}+1}{}$

59. $2-\dfrac{x}{3-\dfrac{2}{x}}$

60. $1+\dfrac{x}{1+\dfrac{1}{x}}$

Thoughts into Words

61. What role does factoring play when simplifying algebraic fractions?

62. Explain in your own words how to multiply two algebraic fractions.

63. What does the concept "least common denominator" mean?

64. Give a step-by-step description of how to add $\dfrac{3x-1}{6}+\dfrac{2x+3}{9}$.

65. Which of the two techniques presented in the text would you use to simplify $\dfrac{\frac{1}{4}+\frac{1}{3}}{\frac{3}{4}-\frac{1}{6}}$?

Which technique would you use to simplify $\dfrac{\frac{3}{8}-\frac{5}{7}}{\frac{7}{9}+\frac{6}{25}}$? Explain your choice for each problem.

Equations Involving Rational Expressions

In Chapter 3 we solved some equations that involved rational expressions with constants as denominators. The following example reviews the general technique used to solve such equations.

EXAMPLE 1   Solve $\dfrac{x-2}{3}+\dfrac{x+1}{4}=\dfrac{1}{6}$.

Solution

$$\frac{x-2}{3}+\frac{x+1}{4}=\frac{1}{6}$$

$$12\left(\frac{x-2}{3}+\frac{x+1}{4}\right)=12\left(\frac{1}{6}\right) \qquad \text{Multiply both sides by 12, which is the}$$
$$\text{LCD of all of the denominators.}$$

$$4(x-2)+3(x+1)=2$$

$$4x-8+3x+3=2$$

$$7x-5=2$$

$$7x=7$$

$$x=1$$

The solution set is $\{1\}$. Check it!   ■

If an equation contains a variable (or variables) in one or more denominators, then we proceed in essentially the same way as in Example 1 above, **except we must avoid any value of the variable that makes a denominator zero**. Consider the following examples.

EXAMPLE 2   Solve $\dfrac{5}{n} + \dfrac{1}{2} = \dfrac{9}{n}$.

Solution   First, we need to realize that **n cannot equal zero**. (Let's indicate this restriction so that it is not forgotten!) Then we can proceed as follows.

$$\frac{5}{n} + \frac{1}{2} = \frac{9}{n}, \qquad n \neq 0$$

$$2n\left(\frac{5}{n} + \frac{1}{2}\right) = 2n\left(\frac{9}{n}\right) \qquad \text{Multiply both sides by the LCD, which is } 2n.$$

$$10 + n = 18$$

$$n = 8$$

The solution set is $\{8\}$. Check it!   ■

EXAMPLE 3   Solve $\dfrac{35 - x}{x} = 7 + \dfrac{3}{x}$.

Solution $$\frac{35 - x}{x} = 7 + \frac{3}{x}, \qquad x \neq 0$$

$$x\left(\frac{35 - x}{x}\right) = x\left(7 + \frac{3}{x}\right) \qquad \text{Multiply both sides by } x.$$

$$35 - x = 7x + 3$$

$$32 = 8x$$

$$4 = x$$

The solution set is $\{4\}$.   ■

EXAMPLE 4   Solve $\dfrac{3}{a - 2} = \dfrac{4}{a + 1}$.

Solution $$\frac{3}{a - 2} = \frac{4}{a + 1}, \qquad a \neq 2 \text{ and } a \neq -1$$

$$(a - 2)(a + 1)\left(\frac{3}{a - 2}\right) = (a - 2)(a + 1)\left(\frac{4}{a + 1}\right) \qquad \begin{array}{l}\text{Multiply both sides by}\\ (a - 2)(a + 1).\end{array}$$

$$3(a + 1) = 4(a - 2)$$

$$3a + 3 = 4a - 8$$

$$11 = a$$

The solution set is $\{11\}$.   ■

Keep in mind that listing the restrictions at the beginning of a problem does not replace *checking* the potential solutions. In Example 4, 11 needs to be checked in the original equation.

EXAMPLE 5   Solve $\dfrac{a}{a-2} + \dfrac{2}{3} = \dfrac{2}{a-2}$.

Solution

$$\frac{a}{a-2} + \frac{2}{3} = \frac{2}{a-2}, \qquad a \neq 2$$

$$3(a-2)\left(\frac{a}{a-2} + \frac{2}{3}\right) = 3(a-2)\left(\frac{2}{a-2}\right) \qquad \text{Multiply both sides by } 3(a-2).$$

$$3a + 2(a-2) = 6$$

$$3a + 2a - 4 = 6$$

$$5a = 10$$

$$a = 2$$

Because our initial restriction was $a \neq 2$, we conclude that this equation *has no solution*. Thus, the solution set is $\varnothing$. ∎

Example 5 demonstrates the importance of recognizing the restrictions that must be made to exclude division by zero. The next example reviews using the *cross-multiplication property of proportions*. If this does not look familiar to you, perhaps a brief look over Section 3.5 would be helpful.

EXAMPLE 6   Solve $\dfrac{x}{7} = \dfrac{4}{x+3}$.

Solution

$$\frac{x}{7} = \frac{4}{x+3}, \qquad x \neq -3$$

$$x(x+3) = 7(4) \qquad \text{cross-multiplication property}$$

$$x^2 + 3x = 28$$

$$x^2 + 3x - 28 = 0$$

$$(x+7)(x-4) = 0$$

$$x + 7 = 0 \qquad \text{or} \qquad x - 4 = 0$$

$$x = -7 \qquad \text{or} \qquad x = 4$$

The solution set is $\{-7, 4\}$. Check these solutions in the original equation. ∎

Let's pause for a moment and look back over Examples 1 through 6. Note that we could also have solved the equation in Example 4 by using the cross-multiplication property of proportions. However, in Examples 1, 2, 3, and 5 we needed to multiply both sides of the equation by the LCD because the equation is not in the form of a proportion.

Problem Solving

The ability to solve equations involving rational expressions broadens our base for solving word problems. We are now ready to tackle some word problems that translate into such equations.

PROBLEM 1   The sum of a number and its reciprocal is $\dfrac{10}{3}$. Find the number.

Solution   Let n represent the number. Then $\dfrac{1}{n}$ represents its reciprocal.

$$n + \frac{1}{n} = \frac{10}{3}, \qquad n \neq 0$$

$$3n\left(n + \frac{1}{n}\right) = 3n\left(\frac{10}{3}\right)$$

$$3n^2 + 3 = 10n$$

$$3n^2 - 10n + 3 = 0$$

$$(3n - 1)(n - 3) = 0$$

$$3n - 1 = 0 \quad \text{or} \quad n - 3 = 0$$

$$3n = 1 \quad \text{or} \quad n = 3$$

$$n = \frac{1}{3} \quad \text{or} \quad n = 3$$

If the number is $\dfrac{1}{3}$, then its reciprocal is $\dfrac{1}{\frac{1}{3}} = 3$. If the number is 3, then its reciprocal is $\dfrac{1}{3}$. ∎

Now let's consider a problem where we use the relationship

$$\frac{\text{dividend}}{\text{divisor}} = \text{quotient} + \frac{\text{remainder}}{\text{divisor}}.$$

PROBLEM 2   The sum of two numbers is 52. If the larger is divided by the smaller, the quotient is 9 and the remainder is 2. Find the numbers.

Solution   Let x represent the smaller number and let y represent the larger number. The problem translates into the following two equations.

sum of 52 $\longrightarrow$ $x + y = 52$

larger divided
by smaller produces $\longrightarrow$ $\dfrac{y}{x} = 9 + \dfrac{2}{x}$
a quotient of 9 and
a remainder of 2

To solve the system let's first rewrite $x + y = 52$ as $y = 52 - x$. Now we can substitute $52 - x$ for y in the second equation.

$$\frac{y}{x} = 9 + \frac{2}{x}$$

$$\frac{52 - x}{x} = 9 + \frac{2}{x}, \qquad x \neq 0$$

$$x\left(\frac{52 - x}{x}\right) = x\left(9 + \frac{2}{x}\right)$$

$$52 - x = 9x + 2$$

$$50 = 10x$$

$$5 = x$$

The smaller number is 5 and the larger number is $52 - 5 = 47$.   ■

Notice that to solve Problem 1 we used one equation and one variable, but to solve Problem 2 we used two equations and two variables. Certainly Problem 2 could also be solved using only one variable and one equation; however, the wording of the problem lends itself to the use of two variables and two equations. Which approach to use is frequently based on personal preference. Don't let our choice of an approach dictate your choice, but be sure that you can handle the problem both ways.

In Chapter 3 we solved several problems using the concept of a proportion. Let's conclude this section by solving a problem of that type.

PROBLEM 3   A sum of $750 is to be divided between two people in the ratio of 2 to 3. How much does each person receive?

Solution   Let d represent the amount of money that one person receives. Then $750 - d$ represents the amount for the other person.

$$\frac{d}{750 - d} = \frac{2}{3}, \qquad d \neq 750$$

$$3d = 2(750 - d)$$

$$3d = 1500 - 2d$$

$$5d = 1500$$

$$d = 300$$

If $d = 300$, then $750 - d$ equals 450. Therefore, one person receives \$300 and the other person receives \$450.

Problem Set 9.5

Solve each of the following equations.

1. $\dfrac{x + 1}{4} + \dfrac{x - 2}{6} = \dfrac{3}{4}$

2. $\dfrac{x + 2}{5} + \dfrac{x - 1}{6} = \dfrac{3}{5}$

3. $\dfrac{x + 3}{2} - \dfrac{x - 4}{7} = 1$

4. $\dfrac{x + 4}{3} - \dfrac{x - 5}{9} = 1$

5. $\dfrac{5}{n} + \dfrac{1}{3} = \dfrac{7}{n}$

6. $\dfrac{3}{n} + \dfrac{1}{6} = \dfrac{11}{3n}$

7. $\dfrac{7}{2x} + \dfrac{3}{5} = \dfrac{2}{3x}$

8. $\dfrac{9}{4x} + \dfrac{1}{3} = \dfrac{5}{2x}$

9. $\dfrac{3}{4x} + \dfrac{5}{6} = \dfrac{4}{3x}$

10. $\dfrac{5}{7x} - \dfrac{5}{6} = \dfrac{1}{6x}$

11. $\dfrac{47 - n}{n} = 8 + \dfrac{2}{n}$

12. $\dfrac{45 - n}{n} = 6 + \dfrac{3}{n}$

13. $\dfrac{n}{65 - n} = 8 + \dfrac{2}{65 - n}$

14. $\dfrac{n}{70 - n} = 7 + \dfrac{6}{70 - n}$

15. $n + \dfrac{1}{n} = \dfrac{17}{4}$

16. $n + \dfrac{1}{n} = \dfrac{37}{6}$

17. $n - \dfrac{2}{n} = \dfrac{23}{5}$

18. $n - \dfrac{3}{n} = \dfrac{26}{3}$

19. $\dfrac{5}{7x - 3} = \dfrac{3}{4x - 5}$

20. $\dfrac{3}{2x - 1} = \dfrac{5}{3x + 2}$

21. $\dfrac{-2}{x - 5} = \dfrac{1}{x + 9}$

22. $\dfrac{5}{2a - 1} = \dfrac{-6}{3a + 2}$

23. $\dfrac{x}{x + 1} - 2 = \dfrac{3}{x - 3}$

24. $\dfrac{x}{x - 2} + 1 = \dfrac{8}{x - 1}$

25. $\dfrac{a}{a + 5} - 2 = \dfrac{3a}{a + 5}$

26. $\dfrac{a}{a - 3} - \dfrac{3}{2} = \dfrac{3}{a - 3}$

27. $\dfrac{5}{x+6} = \dfrac{6}{x-3}$

28. $\dfrac{3}{x-1} = \dfrac{4}{x+2}$

29. $\dfrac{3x-7}{10} = \dfrac{2}{x}$

30. $\dfrac{x}{-4} = \dfrac{3}{12x-25}$

31. $\dfrac{x}{x-6} - 3 = \dfrac{6}{x-6}$

32. $\dfrac{x}{x+1} + 3 = \dfrac{4}{x+1}$

33. $\dfrac{3s}{s+2} + 1 = \dfrac{35}{2(3s+1)}$

34. $\dfrac{s}{2s-1} - 3 = \dfrac{-32}{3(s+5)}$

35. $2 - \dfrac{3x}{x-4} = \dfrac{14}{x+7}$

36. $-1 + \dfrac{2x}{x+3} = \dfrac{-4}{x+4}$

37. $\dfrac{n+6}{27} = \dfrac{1}{n}$

38. $\dfrac{n}{5} = \dfrac{10}{n-5}$

39. $\dfrac{3n}{n-1} - \dfrac{1}{3} = \dfrac{-40}{3n-18}$

40. $\dfrac{n}{n+1} + \dfrac{1}{2} = \dfrac{-2}{n+2}$

41. $\dfrac{-3}{4x+5} = \dfrac{2}{5x-7}$

42. $\dfrac{7}{x+4} = \dfrac{3}{x-8}$

43. $\dfrac{2x}{x-2} + \dfrac{15}{x^2-7x+10} = \dfrac{3}{x-5}$

44. $\dfrac{x}{x-4} - \dfrac{2}{x+3} = \dfrac{20}{x^2-x-12}$

Set up an equation or a system of equations and solve each of the following problems.

45. A sum of $1750 is to be divided between two people in the ratio of 3 to 4. How much does each person receive?

46. A blueprint has a scale of 1 inch represents 5 feet. Find the dimensions of a rectangular room that measures $3\frac{1}{2}$ inches by $5\frac{3}{4}$ inches on the blueprint.

47. One angle of a triangle has a measure of 60° and the measures of the other two angles are in a ratio of 2 to 3. Find the measures of the other two angles.

48. The ratio of the complement of an angle to its supplement is 1 to 4. Find the measure of the angle.

49. The sum of a number and its reciprocal is $\dfrac{53}{14}$. Find the number.

50. The sum of two numbers is 80. If the larger is divided by the smaller, the quotient is 7 and the remainder is 8. Find the numbers.

51. If a home valued at $50,000 is assessed $900 in real estate taxes, at the same rate how much are the taxes on a home valued at $60,000?

52. The ratio of male students to female students at a certain university is 5 to 7. If there is a total of 16,200 students, find the number of male students and the number of female students.

53. Suppose that, together, Laura and Tammy sold $120.75 worth of candy for the annual school fair. If the ratio of Tammy's sales to Laura's sales was 4 to 3, how much did each sell?

54. The total value of a house and a lot is $68,000. If the ratio of the value of the house to the value of the lot is 7 to 1, find the value of the house.

55. The sum of two numbers is 90. If the larger is divided by the smaller, the quotient is 10 and the remainder is 2. Find the numbers.

56. What number must be added to the numerator and denominator of $\frac{2}{5}$ to produce a rational number that is equivalent to $\frac{7}{8}$?

57. A 20-foot board is to be cut into two pieces whose lengths are in the ratio of 7 to 3. Find the lengths of the two pieces.

58. An inheritance of $300,000 is to be divided between a son and the local heart fund in the ratio of 3 to 1. How much money will the son receive?

59. Suppose that in a certain precinct, 1150 people voted in the last presidential election. If the ratio of female voters to male voters was 3 to 2, how many females and how many males voted?

60. The perimeter of a rectangle is 114 centimeters. If the ratio of its width to its length is 7 to 12, find the dimensions of the rectangle.

9.6

More Equations and Applications

Let's begin this section by considering a few more equations that contain rational expressions. We will continue to solve them using the same basic techniques as in the previous section. That is, we will multiply both sides of the equation by the least common denominator of all of the denominators in the equation (with the necessary restrictions to avoid division by zero). Some of the denominators in these problems will require factoring before we can determine a least common denominator.

EXAMPLE 1   Solve $\dfrac{x}{2x - 8} + \dfrac{16}{x^2 - 16} = \dfrac{1}{2}$.

Solution

$$\frac{x}{2x - 8} + \frac{16}{x^2 - 16} = \frac{1}{2}$$

$$\frac{x}{2(x - 4)} + \frac{16}{(x + 4)(x - 4)} = \frac{1}{2}, \qquad x \neq 4 \text{ and } x \neq -4$$

$$2(x - 4)(x + 4)\left(\frac{x}{2(x - 4)} + \frac{16}{(x + 4)(x - 4)}\right) = 2(x + 4)(x - 4)\left(\frac{1}{2}\right)$$

$$x(x + 4) + 2(16) = (x + 4)(x - 4)$$

$$x^2 + 4x + 32 = x^2 - 16$$

$$4x = -48$$

$$x = -12$$

The solution set is $\{-12\}$. Perhaps you should check it!   ∎

In Example 1, notice that the restrictions were not indicated until the denominators were expressed in factored form. It is usually easier to determine the necessary restrictions at this step.

EXAMPLE 2   Solve $\dfrac{3}{n-5} - \dfrac{2}{2n+1} = \dfrac{n+3}{2n^2 - 9n - 5}$.

Solution

$$\frac{3}{n-5} - \frac{2}{2n-1} = \frac{n+3}{2n^2 - 9n - 5}$$

$$\frac{3}{n-5} - \frac{2}{2n+1} = \frac{n+3}{(2n+1)(n-5)}, \qquad n \neq -\frac{1}{2} \text{ and } n \neq 5$$

$$(2n+1)(n-5)\left(\frac{3}{n-5} - \frac{2}{2n+1}\right) = (2n+1)(n-5)\left(\frac{n+3}{(2n+1)(n-5)}\right)$$

$$3(2n+1) - 2(n-5) = n + 3$$

$$6n + 3 - 2n + 10 = n + 3$$

$$4n + 13 = n + 3$$

$$3n = -10$$

$$n = -\frac{10}{3}$$

The solution set is $\left\{ -\dfrac{10}{3} \right\}$.   ∎

EXAMPLE 3   Solve $2 + \dfrac{4}{x-2} = \dfrac{8}{x^2 - 2x}$.

Solution

$$2 + \frac{4}{x-2} = \frac{8}{x^2 - 2x}$$

$$2 + \frac{4}{x-2} = \frac{8}{x(x-2)}, \qquad x \neq 0 \text{ and } x \neq 2$$

$$x(x-2)\left(2 + \frac{4}{x-2}\right) = x(x-2)\left(\frac{8}{x(x-2)}\right)$$

$$2x(x-2) + 4x = 8$$

$$2x^2 - 4x + 4x = 8$$

$$2x^2 = 8$$

$$x^2 = 4$$

$$x^2 - 4 = 0$$

$$(x + 2)(x - 2) = 0$$

$$x + 2 = 0 \qquad \text{or} \qquad x - 2 = 0$$

$$x = -2 \qquad \text{or} \qquad x = 2$$

Since our initial restriction indicated that $x \neq 2$, then the *only solution* is -2. Thus, the solution set is $\{-2\}$. ∎

In Section **4.1**, we discussed using the properties of equality to change the form of various formulas. For example, we considered the simple interest formula "$A = P + Prt$" and changed its form by solving for P as follows.

$$A = P + Prt$$

$$A = P(1 + rt)$$

$$\frac{A}{1 + rt} = P \qquad \text{Multiply both sides by } \frac{1}{1 + rt}$$

If the formula is in the form of a fractional equation, then the techniques of these last two sections are applicable. Consider the following example.

EXAMPLE 4   If the original cost of some business property is C dollars and it is depreciated linearly over N years, its value, V, at the end of T years is given by

$$V = C\left(1 - \frac{T}{N}\right).$$

Solve this formula for N in terms of V, C, and T.

Solution

$$V = C\left(1 - \frac{T}{N}\right)$$

$$V = C - \frac{CT}{N}$$

$$N(V) = N\left(C - \frac{CT}{N}\right) \qquad \text{Multiply both sides by } N.$$

$$NV = NC - CT$$

$$NV - NC = -CT$$

$$N(V - C) = -CT$$

$$N = \frac{-CT}{V - C}$$

$$N = -\frac{CT}{V - C}.$$

∎

Problem Solving

In Section 4.2 we solved some uniform motion problems. The formula $d = rt$ was used in the analysis of the problems and we used guidelines that involve distance relationships. Now let's consider some uniform motion problems for which guidelines that involve either times or rates are appropriate. These problems will generate fractional equations to solve.

PROBLEM 1   An airplane travels 2050 miles in the same time that a car travels 260 miles. If the rate of the plane is 358 miles per hour greater than the rate of the car, find the rate of each.

Solution   Let r represent the rate of the car. Then $r + 358$ represents the rate of the plane. The fact that the times are equal can be a guideline.

$$\underset{\underset{\dfrac{\text{distance of plane}}{\text{rate of plane}}}{\downarrow}}{\text{time of plane}} \quad \overset{\text{equals}}{=} \quad \underset{\underset{\dfrac{\text{distance of car}}{\text{rate of car}}}{\downarrow}}{\text{time of car}}$$

$$\frac{2050}{r + 358} = \frac{260}{r}$$

$$2050r = 260(r + 358)$$

$$2050r = 260r + 93{,}080$$

$$1790r = 93{,}080$$

$$r = 52$$

If $r = 52$, then $r + 358$ equals 410. Thus, the rate of the car is 52 miles per hour and the rate of the plane is 410 miles per hour. ∎

PROBLEM 2   It takes a freight train 2 hours longer to travel 300 miles than it takes an express train to travel 280 miles. The rate of the express train is 20 miles per hour greater than the rate of the freight train. Find the times and rates of both trains.

Solution   Let t represent the time of the express train. Then $t + 2$ represents the time of the freight train. Let's record the information of this problem in a table as follows.

	Distance	*Time*	$r = \dfrac{d}{t}$
Express train	280	t	$\dfrac{280}{t}$
Freight train	300	$t + 2$	$\dfrac{300}{t + 2}$

The fact that the rate of the express train is 20 miles per hour greater than the rate of the freight train can be a guideline.

rate of express   equals   rate of freight train plus 20

$$\frac{280}{t} \qquad = \qquad \frac{300}{t+2} + 20$$

$$t(t+2)\left(\frac{280}{t}\right) = t(t+2)\left(\frac{300}{t+2} + 20\right)$$

$$280(t+2) = 300t + 20t(t+2)$$

$$280t + 560 = 300t + 20t^2 + 40t$$

$$280t + 560 = 340t + 20t^2$$

$$0 = 20t^2 + 60t - 560$$

$$0 = t^2 + 3t - 28$$

$$0 = (t+7)(t-4)$$

$$t + 7 = 0 \qquad \text{or} \qquad t - 4 = 0$$

$$t = -7 \qquad \text{or} \qquad t = 4$$

The negative solution must be discarded, so the time of the express train (t) is 4 hours and the time of the freight train ($t + 2$) is 6 hours. The rate of the express train $\left(\frac{280}{t}\right)$ is $\frac{280}{4} = 70$ miles per hour and the rate of the freight train $\left(\frac{300}{t+2}\right)$ is $\frac{300}{6} = 50$ miles per hour. ∎

REMARK Note that to solve Problem 1 we went directly to a guideline without the use of a table, but for Problem 2 we used a table. Again, remember that this is a personal preference; we are merely exposing you to a variety of techniques.

Uniform motion problems are a special case of a larger group of problems we refer to as **rate-time** problems. For example, if a certain machine can produce 150 items in 10 minutes, then we say that the machine is producing at a rate of $\frac{150}{10} = 15$ items per minute. Likewise, if a person can do a certain job in 3 hours, then, assuming a constant rate of work we say that the person is working at a rate of $\frac{1}{3}$ of the job per hour. In general, if Q is the quantity of something done in t units of time, then the rate, r, is given by $r = \frac{Q}{t}$. We state the rate in terms of *so much quantity per unit of time*. (In uniform motion problems the "quantity" is distance.) Let's consider some examples of rate-time problems.

PROBLEM 3   If Jim can mow a lawn in 50 minutes and his son, Todd, can mow the same lawn in 40 minutes, how long will it take them to mow the lawn if they work together?

Solution   Jim's rate is $\dfrac{1}{50}$ of the lawn per minute and Todd's rate is $\dfrac{1}{40}$ of the lawn per minute.

If we let m represent the number of minutes that they work together, then $\dfrac{1}{m}$ represents their rate when working together. Therefore, since the sum of the individual rates must equal the rate when working together, we can set up and solve the following equation.

Jim's rate   Todd's rate   combined rate

$$\frac{1}{50} + \frac{1}{40} = \frac{1}{m}$$

$$200m\left(\frac{1}{50} + \frac{1}{40}\right) = 200m\left(\frac{1}{m}\right)$$

$$4m + 5m = 200$$

$$9m = 200$$

$$m = \frac{200}{9} = 22\frac{2}{9}$$

It should take them $22\dfrac{2}{9}$ minutes.   ■

PROBLEM 4   Working together, Linda and Kathy can type a term paper in $3\dfrac{3}{5}$ hours. Linda can type the paper by herself in 6 hours. How long would it take Kathy to type the paper by herself?

Solution   Their rate working together is $\dfrac{1}{3\dfrac{3}{5}} = \dfrac{1}{\dfrac{18}{5}} = \dfrac{5}{18}$ of the job per hour and Linda's rate

is $\dfrac{1}{6}$ of the job per hour. If we let h represent the number of hours that it would

take Kathy by herself, then her rate is $\dfrac{1}{h}$ of the job per hour. Thus, we have

Linda's rate   Kathy's rate   combined rate

$$\frac{1}{6} \quad + \quad \frac{1}{h} \quad = \quad \frac{5}{18}.$$

Solving this equation yields

$$18h\left(\frac{1}{6} + \frac{1}{h}\right) = 18h\left(\frac{5}{18}\right)$$

$$3h + 18 = 5h$$

$$18 = 2h$$

$$9 = h.$$

It would take Kathy 9 hours to type the paper by herself.   ■

One final example of this section illustrates another approach that some people find meaningful for rate-time problems. For this approach, think in terms of fractional parts of the job. For example, if a person can do a certain job in 5 hours, then at the end of 2 hours he or she has done $\frac{2}{5}$ of the job. (Again, assume a constant rate of work.) At the end of 4 hours, he or she has finished $\frac{4}{5}$ of the job and, in general, at the end of h hours, he or she has done $\frac{h}{5}$ of the job. Let's see how this works in a problem.

PROBLEM 5

It takes Pat 12 hours to complete a task. After he had been working for 3 hours, he was joined by his brother, Mike, and together they finished the task in 5 hours. How long would it take Mike to do the job by himself?

Solution

Let h represent the number of hours that it would take Mike by himself. Since Pat has been working for 3 hours, he has done $\frac{3}{12} = \frac{1}{4}$ of the job before Mike joins him.

Thus, there is $\frac{3}{4}$ of the original job to be done while working together. We can set up and solve the following equation.

fractional part of the remaining $\frac{3}{4}$ of the job that Pat does

fractional part of the remaining $\frac{3}{4}$ of the job that Mike does

$$\frac{5}{12} + \frac{5}{h} = \frac{3}{4}$$

$$12h\left(\frac{5}{12} + \frac{5}{h}\right) = 12h\left(\frac{3}{4}\right)$$

$$5h + 60 = 9h$$

$$60 = 4h$$

$$15 = h$$

It would take Mike 15 hours to do the entire job by himself. ∎

Problem Set 9.6

Solve each of the following equations.

1. $\dfrac{x}{4x - 4} + \dfrac{5}{x^2 - 1} = \dfrac{1}{4}$

2. $\dfrac{x}{3x - 6} + \dfrac{4}{x^2 - 4} = \dfrac{1}{3}$

3. $3 + \dfrac{6}{t - 3} = \dfrac{6}{t^2 - 3t}$

4. $2 + \dfrac{4}{t - 1} = \dfrac{4}{t^2 - t}$

5. $\dfrac{3}{n - 5} + \dfrac{4}{n + 7} = \dfrac{2n + 11}{n^2 + 2n - 35}$

6. $\dfrac{2}{n + 3} + \dfrac{3}{n - 4} = \dfrac{2n - 1}{n^2 - n - 12}$

7. $\dfrac{5x}{2x + 6} - \dfrac{4}{x^2 - 9} = \dfrac{5}{2}$

8. $\dfrac{3x}{5x + 5} - \dfrac{2}{x^2 - 1} = \dfrac{3}{5}$

9. $1 + \dfrac{1}{n - 1} = \dfrac{1}{n^2 - n}$

10. $3 + \dfrac{9}{n - 3} = \dfrac{27}{n^2 - 3n}$

11. $\dfrac{2}{n - 2} - \dfrac{n}{n + 5} = \dfrac{10n + 15}{n^2 + 3n - 10}$

12. $\dfrac{n}{n + 3} + \dfrac{1}{n - 4} = \dfrac{11 - n}{n^2 - n - 12}$

13. $\dfrac{2}{2x - 3} - \dfrac{2}{10x^2 - 13x - 3} = \dfrac{x}{5x + 1}$

14. $\dfrac{1}{3x + 4} + \dfrac{6}{6x^2 + 5x - 4} = \dfrac{x}{2x - 1}$

15. $\dfrac{2x}{x + 3} - \dfrac{3}{x - 6} = \dfrac{29}{x^2 - 3x - 18}$

16. $\dfrac{x}{x - 4} - \dfrac{2}{x + 8} = \dfrac{63}{x^2 + 4x - 32}$

17. $\dfrac{a}{a - 5} + \dfrac{2}{a - 6} = \dfrac{2}{a^2 - 11a + 30}$

18. $\dfrac{a}{a + 2} + \dfrac{3}{a + 4} = \dfrac{14}{a^2 + 6a + 8}$

19. $\dfrac{-1}{2x - 5} + \dfrac{2x - 4}{4x^2 - 25} = \dfrac{5}{6x + 15}$

20. $\dfrac{-2}{3x + 2} + \dfrac{x - 1}{9x^2 - 4} = \dfrac{3}{12x - 8}$

21. $\dfrac{7y + 2}{12y^2 + 11y - 15} - \dfrac{1}{3y + 5} = \dfrac{2}{4y - 3}$

22. $\dfrac{5y - 4}{6y^2 + y - 12} - \dfrac{2}{2y + 3} = \dfrac{5}{3y - 4}$

23. $\dfrac{2n}{6n^2 + 7n - 3} - \dfrac{n - 3}{3n^2 + 11n - 4} = \dfrac{5}{2n^2 + 11n + 12}$

24. $\dfrac{x + 1}{2x^2 + 7x - 4} - \dfrac{x}{2x^2 - 7x + 3} = \dfrac{1}{x^2 + x - 12}$

25. $\dfrac{1}{2x^2 - x - 1} + \dfrac{3}{2x^2 + x} = \dfrac{2}{x^2 - 1}$

26. $\dfrac{2}{n^2 + 4n} + \dfrac{3}{n^2 - 3n - 28} = \dfrac{5}{n^2 - 6n - 7}$

27. $\dfrac{x+1}{x^3-9x} - \dfrac{1}{2x^2+x-21} = \dfrac{1}{2x^2+13x+21}$

28. $\dfrac{x}{2x^2+5x} - \dfrac{x}{2x^2+7x+5} = \dfrac{2}{x^2+x}$

29. $\dfrac{4t}{4t^2-t-3} + \dfrac{2-3t}{3t^2-t-2} = \dfrac{1}{12t^2+17t+6}$

30. $\dfrac{2t}{2t^2+9t+10} + \dfrac{1-3t}{3t^2+4t-4} = \dfrac{4}{6t^2+11t-10}$

For Problems 31–44, solve each equation for the indicated variable.

31. $y = \dfrac{5}{6}x + \dfrac{2}{9}$ for x

32. $y = \dfrac{3}{4}x - \dfrac{2}{3}$ for x :

33. $\dfrac{-2}{x-4} = \dfrac{5}{y-1}$ for y

34. $\dfrac{7}{y-3} = \dfrac{3}{x+1}$ for y

35. $I = \dfrac{100M}{C}$ for M

36. $V = C\left(1 - \dfrac{T}{N}\right)$ for T

37. $\dfrac{R}{S} = \dfrac{T}{S+T}$ for R

38. $\dfrac{1}{R} = \dfrac{1}{S} + \dfrac{1}{T}$ for R

39. $\dfrac{y-1}{x-3} = \dfrac{b-1}{a-3}$ for y

40. $y = -\dfrac{a}{b}x + \dfrac{c}{d}$ for x

41. $\dfrac{x}{a} + \dfrac{y}{b} = 1$ for y

42. $\dfrac{y-b}{x} = m$ for y

43. $\dfrac{y-1}{x+6} = \dfrac{-2}{3}$ for y

44. $\dfrac{y+5}{x-2} = \dfrac{3}{7}$ for y

Set up an equation or a system of equations and solve each of the following problems.

45. Kent drives his Mazda 270 miles in the same time that Dave drives his Datsun 250 miles. If Kent averages 4 miles per hour faster than Dave, find their rates.

46. Suppose that Wendy rides her bicycle 30 miles in the same time that it takes Kim to ride her bicycle 20 miles. If Wendy rides 5 miles per hour faster than Kim, find the rate of each.

47. An inlet pipe can fill a tank in 10 minutes. A drain can empty the tank in 12 minutes. If the tank is empty and both the pipe and drain are open, how long will it take before the tank overflows?

48. Barry can do a certain job in 3 hours, while it takes Roy 5 hours to do the same job. How long would it take them to do the job working together?

49. Connie can type 600 words in 5 minutes less than it takes Katie to type 600 words. If Connie types at a rate of 20 words per minute faster than Katie types, find the typing rate of each woman.

50. Walt can mow a lawn in 1 hour, while his son, Mike, can mow the same lawn in 50 minutes. One day Mike started mowing the lawn by himself and worked for 30 minutes. Then Walt joined him and they finished the lawn. How long did it take them to finish mowing the lawn after Walt started to help?

51. Plane A can travel 1400 miles in one hour less time than it takes plane B to travel 2000 miles. The rate of plane B is 50 miles per hour faster than the rate of plane A. Find the times and rates of both planes.

52. To travel 60 miles, it takes Sue, riding a moped, 2 hours less time than it takes Ann to travel 50 miles riding a bicycle. Sue travels 10 miles per hour faster than Ann. Find the times and rates of both girls.

53. It takes Amy twice as long to deliver papers as it does Nancy. How long would it take each if they can deliver the papers together in 40 minutes?

54. If two inlet pipes are both open, they can fill a pool in 1 hour and 12 minutes. One of the pipes can fill the pool by itself in 2 hours. How long would it take the other pipe to fill the pool by itself?

55. Dan agreed to mow a vacant lot for $12. It took him an hour longer than what he had anticipated, so he earned $1 per hour less than he originally calculated. How long had he anticipated that it would take him to mow the lot?

56. Last week Al bought some golf balls for $20. The next day they were on sale for $0.50 per ball less and he bought $22.50 worth of balls. If he purchased 5 more balls on the second day than he did on the first day, how many did he buy each day and at what price per ball?

57. Debbie rode her bicycle out into the country for a distance of 24 miles. On the way back, she took a much shorter route of 12 miles and made the return trip in one-half hour less time. If her rate out into the country was 4 miles per hour faster than her rate on the return trip, find both rates.

58. Nick jogs for 10 miles and then walks another 10 miles. He jogs $2\frac{1}{2}$ miles per hour faster than he walks and the entire distance of 20 miles takes 6 hours. Find the rate that he walks and the rate that he jogs.

Chapter 9 Summary

(9.1) Any number that can be written in the form $\frac{a}{b}$, where a and b are integers and $b \neq 0$, is called a **rational number**.

A **rational expression** is defined as the indicated quotient of two polynomials. The following properties pertain to rational numbers and rational expressions.

1. $\dfrac{-a}{b} = \dfrac{a}{-b} = -\dfrac{a}{b}$

2. $\dfrac{-a}{-b} = \dfrac{a}{b}$

3. $\dfrac{a \cdot k}{b \cdot k} = \dfrac{a}{b}$ fundamental principle of fractions

(9.2) Multiplication and division of rational expressions are based upon the following.

$$1.\ \frac{a}{b} \cdot \frac{c}{d} = \frac{ac}{bd} \qquad \text{multiplication}$$

$$2.\ \frac{a}{b} \div \frac{c}{d} = \frac{a}{b} \cdot \frac{d}{c} = \frac{ad}{bc} \qquad \text{division}$$

(9.3) Addition and subtraction of rational expressions are based upon the following.

$$1.\ \frac{a}{b} + \frac{c}{b} = \frac{a+c}{b} \qquad \text{addition}$$

$$2.\ \frac{a}{b} - \frac{c}{b} = \frac{a-c}{b} \qquad \text{subtraction}$$

(9.4) The following basic procedure is used to add or subtract rational expressions.

1. Find the LCD of all denominators.
2. Change each fraction to an equivalent fraction that has the LCD as its denominator.
3. Add or subtract numerators and place this result over the LCD.
4. Look for possibilities to simplify the resulting fraction.

Fractional forms that contain rational numbers or rational expressions in the numerators and/or denominators are called **complex fractions**. The fundamental principle of fractions serves as a basis for simplifying complex fractions.

(9.5) To solve an equation that contains rational expressions, it is often helpful to begin by multiplying both sides of the equation by the LCD of all the denominators in the equation. If an equation contains a variable in one or more denominators, we must be careful to avoid any value of the variable that makes the denominator zero. We can treat some equations that contain rational expressions as proportions. These equations can be solved using the cross-multiplication property.

(9.6) Some formulas contain rational expressions and can be solved for a specific variable using the techniques of Section 9.5. Perhaps it would be helpful to look back over Example 4 of this section.

Chapter 9 Review Problem Set

For Problems 1–6, simplify each of the rational expressions.

1. $\dfrac{26x^2y^3}{39x^4y^2}$

2. $\dfrac{a^2 - 9}{a^2 + 3a}$

3. $\dfrac{n^2 - 3n - 10}{n^2 + n - 2}$

4. $\dfrac{x^4 - 1}{x^3 - x}$

5. $\dfrac{8x^3 - 2x^2 - 3x}{12x^2 - 9x}$

6. $\dfrac{x^4 - 7x^2 - 30}{2x^4 + 7x^2 + 3}$

For Problems 7–10, simplify each complex fraction.

7. $\dfrac{\dfrac{5}{8} - \dfrac{1}{2}}{\dfrac{1}{6} + \dfrac{3}{4}}$

8. $\dfrac{\dfrac{3}{2x} + \dfrac{5}{3y}}{\dfrac{4}{x} - \dfrac{3}{4y}}$

9. $\dfrac{\dfrac{3}{x - 2} - \dfrac{4}{x^2 - 4}}{\dfrac{2}{x + 2} + \dfrac{1}{x - 2}}$

10. $1 - \dfrac{1}{2 - \dfrac{1}{x}}$

For Problems 11–20 perform the indicated operations and express answers in simplest form.

11. $\dfrac{6xy^2}{7y^3} \div \dfrac{15x^2y}{5x^2}$

12. $\dfrac{9ab}{3a + 6} \cdot \dfrac{a^2 - 4a - 12}{a^2 - 6a}$

13. $\dfrac{n^2 + 10n + 25}{n^2 - n} \cdot \dfrac{5n^3 - 3n^2}{5n^2 + 22n - 15}$

14. $\dfrac{x^2 - 2xy - 3y^2}{x^2 + 9y^2} \div \dfrac{2x^2 + xy - y^2}{2x^2 - xy}$

15. $\dfrac{2x + 1}{5} + \dfrac{3x - 2}{4}$

16. $\dfrac{3}{2n} + \dfrac{5}{3n} - \dfrac{1}{9}$

17. $\dfrac{3x}{x + 7} - \dfrac{2}{x}$

18. $\dfrac{10}{x^2 - 5x} + \dfrac{2}{x}$

19. $\dfrac{3}{n^2 - 5n - 36} + \dfrac{2}{n^2 + 3n - 4}$

20. $\dfrac{3}{2y + 3} + \dfrac{5y - 2}{2y^2 - 9y - 18} - \dfrac{1}{y - 6}$

For Problems 21–30, solve each equation.

21. $\dfrac{4x + 5}{3} + \dfrac{2x - 1}{5} = 2$

22. $\dfrac{3}{4x} + \dfrac{4}{5} = \dfrac{9}{10x}$

23. $\dfrac{a}{a - 2} - \dfrac{3}{2} = \dfrac{2}{a - 2}$

24. $\dfrac{4}{5y - 3} = \dfrac{2}{3y + 7}$

25. $n + \dfrac{1}{n} = \dfrac{53}{14}$

26. $\dfrac{1}{2x - 7} + \dfrac{x - 5}{4x^2 - 49} = \dfrac{4}{6x - 21}$

27. $\dfrac{x}{2x + 1} - 1 = \dfrac{-4}{7(x - 2)}$

28. $\dfrac{2x}{-5} = \dfrac{3}{4x - 13}$

29. $\dfrac{2n}{2n^2 + 11n - 21} - \dfrac{n}{n^2 + 5n - 14} = \dfrac{3}{n^2 + 5n - 14}$

30. $\dfrac{2}{t^2 - t - 6} + \dfrac{t + 1}{t^2 + t - 12} = \dfrac{t}{t^2 + 6t + 8}$

31. Solve $\dfrac{y-6}{x+1} = \dfrac{3}{4}$ for y.

32. Solve $\dfrac{x}{a} - \dfrac{y}{b} = 1$ for y.

For Problems 33–38, set up an equation or a system of equations and solve each problem.

33. A sum of $1400 is to be divided between two people in the ratio of $\dfrac{3}{5}$. How much does each person receive?

34. Working together, Dan and Don can mow a lawn in 12 minutes. Don can mow the lawn by himself in 10 minutes less time than it takes Dan by himself. How long does it take each of them to mow the lawn alone?

35. Suppose that car A can travel 250 miles in 3 hours less time than it takes car B to travel 440 miles. The rate of car B is 5 miles per hour faster than car A. Find the rates of both cars.

36. Mark can overhaul an engine in 20 hours and Phil can do the same job by himself in 30 hours. If they both work together for a time and then Mark finishes the job by himself in 5 hours, how long did they work together?

37. Kelly contracted to paint a house for $640. It took him 20 hours longer than he had anticipated, so he earned $1.60 per hour less than he had calculated. How long had he anticipated that it would take him to paint the house?

38. Kent rode his bicycle 66 miles in $4\dfrac{1}{2}$ hours. For the first 40 miles he averaged a certain rate and then for the last 26 miles he reduced his rate by 3 miles per hour. Find his rate for the last 26 miles.

Chapter 10

Quadratic Equations and Inequalities

Solving equations is one of the central themes of this text. Let's pause for a moment and reflect back upon the different types of equations that we have solved.

Type of equation	*Examples*
First-degree equations in one variable	$3x + 2x = x - 4$; $5(x + 4) = 12$; $\dfrac{x + 2}{3} + \dfrac{x - 1}{4} = 2$.
Second-degree equations in one variable *that are factorable*	$x^2 + 5x = 0$; $x^2 + 5x + 6 = 0$; $x^2 - 9 = 0$; $x^2 - 10x + 25 = 0$.
Equations containing rational expressions	$\dfrac{2}{x} + \dfrac{3}{x} = 4$; $\dfrac{5}{a - 1} = \dfrac{6}{a - 2}$; $\dfrac{2}{x^2 - 9} + \dfrac{3}{x + 3} = \dfrac{4}{x - 3}$.
Radical equations	$\sqrt{x} = 2$; $\sqrt{3x - 2} = 5$; $\sqrt{5y + 1} = \sqrt{3y + 4}$.

As the above chart shows, we have solved second-degree equations in one variable, but only those that are factorable. In this chapter we will expand our work to include more general types of second-degree equations as well as inequalities in one variable.

410

10.1
Complex Numbers

Since the square of any real number is nonnegative, a simple equation such as $x^2 = -4$ has no solutions in the set of real numbers. However, we can handle this situation by expanding the set of real numbers into a larger set called the **complex numbers**. This section is devoted to introducing and learning to manipulate complex numbers.

To provide a solution for the equation $x^2 + 1 = 0$, we use the number i, such that

$$i^2 = -1.$$

The number i is not a real number and is often called the **imaginary unit**, but the number i^2 is the real number -1. The imaginary unit i is used to define a complex number as follows.

DEFINITION 10.1

> A **complex number** is any number that can be expressed in the form
>
> $a + bi$,
>
> where a and b are real numbers.

The form $a + bi$ is called the **standard form** of a complex number. The real number a is called the **real part** of the complex number and b is called the **imaginary part**. (Note that b is a real number even though it is called the imaginary part.) The following exemplify this terminology.

1. The number $7 + 5i$ is a complex number that has a real part of 7 and an imaginary part of 5.

2. The number $\frac{2}{3} + i\sqrt{2}$ is a complex number that has a real part of $\frac{2}{3}$ and an imaginary part of $\sqrt{2}$. (It is easy to mistake $\sqrt{2i}$ for $\sqrt{2}\,i$. Thus, it is common to write $i\sqrt{2}$ instead of $\sqrt{2}\,i$ to avoid any difficulties with the radical sign.)

3. The number $-4 - 3i$ can be written in the standard form $-4 + (-3i)$ and therefore is a complex number that has a real part of -4 and an imaginary part of -3. (The form $-4 - 3i$ is often used knowing that it means $-4 + (-3i)$.)

4. The number $-9i$ can be written as $0 + (-9i)$; thus, it is a complex number that has a real part of 0 and an imaginary part of -9. (Complex numbers, such as $-9i$, for which $a = 0$ and $b \neq 0$ are called **pure imaginary numbers**.)

5. The real number 4 can be written as $4 + 0i$ and is thus a complex number that has a real part of 4 and an imaginary part of 0.

From number 5 we see that the set of real numbers is a subset of the set of complex numbers. The following diagram indicates the organizational format of the complex numbers.

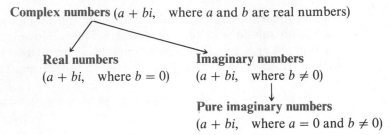

Complex numbers ($a + bi$,   where a and b are real numbers)

Real numbers   **Imaginary numbers**
($a + bi$,   where $b = 0$)   ($a + bi$,   where $b \neq 0$)

Pure imaginary numbers
($a + bi$,   where $a = 0$ and $b \neq 0$)

Two complex numbers $a + bi$ and $c + di$ are said to be **equal** if and only if $a = c$ and $b = d$.

Adding and Subtracting Complex Numbers

To **add complex numbers**, we simply add their real parts and add their imaginary parts. Thus,

$$(a + bi) + (c + di) = (a + c) + (b + d)i.$$

The following examples show addition of two complex numbers.

1. $(4 + 3i) + (5 + 9i) = (4 + 5) + (3 + 9)i = 9 + 12i.$

2. $(-6 + 4i) + (8 - 7i) = (-6 + 8) + (4 - 7)i$
$$= 2 - 3i.$$

3. $\left(\dfrac{1}{2} + \dfrac{3}{4}i\right) + \left(\dfrac{2}{3} + \dfrac{1}{5}i\right) = \left(\dfrac{1}{2} + \dfrac{2}{3}\right) + \left(\dfrac{3}{4} + \dfrac{1}{5}\right)i$
$$= \left(\dfrac{3}{6} + \dfrac{4}{6}\right) + \left(\dfrac{15}{20} + \dfrac{4}{20}\right)i$$
$$= \dfrac{7}{6} + \dfrac{19}{20}i.$$

The set of complex numbers is closed with respect to addition; that is, the sum of two complex numbers is a complex number. Furthermore, the commutative and associative properties of addition hold for all complex numbers. The addition identity element is $0 + 0i$ (or simply the real number 0). The additive inverse of $a + bi$ is $-a - bi$, because

$$(a + bi) + (-a - bi) = 0.$$

To **subtract complex numbers**, $c + di$ from $a + bi$, add the additive inverse of $c + di$. Thus,

$$(a + bi) - (c + di) = (a + bi) + (-c - di)$$
$$= (a - c) + (b - d)i.$$

In other words, we subtract the real parts and subtract the imaginary parts, as in the next examples.

1. $(9 + 8i) - (5 + 3i) = (9 - 5) + (8 - 3)i$
$$= 4 + 5i.$$

2. $(3 - 2i) - (4 - 10i) = (3 - 4) + (-2 - (-10))i$
$$= -1 + 8i.$$

Products and Quotients of Complex Numbers

Since $i^2 = -1$, i is a square root of negative one; so we let $i = \sqrt{-1}$. It should also be evident that $-i$ is a square root of negative one since

$$(-i)^2 = (-i)(-i) = i^2 = -1.$$

Thus, in the set of complex numbers, -1 has two square roots, i and $-i$. We express these symbolically as

$$\sqrt{-1} = i \qquad \text{and} \qquad -\sqrt{-1} = -i.$$

Let us extend our definition so that in the set of complex numbers every negative real number has two square roots. We simply define $\sqrt{-b}$, where b is a positive real number, to be the number whose square is $-b$. Thus,

$$(\sqrt{-b})^2 = -b, \quad \text{for } b > 0.$$

Furthermore, since $(i\sqrt{b})(i\sqrt{b}) = i^2(b) = -1(b) = -b$ we see that

$$\sqrt{-b} = i\sqrt{b}.$$

In other words, a square root of any negative real number can be represented as the product of a real number and the imaginary unit i. Consider the following examples.

$$\sqrt{-4} = i\sqrt{4} = 2i,$$

$$\sqrt{-17} = i\sqrt{17},$$

$$\sqrt{-24} = i\sqrt{24} = i\sqrt{4}\sqrt{6} = 2i\sqrt{6} \qquad \text{Notice that we simplified the radical } \sqrt{24} \text{ to } 2\sqrt{6}.$$

We should also observe that $-\sqrt{-b}$, where $b > 0$, is a square root of $-b$ since

$$(-\sqrt{-b})^2 = (-i\sqrt{b})^2 = i^2(b) = -1(b) = -b.$$

Thus, in the set of complex numbers, $-b$ (where $b > 0$) has two square roots, $i\sqrt{b}$ and $-i\sqrt{b}$. We express these symbolically as

$$\sqrt{-b} = i\sqrt{b} \qquad \text{and} \qquad -\sqrt{-b} = -i\sqrt{b}.$$

We must be very careful with the use of the symbol $\sqrt{-b}$, where $b > 0$. Some relationships that are true in the set of real numbers that involve the square root symbol do not hold if the square root symbol does not represent a real number. For example, $\sqrt{a}\sqrt{b} = \sqrt{ab}$ *does not hold* if a and b are both negative numbers.

Correct $\sqrt{-4}\sqrt{-9} = (2i)(3i) = 6i^2 = 6(-1) = -6.$

Incorrect $\sqrt{-4}\sqrt{-9} = \sqrt{(-4)(-9)} = \sqrt{36} = 6.$

To avoid difficulty with this idea, you should rewrite all expressions of the form $\sqrt{-b}$, where $b > 0$, in the form $i\sqrt{b}$ before doing *any* computations. The following examples further demonstrate this point.

1. $\sqrt{-5}\sqrt{-7} = (i\sqrt{5})(i\sqrt{7}) = i^2\sqrt{35} = (-1)\sqrt{35} = -\sqrt{35}.$

2. $\sqrt{-2}\sqrt{-8} = (i\sqrt{2})(i\sqrt{8}) = i^2\sqrt{16} = (-1)(4) = -4.$

3. $\sqrt{-6}\sqrt{-8} = (i\sqrt{6})(i\sqrt{8}) = i^2\sqrt{48} = (-1)\sqrt{16}\sqrt{3} = -4\sqrt{3}.$

4. $\dfrac{\sqrt{-75}}{\sqrt{-3}} = \dfrac{i\sqrt{75}}{i\sqrt{3}} = \dfrac{\sqrt{75}}{\sqrt{3}} = \sqrt{\dfrac{75}{3}} = \sqrt{25} = 5.$

5. $\dfrac{\sqrt{-48}}{\sqrt{12}} = \dfrac{i\sqrt{48}}{\sqrt{12}} = i\sqrt{\dfrac{48}{12}} = i\sqrt{4} = 2i.$

Since complex numbers have a *binomial form*, we find the *product* of two complex numbers in the same way that we find the product of two binomials. Then by replacing i^2 with -1, we are able to simplify and express the final result in standard form. Consider the following examples.

6. $(2 + 3i)(4 + 5i) = 2(4 + 5i) + 3i(4 + 5i)$
$$= 8 + 10i + 12i + 15i^2$$
$$= 8 + 22i + 15i^2$$
$$= 8 + 22i + 15(-1) = -7 + 22i.$$

7. $(-3 + 6i)(2 - 4i) = -3(2 - 4i) + 6i(2 - 4i)$
$$= -6 + 12i + 12i - 24i^2$$
$$= -6 + 24i - 24(-1)$$
$$= -6 + 24i + 24 = 18 + 24i.$$

8. $(1 - 7i)^2 = (1 - 7i)(1 - 7i)$
$$= 1(1 - 7i) - 7i(1 - 7i)$$
$$= 1 - 7i - 7i + 49i^2$$
$$= 1 - 14i + 49(-1)$$
$$= 1 - 14i - 49$$
$$= -48 - 14i.$$

9. $(2 + 3i)(2 - 3i) = 2(2 - 3i) + 3i(2 - 3i)$
$$= 4 - 6i + 6i - 9i^2$$
$$= 4 - 9(-1)$$
$$= 4 + 9$$
$$= 13.$$

Example 9 illustrates an important situation. The complex numbers $2 + 3i$ and $2 - 3i$ are called conjugates of each other; in general, the two complex numbers $a + bi$ and $a - bi$ are called **conjugates** of each other. *The product of a complex number and its conjugate is always a real number.* This can be shown as follows.

$$
\begin{aligned}
(a + bi)(a - bi) &= a(a - bi) + bi(a - bi) \\
&= a^2 - abi + abi - b^2 i^2 \\
&= a^2 - b^2(-1) \\
&= a^2 + b^2
\end{aligned}
$$

We use conjugates to *simplify expressions* such as $\dfrac{3i}{5 + 2i}$ that *indicate the quotient* of two complex numbers. To eliminate i in the denominator and change the indicated quotient to the standard form of a complex number, we can multiply both the numerator and the denominator by the conjugate of the denominator as follows.

$$
\begin{aligned}
\frac{3i}{5 + 2i} &= \frac{3i(5 - 2i)}{(5 + 2i)(5 - 2i)} \\
&= \frac{15i - 6i^2}{25 - 4i^2} \\
&= \frac{15i - 6(-1)}{25 - 4(-1)} \\
&= \frac{15i + 6}{29} \\
&= \frac{6}{29} + \frac{15}{29}i
\end{aligned}
$$

The following examples further clarify the process of *dividing* complex numbers.

10. $\dfrac{2 - 3i}{4 - 7i} = \dfrac{(2 - 3i)(4 + 7i)}{(4 - 7i)(4 + 7i)}$ $4 + 7i$ is the conjugate of $4 - 7i$.

$$
\begin{aligned}
&= \frac{8 + 14i - 12i - 21i^2}{16 - 49i^2} \\
&= \frac{8 + 2i - 21(-1)}{16 - 49(-1)} \\
&= \frac{8 + 2i + 21}{16 + 49} \\
&= \frac{29 + 2i}{65} \\
&= \frac{29}{65} + \frac{2}{65}i.
\end{aligned}
$$

11. $\dfrac{4 - 5i}{2i} = \dfrac{(4 - 5i)(-2i)}{(2i)(-2i)}$ $-2i$ is the conjugate of $2i$.

$= \dfrac{-8i + 10i^2}{-4i^2}$

$= \dfrac{-8i + 10(-1)}{-4(-1)}$

$= \dfrac{-8i - 10}{4}$

$= -\dfrac{5}{2} - 2i.$

For a problem such as number 11 in which the denominator is a pure imaginary number, we can change to standard form by choosing a multiplier other than the conjugate. Consider the following alternate approach for number 11.

$\dfrac{4 - 5i}{2i} = \dfrac{(4 - 5i)(i)}{(2i)(i)}$

$= \dfrac{4i - 5i^2}{2i^2}$

$= \dfrac{4i - 5(-1)}{2(-1)}$

$= \dfrac{4i + 5}{-2}$

$= -\dfrac{5}{2} - 2i$

Problem Set 10.1

Label each of the following statements true or false.

1. Every complex number is a real number.
2. Every real number is a complex number.
3. The real part of the complex number $6i$ is 0.
4. Every complex number is a pure imaginary number.
5. The sum of two complex numbers is always a complex number.
6. The imaginary part of the complex number 7 is 0.
7. The sum of two complex numbers is sometimes a real number.
8. The sum of two pure imaginary numbers is always a pure imaginary number.

Add or subtract as indicated.

9. $(6 + 3i) + (4 + 5i)$ 10. $(5 + 2i) + (7 + 10i)$

11. $(-8 + 4i) + (2 + 6i)$ 12. $(5 - 8i) + (-7 + 2i)$

13. $(3 + 2i) - (5 + 7i)$

14. $(1 + 3i) - (4 + 9i)$

15. $(-7 + 3i) - (5 - 2i)$

16. $(-8 + 4i) - (9 - 4i)$

17. $(-3 - 10i) + (2 - 13i)$

18. $(-4 - 12i) + (-3 + 16i)$

19. $(4 - 8i) - (8 - 3i)$

20. $(12 - 9i) - (14 - 6i)$

21. $(-1 - i) - (-2 - 4i)$

22. $(-2 - 3i) - (-4 - 14i)$

23. $\left(\frac{3}{2} + \frac{1}{3}i\right) + \left(\frac{1}{6} - \frac{3}{4}i\right)$

24. $\left(\frac{2}{3} - \frac{1}{5}i\right) + \left(\frac{3}{5} - \frac{3}{4}i\right)$

25. $\left(-\frac{5}{9} + \frac{3}{5}i\right) - \left(\frac{4}{3} - \frac{1}{6}i\right)$

26. $\left(\frac{3}{8} - \frac{5}{2}i\right) - \left(\frac{5}{6} + \frac{1}{7}i\right)$

Write each of the following in terms of i and simplify. For example, $\sqrt{-20} = i\sqrt{20} = i\sqrt{4}\sqrt{5} = 2i\sqrt{5}$.

27. $\sqrt{-81}$

28. $\sqrt{-49}$

29. $\sqrt{-14}$

30. $\sqrt{-33}$

31. $\sqrt{-\frac{16}{25}}$

32. $\sqrt{-\frac{64}{36}}$

33. $\sqrt{-18}$

34. $\sqrt{-84}$

35. $\sqrt{-75}$

36. $\sqrt{-63}$

37. $3\sqrt{-28}$

38. $5\sqrt{-72}$

39. $-2\sqrt{-80}$

40. $-6\sqrt{-27}$

41. $12\sqrt{-90}$

42. $9\sqrt{-40}$

Write each of the following in terms of i, perform the indicated operations, and simplify. For example,

$$\sqrt{-3}\sqrt{-8} = (i\sqrt{3})(i\sqrt{8})$$
$$= i^2\sqrt{24}$$
$$= (-1)\sqrt{4}\sqrt{6}$$
$$= -2\sqrt{6}.$$

43. $\sqrt{-4}\sqrt{-16}$

44. $\sqrt{-81}\sqrt{-25}$

45. $\sqrt{-3}\sqrt{-5}$

46. $\sqrt{-7}\sqrt{-10}$

47. $\sqrt{-9}\sqrt{-6}$

48. $\sqrt{-8}\sqrt{-16}$

49. $\sqrt{-15}\sqrt{-5}$

50. $\sqrt{-2}\sqrt{-20}$

51. $\sqrt{-2}\sqrt{-27}$

52. $\sqrt{-3}\sqrt{-15}$

53. $\sqrt{6}\sqrt{-8}$

54. $\sqrt{-75}\sqrt{3}$

55. $\frac{\sqrt{-25}}{\sqrt{-4}}$

56. $\frac{\sqrt{-81}}{\sqrt{-9}}$

57. $\frac{\sqrt{-56}}{\sqrt{-7}}$

58. $\frac{\sqrt{-72}}{\sqrt{-6}}$

59. $\frac{\sqrt{-24}}{\sqrt{6}}$

60. $\frac{\sqrt{-96}}{\sqrt{2}}$

Find each of the following products and express answers in the standard form of a complex number.

61. $(5i)(4i)$

62. $(-6i)(9i)$

63. $(7i)(-6i)$

64. $(-5i)(-12i)$

65. $(3i)(2 - 5i)$

66. $(7i)(-9 + 3i)$

67. $(-6i)(-2 - 7i)$

68. $(-9i)(-4 - 5i)$

69. $(3 + 2i)(5 + 4i)$

70. $(4 + 3i)(6 + i)$ **71.** $(6 - 2i)(7 - i)$ **72.** $(8 - 4i)(7 - 2i)$

73. $(-3 - 2i)(5 + 6i)$ **74.** $(-5 - 3i)(2 - 4i)$ **75.** $(9 + 6i)(-1 - i)$

76. $(10 + 2i)(-2 - i)$ **77.** $(4 + 5i)^2$ **78.** $(5 - 3i)^2$

79. $(-2 - 4i)^2$ **80.** $(-3 - 6i)^2$ **81.** $(6 + 7i)(6 - 7i)$

82. $(5 - 7i)(5 + 7i)$ **83.** $(-1 + 2i)(-1 - 2i)$ **84.** $(-2 - 4i)(-2 + 4i)$

Find each of the following quotients and express answers in the standard form of a complex number.

85. $\dfrac{3i}{2 + 4i}$ **86.** $\dfrac{4i}{5 + 2i}$ **87.** $\dfrac{-2i}{3 - 5i}$ **88.** $\dfrac{-5i}{2 - 4i}$

89. $\dfrac{-2 + 6i}{3i}$ **90.** $\dfrac{-4 - 7i}{6i}$ **91.** $\dfrac{2}{7i}$ **92.** $\dfrac{3}{10i}$

93. $\dfrac{2 + 6i}{1 + 7i}$ **94.** $\dfrac{5 + i}{2 + 9i}$ **95.** $\dfrac{3 + 6i}{4 - 5i}$ **96.** $\dfrac{7 - 3i}{4 - 3i}$

97. $\dfrac{-2 + 7i}{-1 + i}$ **98.** $\dfrac{-3 + 8i}{-2 + i}$ **99.** $\dfrac{-1 - 3i}{-2 - 10i}$ **100.** $\dfrac{-3 - 4i}{-4 - 11i}$

Thoughts into Words

101. Is every real number also a complex number? Explain your answer.

102. Can the product of two nonreal complex numbers be a real number? Explain your answer.

10.2
Quadratic Equations

A second-degree equation in one variable contains the variable with an exponent of two, but no higher power. Such equations are also called **quadratic equations**. The following are examples of quadratic equations.

$$x^2 = 36, \qquad y^2 + 4y = 0, \qquad x^2 + 5x - 2 = 0,$$

$$3n^2 + 2n - 1 = 0, \qquad 5x^2 + x + 2 = 3x^2 - 2x - 1$$

A quadratic equation in the variable x can also be defined as any equation that can be written in the form

$$ax^2 + bx + c = 0,$$

where a, b, and c are real numbers and $a \neq 0$. The form $ax^2 + bx + c = 0$ is called the **standard form** of a quadratic equation.

In previous chapters you solved quadratic equations (the term *quadratic* was not used at that time) by factoring and applying the property, $ab = 0$ if and only if $a = 0$ or $b = 0$. Let's review a few such examples.

EXAMPLE 1   Solve $3n^2 + 14n - 5 = 0$.

Solution

$$3n^2 + 14n - 5 = 0$$

$$(3n - 1)(n + 5) = 0 \qquad \text{Factor the left side.}$$

$$3n - 1 = 0 \quad \text{or} \quad n + 5 = 0 \qquad \begin{array}{l}\text{Apply: } ab = 0 \text{ if and} \\ \text{only if } a = 0 \text{ or } b = 0.\end{array}$$

$$3n = 1 \quad \text{or} \quad n = -5$$

$$n = \frac{1}{3} \quad \text{or} \quad n = -5$$

The solution set is $\left\{-5, \dfrac{1}{3}\right\}$.   ■

EXAMPLE 2   Solve $x^2 + 3kx - 10k^2 = 0$ for x.

Solution

$$x^2 + 3kx - 10k^2 = 0$$

$$(x + 5k)(x - 2k) = 0 \qquad \text{Factor the left side.}$$

$$x + 5k = 0 \quad \text{or} \quad x - 2k = 0 \qquad \begin{array}{l}\text{Apply: } ab = 0 \text{ if and} \\ \text{only if } a = 0 \text{ or } b = 0.\end{array}$$

$$x = -5k \quad \text{or} \quad x = 2k$$

The solution set is $\{-5k, 2k\}$.   ■

EXAMPLE 3   Solve $2\sqrt{x} = x - 8$.

Solution

$$2\sqrt{x} = x - 8$$

$$(2\sqrt{x})^2 = (x - 8)^2 \qquad \text{Square both sides.}$$

$$4x = x^2 - 16x + 64$$

$$0 = x^2 - 20x + 64$$

$$0 = (x - 16)(x - 4) \qquad \text{Factor the right side.}$$

$$x - 16 = 0 \quad \text{or} \quad x - 4 = 0 \qquad \begin{array}{l}\text{Apply: } ab = 0 \text{ if and} \\ \text{only if } a = 0 \text{ or } b = 0.\end{array}$$

$$x = 16 \quad \text{or} \quad x = 4$$

$$\textbf{CHECK} \quad 2\sqrt{x} = x - 8 \qquad\qquad 2\sqrt{x} = x - 8$$

$$2\sqrt{16} \overset{?}{=} 16 - 8 \quad \text{or} \quad 2\sqrt{4} \overset{?}{=} 4 - 8$$

$$2(4) \overset{?}{=} 8 \qquad\qquad 2(2) \overset{?}{=} -4$$

$$8 = 8 \qquad\qquad\quad 4 \neq -4$$

The solution set is $\{16\}$.   ■

We should make two comments about Example 3. First, remember that applying the property, if $a = b$, then $a^n = b^n$, might produce extraneous solutions. Therefore, we *must* check all potential solutions. Secondly, the equation $2\sqrt{x} = x - 8$ is said to be of **quadratic form** because it can be written as $2x^{\frac{1}{2}} = \left(x^{\frac{1}{2}}\right)^2 - 8$. More will be said about the phrase, quadratic form, later.

Let's consider quadratic equations of the form $x^2 = a$, where x is the variable and a is any real number. We can solve $x^2 = a$ as follows.

$$x^2 = a$$

$$x^2 - a = 0$$

$$x^2 - (\sqrt{a})^2 = 0 \qquad\qquad a = (\sqrt{a})^2$$

$$(x - \sqrt{a})(x + \sqrt{a}) = 0 \qquad\qquad \text{Factor the left side.}$$

$$x - \sqrt{a} = 0 \quad \text{or} \quad x + \sqrt{a} = 0 \qquad \begin{array}{l}\text{Apply: } ab = 0 \text{ if and}\\ \text{only if } a = 0 \text{ or } b = 0.\end{array}$$

$$x = \sqrt{a} \quad \text{or} \quad x = -\sqrt{a}.$$

The solutions are $\sqrt{a}$ and $-\sqrt{a}$.

We can state the previous result as a general property and use it to solve certain types of quadratic equations.

PROPERTY 10.1

> For any real number a,
> $$x^2 = a \quad \text{if and only if} \quad x = \sqrt{a} \text{ or } x = -\sqrt{a}.$$
> (The statement $x = \sqrt{a}$ or $x = -\sqrt{a}$ can be written as $x = \pm\sqrt{a}$.)

Property 10.1 along with our knowledge of square roots, makes it very easy to solve quadratic equations of the form $x^2 = a$.

EXAMPLE 4   Solve $x^2 = 45$.

Solution

$$x^2 = 45$$

$$x = \pm\sqrt{45}$$

$$x = \pm 3\sqrt{5} \qquad \sqrt{45} = \sqrt{9}\sqrt{5} = 3\sqrt{5}$$

The solution set is $\{\pm 3\sqrt{5}\}$ ■

EXAMPLE 5   Solve $x^2 = -9$.

Solution

$$x^2 = -9$$

$$x = \pm\sqrt{-9}$$

$$x = \pm 3i. \qquad \sqrt{-9} = i\sqrt{9} = 3i$$

Thus, the solution set is $\{\pm 3i\}$. ∎

EXAMPLE 6   Solve $7n^2 = 12$.

Solution

$$7n^2 = 12$$

$$n^2 = \frac{12}{7}$$

$$n = \pm\sqrt{\frac{12}{7}}$$

$$n = \pm\frac{2\sqrt{21}}{7}. \qquad \sqrt{\frac{12}{7}} = \frac{\sqrt{12}}{\sqrt{7}} \cdot \frac{\sqrt{7}}{\sqrt{7}} = \frac{\sqrt{84}}{7} = \frac{\sqrt{4}\sqrt{21}}{7} = \frac{2\sqrt{21}}{7}$$

The solution set is $\left\{\pm\dfrac{2\sqrt{21}}{7}\right\}$. ∎

EXAMPLE 7   Solve $(3n + 1)^2 = 25$.

Solution

$$(3n + 1)^2 = 25$$

$$(3n + 1) = \pm\sqrt{25}$$

$$3n + 1 = \pm 5$$

$$3n + 1 = 5 \qquad \text{or} \qquad 3n + 1 = -5$$

$$3n = 4 \qquad \text{or} \qquad 3n = -6$$

$$n = \frac{4}{3} \qquad \text{or} \qquad n = -2$$

The solution set is $\left\{-2, \dfrac{4}{3}\right\}$. ∎

EXAMPLE 8   Solve $(x - 3)^2 = -10$.

Solution

$$(x - 3)^2 = -10$$

$$x - 3 = \pm\sqrt{-10}$$

$$x - 3 = \pm i\sqrt{10}$$

$$x = 3 \pm i\sqrt{10}$$

Thus, the solution set is $\{3 \pm i\sqrt{10}\}$. ∎

Sometimes it may be necessary to change the form before we can apply Property 10.1. Let's consider one example to illustrate this idea.

EXAMPLE 9   Solve $3(2x - 3)^2 + 8 = 44$.

Solution

$$3(2x - 3)^2 + 8 = 44$$
$$3(2x - 3)^2 = 36$$
$$(2x - 3)^2 = 12$$
$$2x - 3 = \pm\sqrt{12}$$
$$2x - 3 = \pm 2\sqrt{3}$$
$$2x = 3 \pm 2\sqrt{3}$$
$$x = \frac{3 \pm 2\sqrt{3}}{2}$$

The solution set is $\left\{\dfrac{3 \pm 2\sqrt{3}}{2}\right\}$. ■

Back to the Pythagorean Theorem

Our work with radicals, Property 10.1, and the Pythagorean Theorem form a basis for solving a variety of problems that pertain to right triangles.

PROBLEM 1   Find c in the accompanying figure.

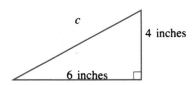

Solution   Apply the Pythagorean Theorem and we obtain

$$c^2 = 6^2 + 4^2$$
$$= 36 + 16$$
$$= 52.$$

Therefore,

$$c = \pm\sqrt{52} = \pm 2\sqrt{13}.$$

Disregard the negative solution and we find that $c = 2\sqrt{13}$ inches or approximately 7.2 inches. ■

There are two special kinds of right triangles that we use extensively in later mathematics courses. An **isosceles right triangle** is a right triangle that has both legs of the same length. Let's consider a problem that involves an isosceles right triangle.

PROBLEM 2   Find the length of each leg of an isosceles right triangle that has a hypotenuse of length 5 meters.

Solution   Let's sketch an isosceles right triangle and let x represent the length of each leg. Then we can apply the Pythagorean Theorem as follows.

$$x^2 + x^2 = 5^2$$
$$2x^2 = 25$$
$$x^2 = \frac{25}{2}$$
$$x = \pm\sqrt{\frac{25}{2}} = \pm\frac{5}{\sqrt{2}} = \pm\frac{5\sqrt{2}}{2}$$

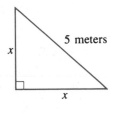

Each leg is $\dfrac{5\sqrt{2}}{2}$ meters or approximately 3.5 meters long.   ∎

The other special kind of right triangle that we use frequently is one that contains acute angles of 30° and 60°. In such a right triangle, which we refer to as a **30°–60° right triangle**, *the side opposite the 30° angle is equal in length to one-half of the length of the hypotenuse*. This relationship, along with the Pythagorean Theorem, provides us with another problem solving technique.

PROBLEM 3   Suppose that the side opposite the 60° angle in a 30°–60° right triangle is 9 feet long. Find the length of the other leg and the length of the hypotenuse.

Solution   Let x represent the length of the side opposite the 30° angle. Then $2x$ represents the length of the hypotenuse, as indicated in the accompanying diagram. Thus, we can apply the Pythagorean Theorem as follows.

$$x^2 + 9^2 = (2x)^2$$
$$x^2 + 81 = 4x^2$$
$$81 = 3x^2$$
$$27 = x^2$$
$$\pm\sqrt{27} = x$$
$$\pm 3\sqrt{3} = x$$

The other leg is $3\sqrt{3}$ feet or approximately 5.2 feet long and the hypotenuse is $2(3\sqrt{3}) = 6\sqrt{3}$ feet or approximately 10.4 feet long.   ∎

Problem Set 10.2

Solve each of the following quadratic equations by factoring and applying the property, $ab = 0$ if and only if $a = 0$ or $b = 0$. If necessary, return to Chapter 3 and review the factoring techniques that we have studied.

1. $x^2 - 9x = 0$ **2.** $x^2 + 5x = 0$ **3.** $x^2 = -3x$

4. $x^2 = 15x$ **5.** $3y^2 + 12y = 0$ **6.** $6y^2 - 24y = 0$

7. $5n^2 - 9n = 0$ **8.** $4n^2 + 13n = 0$

9. $x^2 + x - 30 = 0$ **10.** $x^2 - 8x - 48 = 0$

11. $x^2 - 19x + 84 = 0$ **12.** $x^2 - 21x + 104 = 0$

13. $2x^2 + 19x + 24 = 0$ **14.** $4x^2 + 29x + 30 = 0$

15. $15x^2 + 29x - 14 = 0$ **16.** $24x^2 + x - 10 = 0$

17. $25x^2 - 30x + 9 = 0$ **18.** $16x^2 - 8x + 1 = 0$

19. $6x^2 - 5x - 21 = 0$ **20.** $12x^2 - 4x - 5 = 0$

Solve each of the following radical equations. Don't forget, you *must check* potential solutions.

21. $3\sqrt{x} = x + 2$ **22.** $3\sqrt{2x} = x + 4$ **23.** $\sqrt{2x} = x - 4$

24. $\sqrt{x} = x - 2$ **25.** $\sqrt{3x} + 6 = x$ **26.** $\sqrt{5x} + 10 = x$

Solve each of the following equations for x by factoring and applying the property, $ab = 0$ if and only if $a = 0$ or $b = 0$.

27. $x^2 - 5kx = 0$ **28.** $x^2 + 7kx = 0$

29. $x^2 = 16k^2x$ **30.** $x^2 = 25k^2x$

31. $x^2 - 12kx + 35k^2 = 0$ **32.** $x^2 - 3kx - 18k^2 = 0$

33. $2x^2 + 5kx - 3k^2 = 0$ **34.** $3x^2 - 20kx - 7k^2 = 0$

Use Property 10.1 to help solve each of the following equations.

35. $x^2 = 1$ **36.** $x^2 = 81$ **37.** $x^2 = -36$

38. $x^2 = -49$ **39.** $x^2 = 14$ **40.** $x^2 = 22$

41. $n^2 - 28 = 0$ **42.** $n^2 - 54 = 0$ **43.** $3t^2 = 54$

44. $4t^2 = 108$ **45.** $2t^2 = 7$ **46.** $3t^2 = 8$

47. $15y^2 = 20$ **48.** $14y^2 = 80$ **49.** $10x^2 + 48 = 0$

50. $12x^2 + 50 = 0$ **51.** $24x^2 = 36$ **52.** $12x^2 = 49$

53. $(x - 2)^2 = 9$ **54.** $(x + 1)^2 = 16$ **55.** $(x + 3)^2 = 25$

56. $(x - 2)^2 = 49$ **57.** $(x + 6)^2 = -4$ **58.** $(3x + 1)^2 = 9$

59. $(2x - 3)^2 = 1$ **60.** $(2x + 5)^2 = -4$ **61.** $(n - 4)^2 = 5$

62. $(n - 7)^2 = 6$ **63.** $(t + 5)^2 = 12$ **64.** $(t - 1)^2 = 18$

65. $(3y - 2)^2 = -27$ **66.** $(4y + 5)^2 = 80$

67. $3(x + 7)^2 + 4 = 79$ **68.** $2(x + 6)^2 - 9 = 63$

69. $2(5x - 2)^2 + 5 = 25$ **70.** $3(4x - 1)^2 + 1 = -17$

For Problems 71–76, a and b represent the lengths of the legs of a right triangle, and c represents the length of the hypotenuse. Express answers in simplest radical form and approximate each answer to the nearest tenth of a unit.

71. Find c if $a = 4$ centimeters and $b = 6$ centimeters.

72. Find c if $a = 3$ meters and $b = 7$ meters.

73. Find a if $c = 12$ inches and $b = 8$ inches.

74. Find a if $c = 8$ feet and $b = 6$ feet.

75. Find b if $c = 17$ yards and $a = 15$ yards.

76. Find b if $c = 14$ meters and $a = 12$ meters.

For Problems 77–80, use the following isosceles right triangle. Express answers in simplest radical form.

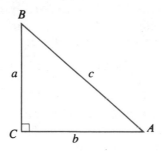

77. If $b = 6$ inches, find c.

78. If $a = 7$ centimeters, find c.

79. If $c = 8$ meters, find a and b.

80. If $c = 9$ feet, find a and b.

For Problems 81–86, use the accompanying figure. Express answers in simplest radical form.

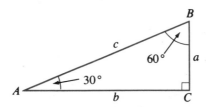

81. If $a = 3$ inches, find b and c.

82. If $a = 6$ feet, find b and c.

83. If $c = 14$ centimeters, find a and b.

84. If $c = 9$ centimeters, find a and b.

85. If $b = 10$ feet, find a and c.

86. If $b = 8$ meters, find a and c.

Miscellaneous Problems

87. Find the length of the hypotenuse (h) of an isosceles right triangle if each leg is s units long. Then use this relationship and redo Problems 77–80.

88. Suppose that the side opposite the 30° angle in a 30°–60° right triangle is s units long. Express the length of the hypotenuse and the length of the other leg in terms of s. Then use these relationships and redo Problems 81–86.

Completing the Square

Thus far we have solved quadratic equations by factoring and applying the property $ab = 0$ if and only if $a = 0$ or $b = 0$, or by applying the property, $x^2 = a$ if and only if $x = \pm\sqrt{a}$. In this section we shall examine another method called **completing the square**, which will give us the power to solve *any* quadratic equation.

A factoring technique we studied in Chapter 3 relied on recognizing **perfect square trinomials**. In each of the following, the perfect square trinomial on the right side is the result of squaring the binomial on the left side.

$$(x + 4)^2 = x^2 + 8x + 16, \qquad (x - 6)^2 = x^2 - 12x + 36,$$

$$(x + 7)^2 = x^2 + 14x + 49, \qquad (x - 9)^2 = x^2 - 18x + 81$$

Notice that in each of the perfect square trinomials *the constant term is equal to the square of one-half of the coefficient of the x-term*. Thus, we can form a perfect square trinomial from an expression of the form $ax^2 + bx$ by adding the constant term $\left(\dfrac{b}{2}\right)^2$. For example, suppose that we want to form a perfect square trinomial from $x^2 + 10x$. Since $\dfrac{1}{2}(10) = 5$ and $5^2 = 25$, the perfect square trinomial $x^2 + 10x + 25$ can be formed. Let's use these ideas to help solve some quadratic equations.

EXAMPLE 1   Solve $x^2 + 10x - 2 = 0$.

Solution

$$x^2 + 10x - 2 = 0$$

$$x^2 + 10x = 2$$

$$x^2 + 10x + 25 = 2 + 25 \qquad \text{We add 25 to the left side to form a perfect square}$$
trinomial; 25 must also be added to the right side.

$$(x + 5)^2 = 27$$

(Now we can proceed as in the last section.)

$$x + 5 = \pm\sqrt{27}$$

$$x + 5 = \pm 3\sqrt{3}$$

$$x = -5 \pm 3\sqrt{3}$$

The solution set is $\{-5 \pm 3\sqrt{3}\}$.   ■

Notice from Example 1 that the method of completing the square to solve a quadratic equation is merely what the name implies. A perfect square trinomial is formed, then the equation can be changed to the necessary form for applying the property, $x^2 = a$ if and only if $x = \pm\sqrt{a}$. Let's consider another example.

EXAMPLE 2   Solve $x^2 + 4x + 7 = 0$.

Solution

$$x^2 + 4x + 7 = 0$$
$$x^2 + 4x = -7$$
$$x^2 + 4x + 4 = -7 + 4 \qquad \text{Add } [\tfrac{1}{2}(4)]^2 = 4 \text{ to both sides.}$$
$$(x + 2)^2 = -3$$
$$x + 2 = \pm\sqrt{-3}$$
$$x + 2 = \pm i\sqrt{3}$$
$$x = -2 \pm i\sqrt{3}$$

The solution set is $\{-2 \pm i\sqrt{3}\}$.   ∎

REMARK Even though we do not show a check for each problem, it is probably a good idea for you to check one now and then. Example 2 is a good one to check since it involves some manipulation with complex numbers.

EXAMPLE 3   Solve $x^2 - 3x + 1 = 0$.

Solution

$$x^2 - 3x + 1 = 0$$
$$x^2 - 3x = -1$$
$$x^2 - 3x + \frac{9}{4} = -1 + \frac{9}{4} \qquad \text{Add } \left[\frac{1}{2}(-3)\right]^2 = \frac{9}{4} \text{ to both sides.}$$
$$\left(x - \frac{3}{2}\right)^2 = \frac{5}{4}$$
$$x - \frac{3}{2} = \pm\sqrt{\frac{5}{4}}$$
$$x - \frac{3}{2} = \pm\frac{\sqrt{5}}{2}$$
$$x = \frac{3}{2} \pm \frac{\sqrt{5}}{2}$$
$$x = \frac{3 \pm \sqrt{5}}{2}$$

The solution set is $\left\{\dfrac{3 \pm \sqrt{5}}{2}\right\}$.   ∎

In Example 3 notice that since the coefficient of the x-term is odd, we are forced into the realm of fractions. The use of common fractions rather than decimals allows us to apply our previous work with radicals.

The relationship for a perfect square trinomial that states that *the constant term is equal to the square of one-half of the coefficient of the x-term* holds only if the coefficient of x^2 is 1. Thus, an adjustment needs to be made when solving quadratic equations that have a coefficient of x^2 other than 1. The next example shows how to make this adjustment.

EXAMPLE 4   Solve $2x^2 + 12x - 5 = 0$.

Solution

$$2x^2 + 12x - 5 = 0$$

$$2x^2 + 12x = 5$$

$$x^2 + 6x = \frac{5}{2} \qquad \text{Multiply both sides by } \frac{1}{2}.$$

$$x^2 + 6x + 9 = \frac{5}{2} + 9 \qquad \text{Add } \left[\frac{1}{2}(6)\right]^2 = 9 \text{ to both sides.}$$

$$x^2 + 6x + 9 = \frac{23}{2}$$

$$(x + 3)^2 = \frac{23}{2}$$

$$x + 3 = \pm\sqrt{\frac{23}{2}}$$

$$x + 3 = \pm\frac{\sqrt{46}}{2} \qquad \sqrt{\frac{23}{2}} = \frac{\sqrt{23}}{\sqrt{2}} \cdot \frac{\sqrt{2}}{\sqrt{2}} = \frac{\sqrt{46}}{2}$$

$$x = -3 \pm \frac{\sqrt{46}}{2}$$

$$x = \frac{-6 \pm \sqrt{46}}{2}$$

The solution set is $\left\{\dfrac{-6 \pm \sqrt{46}}{2}\right\}$.   ■

As we mentioned earlier, we can use the method of completing the square to solve *any* quadratic equation. To illustrate, let's use it to solve an equation that could also be solved by factoring.

EXAMPLE 5   Solve $x^2 - 2x - 8 = 0$ by completing the square.

Solution

$$x^2 - 2x - 8 = 0$$

$$x^2 - 2x = 8$$

$$x^2 - 2x + 1 = 8 + 1 \qquad \text{Add } [\tfrac{1}{2}(-2)]^2 = 1 \text{ to both sides.}$$

$$(x - 1)^2 = 9$$

$$x - 1 = \pm 3$$

$$x - 1 = 3 \quad \text{or} \quad x - 1 = -3$$

$$x = 4 \quad \text{or} \quad x = -2$$

The solution set is $\{-2, 4\}$.   ∎

We make no claim that using the method of completing the square with an equation such as the one in Example 5 is easier than the factoring technique. However, you should recognize that the method of completing the square will work with any quadratic equation.

Problem Set 10.3

Solve each of the following quadratic equations by using (a) the factoring method and (b) the method of completing the square.

1. $x^2 - 4x - 60 = 0$

2. $x^2 + 6x - 16 = 0$

3. $x^2 - 14x = -40$

4. $x^2 - 18x = -72$

5. $x^2 - 5x - 50 = 0$

6. $x^2 + 3x - 18 = 0$

7. $x(x + 7) = 8$

8. $x(x - 1) = 30$

9. $2n^2 - n - 15 = 0$

10. $3n^2 + n - 14 = 0$

11. $3n^2 + 7n - 6 = 0$

12. $2n^2 + 7n - 4 = 0$

13. $n(n + 6) = 160$

14. $n(n - 6) = 216$

Use the method of completing the square to solve each of the following quadratic equations.

15. $x^2 + 4x - 2 = 0$

16. $x^2 + 2x - 1 = 0$

17. $x^2 + 6x - 3 = 0$

18. $x^2 + 8x - 4 = 0$

19. $y^2 - 10y = 1$

20. $y^2 - 6y = -10$

21. $n^2 - 8n + 17 = 0$

22. $n^2 - 4n + 2 = 0$

23. $n(n + 12) = -9$

24. $n(n + 14) = -4$

25. $n^2 + 2n + 6 = 0$

26. $n^2 + n - 1 = 0$

27. $x^2 + 3x - 2 = 0$

28. $x^2 + 5x - 3 = 0$

29. $x^2 + 5x + 1 = 0$

30. $x^2 + 7x + 2 = 0$

31. $y^2 - 7y + 3 = 0$

32. $y^2 - 9y + 30 = 0$

33. $2x^2 + 4x - 3 = 0$

34. $2t^2 - 4t + 1 = 0$

35. $3n^2 - 6n + 5 = 0$

36. $3x^2 + 12x - 2 = 0$

37. $3x^2 + 5x - 1 = 0$

38. $2x^2 + 7x - 3 = 0$

Solve each of the following quadratic equations and use whatever method seems most appropriate.

39. $x^2 + 8x - 48 = 0$

40. $x^2 + 5x - 14 = 0$

41. $2n^2 - 8n = -3$

42. $3x^2 + 6x = 1$

43. $(3x - 1)(2x + 9) = 0$

44. $(5x + 2)(x - 4) = 0$

45. $(x + 2)(x - 7) = 10$

46. $(x - 3)(x + 5) = -7$

47. $(x - 3)^2 = 12$

48. $x^2 = 16x$

49. $3n^2 - 6n + 4 = 0$

50. $2n^2 - 2n - 1 = 0$

51. $n(n + 8) = 240$

52. $t(t - 26) = -160$

53. $3x^2 + 29x = -66$

54. $6x^2 - 13x = 28$

55. $6n^2 + 23n + 21 = 0$

56. $6n^2 + n - 2 = 0$

57. $x^2 + 12x = 4$

58. $x^2 + 6x = -11$

59. $12n^2 - 7n + 1 = 0$

60. $5(x + 2)^2 + 1 = 16$

61. Use the method of completing the square to solve $ax^2 + bx + c = 0$ for x, where a, b, and c are real numbers and $a \neq 0$.

Miscellaneous Problems

Solve each of the following for the indicated variable. Assume that all letters represent positive numbers.

62. $\dfrac{x^2}{a^2} - \dfrac{y^2}{b^2} = 1$ for y

63. $\dfrac{x^2}{a^2} + \dfrac{y^2}{b^2} = 1$ for x

64. $s = \dfrac{1}{2}gt^2$ for t

65. $A = \pi r^2$ for r

Solve each of the following equations for x.

66. $x^2 + 8ax + 15a^2 = 0$

67. $x^2 - 5ax + 6a^2 = 0$

68. $10x^2 - 31ax - 14a^2 = 0$

69. $6x^2 + ax - 2a^2 = 0$

70. $4x^2 + 4bx + b^2 = 0$

71. $9x^2 - 12bx + 4b^2 = 0$

10.4
The Quadratic Formula

As we saw in the last section, the method of completing the square can be used to solve *any* quadratic equation. Furthermore, the equation $ax^2 + bx + c = 0$, where a, b, and c are real numbers with $a \neq 0$, can be used to represent *any* quadratic equation. These two ideas merge to produce the **quadratic formula**, a formula that can be used to solve *any* quadratic equation. The merger is accomplished by using the method of completing the square to solve the equation $ax^2 + bx + c = 0$ as follows.

$$ax^2 + bx + c = 0$$

$$ax^2 + bx = -c$$

$$x^2 + \frac{b}{a}x = -\frac{c}{a} \qquad \text{Multiply both sides by } \frac{1}{a}.$$

$$x^2 + \frac{b}{a}x + \frac{b^2}{4a^2} = -\frac{c}{a} + \frac{b^2}{4a^2} \qquad \textit{Complete the square by adding } \left[\frac{1}{2}\left(\frac{b}{a}\right)\right]^2 = \frac{b^2}{4a^2} \text{ to both sides.}$$

$$\left(x + \frac{b}{2a}\right)^2 = \frac{b^2 - 4ac}{4a^2} \qquad \text{The right side is combined into a single fraction.}$$

$$x + \frac{b}{2a} = \pm\sqrt{\frac{b^2 - 4ac}{4a^2}}$$

$$x + \frac{b}{2a} = \pm\frac{\sqrt{b^2 - 4ac}}{\sqrt{4a^2}}$$

$$x + \frac{b}{2a} = \pm\frac{\sqrt{b^2 - 4ac}}{2a} \qquad \sqrt{4a^2} = |2a| \text{ but } 2a \text{ can be used because of the use of } \pm.$$

$$x + \frac{b}{2a} = \frac{\sqrt{b^2 - 4ac}}{2a} \qquad \text{or} \qquad x + \frac{b}{2a} = -\frac{\sqrt{b^2 - 4ac}}{2a}$$

$$x = -\frac{b}{2a} + \frac{\sqrt{b^2 - 4ac}}{2a} \qquad \text{or} \qquad x = -\frac{b}{2a} - \frac{\sqrt{b^2 - 4ac}}{2a}$$

$$x = \frac{-b + \sqrt{b^2 - 4ac}}{2a} \qquad \text{or} \qquad x = \frac{-b - \sqrt{b^2 - 4ac}}{2a}$$

The quadratic formula is usually stated as follows.

Quadratic Formula

$$x = \frac{-b \pm \sqrt{b^2 - 4ac}}{2a}, \qquad a \neq 0$$

We can use this formula to solve any quadratic equation by expressing the equation in the standard form, $ax^2 + bx + c = 0$, and substituting the values for a, b, and c into the formula. Let's consider some examples.

EXAMPLE 1   Solve $x^2 + 5x + 2 = 0$.

Solution   The given equation is in standard form, so $a = 1, b = 5,$ and $c = 2$. Substituting these values into the formula and simplifying, we obtain

$$x = \frac{-b \pm \sqrt{b^2 - 4ac}}{2a}$$

$$x = \frac{-5 \pm \sqrt{5^2 - 4(1)(2)}}{2(1)}$$

$$x = \frac{-5 \pm \sqrt{25 - 8}}{2}$$

$$x = \frac{-5 \pm \sqrt{17}}{2}.$$

The solution set is $\left\{ \dfrac{-5 \pm \sqrt{17}}{2} \right\}$. ∎

EXAMPLE 2   Solve $x^2 - 2x - 4 = 0$.

Solution   We need to think of $x^2 - 2x - 4 = 0$ as $x^2 + (-2x) + (-4) = 0$ to determine the values $a = 1$, $b = -2$, and $c = -4$. Substitute these values into the quadratic formula and simplify, to obtain

$$x = \frac{-b \pm \sqrt{b^2 - 4ac}}{2a}$$

$$x = \frac{-(-2) \pm \sqrt{(-2)^2 - 4(1)(-4)}}{2(1)}$$

$$x = \frac{2 \pm \sqrt{4 + 16}}{2}$$

$$x = \frac{2 \pm \sqrt{20}}{2}$$

$$x = \frac{2 \pm 2\sqrt{5}}{2}$$

$$= \frac{\cancel{2}(1 \pm \sqrt{5})}{\cancel{2}}.$$

The solution set is $\{1 \pm \sqrt{5}\}$. ∎

EXAMPLE 3   Solve $x^2 - 2x + 19 = 0$.

Solution

$$x^2 - 2x + 19 = 0$$

$$x = \frac{-(-2) \pm \sqrt{(-2)^2 - 4(1)(19)}}{2(1)}$$

$$x = \frac{2 \pm \sqrt{4 - 76}}{2}$$

$$x = \frac{2 \pm \sqrt{-72}}{2}$$

$$x = \frac{2 \pm 6i\sqrt{2}}{2} \qquad \sqrt{-72} = i\sqrt{72} = i\sqrt{36}\sqrt{2} = 6i\sqrt{2}$$

$$x = 1 \pm 3i\sqrt{2}.$$

The solution set is $\{1 \pm 3i\sqrt{2}\}$.   ∎

EXAMPLE 4   Solve $2x^2 + 4x - 3 = 0$.

Solution   Substitute $a = 2$, $b = 4$, and $c = -3$ into the quadratic formula and simplify to obtain

$$x = \frac{-b \pm \sqrt{b^2 - 4ac}}{2a}$$

$$x = \frac{-4 \pm \sqrt{4^2 - 4(2)(-3)}}{2(2)}$$

$$x = \frac{-4 \pm \sqrt{16 + 24}}{4}$$

$$x = \frac{-4 \pm \sqrt{40}}{4}$$

$$x = \frac{-4 \pm 2\sqrt{10}}{4}$$

$$= \frac{-2 \pm \sqrt{10}}{2}.$$

The solution set is $\left\{ \dfrac{-2 \pm \sqrt{10}}{2} \right\}$.   ∎

EXAMPLE 5   Solve $n(3n - 10) = 25$.

Solution   First, we need to change the equation to the standard form $an^2 + bn + c = 0$.

$$n(3n - 10) = 25$$
$$3n^2 - 10n = 25$$
$$3n^2 - 10n - 25 = 0.$$

Now substituting $a = 3$, $b = -10$, and $c = -25$ into the quadratic formula we obtain

$$n = \frac{-b \pm \sqrt{b^2 - 4ac}}{2a}$$

$$n = \frac{-(-10) \pm \sqrt{(-10)^2 - 4(3)(-25)}}{2(3)}$$

$$n = \frac{10 \pm \sqrt{100 + 300}}{2(3)}$$

$$n = \frac{10 \pm \sqrt{400}}{6}$$

$$n = \frac{10 \pm 20}{6}$$

$$n = \frac{10 + 20}{6} \quad \text{or} \quad n = \frac{10 - 20}{6}$$

$$n = 5 \quad \text{or} \quad n = -\frac{5}{3}.$$

The solution set is $\left\{ -\frac{5}{3}, 5 \right\}$.   ■

In Example 5, notice that we used the variable n. The quadratic formula is usually stated in terms of x, but it certainly can be applied to quadratic equations in other variables. Also note in Example 5 that the polynomial $3n^2 - 10n - 25$ can be factored as $(3n + 5)(n - 5)$. Therefore, we could also solve the equation $3n^2 - 10n - 25 = 0$ using the factoring approach. We will give some guidance in the next section as to which approach to use on a particular equation.

Nature of Roots

The quadratic formula makes it easy to determine the nature of the roots of a quadratic equation without completely solving the equation. The number

$$b^2 - 4ac,$$

which appears under the radical sign in the quadratic formula, is called the **discriminant** of the quadratic equation. It is the indicator as to the kind of roots of the equation. For example, suppose that we start to solve the equation $x^2 - 4x + 7 = 0$ as follows.

$$x = \frac{-b \pm \sqrt{b^2 - 4ac}}{2a}$$

$$x = \frac{-(-4) \pm \sqrt{(-4)^2 - 4(1)(7)}}{2(1)}$$

$$x = \frac{4 \pm \sqrt{16 - 28}}{2}$$

$$x = \frac{4 \pm \sqrt{-12}}{2}$$

At this stage we should be able to look ahead and realize that we will obtain two complex solutions for the equation. (By the way, observe that these solutions are complex conjugates.) In other words, the discriminant, -12, is the indicator as to the type of roots that we will obtain.

We make the following general statements relative to the roots of a quadratic equation of the form $ax^2 + bx + c = 0$.

1. If $b^2 - 4ac < 0$, then the equation has two nonreal complex solutions.
2. If $b^2 - 4ac = 0$, then the equation has one real solution.
3. If $b^2 - 4ac > 0$, then the equation has two real solutions.

The following examples illustrate each of these situations. (You may want to solve the equations completely to verify the conclusions.)

Equation	*Discriminant*	*Nature of roots*
$x^2 - 3x + 7 = 0$	$b^2 - 4ac = (-3)^2 - 4(1)(7)$ $= 9 - 28$ $= -19$	two nonreal complex solutions
$9x^2 - 12x + 4 = 0$	$b^2 - 4ac = (-12)^2 - 4(9)(4)$ $= 144 - 144$ $= 0$	one real solution
$2x^2 + 5x - 3 = 0$	$b^2 - 4ac = (5)^2 - 4(2)(-3)$ $= 25 + 24$ $= 49$	two real solutions

There is another very useful relationship that involves the roots of a quadratic equation and the numbers $a, b,$ and c of the general form $ax^2 + bx + c = 0$. Suppose that we let x_1 and x_2 be the two roots generated by the quadratic formula. Thus, we have

$$x_1 = \frac{-b + \sqrt{b^2 - 4ac}}{2a} \qquad \text{and} \qquad x_2 = \frac{-b - \sqrt{b^2 - 4ac}}{2a}.$$

REMARK A point of clarification should be made at this time. Previously, we made the statement that if $b^2 - 4ac = 0$, then the equation has one real solution. Technically, such an equation has two solutions but they are equal. For example, each factor of $(x - 2)(x - 2) = 0$ produces a solution but both solutions are the number 2. This is sometimes referred to as one real solution with a *multiplicity of two*. Using the idea of multiplicity of roots, we can say that every quadratic equation has two roots.

Now let's consider the sum and product of the two roots.

Sum $x_1 + x_2 = \dfrac{-b + \sqrt{b^2 - 4ac}}{2a} + \dfrac{-b - \sqrt{b^2 - 4ac}}{2a} = \dfrac{-2b}{2a} = \boxed{-\dfrac{b}{a}}.$

Product $(x_1)(x_2) = \left(\dfrac{-b + \sqrt{b^2 - 4ac}}{2a}\right)\left(\dfrac{-b - \sqrt{b^2 - 4ac}}{2a}\right) = \dfrac{b^2 - (b^2 - 4ac)}{4a^2}$

$$= \dfrac{b^2 - b^2 + 4ac}{4a^2}$$

$$= \dfrac{4ac}{4a^2} = \boxed{\dfrac{c}{a}}.$$

These relationships provide another way of checking potential solutions when solving quadratic equations. For example, back in Example 3 we solved the equation $x^2 - 2x + 19 = 0$ and obtained solutions of $1 + 3i\sqrt{2}$ and $1 - 3i\sqrt{2}$. Let's check these solutions by using the sum and product relationships.

Check for Example 3

Sum of roots $(1 + 3i\sqrt{2}) + (1 - 3i\sqrt{2}) = 2$ and $-\dfrac{b}{a} = -\dfrac{-2}{1} = 2$

Product of roots $(1 + 3i\sqrt{2})(1 - 3i\sqrt{2}) = 1 - 18i^2 = 1 + 18 = 19$
and $\dfrac{c}{a} = \dfrac{19}{1} = 19$

Likewise, a check for Example 4 is as follows.

Check for Example 4

Sum of roots $\left(\dfrac{-2 + \sqrt{10}}{2}\right) + \left(\dfrac{-2 - \sqrt{10}}{2}\right) = -\dfrac{4}{2} = -2$ and $-\dfrac{b}{a} = -\dfrac{4}{2} = -2$

Product of roots $\left(\dfrac{-2 + \sqrt{10}}{2}\right)\left(\dfrac{-2 - \sqrt{10}}{2}\right) = -\dfrac{6}{4} = -\dfrac{3}{2}$ and $\dfrac{c}{a} = \dfrac{-3}{2} = -\dfrac{3}{2}$

Notice that for both Examples 3 and 4, it was much easier to check by using the sum and product relationships than it would have been by substituting back into the original equation. Don't forget that the values for a, b, and c come from a quadratic equation of the form $ax^2 + bx + c = 0$. Therefore, in Example 5 we must be certain that no errors were made when changing the given equation $n(3n - 10) = 25$ to the form $3n^2 - 10n - 25 = 0$ if we are going to check the potential solutions by using the sum and product relationships.

Problem Set 10.4

For each quadratic equation in Problems 1–10, first use the discriminant to determine whether the equation has two nonreal complex solutions, one real solution with a multiplicity of two, or two real solutions, and then solve the equation.

1. $x^2 + 4x - 21 = 0$

2. $x^2 - 3x - 54 = 0$

3. $9x^2 - 6x + 1 = 0$.

4. $4x^2 + 20x + 25 = 0$

5. $x^2 - 7x + 13 = 0$

6. $2x^2 - x + 5 = 0$

7. $15x^2 + 17x - 4 = 0$

8. $8x^2 + 18x - 5 = 0$

9. $3x^2 + 4x = 2$

10. $2x^2 - 6x = -1$

For Problems 11–50, use the quadratic formula to solve each of the quadratic equations. Check your solutions by using the *sum and product relationships*.

11. $x^2 + 2x - 1 = 0$

12. $x^2 + 4x - 1 = 0$

13. $n^2 + 5n - 3 = 0$

14. $n^2 + 3n - 2 = 0$

15. $a^2 - 8a = 4$

16. $a^2 - 6a = 2$

17. $n^2 + 5n + 8 = 0$

18. $2n^2 - 3n + 5 = 0$

19. $x^2 - 18x + 80 = 0$

20. $x^2 + 19x + 70 = 0$

21. $-y^2 = -9y + 5$

22. $-y^2 + 7y = 4$

23. $2x^2 + x - 4 = 0$

24. $2x^2 + 5x - 2 = 0$

25. $4x^2 + 2x + 1 = 0$

26. $3x^2 - 2x + 5 = 0$

27. $3a^2 - 8a + 2 = 0$

28. $2a^2 - 6a + 1 = 0$

29. $-2n^2 + 3n + 5 = 0$

30. $-3n^2 - 11n + 4 = 0$

31. $3x^2 + 19x + 20 = 0$

32. $2x^2 - 17x + 30 = 0$

33. $36n^2 - 60n + 25 = 0$

34. $9n^2 + 42n + 49 = 0$

35. $4x^2 - 2x = 3$

36. $6x^2 - 4x = 3$

37. $5x^2 - 13x = 0$

38. $7x^2 + 12x = 0$

39. $3x^2 = 5$

40. $4x^2 = 3$

41. $6t^2 + t - 3 = 0$

42. $2t^2 + 6t - 3 = 0$

43. $n^2 + 32n + 252 = 0$

44. $n^2 - 4n - 192 = 0$

45. $12x^2 - 73x + 110 = 0$

46. $6x^2 + 11x - 255 = 0$

47. $-2x^2 + 4x - 3 = 0$

48. $-2x^2 + 6x - 5 = 0$

49. $-6x^2 + 2x + 1 = 0$

50. $-2x^2 + 4x + 1 = 0$

Thoughts into Words

51. Explain the process of *completing the square* to solve a quadratic equation.

52. Explain how to use the quadratic formula to solve $4x = 3x^2 - 2$.

53. How would you solve the equation $x^2 - 4x = 252$? Explain your choice of the method that you would use.

Miscellaneous Problems

 The solution set for $x^2 - 4x - 37 = 0$ is $\{2 \pm \sqrt{41}\}$. With a calculator, we found a rational approximation, to the nearest one-thousandth, for each of these solutions.

$$2 - \sqrt{41} = -4.403 \quad \text{and} \quad 2 + \sqrt{41} = 8.403.$$

Thus, the solution set is $\{-4.403, 8.403\}$, with answers rounded to the nearest one-thousandth. Solve each of the following equations and express solutions to the nearest one-thousandth.

54. $x^2 - 6x - 10 = 0$

55. $x^2 - 16x - 24 = 0$

56. $x^2 + 6x - 44 = 0$

57. $x^2 + 10x - 46 = 0$

58. $x^2 + 8x + 2 = 0$

59. $x^2 + 9x + 3 = 0$

60. $4x^2 - 6x + 1 = 0$

61. $5x^2 - 9x + 1 = 0$

62. $2x^2 - 11x - 5 = 0$

63. $3x^2 - 12x - 10 = 0$

For Problems 64–66, use the discriminant to help solve each problem.

64. Determine k so that the solutions of $x^2 - 2x + k = 0$ are complex but nonreal.

65. Determine k so that $4x^2 - kx + 1 = 0$ has two equal real solutions.

66. Determine k so that $3x^2 - kx - 2 = 0$ has real solutions.

10.5

More Quadratic Equations and Applications

Which method should be used to solve a particular quadratic equation? There is no definite answer to that question; it depends upon the *type* of equation and your personal preference. In the following examples we will state reasons for choosing a specific technique. However, keep in mind that usually this is a decision *you* must make as the need arises. So become familiar with the strengths and weaknesses of each method.

EXAMPLE 1   Solve $2x^2 - 3x - 1 = 0$.

Solution   Because of the leading coefficient of 2 and the constant term of -1, there are very few factoring possibilities to consider. Therefore, with such problems, first try the factoring approach. Unfortunately, this particular polynomial is not factorable using integers. Thus, let's use the quadratic formula to solve the equation.

$$x = \frac{-b \pm \sqrt{b^2 - 4ac}}{2a}$$

$$x = \frac{-(-3) \pm \sqrt{(-3)^2 - 4(2)(-1)}}{2(2)}$$

$$x = \frac{3 \pm \sqrt{9 + 8}}{4}$$

$$x = \frac{3 \pm \sqrt{17}}{4}$$

CHECK We can use the *sum and product-of-roots* relationships for our checking purposes.

Sum of roots $\quad \dfrac{3 + \sqrt{17}}{4} + \dfrac{3 - \sqrt{17}}{4} = \dfrac{6}{4} = \dfrac{3}{2} \quad$ and $\quad -\dfrac{b}{a} = -\dfrac{-3}{2} = \dfrac{3}{2}.$

Product of roots

$$\left(\frac{3 + \sqrt{17}}{4} \right)\left(\frac{3 - \sqrt{17}}{4} \right) = \frac{9 - 17}{16} = -\frac{8}{16} = -\frac{1}{2} \quad \text{and} \quad \frac{c}{a} = \frac{-1}{2} = -\frac{1}{2}.$$

The solution set is $\left\{ \dfrac{3 \pm \sqrt{17}}{4} \right\}.$ ∎

EXAMPLE 2   Solve $\dfrac{3}{n} + \dfrac{10}{n + 6} = 1.$

Solution

$$\frac{3}{n} + \frac{10}{n + 6} = 1, \qquad n \neq 0 \text{ and } n \neq -6.$$

$$n(n + 6)\left(\frac{3}{n} + \frac{10}{n + 6} \right) = 1(n)(n + 6) \qquad \begin{array}{l} \text{Multiply both sides by } n(n + 6), \\ \text{which is the LCD.} \end{array}$$

$$3(n + 6) + 10n - n(n + 6)$$

$$3n + 18 + 10n = n^2 + 6n$$

$$13n + 18 = n^2 + 6n$$

$$0 = n^2 - 7n - 18$$

This is an easy one to consider the possibilities for factoring, and it factors as follows.

$$0 = (n - 9)(n + 2)$$

$$n - 9 = 0 \qquad \text{or} \qquad n + 2 = 0$$

$$n = 9 \qquad \text{or} \qquad n = -2$$

CHECK Substituting 9 and -2 back into the original equation, we obtain

$$\frac{3}{n} + \frac{10}{n+6} = 1 \qquad\qquad \frac{3}{n} + \frac{10}{n+6} = 1$$

$$\frac{3}{9} + \frac{10}{9+6} \overset{?}{=} 1 \qquad\qquad \frac{3}{-2} + \frac{10}{-2+6} \overset{?}{=} 1$$

$$\frac{1}{3} + \frac{10}{15} \overset{?}{=} 1 \qquad \text{or} \qquad -\frac{3}{2} + \frac{10}{4} \overset{?}{=} 1$$

$$\frac{1}{3} + \frac{2}{3} \overset{?}{=} 1 \qquad\qquad -\frac{3}{2} + \frac{5}{2} \overset{?}{=} 1$$

$$1 = 1 \qquad\qquad\qquad \frac{2}{2} = 1$$

The solution set is $\{-2, 9\}$. ∎

We should make two comments about Example 2. First, notice the indication of the initial restrictions $n \neq 0$ and $n \neq -6$. Remember that we need to do this when solving fractional equations. Secondly, the *sum and product-of-roots* relationships were not used for checking purposes in this problem. Those relationships would only check the validity of our work from the step $0 = n^2 - 7n - 18$ to the finish. In other words, an error made in changing the original equation to quadratic form would not be detected by checking the sum and product of potential roots. Thus, with such a problem the only *absolute check* is to substitute the potential solutions back into the *original equation*.

EXAMPLE 3   Solve $x^2 + 22x + 112 = 0$.

Solution   The size of the constant term makes the factoring approach a little cumbersome for this problem. Furthermore, since the leading coefficient is 1 and the coefficient of the x-term is even, the method of completing the square will work rather effectively as follows.

$$x^2 + 22x + 112 = 0$$

$$x^2 + 22x = -112$$

$$x^2 + 22x + 121 = -112 + 121$$

$$(x + 11)^2 = 9$$

$$x + 11 = \pm\sqrt{9}$$

$$x + 11 = \pm 3$$

$$x + 11 = 3 \qquad \text{or} \qquad x + 11 = -3$$

$$x = -8 \qquad \text{or} \qquad x = -14$$

CHECK

Sum of roots $-8 + (-14) = -22$ and $-\dfrac{b}{a} = -22.$

Product of roots $(-8)(-14) = 112$ and $\dfrac{c}{a} = 112.$

The solution set is $\{-14, -8\}$. ∎

EXAMPLE 4   Solve $x^4 - 4x^2 - 96 = 0$.

Solution   An equation such as $x^4 - 4x^2 - 96 = 0$ is not a quadratic equation, but we can solve it using the techniques that we use on quadratic equations. That is to say, we can factor the polynomial and apply the property, $ab = 0$ if and only if $a = 0$ or $b = 0$, as follows.

$$x^4 - 4x^2 - 96 = 0$$
$$(x^2 - 12)(x^2 + 8) = 0$$

$$x^2 - 12 = 0 \qquad \text{or} \qquad x^2 + 8 = 0$$
$$x^2 = 12 \qquad \text{or} \qquad x^2 = -8$$
$$x = \pm\sqrt{12} \qquad \text{or} \qquad x = \pm\sqrt{-8}$$
$$x = \pm 2\sqrt{3} \qquad \text{or} \qquad x = \pm 2i\sqrt{2}$$

The solution set is $\{\pm 2\sqrt{3}, \pm 2i\sqrt{2}\}$. (We will leave the check for this problem for you to do!) ∎

REMARK Another approach to Example 4 would be to substitute y for x^2 and y^2 for x^4. The equation $x^4 - 4x^2 - 96 = 0$ becomes the quadratic equation $y^2 - 4y - 96 = 0$. Thus, we say that $x^4 - 4x^2 - 96 = 0$ is of *quadratic form.* Then we could solve the quadratic equation $y^2 - 4y - 96 = 0$ and use the equation $y = x^2$ to determine the solutions for x.

Applications

Before we conclude this section with some word problems that can be solved using quadratic equations, let's restate the suggestions for solving word problems we made in an earlier chapter.

Suggestions for Solving Word Problems

1. Read the problem carefully and make certain that you understand the meanings of all the words. Be especially alert for any technical terms used in the statement of the problem.

2. Read the problem a second time (perhaps even a third time) to get an

overview of the situation being described and to determine the known facts, as well as what is to be found.

3. Sketch any figure, diagram, or chart that might be helpful in analyzing the problem.

4. Choose a meaningful variable to represent an unknown quantity in the problem (perhaps *l*, if the length of a rectangle is an unknown quantity) and represent any other unknowns in terms of that variable.

5. Look for a *guideline* that you can use to set up an equation. A guideline might be a formula such as "$A = lw$" or a relationship such as "the fractional part of a job done by Bill plus the fractional part of the job done by Mary equals the total job."

6. Form an equation that contains the variable that translates the conditions of the guideline from English to algebra.

7. Solve the equation and use the solutions to determine all facts requested in the problem.

8. Check all answers back into the **original statement of the problem**.

REMARK The problem solving suggestions are stated as if the problems are to be solved using only one variable. However, these same basic suggestions apply when using two variables and two equations. Only the wording of the suggestions would change.

PROBLEM 1   A page for a magazine contains 70 square inches of type. The height of a page is twice the width. If the margin around the type is to be 2 inches uniformly, what are the dimensions of a page?

Solution   Let x represent the width of a page. Then $2x$ represents the height of a page. Now let's draw and label a model of a page.

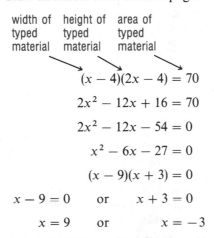

width of   height of   area of
typed   typed   typed
material   material   material

$$(x - 4)(2x - 4) = 70$$

$$2x^2 - 12x + 16 = 70$$

$$2x^2 - 12x - 54 = 0$$

$$x^2 - 6x - 27 = 0$$

$$(x - 9)(x + 3) = 0$$

$$x - 9 = 0 \quad \text{or} \quad x + 3 = 0$$

$$x = 9 \quad \text{or} \quad x = -3$$

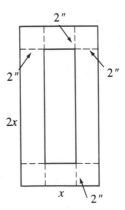

Disregard the negative solution; the page must be 9 inches wide and its height is $2(9) = 18$ inches. ∎

Let's use our knowledge of quadratic equations to analyze some applications of the business world. For example, if P dollars is invested at r rate of interest compounded annually for t years, then the amount of money, A, accumulated at the end of t years is given by the formula

$$A = P(1 + r)^t.$$

This compound interest formula serves as a guideline for the next problem.

PROBLEM 2   Suppose that $100 is invested at a certain rate of interest compounded annually for 2 years. If the accumulated value at the end of 2 years is $121, find the rate of interest.

Solution   Let r represent the rate of interest. Substitute the known values into the compound interest formula to yield

$$A = P(1 + r)^t$$
$$121 = 100(1 + r)^2.$$

Solving this equation, we obtain

$$\frac{121}{100} = (1 + r)^2$$

$$\pm\sqrt{\frac{121}{100}} = (1 + r)$$

$$\pm\frac{11}{10} = 1 + r$$

$$1 + r = \frac{11}{10} \qquad \text{or} \qquad 1 + r = -\frac{11}{10}$$

$$r = -1 + \frac{11}{10} \qquad \text{or} \qquad r = -1 - \frac{11}{10}$$

$$r = \frac{1}{10} \qquad \text{or} \qquad r = -\frac{21}{10}.$$

We must disregard the negative solution, so $r = \frac{1}{10}$ is the only solution. Change $\frac{1}{10}$ to a percent and the rate of interest is 10%.   ■

PROBLEM 3   A businesswoman bought a parcel of land on speculation for $120,000. She subdivided the land into lots and when she had sold all but 18 lots at a profit of $6000 per lot, she regained the entire cost of the land. How many lots were sold and at what price per lot?

Solution   Let x represent the number of lots sold. Then $x + 18$ represents the total number of lots. Therefore, $\dfrac{120{,}000}{x}$ represents the selling price per lot and $\dfrac{120{,}000}{x + 18}$ represents the cost per lot. The following equation represents the situation.

selling price per lot	equals	cost per lot	plus	$6000
$\downarrow$		$\downarrow$		$\downarrow$
$\dfrac{120{,}000}{x}$	$=$	$\dfrac{120{,}000}{x + 18}$	$+$	6000

Solving this equation, we obtain

$$x(x + 18)\left(\frac{120{,}000}{x}\right) = \left(\frac{120{,}000}{x + 18} + 6000\right)(x)(x + 18)$$

$$120{,}000(x + 18) = 120{,}000x + 6000x(x + 18)$$

$$120{,}000x + 2{,}160{,}000 = 120{,}000x + 6000x^2 + 108{,}000x$$

$$0 = 6000x^2 + 108{,}000x - 2{,}160{,}000$$

$$0 = x^2 + 18x - 360.$$

The method of completing the square works very well with this equation.

$$x^2 + 18x = 360$$

$$x^2 + 18x + 81 = 441$$

$$(x + 9)^2 = 441$$

$$x + 9 = \pm\sqrt{441}$$

$$x + 9 = \pm 21$$

$$x + 9 = 21 \quad \text{or} \quad x + 9 = -21$$

$$x = 12 \quad \text{or} \quad x = -30$$

We discard the negative solution; thus, 12 lots were sold at $\dfrac{120{,}000}{x} = \dfrac{120{,}000}{12} =$ $10,000 per lot.   ∎

PROBLEM 4   Barry bought a number of shares of stock for $600. A week later the value of the stock increased $3 per share and he sold all but 10 shares and regained his original investment of $600. How many shares did he sell and at what price per share?

Solution   Let b represent the number of shares that Barry bought, and let s represent the number of shares that he sold. Then $\dfrac{600}{b}$ represents the price per share that he paid for them and $\dfrac{600}{s}$ represents the selling price per share. The problem trans-lates into the following two equations.

number of shares
bought was 10 more $\longrightarrow$ $b = s + 10$
than number sold

selling price per share
was $3 more than purchasing $\longrightarrow$ $\dfrac{600}{s} = \dfrac{600}{b} + 3$
price per share

From the first equation we can substitute $s + 10$ for b in the second equation.

$$\frac{600}{s} = \frac{600}{s + 10} + 3$$

Solving this equation yields

$$s(s + 10)\left(\frac{600}{s}\right) = \left(\frac{600}{s + 10} + 3\right)(s)(s + 10)$$

$$600(s + 10) = 600s + 3s(s + 10)$$

$$600s + 6000 = 600s + 3s^2 + 30s$$

$$0 = 3s^2 + 30s - 6000$$

$$0 = s^2 + 10s - 2000.$$

Use the quadratic formula to obtain

$$s = \frac{-10 \pm \sqrt{10^2 - 4(1)(-2000)}}{2(1)}$$

$$s = \frac{-10 \pm \sqrt{100 + 8000}}{2}$$

$$s = \frac{-10 \pm \sqrt{8100}}{2}$$

$$s = \frac{-10 \pm 90}{2}$$

$$s = \frac{-10 + 90}{2} \qquad \text{or} \qquad s = \frac{-10 - 90}{2}$$

$$s = 40 \qquad \text{or} \qquad s = -50.$$

We discard the negative solution and we know that 40 shares were sold at $\dfrac{600}{s} = \dfrac{600}{40} = \$15$ per share. ∎

This next problem set contains a large variety of word problems. Not only are there some business applications similar to those we discussed in this section, but there are also more problems of the types we discussed back in Chapters 3 and 4. Try to give them your best shot without referring back to examples in earlier chapters.

Problem Set 10.5

Solve each of the following quadratic equations and use the method that seems most appropriate to you.

1. $x^2 - 4x - 6 = 0$

2. $x^2 - 8x - 4 = 0$

3. $3x^2 + 23x - 36 = 0$

4. $n^2 + 22n + 105 = 0$

5. $x^2 - 18x = 9$

6. $x^2 + 20x = 25$

7. $2x^2 - 3x + 4 = 0$

8. $3y^2 - 2y + 1 = 0$

9. $135 + 24n + n^2 = 0$

10. $28 - x - 2x^2 = 0$

11. $(x - 2)(x + 9) = -10$

12. $(x + 3)(2x + 1) = -3$

13. $2x^2 - 4x + 7 = 0$

14. $3x^2 - 2x + 8 = 0$

15. $x^2 - 18x + 15 = 0$

16. $x^2 - 16x + 14 = 0$

17. $20y^2 + 17y - 10 = 0$

18. $12x^2 + 23x - 9 = 0$

19. $4t^2 + 4t - 1 = 0$

20. $5t^2 + 5t - 1 = 0$

Solve each of the following equations.

21. $n + \dfrac{3}{n} = \dfrac{19}{4}$

22. $n - \dfrac{2}{n} = -\dfrac{7}{3}$

23. $\dfrac{3}{x} + \dfrac{7}{x - 1} = 1$

24. $\dfrac{2}{x} + \dfrac{5}{x + 2} = 1$

25. $\dfrac{12}{x - 3} + \dfrac{8}{x} = 14$

26. $\dfrac{16}{x + 5} - \dfrac{12}{x} = -2$

27. $\dfrac{3}{x - 1} - \dfrac{2}{x} = \dfrac{5}{2}$

28. $\dfrac{4}{x + 1} + \dfrac{2}{x} = \dfrac{5}{3}$

29. $\dfrac{6}{x} + \dfrac{40}{x + 5} = 7$

30. $\dfrac{12}{t} + \dfrac{18}{t + 8} = \dfrac{9}{2}$

31. $\dfrac{5}{n - 3} - \dfrac{3}{n + 3} = 1$

32. $\dfrac{3}{t + 2} + \dfrac{4}{t - 2} = 2$

33. $x^4 - 18x^2 + 72 = 0$

34. $x^4 - 21x^2 + 54 = 0$

35. $3x^4 - 35x^2 + 72 = 0$

36. $5x^4 - 32x^2 + 48 = 0$

37. $3x^4 + 17x^2 + 20 = 0$

38. $4x^4 + 11x^2 - 45 = 0$

39. $6x^4 - 29x^2 + 28 = 0$

40. $6x^4 - 31x^2 + 18 = 0$

Set up an equation or a system of equations and solve each of the following problems.

41. Find two consecutive whole numbers such that the sum of their squares is 145.

42. Find two consecutive odd whole numbers such that the sum of their squares is 74.

43. Two positive integers differ by 3, and their product is 108. Find the numbers.

44. Suppose that the sum of two numbers is 20 and the sum of their squares is 232. Find the numbers.

45. Find two numbers such that their sum is 10 and their product is 22.

46. Find two numbers such that their sum is 6 and their product is 7.

47. Suppose that the sum of two whole numbers is 9 and the sum of their reciprocals is $\frac{1}{2}$. Find the numbers.

48. The difference between two whole numbers is 8 and the difference between their reciprocals is $\frac{1}{6}$. Find the two numbers.

49. The sum of the lengths of the two legs of a right triangle is 21 inches. If the length of the hypotenuse is 15 inches, find the length of each leg.

50. The length of a rectangular floor is 1 meter less than twice its width. If a diagonal of the rectangle is 17 meters, find the length and width of the floor.

51. A rectangular plot of ground measuring 12 meters by 20 meters is surrounded by a sidewalk of uniform width. The area of the sidewalk is 68 square meters. Find the width of the walk.

52. A 5-inch-by-7-inch picture is surrounded by a frame of uniform width. The area of the picture and frame together is 80 square inches. Find the width of the frame.

53. The perimeter of a rectangle is 44 inches and its area is 112 square inches. Find the length and width of the rectangle.

54. A rectangular piece of cardboard is 2 units longer than it is wide. From each of its corners a square piece 2 units on a side is cut out. The flaps are then turned up to form an open box that has a volume of 70 cubic units. Find the length and width of the original piece of cardboard.

55. Charlotte traveled 250 miles in one hour more time than it took Lorraine to travel 180 miles. Charlotte drove 5 miles per hour faster than Lorraine. How fast did each one travel?

56. Larry drove 156 miles in one hour more than it took Mike to drive 108 miles. Mike drove at an average rate of 2 miles per hour faster than Larry. How fast did each one travel?

57. On a 570-mile trip, Andy averaged 5 miles per hour faster for the last 240 miles than he did for the first 330 miles. The entire trip took 10 hours. How fast did he travel for the first 330 miles?

58. On a 135-mile bicycle excursion, Maria averaged 5 miles per hour faster for the first 60 miles than she did for the last 75 miles. The entire trip took 8 hours. Find her rate for the first 60 miles.

59. It takes Terry 2 hours longer to do a certain job than it takes Tom. They worked together for 3 hours; then Tom left and Terry finished the job in 1 hour. How long would it take each of them to do the job alone?

60. Suppose that Arlene can mow the entire lawn in 40 minutes less time with the power mower than she can with the push mower. One day the power mower broke down after she had been mowing for 30 minutes. She finished the lawn with the push mower in 20 minutes. How long does it take Arlene to mow the entire lawn with the power mower?

61. A man did a job for $360. It took him 6 hours longer than he expected and therefore he earned $2 per hour less than he anticipated. How long did he expect that it would take to do the job?

62. A group of students agreed to each chip in the same amount to pay for a party that would cost $100. Then they found 5 more students interested in the party and in sharing the expenses. This decreased the amount each had to pay by $1. How many students were involved in the party and how much did they each have to pay?

63. A group of customers agreed to each contribute the same amount to buy their favorite waitress a $100 birthday gift. At the last minute, 2 of the people decided not to chip in. This increased the amount that the remaining people had to pay by $2.50 per person. How many people actually contributed to the gift?

64. A retailer bought a number of special mugs for $48. Two of the mugs were broken in the store, but by selling each of the other mugs $3 above the original cost per mug she made a total profit of $22. How many mugs did she buy and at what price per mug did she sell them?

65. My friend Tony bought a number of shares of stock for $720. A month later the value of the stock increased by $8 per share and he sold all but 20 shares and regained his original investment plus a profit of $80. How many shares did he sell and at what price per share?

66. The formula $D = \dfrac{n(n-3)}{2}$ yields the number of diagonals, D, in a polygon of n sides. Find the number of sides of a polygon that has 54 diagonals.

67. The formula $S = \dfrac{n(n+1)}{2}$ yields the sum, S, of the first n natural numbers $1, 2, 3, 4, \ldots$. How many consecutive natural numbers starting with 1 will give a sum of 1275?

68. At a point 16 yards from the base of a tower, the distance to the top of the tower is 4 yards more than the height of the tower. Find the height of the tower.

 69. Suppose that $500 is invested at a certain rate of interest compounded annually for 2 years. If the accumulated value at the end of 2 years is $594.05, find the rate of interest.

 70. Suppose that $10,000 is invested at a certain rate of interest compounded annually for 2 years. If the accumulated value at the end of 2 years is $12,544, find the rate of interest.

Miscellaneous Problems

Solve each of the following equations.

71. $x - 9\sqrt{x} + 18 = 0$ [*Hint:* Let $y = \sqrt{x}$.]

72. $x - 4\sqrt{x} + 3 = 0$ 73. $x + \sqrt{x} - 2 = 0$

74. $x^{\frac{2}{3}} + x^{\frac{1}{3}} - 6 = 0$ [*Hint:* Let $y = x^{\frac{1}{3}}$.]

75. $6x^{\frac{2}{3}} - 5x^{\frac{1}{3}} - 6 = 0$ 76. $x^{-2} + 4x^{-1} - 12 = 0$

77. $12x^{-2} - 17x^{-1} - 5 = 0$

Quadratic Inequalities

We refer to the equation $ax^2 + bx + c = 0$ as the standard form of a quadratic equation in one variable. Similarly, the following forms express **quadratic inequalities** in one variable.

$$ax^2 + bx + c > 0,$$

$$ax^2 + bx + c \geq 0,$$

$$ax^2 + bx + c < 0,$$

$$ax^2 + bx + c \leq 0$$

We can use the number line very effectively to help solve quadratic inequalities where the quadratic polynomial is factorable. Let's consider some examples to illustrate the procedure.

EXAMPLE 1   Solve and graph the solutions for $x^2 + 2x - 8 > 0$.

Solution   First, let's factor the polynomial.

$$x^2 + 2x - 8 > 0$$

$$(x + 4)(x - 2) > 0$$

On a number line, we shall now indicate that at $x = 2$ and $x = -4$ the product $(x + 4)(x - 2)$ equals zero.

The numbers -4 and 2 divide the number line into three intervals: (1) the numbers less than -4, (2) the numbers between -4 and 2, and (3) the numbers greater than 2. We can choose a **test number** from each of these intervals and see how it affects the signs of the factors $x + 4$ and $x - 2$, and consequently, the sign of the product of these factors. For example, if $x < -4$ (try $x = -5$) then $x + 4$ is negative and $x - 2$ is negative; so their product is positive. If $-4 < x < 2$ (try $x = 0$), then $x + 4$ is positive and $x - 2$ is negative; so their product is negative. If $x > 2$ (try $x = 3$), then $x + 4$ is positive and $x - 2$ is positive; so their product is positive. This information can be conveniently arranged using a number line as follows. Note the open circles at -4 and 2 to indicate that they are not included in the solution set.

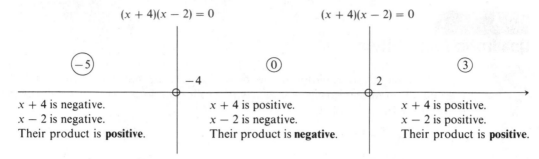

$x + 4$ is negative.
$x - 2$ is negative.
Their product is **positive**.

$x + 4$ is positive.
$x - 2$ is negative.
Their product is **negative**.

$x + 4$ is positive.
$x - 2$ is positive.
Their product is **positive**.

Therefore, the given inequality $x^2 + 2x - 8 > 0$ is satisfied by numbers less than -4 along with numbers greater than 2. The solution set is $\{x \mid x < -4 \text{ or } x > 2\}$ and its graph is as follows.

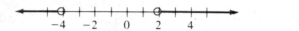

We refer to numbers such as -4 and 2 in the preceding example, where the given polynomial or algebraic expression equals zero or is undefined, as **critical numbers**. Let's consider some additional examples that make use of critical numbers and test numbers.

EXAMPLE 2   Solve and graph the solutions for $x^2 + 2x - 3 \le 0$.

Solution   First, factor the polynomial.

$$x^2 + 2x - 3 \le 0$$

$$(x + 3)(x - 1) \le 0$$

Secondly, locate the values for which $(x + 3)(x - 1)$ equals zero. We put dots at -3 and 1 to remind ourselves that these two numbers are to be included in the solution set since the given statement includes equality. Now let's choose a test number from each of the three intervals and record the sign behavior of the factors $(x + 3)$ and $(x - 1)$.

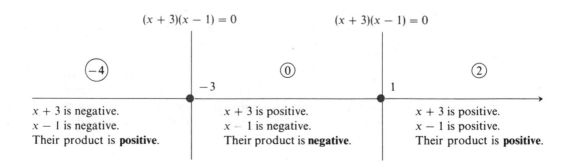

$x + 3$ is negative.
$x - 1$ is negative.
Their product is **positive**.

$x + 3$ is positive.
$x - 1$ is negative.
Their product is **negative**.

$x + 3$ is positive.
$x - 1$ is positive.
Their product is **positive**.

Therefore, the solution set is $\{x \mid x \geq -3 \text{ and } x \leq 1\}$ and it can be graphed as follows.

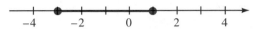

Examples 1 and 2 have indicated a systematic approach for solving quadratic inequalities where the polynomial is factorable. This same type of number line analysis can also be used to solve indicated quotients such as $\dfrac{x+1}{x-5} > 0$.

EXAMPLE 3   Solve and graph the solutions for $\dfrac{x+1}{x-5} > 0$.

Solution   First, indicate that at $x = -1$ the given quotient equals zero, and at $x = 5$ the quotient is undefined. Second, choose test numbers from each of the three intervals and record the sign behavior of $(x + 1)$ and $(x - 5)$.

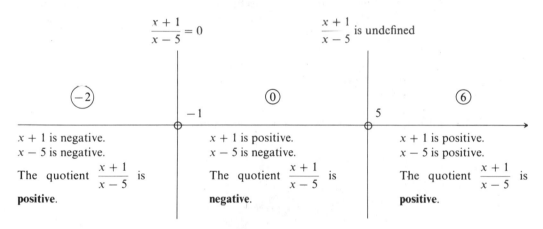

Therefore, the solution set is $\{x \mid x < -1 \text{ or } x > 5\}$ and its graph is as follows.

EXAMPLE 4   Solve $\dfrac{x+2}{x+4} \leq 0$.

Solution   The indicated quotient equals zero at $x = -2$ and is undefined at $x = -4$. (Note that -2 is to be included in the solution set, but -4 is not to be included.) Now let's choose some test numbers and record the sign behavior of $(x + 2)$ and $(x + 4)$.

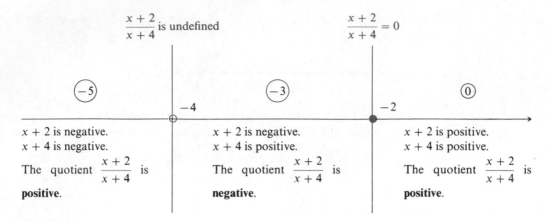

Therefore, the solution set is $\{x \mid x > -4 \text{ and } x \le -2\}$. ■

The final example illustrates that sometimes we need to change the form of the given inequality before we use the number line analysis.

EXAMPLE 5   Solve $\dfrac{x}{x+2} \ge 3$.

Solution   First, let's change the form of the given inequality as follows.

$$\frac{x}{x+2} \ge 3$$

$$\frac{x}{x+2} - 3 \ge 0 \qquad \text{Add } -3 \text{ to both sides.}$$

$$\frac{x - 3(x+2)}{x+2} \ge 0 \qquad \text{Express the left side over a common denominator.}$$

$$\frac{x - 3x - 6}{x+2} \ge 0$$

$$\frac{-2x - 6}{x+2} \ge 0.$$

Now we can proceed as we did with the previous examples. If $x = -3$, then $\dfrac{-2x - 6}{x+2}$ equals zero; and if $x = -2$, then $\dfrac{-2x - 6}{x+2}$ is undefined. Then choosing test numbers we can record the sign behavior of $(-2x - 6)$ and $(x + 2)$.

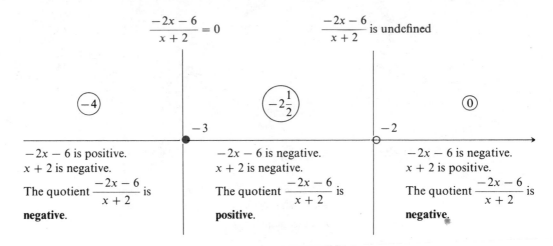

Therefore, the solution set is $\{x \mid x \geq -3 \text{ and } x < -2\}$. (Perhaps you should check a few numbers from this solution set back into the original inequality!) ∎

Problem Set 10.6

Solve each of the following inequalities and graph each solution set on a number line.

1. $(x + 2)(x - 1) > 0$

2. $(x - 2)(x + 3) > 0$

3. $(x + 1)(x + 4) < 0$

4. $(x - 3)(x - 1) < 0$

5. $(2x - 1)(3x + 7) \geq 0$

6. $(3x + 2)(2x - 3) \geq 0$

7. $(x + 2)(4x - 3) \leq 0$

8. $(x - 1)(2x - 7) \leq 0$

9. $(x + 1)(x - 1)(x - 3) > 0$

10. $(x + 2)(x + 1)(x - 2) > 0$

11. $x(x + 2)(x - 4) \leq 0$

12. $x(x + 3)(x - 3) \leq 0$

13. $\dfrac{x + 1}{x - 2} > 0$

14. $\dfrac{x - 1}{x + 2} > 0$

15. $\dfrac{x - 3}{x + 2} < 0$

16. $\dfrac{x + 2}{x - 4} < 0$

17. $\dfrac{2x - 1}{x} \geq 0$

18. $\dfrac{x}{3x + 7} \geq 0$

19. $\dfrac{-x + 2}{x - 1} \leq 0$

20. $\dfrac{3 - x}{x + 4} \leq 0$

Solve each of the following inequalities.

21. $x^2 + 2x - 35 < 0$

22. $x^2 + 3x - 54 < 0$

23. $x^2 - 11x + 28 > 0$

24. $x^2 + 11x + 18 > 0$

25. $3x^2 + 13x - 10 \leq 0$

26. $4x^2 - x - 14 \leq 0$

27. $8x^2 + 22x + 5 \geq 0$

28. $12x^2 - 20x + 3 \geq 0$

29. $x(5x - 36) > 32$

30. $x(7x + 40) < 12$

31. $x^2 - 14x + 49 \geq 0$

32. $(x + 9)^2 \geq 0$

33. $4x^2 + 20x + 25 \leq 0$

34. $9x^2 - 6x + 1 \leq 0$

35. $(x + 1)(x - 3)^2 > 0$

36. $(x - 4)^2(x - 1) \leq 0$

37. $\dfrac{2x}{x+3} > 4$ **38.** $\dfrac{x}{x-1} > 2$ **39.** $\dfrac{x-1}{x-5} \le 2$ **40.** $\dfrac{x+2}{x+4} \le 3$

41. $\dfrac{x+3}{x-3} > -2$ **42.** $\dfrac{x-1}{x-2} < -1$ **43.** $\dfrac{3x+2}{x+4} \le 2$ **44.** $\dfrac{2x-1}{x+2} \ge -1$

45. $\dfrac{x+1}{x-2} < 1$ **46.** $\dfrac{x+3}{x-4} \ge 1$

Thoughts into Words

47. One of our problem solving suggestions is to "look for a *guideline* that can be used to set up an equation." What does this suggestion mean to you?

48. Explain how to solve the inequality $(x+1)(x-2)(x-3) < 0$.

Miscellaneous Problems

49. The product $(x-2)(x+3)$ is positive if both factors are negative *or* if both factors are positive. Therefore, we can solve $(x-2)(x+3) > 0$ as follows.

$$(x-2 < 0 \text{ and } x+3 < 0) \qquad \text{or} \qquad (x-2 > 0 \text{ and } x+3 > 0)$$

$$(x < 2 \text{ and } x < -3) \qquad \text{or} \qquad (x > 2 \text{ and } x > -3)$$

$$x < -3 \qquad \text{or} \qquad x > 2$$

The solution set is $\{x \mid x < -3 \text{ or } x > 2\}$. Use this type of analysis to solve each of the following inequalities.

(a) $(x-2)(x+7) > 0$

(b) $(x-3)(x+9) \ge 0$

(c) $(x+1)(x-6) \le 0$ **(d)** $(x+4)(x-8) < 0$

(e) $\dfrac{x+4}{x-7} > 0$ **(f)** $\dfrac{x-5}{x+8} \le 0$

Chapter 10 Summary

(10.1) A number of the form $a + bi$, where a and b are real numbers and i is the imaginary unit defined by $i = \sqrt{-1}$, is a **complex number**.

Two complex numbers $a + bi$ and $c + di$ are said to be *equal* if and only if $a = c$ and $b = d$.

We describe addition and subtraction of complex numbers as follows.

$$(a + bi) + (c + di) = (a + c) + (b + d)i,$$

$$(a + bi) - (c + di) = (a - c) + (b - d)i$$

We can represent a square root of any negative real number as the product of a real number and the imaginary unit i. That is,

$$\sqrt{-b} = i\sqrt{b}, \quad \text{where } b \text{ is a positive real number.}$$

The product of two complex numbers conforms with the product of two binomials. The **conjugate** of $a + bi$ is $a - bi$. The product of a complex number and its conjugate is a real number. Therefore, conjugates are used to simplify expressions, such as $\dfrac{4 + 3i}{5 - 2i}$, which indicate the quotient of two complex numbers.

(10.2) The **standard form for a quadratic equation** in one variable is

$$ax^2 + bx + c = 0,$$

where a, b, and c are real numbers and $a \neq 0$.

Some quadratic equations can be solved by *factoring* and applying the property, $ab = 0$ if and only if $a = 0$ or $b = 0$.

Don't forget that applying the property, if $a = b$, then $a^n = b^n$, might produce extraneous solutions. Therefore, we *must check* all potential solutions.

We can solve some quadratic equations by applying the property, $x^2 = a$ if and only if $x = \pm\sqrt{a}$.

(10.3) To solve a quadratic equation of the form $x^2 + bx = k$ by **completing the square**, we (1) add $\left(\dfrac{b}{2}\right)^2$ to both sides, (2) factor the left side, and (3) apply the property, $x^2 = a$ if and only if $x = \pm\sqrt{a}$.

(10.4) We can solve any quadratic equation of the form $ax^2 + bx + c = 0$ by the **quadratic formula**, which we usually state as

$$x = \frac{-b \pm \sqrt{b^2 - 4ac}}{2a}.$$

The **discriminant**, $b^2 - 4ac$, can be used to determine the nature of the roots of a quadratic equation as follows.

1. If $b^2 - 4ac < 0$, then the equation has two nonreal complex solutions.
2. If $b^2 - 4ac = 0$, then the equation has two equal real solutions.
3. If $b^2 - 4ac > 0$, then the equation has two unequal real solutions.

If x_1 and x_2 are roots of a quadratic equation, then the following relationships exist.

$$x_1 + x_2 = -\frac{b}{a} \quad \text{and} \quad (x_1)(x_2) = \frac{c}{a}$$

These **sum-and-product relationships** can be used to check potential solutions of quadratic equations.

(10.5) To review the strengths and weaknesses of the three basic methods for solving a quadratic equation (factoring, completing the square, the quadratic formula), go back over the examples in this section.

Keep the following suggestions in mind as you solve word problems.

1. Read the problem carefully.
2. Sketch any figure, diagram, or chart that might help organize and analyze the problem.
3. Choose a meaningful variable.
4. Look for a guideline that can be used to set up an equation.
5. Form an equation that translates the guideline from English to algebra.
6. Solve the equation and use the solutions to determine all facts requested in the problem.
7. Check all answers back into the original statement of the problem.

(10.6) The number line, along with **critical numbers** and **test numbers**, provides a good basis for solving **quadratic inequalities** where the polynomial is factorable. We can use this same basic approach to solve inequalities, such as $\dfrac{3x + 1}{x - 4} > 0$, which indicate quotients.

Chapter 10 Review Problem Set

For Problems 1–8, perform the indicated operations and express the answers in the standard form of a complex number.

1. $(-7 + 3i) + (9 - 5i)$
2. $(4 - 10i) - (7 - 9i)$
3. $5i(3 - 6i)$
4. $(5 - 7i)(6 + 8i)$
5. $(-2 - 3i)(4 - 8i)$
6. $(4 - 3i)(4 + 3i)$
7. $\dfrac{4 + 3i}{6 - 2i}$
8. $\dfrac{-1 - i}{-2 + 5i}$

For Problems 9–12, find the discriminant of each equation and determine whether the equation has (1) two nonreal complex solutions, (2) one real solution with a multiplicity of two, (3) two real solutions. Do not solve the equations.

9. $4x^2 - 20x + 25 = 0$
10. $5x^2 - 7x + 31 = 0$
11. $7x^2 - 2x - 14 = 0$
12. $5x^2 - 2x = 4$

For Problems 13–31, solve each equation.

13. $x^2 - 17x = 0$
14. $(x - 2)^2 = 36$
15. $(2x - 1)^2 = -64$
16. $x^2 - 4x - 21 = 0$
17. $x^2 + 2x - 9 = 0$
18. $x^2 - 6x = -34$
19. $4\sqrt{x} = x - 5$
20. $3n^2 + 10n - 8 = 0$
21. $n^2 - 10n = 200$
22. $3a^2 + a - 5 = 0$

23. $x^2 - x + 3 = 0$

24. $2x^2 - 5x + 6 = 0$

25. $2a^2 + 4a - 5 = 0$

26. $t(t + 5) = 36$

27. $x^2 + 4x + 9 = 0$

28. $(x - 4)(x - 2) = 80$

29. $\dfrac{3}{x} + \dfrac{2}{x + 3} = 1$

30. $2x^4 - 23x^2 + 56 = 0$

31. $\dfrac{3}{n - 2} = \dfrac{n + 5}{4}$

For Problems 32–35, solve each inequality and indicate the solution set on a number line graph.

32. $x^2 + 3x - 10 > 0$

33. $2x^2 + x - 21 \le 0$ |

34. $\dfrac{x - 4}{x + 6} \ge 0$

35. $\dfrac{2x - 1}{x + 1} > 4$

For Problems 36–43, set up an equation or a system of equations and solve each problem.

36. Find two numbers whose sum is 6 and whose product is 2.

37. Sherry bought a number of shares of stock for $250. Six months later the value of the stock increased by $5 per share and she sold all but 5 shares and regained her original investment plus a profit of $50. How many shares did she sell and at what price per share?

38. Dave traveled 270 miles in one hour more time than it took Sandy to travel 260 miles. Sandy drove 7 miles per hour faster than Dave. How fast did each one travel?

39. The area of a square is numerically equal to twice its perimeter. Find the length of a side of the square.

40. Find two consecutive even whole numbers such that the sum of their squares is 164.

41. The perimeter of a rectangle is 38 inches and its area is 84 square inches. Find the length and width of the rectangle.

42. It takes Billy 2 hours longer to do a certain job than it takes Janet. They worked together for 2 hours; then Janet left and Billy finished the job in 1 hour. How long would it take each of them to do the job alone?

43. A company has a rectangular parking lot 40 meters wide and 60 meters long. They plan to increase the area of the lot by 1100 square meters by adding a strip of equal width to one side and one end. Find the width of the strip to be added.

Cumulative Review Problem Set   (Chapters 1–10)

For Problems 1–6, express each radical in simplest radical form.

1. $\sqrt{32xy^3}$

2. $\sqrt{48a^3b^4}$

3. $\dfrac{3\sqrt{2}}{2\sqrt{3}}$

4. $-\dfrac{2\sqrt{5}}{\sqrt{8}}$

5. $\sqrt[3]{48x^5}$

6. $\dfrac{3}{4}\sqrt{108}$

For Problems 7 and 8, rationalize the denominator and simplify.

7. $\dfrac{3}{2\sqrt{5} - 1}$

8. $\dfrac{2\sqrt{3}}{\sqrt{6} + 2\sqrt{2}}$

For Problems 9–18, perform the indicated operations and express answers in simplified form.

9. $(3\sqrt{6})(2\sqrt{8})$

10. $(\sqrt{x} + 2)(\sqrt{x} - 3)$

11. $(2\sqrt{3} + \sqrt{5})(\sqrt{3} - 2\sqrt{5})$

12. $(x - 1)(2x^2 - x + 7)$

13. $\dfrac{9y^2}{x^2 + 12x + 36} \div \dfrac{12y}{x^2 + 6x}$

14. $\dfrac{x^2 - x}{4y} \cdot \dfrac{10xy^2}{2x - 2} \div \dfrac{3x^2 + 3x}{15x^2y^2}$

15. $\dfrac{3}{5x} - \dfrac{2}{3x} + \dfrac{5x}{6}$

16. $\dfrac{5}{x - 9} + \dfrac{4}{x^2 - 3x - 54} - \dfrac{1}{x + 6}$

17. $(20x^2 - 39x + 18) \div (5x$

18. $\dfrac{4x^3 - 5x^2 + 2x - 6}{x^2 - 3x}$

For Problems 19–21, use scientific notation to help perform the indicated operations.

19. $\dfrac{(.00063)(960000)}{(3200)(.0000021)}$

20. $(8000)^{\frac{2}{3}}$

21. $\sqrt{.000009}$

For Problems 22–27, find each of the indicated products or quotients and express answers in standard form.

22. $(3 - 2i)(5 + i)$

23. $(-2 + 5i)(4 - 7i)$

24. $(2 - 6i)^2$

25. $\dfrac{2}{5i}$

26. $\dfrac{3 + i}{2 - 4i}$

27. $\dfrac{-1 - 4i}{-2 + 8i}$

For Problems 28–32, evaluate each of the numerical expressions.

28. $-8^{-\frac{1}{3}}$

29. $(-8)^{\frac{1}{3}}$

30. $\left(\dfrac{2}{5}\right)^{-2}$

31. $\sqrt[4]{16}$

32. $2^0 + 2^{-1} + 2^{-2} + 2^{-3}$

For Problems 33–37, factor each of the algebraic expressions completely.

33. $9x^2 + 12xy + 4y^2$

34. $27x^3 - 64y^3$

35. $4x^4 - 13x^2 + 9$

36. $3x^3 - 30x^2 - 72x$

37. $x^2 - 4xy + 4y^2 - 4$

For Problems 38–57, solve each of the equations. Some of these may have complex numbers as solutions.

38. $16n^2 - 40n + 25 = 0$

39. $x^3 = 8x$

40. $n^2 = -4n - 1$

41. $(y + 2)^2 = -24$

42. $|3x - 2| = |-2x - 4|$

43. $x^3 = 8$

44. $(2x - 7)(x + 4) = 0$

45. $(x - 5)(4x - 1) = 23$

46. $\dfrac{2n - 3}{3} + \dfrac{n + 1}{2} = 3$

47. $(x - 1)(x + 4) = (x + 1)(2x - 5)$

48. $.5(3x + .7) = 20.6$

49. $t^2 - 2t = -4$

50. $4x^2 + 23x - 6 = 0$

51. $x^2 - 4x = 192$

52. $\dfrac{3}{2x - 8} - \dfrac{x - 5}{x^2 - 2x - 8} = \dfrac{7}{x + 2}$

53. $\dfrac{3n}{n^2 + n - 6} + \dfrac{2}{n^2 + 4n + 3} = \dfrac{n}{n^2 - n - 2}$

54. $\sqrt{x + 6} = x$

55. $\sqrt{x + 4} = \sqrt{x - 1} + 1$

56. $x^4 + 5x^2 - 36 = 0$

57. $\dfrac{2}{x - 1} = \dfrac{x + 4}{3}$

For Problems 58–61, solve each equation for the indicated variable.

58. $3x - 5y = 10$ for y

59. $\dfrac{3}{4} = \dfrac{y - 1}{x - 2}$ for y

60. $f = \dfrac{1}{\dfrac{1}{a} + \dfrac{1}{b}}$ for b

61. $V = C\left(1 - \dfrac{T}{N}\right)$ for T

For Problems 62–70, solve each inequality and graph the solution set on a number line.

62. $|-2x - 1| > 3$

63. $|3x + 5| < 2$

64. $(x - 2)(x + 4) > 0$

65. $(x + 1)(2x - 3) < 0$

66. $\dfrac{x - 1}{4} - \dfrac{x + 2}{6} \leq \dfrac{1}{8}$

67. $\dfrac{x - 3}{x - 5} \geq 0$

68. $6x^2 + 13x - 5 > 0$

69. $\dfrac{2x}{x + 3} > 4$

70. $\dfrac{3x + 2}{x + 4} \leq 2$

71. Find the slope of the line determined by the points $(2, 7)$ and $(-1, 2)$.

72. Find the slope of the line determined by the equation $-3x - 5y = 6$.

73. Write the equation of the line that contains the point $(2, -4)$ and has a slope of $-\dfrac{2}{3}$.

74. Write the equation of the line that contains the points $(-2, 6)$ and $(1, -4)$.

75. Solve each of the following systems of equations.

(a) $\begin{pmatrix} 7x + 2y = 18 \\ x - 3y = -4 \end{pmatrix}$

(b) $\begin{pmatrix} 4x - 5y = -39 \\ 3x + 7y = 46 \end{pmatrix}$

76. Graph the equation $y = -2x - 4$.

77. Graph the equation $5x - 3y = -15$.

78. Graph the inequality $x - 3y \le 3$.

For Problems 79–86, set up an equation or a system of equations and solve each problem.

79. One of two complementary angles is 6° less than twice the other angle. Find the measure of each angle.

80. A retailer has some golf shoes that cost him $38 a pair. He wants to sell them at a profit of 60% of the selling price. What price should he charge for the shoes?

81. Dela bought a blouse at a 30% discount sale for $21. What was the original price of the blouse?

82. How many liters of pure alcohol must be added to 20 liters of a 20% alcohol solution to obtain a 50% solution?

83. Find two numbers whose sum is -2 and whose product is -35.

84. The sum of the lengths of the two legs of a right triangle is 9 centimeters. If the length of the hypotenuse is $3\sqrt{5}$ centimeters, find the length of each leg.

85. A 3-by-5-inch picture is surrounded by a frame of uniform width. The area of the picture and frame together is 24 square inches. Find the width of the frame.

86. It takes Kent 2 hours longer to do a certain job than it takes Cindy. They worked together for 2 hours; then Cindy left to go shopping and Kent finished the job in 1 hour. How long would it take each of them to do the job alone?

C h a p t e r 11

Graphing
Techniques

In Chapter 5 we started to show some of the connections between algebraic concepts and their geometric counterparts. This was accomplished by introducing the Cartesian or rectangular coordinate system. Within the framework of this coordinate system we studied algebraic equations of the form $Ax + By = C$ and their straight line geometric representations. In this chapter we will pursue more of the connections between algebra and geometry as well as some more connections between mathematics and the real world. Perhaps at this time it would be helpful for you to go back and look over Sections 5.1 through 5.4.

11.1

Some Graphing Ideas

Let's use some examples to review the graphing ideas presented in Chapter 5 and also to introduce some new graphing techniques.

EXAMPLE 1   Graph the equation $-2x + 3y = -6$.

Solution   From Chapter 5 we know that any equation of the form $Ax + By = C$ produces a straight line graph. Therefore, we can proceed as follows. Let $x = 0$, then

$$-2(0) + 3y = -6$$
$$3y = -6$$
$$y = -2.$$

Thus, $(0, -2)$ is a solution. Let $y = 0$, then

$$-2x + 3(0) = -6$$
$$-2x = -6$$
$$x = 3.$$

Thus, $(3, 0)$ is a solution. Technically, the two points $(0, -2)$ and $(3, 0)$ determine the line, but let's find a third point as a check point. Let $x = -3$, then

$$-2(-3) + 3y = -6$$
$$6 + 3y = -6$$
$$3y = -12$$
$$y = -4.$$

Thus, $(-3, -4)$ is a solution. Plotting the points associated with the three points $(0, -2)$, $(3, 0)$, and $(-3, -4)$, and connecting them with a straight line produces the graph of $-2x + 3y = -6$ in Figure 11.1.

Figure 11.1

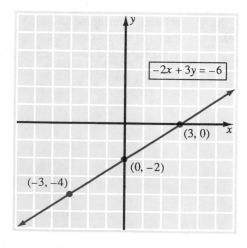

■

The points $(3, 0)$ and $(0, -2)$ in Figure 11.1 are special points. They are the points of the graph that are on the coordinate axes. That is to say, they yield the x-intercept and the y-intercept of the graph. Let's define in general the *intercepts* of a graph.

> The x-coordinates of the points that a graph has in common with the x-axis are called the **x-intercepts** of the graph. (To compute the x-intercepts, let $y = 0$ and solve for x.)
>
> The y-coordinates of the points that a graph has in common with the y-axis are called the **y-intercepts** of the graph. (To compute the y-intercepts, let $x = 0$ and solve for y.)

EXAMPLE 2   Graph $y = (x + 2)(x - 2)$.

Solution   Let's begin by finding the intercepts. If $x = 0$, then

$$y = (0 + 2)(0 - 2)$$

$$y = -4.$$

The point $(0, -4)$ is on the graph. If $y = 0$, then

$$(x + 2)(x - 2) = 0$$

$$x + 2 = 0 \quad \text{or} \quad x - 2 = 0$$

$$x = -2 \quad \text{or} \quad x = 2.$$

The points $(-2, 0)$ and $(2, 0)$ are on the graph. The given equation is in a convenient form for setting up a table of values.

x	y	
0	-4	
-2	0	intercepts
2	0	
1	-3	
-1	-3	other points
3	5	
-3	5	

Plotting these points and connecting them with a smooth curve produces Figure 11.2.

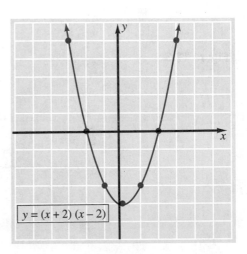

$$y = (x + 2)(x - 2)$$

Figure 11.2   ∎

The curve in Figure 11.2 is called a **parabola** and we will study them in more detail in a later section. However, at this time we do want to emphasize that the parabola in Figure 11.2 is said to be *symmetric with respect to the y-axis*. In other words, the *y*-axis is a line of symmetry. Each half of the curve is a mirror image of the other half through the *y*-axis. Notice in the table of values, that for each ordered pair (x, y), the ordered pair $(-x, y)$ is also a solution. A general test for *y*-axis symmetry can be stated as follows.

---

y-Axis Symmetry

The graph of an equation is symmetric with respect to the *y*-axis if replacing x with $-x$ results in an equivalent equation.

The equation $y = x^2 - 4$ exhibits *y*-axis symmetry because replacing x with $-x$ produces $y = (-x)^2 - 4 = x^2 - 4$. Likewise, the equations $y = -x^2 + 2$, $y = 2x^2 + 5$, and $y = x^4 - x^2$ exhibit *y*-axis symmetry.

EXAMPLE 3   Graph $x = y^2$.

Solution   First, we see that $(0, 0)$ is on the graph and determines both intercepts. Second, the given equation is in a convenient form for setting up a table of values.

x	y	
0	0	intercepts
1	1	
1	-1	
4	2	other points
4	-2	

Plotting these points and connecting them with a smooth curve produces Figure 11.3.

Figure 11.3

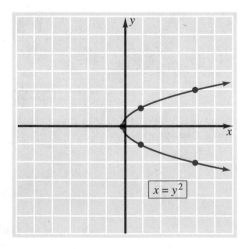

$x = y^2$

The parabola in Figure 11.3 is said to be *symmetric with respect to the x-axis*. Each half of the curve is a mirror image of the other half through the x-axis. Also notice in the table of values, that for each ordered pair (x, y), the ordered pair $(x, -y)$ is a solution. A general test for x-axis symmetry can be stated as follows.

---

x-Axis Symmetry

The graph of an equation is symmetric with respect to the x-axis if replacing y with $-y$ results in an equivalent equation.

The equation $x = y^2$ exhibits x-axis symmetry because replacing y with $-y$ produces $x = (-y)^2 = y^2$. Likewise, the equations $x + y^2 = 5$, $x = y^2 - 5$, and $x = 2y^4 - 3y^2$ exhibit x-axis symmetry.

EXAMPLE 4   Graph $y = \dfrac{1}{x}$.

Solution   First, let's find the intercepts. Let $x = 0$; then $y = \dfrac{1}{x}$ becomes $y = \dfrac{1}{0}$ and $\dfrac{1}{0}$ is undefined. Thus, there is no y-intercept. Let $y = 0$; then $y = \dfrac{1}{x}$ becomes $0 = \dfrac{1}{x}$ and there are no values of x that will satisfy this equation. In other words, this graph has no points on either the x-axis or the y-axis.

Second, let's set up a table of values and keep in mind that neither x nor y can equal zero.

x	y
$\dfrac{1}{2}$	2
1	1
2	$\dfrac{1}{2}$
3	$\dfrac{1}{3}$
$-\dfrac{1}{2}$	-2
-1	-1
-2	$-\dfrac{1}{2}$
-3	$-\dfrac{1}{3}$

In Figure 11.4(a) we have plotted the points associated with the solutions from the table. Since the graph does not intersect either axis, it must consist of two branches. Thus, connecting the points in the first quadrant with a smooth curve and then connecting the points in the third quadrant with a smooth curve, we obtain the graph in Figure 11.4(b).

Figure 11.4

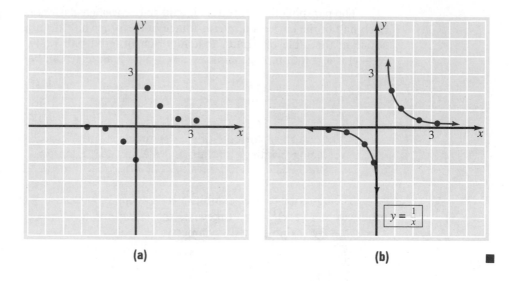

(a)   (b)   ■

The curve in Figure 11.4 is said to be *symmetric with respect to the origin*. Each half of the curve is a mirror image of the other half through the origin. Notice in the table of values, that for each ordered pair (x, y), the ordered pair $(-x, -y)$ is also a solution. A general test for origin symmetry can be stated as follows.

Origin Symmetry

The graph of an equation is symmetric with respect to the origin if replacing x with $-x$ and y with $-y$ results in an equivalent equation.

The equation $y = \dfrac{1}{x}$ exhibits origin symmetry because replacing x with $-x$ and y with $-y$ produces $-y = \dfrac{1}{-x}$ which is equivalent to $y = \dfrac{1}{x}$. $\left(\text{We can multiply both sides of } -y = \dfrac{1}{-x} \text{ by } -1 \text{ and obtain } y = \dfrac{1}{x}.\right)$ Likewise, the equations $xy = 4$, $y = x^3$, and $y = x^5$ exhibit origin symmetry.

Let's pause for a moment and pull together the graphing techniques that have been introduced thus far. Following is a list of graphing suggestions. The order of the suggestions indicates the order in which we usually attack a new graphing problem.

1. Determine the type of symmetry that the equation exhibits.
2. Find the intercepts.
3. Solve the equation for y in terms of x or for x in terms of y if it is not already in such a form.
4. Set up a table of ordered pairs that satisfy the equation. The type of symmetry will affect your choice of values in the table. (This will be illustrated in a moment.)
5. Plot the points associated with the ordered pairs from the table, and connect them with a smooth curve. Then, if appropriate, reflect this part of the curve according to the symmetry shown by the equation.

EXAMPLE 5   Graph $x^2y = -2$.

Solution   Since replacing x with $-x$ produces $(-x)^2y = -2$ or equivalently $x^2y = -2$, the equation exhibits y-axis symmetry. There are no intercepts because neither x nor y can equal 0. Solving the equation for y produces $y = \dfrac{-2}{x^2}$. Since the equation exhibits y-axis symmetry, let's use only positive values for x and then we can reflect across the y-axis.

Let's plot the points determined by the table, connect them with a smooth curve, and reflect this portion of the curve across the y-axis. Figure 11.5 is the result of this process.

x	y
1	-2
2	$-\dfrac{1}{2}$
3	$-\dfrac{2}{9}$
4	$-\dfrac{1}{8}$
$\dfrac{1}{2}$	-8

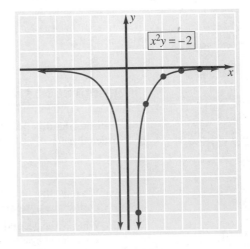

Figure 11.5   ■

EXAMPLE 6   Graph $x = y^3$.

Solution   Since replacing x with $-x$ and y with $-y$ produces $-x = (-y)^3 = -y^3$, which is equivalent to $x = y^3$, the given equation exhibits origin symmetry. If $x = 0$, then $y = 0$; so the origin is a point of the graph. The given equation is in an easy form for deriving a table of values.

x	y
0	0
8	2
1	1
$\dfrac{1}{8}$	$\dfrac{1}{2}$
27	3
64	4

Let's plot the points determined by the table, connect them with a smooth curve, and reflect this portion of the curve through the origin to produce Figure 11.6.

Figure 11.6

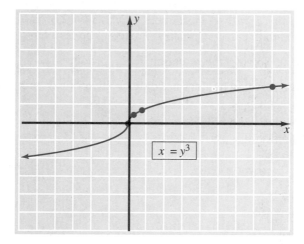

$$x = y^3$$

Problem Set 11.1

1. Indicate the quadrant that contains each of the points represented by the following ordered pairs.

(a) $(-3, -4)$ (b) $(5, -6)$ (c) $(-6, 8)$
(d) $(-1, -6)$ (e) $(4, 10)$ (f) $(9, -1)$

2. (a) What is the first coordinate of any point on the vertical axis in a rectangular coordinate system?

(b) What is the second coordinate of any point on the horizontal axis in a rectangular coordinate system?

3. (a) In which quadrants do the coordinates of a point have the same sign?

(b) In which quadrants do the coordinates of a point have opposite signs?

(c) In which quadrants is the abscissa negative?

(d) In which quadrants is the ordinate negative?

For each of the following points, determine the points that are symmetric with respect to the (a) x-axis, (b) y-axis, and (c) origin (Problems 4–9).

4. $(2, 6)$ **5.** $(-3, 1)$ **6.** $(-2, -4)$ **7.** $(7, -2)$ **8.** $(0, -4)$ **9.** $(5, 0)$

For Problems 10–23, determine the type of symmetry (x-axis, y-axis, origin) possessed by each of the graphs of the following equations. *Do not* sketch the graph.

10. $x^2 + 2y = 4$

11. $-3x + 2y^2 = -4$

12. $x = -y^2 + 5$

13. $y = 4x^2 + 13$

14. $xy = -6$

15. $2x^2y^2 = 5$

16. $2x^2 + 3y^2 = 9$

17. $x^2 - 2x - y^2 = 4$

18. $y = x^2 - 6x - 4$

19. $y = 2x^2 - 7x - 3$

20. $y = x$

21. $y = 2x$

22. $y = x^4 + 4$

23. $y = x^4 - x^2 + 2$

For Problems 24–51, graph each of the equations.

24. $x - y = -1$

25. $x - y = 4$

26. $y = 3x - 6$

27. $y = 2x + 4$

28. $y = -2x + 1$

29. $y = -3x - 1$

30. $y = x^2 - 1$

31. $y = x^2 + 2$

32. $y = (x - 1)(x + 1)$

33. $y = (x + 1)(x + 3)$

34. $y = \dfrac{2}{x^2}$

35. $y = \dfrac{-1}{x^2}$

36. $2x + y = 6$

37. $2x - y = 4$

38. $x = -y^2$

39. $x = y^2 + 2$

40. $xy = -3$

41. $xy = 2$

42. $x^2y = 4$

43. $xy^2 = -4$

44. $y^3 = x^2$

45. $y^2 = x^3$

46. $y = \dfrac{-2}{x^2 + 1}$

47. $y = \dfrac{4}{x^2 + 1}$

48. $x = -y^3$

49. $x = y^2 - 1$

50. $x = (y - 2)^2$

51. $x = -y^3 + 2$

52. The equation $F = \dfrac{9}{5}C + 32$ can be used to convert from degrees in Celsius to degrees in Fahrenheit. Complete the following table and graph the solutions on the set of axes below.

C	0	5	10	15	20	−5	−10	−15	−20	−25
F										

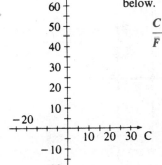

53. At \$0.06 per kilowatt-hour, the equation $A = 0.06t$ determines the amount, A, of an electric bill for t hours. Complete the following table of values.

Hours	t	696	720	740	775	782
Dollars and cents	A					

54. Suppose that a used car dealer determines the selling price of his cars by using a markup of 60% of the cost. Let s represent the selling price and c the cost, and use the following equation.

$$s = c + 0.6c = 1.6c.$$

Complete the following table using the equation $s = 1.6c$.

Dollars	c	250	325	575	895	1095
Dollars	s					

Miscellaneous Problems

55. Use a graphics calculator to graph the equations in Problems 33, 35, 37, 39, 41, 43, 45, 47, 49, and 51.

11.2

Distance Formula and Circles

As we work with the rectangular coordinate system it is sometimes necessary to express the length of certain line segments. In other words, we need to be able to find the *distance between* two points. Let's first consider two specific examples and then develop the general distance formula.

EXAMPLE 1   Find the distance between the points $A(2, 2)$ and $B(5, 2)$ and also between the points $C(-2, 5)$ and $D(-2, -4)$.

Solution   Let's plot the points and draw $\overline{AB}$ as in Figure 11.7. Since $\overline{AB}$ is parallel to the x-axis, its length can be expressed as $|5 - 2|$ or $|2 - 5|$. (The absolute value symbol is used to ensure a nonnegative value.) Thus, the length of $\overline{AB}$ is 3 units. Likewise the length of $\overline{CD}$ is $|5 - (-4)| = |-4 - 5| = 9$ units.

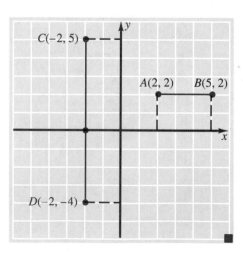

Figure 11.7

EXAMPLE 2   Find the distance between the points $A(2, 3)$ and $B(5, 7)$.

Solution   Let's plot the points and form a right triangle as indicated in Figure 11.8. Notice that the coordinates of point C are $(5, 3)$. Since $\overline{AC}$ is parallel to the horizontal axis, its length is easily determined to be 3 units. Likewise, $\overline{CB}$ is parallel to the vertical axis and its length is 4 units. Let d represent the length of $\overline{AB}$ and apply the Pythagorean Theorem to obtain

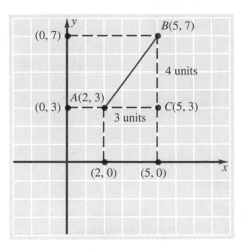

$$d^2 = 3^2 + 4^2$$

$$d^2 = 9 + 16$$

$$d^2 = 25$$

$$d = \pm\sqrt{25} = \pm 5.$$

Figure 11.8

Since *distance between* is a nonnegative value, the length of $\overline{AB}$ is 5 units. ∎

We can use the approach in Example 2 to develop a general distance formula for finding the distance between any two points in a coordinate plane. The development proceeds as follows.

1. Let $P_1(x_1, y_1)$ and $P_2(x_2, y_2)$ represent any two points in a coordinate plane;

2. Form a right triangle as indicated in Figure 11.9. The coordinates of the vertex of the right angle, point R, are (x_2, y_1).

The length of $\overline{P_1R}$ is $|x_2 - x_1|$ and the length of $\overline{RP_2}$ is $|y_2 - y_1|$. (The absolute value symbol is used to ensure a nonnegative value.) Let d represent the length of P_1P_2 and apply the Pythagorean Theorem to obtain

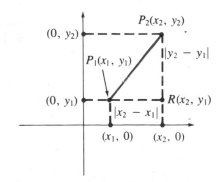

$$d^2 = |x_2 - x_1|^2 + |y_2 - y_1|^2.$$

Since $|a|^2 = a^2$, the **distance formula** can be stated as

Figure 11.9

$$d = \sqrt{(x_2 - x_1)^2 + (y_2 - y_1)^2}.$$

It makes no difference which point you call P_1 or P_2 when using the distance formula. Remember, if you forget the formula, don't panic, but merely form a right triangle and apply the Pythagorean Theorem as we did in Example 2. Let's consider an example that demonstrates the use of the distance formula.

EXAMPLE 3   Verify that the points $(-3, 6)$, $(3, 4)$, and $(1, -2)$ are vertices of an isosceles triangle. (An isosceles triangle has two sides of the same length.)

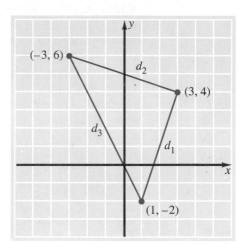

Figure 11.10

Solution   Let's plot the points and draw the triangle (Figure 11.10). Use the distance formula to find the lengths $d_1, d_2,$ and d_3 as follows.

$$d_1 = \sqrt{(3 - 1)^2 + (4 - (-2))^2}$$
$$= \sqrt{2^2 + 6^2} = \sqrt{40} = 2\sqrt{10}$$
$$d_2 = \sqrt{(-3 - 3)^2 + (6 - 4)^2}$$
$$= \sqrt{(-6)^2 + 2^2} = \sqrt{40}$$
$$= 2\sqrt{10}$$
$$d_3 = \sqrt{(-3 - 1)^2 + (6 - (-2))^2}$$
$$= \sqrt{(-4)^2 + 8^2} = \sqrt{80} = 4\sqrt{5}$$

Since $d_1 = d_2$, we know that it is an isosceles triangle.   ∎

Circles

The distance formula, $d = \sqrt{(x_2 - x_1)^2 + (y_2 - y_1)^2}$, when it applies to the definition of a circle produces what is known as the **standard equation of a circle**. We start with a precise definition of a circle.

DEFINITION 11.1

> A **circle** is the set of all points in a plane equidistant from a given fixed point called the **center**. A line segment determined by the center and any point on the circle is called a **radius**.

Let's consider a circle that has a radius of length r and a center at (h, k) on a coordinate system (Figure 11.11). For any point P on the circle with coordinates (x, y), the length of a radius (denoted by r) can be expressed as

$$r = \sqrt{(x - h)^2 + (y - k)^2}.$$

Figure 11.11

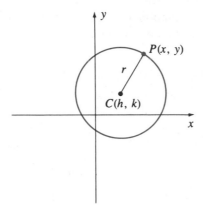

Thus, squaring both sides of the equation, we obtain the **standard form of the equation of a circle**:

$$(x - h)^2 + (y - k)^2 = r^2.$$

We can use the standard form of the equation of a circle to solve two basic kinds of circle problems, namely, (1) given the coordinates of the center and the length of a radius of a circle, find its equation, and (2) given the equation of a circle, find its center and the length of a radius. Let's look at some examples of such problems.

EXAMPLE 4   Write the equation of a circle that has its center at $(3, -5)$ and a radius of length 6 units.

Solution   Let's substitute 3 for h, -5 for k, and 6 for r into the standard form $(x - h)^2 + (y - k)^2 = r^2$ that becomes $(x - 3)^2 + (y + 5)^2 = 6^2$, which we can simplify as follows.

$$(x - 3)^2 + (y + 5)^2 = 6^2$$
$$x^2 - 6x + 9 + y^2 + 10y + 25 = 36$$
$$x^2 + y^2 - 6x + 10y - 2 = 0$$ ∎

Notice in Example 4 that we simplified the equation to the form $x^2 + y^2 + Dx + Ey + F = 0$, where D, E, and F are integers. This is another commonly used form when working with circles.

EXAMPLE 5   Graph $x^2 + y^2 + 4x - 6y + 9 = 0$.

Solution   This equation is of the form $x^2 + y^2 + Dx + Ey + F = 0$, so its graph is a circle.

We can change the given equation into the form $(x - h)^2 + (y - k)^2 = r^2$ by completing the square on x and on y as follows.

$$x^2 + y^2 + 4x - 6y + 9 = 0$$

$$(x^2 + 4x + \underline{\quad}) + (y^2 - 6y + \underline{\quad}) = -9$$

$$(x^2 + 4x + 4) + (y^2 - 6y + 9) = -9 + 4 + 9$$

added 4 to complete the square on x added 9 to complete the square on y added 4 and 9 to compensate for the 4 and 9 added on the left side

$$(x + 2)^2 + (y - 3)^2 = 4$$

$$(x - (-2))^2 + (y - 3)^2 = 2^2$$

         ↑   ↑   ↑

         h k r

The center of the circle is at $(-2, 3)$ and the length of a radius is 2 units (Figure 11.12).

Figure 11.12

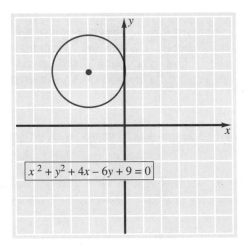

$$x^2 + y^2 + 4x - 6y + 9 = 0$$

As demonstrated by Examples 4 and 5, both forms, $(x - h)^2 + (y - k)^2 = r^2$ and $x^2 + y^2 + Dx + Ey + F = 0$, play an important role when solving problems that deal with circles.

Finally, we need to recognize that the standard form of a circle that has its center at the origin is $x^2 + y^2 = r^2$. This is merely the result of letting $h = 0$ and $k = 0$ in the general standard form.

$$(x - h)^2 + (y - k)^2 = r^2$$

$$(x - 0)^2 + (y - 0)^2 = r^2$$

$$x^2 + y^2 = r^2$$

Thus, by inspection we can recognize that $x^2 + y^2 = 9$ is a circle with its center at the origin; the length of a radius is 3 units. Likewise, the equation of a circle that has its center at the origin and a radius of length 6 units is $x^2 + y^2 = 36$.

Before tackling the next problem, let's review a little plane geometry. A **tangent** to a circle is a line that has one and only one point in common with the circle. This common point is called a point of tangency. Furthermore, any radius drawn to the point of tangency is perpendicular to the tangent line. Now let's consider a problem that uses some ideas about lines from Chapter 5 and the circle concepts of this section.

EXAMPLE 6   Write the equation of the line that is tangent to the circle $x^2 + y^2 = 5$ at the point $(2, 1)$.

Solution   Figure 11.13 is a sketch of the situation presented in the example. We have a circle with its center at the origin and a radius of length $\sqrt{5}$ units. A radius is drawn from the center of the circle to the point $(2, 1)$. The dashed line is the line tangent to the circle at $(2, 1)$. The slope of the radius is $m = \dfrac{y_2 - y_1}{x_2 - x_1} = \dfrac{1 - 0}{2 - 0} = \dfrac{1}{2}$. Therefore, the slope of the tangent line is -2. (Remember that perpendicular lines have slopes that are negative reciprocals of each other.) Now using any point (x, y) on the tangent line we can write the equation of the tangent line as follows.

$$\frac{y - 1}{x - 2} = -2$$

$$\frac{y - 1}{x - 2} = \frac{2}{-1}$$

$$2x - 4 = -y + 1$$

$$2x + y = 5$$

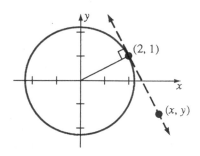

Figure 11.13   ■

Problem Set 11.2

For Problems 1–12, find the distance between each of the pairs of points. Express answers in simplest radical form.

1. $(-2, -1), (7, 11)$ **2.** $(2, 1), (10, 7)$ **3.** $(1, -1), (3, -4)$

4. $(-1, 3), (2, -2)$ **5.** $(6, -4), (9, -7)$ **6.** $(-5, 2), (-1, 6)$

7. $(-3, 3), (0, -3)$ **8.** $(-2, -4), (4, 0)$ **9.** $(1, -6), (-5, -6)$

10. $(-2, 3), (-2, -7)$ **11.** $(1, 7), (4, -2)$ **12.** $(6, 4), (-4, -8)$

13. Verify that the points $(-3, 1), (5, 7)$, and $(8, 3)$ are vertices of a right triangle. [*Hint*: if $a^2 + b^2 = c^2$, then it is a right triangle with the right angle opposite side c.]

14. Verify that the points $(0, 3), (2, -3)$, and $(-4, -5)$ are vertices of an isosceles triangle.

15. Verify that the points $(7, 12)$ and $(11, 18)$ divide the line segment joining $(3, 6)$ and $(15, 24)$ into three segments of equal length.

16. Verify that $(3, 1)$ is the midpoint of the line segment joining $(-2, 6)$ and $(8, -4)$.

Find the center and length of a radius of each of the following circles.

17. $x^2 + y^2 - 2x - 6y - 6 = 0$

18. $x^2 + y^2 + 4x - 12y + 39 = 0$

19. $x^2 + y^2 + 6x + 10y + 18 = 0$

20. $x^2 + y^2 - 10x + 2y + 1 = 0$

21. $x^2 + y^2 = 10$

22. $x^2 + y^2 + 4x + 14y + 50 = 0$

23. $x^2 + y^2 - 16x + 6y + 71 = 0$

24. $x^2 + y^2 = 12$

25. $x^2 + y^2 + 6x - 8y = 0$

26. $x^2 + y^2 - 16x + 30y = 0$

27. $4x^2 + 4y^2 + 4x - 32y + 33 = 0$

28. $9x^2 + 9y^2 - 6x - 12y - 40 = 0$

Write the equation of each of the following circles. Express the final equation in the form $x^2 + y^2 + Dx + Ey + F = 0$.

29. center at $(3, 5)$ and $r = 5$

30. center at $(2, 6)$ and $r = 7$

31. center at $(-4, 1)$ and $r = 8$

32. center at $(-3, 7)$ and $r = 6$

33. center at $(-2, -6)$ and $r = 3\sqrt{2}$

34. center at $(-4, -5)$ and $r = 2\sqrt{3}$

35. center at $(0, 0)$ and $r = 2\sqrt{5}$

36. center at $(0, 0)$ and $r = \sqrt{7}$

37. center at $(5, -8)$ and $r = 4\sqrt{6}$

38. center at $(4, -10)$ and $r = 8\sqrt{2}$

39. Find the equation of the circle that passes through the origin and has its center at $(0, 4)$.

40. Find the equation of the circle that passes through the origin and has its center at $(-6, 0)$.

41. Find the equation of the circle that passes through the origin and has its center at $(-4, 3)$.

42. Find the equation of the circle that passes through the origin and has its center at $(8, -15)$.

43. (a) Find the equation of the line that is tangent to the circle $x^2 + y^2 = 10$ at the point $(1, 3)$.

(b) Find the equation of the line that is tangent to the circle $x^2 + y^2 = 10$ at the point $(-3, 1)$.

(c) Are the tangent lines in parts (a) and (b) perpendicular? Defend your answer.

44. (a) Find the equation of the line that is tangent to the circle $x^2 + y^2 = 5$ at the point $(1, -2)$.

(b) Find the equation of the line that is tangent to the circle $x^2 + y^2 = 5$ at the point $(-1, -2)$.

(c) Are the tangent lines in parts (a) and (b) perpendicular? Defend your answer.

45. (a) Find the center, the length of a radius, and sketch the circle $x^2 + y^2 + 4x - 6y - 4 = 0$.

(b) Find the equation of the line that is tangent to the circle in part (a) at the point $(-1, -1)$.

46. (a) Find the center, the length of a radius, and sketch the circle $x^2 + y^2 - 2x - 8y + 15 = 0$.

(b) Find the equation of the line that is tangent to the circle in part (a) at the point $(2, 3)$.

Miscellaneous Problems

47. By expanding $(x - h)^2 + (y - k)^2 = r^2$, we obtain $x^2 - 2hx + h^2 + y^2 - 2ky + k^2 - r^2 = 0$. Compare this result to the form $x^2 + y^2 + Dx + Ey + F = 0$, and we see that $D = -2h$, $E = -2k$, and $F = h^2 + k^2 - r^2$. Therefore, the center and length of a radius of a circle can be found by using $h = \dfrac{D}{-2}$, $k = \dfrac{E}{-2}$, and $r = \sqrt{h^2 + k^2 - F}$. Use these relationships to find the center and length of a radius of each of the following circles.

(a) $x^2 + y^2 - 2x - 8y + 8 = 0$

(b) $x^2 + y^2 + 4x - 14y + 49 = 0$

(c) $x^2 + y^2 + 12x + 8y - 12 = 0$

(d) $x^2 + y^2 - 16x + 20y + 115 = 0$

(e) $x^2 + y^2 - 12y - 45 = 0$

(f) $x^2 + y^2 + 14x = 0$

 48. Use a graphics calculator to graph each of the following circles.

(a) $x^2 + y^2 = 25$ **(b)** $x^2 + y^2 = 64$

(c) $x^2 + y^2 = 144$ **(d)** $x^2 - 4x + y^2 - 5 = 0$

Graphing Parabolas

Let's begin this section by using the concepts of intercepts and symmetry to help sketch the graph of the equation $y = x^2$.

EXAMPLE 1   Graph $y = x^2$.

Solution   If we replace x with $-x$, the given equation becomes $y = (-x)^2 = x^2$; therefore, we have y-axis symmetry. The origin, $(0, 0)$ is a point of the graph. Now we can set up a table of values that uses nonnegative values for x.

Plot the points determined by the table, connect them with a smooth curve, and reflect that portion of the curve across the y-axis to produce Figure 11.14.

x	y
0	0
1	1
2	4
3	9
$\dfrac{1}{2}$	$\dfrac{1}{4}$

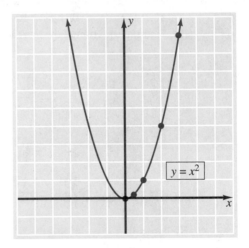

Figure 11.14   ■

The curve in Figure 11.14 is called a **parabola**. The graph of any equation of the form $y = ax^2 + bx + c$, where a, b, and c are real numbers and $a \neq 0$, is a parabola. As we work with parabolas, we will use the vocabulary indicated in Figure 11.15 on page 479.

One way to graph parabolas relies on the ability to find the vertex, to determine whether the parabola opens upward or downward, and to locate two points on opposite sides of the line of symmetry. For some of this information we can compare the parabolas produced by various types of equations such as $y = x^2 + k$, $y = ax^2$, $y = (x - h)^2$, and $y = a(x - h)^2 + k$ to the **basic parabola** produced by the equation $y = x^2$. First, let's consider some equations of the form $y = x^2 + k$, where k is a constant.

Figure 11.15

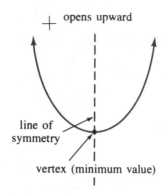

opens upward

line of
symmetry

vertex (minimum value)

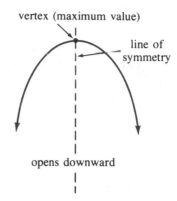

vertex (maximum value)

line of
symmetry

opens downward

EXAMPLE 2   Graph $y = x^2 + 1$.

Solution   Let's set up a table of values to compare y-values for $y = x^2 + 1$ to corresponding y-values for $y = x^2$

x	$y = x^2$	$y = x^2 + 1$
0	0	1
1	1	2
2	4	5
-1	1	2
-2	4	5

It should be evident that y-values for $y = x^2 + 1$ are *one larger* than corresponding y-values for $y = x^2$. For example, if $x = 2$ then $y = 4$ for the equation $y = x^2$, but if $x = 2$ then $y = 5$ for the equation $y = x^2 + 1$. Thus, the graph of $y = x^2 + 1$ is the same as the graph of $y = x^2$ but *moved up* 1 *unit* (Figure 11.16).

Figure 11.16

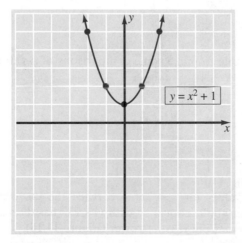

$y = x^2 + 1$

EXAMPLE 3   Graph $y = x^2 - 2$.

Solution   The y-values for $y = x^2 - 2$ are 2 *less than* the corresponding y-values for $y = x^2$ as indicated in the following table.

x	$y = x^2$	$y = x^2 - 2$
0	0	-2
1	1	-1
2	4	2
-1	1	-1
-2	4	2

Thus, the graph of $y = x^2 - 2$ is the same as the graph of $y = x^2$, but *moved down* 2 *units* (Figure 11.17).

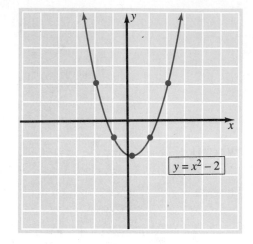

Figure 11.17   ■

In general, the graph of a quadratic equation of the form $y = x^2 + k$ is the same as the graph of $y = x^2$ except moved up or down k units depending on whether k is positive or negative. We say that the graph of $y = x^2 + k$ is a **vertical translation** of the graph of $y = x^2$. Also note that equations of the form $y = x^2 + k$ exhibit y-axis symmetry.

Now let's consider some quadratic equations of the form $y = ax^2$, where a is a nonzero constant.

EXAMPLE 4   Graph $y = 2x^2$.

Solution   Let's again use a table to make some comparisons of y-values.

x	$y = x^2$	$y = 2x^2$
0	0	0
1	1	2
2	4	8
-1	1	2
-2	4	8

Obviously, the y-values for $y = 2x^2$ are *twice* the corresponding y-values for $y = x^2$. Thus, the parabola associated with $y = 2x^2$ has the same vertex (the origin) as the graph of $y = x^2$, but it is narrower (Figure 11.18).

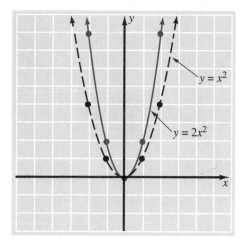

Figure 11.18

EXAMPLE 5   Graph $y = \dfrac{1}{2}x^2$.

Solution   The following table indicates some comparisons of y-values.

x	$y = x^2$	$y = \dfrac{1}{2}x^2$
0	0	0
1	1	$\dfrac{1}{2}$
2	4	2
-1	1	$\dfrac{1}{2}$
-2	4	2

The y-values for $y = \dfrac{1}{2}x^2$ are *one-half* of the corresponding y-values for $y = x^2$. Therefore, the graph of $y = \dfrac{1}{2}x^2$ is wider than the basic parabola (Figure 11.19).

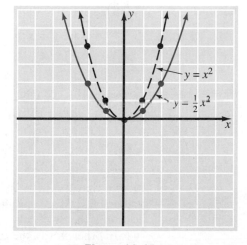

Figure 11.19

EXAMPLE 6   Graph $y = -x^2$

Solution

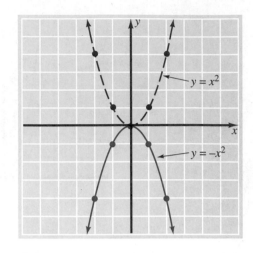

x	$y = x^2$	$y = -x^2$
0	0	0
1	1	-1
2	4	-4
-1	1	-1
-2	4	-4

The y-values for $y = -x^2$ are the *opposites* of the corresponding y-values for $y = x^2$. Thus, the graph of $y = -x^2$ is **a reflection across the x-axis** of the basic parabola (Figure 11.20).

Figure 11.20   ■

In general, the graph of a quadratic equation of the form $y = ax^2$ has its vertex at the origin, exhibits y-axis symmetry, and opens upward if a is positive and downward if a is negative. The parabola is narrower than the basic parabola if $|a| > 1$ and wider if $|a| < 1$.

Let's continue our investigation of quadratic equations by considering those of the form $y = (x - h)^2$, where h is a nonzero constant.

EXAMPLE 7   Graph $y = (x - 2)^2$.

Solution   A fairly extensive table of values illustrates a pattern.

x	$y = x^2$	$y = (x - 2)^2$
-2	4	16
-1	1	9
0	0	4
1	1	1
2	4	0
3	9	1
4	16	4
5	25	9

Notice that $y = (x - 2)^2$ and $y = x^2$ take on the same y-values, *but* for different values of x. More specifically, if $y = x^2$ achieves a certain y-value at x equals a constant, then $y = (x - 2)^2$ achieves the same y-value at x equals the *constant plus two*. In other words, the graph of $y = (x - 2)^2$ is the same as the graph of $y = x^2$ but *moved two units to the right* (Figure 11.21).

Figure 11.21

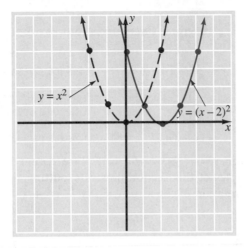

EXAMPLE 8   Graph $y = (x + 3)^2$.

Solution

x	$y = x^2$	$y = (x + 3)^2$
-3	9	0
-2	4	1
-1	1	4
0	0	9
1	1	16
2	4	25
3	9	36

If $y = x^2$ achieves a certain y-value at x equals a constant, then $y = (x + 3)^2$ achieves that same y-value at x equals that *constant minus three*. Therefore, the graph of $y = (x + 3)^2$ is the same as the graph of $y = x^2$ *but moved three units to the left* (Figure 11.22).

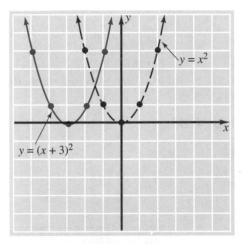

Figure 11.22   ■

In general, the graph of a quadratic equation of the form $y = (x - h)^2$ is the same as the graph of $y = x^2$ but moved to the right h units if h is positive or moved to the left h units if h is negative. For example,

$$y = (x - 4)^2 \qquad\qquad \longrightarrow \text{ moved to the } right \text{ 4 units}$$

$$y = (x + 2)^2 = (x - (-2))^2 \qquad \longrightarrow \text{ moved to the } left \text{ 2 units}$$

We say that the graph of $y = (x - h)^2$ is a **horizontal translation** of the graph of

$y = x^2$. The following diagram summarizes our work thus far in graphing quadratic equations.

$$y = x^2 \longrightarrow y = x^2 + \textcircled{k} \quad \longrightarrow \quad \text{moves the parabola up or down}$$

$$y = x^2 \longrightarrow y = \textcircled{a}x^2 \quad \longrightarrow \quad \text{affects the width and which way the parabola opens}$$

$$\text{basic parabola} \quad y = (x - \textcircled{h})^2 \quad \longrightarrow \quad \text{moves the parabola right or left}$$

Equations of the form $y = x^2 + k$ and $y = ax^2$ are symmetrical about the y-axis. The final two examples of this section illustrate putting these ideas together to graph a quadratic equation of the form $y = a(x - h)^2 + k$.

EXAMPLE 9   Graph $y = 2(x - 3)^2 + 1$.

Solution

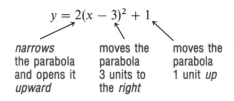

$$y = 2(x - 3)^2 + 1$$

narrows the parabola and opens it upward

moves the parabola 3 units to the *right*

moves the parabola 1 unit *up*

The parabola is drawn in Figure 11.23. Two points in addition to the vertex are located to determine the parabola.

Figure 11.23

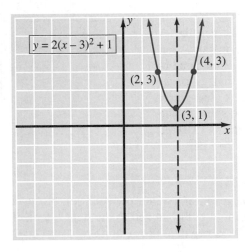

$$y = 2(x - 3)^2 + 1$$

$(2, 3)$

$(4, 3)$

$(3, 1)$

EXAMPLE 10   Graph $y = -\dfrac{1}{2}(x + 1)^2 - 2$.

Solution

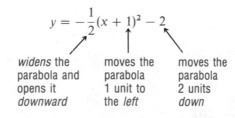

$$y = -\frac{1}{2}(x + 1)^2 - 2$$

widens the parabola and opens it *downward*

moves the parabola 1 unit to the *left*

moves the parabola 2 units *down*

The parabola is drawn in Figure 11.24.

Figure 11.24

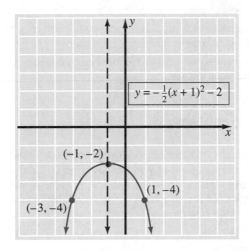

Problem Set 11.3

Graph each of the following parabolas.

1. $y = x^2 + 2$ **2.** $y = x^2 + 3$ **3.** $y = x^2 - 1$

4. $y = x^2 - 5$ **5.** $y = 4x^2$ **6.** $y = 3x^2$

7. $y = -3x^2$ **8.** $y = -4x^2$ **9.** $y = \frac{1}{3}x^2$

10. $y = \frac{1}{4}x^2$ **11.** $y = -\frac{1}{2}x^2$ **12.** $y = -\frac{2}{3}x^2$

13. $y = (x - 1)^2$ **14.** $y = (x - 3)^2$

15. $y = (x + 4)^2$ **16.** $y = (x + 2)^2$

17. $y = 3x^2 + 2$ **18.** $y = 2x^2 + 3$

19. $y = -2x^2 - 2$ **20.** $y = \frac{1}{2}x^2 - 2$

21. $y = (x - 1)^2 - 2$ **22.** $y = (x - 2)^2 + 3$

23. $y = (x + 2)^2 + 1$ **24.** $y = (x + 1)^2 - 4$

25. $y = 3(x - 2)^2 - 4$

26. $y = 2(x + 3)^2 - 1$

27. $y = -(x + 4)^2 + 1$

28. $y = -(x - 1)^2 + 1$

29. $y = -\dfrac{1}{2}(x + 1)^2 - 2$

30. $y = -3(x - 4)^2 - 2$

Thoughts into Words

31. Explain how you would graph the circle $x^2 + y^2 - 8x + 2y + 8 = 0$.

32. Explain how you would graph the parabola $y = -2(x - 1)^2 + 4$.

33. Is it true that if a graph has both x-axis and y-axis symmetry, then it must have origin symmetry? Defend your answer.

34. Is it possible for a graph to have origin symmetry, but not have symmetry with respect to either axis? Defend your answer.

Miscellaneous Problems

35. The points (x, y) and (y, x) are mirror images of each other across the line $y = x$. Therefore, by interchanging x and y in the equation $y = ax^2 + bx + c$, we obtain the equation of its mirror image across the line $y = x$, namely, $x = ay^2 + by + c$. Thus to graph $x = y^2 + 2$, we can first graph $y = x^2 + 2$ and then reflect it across the line $y = x$ as indicated in the following figure.

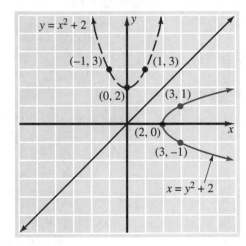

Graph each of the following parabolas.

(a) $x = y^2$

(b) $x = -y^2$

(c) $x = y^2 - 1$

(d) $x = -y^2 + 3$

(e) $x = -2y^2$

(f) $x = 3y^2$

(g) $x = y^2 + 4y + 7$

(h) $x = y^2 - 2y - 3$

11.4

More Parabolas and Some Ellipses

We are now ready to graph quadratic equations of the form $y = ax^2 + bx + c$, where a, b, and c are real numbers and $a \neq 0$. The general approach is one of changing equations of the form $y = ax^2 + bx + c$ to the form $y = a(x - h)^2 + k$. Then we can proceed to graph them as we did in the previous section. The process of *completing the square* serves as the basis for making the change in the form of the equations. Let's consider some examples to illustrate the details.

EXAMPLE 1   Graph $y = x^2 + 6x + 8$.

Solution

$$y = x^2 + 6x + 8$$

Add 9, which is the square of one-half the coefficient of x.

$$y = x^2 + 6x + 9 + 8 - 9 \longleftarrow$$

Subtract 9 to compensate for the 9 that was added.

$$y = (x + 3)^2 - 1$$

$$x^2 + 6x + 9 = (x + 3)^2$$

The graph of $y = (x + 3)^2 - 1$ is the basic parabola moved three units to the left and one unit down (Figure 11.25).

Figure 11.25

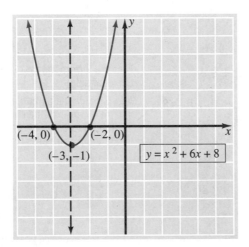

$(-4, 0)$ $(-2, 0)$

$(-3, -1)$

$y = x^2 + 6x + 8$

EXAMPLE 2   Graph $y = x^2 - 3x - 1$.

Solution

$$y = x^2 - 3x - 1$$

$$y = x^2 - 3x + \frac{9}{4} - 1 - \frac{9}{4}$$

Add and subtract $\frac{9}{4}$, which is the square of one-half the coefficient of x.

$$y = \left(x - \frac{3}{2}\right)^2 - \frac{13}{4}$$

The graph of

$$y = \left(x - \frac{3}{2}\right)^2 - \frac{13}{4}$$

is the basic parabola moved $1\frac{1}{2}$ units to the right and $3\frac{1}{4}$ units down (Figure 11.26).

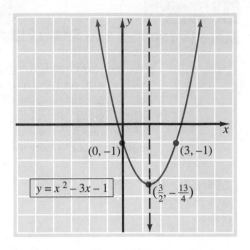

Figure 11.26 ∎

If the coefficient of x^2 is not 1, then a slight adjustment has to be made before we apply the process of completing the square. The next two examples illustrate this situation.

EXAMPLE 3   Graph $y = 2x^2 + 8x + 9$.

Solution

$$y = 2x^2 + 8x + 9$$

$$y = 2(x^2 + 4x) + 9$$ ← Factor a 2 from the first two terms on the right side.

$$y = 2(x^2 + 4x + 4) + 9 - 8$$ ← Add 4 inside the parentheses, which is the square of one-half the coefficient of x. Subtract 8 to compensate for the 4 added inside the parentheses times the factor 2.

$$y = 2(x + 2)^2 + 1$$

See Figure 11.27 for the graph of $y = 2(x + 2)^2 + 1$.

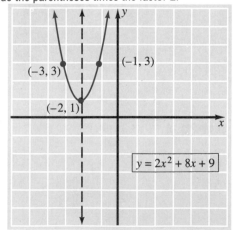

Figure 11.27 ∎

EXAMPLE 4   Graph $y = -3x^2 + 6x - 5$.

Solution

$y = -3x^2 + 6x - 5$

$y = -3(x^2 - 2x) - 5$ Factor -3 from the first two terms on the right side.

$y = -3(x^2 - 2x + 1) - 5 + 3$ Add 1 inside the parentheses to complete the square.
Add 3 to compensate for the 1 added inside the parentheses times the factor of -3.

$y = -3(x - 1)^2 - 2$

The graph of $y = -3(x - 1)^2 - 2$ is drawn in Figure 11.28.

Figure 11.28

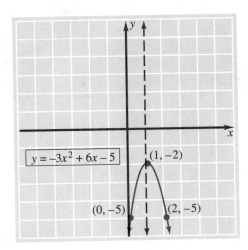

Ellipses

In Section 11.2 we found that the graph of the equation $x^2 + y^2 = 36$ is a circle of radius 6 units with its center at the origin. More generally, it is true that any equation of the form $Ax^2 + By^2 = C$, where $A = B$ and A, B, and C are nonzero constants that have the same sign, is a circle with the center at the origin. For example, $3x^2 + 3y^2 = 12$ is equivalent to $x^2 + y^2 = 4$ (divide both sides of the given equation by 3) and thus it is a circle of radius 2 units with its center at the origin.

The general equation $Ax^2 + By^2 = C$ can be used to describe other geometric figures by changing the restrictions on A and B. For example, if A, B, and C are of the same sign, but $A \neq B$, then the graph of the equation $Ax^2 + By^2 = C$ is an **ellipse**. Let's consider two examples.

EXAMPLE 5   Graph $4x^2 + 25y^2 = 100$.

Solution   Let's find the x- and y-intercepts. Let $x = 0$; then

$$4(0)^2 + 25y^2 = 100$$
$$25y^2 = 100$$
$$y^2 = 4$$
$$y = \pm 2.$$

Thus, the points $(0, 2)$ and $(0, -2)$ are on the graph. Let $y = 0$; then

$$4x^2 + 25(0)^2 = 100$$
$$4x^2 = 100$$
$$x^2 = 25$$
$$x = \pm 5.$$

Thus, the points $(5, 0)$ and $(-5, 0)$ are also on the graph. Plot the four points we have and knowing that it is an ellipse gives us a pretty good sketch of the figure (Figure 11.29).

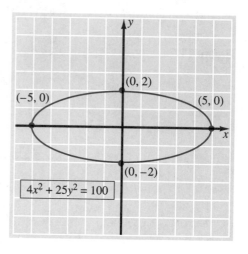

Figure 11.29   ∎

In Figure 11.29 the line segment with endpoints at $(-5, 0)$ and $(5, 0)$ is called the **major axis** of the ellipse. The shorter line segment with endpoints at $(0, -2)$ and $(0, 2)$ is called the **minor axis**. Establishing the endpoints of the major and minor axes provides a basis for sketching an ellipse.

EXAMPLE 6   Graph $9x^2 + 4y^2 = 36$.

Solution   Again let's find the x- and y-intercepts. Let $x = 0$; then

$$9(0)^2 + 4y^2 = 36$$
$$4y^2 = 36$$
$$y^2 = 9$$
$$y = \pm 3.$$

Thus, the points $(0, 3)$ and $(0, -3)$ are on the graph. Let $y = 0$; then

$$9x^2 + 4(0)^2 = 36$$

$$9x^2 = 36$$

$$x^2 = 4$$

$$x = \pm 2.$$

Thus, the points $(2, 0)$ and $(-2, 0)$ are also on the graph. The ellipse is sketched in Figure 11.30.

Figure 11.30

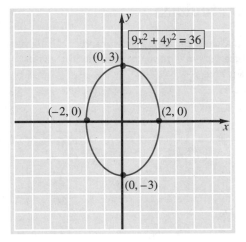

In Figure 11.30 the major axis has endpoints at $(0, -3)$ and $(0, 3)$, and the minor axis has endpoints at $(-2, 0)$ and $(2, 0)$. The ellipses in Figures 11.29 and 11.30 are symmetrical about the x-axis and about the y-axis. In other words, both the x-axis and the y-axis serve as **axes of symmetry**.

Problem Set 11.4

Graph each of the following parabolas.

1. $y = x^2 - 6x + 13$ 2. $y = x^2 - 4x + 7$
3. $y = x^2 + 2x + 6$ 4. $y = x^2 + 8x + 14$
5. $y = x^2 - 5x + 3$ 6. $y = x^2 + 3x + 1$
7. $y = x^2 + 7x + 14$ 8. $y = x^2 - x - 1$
9. $y = 3x^2 - 6x + 5$ 10. $y = 2x^2 + 4x + 7$
11. $y = 4x^2 - 24x + 32$ 12. $y = 3x^2 + 24x + 49$
13. $y = -2x^2 - 4x - 5$ 14. $y = -2x^2 + 8x - 5$
15. $y = -x^2 + 8x - 21$ 16. $y = -x^2 - 6x - 7$

17. $y = 2x^2 - x + 2$

18. $y = 2x^2 + 3x + 1$

19. $y = 3x^2 + 2x + 1$

20. $y = 3x^2 - x - 1$

21. $y = -3x^2 - 7x - 2$

22. $y = -2x^2 + x - 2$

Graph each of the following ellipses.

23. $x^2 + 4y^2 = 36$

24. $x^2 + 4y^2 = 16$

25. $9x^2 + y^2 = 36$

26. $16x^2 + 9y^2 = 144$

27. $4x^2 + 3y^2 = 12$

28. $5x^2 + 4y^2 = 20$

29. $25x^2 + 2y^2 = 50$

30. $12x^2 + y^2 = 36$

Miscellaneous Problems

31. For each of the following parabolas use a graphics calculator to graph the parabola and then use the *trace function* to help estimate the vertex. Then find the vertex by completing the square.

(a) $y = x^2 - 6x + 3$

(b) $y = x^2 - 18x + 66$

(c) $y = -x^2 + 8x - 3$

(d) $y = -x^2 + 24x - 129$

(e) $y = 14x^2 - 7x + 1$

(f) $y = -\dfrac{1}{2}x^2 + 5x - \dfrac{17}{2}$

11.5

Hyperbolas

In Section 11.2 we concluded that the graph of any equation of the form $Ax^2 + By^2 = C$, where $A = B$ and A, B, and C are nonzero constants that have the same sign, is a circle with its center at the origin. Then in Section 11.4 we concluded that if A, B, and C are of the same sign, but $A \neq B$, then the graph of $Ax^2 + By^2 = C$ is an ellipse. Now let's consider the situation where A, B, and C are nonzero constants, and A and B are of unlike signs. In such cases, the graph of an equation of the form $Ax^2 + By^2 = C$ is a **hyperbola**. Let's consider some examples.

EXAMPLE 1   Graph $x^2 - y^2 = 9$.

Solution   If we let $y = 0$, we obtain

$$x^2 - 0^2 = 9$$

$$x^2 = 9$$

$$x = \pm 3.$$

Thus, the points $(3, 0)$ and $(-3, 0)$ are on the graph. If we let $x = 0$, we obtain

$$0^2 - y^2 = 9$$
$$-y^2 = 9$$
$$y^2 = -9.$$

Since $y^2 = -9$ has no real number solutions, there are no points of the y-axis on this graph. That is to say, the graph does not intersect the y-axis. Now let's solve the given equation for y so that we have a more convenient form for finding other solutions.

$$x^2 - y^2 = 9$$
$$-y^2 = 9 - x^2$$
$$y^2 = x^2 - 9$$
$$y = \pm\sqrt{x^2 - 9}$$

Since the radicand, $x^2 - 9$, must be nonnegative, the values we chose for x must be greater than or equal to 3, or less than or equal to -3. With this in mind, we can form the following table of values.

x	y	
3	0	intercepts
-3	0	
4	$\pm\sqrt{7}$	$\approx \pm 2.6$
-4	$\pm\sqrt{7}$	$\approx \pm 2.6$
5	± 4	other points
-5	± 4	

We plot these points and draw the hyperbola as in Figure 11.31. (This graph is also symmetrical about both axes.)

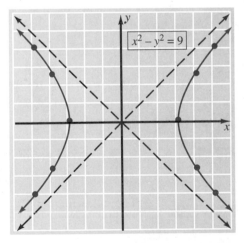

Figure 11.31 ∎

Notice the dashed lines in Figure 11.31, they are called **asymptotes**. Each branch of the hyperbola approaches one of these lines, but does not intersect it. Therefore, the ability to sketch the asymptotes of a hyperbola is very helpful when graphing the hyperbola. Fortunately, the equations of the asymptotes are easy to determine. They can be found by replacing the constant term in the given equation of the hyperbola with 0 and solving for y. (The reason this works will become evident in a later course.) Thus, for the hyperbola in Example 1 we obtain

$$x^2 - y^2 = 0$$

$$y^2 = x^2$$

$$y = \pm x.$$

So, the two lines, $y = x$ and $y = -x$, are the asymptotes indicated by the dashed lines in Figure 11.31.

EXAMPLE 2   Graph $y^2 - 5x^2 = 4$.

Solution   If we let $x = 0$, we obtain

$$y^2 - 5(0)^2 = 4$$

$$y^2 = 4$$

$$y = \pm 2.$$

The points $(0, 2)$ and $(0, -2)$ are on the graph. If we let $y = 0$, we obtain

$$0^2 - 5x^2 = 4$$

$$-5x^2 = 4$$

$$x^2 = -\frac{4}{5}.$$

Since $x^2 = -\dfrac{4}{5}$ has no real number solutions, we know that this hyperbola does not intersect the x-axis. Solving the given equation for y yields

$$y^2 - 5x^2 = 4$$

$$y^2 = 5x^2 + 4$$

$$y = \pm\sqrt{5x^2 + 4}.$$

The following table shows some additional solutions for the equation.

x	y	
0	2	intercepts
0	−2	
1	± 3	other points
−1	± 3	
2	$\pm\sqrt{24}$	$\approx \pm 5$
−2	$\pm\sqrt{24}$	$\approx \pm 5$

The equations of the asymptotes are determined as follows.

$$y^2 - 5x^2 = 0$$
$$y^2 = 5x^2$$
$$y = \pm\sqrt{5}x$$

Sketch the asymptotes and plot the points determined by the table of values to determine the hyperbola in Figure 11.32. (Notice that this hyperbola is also symmetrical about the x-axis and the y-axis.)

Figure 11.32

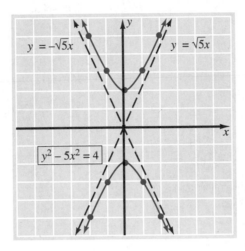

EXAMPLE 3   Graph $4x^2 - 9y^2 = 36$.

Solution   If we let $x = 0$, we obtain

$$4(0)^2 - 9y^2 = 36$$
$$-9y^2 = 36$$
$$y^2 - -4.$$

Since $y^2 = -4$ has no real number solutions, we know that this hyperbola does not intersect the y-axis. If we let $y = 0$, we obtain

$$4x^2 - 9(0)^2 = 36$$
$$4x^2 = 36$$
$$x^2 = 9$$
$$x = \pm 3.$$

Thus, the points $(3, 0)$ and $(-3, 0)$ are on the graph. Now let's solve the equation for y in terms of x and set up a table of values.

$$4x^2 - 9y^2 = 36$$

$$-9y^2 = 36 - 4x^2$$

$$9y^2 = 4x^2 - 36$$

$$y^2 = \frac{4x^2 - 36}{9}$$

$$y = \pm\frac{\sqrt{4x^2 - 36}}{3}$$

x	y	
3	0	intercepts
-3	0	
	$\pm\dfrac{2\sqrt{7}}{3}$	$\approx \pm 1.8$
-4	$\pm\dfrac{2\sqrt{7}}{3}$	$\approx \pm 1.8$
5	$\pm\dfrac{8}{3}$	other points
-5	$\pm\dfrac{8}{3}$	

The equations of the asymptotes are found as follows.

$$4x^2 - 9y^2 = 0$$

$$-9y^2 = -4x^2$$

$$9y^2 = 4x^2$$

$$y^2 = \frac{4x^2}{9}$$

$$y = \pm\frac{2}{3}x$$

Sketch the asymptotes and plot the points determined by the table to determine the hyperbola as shown in Figure 11.33.

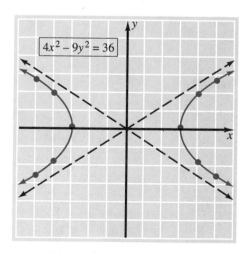

Figure 11.33   ■

As a way of summarizing, let's focus our attention on the continuity pattern used in this chapter. In Section 11.2, we used the definition of a circle to generate a standard form for the equation of a circle. Then, in Sections 11.4 and 11.5, ellipses and hyperbolas were discussed, not from a definition viewpoint, but by considering variations of the general equation of a circle with its center at the origin ($Ax^2 + By^2 = C$, where A, B, and C are of the same sign and $A = B$). In a subsequent mathematics course, parabolas, ellipses, and hyperbolas will also be developed from a definition viewpoint. That is to say, each concept will be first defined and then the definition will be used to generate a standard form of its equation. In such a course you will also study ellipses and hyperbolas that have their centers at points other than the origin.

Parabolas, circles, ellipses, and hyperbolas can be formed by intersecting a plane with a conical surface as shown in Figure 11.34; we often refer to these curves as the **conic sections**. A flashlight produces a "cone of light" that can be cut by the plane of a wall to illustrate the conic sections. Try shining a flashlight against a wall at different angles to produce a circle, an ellipse, a parabola, and one branch of a hyperbola. (You may find it difficult to distinguish between a parabola and a branch of a hyperbola.)

Figure 11.34

    Circle   Ellipse   Parabola   Hyperbola

Problem Set 11.5

For Problems 1–12, graph each of the hyperbolas. Be sure to include the asymptotes and their equations.

1. $x^2 - y^2 = 1$ 2. $x^2 - y^2 = 4$

3. $y^2 - 4x^2 = 9$ 4. $4y^2 - x^2 = 16$

5. $5x^2 - 2y^2 = 20$ 6. $9x^2 - 4y^2 = 9$

7. $y^2 - 16x^2 = 4$ 8. $y^2 - 9x^2 = 16$

9. $-4x^2 + y^2 = -4$ 10. $-9x^2 + y^2 = -36$

11. $25y^2 - 3x^2 = 75$ 12. $16y^2 - 5x^2 = 80$

For Problems 13–22, each equation represents a line, a circle, a parabola, an ellipse, or a hyperbola. First identify each curve and then sketch it.

13. $y = -x^2 - 4x - 1$ 14. $y = -5x - 5$

15. $y^2 - x^2 = -9$ 16. $y = -x^2 - 6x - 10$

17. $y = -2x - 2$ 18. $4x^2 + y^2 = 16$

19. $x^2 + y^2 + 6x + 2y + 6 = 0$ 20. $y^2 - 4x^2 = 4$

21. $x^2 + 9y^2 = 9$ 22. $x^2 + y^2 - 8x + 2y + 13 = 0$

23. The graphs of equations of the form $xy = k$, where k is a nonzero constant, are also hyperbolas, sometimes referred to as rectangular hyperbolas. Graph each of the following.

   (a) $xy = 3$ (b) $xy = 5$

   (c) $xy = -2$ (d) $xy = -4$

24. What is the graph of $xy = 0$? Defend your answer.

25. We have graphed various equations of the form $Ax^2 + By^2 = C$, where C is a nonzero constant. Now graph each of the following.

   (a) $x^2 + y^2 = 0$ (b) $2x^2 + 3y^2 = 0$

   (c) $x^2 - y^2 = 0$ (d) $4y^2 - x^2 = 0$

Thoughts into Words

26. Describe how you would graph the equation $x^2 + 6y^2 = 36$.

27. Describe how you would graph the equation $y = x^2 - 2x + 6$.

28. Describe how you would graph the equation $x^2 - 2y^2 = -8$.

29. Explain the concept of an asymptote and describe how it is used for graphing purposes.

Miscellaneous Problems

30. Use a graphics calculator to graph each of the following ellipses and hyperbolas.

   (a) $2x^2 + y^2 = 16$ (b) $2x^2 - y^2 = 16$ (c) $x^2 - 4y^2 = 36$

   (d) $x^2 + 4y^2 = 36$ (e) $y^2 - x^2 = 64$ (f) $x^2 - y^2 = 64$

   (g) $y^2 - x^2 = 1$ (h) $y^2 + 2y^2 = 1$ (i) $x^2 - y^2 = 121$

   (j) $x^2 + 3y^2 = 110$

Chapter 11 Summary

(11.1) The following suggestions are offered for **graphing an equation** in two variables.

1. Determine the type of symmetry that the equation exhibits.

2. Find the intercepts.

3. Solve the equation for y in terms of x or for x in terms of y if it is not already in such a form.

4. Set up a table of ordered pairs that satisfy the equation. The type of symmetry will affect your choice of values in the table.

5. Plot the points associated with the ordered pairs from the table, and connect them with a smooth curve. Then, if appropriate, reflect this part of the curve according to the symmetry shown by the equation.

(11.2) The distance between any two points (x_1, y_1) and (x_2, y_2) is given by the **distance formula,**

$$d = \sqrt{(x_2 - x_1)^2 + (y_2 - y_1)^2}.$$

The **standard form of the equation of a circle** with its center at (h, k) and a radius of length r is

$$(x - h)^2 + (y - k)^2 = r^2.$$

The standard form of the equation of a circle with its center at the origin and a radius of length r is

$$x^2 + y^2 = r^2.$$

(11.3) and (11.4) The graph of any quadratic equation of the form $y = ax^2 + bx + c$, where a, b, and c are real numbers and $a \neq 0$, is a **parabola.**

The following diagram summarizes the graphing of parabolas.

$y = x^2$ ⟶ $y = x^2 + \text{ⓚ}$ ⟶ moves the parabola up or down

$\quad\quad\quad$ ⟶ $y = \text{ⓐ}x^2$ ⟶ affects the width and which way the parabola opens

↑
basic
parabola ⟶ $y = (x - \text{ⓗ})^2$ ⟶ moves the parabola right or left

The graph of an equation of the form

$$Ax^2 + By^2 = C,$$

where A, B, and C are nonzero constants of the same sign and $A \neq B$, is an **ellipse.**

(11.5) The graph of an equation of the form

$$Ax^2 + By^2 = C,$$

where A, B, and C are nonzero constants and A and B are of unlike signs, is a **hyperbola.**

Circles, ellipses, parabolas, and hyperbolas are often referred to as **conic sections.**

Chapter 11 Review Problem Set

Some of these problems review concepts from Chapter 5.

1. Find the slope of the line determined by each pair of points.
   (a) $(3, 4), (-2, -2)$ (b) $(-2, 3), (4, -1)$
2. Find the slope of each of the following lines.
   (a) $4x + y = 7$ (b) $2x - 7y = 3$
3. Find the lengths of the sides of a triangle whose vertices are at $(2, 3)$, $(5, -1)$, and $(-4, -5)$.

For Problems 4–8, write the equation of each of the lines that satisfy the stated conditions. Express final equations in standard form.

4. containing the points $(-1, 2)$ and $(3, -5)$

5. having a slope of $-\dfrac{3}{7}$ and a y-intercept of 4

6. containing the point $(-1, -6)$ and having a slope of $\dfrac{2}{3}$

7. containing the point $(2, 5)$ and parallel to the line $x - 2y = 4$

8. containing the point $(-2, -6)$ and perpendicular to the line $3x + 2y = 12$

For Problems 9–20, graph each of the equations.

9. $2x - y = 6$ **10.** $y = -2x^2 - 1$

11. $x^2 + y^2 = 1$ **12.** $4x^2 + y^2 = 16$

13. $y = -4x$ **14.** $2x^2 - y^2 = 8$

15. $y = x^2 + 4x - 1$ **16.** $y = 4x^2 - 8x + 2$

17. $xy^2 = -1$ **18.** $3x^2 + 5y^2 = 30$

19. $5y^2 - x^2 = 20$ **20.** $y = -2x^2 + 12x - 16$

For Problems 21 and 22, graph each of the inequalities.

21. $x + 2y \geq 4$ **22.** $2x - 3y \leq 6$

23. Find the center and length of a radius of the circle $x^2 + y^2 + 6x - 8y + 16 = 0$.

24. Find the coordinates of the vertex of the parabola $y = x^2 + 8x + 10$.

25. Find the equations of the asymptotes of the hyperbola $9x^2 - 4y^2 = 72$.

26. Find the length of the major axis of the ellipse $4x^2 + y^2 = 9$.

Chapter 12

Functions

One of the fundamental concepts of mathematics is that of a function. Functions are used to unify mathematics and also to apply mathematics to many real-world problems. Functions provide a means of studying quantities that vary with one another, that is, when a change in one quantity causes a corresponding change in another.

In this chapter we will (1) introduce the basic ideas that pertain to the function concept, (2) review some concepts from Chapter 11 regarding functions, (3) introduce some operations with functions, (4) extend graphing techniques to include the graphing of functions, and (5) discuss some applications of functions.

12.1

Relations and Functions

Mathematically, a function is a special kind of **relation**, so we will begin our discussion with a simple definition of a relation.

DEFINITION 12.1

> A **relation** is a set of ordered pairs.

Thus, a set of ordered pairs such as $\{(1, 2), (3, 7), (8, 14)\}$ is a relation. The set of all first components of the ordered pairs is the **domain** of the relation and the set of all second components is the **range** of the relation. The relation $\{(1, 2), (3, 7), (8, 14)\}$ has a domain of $\{1, 3, 8\}$ and a range of $\{2, 7, 14\}$.

The ordered pairs we refer to in Definition 12.1 may be generated by various means, such as a graph or a chart. However, one of the most common ways of generating ordered pairs is by use of equations. Since the solution set of an equation in two variables is a set of ordered pairs, such an equation describes a relation. Each of the following equations describes a relation between the variables x and y. We have listed *some* of the infinitely many ordered pairs (x, y) of each relation.

1. $x^2 + y^2 = 4$: $(1, \sqrt{3}), (1, -\sqrt{3}), (0, 2), (0, -2)$
2. $y^2 = x^3$: $(0, 0), (1, 1), (1, -1), (4, 8), (4, -8)$
3. $y = x + 2$: $(0, 2), (1, 3), (2, 4), (-1, 1), (5, 7)$
4. $y = \dfrac{1}{x - 1}$: $(0, -1), (2, 1), \left(3, \dfrac{1}{2}\right), \left(-1, -\dfrac{1}{2}\right), \left(-2, -\dfrac{1}{3}\right)$
5. $y = x^2$: $(0, 0), (1, 1), (2, 4), (-1, 1), (-2, 4)$

Now we direct your attention to the ordered pairs associated with equations 3, 4, and 5. Note that in each case no two ordered pairs have the same first component. Such a set of ordered pairs is called a **function**.

DEFINITION 12.2

> A **function** is a relation in which no two ordered pairs have the same first component.

Stated another way, Definition 12.2 means that a function is a relation where each member of the domain is assigned *one and only one* member of the range. Thus, it is easy to determine that each of the following sets of ordered pairs is a function.

$$f = \{(x, y) \mid y = x + 2\},$$

$$g = \left\{(x, y) \,\middle|\, y = \frac{1}{x - 1}\right\},$$

$$h = \{(x, y) \mid y = x^2\}$$

In each case there is one and only one value of y (an element of the range) associated with each value of x (an element of the domain).

Notice that we named the previous functions f, g, and h. It is common to name functions by means of a single letter and the letters f, g, and h are often used. We would suggest more meaningful choices when functions are used to portray real-world situations. For example, if a problem involves a profit function, then naming the function p or even P would seem natural.

The symbol for a function can be used along with a variable that represents an element in the domain to represent the associated element in the range. For example, suppose that we have a function f specified in terms of the variable x. The symbol, $f(x)$, (read "f of x" or "the value of f at x") represents the element in the range associated with the element x from the domain. The function $f = \{(x, y)\,|\,y = x + 2\}$ can be written as $f = \{(x, f(x))\,|\,f(x) = x + 2\}$ and this is usually shortened to read "f is the function determined by the equation $f(x) = x + 2$."

> **REMARK** Be careful with the symbolism $f(x)$. As we stated above, it means the value of the function f at x. It does not mean f times x.

This **function notation** is very convenient when computing and expressing various values of the function. For example, the value of the function $f(x) = 3x - 5$ at $x = 1$ is

$$f(1) = 3(1) - 5 = -2.$$

Likewise, the functional values for $x = 2$, $x = -1$, and $x = 5$ are

$$f(2) = 3(2) - 5 = 1,$$
$$f(-1) = 3(-1) - 5 = -8, \quad \text{and}$$
$$f(5) = 3(5) - 5 = 10.$$

Thus, this function f contains the ordered pairs $(1, -2), (2, 1), (-1, -8), (5, 10)$, and in general all ordered pairs of the form $(x, f(x))$, where $f(x) = 3x - 5$ and x is any real number.

It may be helpful for you to mentally picture the concept of a function in terms of a "function machine" as in Figure 12.1. Each time that a value of x is put

Figure 12.1

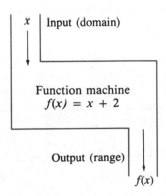

x Input (domain)

Function machine
$f(x) = x + 2$

Output (range)

$f(x)$

into the machine, the equation $f(x) = x + 2$ is used to generate one and only one value for $f(x)$ to be ejected from the machine. For example, if 3 is put into this machine, then $f(3) = 3 + 2 = 5$, and 5 is ejected. Thus, the ordered pair $(3, 5)$ is one element of the function. Now let's look at some examples to help pull together some of the ideas about functions.

EXAMPLE 1   Determine whether the relation $\{(x, y) \mid y^2 = x\}$ is a function and specify its domain and range.

Solution   Because $y^2 = x$ is equivalent to $y = \pm\sqrt{x}$, to each value of x there are assigned *two* values for y. Therefore, this relation is not a function.

The expression $\sqrt{x}$ requires that x be nonnegative; therefore, the domain (D) is

$$D = \{x \mid x \geq 0\}.$$

To each nonnegative real number, the relation assigns two real numbers, $\sqrt{x}$ and $-\sqrt{x}$. Thus, the range (R) is

$$R = \{y \mid y \text{ is a real number}\}.$$ ∎

EXAMPLE 2   For the function $f(x) = x^2$,

    (a) specify its domain,

    (b) determine its range, and

    (c) evaluate $f(-2)$, $f(0)$, and $f(4)$.

Solution   **(a)** Any real number can be squared; therefore, the domain (D) is

$$D = \{x \mid x \text{ is a real number}\}.$$

    (b) Squaring a real number always produces a nonnegative result. Thus, the range (R) is

$$R = \{f(x) \mid f(x) \geq 0\}.$$

    (c) $f(-2) = (-2)^2 = 4,$

        $f(0) = (0)^2 = 0,$

        $f(4) = (4)^2 = 16$ ∎

For our purposes in this text, if the domain of a function is not specifically indicated or determined by a real-world application, then we assume the domain to be all **real number** replacements for the variable, which represents an element in the domain, that will produce **real number** functional values. Consider the following examples.

EXAMPLE 3   Specify the domain for each of the following.

    (a) $f(x) = \dfrac{1}{x - 1}$ **(b)** $f(t) = \dfrac{1}{t^2 - 4}$ **(c)** $f(s) = \sqrt{s - 3}$

Solution

(a) We can replace x with any real number except 1, because 1 makes the denominator zero. Thus, the domain is given by

$$D = \{x \mid x \neq 1\}.$$

(b) We need to eliminate any value of t that will make the denominator zero. Thus, let's solve the equation $t^2 - 4 = 0$.

$$t^2 - 4 = 0$$

$$t^2 = 4$$

$$t = \pm 2$$

The domain is the set

$$D = \{t \mid t \neq -2 \text{ and } t \neq 2\}.$$

(c) The radicand, $s - 3$, must be nonnegative.

$$s - 3 \geq 0$$

$$s \geq 3$$

The domain is the set

$$D = \{s \mid s \geq 3\}.$$ ∎

EXAMPLE 4   If $f(x) = -2x + 7$ and $g(x) = x^2 - 5x + 6$, find $f(3)$, $f(-4)$, $g(2)$, and $g(-1)$.

Solution

$$\underline{f(x) = -2x + 7} \qquad\qquad \underline{g(x) = x^2 - 5x + 6}$$

$$f(3) = -2(3) + 7 = 1, \qquad\qquad g(2) = 2^2 - 5(2) + 6 = 0,$$
$$f(-4) = -2(-4) + 7 = 15 \qquad\qquad g(-1) = (-1)^2 - 5(-1) + 6 = 12$$ ∎

In Example 4, notice that we are working with two different functions in the same problem. Thus, different names, f and g, are used.

The quotient $\dfrac{f(a + h) - f(a)}{h}$ is often called a **difference quotient** and we use it extensively with functions when studying the limit concept in calculus. The next two examples show how we found the difference quotient for two specific functions.

EXAMPLE 5   If $f(x) = 3x - 5$, find $\dfrac{f(a + h) - f(a)}{h}$.

Solution

$$f(a + h) = 3(a + h) - 5$$

$$= 3a + 3h - 5$$

and

$$f(a) = 3a - 5$$

Therefore,

$$f(a + h) - f(a) = (3a + 3h - 5) - (3a - 5)$$
$$= 3a + 3h - 5 - 3a + 5$$
$$= 3h$$

and

$$\frac{f(a + h) - f(a)}{h} = \frac{3h}{h} = 3.$$ ■

EXAMPLE 6   If $f(x) = x^2 + 2x - 3$, find $\dfrac{f(a + h) - f(a)}{h}$.

Solution

$$f(a - h) = (a + h)^2 + 2(a + h) - 3$$
$$= a^2 + 2ah + h^2 + 2a + 2h - 3,$$

and

$$f(a) = a^2 + 2a - 3$$

Therefore,

$$f(a + h) - f(a) = (a^2 + 2ah + h^2 + 2a + 2h - 3) - (a^2 + 2a - 3)$$
$$= a^2 + 2ah + h^2 + 2a + 2h - 3 - a^2 - 2a + 3$$
$$= 2ah + h^2 + 2h,$$

and

$$\frac{f(a + h) - f(a)}{h} = \frac{2ah + h^2 + 2h}{h}$$
$$= \frac{\cancel{h}(2a + h + 2)}{\cancel{h}}$$
$$= 2a + h + 2.$$ ■

Functions and functional notation provide the basis for describing many real-world relationships. The next example illustrates this point.

EXAMPLE 7   Suppose a factory determines that the overhead for producing a quantity of a certain item is $500 and the cost for each item is $25. Express the total expenses as a function of the number of items produced and compute the expenses for producing 12, 25, 50, 75, and 100 items.

Solution   Let n represent the number of items produced. Then $25n + 500$ represents the total expenses. Let's use E to represent the *expense function*, so that we have

$$E(n) = 25n + 500, \quad \text{where } n \text{ is a whole number,}$$

from which we obtain

$$E(12) = 25(12) + 500 = 800,$$

$$E(25) = 25(25) + 500 = 1125,$$

$$E(50) = 25(50) + 500 = 1750,$$

$$E(75) = 25(75) + 500 = 2375,$$

$$E(100) = 25(100) + 500 = 3000.$$

So the total expenses for producing 12, 25, 50, 75, and 100 items are $800, $1125, $1750, $2375, and $3000, respectively.   ■

Problem Set 12.1

For Problems 1–10, specify the domain and the range for each relation. Also state whether or not the relation is a function.

1. $\{(1, 5), (2, 8), (3, 11), (4, 14)\}$
2. $\{(0, 0), (2, 10), (4, 20), (6, 30), (8, 40)\}$
3. $\{(0, 5), (0, -5), (1, 2\sqrt{6}), (1, -2\sqrt{6})\}$
4. $\{(1, 1), (1, 2), (1, -1), (1, -2), (1, 3)\}$
5. $\{(1, 2), (2, 5), (3, 10), (4, 17), (5, 26)\}$
6. $\{(-1, 5), (0, 1), (1, -3), (2, -7)\}$
7. $\{(x, y) \mid 5x - 2y = 6\}$
8. $\{(x, y) \mid y = -3x\}$
9. $\{(x, y) \mid x^2 = y^3\}$
10. $\{(x, y) \mid x^2 - y^2 = 16\}$

For Problems 11–36, specify the domain for each of the functions.

11. $f(x) = 7x - 2$

12. $f(x) = x^2 + 1$

13. $f(x) = \dfrac{1}{x - 1}$

14. $f(x) = \dfrac{-3}{x + 4}$

15. $g(x) = \dfrac{3x}{4x - 3}$

16. $g(x) = \dfrac{5x}{2x + 7}$

17. $h(x) = \dfrac{2}{(x + 1)(x - 4)}$

18. $h(x) = \dfrac{-3}{(x - 6)(2x + 1)}$

19. $f(x) = \dfrac{14}{x^2 + 3x - 40}$

20. $f(x) = \dfrac{7}{x^2 - 8x - 20}$

21. $f(x) = \dfrac{-4}{x^2 + 6x}$

22. $f(x) = \dfrac{9}{x^2 - 12x}$

23. $f(t) = \dfrac{4}{t^2 + 9}$

24. $f(t) = \dfrac{8}{t^2 + 1}$

25. $f(t) = \dfrac{3t}{t^2 - 4}$

26. $f(t) = \dfrac{-2t}{t^2 - 25}$

27. $h(x) = \sqrt{x + 4}$

28. $h(x) = \sqrt{5x - 3}$

29. $f(s) = \sqrt{4s - 5}$

30. $f(x) = \sqrt{s - 2} + 5$

31. $f(x) = \sqrt{x^2 - 16}$

32. $f(x) = \sqrt{x^2 - 49}$

33. $f(x) = \sqrt{x^2 - 3x - 18}$

34. $f(x) = \sqrt{x^2 + 4x - 32}$

35. $f(x) = \sqrt{1 - x^2}$

36. $f(x) = \sqrt{9 - x^2}$

37. If $f(x) = 5x - 2$, find $f(0)$, $f(2)$, $f(-1)$, and $f(-4)$.

38. If $f(x) = -3x - 4$, find $f(-2)$, $f(-1)$, $f(3)$, and $f(5)$.

39. If $f(x) = \dfrac{1}{2}x - \dfrac{3}{4}$, find $f(-2)$, $f(0)$, $f\left(\dfrac{1}{2}\right)$, and $f\left(\dfrac{2}{3}\right)$.

40. If $g(x) = x^2 + 3x - 1$, find $g(1)$, $g(-1)$, $g(3)$, and $g(-4)$.

41. If $g(x) = 2x^2 - 5x - 7$, find $g(-1)$, $g(2)$, $g(-3)$, and $g(4)$.

42. If $h(x) = -x^2 - 3$, find $h(1)$, $h(-1)$, $h(-3)$, and $h(5)$.

43. If $h(x) = -2x^2 - x + 4$, find $h(-2)$, $h(-3)$, $h(4)$, and $h(5)$.

44. If $f(x) = \sqrt{x - 1}$, find $f(1)$, $f(5)$, $f(13)$, *and* $f(26)$.

45. If $f(x) = \sqrt{2x + 1}$, find $f(3)$, $f(4)$, $f(10)$, and $f(12)$.

46. If $f(x) = \dfrac{3}{x - 2}$, find $f(3)$, $f(0)$, $f(-1)$, and $f(-5)$.

47. If $f(x) = \dfrac{-4}{x + 3}$, find $f(1)$, $f(-1)$, $f(3)$, and (-6).

48. If $f(x) = 2x^2 - 7$ and $g(x) = x^2 + x - 1$, find $f(-2)$, $f(3)$, $g(-4)$, and $g(5)$.

49. If $f(x) = 5x^2 - 2x + 3$ and $g(x) = -x^2 + 4x - 5$, find $f(-2)$, $f(3)$, $g(-4)$, and $g(6)$.

50. If $f(x) = |3x - 2|$ and $g(x) = |x| + 2$, find $f(1)$, $f(-1)$, $g(2)$, and $g(-3)$.

51. If $f(x) = 3|x| - 1$ and $g(x) = -|x| + 1$, find $f(-2)$, $f(3)$, $g(-4)$, and $g(5)$.

For Problems 52–59, find $\dfrac{f(a + h) - f(a)}{h}$ for each of the given functions.

52. $f(x) = 5x - 4$

53. $f(x) = -3x + 6$

54. $f(x) = x^2 + 5$

55. $f(x) = -x^2 - 1$

56. $f(x) = x^2 - 3x + 7$

57. $f(x) = 2x^2 - x + 8$

58. $f(x) = -3x^2 + 4x - 1$

59. $f(x) = -4x^2 - 7x - 9$

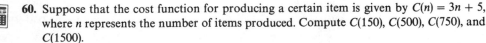

60. Suppose that the cost function for producing a certain item is given by $C(n) = 3n + 5$, where n represents the number of items produced. Compute $C(150)$, $C(500)$, $C(750)$, and $C(1500)$.

61. The height of a projectile fired vertically into the air (neglecting air resistance) at an initial velocity of 64 feet per second is a function of the time (t) and is given by the equation

$$h(t) = 64t - 16t^2.$$

Compute $h(1)$, $h(2)$, $h(3)$, and $h(4)$.

 62. The profit function for selling n items is given by $P(n) = -n^2 + 500n - 61,500$. Compute $P(200)$, $P(230)$, $P(250)$, and $P(260)$.

 63. A car rental agency charges $50 per day plus $0.32 a mile. Therefore, the daily charge for renting a car is a function of the number of miles traveled (m) and can be expressed as $C(m) = 50 + 0.32m$. Compute $C(75)$, $C(150)$, $C(225)$, and $C(650)$.

 64. The equation $A(r) = \pi r^2$ expresses the area of a circular region as a function of the length of a radius (r). Use 3.14 as an approximation for π and compute $A(2)$, $A(3)$, $A(12)$, and $A(17)$.

 65. The equation $I(r) = 500r$ expresses the amount of simple interest earned by an investment of $500 for one year as a function of the rate of interest (r). Compute $I(0.11)$, $I(0.12)$, $I(0.135)$, and $I(0.15)$.

Thoughts into Words

66. How would you distinguish between a relation and a function?

67. Are all functions also relations? Are all relations also functions? Defend your answers.

68. What does it mean to say that the domain of a function may be restricted if the function represents a real-world situation?

12.2

Special Functions and Problem Solving

We have used phrases such as "the graph of the solution set of the equation $y = x - 1$," or simply "the graph of the equation $y = x - 1$" is a line that contains the points $(0, -1)$ and $(1, 0)$. Because the equation $y = x - 1$ (which can be written as $f(x) = x - 1$) can be used to specify a function, the line previously referred to is also called the **graph of the function specified by the equation** or simply the **graph of the function**. Generally speaking, the graph of any equation that determines a function is also called the graph of the function. Thus, the graphing techniques discussed in Chapter 11 will continue to play an important role as we graph functions.

As we use the function concept in our study of mathematics, it is helpful to classify certain types of functions and become familiar with their equations, characteristics, and graphs. In this section we will discuss two special types of functions—**linear** and **quadratic functions**. These functions are merely an outgrowth of our earlier study of linear and quadratic equations.

Linear Functions

Any function that can be written in the form

$$f(x) = ax + b,$$

where a and b are real numbers, is called a **linear function**. The following are examples of linear functions.

$$f(x) = -3x + 6, \qquad f(x) = 2x + 4, \qquad f(x) = -\frac{1}{2}x - \frac{3}{4}$$

Graphing linear functions is quite easy because the graph of every linear function is a straight line. Therefore, all we need to do is determine two points of the graph and draw the line determined by those two points. You may want to continue using a third point as a check point.

EXAMPLE 1   Graph the function $f(x) = -3x + 6$.

Solution   Because $f(0) = 6$, the point $(0, 6)$ is on the graph. Likewise, because $f(1) = 3$, the point $(1, 3)$ is on the graph. Plot these two points and draw the line determined by the two points to produce Figure 12.2.

Figure 12.2

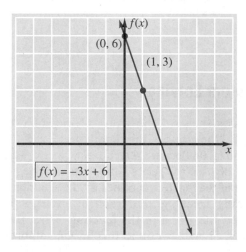

REMARK Note that in Figure 12.2 we labeled the vertical axis $f(x)$. We could also label it y, since $f(x) = -3x + 6$ and $y = -3x + 6$ mean the same thing. We will continue to use the $f(x)$ label in this chapter to help you adjust to the function notation.

EXAMPLE 2   Graph the function $f(x) = x$.

Solution   The equation $f(x) = x$ can be written as $f(x) = 1x + 0$; thus, it is a linear function. Since $f(0) = 0$ and $f(2) = 2$, the points $(0, 0)$ and $(2, 2)$ determine the line in Figure 12.3. The function $f(x) = x$ is often called the **identity function**.

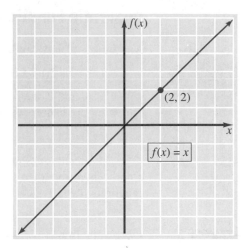

Figure 12.3   ■

As you use function notation to graph functions, it is often helpful to think of the ordinate of every point on the graph as the value of the function at a specific value of x. Geometrically, this functional value is the directed distance of the point from the x-axis as illustrated in Figure 12.4, with the function $f(x) = 2x - 4$. For example, consider the graph of the function $f(x) = 2$. The function $f(x) = 2$ means that every functional value is 2, or geometrically, that every point on the graph is 2 units above the x-axis. Thus the graph is the horizontal line shown in Figure 12.5.

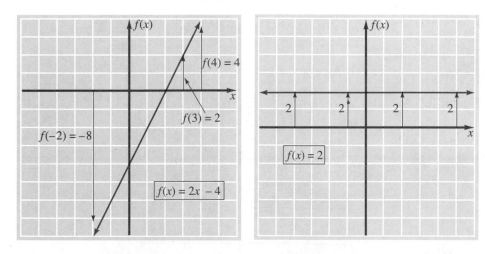

Figure 12.4   **Figure 12.5**

Any linear function of the form $f(x) = ax + b$, where $a = 0$, is called a **constant function** and its graph is a *horizontal line*.

Quadratic Functions

Any function that can be written in the form

$$f(x) = ax^2 + bx + c,$$

where a, b, and c are real numbers with $a \neq 0$, is called a **quadratic function**. The following are examples of quadratic functions.

$$f(x) = 3x^2, \qquad f(x) = -2x^2 + 5x, \quad f(x) = 4x^2 - 7x + 1$$

The techniques we discussed in Chapter 11 relative to graphing quadratic equations of the form $y = ax^2 + bx + c$ provide the basis for graphing quadratic functions. Let's review some work from Chapter 11 with an example.

EXAMPLE 3   Graph the function $f(x) = 2x^2 - 4x + 5$.

Solution

$$f(x) = 2x^2 - 4x + 5$$
$$= 2(x^2 - 2x + \underline{\qquad}) + 5$$
$$= 2(x^2 - 2x + 1) + 5 - 2$$
$$= 2(x - 1)^2 + 3$$

Recall the process of completing the square!

From this form we can obtain the following information about the parabola.

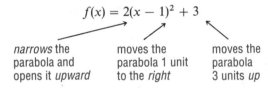

$$f(x) = 2(x - 1)^2 + 3$$

narrows the parabola and opens it *upward*

moves the parabola 1 unit to the *right*

moves the parabola 3 units *up*

Thus, the parabola can be drawn as in Figure 12.6.

Figure 12.6

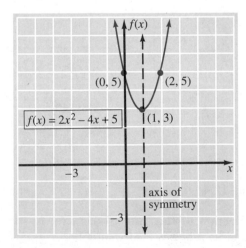

In general, if we complete the square on

$$f(x) = ax^2 + bx + c,$$

we obtain

$$f(x) = a\left(x^2 + \frac{b}{a}x + \underline{\quad} \right) + c$$

$$= a\left(x^2 + \frac{b}{a}x + \frac{b^2}{4a^2} \right) + c - \frac{b^2}{4a}$$

$$= a\left(x + \frac{b}{2a} \right)^2 + \frac{4ac - b^2}{4a}.$$

Therefore, the parabola associated with $f(x) = ax^2 + bx + c$ has its vertex at $\left(-\frac{b}{2a}, \frac{4ac - b^2}{4a} \right)$ and the equation of its axis of symmetry is $x = -\frac{b}{2a}$. These facts are illustrated in Figure 12.7.

Figure 12.7

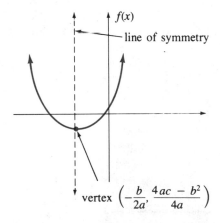

By using the information from Figure 12.7 we now have another way of graphing quadratic functions of the form $f(x) = ax^2 + bx + c$, shown by the following steps.

1. Determine whether the parabola opens upward (if $a > 0$) or downward (if $a < 0$).

2. Find $-\frac{b}{2a}$, which is the x-coordinate of the vertex.

3. Find $f\left(-\frac{b}{2a} \right)$, which is the y-coordinate of the vertex. $\left(\text{You could also find the } y\text{-coordinate by evaluating } \frac{4ac - b^2}{4a}. \right)$

4. Locate another point on the parabola and also locate its image across the line of symmetry, $x = -\dfrac{b}{2a}$.

The three points in Steps 2, 3, and 4 should determine the general shape of the parabola. Let's try two examples and use these steps.

EXAMPLE 4   Graph $f(x) = 3x^2 - 6x + 5$.

Solution

Step 1. Because $a = 3$, the parabola opens upward.

Step 2. $-\dfrac{b}{2a} = -\dfrac{-6}{6} = 1.$

Step 3. $f\left(-\dfrac{b}{2a}\right) = f(1) = 3 - 6 + 5 = 2$. Thus, the vertex is at $(1, 2)$.

Step 4. Letting $x = 2$, we obtain $f(2) = 12 - 12 + 5 = 5$. Thus $(2, 5)$ is on the graph and so is its reflection $(0, 5)$ across the line of symmetry $x = 1$.

The three points $(1, 2)$, $(2, 5)$, and $(0, 5)$ are used to graph the parabola in Figure 12.8.

Figure 12.8

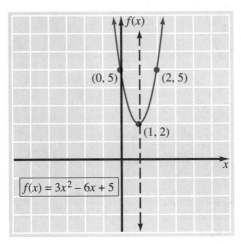

$(0, 5)$ $(2, 5)$

$(1, 2)$

$f(x) = 3x^2 - 6x + 5$

EXAMPLE 5   Graph $f(x) = -x^2 - 4x - 7$.

Solution

Step 1. Since $a = -1$, the parabola opens downward.

Step 2. $-\dfrac{b}{2a} = -\dfrac{-4}{-2} = -2.$

Step 3. $f\left(-\dfrac{b}{2a}\right) = f(-2) = -(-2)^2 - 4(-2) - 7 = -3.$

Step 4. Letting $x = 0$, we obtain $f(0) = -7$. Thus, $(0, -7)$ is on the graph and so is its reflection $(-4, -7)$ across the line of symmetry $x = -2$.

The three points $(-2, \quad 3), (0, -7)$ and $(-4, -7)$ are used to draw the parabola in Figure 12.9.

Figure 12.9

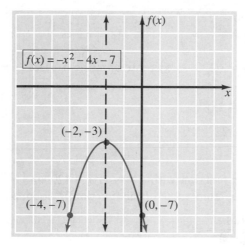

In summary, to graph a quadratic function we basically have two methods.

1. We can express the function in the form $f(x) = a(x - h)^2 + k$ and use the values of a, h, and k to determine the parabola; *or*

2. We can express the function in the form $f(x) = ax^2 + bx + c$ and use the approach demonstrated in Examples 4 and 5.

Problem Solving

As we have seen, the vertex of the graph of a quadratic function is either the lowest or the highest point on the graph. Thus, the vocabulary *minimum value* or *maximum value* of a function is often used in applications of the parabola. The x-value of the vertex indicates where the minimum or maximum occurs and $f(x)$ yields the minimum or maximum value of the function. Let's consider some examples that use these ideas.

EXAMPLE 6   A farmer has 120 rods of fencing and wants to enclose a rectangular plot of land that requires fencing on only three sides, since it is bounded by a river on one side. Find the length and width of the plot that will maximize the area.

Solution   Let x represent the width; then $120 - 2x$ represents the length, as indicated in Figure 12.10.

Figure 12.10

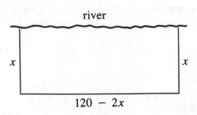

The function $A(x) = x(120 - 2x)$ represents the area of the plot in terms of the width x. Since

$$A(x) = x(120 - 2x)$$

$$= 120x - 2x^2$$

$$= -2x^2 + 120x,$$

we have a quadratic function with $a = -2$, $b = 120$, and $c = 0$. Therefore, the x-value where the maximum value of the function is obtained is

$$-\frac{b}{2a} = -\frac{120}{2(-2)} = 30.$$

If $x = 30$, then $120 - 2x = 120 - 2(30) = 60$.
Thus, the farmer should make the plot 30 rods wide and 60 rods long to maximize the area at $(30)(60) = 1800$ square rods. ∎

EXAMPLE 7   Find two numbers whose sum is 30, such that the sum of their squares is a minimum.

Solution   Let x represent one of the numbers; then $30 - x$ represents the other number. By expressing the sum of the squares as a function of x we obtain

$$f(x) = x^2 + (30 - x)^2,$$

which can be simplified to

$$f(x) = x^2 + 900 - 60x + x^2$$

$$= 2x^2 - 60x + 900.$$

This is a quadratic function with $a = 2$, $b = -60$, and $c = 900$. Therefore, the x-value where the minimum occurs is

$$-\frac{b}{2a} = -\frac{-60}{4} = 15.$$

If $x = 15$, then $30 - x = 30 - (15) = 15$. Thus, the two numbers should both be 15. ∎

EXAMPLE 8   A golf pro-shop operator finds that she can sell 30 sets of golf clubs at $500 per set in a year. Furthermore, she predicts that for each $25 decrease in price, three extra sets of golf clubs could be sold. At what price should she sell the clubs to maximize gross income?

Solution   Sometimes in analyzing such a problem it helps to start setting up a table as follows.

	Number of sets	*Price per set*	*Income*
3 additional sets can be sold for → a $25 decrease in price	30	$500	$15,000
	33	$475	$15,675
	36	$450	$16,200

Let x represent the number of $25 decreases in price. Then we can express the income as a function of x as follows.

$$f(x) = (30 + 3x)(500 - 25x)$$

$$\uparrow \qquad\qquad \uparrow$$

number of sets   price per set

Simplifying this, we obtain

$$f(x) = 15,000 - 750x + 1500x - 75x^2$$

$$= -75x^2 + 750x + 15,000.$$

Completing the square we obtain

$$f(x) = -75x^2 + 750x + 15,000$$

$$= -75(x^2 - 10x + \underline{\qquad}) + 15,000$$

$$= -75(x^2 - 10x + 25) + 15,000 + 1875$$

$$= -75(x - 5)^2 + 16,875.$$

From this form we know that the vertex of the parabola is at (5,16875). So 5 decreases of $25 each, that is, a $125 reduction in price, will give a maximum income of $16,875. The golf clubs should be sold at $375 per set.   ∎

Problem Set 12.2

For Problems 1–30, graph each of the linear and quadratic functions.

1. $f(x) = 2x - 4$ **2.** $f(x) = 3x + 3$

3. $f(x) = -2x^2$ **4.** $f(x) = -4x^2$

5. $f(x) = -3x$ **6.** $f(x) = -4x$

7. $f(x) = -(x + 1)^2 - 2$ **8.** $f(x) = -(x - 2)^2 + 4$

9. $f(x) = -x + 3$

10. $f(x) = -2x - 4$

11. $f(x) = x^2 + 2x - 2$

12. $f(x) = x^2 - 4x - 1$

13. $f(x) = -x^2 + 6x - 8$

14. $f(x) = -x^2 - 8x - 15$

15. $f(x) = -3$

16. $f(x) = 1$

17. $f(x) = 2x^2 - 20x + 52$

18. $f(x) = 2x^2 + 12x + 14$

19. $f(x) = -3x^2 + 6x$

20. $f(x) = -4x^2 - 8x$

21. $f(x) = x^2 - x + 2$

22. $f(x) = x^2 + 3x + 2$

23. $f(x) = 2x^2 + 10x + 11$

24. $f(x) = 2x^2 - 10x + 15$

25. $f(x) = -2x^2 - 1$

26. $f(x) = -3x^2 + 2$

27. $f(x) = -3x^2 + 12x - 7$

28. $f(x) = -3x^2 - 18x - 23$

29. $f(x) = -2x^2 + 14x - 25$

30. $f(x) = -2x^2 - 10x - 14$

31. Suppose that the cost function for a particular item is given by the equation $C(x) = 2x^2 - 320x + 12{,}920$, where x represents the number of items. How many items should be produced to minimize the cost?

32. Suppose that the equation $p(x) = -2x^2 + 280x - 1000$, where x represents the number of items sold, describes the profit function for a certain business. How many items should be sold to maximize the profit?

33. Find two numbers whose sum is 30, such that the sum of the square of one number plus ten times the other number is a minimum.

34. The height of a projectile fired vertically into the air (neglecting air resistance) at an initial velocity of 96 feet per second is a function of the time and is given by the equation $f(x) = 96x - 16x^2$, where x represents the time. Find the highest point reached by the projectile.

35. Two hundred and forty meters of fencing is available to enclose a rectangular playground. What should be the dimensions of the playground to maximize the area?

36. Find two numbers whose sum is 50 and whose product is a maximum.

37. A Cable TV company has 1000 subscribers who each pay $15 per month. Based on a survey, they feel that for each decrease of $0.25 on the monthly rate, they could obtain 20 additional subscribers. At what rate will maximum revenue be obtained and how many subscribers will it take at that rate?

38. A motel advertises that they will provide dinner, a dance, and drinks at $50 per couple for a New Year's Eve party. They must have a guarantee of 30 couples. Furthermore, they will agree that for each couple in excess of 30, they will reduce the price per couple for all attending by $0.50. How many couples will it take to maximize the motel's revenue?

39. Find two numbers whose difference is 30, such that the sum of the square of one number plus eight times the other number is a minimum.

40. Find the maximum product of two numbers whose sum is 40. What are the numbers?

41. Find the minimum product of two numbers whose difference is 10. What are the numbers?

12.3
Graphing Functions

To graph a new function, that is, one we are unfamiliar with, we can use some of the graphing suggestions offered in Chapter 11. Let's restate those suggestions in terms of function vocabulary and notation. Pay special attention to steps 2 and 3 where we have restated the concepts of intercepts and symmetry using function notation.

1. Determine the domain of the function.
2. Determine any types of symmetry that the equation possesses. If $f(-x) = f(x)$, then the function exhibits y-axis symmetry. If $f(-x) = -f(x)$, then the function exhibits origin symmetry. (Note that the definition of a function rules out the possibility that the graph of a function has x-axis symmetry.)
3. Find the y-intercept (we are labeling the y-axis with $f(x)$) by evaluating $f(0)$. Find the x-intercept by finding the value(s) of x such that $f(x) = 0$.
4. Set up a table of ordered pairs that satisfy the equation. The type of symmetry and the domain will affect your choice of values of x in the table.
5. Plot the points associated with the ordered pairs and connect them with a smooth curve. Then, if appropriate, reflect this part of the curve according to any symmetries possessed by the graph.

EXAMPLE 1   Graph $f(x) = \sqrt{x}$.

Solution   Since the radicand must be nonnegative, the domain is the set of nonnegative real numbers. Since $x \geq 0$, $f(-x)$ is not a real number; so there is no symmetry for this graph. We see that $f(0) = 0$; so both intercepts are 0. That is to say, the origin $(0, 0)$ is a point of the graph. Now let's set up a table of values keeping in mind that $x \geq 0$. Plotting these points and connecting them with a smooth curve produces Figure 12.11.

x	$f(x)$
0	0
1	1
4	2
9	3

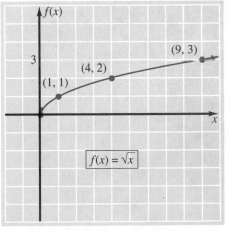

Figure 12.11

EXAMPLE 2   Graph $f(x) = x^3$.

Solution   Since any real number can be cubed, the domain of the function is the set of all real numbers. Since $f(-x) = (-x)^3 = -x^3$ and $-f(x) = -x^3$, we have $f(-x) = -f(x)$ and therefore the graph of $f(x) = x^3$ has origin symmetry. Because $f(0) = 0^3 = 0$, the origin is on the graph. Because of origin symmetry, we can limit the table of values to positive values of x.

x	$f(x)$
1	1
$\dfrac{1}{2}$	$\dfrac{1}{8}$
1	1
$\dfrac{3}{2}$	$\dfrac{27}{8}$
2	8

Plotting the points determined by the table along with the origin, and connecting them with a smooth curve produces Figure 12.12(a). Then reflecting this portion of the curve through the origin produces the complete graph as shown in Figure 12.12(b).

Figure 12.12

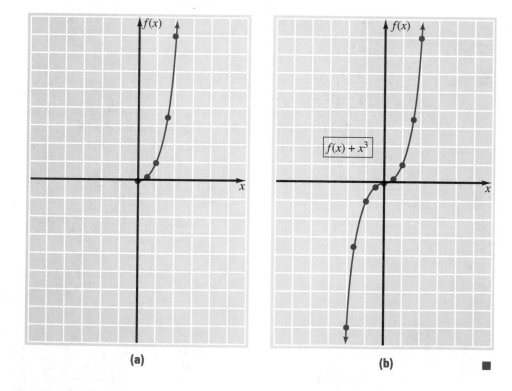

(a)   (b)

Sometimes a new function is defined in terms of old functions. In such cases, the definition plays an important role in the study of the new function. Consider the following example.

EXAMPLE 3   Graph the function $f(x) = |x|$.

Solution   The concept of absolute value is defined for all real numbers as

$$|x| = x \quad \text{if } x \geq 0,$$
$$|x| = -x \quad \text{if } x < 0.$$

Therefore, we can express the absolute value function as

$$f(x) = |x| = \begin{cases} x & \text{if } x \geq 0 \\ -x & \text{if } x < 0 \end{cases}.$$

The graph of $f(x) = x$ for $x \geq 0$ is the ray in the first quadrant and the graph of $f(x) = -x$ for $x < 0$ is the half-line in the second quadrant as indicated in Figure 12.13.

Figure 12.13

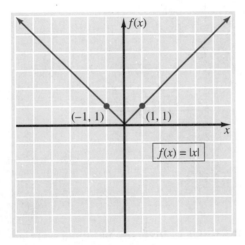

REMARK 1 Note in Example 3 that the equation $f(x) = |x|$ does exhibit y-axis symmetry because $f(-x) = |-x| = |x|$. Even though we did not use the symmetry idea to sketch the curve, we should recognize that the symmetry does exist.

REMARK 2 At this time it might be helpful to go back and look over Section 11.3. We will be referring back to that section frequently in the next few pages.

Translations of the Basic Curves

In Section 11.3 we graphed various vertical and horizontal translations of the basic parabola $y = x^2$. For example, the graph of $y = x^2 + 1$ is the graph of $y = x^2$ moved up one unit. Likewise, the graph of $y = x^2 - 2$ is the graph of $y = x^2$ moved down two units. Now let's describe in general the concept of a vertical translation.

Vertical Translation

The graph of $y = f(x) + k$ is the graph of $y = f(x)$ shifted k units upward if $k > 0$ or shifted $|k|$ units downward if $k < 0$.

In Figure 12.14 the graph of $f(x) = |x| + 2$ is obtained by shifting the graph of $f(x) = |x|$ upward two units and the graph of $f(x) = |x| - 4$ is obtained by shifting the graph of $f(x) = |x|$ downward four units. (Remember that $f(x) = |x| - 4$ can be written as $f(x) = |x| + (-4)$.)

Figure 12.14

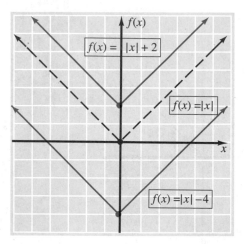

Horizontal translations of the basic parabola were also graphed in Section 11.3. For example, the graph of $y = (x - 2)^2$ is the graph of $y = x^2$ shifted two units to the right and the graph of $y = (x + 3)^2$ is the graph of $y = x^2$ shifted three units to the left. The general concept of a horizontal translation can be described as follows.

Horizontal Translation

The graph of $y = f(x - h)$ is the graph of $y = f(x)$ shifted h units to the right if $h > 0$ or shifted $|h|$ units to the left if $h < 0$.

In Figure 12.15 the graph of $f(x) = (x - 3)^3$ is obtained by shifting the graph of $f(x) = x^3$ three units to the right. Likewise, in Figure 12.15 the graph of $f(x) = (x + 2)^3$ is obtained by shifting the graph of $f(x) = x^3$ two units to the left. (Remember that $f(x) = (x + 2)^3$ can be written as $f(x) = (x - (-2))^3$.)

Figure 12.15

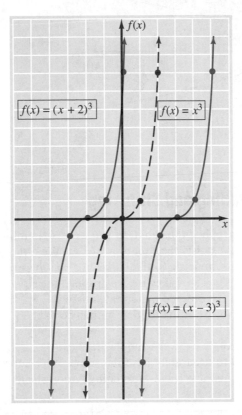

Reflections Through the x-axis

In Section 11.3 the graph of $y = -x^2$ was obtained by reflecting the graph of $y = x^2$ through the x-axis. The general concept of an x-axis reflection can be described as follows.

> **x-axis Reflection**
>
> The graph of $y = -f(x)$ is the graph of $y = f(x)$ reflected through the x-axis.

In Figure 12.16 the graph of $f(x) = -\sqrt{x}$ is obtained by reflecting the graph of $f(x) = \sqrt{x}$ through the x-axis.

Figure 12.16

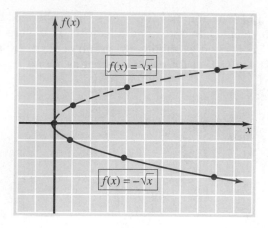

Vertical Stretching and Shrinking

In Section 11.3 we graphed the equation $y = 2x^2$ by doubling the y-coordinates of the ordered pairs that satisfy the equation $y = x^2$. We obtained a parabola with its vertex at the origin, symmetric to the y-axis but *narrower* than the basic parabola, which is the graph of $y = x^2$. Likewise, we graphed the equation $y = \frac{1}{2}x^2$ by halving the y-coordinates of the ordered pairs that satisfy $y = x^2$. In this case, we obtained a parabola with its vertex at the origin, symmetric to the y-axis but *wider* than the basic parabola.

The concepts of "narrower" and "wider" can be used to describe parabolas but cannot be used to accurately describe some other curves. Instead we use the more general concepts of vertical "stretching" and "shrinking."

Vertical Stretching and Shrinking

The graph of $y = cf(x)$ is obtained from the graph of $y = f(x)$ by multiplying the y-coordinates of $y = f(x)$ by c. If $c > 1$, the graph is said to be stretched by a factor of c, and if $0 < c < 1$, the graph is said to be shrunk by a factor of c.

In Figure 12.17 the graph of $f(x) = 2\sqrt{x}$ is obtained by doubling the y-coordinates of points on the graph of $f(x) = \sqrt{x}$. Likewise, in Figure 12.17, the graph of $f(x) = \frac{1}{2}\sqrt{x}$ is obtained by halving the y-coordinates of points on the graph of $f(x) = \sqrt{x}$.

Finally, let's consider the graph of a function that involves a stretching, a reflection, a horizontal translation, and a vertical translation of the basic absolute value curve.

Figure 12.17

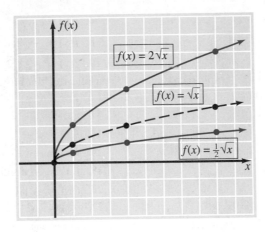

EXAMPLE 4   Graph $f(x) = -2|x - 3| + 1$.

Solution   This is the basic absolute value curve stretched by a factor of two, reflected through the x-axis, shifted three units to the right, and shifted one unit upward. To sketch the graph we locate the point $(3, 1)$, and then determine a point on each of the rays. The graph is shown in Figure 12.18.

Figure 12.18

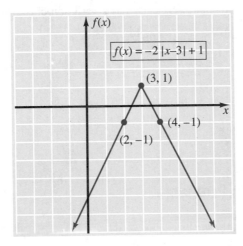

Problem Set 12.3

For Problems 1–10, graph each curve as a variation of the basic square root curve determined by $f(x) = \sqrt{x}$.

1. $f(x) = \sqrt{x} - 1$

2. $f(x) = \sqrt{x} + 3$

3. $f(x) = \sqrt{x + 2}$

4. $f(x) = \sqrt{x - 1}$

5. $f(x) = -\sqrt{x-2}$

6. $f(x) = -\sqrt{x}$

7. $f(x) = \sqrt{x+3} + 1$

8. $f(x) = -\sqrt{x-2} - 2$

9. $f(x) = 3\sqrt{x}$

10. $f(x) = -2\sqrt{x}$

For Problems 11–22, graph each curve as a variation of the basic absolute value graph determined by $f(x) = |x|$.

11. $f(x) = |x| - 1$

12. $f(x) = |x| + 2$

13. $f(x) = |x + 2|$

14. $f(x) = |x - 1|$

15. $f(x) = -|x - 2|$

16. $f(x) = -|x + 3|$

17. $f(x) = -|x| + 3$

18. $f(x) = -|x| - 2$

19. $f(x) = 2|x|$

20. $f(x) = -3|x|$

21. $f(x) = 3|x - 2| + 1$

22. $f(x) = 2|x + 2| - 1$

For Problems 23–30, graph each curve as a variation of the basic cubic curve determined by $f(x) = x^3$.

23. $f(x) = x^3 - 1$

24. $f(x) = x^3 + 2$

25. $f(x) = (x + 2)^3$

26. $f(x) = (x - 3)^3$

27. $f(x) = -x^3$

28. $f(x) = -2x^3$

29. $f(x) = (x - 1)^3 - 2$

30. $f(x) = -(x + 3)^3 + 1$

31. **(a)** Graph $f(x) = \dfrac{1}{x}$.

(b) Graph each of the following curves as a variation of the curve from part (a).

(i) $f(x) = \dfrac{1}{x} - 2$

(ii) $f(x) = \dfrac{1}{x} + 3$

(iii) $f(x) = \dfrac{1}{x + 1}$

(iv) $f(x) = \dfrac{1}{x - 2}$

(v) $f(x) = -\dfrac{1}{x}$

(vi) $f(x) = \dfrac{3}{x}$

Miscellaneous Problems

Use a graphics calculator for Problems 32–34.

32. Graph $f(x) = x^3$, $f(x) = 4x^3$, $f(x) = \dfrac{1}{2}x^3$ and $f(x) = \dfrac{1}{5}x^3$ on the same set of axes.

33. Graph $f(x) = \dfrac{1}{x^2}$, $f(x) = \dfrac{1}{x^2} + 2$, $f(x) = \dfrac{1}{(x + 2)^2}$, and $f(x) = -\dfrac{1}{x^2}$ on the same set of axes.

34. Graph $f(x) = (x + 1)^3$ and $f(x) = x^3 + 3x^2 + 3x + 1$ on the same set of axes. What does this result mean? Now use a graphic approach to show that $(x + 2)^4 = x^4 + 8x^3 + 24x^2 + 32x + 16$.

12.4

Operations With Functions

In subsequent mathematics courses, it is common to encounter functions that are defined in terms of sums, differences, products, and quotients of simpler functions. For example, if $h(x) = x^2 + \sqrt{x - 1}$, then we may consider the function h as the sum of f and g, where $f(x) = x^2$ and $g(x) = \sqrt{x - 1}$. In general, *if f and g are functions and D is the intersection of their domains*, then the following definitions can be made.

Sum: $\qquad\qquad (f + g)(x) = f(x) + g(x);$

Difference: $\qquad (f - g)(x) = f(x) - g(x);$

Product: $\qquad\quad (f \cdot g)(x) = f(x) \cdot g(x);$

Quotient: $\qquad \left(\dfrac{f}{g}\right)(x) = \dfrac{f(x)}{g(x)}, \qquad g(x) \neq 0.$

EXAMPLE 1   If $f(x) = 3x - 1$ and $g(x) = x^2 - x - 2$, find **(a)** $(f + g)(x)$; **(b)** $(f - g)(x)$; **(c)** $(f \cdot g)(x)$; and **(d)** $(f/g)(x)$. Determine the domain of each.

Solution

(a) $(f + g)(x) = f(x) + g(x) = (3x - 1) + (x^2 - x - 2) = x^2 + 2x - 3.$

(b) $(f - g)(x) = f(x) - g(x)$

$\qquad = (3x - 1) - (x^2 - x - 2)$

$\qquad = 3x - 1 - x^2 + x + 2$

$\qquad = -x^2 + 4x + 1.$

(c) $(f \cdot g)(x) = f(x) \cdot g(x)$

$\qquad = (3x - 1)(x^2 - x - 2)$

$\qquad = 3x^3 - 3x^2 - 6x - x^2 + x + 2$

$\qquad = 3x^3 - 4x^2 - 5x + 2.$

(d) $\left(\dfrac{f}{g}\right)(x) = \dfrac{f(x)}{g(x)} = \dfrac{3x - 1}{x^2 - x - 2}.$

The domain of both f and g is the set of all real numbers. Therefore, the domain of $f + g$, $f - g$, and $f \cdot g$ is the set of all real numbers. For f/g, the denominator $x^2 - x - 2$ cannot equal zero. Solving $x^2 - x - 2 = 0$ produces

$(x - 2)(x + 1) = 0$

$x - 2 = 0 \quad \text{or} \quad x + 1 = 0$

$x = 2 \quad \text{or} \qquad x = -1.$

Therefore, the domain for f/g is the set of all real numbers except 2 and -1. ∎

The Composition of Functions

Besides adding, subtracting, multiplying, and dividing functions, there is another important operation called **composition**. The composition of two functions can be defined as follows.

DEFINITION 12.3

The **composition** of functions f and g is defined by

$$(f \circ g)(x) = f(g(x)),$$

for all x in the domain of g such that $g(x)$ is in the domain of f.

The left side, $(f \circ g)(x)$, of the equation in Definition 12.3 can be read as "the composition of f and g" and the right side, $f(g(x))$, can be read as "f of g of x." It may also be helpful for you to mentally picture Definition 12.3 as two function machines *hooked together* to produce another function (often called a **composite function**) as illustrated in Figure 12.19. Notice that what comes out of the function g is substituted into the function f. Thus, composition is sometimes called the substitution of functions.

Figure 12.19

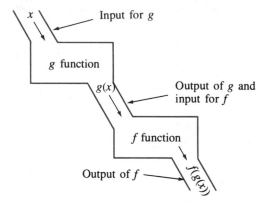

Figure 12.19 also vividly illustrates the fact that $f \circ g$ is defined *for all x in the domain of g such that g(x) is in the domain of f.* In other words, what comes out of g must be capable of being fed into f. Let's consider some examples.

EXAMPLE 2   If $f(x) = x^2$ and $g(x) = x - 3$, find $(f \circ g)(x)$ and determine its domain.

Solution   Applying Definition 12.3 we obtain

$$(f \circ g)(x) = f(g(x))$$
$$= f(x - 3)$$
$$= (x - 3)^2.$$

Because g and f are both defined for all real numbers, so is $f \circ g$.   ∎

EXAMPLE 3   If $f(x) = \sqrt{x}$ and $g(x) = x - 4$, find $(f \circ g)(x)$ and determine its domain.

Solution   Applying Definition 12.3 we obtain

$$(f \circ g)(x) = f(g(x))$$
$$= f(x - 4)$$
$$= \sqrt{x - 4}.$$

The domain of g is all real numbers but the domain of f is only the nonnegative real numbers. Thus $g(x)$, which is $x - 4$, has to be nonnegative. So

$$x - 4 \geq 0$$
$$x \geq 4,$$

and the domain of $f \circ g$ is $D = \{x \mid x \geq 4\}$.   ∎

Definition 12.3 with f and g interchanged, defines the composition of g and f as $(g \circ f)(x) = g(f(x))$.

EXAMPLE 4   If $f(x) = x^2$ and $g(x) = x - 3$, find $(g \circ f)(x)$ and determine its domain.

Solution
$$(g \circ f)(x) = g(f(x))$$
$$= g(x^2)$$
$$= x^2 - 3$$

Since f and g are both defined for all real numbers, the domain of $g \circ f$ is the set of all real numbers.   ∎

The results of Examples 2 and 4 demonstrate an important idea, namely, that the composition of functions is *not a commutative operation*. In other words, it is not true that $f \circ g = g \circ f$ for all functions f and g. However, as we will see in the next section, there is a special class of functions where $f \circ g = g \circ f$.

EXAMPLE 5   If $f(x) = 2x + 3$ and $g(x) = \sqrt{x - 1}$, determine each of the following.
(a) $(f \circ g)(x)$ (b) $(g \circ f)(x)$ (c) $(f \circ g)(5)$ (d) $(g \circ f)(7)$

Solution
(a) $(f \circ g)(x) = f(g(x))$
$$= f(\sqrt{x - 1})$$
$$= 2\sqrt{x - 1} + 3 \qquad D = \{x \mid x \geq 1\}$$

(b) $(g \circ f)(x) = g(f(x))$

$$= g(2x + 3)$$
$$= \sqrt{2x + 3 - 1}$$
$$= \sqrt{2x + 2} \qquad D = \{x \mid x \geq -1\}$$

(c) By using the composite function formed in part (a), we obtain

$$(f \circ g)(5) = 2\sqrt{5 - 1} + 3$$
$$= 2\sqrt{4} + 3$$
$$= 2(2) + 3 = 7.$$

(d) By using the composite function formed in part (b) we obtain

$$(g \circ f)(7) = \sqrt{2(7) + 2} = \sqrt{16} = 4.$$ ∎

Our final example of this section presents a situation in which you must be very careful when determining the domain of a composite function. Remember that the domain of the composite function $(f \circ g)(x)$ is the set of all x in the domain of g such that $g(x)$ is in the domain of f.

EXAMPLE 6   If $f(x) = \dfrac{2}{x - 1}$ and $g(x) = \dfrac{1}{x}$, find $(f \circ g)(x)$ and determine its domain.

Solution $(f \circ g)(x) = f(g(x))$

$$= f\left(\frac{1}{x}\right)$$

$$= \frac{2}{\dfrac{1}{x} - 1} = \frac{2}{\dfrac{1 - x}{x}}$$

$$= \frac{2x}{1 - x}.$$

The domain of g is all real numbers except zero, and the domain of f is all real numbers except one. Since $g(x)$, which is $1/x$, cannot equal 1,

$$\frac{1}{x} \neq 1$$

$$x \neq 1.$$

Therefore, the domain of $f \circ g$ is $D = \{x \mid x \neq 0 \text{ and } x \neq 1\}$. ∎

Problem Set 12.4

For Problems 1–8, find $f + g$, $f - g$, $f \cdot g$, and $\dfrac{f}{g}$.

1. $f(x) = 3x - 4$, $g(x) = 5x + 2$
2. $f(x) = -6x - 1$, $g(x) = -8x + 7$
3. $f(x) = x^2 - 6x + 4$, $g(x) = -x - 1$
4. $f(x) = 2x^2 - 3x + 5$, $g(x) = x^2 - 4$
5. $f(x) = x^2 - x - 1$, $g(x) = x^2 + 4x - 5$
6. $f(x) = x^2 - 2x - 24$, $g(x) = x^2 - x - 30$
7. $f(x) = \sqrt{x - 1}$, $g(x) = \sqrt{x}$
8. $f(x) = \sqrt{x + 2}$, $g(x) = \sqrt{3x - 1}$

For Problems 9–20, determine the indicated functional values.

9. If $f(x) = 9x - 2$ and $g(x) = -4x + 6$, find $(f \circ g)(-2)$ and $(g \circ f)(4)$.
10. If $f(x) = -2x - 6$ and $g(x) = 3x + 10$, find $(f \circ g)(5)$ and $(g \circ f)(-3)$.
11. If $f(x) = 4x^2 - 1$ and $g(x) = 4x + 5$, find $(f \circ g)(1)$ and $(g \circ f)(4)$.
12. If $f(x) = -5x + 2$ and $g(x) = -3x^2 + 4$, find $(f \circ g)(-2)$ and $(g \circ f)(-1)$.
13. If $f(x) = \dfrac{1}{x}$ and $g(x) = \dfrac{2}{x - 1}$, find $(f \circ g)(2)$ and $(g \circ f)(-1)$.
14. If $f(x) = \dfrac{2}{x - 1}$ and $g(x) = -\dfrac{3}{x}$, find $(f \circ g)(1)$ and $(g \circ f(-1)$.
15. If $f(x) = \dfrac{1}{x - 2}$ and $g(x) = \dfrac{4}{x - 1}$, find $(f \circ g)(3)$ and $(g \circ f)(2)$.
16. If $f(x) = \sqrt{x + 6}$ and $g(x) = 3x - 1$, find $(f \circ g)(-2)$ and $(g \circ f)(-2)$,
17. If $f(x) = \sqrt{3x - 2}$ and $g(x) = -x + 4$, find $(f \circ g)(1)$ and $(g \circ f)(6)$.
18. If $f(x) = -5x + 1$ and $g(x) = \sqrt{4x + 1}$, find $(f \circ g)(6)$ and $(g \circ f)(-1)$.
19. If $f(x) = |4x - 5|$ and $g(x) = x^3$, find $(f \circ g)(-2)$ and $(g \circ f)(2)$.
20. If $f(x) = -x^3$ and $g(x) = |2x + 4|$, find $(f \circ g)(-1)$ and $(g \circ f)(-3)$.

For Problems 21–38, determine $(f \circ g)(x)$ and $(g \circ f)(x)$ for each pair of functions. Also specify the domain of $(f \circ g)(x)$ and $(g \circ f)(x)$.

21. $f(x) = 3x$ and $g(x) = 5x - 1$
22. $f(x) = 4x - 3$ and $q(x) = -2x$
23. $f(x) = -2x + 1$ and $g(x) = 7x + 4$
24. $f(x) = 6x - 5$ and $g(x) = -x + 6$
25. $f(x) = 3x + 2$ and $g(x) = x^2 + 3$
26. $f(x) = -2x + 4$ and $g(x) = 2x^2 - 1$
27. $f(x) = 2x^2 - x + 2$ and $g(x) = -x + 3$
28. $f(x) = 3x^2 - 2x - 4$ and $g(x) = -2x + 1$
29. $f(x) = \dfrac{3}{x}$ and $g(x) = 4x - 9$

30. $f(x) = -\dfrac{2}{x}$ and $g(x) = -3x + 6$

31. $f(x) = \sqrt{x + 1}$ and $g(x) = 5x + 3$

32. $f(x) = 7x - 2$ and $g(x) = \sqrt{2x - 1}$

33. $f(x) = \dfrac{1}{x}$ and $g(x) = \dfrac{1}{x - 4}$

34. $f(x) = \dfrac{2}{x + 3}$ and $g(x) = -\dfrac{3}{x}$

35. $f(x) = \sqrt{x}$ and $g(x) = \dfrac{4}{x}$

36. $f(x) = \dfrac{2}{x}$ and $g(x) = |x|$

37. $f(x) = \dfrac{3}{2x}$ and $g(x) = \dfrac{1}{x + 1}$

38. $f(x) = \dfrac{4}{x - 2}$ and $g(x) = \dfrac{3}{4x}$

For Problems 39–46, show that $(f \circ g)(x) = x$ and $(g \circ f)(x) = x$ for each pair of functions.

39. $f(x) = 3x$ and $g(x) = \dfrac{1}{3}x$

40. $f(x) = -2x$ and $g(x) = -\dfrac{1}{2}x$

41. $f(x) = 4x + 2$ and $g(x) = \dfrac{x - 2}{4}$

42. $f(x) = 3x - 7$ and $g(x) = \dfrac{x + 7}{3}$

43. $f(x) = \dfrac{1}{2}x + \dfrac{3}{4}$ and $g(x) = \dfrac{4x - 3}{2}$

44. $f(x) = \dfrac{2}{3}x - \dfrac{1}{5}$ and $g(x) = \dfrac{3}{2}x + \dfrac{3}{10}$

45. $f(x) = -\dfrac{1}{4}x - \dfrac{1}{2}$ and $g(x) = -4x - 2$

46. $f(x) = -\dfrac{3}{4}x + \dfrac{1}{3}$ and $g(x) = -\dfrac{4}{3}x + \dfrac{4}{9}$

Thoughts into Words

47. What does it mean to say that the *composition of functions* is not a commutative operation?

48. Is *addition of functions* a commutative operation? Defend your answer.

49. How would you go about graphing the function $f(x) = -\sqrt{x - 5} + 4$?

50. How would you go about graphing the function $f(x) = \dfrac{1}{x^2}$?

12.5

Inverse Functions

Graphically, the distinction between a relation and a function can be easily recognized. In Figure 12.20 we have sketched four graphs. Which of these are graphs of functions and which are graphs of relations that are not functions? Think in terms of "to each member of the domain there is assigned one and only one member of the range"; this is the basis for what is known as the **vertical line test for functions**. Because each value of x produces only one value of $f(x)$, any vertical line drawn through a **graph of a function must not intersect the graph in more than one point**. Therefore, parts (a) and (c) of Figure 12.20 are graphs of functions and parts (b) and (d) are graphs of relations that are not functions.

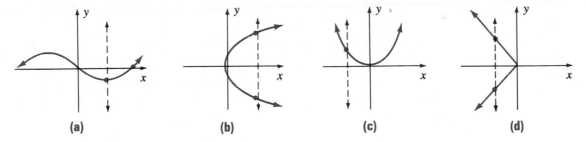

(a)   (b)   (c)   (d)

Figure 12.20

There is also a useful distinction made between two basic types of functions. Consider the graphs of the two functions $f(x) = 2x - 1$ and $f(x) = x^2$ in Figure 12.21. In part (a) any *horizontal line* will intersect the graph in no more than one

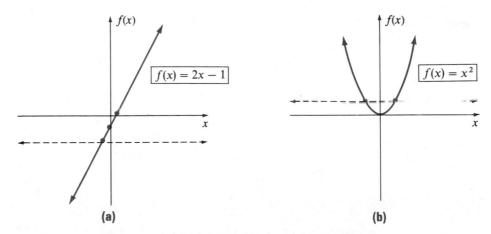

(a)   (b)

Figure 12.21

point. Therefore, every value of $f(x)$ has only one value of x associated with it. Any function that has the additional property of having only one value of x associated with each value of $f(x)$ is called a **one-to-one function**. The function $f(x) = x^2$ is not a one-to-one function because the horizontal line in part (b) of Figure 12.21 intersects the parabola in two points.

In terms of ordered pairs, a one-to-one function does not contain any ordered pairs that have the same second component. For example,

$$f = \{(1, 3), (2, 6), (4, 12)\}$$

is a one-to-one function, but

$$g = \{(1, 2), (2, 5), (-2, 5)\}$$

is not a one-to-one function.

If the components of each ordered pair of a given one-to-one function are interchanged, the resulting function and the given function are called **inverses** of each other. Thus,

$$\{(1, 3), (2, 6), (4, 12)\} \quad \text{and} \quad \{(3, 1), (6, 2), (12, 4)\}$$

are **inverse functions**. The inverse of a function f is denoted by f^{-1} (read "f inverse" or "the inverse of f"). If (a, b) is an ordered pair of f, then (b, a) is an ordered pair of f^{-1}. The domain and range of f^{-1} are the range and domain, respectively, of f.

REMARK Do not confuse the -1 in f^{-1} with a negative exponent. The symbol f^{-1} does not mean $\dfrac{1}{f^1}$, but refers to the inverse function of function f.

Graphically, two functions that are inverses of each other are mirror images with reference to the line $y = x$. This is due to the fact that ordered pairs (a, b) and (b, a) are mirror images with respect to the line $y = x$ as illustrated in Figure 12.22.

Figure 12.22

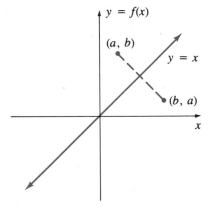

Therefore, if we know the graph of a function f, as in Figure 12.23(a) then we can determine the graph of f^{-1} by reflecting f across the line $y = x$ (Figure 12.23(b)).

Figure 12.23

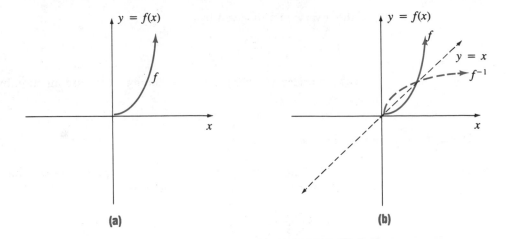

(a)   (b)

Another useful way of viewing inverse functions is in terms of composition. Basically, inverse functions *undo* each other and this can be more formally stated as follows. If f and g are inverses of each other, then

1. $(f \circ g)(x) = f(g(x)) = x$ for all x in domain of g; and

2. $(g \circ f)(x) = g(f(x)) = x$ for all x in domain of f.

As we will see in a moment, this relationship of inverse functions can be used to verify whether two functions are indeed inverses of each other.

Finding Inverse Functions

The idea of inverse functions *undoing each other* provides the basis for a rather informal approach to finding the inverse of a function. Consider the function

$$f(x) = 2x + 1.$$

To each x this function assigns *twice x plus 1*. To *undo* this function, we could *subtract* 1 *and divide by* 2. So, the inverse should be

$$f^{-1}(x) = \frac{x-1}{2}.$$

Now let's verify that f and f^{-1} are inverses of each other.

$$(f \circ f^{-1})(x) = f(f^{-1}(x)) = f\left(\frac{x-1}{2}\right) = 2\left(\frac{x-1}{2}\right) + 1 = x$$

and

$$(f^{-1} \circ f)(x) = f^{-1}(f(x)) = f^{-1}(2x + 1) = \frac{2x + 1 - 1}{2} = x$$

Thus, the inverse of f is given by

$$f^{-1}(x) = \frac{x-1}{2}.$$

Let's consider another example of finding an inverse function by the *undoing* process.

EXAMPLE 1   Find the inverse of $f(x) = 3x - 5$.

Solution   To each x, the function f assigns *three times x minus 5*. To *undo* this we can *add 5 and then divide by 3*. So, the inverse should be

$$f^{-1}(x) = \frac{x+5}{3}.$$

To verify that f and f^{-1} are inverses we can show that

$$(f \circ f^{-1})(x) = f(f^{-1}(x)) = f\left(\frac{x+5}{3}\right)$$

$$= 3\left(\frac{x+5}{3}\right) - 5 = x$$

and

$$(f^{-1} \circ f)(x) = f^{-1}(f(x)) = f^{-1}(3x - 5)$$

$$= \frac{3x - 5 + 5}{3} = x.$$

Thus, f and f^{-1} are inverses and we can write

$$f^{-1}(x) = \frac{x+5}{3}. \qquad \blacksquare$$

This informal approach may not work very well with more complex functions, but it does emphasize how inverse functions are related to each other. A more formal and systematic technique for finding the inverse of a function can be described as follows.

1. Replace the symbol $f(x)$ by y.
2. Interchange x and y.
3. Solve the equation for y in terms of x.
4. Replace y by the symbol $f^{-1}(x)$.

Now let's use two examples to illustrate this technique.

EXAMPLE 2   Find the inverse of $f(x) = -3x + 11$.

Solution

Replacing $f(x)$ by y, the given equation becomes

$$y = -3x + 11.$$

Interchanging x and y produces

$$x = -3y + 11.$$

Now, solving for y we obtain

$$x = -3y + 11$$

$$3y = -x + 11$$

$$y = \frac{-x + 11}{3}.$$

Finally, replacing y by $f^{-1}(x)$ we can express the inverse function as

$$f^{-1}(x) = \frac{-x + 11}{3}.$$

∎

EXAMPLE 3

Find the inverse of $f(x) - \frac{3}{2}x - \frac{1}{4}$.

Solution

Replacing $f(x)$ by y, the given equation becomes

$$y = \frac{3}{2}x - \frac{1}{4}.$$

Interchanging x and y produces

$$x = \frac{3}{2}y - \frac{1}{4}.$$

Now, solving for y we obtain

$$x = \frac{3}{2}y - \frac{1}{4}$$

$$4x = 6y - 1$$

$$4x + 1 = 6y$$

$$\frac{4x + 1}{6} = y$$

$$\frac{2}{3}x + \frac{1}{6} = y.$$

Finally, replacing y by $f^{-1}(x)$ we can express the inverse function as

$$f^{-1}(x) = \frac{2}{3}x + \frac{1}{6}.$$

∎

For both Examples 2 and 3 you should be able to show that $(f \circ f^{-1})(x) = x$ and $(f^{-1} \circ f)(x) = x$.

Problem Set 12.5

For Problems 1–8, identify each graph as (a) the graph of a function or (b) the graph of a relation that is not a function. Use the vertical line test.

1.

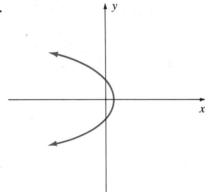

2.

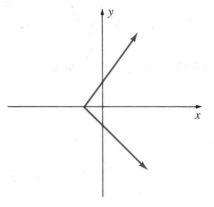

3.

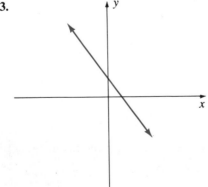

4.

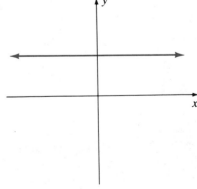

5.

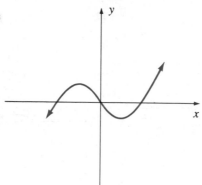

Function

6.

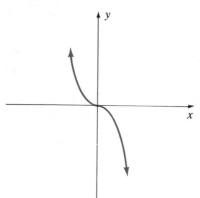

Function

7.

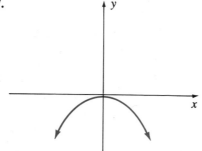

8.

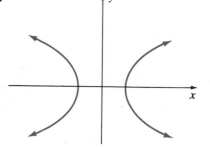

For Problems 9–16, identify each graph as (a) the graph of a one-to-one function or (b) the graph of a function that is not one-to-one. Use the horizontal line test.

9.

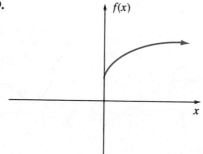

10.

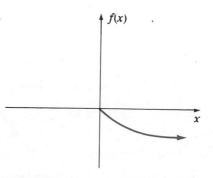

11.

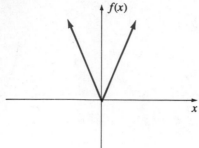

12.

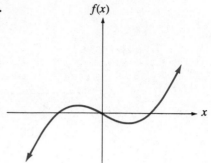

13.

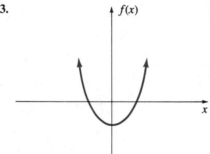

14.

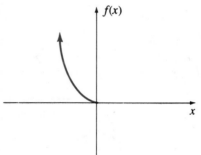

15.

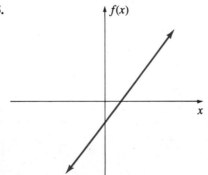

One-to-one function

16.

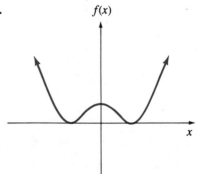

Not a one-to-one function

For Problems 17–20, (a) list the domain and range of the given function, (b) form the inverse function, and (c) list the domain and range of the inverse function.

17. $f = \{(1, 3), (2, 6), (3, 11), (4, 18)\}$ **18.** $f = \{(0, -4), (1, -3), (4, -2)\}$

19. $f = \{(-2, -1), (-1, 1), (0, 5), (5, 10)\}$ **20.** $f = \{(-1, 1), (-2, 4), (1, 9), (2, 12)\}$

For Problems 21–30, find the inverse of the given function by using the "undoing process" and then verify that $(f \circ f^{-1})(x) = x$ and $(f^{-1} \circ f)(x) = x$.

21. $f(x) = 5x - 4$ **22.** $f(x) = 7x + 9$

23. $f(x) = -2x + 1$ **24.** $f(x) = -4x - 3$

25. $f(x) = \dfrac{4}{5}x$ **26.** $f(x) = -\dfrac{2}{3}x$

27. $f(x) = \dfrac{1}{2}x + 4$ **28.** $f(x) = \dfrac{3}{4}x - 2$

29. $f(x) = \dfrac{1}{3}x - \dfrac{2}{5}$ **30.** $f(x) = \dfrac{2}{5}x + \dfrac{1}{3}$

For Problems 31–40, find the inverse of the given function by using the process illustrated in Examples 2 and 3 of this section and then verify that $(f \circ f^{-1})(x) = x$ and $(f^{-1} \circ f)(x) = x$.

31. $f(x) = 9x + 4$ **32.** $f(x) = 8x - 5$

33. $f(x) = -5x - 4$ **34.** $f(x) = -6x + 2$

35. $f(x) = -\dfrac{2}{3}x + 7$ **36.** $f(x) = -\dfrac{3}{5}x + 1$

37. $f(x) = \dfrac{4}{3}x - \dfrac{1}{4}$ **38.** $f(x) = \dfrac{5}{2}x + \dfrac{2}{7}$

39. $f(x) = -\dfrac{3}{7}x - \dfrac{2}{3}$ **40.** $f(x) = -\dfrac{3}{5}x + \dfrac{3}{4}$

For Problems 41–50, (a) find the inverse of the given function, and (b) graph the given function and its inverse on the same set of axes.

41. $f(x) = 4x$ **42.** $f(x) = \dfrac{2}{5}x$

43. $f(x) = -\dfrac{1}{3}x$ **44.** $f(x) = -6x$

45. $f(x) = 3x - 3$ **46.** $f(x) = 2x + 2$

47. $f(x) = -2x - 4$ **48.** $f(x) = -3x + 9$

49. $f(x) = x^2, \quad x \geq 0$ **50.** $f(x) = x^2 + 2, \quad x \geq 0$

Miscellaneous Problems

51. Explain why every nonconstant linear function has an inverse.

52. The composition idea can also be used to find the inverse of a function. For example, to find the inverse of $f(x) = 5x + 3$, we could proceed as follows.

$$f(f^{-1}(x)) = 5(f^{-1}(x)) + 3 \quad \text{and} \quad f(f^{-1}(x)) = x.$$

Therefore, equating the two expressions for $f(f^{-1}(x))$ we obtain

$$5(f^{-1}(x)) + 3 = x$$

$$5(f^{-1}(x)) = x - 3$$

$$f^{-1}(x) = \frac{x - 3}{5}.$$

Use this approach to find the inverse of each of the following functions.

(a) $f(x) = 2x + 1$ (b) $f(x) = 3x - 2$

(c) $f(x) = -4x + 5$ (d) $f(x) = -x + 1$

(e) $f(x) = 2x$ (f) $f(x) = -5x$

 53. Use a graphics calculator and graph $f(x) = x$, $g(x) = 2x + 1$, and $g^{-1}(x) = \frac{1}{2}x - \frac{1}{2}$ on the same set of axes. The graphs of g and g^{-1} should be reflections of each other through the line $f(x) = x$.

 54. Use a graphics calculator and graph $f(x) = x$, $g(x) = -3x + 2$, and $g^{-1}(x) = -\frac{1}{3}x + \frac{2}{3}$ on the same set of axes. The graphs of g and g^{-1} should be reflections of each other through the line $f(x) = x$.

12.6

Direct and Inverse Variations

"The distance a car travels at a fixed rate *varies directly* as the time." "At a constant temperature, the volume of an enclosed gas *varies inversely* as the pressure." Such statements illustrate two basic types of functional relationships, called **direct** and **inverse variation**, which are widely used, especially in the physical sciences. These relationships can be expressed by equations that specify functions. The purpose of this section is to investigate these special functions.

The statement *y varies directly as x* means

$$y = kx$$

where k is a nonzero constant, called the **constant of variation**. The phrase "*y is directly proportional to x*," is also used to indicate direct variation; k is then referred to as the **constant of proportionality**.

> **REMARK** Notice that the equation $y = kx$ defines a function and could be written as $f(x) = kx$ by using function notation. However, in this section it is more convenient to avoid the function notation and use variables that are meaningful in terms of the physical entities involved in the problem.

Statements that indicate direct variation may also involve powers of x. For example, "y varies directly as the square of x" can be written as

$y = kx^2$.

In general, "y varies directly as the nth power of $x(n > 0)$" means

$y = kx^n$.

There are basically three types of problems that deal with direct variation, namely, (1) to translate an English statement into an equation that expresses the direct variation, (2) to find the constant of variation from given values of the variables, and (3) to find additional values of the variables once the constant of variation has been determined. Let's consider an example of each of these types of problems.

EXAMPLE 1   Translate the statement "the tension on a spring varies directly as the distance it is stretched" into an equation and use k as the constant of variation.

Solution   Letting t represent the tension and d the distance, the equation becomes

$t = kd$.   ■

EXAMPLE 2   If A varies directly as the square of s, and $A = 28$ when $s = 2$, find the constant of variation.

Solution   Since A varies directly as the square of s, we have

$A = ks^2$.

Substitute $A = 28$ and $s = 2$, to obtain

$28 = k(2)^2$.

Solving this equation for k yields

$28 = 4k$

$7 - k$,

The constant of variation is 7.   ■

EXAMPLE 3   If y is directly proportional to x and if $y = 6$ when $x = 9$, find the value of y when $x = 24$.

Solution   The statement "y is directly proportional to x" translates into

$y = kx$.

Letting $y = 6$ and $x = 9$, the constant of variation becomes

$$6 = k(9)$$

$$6 = 9k$$

$$\frac{6}{9} = k$$

$$\frac{2}{3} = k.$$

So, the specific equation is $y = \frac{2}{3}x$. Now, letting $x = 24$, we obtain

$$y = \frac{2}{3}(24) = 16.$$

The required value of y is 16.   ■

Inverse Variation

We define the second basic type of variation, called **inverse variation**, as follows. The statement *y varies inversely as x* means

$$y = \frac{k}{x}$$

where k is a nonzero constant; again we refer to it as the constant of variation. The phrase "y is inversely proportional to x" is also used to express inverse variation. As with direct variation, statements that indicate inverse variation may involve powers of x. For example, "y varies inversely as the square of x" can be written as

$$y = \frac{k}{x^2}.$$

In general, "y varies inversely as the nth power of $x(n > 0)$" means

$$y = \frac{k}{x^n}.$$

The following examples illustrate the three basic kinds of problems that we run across which involve inverse variation.

EXAMPLE 4   Translate the statement "the length of a rectangle of a fixed area varies inversely as the width" into an equation that uses k as the constant of variation.

Solution   Let l represent the length and w the width and the equation is

$$l = \frac{k}{w}.$$ ■

EXAMPLE 5   If y is inversely proportional to x and $y = 4$ when $x = 12$, find the constant of variation.

Solution   Since y is inversely proportional to x, we have

$$y = \frac{k}{x}.$$

Substituting $y = 4$ and $x = 12$, we obtain

$$4 = \frac{k}{12}.$$

Solving this equation for k yields

$$k = 48.$$

The constant of variation is 48. ■

EXAMPLE 6   Suppose the number of days it takes to complete a construction job varies inversely as the number of people assigned to the job. If it takes 7 people 8 days to do the job, how long would it take 14 people to complete the job?

Solution   Let d represent the number of days and p the number of people. The phrase "number of days ... varies inversely as the number of people" translates into

$$d = \frac{k}{p}.$$

Let $d = 8$ when $p = 7$ and the constant of variation becomes

$$8 = \frac{k}{7}$$

$$k = 56.$$

So, the specific equation is

$$d = \frac{56}{p}.$$

Now, let $p = 14$, to obtain

$$d = \frac{56}{14}$$

$$d = 4.$$

It should take 14 people 4 days to complete the job. ■

The terms *direct* and *inverse*, as applied to variation, refer to the relative behavior of the variables involved in the equation. That is to say, in direct variation

$(y = kx)$ an assignment of *increasing absolute values* for x produces *increasing absolute values* for y. Whereas, in inverse variation $\left(y = \dfrac{k}{x} \right)$ an assignment of *increasing absolute values* for x produces *decreasing absolute values* for y.

Joint Variation

Variation may involve more than two variables. The following table illustrates some variation statements and their equivalent algebraic equations that use k as the constant of variation.

Variation statement	Algebraic equation
1. y varies jointly as x and z	$y = kxz$
2. y varies jointly as x, z, and w	$y = kxzw$
3. V varies jointly as h and the square of r	$V = khr^2$
4. h varies directly as V and inversely as w	$h = \dfrac{kV}{w}$
5. y is directly proportional to x and inversely proportional to the square of z	$y = \dfrac{kx}{z^2}$
6. y varies jointly as w and z, and inversely as x	$y = \dfrac{kwz}{x}$

Statements 1, 2, and 3 illustrate the concept of **joint variation**. Statements 4 and 5 show that both direct and inverse variation may occur in the same problem. Statement 6 combines joint variation with inverse variation.

The two final examples illustrate problems that involve some of these possible variation situations.

EXAMPLE 7   The length of a rectangular box with a fixed height varies directly as the volume and inversely as the width. If the length is 12 centimeters when the volume is 960 cubic centimeters and the width is 8 centimeters, find the length when the volume is 700 centimeters and the width is 5 centimeters.

Solution   Use l for length, V for volume, and w for width and the phrase "length varies directly as the volume and inversely as the width" translates into

$$l = \frac{kV}{w}.$$

Substitute $l = 12$, $V = 960$, and $w = 8$ and the constant of variation becomes

$$12 = \frac{k(960)}{8}$$

$$12 = 120k$$

$$\frac{1}{10} = k.$$

So the specific equation is

$$l = \frac{\frac{1}{10}V}{w} = \frac{V}{10w}.$$

Now, let $V = 700$ and $w = 5$ to obtain

$$l = \frac{700}{10(5)} = \frac{700}{50} = 14.$$

The length is 14 centimeters.   ■

EXAMPLE 8   Suppose that y varies jointly as x and z, and inversely as w. If $y = 154$ when $x = 6$, $z = 11$, and $w = 3$, find the constant of variation.

Solution   The statement "y varies jointly as x and z, and inversely as w" translates into

$$y = \frac{kxz}{w}.$$

Substitute $y = 154$, $x = 6$, $z = 11$, and $w = 3$ to obtain

$$154 = \frac{k(6)(11)}{3}$$

$$154 = 22k$$

$$7 = k.$$

The constant of variation is 7.   ■

Problem Set 12.6

For Problems 1–10, translate each statement of variation into an equation and use k as the constant of variation.

1. y varies inversely as the square of x.

2. y varies directly as the cube of x.

3. C varies directly as g and inversely as the cube of t.

4. V varies jointly as l and w.

5. The volume (V) of a sphere is directly proportional to the cube of its radius (r).

6. At a constant temperature, the volume (V) of a gas varies inversely as the pressure (P).

7. The surface area (S) of a cube varies directly as the square of the length of an edge (e).

8. The intensity of illumination (I) received from a source of light is inversely proportional to the square of the distance (d) from the source.

9. The volume (V) of a cone varies jointly as its height and the square of its radius.

10. The volume (V) of a gas varies directly as the absolute temperature (T) and inversely as the pressure (P).

For Problems 11–24, find the constant of variation for each of the stated conditions.

11. y varies directly as x, and $y = 8$ when $x = 12$.

12. y varies directly as x, and $y = 60$ when $x = 24$.

13. y varies directly as the square of x, and $y = -144$ when $x = 6$.

14. y varies directly as the cube of x, and $y = 48$ when $x = -2$.

15. V varies jointly as B and h, and $V = 96$ when $B = 24$ and $h = 12$.

16. A varies jointly as b and h, and $A = 72$ when $b = 16$ and $h = 9$.

17. y varies inversely as x, and $y = -4$ when $x = \dfrac{1}{2}$.

18. y varies inversely as x, and $y = -6$ when $x = \dfrac{4}{3}$.

19. r varies inversely as the square of t, and $r = \dfrac{1}{8}$ when $t = 4$.

20. r varies inversely as the cube of t, and $r = \dfrac{1}{16}$ when $t = 4$.

21. y varies directly as x and inversely as z, and $y = 45$ when $x = 18$ and $z = 2$.

22. y varies directly as x and inversely as z, and $y = 24$ when $x = 36$ and $z = 18$.

23. y is directly proportional to x and inversely proportional to the square of z, and $y = 81$ when $x = 36$ and $z = 2$.

24. y is directly proportional to the square of x and inversely proportional to the cube of z, and $y = 4\dfrac{1}{2}$ when $x = 6$ and $z = 4$.

Solve each of the following problems.

25. If y is directly proportional to x, and $y = 36$ when $x = 48$, find the value of y when $x = 12$.

26. If y is directly proportional to x, and $y = 42$ when $x = 28$, find the value of y when $x = 38$.

27. If y is inversely proportional to x, and $y = \dfrac{1}{9}$ when $x = 12$, find the value of y when $x = 8$.

28. If y is inversely proportional to x, and $y = \dfrac{1}{35}$ when $x = 14$, find the value of y when $x = 16$.

29. If A varies jointly as b and h, and $A = 60$ when $b = 12$ and $h = 10$, find A when $b = 16$ and $h = 14$.

30. If V varies jointly as B and h, and $V = 51$ when $B = 17$ and $h = 9$, find V when $B - 19$ and $h = 12$.

31. The volume of a gas at a constant temperature varies inversely as the pressure. What is the volume of a gas under pressure of 25 pounds if the gas occupies 15 cubic centimeters under a pressure of 20 pounds?

32. The time required for a car to travel a certain distance varies inversely as the rate at which it travels. If it takes 4 hours at 50 miles per hour to travel the distance, how long will it take at 40 miles per hour?

33. The volume (V) of a gas varies directly as the temperature (T) and inversely as the pressure (P). If $V = 48$ when $T = 320$ and $P = 20$, find V when $T = 280$ and $P = 30$.

34. The distance that a freely falling body falls varies directly as the square of the time it falls. If a body falls 144 feet in 3 seconds, how far will it fall in 5 seconds?

35. The period (the time required for one complete oscillation) of a simple pendulum varies directly as the square root of its length. If a pendulum 12 feet long has a period of 4 seconds, find the period of a pendulum of length 3 feet.

36. The simple interest earned by a certain amount of money varies jointly as the rate of interest and the time (in years) that the money is invested. If \$120 is earned for the money invested at 12% for 2 years, how much is earned if the money is invested at 14% for 3 years?

37. The electrical resistance of a wire varies directly as its length and inversely as the square of its diameter. If the resistance of 200 meters of wire that has a diameter of $\frac{1}{2}$ centimeter is 1.5 ohms, find the resistance of 400 meters of wire with a diameter of $\frac{1}{4}$ centimeter.

38. The volume of a cylinder varies jointly as its altitude and the square of the radius of its base. If the volume of a cylinder is 1386 cubic centimeters when the radius of the base is 7 centimeters and its altitude is 9 centimeters, find the volume of a cylinder that has a base of radius 14 centimeters; the altitude of the cylinder is 5 centimeters.

Miscellaneous Problems

 In the previous problems numbers were chosen to make computations reasonable without the use of a calculator. However, many times variation-type problems involve messy computations and the calculator becomes a very useful tool. Use your calculator to help solve the following problems.

39. The simple interest earned by a certain amount of money varies jointly as the rate of interest and the time (in years) that the money is invested.

    (a) If some money invested at 11% for 2 years earns \$385, how much would the same amount earn at 12% for 1 year?

    (b) If some money invested at 12% for 3 years earns \$819, how much would the same amount earn at 14% for 2 years?

(c) If some money invested at 14% for 4 years earns $1960, how much would the same amount earn at 15% for 2 years?

40. The period (the time required for one complete oscillation) of a simple pendulum varies directly as the square root of its length. If a pendulum 9 inches long has a period of 2.4 seconds, find the period of a pendulum of length 12 inches. Express your answer to the nearest one-tenth of a second.

41. The volume of a cylinder varies jointly as its altitude and the square of the radius of its base. If the volume of a cylinder is 549.5 cubic meters when the radius of the base is 5 meters and its altitude is 7 meters, find the volume of a cylinder that has a base of a radius 9 meters and an altitude of 14 meters.

42. If y is directly proportional to x and inversely proportional to the square of z, and if $y = 0.336$ when $x = 6$ and $z = 5$, find the constant of variation.

43. If y is inversely proportional to the square root of x, and $y = 0.08$ when $x = 225$, find y when $x = 625$.

Chapter 12 Summary

(12.1) A **relation** is a set of ordered pairs; a **function** is a relation in which no two ordered pairs have the same first component. The **domain** of a relation (or function) is the set of all first components and the **range** is the set of all second components.

Single symbols such as f, g, and h are commonly used to name functions. The symbol $f(x)$ represents the element in the range associated with x from the domain. Thus, if $f(x) = 3x + 7$, then $f(1) = 3(1) + 7 = 10$.

(12.2) A summary of the three special types of functions follows.

Equation	Type of function	Graph
1. $f(x) = b$, where b is a real number	constant	horizontal line
2. $f(x) = ax + b$, where a and b are real numbers and $a \neq 0$	linear	straight line
3. $f(x) = ax^2 + bx + c$, where a, b, and c are real numbers and $a \neq 0$	quadratic	parabola

We can solve some applications that involve maximum or minimum values with our knowledge of parabolas that are generated by quadratic functions.

(12.3) We list the following suggestions for graphing functions you are unfamiliar with.

1. Determine the domain of the function.

2. Determine the type of symmetry exhibited by the equation.

3. Find the intercepts.

4. Set up a table of values that satisfy the equation.

5. Plot the points associated with the ordered pairs and connect them with a smooth curve. Then, if appropriate, reflect this part of the curve according to any symmetry possessed by the graph.

Variations of basic curves can be formed by translations, reflections, and stretching and shrinking transformations. Each of these transformations can be described as follows.

Vertical translation   The graph of $y = f(x) + k$ is the graph of $y = f(x)$ shifted k units upward if $k > 0$ or shifted $|k|$ units downward if $k < 0$.

Horizontal translation   The graph of $y = f(x - h)$ is the graph of $y = f(x)$ shifted h units to the right if $h > 0$ or shifted $|h|$ units to the left if $h < 0$.

x-axis reflection   The graph of $y = -f(x)$ is the graph of $y = f(x)$ reflected through the x-axis.

Vertical Stretching and Shrinking   The graph of $y = cf(x)$ is obtained from the graph of $y = f(x)$ by multiplying the y-coordinates of $y = f(x)$ by c. If $c > 1$, the graph is said to be stretched by a factor of c, and if $0 < c < 1$, the graph is said to be shrunk by a factor of c.

(12.4) The basic operations of addition, subtraction, multiplication, and division of functions are defined as follows.

Sum: $\qquad\qquad (f + g)(x) = f(x) + g(x);$

Difference: $\qquad (f - g)(x) = f(x) - g(x);$

Product: $\qquad\quad (f \cdot g)(x) = f(x) \cdot g(x);$

Quotient: $\qquad\quad \left(\dfrac{f}{g}\right)(x) = \dfrac{f(x)}{g(x)}, \qquad g(x) \neq 0.$

The **composition** of two functions f and g is defined by

$$(f \circ g)(x) = f(g(x))$$

for all x in the domain of g such that $g(x)$ is in the domain of f.

(12.5) A one-to-one function is a function such that no two ordered pairs have the same second component.

If the components of each ordered pair of a given one-to-one function are interchanged, the resulting function and the given function are **inverses** of each other. The inverse of a function f is denoted by f^{-1}.

Graphically, two functions that are inverses of each other are mirror images with reference to the line $y = x$.

We can show that two functions f and f^{-1} are inverses of each other by verifying that

1. $(f^{-1} \circ f)(x) = x$ for all x in the domain of f,
2. $(f \circ f^{-1})(x) = x$ for all x in the domain of f^{-1}.

A technique for finding the inverse of a function follows.

1. Let $y = f(x)$.

2. Interchange x and y.

3. Solve the equation for y in terms of x.

4. $f^{-1}(x)$ is determined by the final equation.

(12.6) The equation $y = kx$ (k is a nonzero constant) defines a function called a **direct variation**. The equation $y = \dfrac{k}{x}$ defines a function called **inverse variation**. In both cases, k is called the **constant of variation**.

Chapter 12 Review Problem Set

For Problems 1–4, specify the domain of each function.

1. $f = \{(1, 3), (2, 5), (4, 9)\}$

2. $f(x) = \dfrac{4}{x - 5}$

3. $f(x) = \dfrac{3}{x^2 + 4x}$

4. $f(x) = \sqrt{x^2 - 25}$

5. If $f(x) = x^2 - 2x - 1$, find $f(2)$, $f(-3)$, and $f(a)$.

6. If $f(x) = 2x^2 + x - 7$, find $\dfrac{f(a + h) - f(a)}{h}$.

For Problems 7–16, graph each of the functions.

7. $f(x) = 4$

8. $f(x) = -3x + 2$

9. $f(x) = x^2 + 2x + 2$

10. $f(x) = |x| + 4$

11. $f(x) = -|x - 2|$

12. $f(x) = \sqrt{x - 2} - 3$

13. $f(x) = \dfrac{1}{x^2}$

14. $f(x) = -\dfrac{1}{2}x^2$

15. $f(x) = -3x^2 + 6x - 2$

16. $f(x) = -\sqrt{x + 1} - 2$

17. Find the coordinates of the vertex and the equation of the line of symmetry for each of the following parabolas.

  (a) $f(x) = x^2 + 10x - 3$

  (b) $f(x) = -2x^2 - 14x + 9$

For Problems 18–20, determine $(f \circ g)(x)$ and $(g \circ f)(x)$ for each pair of functions.

18. $f(x) = 2x - 3$ and $g(x) = 3x - 4$

19. $f(x) = x - 4$ and $g(x) = x^2 - 2x + 3$

20. $f(x) = x^2 - 5$ and $g(x) = -2x + 5$

For Problems 21–23, find the inverse (f^{-1}) of the given function.

21. $f(x) = 6x - 1$

22. $f(x) = \dfrac{2}{3}x + 7$

23. $f(x) = -\dfrac{3}{5}x - \dfrac{2}{7}$

24. If y varies directly as x and inversely as z, and if $y = 21$ when $x = 14$ and $z = 6$, find the constant of variation.

25. If y varies jointly as x and the square root of z, and if $y = 60$ when $x = 2$ and $z = 9$, find y when $x = 3$ and $z = 16$.

26. The weight of a body above the surface of the earth varies inversely as the square of its distance from the center of the earth. Assume that the radius of the earth is 4000 miles. How much would a man weigh 1000 miles above the earth's surface if he weighs 200 pounds on the surface?

27. Find two numbers whose sum is 40 and whose product is a maximum.

28. Find two numbers whose sum is 50 such that the square of one number plus six times the other number is a minimum.

29. Suppose that 50 students are able to raise $250 for a party when each one contributes $5. Furthermore, they figure that for each additional student they can find to contribute, the cost per student will decrease by a nickel. How many additional students do they need to maximize the amount they will have for a party?

30. The surface area of a cube varies directly as the square of the length of an edge. If the surface area of a cube having edges 8 inches long is 384 square inches, find the surface area of a cube having edges 10 inches long.

Cumulative Review Problem Set (Chapters 1–12)

For Problems 1–5, evaluate each algebraic expression for the given values of the variables.

1. $-5(x - 1) - 3(2x + 4) + 3(3x - 1)$ for $x = -2$

2. $\dfrac{14a^3b^2}{7a^2b}$ for $a = -1$ and $b = 4$

3. $\dfrac{2}{n} - \dfrac{3}{2n} + \dfrac{5}{3n}$ for $n = 4$

4. $-4\sqrt{2x - y} + 5\sqrt{3x + y}$ for $x = 16$ and $y = 16$

5. $\dfrac{3}{x - 2} - \dfrac{5}{x + 3}$ for $x = 3$

For Problems 6–15, perform the indicated operations and express answers in simplified form.

6. $(-5\sqrt{6})(3\sqrt{12})$

7. $(2\sqrt{x} - 3)(\sqrt{x} + 4)$

8. $(3\sqrt{2} - \sqrt{6})(\sqrt{2} + 4\sqrt{6})$

9. $(2x - 1)(x^2 + 6x - 4)$

10. $\dfrac{x^2 - x}{x + 5} \cdot \dfrac{x^2 + 5x + 4}{x^4 - x^2}$

11. $\dfrac{16x^2y}{24xy^3} \div \dfrac{9xy}{8x^2y^2}$

12. $\dfrac{x + 3}{10} + \dfrac{2x + 1}{15} - \dfrac{x - 2}{18}$

13. $\dfrac{7}{12ab} - \dfrac{11}{15a^2}$

14. $\dfrac{8}{x^2 - 4x} + \dfrac{2}{x}$

15. $(8x^3 - 6x^2 - 15x + 4) \div (4x - 1)$

For Problems 16–19, simplify each of the complex fractions.

16. $\dfrac{\dfrac{5}{x^2} - \dfrac{3}{x}}{\dfrac{1}{y} + \dfrac{2}{y^2}}$

17. $\dfrac{\dfrac{2}{x} - 3}{\dfrac{3}{y} + 4}$

18. $\dfrac{2 - \dfrac{1}{n + 2}}{3 + \dfrac{4}{n + 3}}$

19. $\dfrac{3a - 1}{2 - \dfrac{1}{a}}$

For Problems 20–25, factor each of the algebraic expressions completely.

20. $20x^2 + 7x - 6$

21. $16x^3 + 54$

22. $4x^4 - 25x^2 + 36$

23. $12x^3 - 52x^2 - 40x$

24. $xy - 6x + 3y - 18$

25. $10 + 9x - 9x^2$

For Problems 26–33, evaluate each of the numerical expressions.

26. $\left(\dfrac{2}{3}\right)^{-4}$

27. $\dfrac{3}{\left(\dfrac{4}{3}\right)^{-1}}$

28. $\sqrt[3]{-\dfrac{27}{64}}$

29. $-\sqrt{.09}$

30. $(27)^{-\frac{4}{3}}$

31. $4^0 + 4^{-1} + 4^{-2}$

32. $\left(\dfrac{3^{-1}}{2^{-3}}\right)^{-2}$

33. $(2^{-3} - 3^{-2})^{-1}$

For Problems 34–36, find the indicated products and quotients and express final answers with positive integral exponents only.

34. $(-3x^{-1}y^2)(4x^{-2}y^{-3})$

35. $\dfrac{48x^{-4}y^2}{6xy}$

36. $\left(\dfrac{27a^{-4}b^{-3}}{-3a^{-1}b^{-4}}\right)^{-1}$

For Problems 37–44, express each radical expression in simplest radical form.

37. $\sqrt{80}$

38. $-2\sqrt{54}$

39. $\sqrt{\dfrac{75}{81}}$

40. $\dfrac{4\sqrt{6}}{3\sqrt{8}}$

41. $\sqrt[3]{56}$

42. $\dfrac{\sqrt[3]{3}}{\sqrt[3]{4}}$

43. $4\sqrt{52x^3y^2}$

44. $\sqrt{\dfrac{2x}{3y}}$

For Problems 45–47, use the distributive property to help simplify each of the following.

45. $-3\sqrt{24} + 6\sqrt{54} - \sqrt{6}$

46. $\dfrac{\sqrt{8}}{3} - \dfrac{3\sqrt{18}}{4} - \dfrac{5\sqrt{50}}{2}$

47. $8\sqrt[3]{3} - 6\sqrt[3]{24} - 4\sqrt[3]{81}$

For Problems 48 and 49, rationalize the denominator and simplify

48. $\dfrac{\sqrt{3}}{\sqrt{6} - 2\sqrt{2}}$

49. $\dfrac{3\sqrt{5} - \sqrt{3}}{2\sqrt{3} + \sqrt{7}}$

For Problems 50–52, use scientific notation to help perform the indicated operations.

50. $\dfrac{(.00016)(300)(.028)}{.064}$

51. $\dfrac{.00072}{.0000024}$

52. $\sqrt{.00000009}$

For Problems 53–56, find each of the indicated products or quotients and express answers in standard form.

53. $(5 - 2i)(4 + 6i)$

54. $(-3 - i)(5 - 2i)$

55. $\dfrac{5}{4i}$

56. $\dfrac{-1 + 6i}{7 - 2i}$

57. Find the slope of the line determined by the points $(2, -3)$ and $(-1, 7)$.

58. Find the slope of the line determined by the equation $4x - 7y = 9$.

59. Find the length of the line segment whose endpoints are $(4, 5)$ and $(-2, 1)$.

60. Write the equation of the line that contains the points $(3, -1)$ and $(7, 4)$.

61. Write the equation of the line that is perpendicular to the line $3x - 4y = 6$ and contains the point $(-3, -2)$.

62. Find the center and the length of a radius of the circle $x^2 + 4x + y^2 - 12y + 31 = 0$.

63. Find the coordinates of the vertex of the parabola $y = x^2 + 10x + 21$.

64. Find the length of the major axis of the ellipse $x^2 + 4y^2 = 16$.

For Problems 65–70, graph each of the equations.

65. $-x + 2y = -4$

66. $x^2 + y^2 = 9$

67. $x^2 - y^2 = 9$

68. $x^2 + 2y^2 = 8$

69. $y = -3x$

70. $x^2 y = 4$

For Problems 71–76, graph each of the functions.

71. $f(x) = -2x - 4$

72. $f(x) = -2x^2 - 2$

73. $f(x) = x^2 - 2x - 2$

74. $f(x) = \sqrt{x + 1} + 2$

75. $f(x) = 2x^2 + 8x + 9$

76. $f(x) = -|x - 2| + 1$

77. If $f(x) = x - 3$ and $g(x) = 2x^2 - x - 1$, find $(g \circ f)(x)$ and $(f \circ g)(x)$.

78. Find the inverse (f^{-1}) of $f(x) = 3x - 7$.

79. Find the inverse of $f(x) = -\dfrac{1}{2}x + \dfrac{2}{3}$.

80. Find the constant of variation if y varies directly as x, and $y = 2$ when $x = -\dfrac{2}{3}$.

81. If y is inversely proportional to the square of x, and $y = 4$ when $x = 3$, find y when $x = 6$.

82. The volume of a gas at a constant temperature varies inversely as the pressure. What is the volume of a gas under a pressure of 25 pounds if the gas occupies 15 cubic centimeters under a pressure of 20 pounds?

83. If A varies jointly as b and h, and $A = 130$ when $b = 20$ and $h = 13$, find A when $b = 14$ and $h = 17$.

84. If V varies jointly as B and h, and $V = 150$ when $B = 18$ and $h = 25$, find V when $B = 19$ and $h = 27$.

For Problems 85–105, solve each of the equations.

85. $3(2x - 1) - 2(5x + 1) = 4(3x + 4)$

86. $n + \dfrac{3n - 1}{9} - 4 = \dfrac{3n + 1}{3}$

87. $.92 + .9(x - .3) = 2x - 5.95$

88. $|4x - 1| = 11$

89. $3x^2 = 7x$

90. $x^3 - 36x = 0$

91. $30x^2 + 13x - 10 = 0$

92. $8x^3 + 12x^2 - 36x = 0$

93. $x^4 + 8x^2 - 9 = 0$

94. $(n + 4)(n - 6) = 11$

95. $2 - \dfrac{3x}{x - 4} = \dfrac{14}{x + 7}$

96. $\dfrac{2n}{6n^2 + 7n - 3} - \dfrac{n - 3}{3n^2 + 11n - 4} = \dfrac{5}{2n^2 + 11n + 12}$

97. $\sqrt{3y} - y = -6$

98. $\sqrt{x + 19} - \sqrt{x + 28} = -1$

99. $(3x - 1)^2 = 45$

100. $(2x + 5)^2 = -32$

101. $2x^2 - 3x + 4 = 0$

102. $3n^2 - 6n + 2 = 0$

103. $\dfrac{5}{n - 3} - \dfrac{3}{n + 3} = 1$

104. $12x^4 - 19x^2 + 5 = 0$

105. $2x^2 + 5x + 5 = 0$

For Problems 106–115, solve each of the inequalities.

106. $-5(y - 1) + 3 > 3y - 4 - 4y$

107. $.06x + .08(250 - x) \geq 19$

108. $|5x - 2| > 13$

109. $|6x + 2| < 8$

110. $\dfrac{x - 2}{5} - \dfrac{3x - 1}{4} \leq \dfrac{3}{10}$

111. $(x - 2)(x + 4) \leq 0$

112. $(3x - 1)(x - 4) > 0$

113. $x(x + 5) < 24$

114. $\dfrac{x - 3}{x - 7} \geq 0$

115. $\dfrac{2x}{x + 3} > 4$

For Problems 116–120, solve each of the systems of equations.

116. $\begin{pmatrix} 4x - 3y = 18 \\ 3x - 2y = 15 \end{pmatrix}$

117. $\begin{pmatrix} y = \dfrac{2}{5}x - 1 \\ 3x + 5y = 4 \end{pmatrix}$

118. $\begin{pmatrix} \dfrac{x}{2} - \dfrac{y}{3} = 1 \\ \dfrac{2x}{5} + \dfrac{y}{2} = 2 \end{pmatrix}$

119. $\begin{pmatrix} x + y = 650 \\ .05x + .06y = 36.5 \end{pmatrix}$

120. $\begin{pmatrix} 5x + 2y = -32 \\ x = 4y + 20 \end{pmatrix}$

For Problems 121–135, set up an equation, an inequality, or a system of equations to help solve each problem.

121. Find three consecutive odd integers whose sum is 57.

122. Suppose that Eric has a collection of 63 coins consisting of nickels, dimes, and quarters. The number of dimes is 6 more than the number of nickels and the number of quarters is one more than twice the number of nickels. How many coins of each kind are in the collection?

123. One of two supplementary angles is 4° more than one-third of the other angle. Find the measure of each of the angles.

124. If a ring costs a jeweler $300, at what price should it be sold to make a profit of 50% on the selling price?

125. Last year Beth invested a certain amount of money at 8% and $300 more than that amount at 9%. Her total yearly interest was $316. How much did she invest at each rate?

126. Two trains leave the same depot at the same time, one traveling east and the other traveling west. At the end of $4\frac{1}{2}$ hours they are 639 miles apart. If the rate of the train traveling east is 10 miles per hour faster than the other train, find their rates.

127. Suppose that a 10-quart radiator contains a 50% solution of antifreeze. How much needs to be drained out and replaced with pure antifreeze to obtain a 70% antifreeze solution?

128. Sam shot rounds of 70, 73, and 76 on the first three days of a golf tournament. What must he shoot on the fourth day of the tournament to average 72 or less for the four days?

129. The cube of a number equals nine times the same number. Find the number.

130. A strip of uniform width is to be cut off of both sides and both ends of a sheet of paper that is 8 inches by 14 inches to reduce the size of the paper to an area of 72 square inches. Find the width of the strip.

131. A sum of $2450 is to be divided between two people in the ratio of 3 to 4. How much does each person receive?

132. Working together Sue and Dean can complete a task in $1\frac{1}{5}$ hours. Dean can do the task by himself in 2 hours. How long would it take Sue to complete the task by herself?

133. Dudley bought a number of shares of stock for $300. A month later he sold all but 10 shares at a profit of $5 per share and regained his original investment of $300. How many shares did he originally buy and at what price per share?

134. The units digit of a two-digit number is one more than twice the tens digit. The sum of the digits is 10. Find the number.

135. The sum of the two smallest angles of a triangle is 40° less than the other angle. The sum of the smallest and largest angle is twice the other angle. Find the measures of the three angles of the triangle.

Chapter 13

Exponential and Logarithmic Functions

This chapter will extend the meaning of an exponent, and introduce the concept of a logarithm. We will (1) work with some exponential functions, (2) work with some logarithmic functions, and (3) use the concepts of exponent and logarithm to expand our capabilities for solving problems. Your calculator will be a valuable tool throughout this chapter.

13.1

Exponential Functions

In Chapter 2 the expression b^n was defined to mean n factors of b, where n is any positive integer and b is any real number. For example,

$$4^3 = 4 \cdot 4 \cdot 4 = 64,$$

$$\left(\frac{1}{2}\right)^4 = \left(\frac{1}{2}\right)\left(\frac{1}{2}\right)\left(\frac{1}{2}\right)\left(\frac{1}{2}\right) = \frac{1}{16},$$

$$(-0.3)^2 = (-0.3)(-0.3) = 0.09.$$

In Chapter 8 by defining $b^0 = 1$ and $b^{-n} = \dfrac{1}{b^n}$, where n is any positive integer and b is any nonzero real number, the concept of an exponent was extended to include all integers. For example,

$$2^{-3} = \frac{1}{2^3} = \frac{1}{2 \cdot 2 \cdot 2} = \frac{1}{8}, \qquad \left(\frac{1}{3}\right)^{-2} = \frac{1}{\left(\frac{1}{3}\right)^2} = \frac{1}{\frac{1}{9}} = 9,$$

$$(0.4)^{-1} = \frac{1}{(0.4)^1} = \frac{1}{0.4} = 2.5, \qquad (-0.98)^0 = 1.$$

Chapter 8 also provided for the use of all rational numbers as exponents by defining $b^{\frac{m}{n}} = \sqrt[n]{b^m}$, where n is a positive integer greater than one and b is a real number such that $\sqrt[n]{b}$ exists. For example,

$$8^{\frac{2}{3}} = \sqrt[3]{8^2} = \sqrt[3]{64} = 4, \qquad 16^{\frac{1}{4}} = \sqrt[4]{16^1} = 2, \qquad 32^{-\frac{1}{5}} = \frac{1}{32^{\frac{1}{5}}} = \frac{1}{\sqrt[5]{32}} = \frac{1}{2}.$$

To formally extend the concept of an exponent to include the use of irrational numbers requires some ideas from calculus and is therefore beyond the scope of this text. However, here's a glance at the general idea involved. Consider the number $2^{\sqrt{3}}$. By using the nonterminating and nonrepeating decimal representation $1.73205\ldots$ for $\sqrt{3}$, form the sequence of numbers 2^1, $2^{1.7}$, $2^{1.73}$, $2^{1.732}$, $2^{1.7320}$, $2^{1.73205}\ldots$. It would seem reasonable that each successive power gets closer to $2^{\sqrt{3}}$. This is precisely what happens if b^n, where n is irrational, is properly defined by using the concept of a limit.

So from now on we can use any real number as an exponent and the basic properties stated in Chapter 8 can be extended to include all real numbers as exponents. Let's restate those properties at this time with the restriction that the bases a and b are to be positive numbers (to avoid expressions such as $(-4)^{\frac{1}{2}}$, which do not represent real numbers).

PROPERTY 13.1

> If a and b are positive real numbers and m and n are any real numbers, then
>
> 1. $b^n \cdot b^m = b^{n+m}$ product of two powers
> 2. $(b^n)^m = b^{mn}$ power of a power
> 3. $(ab)^n = a^n b^n$ power of a product
> 4. $\left(\dfrac{a}{b}\right)^n = \dfrac{a^n}{b^n}$ power of a quotient
> 5. $\dfrac{b^n}{b^m} = b^{n-m}$ quotient of two powers

Another property that can be used to solve certain types of equations involving exponents can be stated as follows.

PROPERTY 13.2

> If $b > 0$, $b \neq 1$, and m and n are real numbers, then
>
> $b^n = b^m$ if and only if $n = m$.

The following examples illustrate the use of Property 13.2.

EXAMPLE 1   Solve $2^x = 32$.

Solution

$$2^x = 32$$
$$2^x = 2^5 \qquad 32 = 2^5$$
$$x = 5 \qquad \text{Property 13.2}$$

The solution set is $\{5\}$. ∎

EXAMPLE 2   Solve $3^{2x} = \dfrac{1}{9}$.

Solution

$$3^{2x} = \frac{1}{9} = \frac{1}{3^2}$$
$$3^{2x} = 3^{-2}$$
$$2x = -2 \qquad \text{Property 13.2}$$
$$x = -1$$

The solution set is $\{-1\}$. ∎

EXAMPLE 3   Solve $\left(\dfrac{1}{5}\right)^{x-2} = \dfrac{1}{125}$.

Solution

$$\left(\frac{1}{5}\right)^{x-2} = \frac{1}{125} = \left(\frac{1}{5}\right)^3$$

$$x - 2 = 3 \qquad \text{Property 13.2}$$

$$x = 5$$

The solution set is $\{5\}$. ∎

EXAMPLE 4   Solve $8^x = 32$.

Solution

$$8^x = 32$$

$$(2^3)^x = 2^5 \qquad 8 = 2^3$$

$$2^{3x} = 2^5$$

$$3x = 5 \qquad \text{Property 13.2}$$

$$x = \frac{5}{3}$$

The solution set is $\left\{\dfrac{5}{3}\right\}$. ∎

EXAMPLE 5   Solve $(3^{x+1})(9^{x-2}) = 27$.

Solution

$$(3^{x+1})(9^{x-2}) = 27$$

$$(3^{x+1})(3^2)^{x-2} = 3^3$$

$$(3^{x+1})(3^{2x-4}) = 3^3$$

$$3^{3x-3} = 3^3$$

$$3x - 3 = 3 \qquad \text{Property 13.2}$$

$$3x = 6$$

$$x = 2$$

The solution set is $\{2\}$. ∎

If b is any positive number, then the expression b^x designates exactly one real number for every real value of x. Thus, the equation $f(x) = b^x$ defines a function whose domain is the set of real numbers. Furthermore, if we place the additional restriction $b \neq 1$, then any equation of the form $f(x) = b^x$ describes a one-to-one function and is called an **exponential function**. This leads to the following definition.

DEFINITION 13.1

> If $b > 0$ and $b \neq 1$, then the function f defined by
>
> $$f(x) = b^x,$$
>
> where x is any real number, is called the **exponential function with base b**.

REMARK The function $f(x) = 1^x$ is a constant function whose graph is a horizontal line, and therefore it is not a one-to-one function. Remember from Chapter 2 that one-to-one functions have inverses; this becomes a key issue in a later section.

Now let's consider graphing some exponential functions.

EXAMPLE 6   Graph the function $f(x) = 2^x$.

Solution   First, let's set up a table of values.

x	$f(x) = 2^x$
-2	$\dfrac{1}{4}$
-1	$\dfrac{1}{2}$
0	1
1	2
2	4
3	8

Plot these points and connect them with a smooth curve to produce Figure 13.1.

Figure 13.1

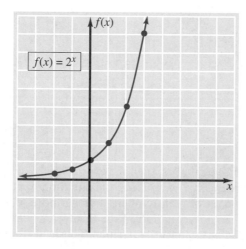

$f(x) = 2^x$

EXAMPLE 7   Graph $f(x) = \left(\dfrac{1}{2}\right)^x$.

Solution   Again, let's set up a table of values.

x	$f(x) = \left(\dfrac{1}{2}\right)^x$
-2	4
-1	2
0	1
1	$\dfrac{1}{2}$
2	$\dfrac{1}{4}$
3	$\dfrac{1}{8}$

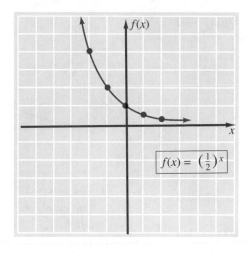

Figure 13.2

Plot these points and connect them with a smooth curve to produce Figure 13.2.   ∎

In the tables for Examples 6 and 7 we chose integral values for x to keep the computation simple. However, with the use of a calculator we could easily acquire functional values by using nonintegral exponents. Consider the following additional values for each of the tables.

$f(x) = 2^x$		$f(x) = \left(\dfrac{1}{2}\right)^x$	
$f(0.5) \approx 1.41$	$f(-0.5) \approx 0.71$	$f(0.7) \approx 0.62$	$f(-0.8) \approx 1.74$
$f(1.7) \approx 3.25$	$f(-2.6) \approx 0.16$	$f(2.3) \approx 0.20$	$f(-2.1) \approx 4.29$

Use your calculator to check these results. Also, it would be worthwhile for you to go back and see that the points determined do *fit the graphs* in Figures 13.1 and 13.2.

The graphs in Figures 13.1 and 13.2 illustrate a *general behavior* pattern of exponential functions. That is to say, if $b > 1$, then the graph of $f(x) = b^x$ *goes up to the right* and the function is called an **increasing function**. If $0 < b < 1$, then the graph of $f(x) = b^x$ *goes down to the right* and the function is called a **decreasing function**. These facts are illustrated in Figure 13.3. Notice that because $b^0 = 1$ for any $b > 0$, all graphs of $f(x) = b^x$ contain the point $(0, 1)$.

As you graph exponential functions, don't forget your previous graphing experiences.

Figure 13.3

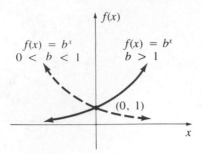

1. The graph of $f(x) = 2^x - 4$ is the graph of $f(x) = 2^x$ *moved down four units.*
2. The graph of $f(x) = 2^{x+3}$ is the graph of $f(x) = 2^x$ *moved three units to the left.*
3. The graph of $f(x) = -2^x$ is the graph of $f(x) = 2^x$ *reflected across the x-axis.*

Problem Set 13.1

Solve each of the following equations.

1. $3^x = 27$ **2.** $2^x = 64$ **3.** $2^{2x} = 16$ **4.** $3^{2x} = 81$

5. $\left(\dfrac{1}{4}\right)^x = \dfrac{1}{256}$ **6.** $\left(\dfrac{1}{2}\right)^x = \dfrac{1}{128}$ **7.** $5^{x+2} = 125$ **8.** $4^{x-3} = 16$

9. $3^{-x} = \dfrac{1}{243}$ **10.** $5^{-x} = \dfrac{1}{25}$ **11.** $6^{3x-1} = 36$ **12.** $2^{2x+3} = 32$

13. $4^x = 8$ **14.** $16^x = 64$ **15.** $8^{2x} = 32$ **16.** $9^{3x} = 27$

17. $\left(\dfrac{1}{2}\right)^{2x} = 64$ **18.** $\left(\dfrac{1}{3}\right)^{5x} = 243$ **19.** $\left(\dfrac{3}{4}\right)^x = \dfrac{64}{27}$ **20.** $\left(\dfrac{2}{3}\right)^x = \dfrac{9}{4}$

21. $9^{4x-2} = \dfrac{1}{81}$ **22.** $8^{3x+2} = \dfrac{1}{16}$ **23.** $6^{2x} + 3 = 39$ **24.** $5^{2x} - 2 = 123$

25. $10^x = .1$ **26.** $10^x = .0001$ **27.** $32^x = \dfrac{1}{4}$ **28.** $9^x = \dfrac{1}{27}$

29. $(2^{x+1})(2^x) = 64$ **30.** $(2^{2x-1})(2^{x+2}) = 32$ **31.** $(27)(3^x) = 9^x$

32. $(3^x)(3^{5x}) = 81$ **33.** $(4^x)(16^{3x-1}) = 8$ **34.** $(8^{2x})(4^{2x-1}) = 16$

Graph each of the following exponential functions.

35. $f(x) = 4^x$ **36.** $f(x) = 3^x$

37. $f(x) = 6^x$ **38.** $f(x) = 5^x$

39. $f(x) = \left(\dfrac{1}{3}\right)^x$ **40.** $f(x) = \left(\dfrac{1}{4}\right)^x$

41. $f(x) = \left(\dfrac{3}{4}\right)^x$ **42.** $f(x) = \left(\dfrac{2}{3}\right)^x$

43. $f(x) = 2^{x-2}$ **44.** $f(x) = 2^{x+1}$

45. $f(x) = 3^{-x}$ **46.** $f(x) = 2^{-x}$

47. $f(x) = 3^{2x}$ **48.** $f(x) = 2^{2x}$

49. $f(x) = 3^x - 2$ **50.** $f(x) = 2^x + 1$

51. $f(x) = 2^{-x-2}$ **52.** $f(x) = 3^{-x+1}$

Thoughts into Words

53. Why is the base of an exponential function restricted to positive numbers not including one?

54. Explain how you would solve the equation $(2^{x+1})(8^{2x-3}) = 64$.

Miscellaneous Problems

For Problems 55–57, use a graphics calculator.

55. Graph $f(x) = 6^x$, $f(x) = 6^{x+3}$, $f(x) = 6^x - 3$, and $f(x) = -6^x$ on the same set of axes.

56. Graph $f(x) = \left(\dfrac{1}{2}\right)^x$, $f(x) = \left(\dfrac{1}{3}\right)^x$, $f(x) = \left(\dfrac{1}{4}\right)^x$, and $f(x) = \left(\dfrac{1}{5}\right)^x$ on the same set of axes.

57. Graph $f(x) = (-2)^x$. Explain your result.

13.2

Applications of Exponential Functions

Equations that describe exponential functions can represent many real-world situations that exhibit growth or decay. For example, suppose that an economist predicts an annual inflation rate of 5% for the next 10 years. This means an item that presently costs $8 will cost $8(105\%) = 8(1.05) = \$8.40$ a year from now. The same item will cost $(105\%)[8(105\%)] = 8(1.05)^2 = \$8.82$ in 2 years. In general, the equation

$$P = P_0(1.05)^t$$

yields the predicted price P of an item in t years at the annual inflation rate of 5%, where that item presently costs P_0. By using this equation, we can look at some future prices based on the prediction of a 5% inflation rate. For example,

1. A $3.27 container of hot cocoa mix will cost $3.27(1.05)^3 = \$3.79$ in 3 years;

2. A $4.07 jar of coffee will cost $4.07(1.05)^5 = \$5.19$ in 5 years;

3. A $9500 car will cost $9500(1.05)^7 = \$13,367$ (rounded to the nearest dollar) in 7 years.

Compound Interest

Compound interest provides another illustration of exponential growth. Suppose that $500 (called the **principal**) is invested at an interest rate of 8% **compounded annually**. The interest earned the first year is $500(0.08) = $40 and this amount is added to the original $500 to form a new principal of $540 for the second year. The interest earned during the second year is $540(0.08) = $43.20 and this amount is added to $540 to form a new principal of $583.20 for the third year. Each year a new principal is formed by reinvesting the interest earned that year.

In general, suppose that a sum of money P (called the principal) is invested at an interest rate of r percent compounded annually. The interest earned the first year is Pr and the new principal for the second year is $P + Pr$ or $P(1 + r)$. Note that the new principal for the second year can be found by multiplying the original principal P by $(1 + r)$. In a like fashion, the new principal for the third year can be found by multiplying the previous principal $P(1 + r)$ by $(1 + r)$, thus obtaining $P(1 + r)^2$. If this process is continued, then after t years the *total amount of money accumulated*, A, is given by

$$A = P(1 + r)^t.$$

Consider the following examples of investments made at a certain rate of interest compounded annually.

1. $750 invested for 5 years at 9% compounded annually produces

$$A = \$750(1.09)^5 = \$1153.97.$$

2. $1000 invested for 10 years at 11% compounded annually produces

$$A = \$1000(1.11)^{10} = \$2839.42.$$

3. $5000 invested for 20 years at 12% compounded annually produces

$$A = \$5000(1.12)^{20} = \$48,231.47.$$

If we invest money at a certain rate of interest to be *compounded* more than once a year, then we can adjust the basic formula, $A = P(1 + r)^t$, according to the number of compounding periods in the year. For example, for **compounding semiannually**, the formula becomes $A = P\left(1 + \dfrac{r}{2}\right)^{2t}$ and for **compounding quarterly**, the formula becomes $A = P\left(1 + \dfrac{r}{4}\right)^{4t}$. In general, we have the following formula for which n represents the number of compounding periods in a year.

$$A = P\left(1 + \frac{r}{n}\right)^{nt}$$

The following examples should clarify the use of this formula.

1. $750 invested for 5 years at 9% compounded semiannually produces

$$A = \$750\left(1 + \frac{0.09}{2}\right)^{2(5)} = \$750(1.045)^{10} = \$1164.73.$$

2. $1000 invested for 10 years at 11% compounded quarterly produces

$$A = \$1000\left(1 + \frac{0.11}{4}\right)^{4(10)} = \$1000(1.0275)^{40} = \$2959.87.$$

3. $5000 invested for 20 years at 12% compounded monthly produces

$$A = \$5000\left(1 + \frac{0.12}{12}\right)^{12(20)} = \$5000(1.01)^{240} = \$54,462.77.$$

You may find it interesting to compare these results with those obtained earlier for compounding annually.

Exponential Decay

Suppose it is estimated that the value of a car depreciates 15% per year for the first five years. Thus a car that costs $9500 will be worth $9500(100\% - 15\%) = 9500(85\%) = 9500(0.85) = \$8075$ in one year. In two years the value of the car will have depreciated to $9500(0.85)^2 = \$6864$(nearest dollar). The equation

$$V = V_0(0.85)^t$$

yields the value V of a car in t years at the annual depreciation rate of 15%, where the car initially cost V_0. By using this equation, we can estimate some car values to the nearest dollar, as follows.

1. A $6900 car will be worth $6900(0.85)^3 = \$4237$ in 3 years.

2. A $10,900 car will be worth $10,900(0.85)^4 = \$5690$ in 4 years.

3. A $13,000 car will be worth $\$13,000(0.85)^5 = \$5768$ in 5 years.

Another example of exponential decay is associated with radioactive substances. We can describe the rate of decay exponentially, based on the half-life of a substance. The *half-life* of a radioactive substance is the amount of time that it takes for one-half of an initial amount of the substance to disappear as the result of decay. For example, suppose that we have 200 grams of a certain substance that has a half-life of 5 days. After 5 days, $200\left(\frac{1}{2}\right) = 100$ grams remain. After 10 days, $200\left(\frac{1}{2}\right)^2 = 50$ grams remain. After 15 days, $200\left(\frac{1}{2}\right)^3 = 25$ grams remain. In general, after t days, $200\left(\frac{1}{2}\right)^{\frac{t}{5}}$ grams remain.

This discussion leads us to the following half-life formula. Suppose there is an initial amount, Q_0, of a radioactive substance with a half-life of h. The amount of substance remaining, Q, after a time period of t, is given by the formula

$$Q = Q_0 \left(\frac{1}{2}\right)^{\frac{t}{h}}.$$

The units of measure for t and h must be the same.

EXAMPLE 1   Barium-140 has a half-life of 13 days. If there are 500 milligrams of barium initially, how many milligrams remain after 26 days? After 100 days?

Solution   Using $Q_0 = 500$ and $h = 13$, the half-life formula becomes

$$Q = 500 \left(\frac{1}{2}\right)^{\frac{t}{13}}.$$

If $t = 26$, then

$$Q = 500 \left(\frac{1}{2}\right)^{\frac{26}{13}}$$

$$= 500 \left(\frac{1}{2}\right)^{2}$$

$$= 500 \left(\frac{1}{4}\right)$$

$$= 125.$$

So, 125 milligrams remain after 26 days. If $t = 100$, then

$$Q = 500 \left(\frac{1}{2}\right)^{\frac{100}{13}}$$

$$= 500(.5)^{\frac{100}{13}}$$

$$= 2.4 \qquad \text{to the nearest tenth of a milligram.}$$

So, approximately 2.4 milligrams remain after 100 days.   ■

REMARK In the preface of this book we stated that the calculator is "useful at times, unnecessary at other times." Example 1 clearly demonstrates this point. We solved the first part of the problem very easily without the calculator but it certainly was helpful for the second part of the problem.

The Number *e*

An interesting situation occurs if we consider the compound interest formula for $P = \$1$, $r = 100\%$, and $t = 1$ year. The formula becomes $A = 1\left(1 + \dfrac{1}{n}\right)^n$. The following table shows some values, rounded to eight decimal places, of $\left(1 + \dfrac{1}{n}\right)^n$ for different values of *n*.

n	$\left(1 + \dfrac{1}{n}\right)^n$
1	2.00000000
10	2.59374246
100	2.70481383
1000	2.71692393
10,000	2.71814593
100,000	2.71826824
1,000,000	2.71828047
10,000,000	2.71828169
100,000,000	2.71828181
1,000,000,000	2.71828183

The table suggests that as *n* increases, the value of $\left(1 + \dfrac{1}{n}\right)^n$ gets closer and closer to some fixed number. This does happen and the fixed number is called *e*. To five decimal places, $e = 2.71828$.

The function defined by the equation $f(x) = e^x$ is the **natural exponential function**. It has a great many real-world applications; some we will look at in a moment. First, however, let's get a picture of the natural exponential function.

Since $2 < e < 3$, the graph of $f(x) = e^x$ must fall between the graphs of $f(x) = 2^x$ and $f(x) = 3^x$. To be more specific, let's use our calculator to determine a table of values. Use the $\boxed{e^x}$ key, and round the results to the nearest tenth to obtain the following table. Plot the points determined by this table and connect them with a smooth curve to produce Figure 13.4.

x	$f(x) = e^x$
0	1.0
1	2.7
2	7.4
-1	0.4
-2	0.1

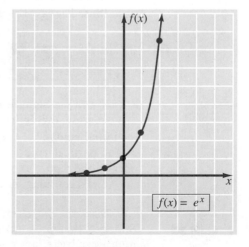

Figure 13.4

Back to Compound Interest

Let's return to the concept of compound interest. If the number of compounding periods in a year is increased indefinitely, we arrive at the concept of **compounding continuously**. Mathematically, this can be accomplished by applying the limit concept to the expression $P\left(1 + \dfrac{r}{n}\right)^{nt}$. We will not show the details here, but the following result is obtained. The formula

$$A = Pe^{rt}$$

yields the accumulated value, A, of a sum of money, P, that has been invested for t years at a rate of r percent compounded continuously. The following examples show the use of the formula.

1. $750 invested for 5 years at 9% compounded continuously produces

$$A = 750e^{(.09)(5)} = 750e^{.45} = \$1176.23.$$

2. $1000 invested for 10 years at 11% compounded continuously produces

$$A = 1000e^{(.11)(10)} = 1000e^{1.1} = \$3004.17.$$

3. $5000 invested for 20 years at 12% compounded continuously produces

$$A = 5000e^{(.12)(20)} = 5000e^{2.4} = \$55,115.88.$$

Again you may find it interesting to compare these results with those you obtained earlier when using a different number of compounding periods.

The ideas behind compounding continuously carry over to other growth situations. The law of exponential growth

$$Q(t) = Q_0e^{kt}$$

is used as a mathematical model for numerous growth-and-decay applications. In this equation, $Q(t)$ represents the quantity of a given substance at any time t; Q_0 is the initial amount of the substance (when $t = 0$); and k is a constant that depends on the particular application. If $k < 0$, then $Q(t)$ decreases as t increases, and we refer to the model as the **law of decay**.

Let's consider some growth-and-decay applications.

EXAMPLE 2   Suppose that in a certain culture the equation $Q(t) = 15000e^{.3t}$ expresses the number of bacteria present as a function of the time t, where t is expressed in hours. Find (a) the initial number of bacteria, and (b) the number of bacteria after 3 hours.

Solution   (a) The initial number of bacteria is produced when $t = 0$.

$$Q(0) = 15000e^{.3(0)}$$

$$= 15000e^0$$

$$= 15000 \qquad e^0 = 1$$

(b) $Q(3) = 15000e^{.3(3)}$

$$= 15000e^{.9}$$

$$= 36894 \qquad \text{nearest whole number}$$

Therefore, there should be approximately 36,894 bacteria present after 3 hours.   ■

EXAMPLE 3   Suppose the number of bacteria present in a certain culture after t minutes is given by the equation $Q(t) = Q_0 e^{.05t}$, where Q_0 represents the initial number of bacteria. If 5000 bacteria are present after 20 minutes, how many bacteria were present initially?

Solution   If 5000 bacteria are present after 20 minutes, then $Q(20) = 5000$.

$$5000 = Q_0 e^{.05(20)}$$

$$5000 = Q_0 e^1$$

$$\frac{5000}{e} = Q_0$$

$$1839 = Q_0 \qquad \text{nearest whole number}$$

Therefore, there were approximately 1839 bacteria present initially.   ■

EXAMPLE 4   The number of grams Q of a certain radioactive substance present after t seconds is given by $Q(t) = 200e^{-.3t}$. How many grams remain after 7 seconds?

Solution   We need to evaluate $Q(7)$.

$$Q(7) = 200e^{-.3(7)}$$

$$= 200e^{-2.1}$$

$$= 24 \qquad \text{nearest whole number}$$

Thus, approximately 24 grams remain after 7 seconds.   ■

Problem Set 13.2

1. Assuming that the rate of inflation is 4% per year, the equation $P = P_0(1.04)^t$ yields the predicted price P of an item in t years that presently costs P_0. Find the predicted price of each of the following items for the indicated years ahead.

   (a) \$0.55 can of soup in 3 years

   (b) \$3.43 container of cocoa mix in 5 years

   (c) \$1.76 jar of coffee creamer in 4 years

   (d) \$0.44 can of beans and bacon in 10 years

   (e) \$9000 car in 5 years (nearest dollar)

   (f) \$50,000 house in 8 years (nearest dollar)

   (g) \$500 TV set in 7 years (nearest dollar)

2. Suppose it is estimated that the value of a car depreciates 30% per year for the first 5 years. The equation $A = P_0(.7)^t$ yields the value (A) of a car after t years if the original price is P_0. Find the value (to the nearest dollar) of each of the following priced cars after the indicated time.

   (a) \$9000 car after 4 years

   (b) \$5295 car after 2 years

   (c) \$6395 car after 5 years

   (d) \$15,595 car after 3 years

For Problems 3–14, use the formula $A = P\left(1 + \dfrac{r}{n}\right)^{nt}$ to find the total amount of money accumulated at the end of the indicated time period for each of the following investments.

   3. \$200 for 6 years at 6% compounded annually

   4. \$250 for 5 years at 7% compounded annually

   5. \$500 for 7 years at 8% compounded semiannually

   6. \$750 for 8 years at 8% compounded semiannually

   7. \$800 for 9 years at 9% compounded quarterly

   8. \$1200 for 10 years at 10% compounded quarterly

   9. \$1500 for 5 years at 12% compounded monthly

   10. \$2000 for 10 years at 9% compounded monthly

   11. \$5000 for 15 years at 8.5% compounded annually

   12. \$7500 for 20 years at 9.5% compounded semiannually

   13. \$8000 for 10 years at 10.5% compounded quarterly

   14. \$10,000 for 25 years at 9.25% compounded monthly

For Problems 15–23, use the formula $A = Pe^{rt}$ to find the total amount of money accumulated at the end of the indicated time period by compounding continuously.

15. \$400 for 5 years at 7%	16. \$500 for 7 years at 6%
17. \$750 for 8 years at 8%	18. \$1000 for 10 years at 9%
19. \$2000 for 15 years at 10%	20. \$5000 for 20 years at 11%

21. $7500 for 10 years at 8.5%

22. $10,000 for 25 years at 9.25%

23. $15,000 for 10 years at 7.75%

24. Complete the following chart that illustrates what happens to $1000 invested at various rates of interest for different lengths of time but always compounded continuously. Round your answers to the nearest dollar.

$1000 compounded continuously

	8%	10%	12%	14%
5 years				
10 years				
15 years				
20 years				
25 years				

25. Complete the following chart that illustrates what happens to $1000 invested at 12% for different lengths of time and different numbers of compounding periods. Round all of your answers to the nearest dollar.

$1000 at 12%

	1 year	5 years	10 years	20 years
Compounded annually				
Compounded semiannually				
Compounded quarterly				
Compounded monthly				
Compounded continuously				

26. Complete the following chart that illustrates what happens to $1000 in 10 years based on different rates of interest and different numbers of compounding periods. Round your answers to the nearest dollar.

$1000 for 10 years

	8%	10%	12%	14%
Compounded annually				
Compounded semiannually				
Compounded quarterly				
Compounded monthly				
Compounded continuously				

27. Suppose that Nora invested $500 at 8.25% compounded annually for 5 years and Patti invested $500 at 8% compounded quarterly for 5 years. At the end of 5 years, who will have the most money and by how much?

28. Two years ago Daniel invested some money at 8% interest compounded annually. Today it is worth $758.16. How much did he invest two years ago?

29. What rate of interest (nearest hundredth of a percent) is needed so that an investment of $2500 will yield $3000 in 2 years if the money is compounded annually?

30. Suppose that a certain radioactive substance has a half-life of 20 years. If there are presently 2500 milligrams of the substance, how much, to the nearest milligram, will remain after 40 years? After 50 years?

31. Strontium-90 has a half-life of 29 years. If there are 400 grams of strontium initially, how much, to the nearest gram, will remain after 87 years? After 100 years?

32. The half-life of radium is approximately 1600 years. If the present amount of radium in a certain location is 500 grams, how much will remain after 800 years? Express your answer to the nearest gram.

For Problems 33–38, graph each of the exponential functions.

33. $f(x) = e^x + 1$ 34. $f(x) = e^x - 2$ 35. $f(x) = 2e^x$

36. $f(x) = -e^x$ 37. $f(x) = e^{2x}$ 38. $f(x) = e^{-x}$

For Problems 39–44, express your answers to the nearest whole number.

39. Suppose that in a certain culture, the equation $Q(t) = 1000e^{.4t}$ expresses the number of bacteria present as a function of the time t, where t is expressed in hours. How many bacteria are present at the end of 2 hours? 3 hours? 5 hours?

40. The number of bacteria present at a given time under certain conditions is given by the equation $Q = 5000e^{0.05t}$, where t is expressed in minutes. How many bacteria are present at the end of 10 minutes? 30 minutes? 1 hour?

41. The number of bacteria present in a certain culture after t hours is given by the equation $Q = Q_0e^{0.3t}$, where Q_0 represents the initial number of bacteria. If 6640 bacteria are present after 4 hours, how many bacteria were present initially?

42. The number of grams Q of a certain radioactive substance present after t seconds is given by the equation $Q = 1500e^{-0.4t}$. How many grams remain after 5 seconds? 10 seconds? 20 seconds?

43. Suppose that the present population of a city is 75,000. Using the equation $P(t) = 75000e^{.01t}$ to estimate future growth, estimate the population

    (a) 10 years from now   (b) 15 years from now, and   (c) 25 years from now.

44. Suppose that the present population of a city is 150,000. Use the equation $P(t) = 150000e^{.032t}$ to estimate future growth. Estimate the population

    (a) 10 years from now,   (b) 20 years from now, and   (c) 30 years from now.

45. The atmospheric pressure, measured in pounds per square inch, is a function of the altitude above sea level. The equation $P(a) = 14.7e^{-.21a}$, where a is the altitude measured in miles, can be used to approximate atmospheric pressure. Find the atmospheric pressure at each of the following locations. Express each answer to the nearest tenth of a pound per square inch.

    (a) Mount McKinley in Alaska—altitude of 3.85 miles;

    (b) Denver, Colorado—the "mile-high" city;

    (c) Asheville, North Carolina—altitude of 1985 feet (5280 feet = 1 mile);

    (d) Phoenix, Arizona—altitude of 1090 feet.

Thoughts into Words

46. Explain the difference between simple interest and compound interest.

47. Suppose that a certain radioactive substance has a half-life of 5000 years. What does this mean?

Miscellaneous Problems

 48. Use a graphics calculator to check your graphs for Problems 33–38.

13.3

Logarithms

In Sections 13.1 and 13.2, (1) we learned about exponential expressions of the form b^n, where b is any positive real number and n is any real number, (2) we used exponential expressions of the form b^n to define exponential functions, and (3) we used exponential functions to help solve problems. In the next three sections we will follow the same basic pattern with respect to a new concept, that of a **logarithm**. Let's begin with the following definition.

DEFINITION 13.2

> If r is any positive real number, then the unique exponent t such that $b^t = r$ is called the **logarithm of r with base b** and is denoted by $\log_b r$.

According to Definition 13.2, the logarithm of 8 base 2 is the exponent t such that $2^t = 8$; thus, we can write $\log_2 8 = 3$. Likewise, we can write $\log_{10} 100 = 2$ because $10^2 = 100$. In general, we can remember Definition 13.2 in terms of the statement

> $\log_b r = t$ is equivalent to $b^t = r$.

Thus, we can easily switch back and forth between exponential and logarithmic forms of equations, as the next examples illustrate.

$$\log_3 81 = 4 \text{ is equivalent to } 3^4 = 81,$$

$$\log_{10} 100 = 2 \text{ is equivalent to } 10^2 = 100,$$

$$\log_{10} 0.001 = -3 \text{ is equivalent to } 10^{-3} = 0.001,$$

$$\log_2 128 = 7 \text{ is equivalent to } 2^7 = 128,$$

$$2^4 = 16 \text{ is equivalent to } \log_2 16 = 4,$$

$$5^2 = 25 \text{ is equivalent to } \log_5 25 = 2,$$

$$\left(\frac{1}{2}\right)^4 = \frac{1}{16} \text{ is equivalent to } \log_{\frac{1}{2}}\left(\frac{1}{16}\right) = 4,$$

$$10^{-2} = 0.01 \text{ is equivalent to } \log_{10} 0.01 = -2$$

We can conveniently calculate some logarithms by changing to exponential form as in the next examples.

EXAMPLE 1   Evaluate $\log_4 64$.

Solution   Let $\log_4 64 = x$. Then by switching to exponential form we have $4^x = 64$, which we can solve as we did back in Section 13.1.

$$4^x = 64$$

$$4^x = 4^3$$

$$x = 3$$

Therefore, we can write $\log_4 64 = 3$.   ∎

EXAMPLE 2   Evaluate $\log_{10} 0.1$.

Solution   Let $\log_{10} 0.1 = x$. Then by switching to exponential form we have $10^x = 0.1$, which we can solve as follows.

$$10^x = 0.1$$

$$10^x = \frac{1}{10}$$

$$10^x = 10^{-1}$$

$$x = -1$$

Thus, we obtain $\log_{10} 0.1 = -1$ ∎

The link between logarithms and exponents also provides the basis for solving some equations that involve logarithms, as the next two examples illustrate.

EXAMPLE 3   Solve $\log_8 x = \frac{2}{3}$.

Solution $\log_8 x = \frac{2}{3}$

$8^{\frac{2}{3}} = x$ by switching to exponential form

$\sqrt[3]{8^2} = x$

$$(\sqrt[3]{8})^2 = x$$

$$4 = x$$

The solution set is $\{4\}$.   ■

EXAMPLE 4   Solve $\log_b 1000 = 3$.

Solution

$$\log_b 1000 = 3$$

$$b^3 = 1000$$

$$b = 10$$

The solution set is $\{10\}$.   ■

Properties of Logarithms

There are some properties of logarithms that are a direct consequence of Definition 13.2 and our knowledge of exponents. For example, by writing the exponential equations $b^1 = b$ and $b^0 = 1$ in logarithmic form, the following property is obtained.

PROPERTY 13.3

For $b > 0$ and $b \neq 1$,
1. $\log_b b = 1$
2. $\log_b 1 = 0$.

Thus, we can write

$$\log_{10} 10 = 1,$$

$$\log_2 2 = 1,$$

$$\log_{10} 1 = 0,$$

$$\log_5 1 = 0.$$

By Definition 13.2 $\log_b r$ is the exponent t such that $b^t = r$. Therefore, raising b to the $\log_b r$ power must produce r. We state this fact in Property 13.4.

PROPERTY 13.4

For $b > 0$, $b \neq 1$, and $r > 0$
$$b^{\log_b r} = r.$$

The following examples illustrate Property 13.4.

$$10^{\log_{10} 19} = 19,$$

$$2^{\log_2 14} = 14,$$

$$e^{\log_e 5} = 5.$$

Because a logarithm is by definition an exponent, it would seem reasonable to predict that there are some properties of logarithms that correspond to the basic exponential properties. This is an accurate prediction; these properties provide a basis for computational work with logarithms. Let's state the first of these properties and show how it can be verified by using our knowledge of exponents.

PROPERTY 13.5

> For positive real numbers b, r, and s where $b \neq 1$,
>
> $$\log_b rs = \log_b r + \log_b s.$$

To verify Property 13.5 we can proceed as follows. Let $m = \log_b r$ and $n = \log_b s$. Change each of these equations to exponential form.

$$m = \log_b r \quad \text{becomes} \quad r = b^m,$$

$$n = \log_b s \quad \text{becomes} \quad s = b^n$$

Thus, the product rs becomes

$$rs = b^m \cdot b^n = b^{m+n}.$$

Now, by changing $rs = b^{m+n}$ back to logarithmic form, we obtain

$$\log_b rs = m + n.$$

Replacing m with $\log_b r$ and n with $\log_b s$ yields

$$\log_b rs = \log_b r + \log_b s.$$

The following three examples demonstrate a use of Property 13.5.

EXAMPLE 5   If $\log_2 5 = 2.3222$ and $\log_2 3 = 1.5850$, evaluate $\log_2 15$.

Solution   Because $15 = 5 \cdot 3$, we can apply Property 13.5 as follows.

$$\log_2 15 = \log_2(5 \cdot 3)$$

$$= \log_2 5 + \log_2 3$$

$$= 2.3222 + 1.5850 = 3.9072 \qquad \blacksquare$$

EXAMPLE 6   If $\log_{10} 178 = 2.2504$ and $\log_{10} 89 = 1.9494$, evaluate $\log_{10}(178 \cdot 89)$.

Solution $$\log_{10}(178 \cdot 89) = \log_{10} 178 + \log_{10} 89$$

$$= 2.2504 + 1.9494 = 4.1998 \qquad \blacksquare$$

EXAMPLE 7 If $\log_3 8 = 1.8928$, then evaluate $\log_3 72$.

Solution
$$\log_3 72 = \log_3(9 \cdot 8)$$
$$= \log_3 9 + \log_3 8$$
$$= 2 + 1.8928 \qquad \log_3 9 = 2 \text{ because } 3^2 = 9$$
$$= 3.8928 \qquad\qquad\qquad\blacksquare$$

Since $\dfrac{b^m}{b^n} = b^{m-n}$, we would expect a corresponding property pertaining to logarithms. There is such a property, Property 13.6.

PROPERTY 13.6

For positive numbers b, r, and s, where $b \neq 1$,
$$\log_b\left(\frac{r}{s}\right) = \log_b r - \log_b s.$$

This property can be verified by using an approach similar to the one we used to verify Property 13.5. We leave it for you to do in an exercise in the next problem set.

We can use Property 13.6 to change a division problem into a subtraction problem as in the next two examples.

EXAMPLE 8 If $\log_5 36 = 2.2265$ and $\log_5 4 = 0.8614$, evaluate $\log_5 9$.

Solution Since $9 = \dfrac{36}{4}$, we can use Property 13.6 as follows.
$$\log_5 9 = \log_5\left(\frac{36}{4}\right)$$
$$= \log_5 36 - \log_5 4$$
$$= 2.2265 - 0.8614 = 1.3651 \qquad\blacksquare$$

EXAMPLE 9 Evaluate $\log_{10}\left(\dfrac{379}{86}\right)$ given that $\log_{10} 379 = 2.5786$ and $\log_{10} 86 = 1.9345$.

Solution
$$\log_{10}\left(\frac{379}{86}\right) = \log_{10} 379 - \log_{10} 86$$
$$= 2.5786 - 1.9345$$
$$= 0.6441 \qquad\blacksquare$$

The next property of logarithms provides the basis for evaluating expressions such as $3^{\sqrt{2}}$, $(\sqrt{5})^{\frac{2}{3}}$, and $(0.076)^{\frac{3}{4}}$. Here follows the property, a basis for its justification, and illustrations of its use.

PROPERTY 13.7

> If r is a positive real number, b is a positive real number other than 1, and p is any real number, then
>
> $$\log_b r^p = p(\log_b r).$$

As you might expect, the exponential property $(b^n)^m = b^{mn}$ plays an important role in the verification of Property 13.7. This is an exercise for you in the next problem set. Let's look at some uses of Property 13.7.

EXAMPLE 10   Evaluate $\log_2 22^{\frac{1}{3}}$ given that $\log_2 22 = 4.4598$.

Solution

$$\log_2 22^{\frac{1}{3}} = \frac{1}{3}\log_2 22 \qquad \text{Property 13.7}$$

$$= \frac{1}{3}(4.4598)$$

$$= 1.4866 \qquad\qquad\qquad\qquad\qquad \blacksquare$$

EXAMPLE 11   Evaluate $\log_{10}(8540)^{\frac{3}{5}}$ given that $\log_{10} 8540 = 3.9315$.

Solution

$$\log_{10}(8540)^{\frac{3}{5}} = \frac{3}{5}\log_{10} 8540 \qquad \text{Property 13.7}$$

$$= \frac{3}{5}(3.9315)$$

$$= 2.3589 \qquad\qquad\qquad\qquad\qquad \blacksquare$$

Working together, the properties of logarithms allow us to change the forms of various logarithmic expressions. For example, an expression such as $\log_b \sqrt{\dfrac{xy}{z}}$ can be rewritten in terms of sums and differences of simpler logarithmic quantities as follows.

$$\log_b \sqrt{\frac{xy}{z}} = \log_b \left(\frac{xy}{z}\right)^{\frac{1}{2}}$$

$$= \frac{1}{2}\log_b \left(\frac{xy}{z}\right) \qquad\qquad \text{Property 13.7}$$

$$= \frac{1}{2}(\log_b xy - \log_b z) \qquad \text{Property 13.6}$$

$$= \frac{1}{2}(\log_b x + \log_b y - \log_b z) \qquad \text{Property 13.5}$$

Sometimes we need to change from an indicated sum or difference of logarithmic quantities to an indicated product or quotient. This is especially helpful when solving certain kinds of equations that involve logarithms. Note in these next two examples how we can use the properties, along with the process of changing from logarithmic form to exponential form, to solve some equations.

EXAMPLE 12   Solve $\log_{10} x + \log_{10}(x + 9) = 1$.

Solution

$$\log_{10} x + \log_{10}(x + 9) = 1$$

$$\log_{10}[x(x + 9)] = 1 \qquad \text{Property 13.5}$$

$$10^1 = x(x + 9) \qquad \text{change to exponential form}$$

$$10 = x^2 + 9x$$

$$0 = x^2 + 9x - 10$$

$$0 = (x + 10)(x - 1)$$

$$x + 10 = 0 \qquad \text{or} \qquad x - 1 = 0$$

$$x = -10 \qquad \text{or} \qquad x = 1$$

Since logarithms are defined only for positive numbers, x and $x + 9$ have to be positive. Therefore, the solution of -10 must be discarded. The solution set is $\{1\}$.

■

EXAMPLE 13   Solve $\log_5(x + 4) - \log_5 x = 2$.

Solution

$$\log_5(x + 4) - \log_5 x = 2$$

$$\log_5\left(\frac{x + 4}{x}\right) = 2 \qquad \text{Property 13.6}$$

$$5^2 = \frac{x + 4}{x} \qquad \text{change to exponential form}$$

$$25 = \frac{x + 4}{x}$$

$$25x = x + 4$$

$$24x = 4$$

$$x = \frac{4}{24} = \frac{1}{6}$$

The solution set is $\left\{\frac{1}{6}\right\}$.

■

Problem Set 13.3

Write each of the following in logarithmic form. For example, $2^3 = 8$ becomes $\log_2 8 = 3$ in logarithmic form.

1. $2^7 = 128$ **2.** $3^3 = 27$ **3.** $5^3 = 125$ **4.** $2^6 = 64$

5. $10^3 = 1000$ **6.** $10^1 = 10$ **7.** $2^{-2} = \dfrac{1}{4}$ **8.** $3^{-4} = \dfrac{1}{81}$

9. $10^{-1} = .1$ **10.** $10^{-2} = .01$

Write each of the following in exponential form. For example, $\log_2 8 = 3$ becomes $2^3 = 8$ in exponential form.

11. $\log_3 81 = 4$ **12.** $\log_2 256 = 8$

13. $\log_4 64 = 3$ **14.** $\log_5 25 = 2$

15. $\log_{10} 10000 = 4$ **16.** $\log_{10} 100000 = 5$

17. $\log_2\left(\dfrac{1}{16}\right) = -4$ **18.** $\log_5\left(\dfrac{1}{125}\right) = -3$

19. $\log_{10} .001 = -3$ **20.** $\log_{10} .000001 = -6$

Evaluate each of the following.

21. $\log_2 16$ **22.** $\log_3 9$ **23.** $\log_3 81$ **24.** $\log_2 512$

25. $\log_6 216$ **26.** $\log_4 256$ **27.** $\log_7 \sqrt{7}$ **28.** $\log_2 \sqrt[3]{2}$

29. $\log_{10} 1$ **30.** $\log_{10} 10$ **31.** $\log_{10} .1$ **32.** $\log_{10} .0001$

33. $10^{\log_{10} 5}$ **34.** $10^{\log_{10} 14}$ **35.** $\log_2\left(\dfrac{1}{32}\right)$ **36.** $\log_5\left(\dfrac{1}{25}\right)$

37. $\log_5(\log_2 32)$ **38.** $\log_2(\log_4 16)$ **39.** $\log_{10}(\log_7 7)$ **40.** $\log_2(\log_5 5)$

Solve each of the following equations.

41. $\log_7 x = 2$ **42.** $\log_2 x = 5$ **43.** $\log_8 x = \dfrac{4}{3}$ **44.** $\log_{16} x = \dfrac{3}{2}$

45. $\log_9 x = \dfrac{3}{2}$ **46.** $\log_8 x = -\dfrac{2}{3}$ **47.** $\log_4 x = -\dfrac{3}{2}$ **48.** $\log_9 x = -\dfrac{5}{2}$

49. $\log_x 2 = \dfrac{1}{2}$ **50.** $\log_x 3 = \dfrac{1}{2}$

Given that $\log_2 5 = 2.3219$ and $\log_2 7 = 2.8074$, evaluate each of the following by using Properties 13.5–13.7.

51. $\log_2 35$ **52.** $\log_2(\tfrac{7}{5})$ **53.** $\log_2 125$

54. $\log_2 49$ **55.** $\log_2 \sqrt{7}$ **56.** $\log_2 \sqrt[3]{5}$

57. $\log_2 175$ **58.** $\log_2 56$ **59.** $\log_2 80$

Given that $\log_8 5 = .7740$ and $\log_8 11 = 1.1531$, evaluate each of the following using Properties 13.5–13.7.

60. $\log_8 55$ **61.** $\log_8\left(\tfrac{5}{11}\right)$ **62.** $\log_8 25$

63. $\log_8 \sqrt{11}$ **64.** $\log_8(5)^{\frac{2}{3}}$ **65.** $\log_8 88$

66. $\log_8 320$ **67.** $\log_8\left(\tfrac{25}{11}\right)$ **68.** $\log_8\left(\tfrac{121}{25}\right)$

Express each of the following as the sum or difference of simpler logarithmic quantities. Assume that all variables represent positive real numbers. For example,

$$\log_b \frac{x^3}{y^2} = \log_b x^3 - \log_b y^2$$

$$= 3\log_b x - 2\log_b y$$

69. $\log_b xyz$ **70.** $\log_b 5x$ **71.** $\log_b\left(\dfrac{y}{z}\right)$ **72.** $\log_b\left(\dfrac{x^2}{y}\right)$

73. $\log_b y^3 z^4$ **74.** $\log_b x^2 y^3$ **75.** $\log_b\left(\dfrac{x^{\frac{1}{2}}y^{\frac{1}{3}}}{z^4}\right)$ **76.** $\log_b x^{\frac{2}{3}}y^{\frac{3}{4}}$

77. $\log_b \sqrt[3]{x^2 z}$ **78.** $\log_b \sqrt{xy}$ **79.** $\log_b\left(x\sqrt{\dfrac{x}{y}}\right)$ **80.** $\log_b \sqrt{\dfrac{x}{y}}$

For Problems 81–92, solve each of the equations.

81. $\log_3 x + \log_3 4 = 2$ **82.** $\log_7 5 + \log_7 x = 1$

83. $\log_{10} x + \log_{10}(x - 21) = 2$ **84.** $\log_{10} x + \log_{10}(x - 3) = 1$

85. $\log_2 x + \log_2 (x - 3) = 2$ **86.** $\log_3 x + \log_3 (x - 2) = 1$

87. $\log_{10}(2x - 1) - \log_{10}(x - 2) = 1$ **88.** $\log_{10}(9x - 2) = 1 + \log_{10}(x - 4)$

89. $\log_5 (3x - 2) = 1 + \log_5 (x - 4)$ **90.** $\log_6 x + \log_6 (x + 5) = 2$

91. $\log_8 (x + 7) + \log_8 x = 1$ **92.** $\log_6 (x + 1) + \log_6 (x - 4) = 2$

93. Verify Property 13.6   **94.** Verify Property 13.7

Thoughts into Words

95. How would you explain the concept of a logarithm to someone who has never studied algebra?

96. Explain, without using Property 13.4, why $4^{\log_4 9}$ equals 9.

13.4
Logarithmic Functions

We can now use the concept of a logarithm to define a new function as follows.

DEFINITION 13.3

If $b > 0$ and $b \neq 1$, then the function f defined by

$$f(x) = \log_b x,$$

where x is any positive real number, is called the **logarithmic function with base b**.

We can obtain the graph of a specific logarithmic function in various ways.

For example, we can change the equation $y = \log_2 x$ to the exponential equation $2^y = x$, where we can determine a table of values. The next set of exercises asks you to graph some logarithmic functions with this approach.

We can obtain the graph of a logarithmic function by setting up a table of values directly from the logarithmic equation. Example 1 illustrates this approach.

EXAMPLE 1   Graph $f(x) = \log_2 x$.

Solution   Let's choose some values for x where the corresponding values for $\log_2 x$ are easily determined. (Remember that logarithms are only defined for the positive real numbers.)

x	$f(x)$
$\frac{1}{8}$	-3
$\frac{1}{4}$	-2
$\frac{1}{2}$	-1
1	0
2	1
4	2
8	3

$\log_2 \dfrac{1}{8} = -3$ because $2^{-3} = \dfrac{1}{2^3} = \dfrac{1}{8}$

$\log_2 1 = 0$ because $2^0 = 1$

Plot these points and connect them with a smooth curve to produce Figure 13.5.

Figure 13.5

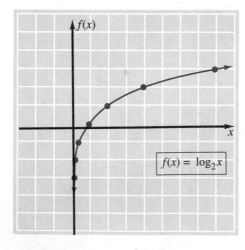

$f(x) = \log_2 x$

Suppose that we consider two functions f and g as follows.

$f(x) = b^x$ *Domain*: all real numbers
                            Range: positive real numbers.

$g(x) = \log_b x$ *Domain*: positive real numbers
                            Range: all real numbers.

Furthermore, suppose that we consider the composition of f and g, and the composition of g and f.

$$(f \circ g)(x) = f(g(x)) = f(\log_b x) = b^{\log_b x} = x,$$
$$(g \circ f)(x) = g(f(x)) = g(b^x) = \log_b b^x = x \log_b b = x(1) = x$$

Therefore, because the domain of f is the range of g, the range of f is the domain of g, $f(g(x)) = x$, and $g(f(x)) = x$, *the two functions f and g are inverses of each other.*

Remember also from Chapter 12 that the graphs of a function and its inverse are reflections of each other through the line $y = x$. Thus, the graph of a logarithmic function can also be determined by reflecting the graph of its inverse exponential function through the line $y = x$. We see this in Figure 13.6 where the graph of $y = 2^x$ has been reflected across the line $y = x$ to produce the graph of $y = \log_2 x$.

Figure 13.6

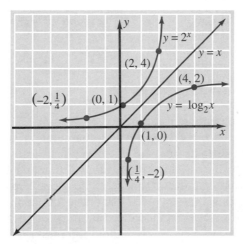

The *general behavior* patterns of exponential functions were illustrated by two graphs back in Figure 13.3. We can now reflect each of those graphs through the line $y = x$ and observe the general behavior patterns of logarithmic functions as shown in Figure 13.7.

Figure 13.7

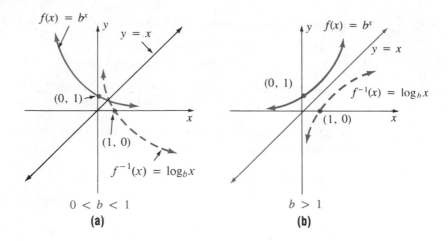

$$0 < b < 1$$
(a)

$$b > 1$$
(b)

Finally, when graphing logarithmic functions, don't forget about variations of the basic curves.

1. The graph of $f(x) = 3 + \log_2 x$ is the graph of $f(x) = \log_2 x$ *moved up three units.* (Since $\log_2 x + 3$ is apt to be confused with $\log_2(x + 3)$, we commonly write $3 + \log_2 x$.)

2. The graph of $f(x) = \log_2(x - 4)$ is the graph of $f(x) = \log_2 x$ *moved four units to the right.*

3. The graph of $f(x) = -\log_2 x$ is the graph of $f(x) = \log_2 x$ *reflected across the x-axis.*

Common Logarithms—Base 10

The properties of logarithms we discussed in Section 13.3 are true for any valid base. For example, since the Hindu-Arabic numeration system that we use is a base-10 system, logarithms to base 10 have historically been used for computational purposes. Base-10 logarithms are called **common logarithms**.

Originally, common logarithms were developed to assist in complicated numerical calculations that involved products, quotients, and powers of real numbers. Today they are seldom used for that purpose because the calculator and computer can much more effectively handle the messy computational problems. However, common logarithms do still occur in applications; they are deserving of our attention.

REMARK In Appendix C we have included a short discussion relative to the computational aspects of common logarithms. You may find it interesting to browse through this material. It should enhance your appreciation of the calculator.

As we know from earlier work, the definition of a logarithm provides the basis for evaluating $\log_{10} x$ for values of x that are integral powers of 10. Consider the following examples.

$\log_{10} 1000 = 3$ because $10^3 = 1000$,

$\log_{10} 100 = 2$ because $10^2 = 100$,

$\log_{10} 10 = 1$ because $10^1 = 10$,

$\log_{10} 1 = 0$ because $10^0 = 1$,

$\log_{10} 0.1 = -1$ because $10^{-1} = \dfrac{1}{10} = 0.1$,

$\log_{10} 0.01 = -2$ because $10^{-2} = \dfrac{1}{10^2} = 0.01$,

$\log_{10} 0.001 = -3$ because $10^{-3} = \dfrac{1}{10^3} = 0.001$

When working exclusively with base 10 logarithms, it is customary to omit writing the numeral 10 to designate the base. Thus, the expression $\log_{10} x$ is written as $\log x$ and a statement such as $\log_{10} 1000 = 3$ becomes $\log 1000 = 3$. We will follow this practice from now on in this chapter, but don't forget that the base is understood to be 10.

$$\log_{10} x = \log x$$

To find the common logarithm of a positive number that is not an integral power of 10, we can use an appropriately equipped calculator or a table such as the one that appears inside the back cover of this text. Using a calculator equipped with a common logarithm function (ordinarily a key labeled $\boxed{\log}$ is used), we obtained the following results rounded to four decimal places.

$\log 1.75 = 0.2430$,

$\log 23.8 = 1.3766$,

$\log 134 = 2.1271$,   (Be sure that you can use a calculator and obtain these results.)

$\log 0.192 = -0.7167$,

$\log 0.0246 = -1.6091$

In order to use logarithms to solve problems we sometimes need to be able to determine a number when the logarithm of the number is known. That is to say, we may need to determine x if $\log x$ is known. Let's consider an example.

EXAMPLE 2   Find x if $\log x = .2430$.

Solution   If $\log x = .2430$, then by changing to exponential form we have $10^{.2430} = x$. Therefore, using the $\boxed{10^x}$ key we can find x.

$$x = 10^{.2430} \approx 1.749846689$$

Therefore, $x = 1.7498$, rounded to five significant digits.   ∎

Be sure that you can use your calculator to obtain the following results. We have rounded the values for x to 5 significant digits.

If $\log x = .7629$, then $x = 10^{.7629} = 5.7930$.

If $\log x = 1.4825$, then $x = 10^{1.4825} = 30.374$.

If $\log x = 4.0214$, then $x = 10^{4.0214} = 10505$.

If $\log x = -1.5162$, then $x = 10^{-1.5162} = .030465$.

If $\log x = -3.8921$, then $x = 10^{-3.8921} = .00012820$.

The **common logarithmic function** is defined by the equation $f(x) = \log x$. It should now be a simple matter to set up a table of values and sketch the function. We will have you do this in the next set of exercises. Remember that $f(x) = 10^x$ and $g(x) = \log x$ are inverses of each other. Therefore, we could also get the graph of $g(x) = \log x$ by reflecting the exponential curve $f(x) = 10^x$ across the line $y = x$.

Natural Logarithms—Base e

In many practical applications of logarithms, the number e (remember $e \approx 2.71828$) is used as a base. Logarithms with a base of e are called **natural logarithms** and the symbol $\ln x$ is commonly used instead of $\log_e x$.

$$\boxed{\log_e x = \ln x}$$

Natural logarithms can be found with an appropriately equipped calculator or with a table of natural logarithms. (A table of natural logarithms is provided in Appendix D.) Using a calculator with a natural logarithm function (ordinarily a key labeled $\boxed{\ln x}$), we can obtain the following results rounded to four decimal places.

$\ln 3.21 = 1.1663$,

$\ln 47.28 = 3.8561$,

$\ln 842 = 6.7358$,

$\ln 0.21 = -1.5606$,

$\ln 0.0046 = -5.3817$,

$\ln 10 = 2.3026$

Be sure that you can use your calculator to obtain these results. Keep in mind the significance of a statement such as $\ln 3.21 = 1.1663$. By changing to exponential form we are claiming that e raised to the 1.1663 power is approximately 3.21. Using a calculator we obtain $e^{1.1663} = 3.210093293$.

Let's do a few more problems and find x when given $\ln x$. Be sure that you agree with these results.

If $\ln x = 2.4156$, then $x = e^{2.4156} = 11.196$.

If $\ln x = .9847$, then $x = e^{.9847} = 2.6770$.

If $\ln x = 4.1482$, then $x = e^{4.1482} = 63.320$.

If $\ln x = 1.7654$, then $x = e^{-1.7654} = .17112$.

The **natural logarithmic function** is defined by the equation $f(x) = \ln x$. It is the inverse of the natural exponential function $f(x) = e^x$. Thus, one way to graph $f(x) = \ln x$ is to reflect the graph of $f(x) = e^x$ across the line $y = x$. We will have you do this in the next set of problems.

Problem Set 13.4

For Problems 1–10, use a calculator to find each **common logarithm**. Express answers to four decimal places.

1. log 7.24	**2.** log 2.05	**3.** log 52.23
4. log 825.8	**5.** log 3214.1	**6.** log 14,189
7. log 0.729	**8.** log 0.04376	**9.** log 0.00034

10. log 0.000069

For Problems 11–20, use your calculator to find x when given $\log x$. Express answers to five significant digits.

11. $\log x = 2.6143$	**12.** $\log x = 1.5263$	**13.** $\log x = 4.9547$
14. $\log x = 3.9335$	**15.** $\log x = 1.9006$	**16.** $\log x = 0.5517$
17. $\log x = -1.3148$	**18.** $\log x = -0.1452$	**19.** $\log x = -2.1928$

20. $\log x = -2.6542$

For Problems 21–30, use your calculator to find each **natural logarithm**. Express answers to four decimal places.

21. ln 5	**22.** ln 18	**23.** ln 32.6	**24.** ln 79.5
25. ln 430	**26.** ln 371.8	**27.** ln 0.46	**28.** ln 0.524
29. ln 0.0314	**30.** ln 0.008142		

For Problems 31–40, use your calculator to find x when given $\ln x$. Express answers to five significant digits.

31. $\ln x = 0.4721$	**32.** $\ln x = 0.9413$	**33.** $\ln x = 1.1425$
34. $\ln x = 2.7619$	**35.** $\ln x = 4.6873$	**36.** $\ln x = 3.0259$
37. $\ln x = -0.7284$	**38.** $\ln x = -1.6246$	**39.** $\ln x = -3.3244$

40. $\ln x = -2.3745$

41. (a) Complete the following table and then graph $f(x) = \log x$. (Express the values for $\log x$ to the nearest tenth.)

x	0.1	0.5	1	2	4	8	10
$\log x$							

(b) Complete the following table and express values for 10^x to the nearest tenth.

x	-1	$-.3$	0	.3	.6	.9	1
10^x							

Then graph $f(x) = 10^x$ and reflect it across the line $y = x$ to produce the graph for $f(x) = \log x$.

42. (a) Complete the following table and then graph $f(x) = \ln x$. (Express the values for $\ln x$ to the nearest tenth.)

x	0.1	0.5	1	2	4	8	10
$\ln x$							

(b) Complete the following table and express values for e^x to the nearest tenth.

x	-2.3	$-.7$	0	.7	1.4	2.1	2.3
e^x							

Then graph $f(x) = e^x$ and reflect it across the line $y = x$ to produce the graph for $f(x) = \ln x$.

43. Graph $y = \log_{\frac{1}{2}} x$ by graphing $\left(\dfrac{1}{2}\right)^y = x$.

44. Graph $y = \log_2 x$ by graphing $2^y = x$.

45. Graph $f(x) = \log_3 x$ by reflecting the graph of $g(x) = 3^x$ across the line $y = x$.

46. Graph $f(x) = \log_4 x$ by reflecting the graph of $g(x) = 4^x$ across the line $y = x$.

For Problems 47–53, graph each of the functions. Remember that the graph of $f(x) = \log_2 x$ is given in Figure 13.5.

47. $f(x) = 3 + \log_2 x$ **48.** $f(x) = -2 + \log_2 x$

49. $f(x) = \log_2(x + 3)$ **50.** $f(x) = \log_2(x - 2)$

51. $f(x) = \log_2 2x$ **52.** $f(x) = -\log_2 x$

53. $f(x) = 2\log_2 x$

For Problems 54–61, perform the following calculations and express answers to the nearest hundredth. (These calculations are in preparation for our work in the next section.)

54. $\dfrac{\log 7}{\log 3}$ **55.** $\dfrac{\ln 2}{\ln 7}$ **56.** $\dfrac{2\ln 3}{\ln 8}$

57. $\dfrac{\ln 5}{2\ln 3}$ **58.** $\dfrac{\ln 3}{0.04}$ **59.** $\dfrac{\ln 2}{0.03}$

60. $\dfrac{\log 2}{5\log 1.02}$ **61.** $\dfrac{\log 5}{3\log 1.07}$

Miscellaneous Problems

For Problems 62–65, use a graphics calculator.

62. Graph $f(x) = x$, $f(x) = e^x$, and $f(x) = \ln x$ on the same set of axes. The graphs of $f(x) = e^x$ and $f(x) = \ln x$ should be reflections of each other through the line $f(x) = x$.

63. Graph $f(x) = x$, $f(x) = 10^x$, and $f(x) = \log x$ on the same set of axes.

64. Graph $f(x) = \ln x$, $f(x) = 2 \ln x$, $f(x) = 3 \ln x$ and $f(x) = -2 \ln x$ on the same set of axes.

65. Graph $f(x) = \log x$, $f(x) = 2 + \log x$, and $f(x) = -3 + \log x$ on the same set of axes.

13.5

Exponential Equations, Logarithmic Equations, and Problem Solving

In Section 13.1 we solved exponential equations such as $3^x = 81$ when we expressed both sides of the equation as a power of 3 and then applied the property "if $b^n = b^m$, then $n = m$." However, if we try to use this same approach with an equation such as $3^x = 5$, we face the difficulty of expressing 5 as a power of 3. We can solve this type of problem by using the properties of logarithms and the following property of equality.

PROPERTY 13.8

> If $x > 0$, $y > 0$, and $b \neq 1$, then
>
> $\qquad x = y$ if and only if $\log_b x = \log_b y$.

Property 13.8 is stated in terms of any valid base b; however, for most applications either common logarithms (base 10) or natural logarithms (base e) are used. Let's consider some examples.

EXAMPLE 1   Solve $3^x = 5$ to the nearest hundredth.

Solution   By using common logarithms we can proceed as follows.

$$3^x = 5$$
$$\log 3^x = \log 5 \qquad \text{Property 13.8}$$
$$x \log 3 = \log 5 \qquad \log r^p = p \log r$$
$$x = \frac{\log 5}{\log 3}$$
$$x = 1.46 \qquad \text{nearest hundredth}$$

CHECK Since $3^{1.46} \approx 4.972754647$, we say that, to the nearest hundredth, the solution set for $3^x = 5$ is $\{1.46\}$. ∎

EXAMPLE 2 Solve $e^{x+1} = 5$ to the nearest hundredth.

Solution Since the base of e is used in the exponential expression, let's use natural logarithms to help solve this equation.

$$e^{x+1} = 5$$

$$\ln e^{x+1} = \ln 5 \qquad \text{Property 13.8}$$

$$(x + 1)\ln e = \ln 5 \qquad \ln r^2 = p \ln r$$

$$(x + 1)(1) = \ln 5 \qquad \ln e = 1$$

$$x = \ln 5 - 1$$

$$x \approx 0.609437912$$

$$x = 0.61 \qquad \text{nearest hundredth}$$

The solution set is $\{0.61\}$. Check it! ∎

Logarithmic Equations

In Example 12 of Section 13.3 we solved the logarithmic equation

$$\log_{10} x + \log_{10}(x + 9) = 1$$

by simplifying the left side of the equation to $\log_{10}[x(x + 9)]$ and then changing the equation to exponential form to complete the solution. Now, using Property 13.8, we can solve such a logarithmic equation another way and also expand our equation solving capabilities. Let's consider some examples.

EXAMPLE 3 Solve $\log x + \log(x - 15) = 2$.

Solution Since $\log 100 = 2$, the given equation becomes

$$\log x + \log(x - 15) = \log 100.$$

Now, simplify the left side, apply Property 13.8 and proceed as follows.

$$\log(x)(x - 15) = \log 100$$

$$x(x - 15) = 100$$

$$x^2 - 15x - 100 = 0$$

$$(x - 20)(x + 5) = 0$$

$$x - 20 = 0 \qquad \text{or} \qquad x + 5 = \quad 0$$

$$x = 20 \qquad \text{or} \qquad x = -5$$

The domain of a logarithmic function must contain only positive numbers, so x and $x - 15$ must be positive in this problem. Therefore, we discard the solution of -5; the solution set is $\{20\}$. ■

EXAMPLE 4   Solve $\ln(x + 2) = \ln(x - 4) + \ln 3$.

Solution

$$\ln(x + 2) = \ln(x - 4) + \ln 3$$

$$\ln(x + 2) = \ln[3(x - 4)]$$

$$x + 2 = 3(x - 4)$$

$$x + 2 = 3x - 12$$

$$14 = 2x$$

$$7 = x$$

The solution set is $\{7\}$. ■

Problem Solving

In Section 13.2 we used the compound interest formula

$$A = P\left(1 + \frac{r}{n}\right)^{nt}$$

to determine the amount of money (A) accumulated at the end of t years if P dollars is invested at r rate of interest compounded n times per year. Now let's use this formula to solve other types of problems that deal with compound interest.

EXAMPLE 5   How long will $500 take to double itself if invested at 12% interest compounded quarterly?

Solution   To *double itself* means that the $500 will grow into $1000. Thus,

$$1000 = 500\left(1 + \frac{0.12}{4}\right)^{4t}$$

$$= 500(1 + 0.03)^{4t}$$

$$= 500(1.03)^{4t}.$$

Multiply both sides of $1000 = 500(1.03)^{4t}$ by $\dfrac{1}{500}$ to yield

$$2 = (1.03)^{4t}.$$

Therefore,

$$\log 2 = \log(1.03)^{4t} \qquad \text{Property 13.8}$$

$$= 4t \log 1.03. \qquad \log r^p = p \log r$$

Solve for t to obtain

$$\log 2 = 4t \log 1.03$$

$$\frac{\log 2}{\log 1.03} = 4t$$

$$\frac{\log 2}{4 \log 1.03} = t \qquad \text{Multiply both sides by } \frac{1}{4}.$$

$$t \approx 5.862443063$$

$$t = 5.9. \qquad \text{nearest tenth}$$

Therefore, we are claiming that $500 invested at 12% interest compounded quarterly will double itself in approximately 5.9 years.

CHECK $500 invested at 12% compounded quarterly for 5.9 years will produce

$$A = \$500\left(1 + \frac{0.12}{4}\right)^{4(5.9)}$$

$$= \$500(1.03)^{23.6}$$

$$= \$1004.45. \qquad \blacksquare$$

In Section 13.2, we also used the formula $A = Pe^{rt}$ when money was to be compounded continuously. At this time, with the help of natural logarithms, we can extend our use of this formula.

EXAMPLE 6   How long will it take $100 to triple itself if it is invested at 8% interest compounded continuously?

Solution   To *triple itself* means that the $100 will grow into $300. Thus, using the formula for interest that is compounded continuously, we can proceed as follows.

$$A = Pe^{rt}$$

$$\$300 = \$100e^{(0.08)t}$$

$$3 = e^{0.08t}$$

$$\ln 3 = \ln e^{0.08t} \qquad \text{Property 13.8}$$

$$\ln 3 = 0.08t \ln e \qquad \ln r^p = p \ln r$$

$$\ln 3 = 0.08t \qquad \ln e = 1$$

$$\frac{\ln 3}{0.08} = t$$

$$t \approx 13.73265361$$

$$t = 13.7 \qquad\qquad \text{nearest tenth}$$

Therefore, in approximately 13.7 years, $100 will triple itself at 8% interest compounded continuously.

CHECK $100 invested at 8% compounded continuously for 13.7 years produces

$$A = Pe^{rt}$$

$$= \$100e^{0.08(13.7)}$$

$$= \$100e^{1.096}$$

$$\approx \$299.22.$$ ∎

Seismologists use the Richter scale to measure and report the magnitude of earthquakes. The equation

$$R = \log\frac{I}{I_0} \qquad\qquad \text{R is called a Richter number.}$$

compares the intensity I of an earthquake to a minimal or reference intensity I_0. The reference intensity is the smallest earth movement that can be recorded on a seismograph. Suppose that the intensity of an earthquake was determined to be 50,000 times the reference intensity. In this case, $I = 50{,}000\ I_0$ and the Richter number would be calculated as follows.

$$R = \log\frac{50{,}000\ I_0}{I_0}$$

$$R = \log 50000$$

$$R \approx 4.698970004$$

Thus, a Richter number of 4.7 would be reported. Let's consider two more examples that involve Richter numbers.

EXAMPLE 7 An earthquake in San Francisco in 1989 was reported to have a Richter number of 6.9. How did its intensity compare to the reference intensity?

Solution

$$6.9 = \log \frac{I}{I_0}$$

$$10^{6.9} = \frac{I}{I_0}$$

$$I = (10^{6.9})(I_0)$$

$$I \approx 7943282\, I_0$$

So its intensity was a little less than 8 million times the reference intensity. ∎

EXAMPLE 8   An earthquake in Iran in 1990 had a Richter number of 7.7. Compare the intensity level of that earthquake to the one in San Francisco (Example 7).

Solution   From Example 7 we have $I = (10^{6.9})(I_0)$ for the earthquake in San Francisco. Then using a Richter number of 7.7 we obtain $I = (10^{7.7})(I_0)$ for the earthquake in Iran. Therefore, by comparison,

$$\frac{(10^{7.7})(I_0)}{(10^{6.9})(I_0)} = 10^{7.7-6.9} = 10^{0.8} \approx 6.3.$$

The earthquake in Iran was about 6 times as intense as the one in San Francisco. ∎

Logarithms with Bases Other Than 10 or *e*

Now let's use either common or natural logarithms to evaluate logarithms that have bases other than 10 or *e*. Consider the following example.

EXAMPLE 9   Evaluate $\log_3 41$.

Solution   Let $x = \log_3 41$. Change to exponential form to obtain

$$3^x = 41.$$

Now we can apply Property 13.8 and proceed as follows.

$$\log 3^x = \log 41$$
$$x \log 3 = \log 41$$
$$x = \frac{\log 41}{\log 3}$$
$$x = 3.3802 \qquad \text{rounded to four decimal places}$$

∎

Using the method of Example 9 to evaluate $\log_a r$ produces the following formula, which we often refer to as the **change-of-base** formula for logarithms.

PROPERTY 13.9

If a, b, and r are positive numbers with $a \neq 1$ and $b \neq 1$, then

$$\log_a r = \frac{\log_b r}{\log_b a}.$$

By using Property 13.9 we can easily determine a relationship between logarithms of a different base. For example, suppose that in Property 13.9 we let $a = 10$ and $b = e$. Then

$$\log_a r = \frac{\log_b r}{\log_b a}$$

becomes

$$\log_{10} r = \frac{\log_e r}{\log_e 10},$$

which can be written as

$$\log_e r = (\log_e 10)(\log_{10} r).$$

Since $\log_e 10 = 2.3026$, rounded to four decimal places, we have

$$\log_e r = (2.3026)(\log_{10} r).$$

Thus, the natural logarithm of any positive number is approximately equal to the common logarithm of the number times 2.3026.

Problem Set 13.5

For Problems 1–14, solve each exponential equation and express solutions to the nearest hundredth.

1. $3^x = 32$ **2.** $2^x = 40$ **3.** $4^x = 21$

4. $5^x = 73$ **5.** $3^{x-2} = 11$ **6.** $2^{x+1} = 7$

7. $5^{3x+1} = 9$ **8.** $7^{2x-1} = 35$ **9.** $e^x = 5.4$

10. $e^x = 45$ **11.** $e^{x-2} = 13.1$ **12.** $e^{x-1} = 8.2$

13. $3e^x = 35.1$ **14.** $4e^x - 2 = 26$

For Problems 15–22, solve each logarithmic equation.

15. $\log x + \log(x + 21) = 2$ **16.** $\log x + \log(x + 3) = 1$

17. $\log(3x - 1) = 1 + \log(5x - 2)$ **18.** $\log(2x - 1) - \log(x - 3) = 1$

19. $\log(x + 2) - \log(2x + 1) = \log x$ **20.** $\log(x + 1) - \log(x + 2) = \log\dfrac{1}{x}$

21. $\ln(2t + 5) = \ln 3 + \ln(t - 1)$ **22.** $\ln(3t - 4) - \ln(t + 1) = \ln 2$

For Problems 23–32, approximate each of the following logarithms to three decimal places.

23. $\log_2 23$

24. $\log_3 32$

25. $\log_6 .214$

26. $\log_5 1.4$

27. $\log_7 421$

28. $\log_8 514$

29. $\log_9 .0017$

30. $\log_4 .00013$

31. $\log_3 720$

32. $\log_2 896$

For Problems 33–41, solve each problem and express answers to the nearest tenth.

33. How long will it take $750 to be worth $1000 if it is invested at 12% interest compounded quarterly?

34. How long will it take $1000 to double itself if it is invested at 9% interest compounded semiannually?

35. How long will it take $2000 to double itself if it is invested at 13% interest compounded continuously?

36. How long will it take $500 to triple itself if it is invested at 9% interest compounded continuously?

37. For a certain strain of bacteria, the number present after t hours is given by the equation $Q = Q_0 e^{0.34t}$, where Q_0 represents the initial number of bacteria. How long will it take 400 bacteria to increase to 4000 bacteria?

38. A piece of machinery valued at $30,000 depreciates at a rate of 10% yearly. How long will it take until it has a value of $15,000?

39. The number of grams of a certain radioactive substance present after t hours is given by the equation $Q = Q_0 e^{-0.45t}$, where Q_0 represents the initial number of grams. How long would it take 2500 grams to be reduced to 1250 grams?

40. For a certain culture the equation $Q(t) = Q_0 e^{0.4t}$, where Q_0 is an initial number of bacteria and t is time measured in hours, yields the number of bacteria as a function of time. How long will it take 500 bacteria to increase to 2000?

41. Suppose that the equation $P(t) = P_0 e^{0.02t}$, where P_0 represents an initial population and t is the time in years, is used to predict population growth. How long would it take a city of 50,000 to double its population?

Solve each of the Problems 42–46.

42. The equation $P(a) = 14.7 e^{-0.21a}$, where a is the altitude above sea level measured in miles, yields the atmospheric pressure in pounds per square inch. If the atmospheric pressure at Cheyenne, Wyoming is approximately 11.53 pounds per square inch, find that city's altitude above sea level. Express your answer to the nearest hundred feet.

43. An earthquake in Los Angeles in 1971 had an intensity of approximately five million times the reference intensity. What was the Richter number associated with that earthquake?

44. An earthquake in San Francisco in 1906 was reported to have a Richter number of 8.3. How did its intensity compare to the reference intensity?

45. Calculate how many times more intense an earthquake with a Richter number of 7.3 is than an earthquake with a Richter number of 6.4.

46. Calculate how many times more intense an earthquake with a Richter number of 8.9 is than an earthquake with a Richter number of 6.2.

Thoughts into Words

47. Explain the concept of a Richter number.

48. Explain how to determine $\log_3 746$ without using Property 13.9.

Chapter 13 Summary

(13.1) If a and b are positive real numbers, and m and n are any real numbers, then

1. $b^n \cdot b^m = b^{n+m}$ product of two powers

2. $(b^n)^m = b^{mn}$ power of a power

3. $(ab)^n = a^n b^n$ power of a product

4. $\left(\dfrac{a}{b}\right)^n = \dfrac{a^n}{b^n}$ power of a quotient

5. $\dfrac{b^n}{b^m} = b^{n-m}$ quotient of two powers

If $b > 0$, $b \neq 1$, and m and n are real numbers, then

$$b^n = b^m \quad \text{if and only if } n = m.$$

A function defined by an equation of the form

$$f(x) = b^x, \qquad b > 0 \text{ and } b \neq 1$$

is called an **exponential function**.

(13.2) A general formula for any principal, P, being compounded n times per year for any number of years (t) at a rate of r percent is

$$A = P\left(1 + \frac{r}{n}\right)^{nt}$$

where A represents the total amount of money accumulated at the end of the t years. The value of $\left(1 + \dfrac{1}{n}\right)^n$, as n gets infinitely large, approaches the number e, where e equals 2.71828 to five decimal places.

The formula

$$A = Pe^{rt}$$

yields the accumulated value, A, of a sum of money, P, that has been invested for t years at a rate of r percent **compounded continuously**.

The equation

$$Q(t) = Q_0 e^{kt}$$

is used as a mathematical model for many growth-and-decay applications.

(13.3) If r is any positive real number, then the unique exponent t such that $b^t = r$ is called the **logarithm of r with base b** and is denoted by $\log_b r$.

For $b > 0$ and $b \neq 1$, and $r > 0$,

1. $\log_b b = 1$,
2. $\log_b 1 = 0$,
3. $r = b^{\log_b r}$.

The following properties of logarithms are derived from the definition of a logarithm and the properties of exponents.

For positive real numbers b, r, and s where $b \neq 1$,

1. $\log_b rs = \log_b r + \log_b s$.
2. $\log_b \left(\dfrac{r}{s}\right) = \log_b r - \log_b s$.
3. $\log_b r^p = p \log_b r$. p is any real number.

(13.4) A function defined by an equation of the form

$$f(x) = \log_b x, \qquad b > 0 \text{ and } b \neq 1$$

is called a **logarithmic function**. The equation $y = \log_b x$ is equivalent to $x = b^y$. The two functions $f(x) = b^x$ and $g(x) = \log_b x$ are inverses of each other.

Logarithms with a base of 10 are called **common logarithms**. The expression $\log_{10} x$ is commonly written as $\log x$.

Natural logarithms are logarithms that have a base of e, where e is an irrational number whose decimal approximation to eight digits is 2.7182818. Natural logarithms are denoted by $\log_e x$ or **$\ln x$**.

(13.5) The properties of equality along with the properties of exponents and logarithms merge to help us solve a variety of exponential and logarithmic equations. Due to these properties we can now solve problems that deal with various applications including compound interest and growth problems.

The formula

$$\log_a r = \frac{\log_b r}{\log_b a}$$

is often called the **change-of-base** formula.

Chapter 13 Review Problem Set

For Problems 1–8, evaluate each expression without using a calculator.

1. $\log_2 128$
2. $\log_4 64$
3. $\log 10{,}000$
4. $\log .001$
5. $\ln e^2$
6. $5^{\log_5 13}$
7. $\log(\log_3 3)$
8. $\log_2 \left(\dfrac{1}{4}\right)$

For Problems 9–16, solve each equation without using your calculator.

9. $2^x = \dfrac{1}{16}$

10. $3^x - 4 = 23$

11. $16^x = \dfrac{1}{8}$

12. $\log_8 x = \dfrac{2}{3}$

13. $\log_x 3 = \dfrac{1}{2}$

14. $\log_5 2 + \log_5(3x + 1) = 1$

15. $\log_2(x + 5) - \log_2 x = 1$

16. $\log_2 x + \log_2(x - 4) = 5$

For Problems 17–20, use your calculator to find each logarithm. Express answers to four decimal places.

17. $\log 73.14$

18. $\ln 114.2$

19. $\ln 0.014$

20. $\log 0.00235$

For Problems 21–24, use your calculator to find x when given $\log x$ or $\ln x$. Express answers to five significant digits.

21. $\ln x = 0.1724$

22. $\log x = 3.4215$

23. $\log x = -1.8765$

24. $\ln x = -2.5614$

For Problems 25–28, use your calculator to help solve each equation. Express solutions to the nearest hundredth.

25. $3^x = 42$

26. $2e^x = 14$

27. $2^{x+1} = 79$

28. $e^{x-2} = 37$

For Problems 29–32, graph each of the functions.

29. $f(x) = 2^x - 3$

30. $f(x) = -3^x$

31. $f(x) = -1 + \log_2 x$

32. $f(x) = \log_2(x + 1)$

33. Approximate a value for $\log_5 97$ to three decimal places.

34. Suppose that $800 is invested at 9% interest compounded annually. How much money has accumulated at the end of 15 years?

35. If $2500 is invested at 10% interest compounded quarterly, how much money has accumulated at the end of 12 years?

36. If $3500 is invested at 8% interest compounded continuously, how much money will accumulate in 6 years?

37. Suppose that a certain radioactive substance has a half-life of 40 days. If there are presently 750 grams of the substance, how much, to the nearest gram, will remain after 100 days?

38. How long will it take $100 to double itself if it is invested at 14% interest compounded annually?

39. How long will it take $1000 to be worth $3500 if it is invested at 10.5% interest compounded quarterly?

40. Suppose that the present population of a city is 50,000. Use the equation $P(t) = P_0 e^{0.02t}$, where P_0 represents an initial population, to estimate future populations. Estimate the population of that city in 10 years, 15 years, and 20 years.

41. The number of bacteria present in a certain culture after t hours is given by the equation

$Q = Q_0 e^{0.29t}$, where Q_0 represents the initial number of bacteria. How long will it take 500 bacteria to increase to 2000 bacteria?

42. An earthquake in Mexico City in 1985 had an intensity level about 125,000,000 times the reference intensity. Find the Richter number for that earthquake.

Chapter 14

Using Matrices and Determinants to Solve Linear Systems

In Chapter 5 we solved systems of two linear equations in two variables using the techniques of substitution and elimination-by-addition. In this chapter we will work with systems of three linear equations in three variables and introduce some techniques that can be used with larger systems.

Systems of Three Linear Equations in Three Variables

Let's begin this section by introducing a different format for solving systems of linear equations using the elimination-by-addition method. This format involves the replacement of systems of equations with simpler equivalent systems until we obtain a system whereby we can easily extract the solutions. *Equivalent systems of equations are systems that have exactly the same solution set.* We can apply the following operations or transformations to a system of equations to produce an equivalent system.

1. Any two equations of the system can be interchanged.
2. Both sides of an equation of the system can be multiplied by any nonzero real number.
3. Any equation of the system can be replaced by the *sum* of that equation and a nonzero multiple of another equation.

Now let's see how to apply these operations to solve a system of two linear equations in two unknowns.

EXAMPLE 1   Solve the system $\begin{pmatrix} 3x + 2y = 1 \\ 5x - 2y = 23 \end{pmatrix}$. $\qquad$ (1)
$\qquad$ (2)

Solution   Let's replace equation (2) with an equation we form by multiplying equation (1) by 1 and then adding that result to equation (2).

$$\begin{pmatrix} 3x + 2y = 1 \\ 8x = 24 \end{pmatrix} \qquad\qquad (3)$$
$$(4)$$

From equation (4) we can easily obtain the value of x.

$$8x = 24$$
$$x = 3$$

Then we can substitute 3 for x in equation (3).

$$3x + 2y = 1$$
$$3(3) + 2y = 1$$
$$2y = -8$$
$$y = -4$$

The solution set is $\{(3, -4)\}$. Check it! $\qquad\blacksquare$

EXAMPLE 2   Solve the system $\begin{pmatrix} x + 5y = & -2 \\ 3x - 4y = & -25 \end{pmatrix}$.

$$\quad\quad\quad\quad\quad\quad\quad\quad\quad (1)$$
$$\quad\quad\quad\quad\quad\quad\quad\quad\quad (2)$$

Solution   Let's replace equation (2) with an equation we form by multiplying equation (1) by -3 and then adding that result to equation (2).

$$\begin{pmatrix} x + 5y = & -2 \\ -19y = & -19 \end{pmatrix}$$

$$\quad\quad\quad\quad\quad\quad\quad\quad\quad (3)$$
$$\quad\quad\quad\quad\quad\quad\quad\quad\quad (4)$$

From equation (4) we can obtain the value of y.

$$-19y = -19$$

$$y = 1$$

Now we can substitute 1 for y in equation (3).

$$x + 5y = -2$$

$$x + 5(1) = -2$$

$$x = -7$$

The solution set is $\{(-7, 1)\}$. ∎

Notice that our objective has been to produce an equivalent system of equations whereby one of the variables can be *eliminated* from one equation. We accomplish this by multiplying one equation of the system by an appropriate number and then *adding* that result to the other equation. Thus, the method is called **elimination-by-addition**. Let's look at another example.

EXAMPLE 3   Solve the system $\begin{pmatrix} 2x + 5y = & 4 \\ 5x - 7y = & -29 \end{pmatrix}$.

$$\quad\quad\quad\quad\quad\quad\quad\quad\quad (1)$$
$$\quad\quad\quad\quad\quad\quad\quad\quad\quad (2)$$

Solution   Let's form an equivalent system where the second equation has no x-term. First, we can multiply equation (2) by -2.

$$\begin{pmatrix} 2x + 5y = & 4 \\ -10x + 14y = & 58 \end{pmatrix}$$

$$\quad\quad\quad\quad\quad\quad\quad\quad\quad (3)$$
$$\quad\quad\quad\quad\quad\quad\quad\quad\quad (4)$$

Now we can replace equation (4) with an equation that we form by multiplying equation (3) by 5 and then adding that result to equation (4).

$$\begin{pmatrix} 2x + 5y = & 4 \\ 39y = & 78 \end{pmatrix}$$

$$\quad\quad\quad\quad\quad\quad\quad\quad\quad (5)$$
$$\quad\quad\quad\quad\quad\quad\quad\quad\quad (6)$$

From equation (6) we can find the value of y.

$$39y = 78$$

$$y = 2$$

Now we can substitute 2 for y in equation (5).

$$2x + 5y = 4$$
$$2x + 5(2) = 4$$
$$2x = -6$$
$$x = -3$$

The solution set is $\{(-3, 2)\}$. ■

Systems of Three Linear Equations in Three Variables

Consider a linear equation in three variables x, y, and z, such as $3x - 2y + z = 7$. Any **ordered triple** (x, y, z) that makes the equation a true numerical statement is said to be a solution of the equation. For example, the ordered triple $(2, 1, 3)$ is a solution because $3(2) - 2(1) + 3 = 7$. However, the ordered triple $(5, 2, 4)$ is not a solution because $3(5) - 2(2) + 4 \neq 7$. There are infinitely many solutions in the solution set.

> **REMARK** The concept of a *linear* equation is generalized to include equations of more than two variables. Thus, an equation such as $5x - 2y + 9z = 8$ is called a linear equation in three variables; the equation $5x - 7y + 2z - 11w = 1$ is called a linear equation in four variables, and so on.

To *solve* a system of three linear equations in three variables, such as

$$\begin{pmatrix} 3x - y + 2z = 13 \\ 4x + 2y + 5z = 30 \\ 5x - 3y - z = 3 \end{pmatrix}$$

means to find all of the ordered triples that satisfy all three equations. In other words, the solution set of the system is the intersection of the solution sets of all three equations in the system.

The graph of a linear equation in three variables is a **plane**, not a line. In fact, graphing equations in three variables requires the use of a three-dimensional coordinate system. Thus, using a graphing approach to solve systems of three linear equations in three variables is not at all practical. However, a simple graphic analysis does provide us with some direction as to what we can expect as we begin solving such systems.

In general, since each linear equation in three variables produces a plane, a system of three such equations produces three planes. There are various ways that three planes can be related. For example, they may be mutually parallel, or two of the planes may be parallel and the third one intersect each of the two. (You may want to analyze all of the other possibilities for the three planes!) However, for our purposes at this time we need to realize that from a solution set viewpoint, a system of three linear equations in three variables produces one of the following possibilities.

1. There is *one ordered triple* that satisfies all three equations. The three planes have a common point of intersection as indicated in Figure 14.1.

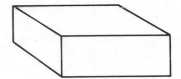

Figure 14.1

2. There are *infinitely many* ordered triples in the solution set, all of which are coordinates of points on a line common to the planes. This can happen if the three planes have a common line of intersection (Figure 14.2(a)) or if two of the planes coincide and the third plane intersects them (Figure 14.2(b)).

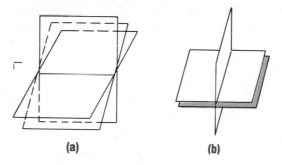

(a)   (b)

Figure 14.2

3. There are *infinitely many* ordered triples in the solution set, all of which are coordinates of points on a plane. This can happen if the three planes coincide, as illustrated in Figure 14.3.

Figure 14.3

4. The solution set is *empty*; it is $\emptyset$. This can happen in various ways, as we see in Figure 14.4. Notice that in each situation there are no points common to all three planes.

Figure 14.4

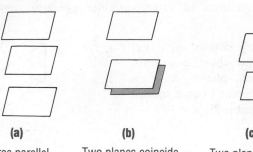 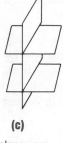 

(a)	(b)	(c)	(d)
Three parallel planes.	Two planes coincide, and the third one is parallel to the coinciding planes.	Two planes are parallel and the third one intersects them in parallel lines.	No two planes are parallel, but two of them intersect in a line which is parallel to the third plane.

Now that we know what possibilities exist, let's consider finding the solution sets for some systems. Our approach will be the elimination-by-addition method, whereby systems are replaced with equivalent systems until a system is obtained where we can easily determine the solution set. Let's start with an example that allows us to determine the solution set without changing to another equivalent system.

EXAMPLE 4   Solve the system $\begin{pmatrix} 2x - 3y + 5z = -5 \\ 2y - 3z = 4 \\ 4z = -8 \end{pmatrix}$.

$\qquad$ (1)
$\qquad$ (2)
$\qquad$ (3)

Solution   From equation (3) we can find the value of z.

$$4z = -8$$

$$z = -2$$

Now we can substitute -2 for z in equation (2).

$$2y - 3z = 4$$

$$2y - 3(-2) = 4$$

$$2y + 6 = 4$$

$$2y = -2$$

$$y = -1$$

Finally, we can substitute -2 for z and -1 for y in equation (1).

$$2x - 3y + 5z = -5$$

$$2x - 3(-1) + 5(-2) = -5$$

$$2x + 3 - 10 = -5$$

$$2x - 7 = -5$$
$$2x = 2$$
$$x = 1$$

The solution set is $\{(1, -1, -2)\}$.   ∎

Notice the format of the equations in the system of Example 4. The first equation contains all three variables, the second equation has only two variables, and the third equation has only one variable. This allowed us to solve the third equation and then to use "back substitution" to find the values of the other variables. Let's consider another example where we have to make one replacement of an equivalent system.

EXAMPLE 5

Solve the system $\begin{pmatrix} 3x + 2y - 7z = -34 \\ y + 5z = 21 \\ 3y - 2z = -22 \end{pmatrix}$.   (1) (2) (3)

Solution

Let's replace equation (3) with an equation we form by multiplying equation (2) by -3 and then adding that result to equation (3).

$\begin{pmatrix} 3x + 2y - 7z = -34 \\ y + 5z = 21 \\ -17z = -85 \end{pmatrix}$ (4) (5) (6)

From equation (6) we can find the value of z.

$$-17z = -85$$
$$z = 5$$

Now we can substitute 5 for z in equation (5).

$$y + 5z = 21$$
$$y + 5(5) = 21$$
$$y = -4$$

Finally, we can substitute 5 for z and -4 for y in equation (4).

$$3x + 2y - 7z = -34$$
$$3x + 2(-4) - 7(5) = -34$$
$$3x - 8 - 35 = -34$$
$$3x - 43 = -34$$
$$3x = 9$$
$$x = 3$$

The solution set is $\{(3, -4, 5)\}$.   ∎

Now let's consider some examples where we have to make more than one replacement of equivalent systems.

EXAMPLE 6

Solve the system $\begin{pmatrix} x - y + 4z = -29 \\ 3x - 2y - z = -6 \\ 2x - 5y + 6z = -55 \end{pmatrix}$.

$$\quad (1)$$
$$\quad (2)$$
$$\quad (3)$$

Solution

Let's replace equation (2) with an equation we form by multiplying equation (1) by -3 and then adding that result to equation (2). Let's also replace equation (3) with an equation we form by multiplying equation (1) by -2 and then adding that result to equation (3).

$$\begin{pmatrix} x - y\ + 4z = -29 \\ y - 13z =\ \ \ 81 \\ -3y\ - 2z =\ \ \ \ 3 \end{pmatrix}$$

$$\quad (4)$$
$$\quad (5)$$
$$\quad (6)$$

Now let's replace equation (6) with an equation we form by multiplying equation (5) by 3 and then adding that result to equation (6).

$$\begin{pmatrix} x - y\ + 4z = -29 \\ y - 13z =\ \ \ 81 \\ -41z =\ \ 246 \end{pmatrix}$$

$$\quad (7)$$
$$\quad (8)$$
$$\quad (9)$$

From equation (9) we can determine the value of z.

$$-41z = 246$$

$$z = -6$$

Now we can substitute -6 for z in equation (8).

$$y - 13z = 81$$

$$y - 13(-6) = 81$$

$$y + 78 = 81$$

$$y =\ \ 3$$

Finally, we can substitute -6 for z and 3 for y in equation (7).

$$x - y + 4z = -29$$

$$x - 3 + 4(-6) = -29$$

$$x - 3 - 24 = -29$$

$$x - 27 = -29$$

$$x =\ \ -2$$

The solution set is $\{(-2, 3, -6)\}$.   ∎

EXAMPLE 7   Solve the system $\begin{pmatrix} 3x - 4y + z = 14 \\ 5x + 3y - 2z = 27 \\ 7x - 9y + 4z = 31 \end{pmatrix}$.
$$\begin{align} &(1)\\ &(2)\\ &(3) \end{align}$$

Solution   A glance at the coefficients in the system indicates that eliminating the z-terms from equations (2) and (3) would be easy to do. Let's replace equation (2) with an equation we form by multiplying equation (1) by 2 and then adding that result to equation (2). Also let's replace equation (3) with an equation we form by multiplying equation (1) by -4 and then adding that result to equation (3).

$$\begin{pmatrix} 3x - 4y + z = 14 \\ 11x - 5y = 55 \\ -5x + 7y = -25 \end{pmatrix} \quad \begin{matrix} (4)\\ (5)\\ (6) \end{matrix}$$

Now let's eliminate the y-terms from equations (5) and (6). First, let's multiply equation (6) by 5.

$$\begin{pmatrix} 3x - 4y + z = 14 \\ 11x - 5y = 55 \\ -25x + 35y = -125 \end{pmatrix} \quad \begin{matrix} (7)\\ (8)\\ (9) \end{matrix}$$

Now we can replace equation (9) with an equation we form by multiplying equation (8) by 7 and then adding that result to equation (9).

$$\begin{pmatrix} 3x - 4y + z = 14 \\ 11x - 5y = 55 \\ 52x = 260 \end{pmatrix} \quad \begin{matrix} (10)\\ (11)\\ (12) \end{matrix}$$

From equation (12) we can determine the value of x.

$$52x = 260$$
$$x = 5$$

Now we can substitute 5 for x in equation (11).

$$11x - 5y = 55$$
$$11(5) - 5y = 55$$
$$-5y = 0$$
$$y = 0$$

Finally, we can substitute 5 for x and 0 for y in equation (10).

$$3x - 4y + z = 14$$
$$3(5) - 4(0) + z = 14$$
$$15 - 0 + z = 14$$
$$z = -1$$

The solution set is $\{(5, 0, -1)\}$.   ■

EXAMPLE 8   Solve the system $\begin{pmatrix} x - 2y + 3z = 1 \\ 3x - 5y - 2z = 4 \\ 2x - 4y + 6z = 7 \end{pmatrix}$.

$$\begin{align} &(1) \\ &(2) \\ &(3) \end{align}$$

Solution   A glance at the coefficients indicates that it should be easy to eliminate the x-terms from equations (2) and (3). We can replace equation (2) with an equation we form by multiplying equation (1) by -3 and then adding that result to equation (2). Likewise, we can replace equation (3) with an equation we form by multiplying equation (1) by -2 and then adding that result to equation (3).

$$\begin{pmatrix} x - 2y + 3z = 1 \\ y - 11z = 1 \\ 0 + 0 + 0 = 5 \end{pmatrix}$$

$$\begin{align} &(4) \\ &(5) \\ &(6) \end{align}$$

The false statement, $0 = 5$, indicates that the system is inconsistent and therefore the solution set is $\varnothing$. (If you were to graph this system, equations (1) and (3) would produce parallel planes. Thus, this is the situation depicted back in Figure 14.4(c)). ∎

EXAMPLE 9   Solve the system $\begin{pmatrix} 2x - y + 4z = 1 \\ 3x + 2y - z = -5 \\ 5x - 6y + 17z = -1 \end{pmatrix}$.

$$\begin{align} &(1) \\ &(2) \\ &(3) \end{align}$$

Solution   A glance at the coefficients indicates that it is easy to eliminate the y-terms from equations (2) and (3). We can replace equation (2) with an equation we form by multiplying equation (1) by 2 and then adding that result to equation (2). Likewise, we can replace equation (3) with an equation we form by multiplying equation (1) by -6 and then adding that result to equation (3).

$$\begin{pmatrix} 2x - y + 4z = 1 \\ 7x + 7z = 7 \\ -7x - 7z = -7 \end{pmatrix}$$

$$\begin{align} &(4) \\ &(5) \\ &(6) \end{align}$$

Now let's replace equation (6) with an equation we form by multiplying equation (5) by 1 and then adding that result to equation (6).

$$\begin{pmatrix} 2x - y + 4z = 1 \\ 7x + 7z = 7 \\ 0 + 0 = 0 \end{pmatrix}$$

$$\begin{align} &(7) \\ &(8) \\ &(9) \end{align}$$

The true numerical statement, $0 + 0 = 0$, indicates that the system has *infinitely many solutions*. (The graph of this system is shown in Figure 14.2(a).) ∎

REMARK   It can be shown that the solutions for the system in Example 9 are of the form $(t, 3 - 2t, 1 - t)$, where t is any real number. For example, if we let

$t = 2$, then we get the ordered triple $(2, -1, -1)$ and this triple will satisfy all three of the original equations. For our purposes in this text we shall simply indicate that such a system has *infinitely many solutions*.

Problem Set 14.1

Solve each of the following systems. If the solution set is $\varnothing$ or if it contains infinitely many solutions, then so indicate.

1. $\begin{pmatrix} 2x + 3y = -1 \\ 5x - 3y = 29 \end{pmatrix}$

2. $\begin{pmatrix} 3x - 4y = -30 \\ 7x + 4y = 10 \end{pmatrix}$

3. $\begin{pmatrix} 6x - 7y = 15 \\ 6x + 5y = -21 \end{pmatrix}$

4. $\begin{pmatrix} 5x + 2y = -4 \\ 5x - 3y = 6 \end{pmatrix}$

5. $\begin{pmatrix} x - 2y = -12 \\ 2x + 9y = 2 \end{pmatrix}$

6. $\begin{pmatrix} x - 4y = 29 \\ 3x + 2y = -11 \end{pmatrix}$

7. $\begin{pmatrix} 4x + 7y = -16 \\ 6x - y = -24 \end{pmatrix}$

8. $\begin{pmatrix} 6x + 7y = 17 \\ 3x + y = -4 \end{pmatrix}$

9. $\begin{pmatrix} 3x - 2y = 5 \\ 2x + 5y = -3 \end{pmatrix}$

10. $\begin{pmatrix} 4x + 3y = -4 \\ 3x - 7y = 34 \end{pmatrix}$

11. $\begin{pmatrix} 7x - 2y = 4 \\ 7x - 2y = 9 \end{pmatrix}$

12. $\begin{pmatrix} 5x - y = 6 \\ 10x - 2y = 12 \end{pmatrix}$

13. $\begin{pmatrix} x + 2y - 3z = 2 \\ 3y - z = 13 \\ 3y + 5z = 25 \end{pmatrix}$

14. $\begin{pmatrix} 2x + 3y - 4z = -10 \\ 2y + 3z = 16 \\ 2y - 5z = -16 \end{pmatrix}$

15. $\begin{pmatrix} 3x + 2y - 2z = 14 \\ x - 6z = 16 \\ 2x + 5z = -2 \end{pmatrix}$

16. $\begin{pmatrix} 3x + 2y - z = -11 \\ 2x - 3y = -1 \\ 4x + 5y = -13 \end{pmatrix}$

17. $\begin{pmatrix} 2x - y + z = 0 \\ 3x - 2y + 4z = 11 \\ 5x + y - 6z = -32 \end{pmatrix}$

18. $\begin{pmatrix} x - 2y + 3z = 7 \\ 2x + y + 5z = 17 \\ 3x - 4y - 2z = 1 \end{pmatrix}$

19. $\begin{pmatrix} 4x - y + z - 5 \\ 3x + y + 2z = 4 \\ x - 2y - z = 1 \end{pmatrix}$

20. $\begin{pmatrix} 2x - y + 3z - -14 \\ 4x + 2y - z = 12 \\ 6x - 3y + 4z = -22 \end{pmatrix}$

21. $\begin{pmatrix} x - y + 2z = 4 \\ 2x - 2y + 4z = 7 \\ 3x - 3y + 6z = 1 \end{pmatrix}$

22. $\begin{pmatrix} x + y - z = 2 \\ 3x - 4y + 2z = 5 \\ 2x + 2y - 2z = 7 \end{pmatrix}$

23. $\begin{pmatrix} x - 2y + z = -4 \\ 2x + 4y - 3z = -1 \\ -3x - 6y + 7z = 4 \end{pmatrix}$

24. $\begin{pmatrix} 2x - y + 3z = 1 \\ 4x + 7y - z = 7 \\ x + 4y - 2z = 3 \end{pmatrix}$

25. $\begin{pmatrix} 3x - 2y + 4z = 6 \\ 9x + 4y - z = 0 \\ 6x - 8y - 3z = 3 \end{pmatrix}$

26. $\begin{pmatrix} 2x - y + 3z = 0 \\ 3x + 2y - 4z = 0 \\ 5x - 3y + 2z = 0 \end{pmatrix}$

27. $\begin{pmatrix} 3x - y + 4z = 9 \\ 3x + 2y - 8z = -12 \\ 9x + 5y - 12z = -23 \end{pmatrix}$

28. $\begin{pmatrix} 5x - 3y + z = 1 \\ 2x - 5y = -2 \\ 3x - 2y - 4z = -27 \end{pmatrix}$

29. $\begin{pmatrix} 4x - y + 3z = -12 \\ 2x + 3y - z = 8 \\ 6x + y + 2z = -8 \end{pmatrix}$

30. $\begin{pmatrix} x + 3y - 2z = 19 \\ 3x - y - z = 7 \\ -2x + 5y + z = 2 \end{pmatrix}$

31. $\begin{pmatrix} x + y + z = 1 \\ 2x - 3y + 6z = 1 \\ -x + y + z = 0 \end{pmatrix}$

32. $\begin{pmatrix} 3x + 2y - 2z = -2 \\ x - 3y + 4z = -13 \\ -2x + 5y + 6z = 29 \end{pmatrix}$

For Problems 33–42, solve each problem by setting up and solving a system of three linear equations in three variables.

33. The sum of the digits of a three-digit number is 14. The number is 14 larger than twenty times the tens digit. The sum of the tens digit and the units digit is 12 larger than the hundreds digit. Find the number.

34. The sum of the digits of a three-digit number is 13. The sum of the hundreds digit and the tens digit is one less than the units digit. The sum of three times the hundreds digit and four times the units digit is 26 more than twice the tens digit. Find the number.

35. Two bottles of catsup, 2 jars of peanut butter, and 1 jar of pickles cost $4.20. Three bottles of catsup, 4 jars of peanut butter, and 2 jars of pickles cost $7.70. Four bottles of catsup, 3 jars of peanut butter, and 5 jars of pickles cost $9.80. Find the cost per bottle of catsup and per jar for peanut butter and pickles.

36. Five pounds of potatoes, 1 pound of onions, and 2 pounds of apples cost $1.26. Two pounds of potatoes, 3 pounds of onions, and 4 pounds of apples cost $1.88. Three pounds of potatoes, 4 pounds of onions, and 1 pound of apples cost $1.24. Find the price per pound for each item.

37. The sum of three numbers is 20. The sum of the first and third numbers is 2 more than twice the second number. The third number minus the first yields three times the second number. Find the numbers.

38. The sum of three numbers is 40. The third number is 10 less than the sum of the first two numbers. The second number is 1 larger than the first. Find the numbers.

39. The sum of the measures of the angles of a triangle is 180°. The largest angle is twice the smallest angle. The sum of the smallest and the largest angle is twice the other angle. Find the measure of each angle.

40. A box contains $2 in nickels, dimes, and quarters. There are 19 coins in all with twice as many nickels as dimes. How many coins of each kind are there?

41. Part of $3000 is invested at 12%, another part at 13%, and the remainder at 14%. The total yearly income from the three investments is $400. The sum of the amounts invested at 12% and 13% equals the amount invested at 14%. Determine how much is invested at each rate.

42. The perimeter of a triangle is 45 centimeters. The longest side is 4 centimeters less than

twice the shortest side. The sum of the lengths of the shortest and longest sides is 7 centimeters less than three times the length of the remaining side. Find the lengths of all three sides of the triangle.

14.2

A Matrix Approach to Solving Systems

In the previous section we adopted a specific format for using the elimination-by-addition method to solve systems of linear equations. This format leads to our next technique, that of using matrices to solve systems of linear equations.

A **matrix** is simply an array of numbers arranged in horizontal rows and vertical columns. For example, the matrix

$$\text{2 rows} \left.\begin{array}{c}\longrightarrow \\ \longrightarrow\end{array}\right. \begin{bmatrix} 2 & 1 & -4 \\ 5 & -7 & 6 \end{bmatrix}$$
$$\uparrow \quad \uparrow \quad \uparrow$$
$$\text{3 columns}$$

has 2 rows and 3 columns, which we refer to as a 2×3 (read "two-by-three") matrix. Some additional examples of matrices (*matrices* is the plural of matrix) are as follows.

$$\overset{3 \times 2}{\begin{bmatrix} 3 & 2 \\ -1 & 4 \\ 5 & 7 \end{bmatrix}} \quad \overset{2 \times 2}{\begin{bmatrix} 4 & 1 \\ 0 & -5 \end{bmatrix}} \quad \overset{1 \times 4}{\begin{bmatrix} 1 & 2 & 6 & 8 \end{bmatrix}} \quad \overset{5 \times 1}{\begin{bmatrix} 3 \\ 7 \\ 10 \\ 2 \\ -4 \end{bmatrix}}$$

In general, *a* matrix of *m* rows and *n* columns is called a matrix of dimension $m \times n$.

With every system of linear equations we can associate a matrix that consists of the coefficients and constant terms. For example, with the system

$$\begin{pmatrix} x - 3y = -17 \\ 2x + 7y = 31 \end{pmatrix}$$

we can associate the matrix

$$\begin{bmatrix} 1 & -3 & \vdots & -17 \\ 2 & 7 & \vdots & 31 \end{bmatrix},$$

which is called the **augmented matrix** of the system. The dashed line separates the coefficients from the constant terms; technically it is not necessary.

Since augmented matrices represent systems of equations, we can operate with them as we do with systems of equations. Our previous work with systems of equations was based on the following properties.

1. Any two equations of a system may be interchanged.

Example By interchanging the two equations, the system $\begin{pmatrix} 2x - 5y = 9 \\ x + 3y = 4 \end{pmatrix}$ is equivalent to the system $\begin{pmatrix} x + 3y = 4 \\ 2x - 5y = 9 \end{pmatrix}$.

2. Any equation of the system may be multiplied by a nonzero constant.

Example By multiplying the top equation by -2, the system

$$\begin{pmatrix} x + 3y = 4 \\ 2x - 5y = 9 \end{pmatrix}$$

is equivalent to the system $\begin{pmatrix} -2x - 6y = -8 \\ 2x - 5y = 9 \end{pmatrix}$.

3. Any equation of the system can be replaced by adding a nonzero multiple of another equation to that equation.

Example By adding -2 times the first equation to the second equation, the system $\begin{pmatrix} x + 3y = 4 \\ 2x - 5y = 9 \end{pmatrix}$ is equivalent to the system $\begin{pmatrix} x + 3y = 4 \\ -11y = 1 \end{pmatrix}$.

Each of the properties geared to solving a system of equations produces a corresponding property of the augmented matrix of the system. For example, exchanging two equations of a system corresponds to exchanging two rows of the augmented matrix that represents the system. We usually refer to these properties as *elementary row operations* and we can state them as follows.

Elementary Row Operations

1. Any two rows of an augmented matrix can be interchanged.

2. Any row can be multiplied by a nonzero constant.

3. Any row of the augmented matrix can be replaced by adding a nonzero multiple of another row to that row.

Using the elementary row operations on an augmented matrix provides a basis for solving systems of linear equations. Study the following examples very carefully; keep in mind that the general scheme, called **Gaussian elimination**, is one of using elementary row operations on a matrix to continue replacing a system of equations with an equivalent system until a system is obtained where the solutions are easily determined. We will use a format similar to the one we used in the previous section except that we will represent systems of equations by matrices.

EXAMPLE 1   Solve the system $\left(\begin{matrix} x - 3y = -17 \\ 2x + 7y = 31 \end{matrix} \right)$.

Solution   The augmented matrix of the system is

$$\begin{bmatrix} 1 & -3 & \vdots & -17 \\ 2 & 7 & \vdots & 31 \end{bmatrix}.$$

We can multiply row one by -2 and add this result to row two to produce a new row two.

$$\begin{bmatrix} 1 & -3 & \vdots & -17 \\ 0 & 13 & \vdots & 65 \end{bmatrix}$$

This matrix represents the system

$$\left(\begin{matrix} x - 3y = -17 \\ 13y = 65 \end{matrix} \right).$$

From the last equation we can determine the value of y.

$$13y = 65$$

$$y = 5$$

Now we can substitute 5 for y in the equation $x - 3y = -17$.

$$x - 3(5) = -17$$

$$x - 15 = -17$$

$$x = -2$$

The solution set is $\{(-2, 5)\}$.   ■

EXAMPLE 2   Solve the system $\left(\begin{matrix} 3x + 2y = 3 \\ 30x - 6y = 17 \end{matrix} \right)$.

Solution   The augmented matrix of the system is

$$\begin{bmatrix} 3 & 2 & \vdots & 3 \\ 30 & -6 & \vdots & 17 \end{bmatrix}.$$

We can multiply row one by -10 and add this result to row two to produce a new row two.

$$\begin{bmatrix} 3 & 2 & \vdots & 3 \\ 0 & -26 & \vdots & -13 \end{bmatrix}$$

This matrix represents the system

$$\left(\begin{matrix} 3x + 2y = 3 \\ -26y = -13 \end{matrix} \right).$$

From the last equation we can determine the value of y.

$$-26y = -13$$

$$y = \frac{-13}{-26} = \frac{1}{2}$$

Now we can substitute $\frac{1}{2}$ for y in the equation $3x + 2y = 3$.

$$3x + 2\left(\frac{1}{2}\right) = 3$$

$$3x + 1 = 3$$

$$3x = 2$$

$$x = \frac{2}{3}$$

The solution set is $\left\{\left(\frac{2}{3}, \frac{1}{2}\right)\right\}$.   ■

It should be evident from Examples 1 and 2 that the matrix approach does not provide us with much extra power for solving systems of two linear equations in two unknowns. However, as the systems become larger, the compactness of the matrix approach becomes more convenient.

Suppose that we had the following system to solve.

$$\begin{pmatrix} 5x - 3y - 2z = & -1 \\ 4y + 7z = & 3 \\ 4z = & -12 \end{pmatrix}$$

It would be easy to determine z from the last equation and then substitute that value in the second equation to determine y. The values for y and z could then be substituted in the first equation to determine x. In other words, the form of this system makes it convenient to solve. We say that the system is in **triangular form**; notice the location of zeros in its augmented matrix

$$\begin{bmatrix} 5 & -3 & -2 & \vdots & -1 \\ 0 & 4 & 7 & \vdots & 3 \\ 0 & 0 & 4 & \vdots & -12 \end{bmatrix}.$$

EXAMPLE 3   Solve the system $\begin{pmatrix} 2x - 3y - z = -2 \\ x - 2y + 3z = 9 \\ 3x + y - 5z = -8 \end{pmatrix}$.

Solution

The augmented matrix of the system is

$$\begin{bmatrix} 2 & -3 & -1 & \vdots & -2 \\ 1 & -2 & 3 & \vdots & 9 \\ 3 & 1 & -5 & \vdots & -8 \end{bmatrix}.$$

Let's begin by interchanging the top two rows.

$$\begin{bmatrix} 1 & -2 & 3 & \vdots & 9 \\ 2 & -3 & -1 & \vdots & -2 \\ 3 & 1 & -5 & \vdots & -8 \end{bmatrix}$$

Now we can multiply row one by -2 and add this result to row two to produce a new row two. Also we can multiply row one by -3 and add this result to row three to produce a new row three.

$$\begin{bmatrix} 1 & -2 & 3 & \vdots & 9 \\ 0 & 1 & -7 & \vdots & -20 \\ 0 & 7 & -14 & \vdots & -35 \end{bmatrix}$$

Now we can multiply row two by -7 and add this result to row three to produce a new row three.

$$\begin{bmatrix} 1 & -2 & 3 & \vdots & 9 \\ 0 & 1 & -7 & \vdots & -20 \\ 0 & 0 & 35 & \vdots & 105 \end{bmatrix}$$

This last matrix represents the system $\begin{pmatrix} x - 2y + 3z = & 9 \\ y - 7z = & -20 \\ 35z = & 105 \end{pmatrix}$, which is in **triangular form**. We can use the third equation to determine the value of z.

$$35z = 105$$

$$z = 3$$

Now we can substitute 3 for z in the second equation.

$$y - 7z = -20$$

$$y - 7(3) = -20$$

$$y - 21 = -20$$

$$y = \quad 1$$

Finally, we can substitute 3 for z and 1 for y in the first equation.

$$x - 2y + 3z = 9$$
$$x - 2(1) + 3(3) = 9$$
$$x - 2 + 9 = 9$$
$$x + 7 = 9$$
$$x = 2$$

The solution set is $\{(2, 1, 3)\}$.   ∎

EXAMPLE 4   Solve the system $\begin{pmatrix} x + 3y - 4z = & 29 \\ 2x + 4y - 5z = & 37 \\ 5x - y + 3z = & -20 \end{pmatrix}$.

Solution   The augmented matrix is

$$\begin{bmatrix} 1 & 3 & -4 & \vdots & 29 \\ 2 & 4 & -5 & \vdots & 37 \\ 5 & -1 & 3 & \vdots & -20 \end{bmatrix}.$$

Now we can get zeros in the first column beneath the 1 by multiplying -2 times row 1 and adding the result to row 2 and multiplying -5 times row 1 and adding the result to row 3.

$$\begin{bmatrix} 1 & 3 & -4 & \vdots & 29 \\ 0 & -2 & 3 & \vdots & -21 \\ 0 & -16 & 23 & \vdots & -165 \end{bmatrix}$$

Finally, we can multiply row 2 by -8 and add this result to row 3.

$$\begin{bmatrix} 1 & 3 & -4 & \vdots & 29 \\ 0 & -2 & 3 & \vdots & -21 \\ 0 & 0 & -1 & \vdots & 3 \end{bmatrix}$$

The last row represents the equation $-z = 3$ from which we obtain $z = -3$. Now we can substitute -3 for z in the equation represented by the second row.

$$-2y + 3z = -21$$
$$-2y + 3(-3) = -21$$
$$-2y - 9 = -21$$
$$-2y = -12$$
$$y = 6$$

Now we can substitute -3 for z and 6 for y in the equation represented by the first row.

$$x + 3y - 4z = 29$$
$$x + 3(6) - 4(-3) = 29$$
$$x + 18 + 12 = 29$$
$$x + 30 = 29$$
$$x = -1$$

The solution set of the original system is $\{(-1, 6, -3)\}$. ∎

EXAMPLE 5   Solve the system $\begin{pmatrix} x - 2y + 3z = 3 \\ 5x - 9y + 4z = 2 \\ 2x - 4y + 6z = -1 \end{pmatrix}$.

Solution   The augmented matrix of the system is

$$\begin{bmatrix} 1 & -2 & 3 & \vdots & 3 \\ 5 & -9 & 4 & \vdots & 2 \\ 2 & -4 & 6 & \vdots & -1 \end{bmatrix}.$$

We can get zeros below the 1 in the first column by multiplying -5 times row 1 and adding the result to row 2 and multiplying -2 times row 1 and adding the result to row 3.

$$\begin{bmatrix} 1 & -2 & 3 & \vdots & 3 \\ 0 & 1 & -11 & \vdots & -13 \\ 0 & 0 & 0 & \vdots & -7 \end{bmatrix}$$

At this point we can stop, because the bottom row of the matrix represents the statement $0(x) + 0(y) + 0(z) = -7$, which is obviously a false statement for all values of x, y, and z. Thus, the original system is *inconsistent*; its solution set is $\varnothing$. ∎

EXAMPLE 6   Solve the system $\begin{pmatrix} x + 2y + 2z = 9 \\ x + 3y - 4z = 5 \\ 2x + 5y - 2z = 14 \end{pmatrix}$.

Solution   The augmented matrix of the system is

$$\begin{bmatrix} 1 & 2 & 2 & \vdots & 9 \\ 1 & 3 & -4 & \vdots & 5 \\ 2 & 5 & -2 & \vdots & 14 \end{bmatrix}.$$

We can get zeros in the first column below the 1 in the upper left-hand corner by multiplying -1 times row 1 and adding the result to row 2, and multiplying -2 times row 1 and adding the result to row 3.

$$\begin{bmatrix} 1 & 2 & 2 & \vdots & 9 \\ 0 & 1 & -6 & \vdots & -4 \\ 0 & 1 & -6 & \vdots & -4 \end{bmatrix}$$

Now multiply row 2 by -1 and add this result to row 3.

$$\begin{bmatrix} 1 & 2 & 2 & \vdots & 9 \\ 0 & 1 & -6 & \vdots & -4 \\ 0 & 0 & 0 & \vdots & 0 \end{bmatrix}$$

The bottom row of zeros represents the statement $0(x) + 0(y) + 0(z) = 0$, which is true for all values of x, y, and z. Thus, the original system is a *dependent system*, and its solution set consists of infinitely many ordered triples. ■

Problem Set 14.2

Solve each of the following systems and use matrices as we did in the examples of this section.

1. $\begin{pmatrix} x - 2y = 14 \\ 4x + 5y = 4 \end{pmatrix}$

2. $\begin{pmatrix} x + 5y = -3 \\ 3x - 2y = -26 \end{pmatrix}$

3. $\begin{pmatrix} 3x + 7y = -40 \\ x + 4y = -20 \end{pmatrix}$

4. $\begin{pmatrix} 7x - 9y = 53 \\ x - 3y = 11 \end{pmatrix}$

5. $\begin{pmatrix} x - 3y = 4 \\ 4x - 5y = 3 \end{pmatrix}$

6. $\begin{pmatrix} x + 3y = 7 \\ 2x - 4y = 9 \end{pmatrix}$

7. $\begin{pmatrix} 6x + 7y = -15 \\ 4x - 9y = 31 \end{pmatrix}$

8. $\begin{pmatrix} 5x - 3y = -16 \\ 6x + 5y = -2 \end{pmatrix}$

9. $\begin{pmatrix} x + 3y - 4z = 5 \\ -2x - 5y + z = 9 \\ 7x - y - z = -2 \end{pmatrix}$

10. $\begin{pmatrix} x - y + 5z = -2 \\ -3x + 2y + z = 17 \\ 4x - 5y - 3z = -36 \end{pmatrix}$

11. $\begin{pmatrix} x - 2y - 3z = -11 \\ 2x - 3y + z = 7 \\ -3x - 5y + 7z = 14 \end{pmatrix}$

12. $\begin{pmatrix} x + y + 3z = -8 \\ 3x + 2y - 5z = 19 \\ 5x - y - 4z = 23 \end{pmatrix}$

13. $\begin{pmatrix} y + 3z = -3 \\ 2x - 5z = 18 \\ 3x - y + 2z = 5 \end{pmatrix}$

14. $\begin{pmatrix} x - z = -1 \\ -2x + y + 3z = 4 \\ 3x - 4y = 31 \end{pmatrix}$

15. $\begin{pmatrix} -x - 5y + 2z = -5 \\ 3x + 14y - z = 13 \\ 4x - 3y + 5z = -26 \end{pmatrix}$

16. $\begin{pmatrix} -x - 3y + 4z = 3 \\ 3x + 8y - z = 27 \\ 5x - y + 2z = -5 \end{pmatrix}$

17. $\begin{pmatrix} x + 2y - z = -5 \\ 3x + 4y + 2z = -8 \\ -2x - y + 5z = 10 \end{pmatrix}$

18. $\begin{pmatrix} x - 3y + 2z = 0 \\ 2x - 4y - 3z = 19 \\ -3x - y + z = -11 \end{pmatrix}$

19. $\begin{pmatrix} -3x + 2y - z = 12 \\ 5x + 2y - 3z = 6 \\ x - y + 5z = -10 \end{pmatrix}$

20. $\begin{pmatrix} -2x - 3y + 5z = 15 \\ 4x - y + 2z = -4 \\ x + y - 3z = -7 \end{pmatrix}$

21. $\begin{pmatrix} -2x + 5y - z = -1 \\ 4x + y - 5z = 23 \\ x - 2y + 3z = -7 \end{pmatrix}$

22. $\begin{pmatrix} 2x + 5y + z = 1 \\ x + 2y - 3z = -13 \\ 3x - y - 2z = -4 \end{pmatrix}$

23. $\begin{pmatrix} x + 2y - 5z = -1 \\ 2x + 3y - 2z = 2 \\ 3x + 5y - 7z = 4 \end{pmatrix}$

24. $\begin{pmatrix} x - y + 2z = 1 \\ -3x + 4y - z = 4 \\ -x + 2y + 3z = 6 \end{pmatrix}$

Miscellaneous Problems

25. Solve the system $\begin{pmatrix} x - 3y - 2z + w = -3 \\ -2x + 7y + z - 2w = -1 \\ 3x - 7y - 3z + 3w = -5 \\ 5x + y + 4z - 2w = 18 \end{pmatrix}$.

26. Solve the system $\begin{pmatrix} x - 2y + 2z - w = -2 \\ -3x + 5y - z - 3w = 2 \\ 2x + 3y + 3z + 5w = -9 \\ 4x - y - z - 2w = 8 \end{pmatrix}$.

14.3

Determinants

A **square matrix** is one that has the same number of rows as columns. Associated with each square matrix that has real number entries is a real number called the **determinant** of the matrix. For a 2×2 matrix

$$\begin{bmatrix} a_1 & b_1 \\ a_2 & b_2 \end{bmatrix},$$

the determinant is written as

$$\begin{vmatrix} a_1 & b_1 \\ a_2 & b_2 \end{vmatrix}$$

and defined by

$$\begin{vmatrix} a_1 & b_1 \\ a_2 & b_2 \end{vmatrix} = a_1 b_2 - a_2 b_1. \tag{1}$$

Notice that a determinant is simply a number and the determinant notation used on the left side of equation (1) is a way of expressing the number on the right side.

EXAMPLE 1   Find the determinant of the matrix $\begin{bmatrix} 3 & -2 \\ 5 & 8 \end{bmatrix}$.

Solution   In this case, $a_1 = 3$, $b_1 = -2$, $a_2 = 5$, and $b_2 = 8$. Thus, we have

$$\begin{vmatrix} 3 & -2 \\ 5 & 8 \end{vmatrix} = 3(8) - 5(-2) = 24 + 10 = 34.$$ ∎

Finding the determinant of a square matrix is commonly called *evaluating the determinant* and the matrix notation is sometimes omitted.

EXAMPLE 2   Evaluate $\begin{vmatrix} -3 & 5 \\ 1 & 2 \end{vmatrix}$.

Solution $$\begin{vmatrix} -3 & 5 \\ 1 & 2 \end{vmatrix} = -3(2) - 1(5) = -11.$$ ∎

Cramer's Rule

Determinants provide the basis for another method of solving linear systems. Consider the system

$$\left(\begin{matrix} a_1 x + b_1 y = c_1 \\ a_2 x + b_2 y = c_2 \end{matrix} \right). \qquad \begin{matrix} (1) \\ (2) \end{matrix}$$

We shall solve this system by using the elimination method; observe that our solutions can be conveniently written in determinant form. To solve for x we can multiply equation (1) by b_2 and equation (2) by $-b_1$ and then add.

$$a_1 b_2 x + b_1 b_2 y = c_1 b_2$$

$$-a_2 b_1 x - b_1 b_2 y = -c_2 b_1$$

$$\overline{a_1 b_2 x - a_2 b_1 x = c_1 b_2 - c_2 b_1}$$

$$(a_1 b_2 - a_2 b_1)x = c_1 b_2 - c_2 b_1$$

$$x = \frac{c_1 b_2 - c_2 b_1}{a_1 b_2 - a_2 b_1}. \qquad \text{if } a_1 b_2 - a_2 b_1 \neq 0$$

To solve for y we can multiply equation (1) by $-a_2$ and equation (2) by a_1 and add.

$$-a_1 a_2 x - a_2 b_1 y = -a_2 c_1$$

$$a_1 a_2 x + a_1 b_2 y = a_1 c_2$$

$$\overline{ a_1 b_2 y - a_2 b_1 y = a_1 c_2 - a_2 c_1}$$

$$(a_1 b_2 - a_2 b_1)y = a_1 c_2 - a_2 c_1$$

$$y = \frac{a_1 c_2 - a_2 c_1}{a_1 b_2 - a_2 b_1}. \qquad \text{if } a_1 b_2 - a_2 b_1 \neq 0$$

We can express the solutions for x and y in determinant form as follows.

$$x = \frac{c_1 b_2 - c_2 b_1}{a_1 b_2 - a_2 b_1} = \frac{\begin{vmatrix} c_1 & b_1 \\ c_2 & b_2 \end{vmatrix}}{\begin{vmatrix} a_1 & b_1 \\ a_2 & b_2 \end{vmatrix}} \qquad\qquad y = \frac{a_1 c_2 - a_2 c_1}{a_1 b_2 - a_2 b_1} = \frac{\begin{vmatrix} a_1 & c_1 \\ a_2 & c_2 \end{vmatrix}}{\begin{vmatrix} a_1 & b_1 \\ a_2 & b_2 \end{vmatrix}}$$

For convenience, we shall denote the three determinants in the solution as

$$\begin{vmatrix} a_1 & b_1 \\ a_2 & b_2 \end{vmatrix} = D, \qquad \begin{vmatrix} c_1 & b_1 \\ c_2 & b_2 \end{vmatrix} = D_x, \qquad \begin{vmatrix} a_1 & c_1 \\ a_2 & c_2 \end{vmatrix} = D_y.$$

Notice that the elements of D are the coefficients of the variables in the given system. In D_x, we obtain the elements by replacing the coefficients of x with the respective constants. In D_y, we replace the coefficients of y with the respective constants. This method of using determinants to solve a system of two linear equations in two variables is called **Cramer's Rule** and can be stated as follows.

Cramer's Rule

Given the system

$$\begin{pmatrix} a_1 x + b_1 y = c_1 \\ a_2 x + b_2 y = c_2 \end{pmatrix} \quad \text{with } a_1 b_2 - a_2 b_1 \neq 0,$$

then

$$x = \frac{\begin{vmatrix} c_1 & b_1 \\ c_2 & b_2 \end{vmatrix}}{\begin{vmatrix} a_1 & b_1 \\ a_2 & b_2 \end{vmatrix}} = \frac{D_x}{D} \quad \text{and} \quad y = \frac{\begin{vmatrix} a_1 & c_1 \\ a_2 & c_2 \end{vmatrix}}{\begin{vmatrix} a_1 & b_1 \\ a_2 & b_2 \end{vmatrix}} = \frac{D_y}{D}.$$

Let's use Cramer's Rule to solve some systems.

EXAMPLE 3   Solve the system $\begin{pmatrix} x + 2y = 11 \\ 2x - y = 2 \end{pmatrix}$.

Solution   Let's find D, D_x, and D_y.

$$D = \begin{vmatrix} 1 & 2 \\ 2 & -1 \end{vmatrix} = -1 - 4 = -5,$$

$$D_x = \begin{vmatrix} 11 & 2 \\ 2 & -1 \end{vmatrix} = -11 - 4 = -15,$$

$$D_y = \begin{vmatrix} 1 & 11 \\ 2 & 2 \end{vmatrix} = 2 - 22 = -20$$

Thus, we have

$$x = \frac{D_x}{D} = \frac{-15}{-5} = 3,$$

$$y = \frac{D_y}{D} = \frac{-20}{-5} = 4.$$

The solution set is $\{(3, 4)\}$ which can be verified, as always, by substituting back into the original equations.   ■

REMARK Notice that Cramer's Rule has a restriction, namely, $a_1 b_2 - a_2 b_1 \neq 0$; that is, $D \neq 0$. Thus, it is a good idea to find D first. Then if $D = 0$, Cramer's Rule does not apply and you can revert to one of the other methods to determine whether the solution set is empty or has infinitely many solutions.

EXAMPLE 4   Solve the system $\begin{pmatrix} 2x - 3y = -8 \\ 3x + 5y = 7 \end{pmatrix}$.

Solution
$$D = \begin{vmatrix} 2 & -3 \\ 3 & 5 \end{vmatrix} = 10 - (-9) = 19,$$

$$D_x = \begin{vmatrix} -8 & -3 \\ 7 & 5 \end{vmatrix} = -40 - (-21) = -19,$$

$$D_y = \begin{vmatrix} 2 & -8 \\ 3 & 7 \end{vmatrix} = 14 - (-24) = 38$$

Thus, we obtain

$$x = \frac{D_x}{D} = \frac{-19}{19} = -1 \quad \text{and} \quad y = \frac{D_y}{D} = \frac{38}{19} = 2.$$

The solution set is $\{(-1, 2)\}$.   ■

EXAMPLE 5   Solve the system $\begin{pmatrix} y = -2x - 2 \\ 4x - 5y = 17 \end{pmatrix}$.

Solution   First, we must change the form of the first equation so that the system fits the form given in Cramer's Rule. The equation $y = -2x - 2$ can be written as $2x + y = -2$. The system now becomes

$$\begin{pmatrix} 2x + \ y = -2 \\ 4x - 5y = \ 17 \end{pmatrix}$$

and we can proceed as before.

$$D = \begin{vmatrix} 2 & 1 \\ 4 & -5 \end{vmatrix} = -10 - 4 = -14,$$

$$D_x = \begin{vmatrix} -2 & 1 \\ 17 & -5 \end{vmatrix} = 10 - 17 = -7,$$

$$D_y = \begin{vmatrix} 2 & -2 \\ 4 & 17 \end{vmatrix} = 34 - (-8) = 42$$

Thus, the solutions are

$$x = \frac{D_x}{D} = \frac{-7}{-14} = \frac{1}{2} \quad \text{and} \quad y = \frac{D_y}{D} = \frac{42}{-14} = -3.$$

The solution set is $\left\{\left(\dfrac{1}{2}, -3\right)\right\}$.   ■

Problem Set 14.3

Evaluate each of the following determinants.

1. $\begin{vmatrix} 6 & 2 \\ 4 & 3 \end{vmatrix}$

2. $\begin{vmatrix} 7 & 6 \\ 2 & 5 \end{vmatrix}$

3. $\begin{vmatrix} 4 & 7 \\ 8 & 2 \end{vmatrix}$

4. $\begin{vmatrix} 3 & 9 \\ 6 & 4 \end{vmatrix}$

5. $\begin{vmatrix} -3 & 2 \\ 7 & 5 \end{vmatrix}$

6. $\begin{vmatrix} 5 & 1 \\ 8 & -4 \end{vmatrix}$

7. $\begin{vmatrix} 8 & -3 \\ 6 & 4 \end{vmatrix}$

8. $\begin{vmatrix} 5 & 9 \\ -3 & 6 \end{vmatrix}$

9. $\begin{vmatrix} -3 & 2 \\ 5 & -6 \end{vmatrix}$

10. $\begin{vmatrix} -2 & 4 \\ 9 & -7 \end{vmatrix}$

11. $\begin{vmatrix} 3 & -3 \\ -6 & 8 \end{vmatrix}$

12. $\begin{vmatrix} 6 & -5 \\ -8 & 12 \end{vmatrix}$

13. $\begin{vmatrix} -7 & -2 \\ -2 & 4 \end{vmatrix}$

14. $\begin{vmatrix} 6 & -1 \\ -8 & -3 \end{vmatrix}$

15. $\begin{vmatrix} -2 & -3 \\ -4 & -5 \end{vmatrix}$

16. $\begin{vmatrix} -9 & -7 \\ -6 & -4 \end{vmatrix}$

17. $\begin{vmatrix} \dfrac{1}{4} & -2 \\ \dfrac{3}{2} & 8 \end{vmatrix}$

18. $\begin{vmatrix} -\dfrac{2}{3} & 10 \\ -\dfrac{1}{2} & 6 \end{vmatrix}$

19. $\begin{vmatrix} \dfrac{3}{2} & -\dfrac{1}{2} \\ 1 & -\dfrac{2}{5} \\ 2 & \end{vmatrix}$

20. $\begin{vmatrix} -\dfrac{1}{4} & \dfrac{1}{3} \\ \dfrac{3}{4} & \dfrac{2}{3} \end{vmatrix}$

Use Cramer's Rule to find the solution set for each of the following systems.

21. $\begin{pmatrix} 2x + y = 14 \\ 3x - y = 1 \end{pmatrix}$

22. $\begin{pmatrix} 4x - y = 11 \\ 2x + 3y = 23 \end{pmatrix}$

23. $\begin{pmatrix} -x + 3y = 17 \\ 4x - 5y = -33 \end{pmatrix}$

24. $\begin{pmatrix} 5x + 2y = -15 \\ 7x - 3y = 37 \end{pmatrix}$

25. $\begin{pmatrix} 9x + 5y = -8 \\ 7x - 4y = -22 \end{pmatrix}$

26. $\begin{pmatrix} 8x - 11y = 3 \\ -x + 4y = -3 \end{pmatrix}$

27. $\begin{pmatrix} x + 5y = 4 \\ 3x + 15y = -1 \end{pmatrix}$

28. $\begin{pmatrix} 4x - 7y = 0 \\ 7x + 2y = 0 \end{pmatrix}$

29. $\begin{pmatrix} 6x - y = 0 \\ 5x + 4y = 29 \end{pmatrix}$

30. $\begin{pmatrix} 3x - 4y = 2 \\ 9x - 12y = 6 \end{pmatrix}$

31. $\begin{pmatrix} -4x + 3y = 3 \\ 4x - 6y = -5 \end{pmatrix}$

32. $\begin{pmatrix} x - 2y = -1 \\ x = -6y + 5 \end{pmatrix}$

33. $\begin{pmatrix} 6x - 5y = 1 \\ 4x + 7y = 2 \end{pmatrix}$

34. $\begin{pmatrix} y = 3x + 5 \\ y = 6x + 6 \end{pmatrix}$

35. $\begin{pmatrix} 7x + 2y = -1 \\ y = -x + 2 \end{pmatrix}$

36. $\begin{pmatrix} 9x - y = -2 \\ y = 4 - 8x \end{pmatrix}$

37. $\begin{pmatrix} -\dfrac{2}{3}x + \dfrac{1}{2}y = -7 \\ \dfrac{1}{3}x - \dfrac{3}{2}y = 6 \end{pmatrix}$

38. $\begin{pmatrix} \dfrac{1}{2}x + \dfrac{2}{3}y = -6 \\ \dfrac{1}{4}x - \dfrac{1}{3}y = -1 \end{pmatrix}$

39. $\begin{pmatrix} x + \dfrac{2}{3}y = -6 \\ -\dfrac{1}{4}x + 3y = -8 \end{pmatrix}$

40. $\begin{pmatrix} 3x - \dfrac{1}{2}y = 6 \\ -2x + \dfrac{1}{3}y = -4 \end{pmatrix}$

Thoughts into Words

41. Why does the solution set for the system $\begin{pmatrix} x = 2y + 4 \\ 3x - 6y = 12 \end{pmatrix}$ contain infinitely many ordered pairs as solutions? Find at least five of these ordered pairs.

42. Explain how you would solve the system $\begin{pmatrix} 2x + 4y = 10 \\ 5x - 7y = -2 \end{pmatrix}$.

43. Explain the difference between a matrix and a determinant.

Miscellaneous Problems

44. Verify each of the following. The variables represent real numbers.

(a) $\begin{vmatrix} a & b \\ a & b \end{vmatrix} = 0$

(b) $\begin{vmatrix} a & a \\ b & b \end{vmatrix} = 0$

(c) $\begin{vmatrix} a & b \\ c & d \end{vmatrix} = -\begin{vmatrix} b & a \\ d & c \end{vmatrix}$

(d) $\begin{vmatrix} a & b \\ c & d \end{vmatrix} = -\begin{vmatrix} c & d \\ a & b \end{vmatrix}$

(e) $k\begin{vmatrix} a & b \\ c & d \end{vmatrix} = \begin{vmatrix} ka & b \\ kc & d \end{vmatrix}$

(f) $k\begin{vmatrix} a & b \\ c & d \end{vmatrix} = \begin{vmatrix} ka & kb \\ c & d \end{vmatrix}$

14.4

3 x 3 Determinants and Cramer's Rule

This section will extend the concept of a determinant to include 3×3 determinants and then also extend the use of determinants to solve systems of three linear equations in three variables.

For a 3×3 matrix

$$\begin{bmatrix} a_1 & b_1 & c_1 \\ a_2 & b_2 & c_2 \\ a_3 & b_3 & c_3 \end{bmatrix},$$

the determinant is written as

$$\begin{vmatrix} a_1 & b_1 & c_1 \\ a_2 & b_2 & c_2 \\ a_3 & b_3 & c_3 \end{vmatrix}$$

and defined by

$$\begin{vmatrix} a_1 & b_1 & c_1 \\ a_2 & b_2 & c_2 \\ a_3 & b_3 & c_3 \end{vmatrix} = a_1 b_2 c_3 + b_1 c_2 a_3 + c_1 a_2 b_3 - a_3 b_2 c_1 - b_3 c_2 a_1 - c_3 a_2 b_1. \quad (1)$$

It is evident that the definition given by equation (1) is a bit complicated to be very useful in practice. Fortunately, there is a method, called **expansion of a determinant by minors**, that we can use to calculate such a determinant.

The **minor** of an element in a determinant is the determinant that remains after deleting the row and column in which the element appears. For example, consider the determinant of equation (1).

The minor of a_1 is $\begin{vmatrix} b_2 & c_2 \\ b_3 & c_3 \end{vmatrix}$,

The minor of a_2 is $\begin{vmatrix} b_1 & c_1 \\ b_3 & c_3 \end{vmatrix}$.

The minor of a_3 is $\begin{vmatrix} b_1 & c_1 \\ b_2 & c_2 \end{vmatrix}$

Now let's consider the terms, in pairs, of the right side of equation (1) and show the tie-up with minors.

$$a_1 b_2 c_3 - b_3 c_2 a_1 = a_1(b_2 c_3 - b_3 c_2).$$

$$= a_1 \begin{vmatrix} b_2 & c_2 \\ b_3 & c_3 \end{vmatrix},$$

$$c_1 a_2 b_3 - c_3 a_2 b_1 = -(c_3 a_2 b_1 - c_1 a_2 b_3)$$

$$= -a_2(b_1 c_3 - b_3 c_1).$$

$$= -a_2 \begin{vmatrix} b_1 & c_1 \\ b_3 & c_3 \end{vmatrix},$$

$$b_1 c_2 a_3 - a_3 b_2 c_1 = a_3(b_1 c_2 - b_2 c_1).$$

$$= a_3 \begin{vmatrix} b_1 & c_1 \\ b_2 & c_2 \end{vmatrix}$$

Therefore, we have

$$\begin{vmatrix} a_1 & b_1 & c_1 \\ a_2 & b_2 & c_2 \\ a_3 & b_3 & c_3 \end{vmatrix} = a_1 \begin{vmatrix} b_2 & c_2 \\ b_3 & c_3 \end{vmatrix} - a_2 \begin{vmatrix} b_1 & c_1 \\ b_3 & c_3 \end{vmatrix} + a_3 \begin{vmatrix} b_1 & c_1 \\ b_2 & c_2 \end{vmatrix},$$

and this is called the **expansion of the determinant by minors about the first column**.

EXAMPLE 1   Evaluate $\begin{vmatrix} 1 & 2 & -1 \\ 3 & 1 & -2 \\ 2 & 4 & 3 \end{vmatrix}$ by expanding by minors about the first column.

Solution
$$\begin{vmatrix} 1 & 2 & -1 \\ 3 & 1 & -2 \\ 2 & 4 & 3 \end{vmatrix} = 1 \begin{vmatrix} 1 & -2 \\ 4 & 3 \end{vmatrix} - 3 \begin{vmatrix} 2 & -1 \\ 4 & 3 \end{vmatrix} + 2 \begin{vmatrix} 2 & -1 \\ 1 & -2 \end{vmatrix}$$

$$= 1(3 - (-8)) - 3(6 - (-4)) + 2(-4 - (-1))$$

$$= 1(11) - 3(10) + 2(-3) = -25 \qquad \blacksquare$$

It is possible to expand a determinant by minors about *any row* or *any column.* To help determine the signs of the terms in the expansion, the following *sign array* is very useful.

$$+ - +$$
$$- + -$$
$$+ - +$$

For example, let's expand the determinant in Example 1 by minors about the *second row.* The second row in the sign array is $- + -$. Therefore,

$$\begin{vmatrix} 1 & 2 & -1 \\ 3 & 1 & -2 \\ 2 & 4 & 3 \end{vmatrix} = -3 \begin{vmatrix} 2 & -1 \\ 4 & 3 \end{vmatrix} + 1 \begin{vmatrix} 1 & -1 \\ 2 & 3 \end{vmatrix} - (-2) \begin{vmatrix} 1 & 2 \\ 2 & 4 \end{vmatrix}$$

$$= -3(6 - (-4)) + 1(3 - (-2)) + 2(4 - 4)$$

$$= -3(10) + 1(5) + 2(0)$$

$$= -25.$$

Your decision as to which row or column to use for expanding a particular determinant by minors may depend upon the numbers involved in the determinant. A row or column with one or more zeros is frequently a good choice, as the next example illustrates.

EXAMPLE 2   Evaluate $\begin{vmatrix} 3 & -1 & 4 \\ 5 & 2 & 0 \\ -2 & 6 & 0 \end{vmatrix}$.

Solution   Since the third column has two zeros, we shall expand about it.

$$\begin{vmatrix} 3 & -1 & 4 \\ 5 & 2 & 0 \\ -2 & 6 & 0 \end{vmatrix} = 4 \begin{vmatrix} 5 & 2 \\ -2 & 6 \end{vmatrix} - 0 \begin{vmatrix} 3 & -1 \\ -2 & 6 \end{vmatrix} + 0 \begin{vmatrix} 3 & -1 \\ 5 & 2 \end{vmatrix}$$

$$= 4(30 - (-4)) - 0 + 0 = 136$$

(Notice that because of the zeros there is no need to evaluate the last two minors.) ∎

REMARK 1 The expansion-by-minors method can be extended to determinants of size 4×4, 5×5, and so on. However, it should be obvious that it becomes increasingly more tedious with "bigger" determinants. Fortunately, the computer handles the calculation of such determinants with a different technique.

REMARK 2 There is another method for evaluating 3×3 determinants. This method is demonstrated in Problem 34 of the next problem set. If you choose to use that method, keep in mind *that it only works for 3 × 3 determinants.*

Without showing all of the details, we will simply state that *Cramer's Rule* also applies to solving systems of three linear equations in three variables. It can be stated as follows.

Cramer's Rule

Given the system

$$\begin{cases} a_1x + b_1y + c_1z = d_1 \\ a_2x + b_2y + c_2z = d_2 \\ a_3x + b_3y + c_3z = d_3 \end{cases}$$

with $D = \begin{vmatrix} a_1 & b_1 & c_1 \\ a_2 & b_2 & c_2 \\ a_3 & b_3 & c_3 \end{vmatrix} \neq 0,$ $D_x = \begin{vmatrix} d_1 & b_1 & c_1 \\ d_2 & b_2 & c_2 \\ d_3 & b_3 & c_3 \end{vmatrix},$

$$D_y = \begin{vmatrix} a_1 & d_1 & c_1 \\ a_2 & d_2 & c_2 \\ a_3 & d_3 & c_3 \end{vmatrix}, \qquad D_z = \begin{vmatrix} a_1 & b_1 & d_1 \\ a_2 & b_2 & d_2 \\ a_3 & b_3 & d_3 \end{vmatrix},$$

then $x = \dfrac{D_x}{D}$, $y = \dfrac{D_y}{D}$, and $z = \dfrac{D_z}{D}$.

Notice that the elements of D are the coefficients of the variables in the given system. Then D_x, D_y, and D_z are formed by replacing the elements in the x, y, or z column, respectively, by the constants of the system d_1, d_2, and d_3. Again, note the restriction $D \neq 0$. As before, if $D = 0$, then Cramer's Rule does not apply and you can use the elimination method to determine whether the system has *no solution* or *infinitely many* solutions.

EXAMPLE 3

Use Cramer's Rule to solve the system $\begin{cases} x - 2y + z = -4 \\ 2x + y - z = 5 \\ 3x + 2y + 4z = 3 \end{cases}$.

Solution

To find D, let's expand about *row* 1.

$$D = \begin{vmatrix} 1 & -2 & 1 \\ 2 & 1 & -1 \\ 3 & 2 & 4 \end{vmatrix} = 1\begin{vmatrix} 1 & -1 \\ 2 & 4 \end{vmatrix} - (-2)\begin{vmatrix} 2 & -1 \\ 3 & 4 \end{vmatrix} + 1\begin{vmatrix} 2 & 1 \\ 3 & 2 \end{vmatrix}$$

$$= 1(4 - (-2)) + 2(8 - (-3)) + 1(4 - 3)$$

$$= 1(6) + 2(11) + 1(1) = 29$$

To find D_x, let's expand about *column 3*.

$$D_x = \begin{vmatrix} -4 & -2 & 1 \\ 5 & 1 & -1 \\ 3 & 2 & 4 \end{vmatrix} = 1 \begin{vmatrix} 5 & 1 \\ 3 & 2 \end{vmatrix} - (-1) \begin{vmatrix} -4 & -2 \\ 3 & 2 \end{vmatrix} + 4 \begin{vmatrix} -4 & -2 \\ 5 & 1 \end{vmatrix}$$

$$= 1(10 - 3) + 1(-8 - (-6)) + 4(-4 - (-10))$$

$$= 1(7) + 1(-2) + 4(6)$$

$$= 29$$

To find D_y, let's expand about *row 1*.

$$D_y = \begin{vmatrix} 1 & -4 & 1 \\ 2 & 5 & -1 \\ 3 & 3 & 4 \end{vmatrix} = 1 \begin{vmatrix} 5 & -1 \\ 3 & 4 \end{vmatrix} - (-4) \begin{vmatrix} 2 & -1 \\ 3 & 4 \end{vmatrix} + 1 \begin{vmatrix} 2 & 5 \\ 3 & 3 \end{vmatrix}$$

$$= 1(20 - (-3)) + 4(8 - (-3)) + 1(6 - 15)$$

$$= 1(23) + 4(11) + 1(-9)$$

$$= 58$$

To find D_z, let's expand about *column 1*.

$$D_z = \begin{vmatrix} 1 & -2 & -4 \\ 2 & 1 & 5 \\ 3 & 2 & 3 \end{vmatrix} = 1 \begin{vmatrix} 1 & 5 \\ 2 & 3 \end{vmatrix} - 2 \begin{vmatrix} -2 & -4 \\ 2 & 3 \end{vmatrix} + 3 \begin{vmatrix} -2 & -4 \\ 1 & 5 \end{vmatrix}$$

$$= 1(3 - 10) - 2(-6 - (-8)) + 3(-10 - (-4))$$

$$= 1(-7) - 2(2) + 3(-6)$$

$$= -29$$

Thus,

$$x = \frac{D_x}{D} = \frac{29}{29} = 1,$$

$$y = \frac{D_y}{D} = \frac{58}{29} = 2,$$

$$z = \frac{D_z}{D} = \frac{-29}{29} = -1.$$

The solution set is $\{(1, 2, -1)\}$. (Be sure to check it!) ■

EXAMPLE 4 Use Cramer's Rule to solve the system $\begin{pmatrix} 2x - y + 3z = -17 \\ 3y + z = 5 \\ x - 2y - z = -3 \end{pmatrix}$.

Solution To find D, let's expand about column 1.

$$D = \begin{vmatrix} 2 & -1 & 3 \\ 0 & 3 & 1 \\ 1 & -2 & -1 \end{vmatrix} = 2 \begin{vmatrix} 3 & 1 \\ -2 & -1 \end{vmatrix} - 0 \begin{vmatrix} -1 & 3 \\ -2 & -1 \end{vmatrix} + 1 \begin{vmatrix} -1 & 3 \\ 3 & 1 \end{vmatrix}$$

$$= 2(-3 - (-2)) - 0 + 1(-1 - 9)$$

$$= 2(-1) - 0 - 10 = -12$$

To find D_x, let's expand about column 3.

$$D_x = \begin{vmatrix} -17 & -1 & 3 \\ 5 & 3 & 1 \\ -3 & -2 & -1 \end{vmatrix}$$

$$= 3 \begin{vmatrix} 5 & 3 \\ -3 & -2 \end{vmatrix} - 1 \begin{vmatrix} -17 & -1 \\ -3 & -2 \end{vmatrix} + (-1) \begin{vmatrix} -17 & -1 \\ 5 & 3 \end{vmatrix}$$

$$= 3(-10 - (-9)) - 1(34 - 3) - 1(-51 - (-5))$$

$$= 3(-1) - 1(31) - 1(-46) = 12$$

To find D_y, let's expand about column 1.

$$D_y = \begin{vmatrix} 2 & -17 & 3 \\ 0 & 5 & 1 \\ 1 & -3 & -1 \end{vmatrix} = 2 \begin{vmatrix} 5 & 1 \\ -3 & -1 \end{vmatrix} - 0 \begin{vmatrix} -17 & 3 \\ -3 & -1 \end{vmatrix} + 1 \begin{vmatrix} -17 & 3 \\ 5 & 1 \end{vmatrix}$$

$$= 2(-5 - (-3)) - 0 + 1(-17 - 15)$$

$$= 2(-2) - 0 + 1(-32) = -36$$

To find D_z, let's expand about column 1.

$$D_z = \begin{vmatrix} 2 & -1 & -17 \\ 0 & 3 & 5 \\ 1 & -2 & -3 \end{vmatrix} = 2 \begin{vmatrix} 3 & 5 \\ -2 & -3 \end{vmatrix} - 0 \begin{vmatrix} -1 & -17 \\ -2 & -3 \end{vmatrix} + 1 \begin{vmatrix} -1 & -17 \\ 3 & 5 \end{vmatrix}$$

$$= 2(-9 - (-10)) - 0 + 1(-5 - (-51))$$

$$= 2(1) - 0 + 1(46) = 48$$

Thus,

$$x = \frac{D_x}{D} = \frac{12}{-12} = -1, \qquad y = \frac{D_y}{D} = \frac{-36}{-12} = 3, \qquad z = \frac{D_z}{D} = \frac{48}{-12} = -4.$$

The solution set is $\{(-1, 3, -4)\}$. ∎

Problem Set 14.4

Use expansion-by-minors to evaluate each of the following determinants.

1. $\begin{vmatrix} 2 & 7 & 5 \\ 1 & -1 & 1 \\ -4 & 3 & 2 \end{vmatrix}$

2. $\begin{vmatrix} 2 & 4 & 1 \\ -1 & 5 & 1 \\ -3 & 6 & 2 \end{vmatrix}$

3. $\begin{vmatrix} 3 & -2 & 1 \\ 2 & 1 & 4 \\ -1 & 3 & 5 \end{vmatrix}$

4. $\begin{vmatrix} 1 & -1 & 2 \\ 2 & 1 & 3 \\ -1 & -2 & 1 \end{vmatrix}$

5. $\begin{vmatrix} -3 & -2 & 1 \\ 5 & 0 & 6 \\ 2 & 1 & -4 \end{vmatrix}$

6. $\begin{vmatrix} -5 & 1 & -1 \\ 3 & 4 & 2 \\ 0 & 2 & -3 \end{vmatrix}$

7. $\begin{vmatrix} 3 & -4 & -2 \\ 5 & -2 & 1 \\ 1 & 0 & 0 \end{vmatrix}$

8. $\begin{vmatrix} -6 & 5 & 3 \\ 2 & 0 & -1 \\ 4 & 0 & 7 \end{vmatrix}$

9. $\begin{vmatrix} 4 & -2 & 7 \\ 1 & -1 & 6 \\ 3 & 5 & -2 \end{vmatrix}$

10. $\begin{vmatrix} -5 & 2 & 6 \\ 1 & -1 & 3 \\ 4 & -2 & -4 \end{vmatrix}$

Use Cramer's Rule to find the solution set for each of the following systems.

11. $\begin{pmatrix} 2x - y + 3z = -10 \\ x + 2y - 3z = 2 \\ 3x - 2y + 5z = -16 \end{pmatrix}$

12. $\begin{pmatrix} -x + y - z = 1 \\ 2x + 3y - 4z = 10 \\ -3x - y + z = -5 \end{pmatrix}$

13. $\begin{pmatrix} x - y + 2z = -8 \\ 2x + 3y - 4z = 18 \\ -x + 2y - z = 7 \end{pmatrix}$

14. $\begin{pmatrix} x - 2y + z = 3 \\ 3x + 2y + z = -3 \\ 2x - 3y - 3z = -5 \end{pmatrix}$

15. $\begin{pmatrix} 3x - 2y - 3z = -5 \\ x + 2y + 3z = -3 \\ -x + 4y - 6z = 8 \end{pmatrix}$

16. $\begin{pmatrix} 2x - 3y + 3z = -3 \\ -2x + 5y - 3z = 5 \\ 3x - y + 6z = -1 \end{pmatrix}$

17. $\begin{pmatrix} -x + y + z = -1 \\ x - 2y + 5z = -4 \\ 3x + 4y - 6z = -1 \end{pmatrix}$

18. $\begin{pmatrix} x - 2y + 3z = 1 \\ 2x + y + z = 4 \\ 4x - 3y + 7z = 6 \end{pmatrix}$

19. $\begin{pmatrix} x - y + 2z = 4 \\ 3x - 2y + 4z = 6 \\ 2x - 2y + 4z = -1 \end{pmatrix}$

20. $\begin{pmatrix} -x - 2y + z = 8 \\ 3x + y - z = 5 \\ 5x - y + 4z = 33 \end{pmatrix}$

21. $\begin{pmatrix} 2x - y + 3z = -5 \\ 3x + 4y - 2z = -25 \\ -x + z = 6 \end{pmatrix}$

22. $\begin{pmatrix} 3x - 2y + z = 11 \\ 5x + 3y = 17 \\ x + y - 2z = 6 \end{pmatrix}$

23. $\begin{pmatrix} 2y - z = 10 \\ 3x + 4y = 6 \\ x - y + z = -9 \end{pmatrix}$

24. $\begin{pmatrix} 6x - 5y + 2z = 7 \\ 2x + 3y - 4z = -21 \\ 2y + 3z = 10 \end{pmatrix}$

25. $\begin{pmatrix} -2x + 5y - 3z = -1 \\ 2x - 7y + 3z = 1 \\ 4x - y - 6z = -6 \end{pmatrix}$

26. $\begin{pmatrix} 7x - 2y + 3z = -4 \\ 5x + 2y - 3z = 4 \\ -3x - 6y + 12z = -13 \end{pmatrix}$

27. $\begin{pmatrix} -x - y + 5z = 4 \\ x + y - 7z = -6 \\ 2x + 3y + 4z = 13 \end{pmatrix}$

28. $\begin{pmatrix} x + 7y - z = -1 \\ -x - 9y + z = 3 \\ 3x + 4y - 6z = 5 \end{pmatrix}$

29. $\begin{pmatrix} 5x - y + 2z = 10 \\ 7x + 2y - 2z = -4 \\ -3x - y + 4z = 1 \end{pmatrix}$

30. $\begin{pmatrix} 4x - y - 3z = -12 \\ 5x + y + 6z = 4 \\ 6x - y - 3z = -14 \end{pmatrix}$

Miscellaneous Problems

31. Evaluate the following determinant by expanding about the *second column*.

$$\begin{vmatrix} a & e & a \\ b & f & b \\ c & g & c \end{vmatrix}$$

Make a conjecture about determinants that contain two identical columns.

32. Show that $\begin{vmatrix} 1 & -1 & 2 \\ 2 & 3 & -1 \\ -1 & 2 & 4 \end{vmatrix} = -\begin{vmatrix} -1 & 1 & 2 \\ 3 & 2 & -1 \\ 2 & -1 & 4 \end{vmatrix}$.

Make a conjecture about the result of interchanging two columns of a determinant.

33. (a) Show that $\begin{vmatrix} 2 & 1 & 2 \\ 4 & -1 & -2 \\ 6 & 3 & 1 \end{vmatrix} = 2\begin{vmatrix} 1 & 1 & 2 \\ 2 & -1 & -2 \\ 3 & 3 & 1 \end{vmatrix}$.

Make a conjecture about the result of factoring a common factor from each element of a column in a determinant.

(b) Use your conjecture from part (a) to help evaluate the following determinant.

$$\begin{vmatrix} 2 & 4 & -1 \\ -3 & -4 & -2 \\ 5 & 4 & 3 \end{vmatrix}.$$

34. We can describe another technique for evaluating 3×3 determinants as follows. First, let's write the given determinant with its first two columns repeated on the right.

$$\begin{vmatrix} a_1 & b_1 & c_1 \\ a_2 & b_2 & c_2 \\ a_3 & b_3 & c_3 \end{vmatrix} \begin{matrix} a_1 & b_1 \\ a_2 & b_2 \\ a_3 & b_3 \end{matrix}$$

Then we can add the three products shown with $+$, and subtract the three products shown with $-$.

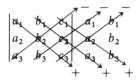

(**a**) Be sure that the previous description will produce equation (1) on page 629.

(**b**) Use this technique to do Problems 1–10.

Chapter 14 Summary

(**14.1**) Solving **a system of three linear equations in three variables** produces one of the following results.

1. There is **one ordered triple** that satisfies all three equations.
2. There are **infinitely many ordered triples** in the solution set, all of which are coordinates of points on a line common to the planes.
3. There are **infinitely many ordered triples** in the solution set, all of which are coordinates on a plane.
4. The solution set is empty; it is $\varnothing$.

(**14.2**) A **matrix** is an array of numbers arranged in horizontal rows and vertical columns. A matrix of m rows and n columns is called an $m \times n$ ("m-by-n") matrix. The **augmented matrix** of the system

$$\begin{pmatrix} 5x - 2y - z = 4 \\ 3x - y - z = 7 \\ 2x + 3y + 7z = 9 \end{pmatrix}$$

is

$$\begin{bmatrix} 5 & 2 & -1 & \vdots & 4 \\ 3 & -1 & -1 & \vdots & 7 \\ 2 & 3 & 7 & \vdots & 9 \end{bmatrix}.$$

The following **elementary row operations** provide the basis for transforming matrices.

1. Any two rows of an augmented matrix can be interchanged.
2. Any row can be multiplied by a nonzero constant.
3. Any row can be replaced by adding a nonzero multiple of another row to that row.

Transforming an augmented matrix to **triangular form** and then using back-substitution provides a systematic technique for solving systems of linear equations.

(14.3) A rectangular array of numbers is called a **matrix**. A **square matrix** has the same number of rows as columns. For a 2×2 matrix

$$\begin{bmatrix} a_1 & b_1 \\ a_2 & b_2 \end{bmatrix},$$

the **determinant** of the matrix is written as

$$\begin{vmatrix} a_1 & b_1 \\ a_2 & b_2 \end{vmatrix}$$

and defined by

$$\begin{vmatrix} a_1 & b_1 \\ a_2 & b_2 \end{vmatrix} = a_1 b_2 - a_2 b_1.$$

Cramer's Rule for solving a system of two linear equations in two variables is stated as follows.

Given the system $\begin{pmatrix} a_1 x + b_1 y = c_1 \\ a_2 x + b_2 y = c_2 \end{pmatrix}$

with

$$D = \begin{vmatrix} a_1 & b_1 \\ a_2 & b_2 \end{vmatrix} \neq 0, \qquad D_x = \begin{vmatrix} c_1 & b_1 \\ c_2 & b_2 \end{vmatrix}, \qquad D_y = \begin{vmatrix} a_1 & c_1 \\ a_2 & c_2 \end{vmatrix},$$

then

$$x = \frac{D_x}{D} \qquad \text{and} \qquad y = \frac{D_y}{D}.$$

(14.4) A 3×3 determinant is defined by

$$\begin{vmatrix} a_1 & b_1 & c_1 \\ a_2 & b_2 & c_2 \\ a_3 & b_3 & c_3 \end{vmatrix} = a_1 b_2 c_3 + b_1 c_2 a_3 + c_1 a_2 b_3 - a_3 b_2 c_1 - b_3 c_2 a_1 - c_3 a_2 b_1.$$

The **minor** of an element in a determinant is the determinant that remains after deleting the row and column in which the element appears. A determinant can be evaluated by **expansion-by-minors** of the elements of any row or any column.

Cramer's Rule for solving a system of three linear equations in three variables is stated as follows.

Given the system $\begin{pmatrix} a_1 x + b_1 y + c_1 z = d_1 \\ a_2 x + b_2 y + c_2 z = d_2 \\ a_3 x + b_3 y + c_3 z = d_3 \end{pmatrix}$

with

$$D = \begin{vmatrix} a_1 & b_1 & c_1 \\ a_2 & b_2 & c_2 \\ a_3 & b_3 & c_3 \end{vmatrix} \neq 0, \qquad D_x = \begin{vmatrix} d_1 & b_1 & c_1 \\ d_2 & b_2 & c_2 \\ d_3 & b_3 & c_3 \end{vmatrix},$$

$$D_y = \begin{vmatrix} a_1 & d_1 & c_1 \\ a_2 & d_2 & c_2 \\ a_3 & d_3 & c_3 \end{vmatrix}, \qquad D_z = \begin{vmatrix} a_1 & b_1 & d_1 \\ a_2 & b_2 & d_2 \\ a_3 & b_3 & d_3 \end{vmatrix},$$

then $x = \dfrac{D_x}{D}$, $y = \dfrac{D_y}{D}$, and $z = \dfrac{D_z}{D}$.

Chapter 14 Review Problem Set

For Problems 1–4, solve each system of equations using (a) the elimination-by-addition method, (b) a matrix approach, and (c) Cramer's Rule.

1. $\begin{pmatrix} 3x - 2y = -6 \\ 2x + 5y = 34 \end{pmatrix}$

2. $\begin{pmatrix} x + 4y = 25 \\ y = -3x - 2 \end{pmatrix}$

3. $\begin{pmatrix} x = 5y - 49 \\ 4x + 3y = -12 \end{pmatrix}$

4. $\begin{pmatrix} x - 6y = 7 \\ 3x + 5y = 9 \end{pmatrix}$

For Problems 5–14, solve each system using the method that seems most appropriate to you.

5. $\begin{pmatrix} x - 3y = 25 \\ -3x + 2y = -26 \end{pmatrix}$

6. $\begin{pmatrix} 5x - 7y = -66 \\ x + 4y = 30 \end{pmatrix}$

7. $\begin{pmatrix} 4x + 3y = -9 \\ 3x - 5y = 15 \end{pmatrix}$

8. $\begin{pmatrix} 2x + 5y = 47 \\ 4x - 7y = -25 \end{pmatrix}$

9. $\begin{pmatrix} 7x - 3y = 25 \\ y = 3x - 9 \end{pmatrix}$

10. $\begin{pmatrix} x = -4 - 5y \\ y = 4x + 16 \end{pmatrix}$

11. $\begin{pmatrix} \dfrac{1}{2}x + \dfrac{2}{3}y = 6 \\ \dfrac{3}{4}x - \dfrac{5}{6}y = -24 \end{pmatrix}$

12. $\begin{pmatrix} \dfrac{3}{4}x - \dfrac{1}{2}y = 14 \\ \dfrac{5}{12}x + \dfrac{3}{4}y = 16 \end{pmatrix}$

13. $\begin{pmatrix} 6x - 4y = 7 \\ 9x + 8y = 0 \end{pmatrix}$

14. $\begin{pmatrix} 4x - 5y = -5 \\ 6x - 10y = -9 \end{pmatrix}$

For Problems 15–18, evaluate each of the determinants.

15. $\begin{vmatrix} 3 & 5 \\ 6 & 4 \end{vmatrix}$

16. $\begin{vmatrix} -1 & -5 \\ 4 & 9 \end{vmatrix}$

17. $\begin{vmatrix} 4 & -1 & -3 \\ 2 & 1 & 4 \\ -3 & 2 & 2 \end{vmatrix}$

18. $\begin{vmatrix} 5 & 3 & 4 \\ -2 & 0 & 1 \\ -1 & -2 & 6 \end{vmatrix}$

For Problems 19 and 20, solve each system of equations using (a) the elimination method, (b) a matrix approach, and (c) Cramer's Rule.

19. $\begin{pmatrix} x - 2y + 4z = -14 \\ 3x - 5y + z = 20 \\ -2x + y - 5z = 22 \end{pmatrix}$

20. $\begin{pmatrix} x + 3y - 2z = 28 \\ 2x - 8y + 3z = -63 \\ 3x + 8y - 5z = 72 \end{pmatrix}$

For Problems 21–24, solve each system using the method that seems most appropriate to you.

21. $\begin{pmatrix} x + y - z = -2 \\ 2x - 3y + 4z = 17 \\ -3x + 2y + 5z = -7 \end{pmatrix}$

22. $\begin{pmatrix} -x - y + z = -3 \\ 3x + 2y - 4z = 12 \\ 5x + y + 2z = 5 \end{pmatrix}$

23. $\begin{pmatrix} 3x + y - z = -6 \\ 3x + 2y + 3z = 9 \\ 6x - 2y + 2z = 8 \end{pmatrix}$

24. $\begin{pmatrix} x - 3y + z = 2 \\ 2x - 5y - 3z = 22 \\ -4x + 3y + 5z = -26 \end{pmatrix}$

Cumulative Review Problem Set (Chapters 1–14)

For Problems 1–10, evaluate each expression without using a calculator.

1. $\dfrac{2}{\left(\frac{1}{2}\right)^{-1}}$

2. -2^3

3. $\left(\dfrac{2}{3}\right)^{-3}$

4. $\log .001$

5. $\ln e^3$

6. $(2^{-1} + 3^{-1})^{-1}$

7. $\sqrt{\dfrac{9}{64}}$

8. $-\sqrt[3]{-8}$

9. $\log_3 81$

10. $2^0 + 2^{-2} + 2^{-4}$

For Problems 11–25, solve each of the equations.

11. $-3(2x - 1) = 2(4x - 1)$

12. $|3x - 2| = 13$

13. $5x^2 = -9x$

14. $9^{2x-1} = \dfrac{1}{3}$

15. $\log_3 x = 2$

16. $\ln(x + 1) = \ln(x - 2) + \ln 2$

17. $(2x + 3)(x - 1) = (2x + 3)(x - 4)$

18. $(5x - 2)(3x + 4) = 0$

19. $2x^2 - x + 4 = 0$

20. $\sqrt{x - 1} = x - 1$

21. $\dfrac{3}{x - 2} = \dfrac{4}{x + 3}$

22. $(x - 3)(2x + 1) = 9$

23. $15x^3 + x^2 - 2x = 0$

24. $.04x + .05(900 - x) = 42$

25. $(2x - 3)^2 = 24$

For Problems 26–29, solve each of the systems of equations.

26. $\begin{pmatrix} 4x - y = -1 \\ y = 15 - 3x \end{pmatrix}$

27. $\begin{pmatrix} 5x + 3y = 7 \\ 2x - 7y = -30 \end{pmatrix}$

28. $\begin{pmatrix} x + y = 800 \\ .03x + .04y = 30 \end{pmatrix}$

29. $\begin{pmatrix} x - 2y + 3z = -14 \\ 2x + y - z = 3 \\ 5x - 3y - 7z = 10 \end{pmatrix}$

For Problems 30–37, solve each of the inequalities.

30. $-3(x - 1) > 4x - 3 - 2x$

31. $\dfrac{x - 1}{2} - \dfrac{x + 3}{4} \le 1$

32. $(2x + 1)(x - 5) \ge 0$

33. $|3x - 1| < 8$

34. $|5x + 2| > 12$

35. $\dfrac{x + 2}{x - 3} > 0$

36. $\dfrac{x}{x - 2} \le 10$

37. $x(x + 2) < 35$

For Problems 38–44, perform the indicated operations and express answers in simplified form.

38. $(3\sqrt{8})(-6\sqrt{5})$

39. $(3\sqrt{2} + \sqrt{3})(\sqrt{2} - 2\sqrt{3})$

40. $\dfrac{x + 1}{3} + \dfrac{x - 1}{4} - \dfrac{x + 6}{6}$

41. $\dfrac{9}{14x^2 y} - \dfrac{7}{10x}$

42. $\dfrac{15xy^2}{13y} \div \dfrac{12x^2 y^2}{5y}$

43. $(2x^3 + 9x^2 - 26x + 12) \div (2x - 3)$

44. $\dfrac{3x^2 + x - 2}{x^3 + 4x^2 - 45x} \cdot \dfrac{7x^2 + 63x}{9x^2 - 4}$

For Problems 45–50, factor each algebraic expression completely.

45. $3x^3 - 27x$

46. $24n^2 + 14n - 5$

47. $x^3 - 27$

48. $xy + 3x - 2y - 6$

49. $3x^3 + 192$

50. $9x^4 + 5x^2 - 4$

For Problems 51–56, express each radical expression in simplest radical form.

51. $\sqrt{56}$

52. $-3\sqrt{72}$

53. $\sqrt[3]{72}$

54. $\dfrac{3\sqrt{2}}{7\sqrt{3}}$

55. $\dfrac{12\sqrt{6}}{6\sqrt{3}}$

56. $\dfrac{\sqrt{3}}{2\sqrt{3} + \sqrt{2}}$

For Problems 57–60, find each of the indicated products or quotients and express answers in the standard form of a complex number.

57. $(7i)(-5i)$

58. $\dfrac{3}{7i}$

59. $(3 - 6i)(4 + 2i)$

60. $\dfrac{-2 + i}{5 - 2i}$

61. Find the slope of the line determined by the points $(-1, 6)$ and $(2, -4)$.

62. Find the slope of the line determined by the equation $3y - 5x = 7$.

63. Find the length of the line segment whose endpoints are $(2, 4)$ and $(-3, -2)$.

64. Write the equation of the line that is perpendicular to the line $5x - 3y = 9$ and contains the point $(4, -2)$.

65. Write the equation of the line that is parallel to the line $x - 6y = 1$ and contains the point $(-2, -7)$.

66. Write the equation of the line that contains the points $(-1, 8)$ and $(4, 5)$.

67. Find the center and the length of a radius of the circle $x^2 + y^2 - 10x + 12y + 25 = 0$.

68. Find the vertex of the parabola $y = x^2 + 2x + 4$.

69. Find the length of the major axis of the ellipse $4x^2 + y^2 = 49$.

70. Find the equations of the asymptotes for the hyperbola $y^2 - 4x^2 = 16$.

71. Evaluate the determinant $\begin{vmatrix} 3 & 0 & -4 \\ 5 & 2 & 1 \\ -6 & 0 & -7 \end{vmatrix}$.

72. Find the inverse of the function $f(x) = -4x + 6$.

73. If $f(x) = 7x + 2$, find $\dfrac{f(a + h) - f(a)}{h}$.

74. If $f(x) = -x^2 + 6x - 1$, find $f(-2)$.

75 Find the constant of variation if y varies directly as x, and $y = 2$ when $x = -\dfrac{1}{2}$.

76. If y is inversely proportional to x, and $y = 6$ when $x = \dfrac{1}{2}$, find y when $x = 4$.

For Problems 77–80, graph each of the equations.

77. $2x - y = -4$ **78.** $y^2 - x^2 = 3$ **79.** $x^2 + 2y^2 = 10$ **80.** $x = y^2 - 3$

For Problems 81–86, graph each of the functions.

81. $f(x) = -x^2 - 1$ **82.** $f(x) = -x - 1$ **83.** $f(x) = 3x^2 - 12x + 13$

84. $f(x) = -\sqrt{x + 2}$ **85.** $f(x) = 2|x + 2| + 1$ **86.** $f(x) = -(x - 1)^3$

For Problems 87–94, set up an equation or a system of equations to help solve each problem.

87. One of two complementary angles is $12°$ less than twice the other angle. Find the measure of each angle.

88. The largest angle of a triangle is $7°$ more than twice the smallest angle. The sum of the smallest and largest angle is $6°$ less than twice the other angle. Find the measure of each angle of the triangle.

89. Tanya has 130 coins consisting of dimes and quarters totaling $20.50. How many coins of each kind does she have?

90. How many pints of pure alcohol must be added to 12 pints of a 10% alcohol solution to raise it to a 20% solution?

91. Working together Marty and Nana can complete a task in 1 hour and 20 minutes. Marty can do the task by himself in 4 hours. How long would it take Nana to complete the task by herself?

92. To travel 300 miles, it takes a freight train 2 hours more than it does an express train to travel 280 miles. The rate of the express train is 20 miles per hour greater than the rate of the freight train. Find the rates of both trains.

93. Last year Maria invested $13,000, part of it at 6% and the balance of it at 7%. She received $860 in interest from the two investments. How much did she invest at each rate?

94. Caleb bought a pair of walking shoes at a 20% discount sale for $64. What was the original price of the shoes?

For Problems 95–100, use your calculator to help solve each problem.

95. Find a value, rounded to four decimal places, for $\log_5 87$.

96. Solve the equation $4^x = 127$ to the nearest hundredth.

97. Solve the equation $e^x = 78$ to the nearest hundredth.

98. How long will it take $500 to earn $500 in interest if it is invested at 7% interest compounded annually? Express the answer to the nearest tenth of a year.

99. Suppose that a certain radioactive substance has a half-life of 150 days. If there are presently 2500 grams of the substance, how much, to the nearest gram, will remain after 110 days?

100. If $3500 is invested at 8% interest compounded quarterly, how much money is accumulated after 10 years? Express the answer to the nearest dollar.

Chapter 15

Sequences and Series

15.1 **Arithmetic Sequences**

15.2 **Arithmetic Series**

15.3 **Geometric Sequences and Series**

15.4 **Infinite Geometric Series**

15.5 **Binomial Expansions**

Suppose that Math University had an enrollment of 7500 students in 1987 and each year thereafter through 1991 the enrollment increased by 500 students. The numbers 7500, 8000, 8500, 9000, 9500 represent the enrollment figures for the years 1987 through 1991. This list of numbers, where there is a constant difference of 500 between any two successive numbers in the list, is called an arithmetic sequence.

Suppose a woman's present yearly salary

is $20,000 and she expects a 10% raise for each of the next four years. The numbers 20,000, 22,000, 24,200, 26,620, 29,282 represent her present salary and her salary for each of the next four years. This list of numbers, where each number after the first one is 1.1 times the previous number in the list, is called a geometric sequence.

Arithmetic and geometric sequences are the center of our attention in this chapter.

15.1

Arithmetic Sequences

An **infinite sequence** is a function whose domain is the set of positive integers. For example, consider the function defined by the equation

$$f(n) = 3n + 2$$

where the domain is the set of positive integers. Furthermore, let's substitute the numbers of the domain, in order, starting with 1. The resulting ordered pairs can be listed as

(1, 5),   (2, 8),   (3, 11),   (4, 14),   (5, 17),

and so on. Since we have agreed to use the domain of positive integers, in order starting with 1, there is no need to use ordered pairs. We can simply express the infinite sequence as

5, 8, 11, 14, 17, ...

Frequently, the letter a is used to represent *sequential functions* and the functional value at n is written as a_n (read "a sub n") instead of $a(n)$. The sequence is then expressed as

$$a_1, a_2, a_3, a_4, \ldots, a_n, \ldots,$$

where a_1 is the *first term*, a_2 the *second term*, a_3 the *third term*, and so on. The expression a_n which defines the sequence, is called the **general term** of the sequence. Knowing the general term of a sequence allows us to find as many terms of the sequence as needed and also to find any specific terms. Consider the following examples.

EXAMPLE 1   Find the first five terms of each of the following sequences.

(a) $a_n = n^2 + 1$.   (b) $a_n = 2^n$.

Solution   (a) The first five terms are found by replacing n with 1, 2, 3, 4, and 5.

$$a_1 = 1^2 + 1 = 2,$$

$$a_2 = 2^2 + 1 = 5,$$

$$a_3 = 3^2 + 1 = 10,$$

$$a_4 = 4^2 + 1 = 17,$$

$$a_5 = 5^2 + 1 = 26$$

Thus, the first five terms are 2, 5, 10, 17, and 26.

(b) $a_1 = 2^1 = 2,$

$a_2 = 2^2 = 4,$

$a_3 = 2^3 = 8,$

$a_4 = 2^4 = 16,$

$a_5 = 2^5 = 32$

The first five terms are 2, 4, 8, 16, and 32.   ∎

EXAMPLE 2   Find the 12th and 25th terms of the sequence $a_n = 5n - 1$.

Solution $a_{12} = 5(12) - 1 = 59,$

$a_{25} = 5(25) - 1 = 124$

The 12th term is 59 and the 25th term is 124.   ∎

An **arithmetic sequence** (also called an *arithmetic progression*) is a sequence where there is a **common difference** between successive terms. The following are examples of arithmetic sequences.

1. 1, 4, 7, 10, 13,...
2. 6, 11, 16, 21, 26,...
3. 14, 25, 36, 47, 58,...
4. 4, 2, 0, −2, −4,...
5. −1, −7, −13, −19, −25,...

The common difference in number 1 of the previous list is 3. That is to say, $4 - 1 = 3, 7 - 4 = 3, 10 - 7 = 3, 13 - 10 = 3$, and so on. The common differences for numbers 2, 3, 4, and 5 are 5, 11, −2, and −6, respectively. It is sometimes stated that arithmetic sequences exhibit *constant growth*. This is an accurate description if we keep in mind that the *growth* may be in a negative direction as illustrated by numbers 4 and 5 above.

In a more general setting we say that the sequence

$$a_1, a_2, a_3, a_4, \ldots, a_n, \ldots$$

is an arithmetic sequence if and only if there is a real number d such that

$$a_{k+1} - a_k = d \tag{1}$$

for every positive integer k. The number d is called the **common difference.**

From equation (1) we see that $a_{k+1} = a_k + d$. In other words, we can generate an arithmetic sequence that has a common difference of d by starting with a first term of a_1 and then simply adding d to each successive term as follows.

first term a_1

second term $a_1 + d$

third term $a_1 + 2d$ $(a_1 + d) + d = a_1 + 2d$

fourth term $a_1 + 3d$ $(a_1 + 2d) + d = a_1 + 3d$

$\vdots$

nth term $a_1 + (n - 1)d$

Thus, the *general term of an arithmetic sequence* is given by

$$a_n = a_1 + (n - 1)d$$

where a_1 is the first term and d the common difference. This general term formula provides the basis for doing a variety of problems that involve arithmetic sequences.

EXAMPLE 3   Find the general term for each of the following arithmetic sequences.

          (a) 1, 5, 9, 13,....   (b) 5, 2, -1, -4,....

Solution   (a) The common difference is 4 and the first term is 1. Substituting these values into $a_n = a_1 + (n - 1)d$ and simplifying, we obtain

$$a_n = a_1 + (n - 1)d$$
$$= 1 + (n - 1)4$$
$$= 1 + 4n - 4$$
$$= 4n - 3.$$

(Perhaps you should verify that the general term $a_n = 4n - 3$ does produce the sequence 1, 5, 9, 13,....)

(b) Since the first term is 5 and the common difference is -3, we obtain

$$a_n = a_1 + (n - 1)d$$
$$= 5 + (n - 1)(-3)$$
$$= 5 - 3n + 3 = -3n + 8.$$ ■

EXAMPLE 4   Find the 50th term of the arithmetic sequence 2, 6, 10, 14,....

Solution   Certainly, we could simply continue to write the terms of the given sequence until the 50th term is reached. However, let's use the general term formula, $a_n = a_1 + (n-1)d$, to find the 50th term without determining all of the other terms.

$$a_{50} = 2 + (50 - 1)4$$
$$= 2 + 49(4) = 2 + 196 = 198. \qquad \blacksquare$$

EXAMPLE 5   Find the first term of the arithmetic sequence where the 3rd term is 13 and the 10th term is 62.

Solution   Using $a_n = a_1 + (n-1)d$ with $a_3 = 13$ (the third term is 13) and $a_{10} = 62$ (the 10th term is 62) we have

$$13 = a_1 + (3-1)d = a_1 + 2d,$$
$$62 = a_1 + (10-1)d = a_1 + 9d.$$

Solve the system of equations

$$\begin{pmatrix} a_1 + 2d = 13 \\ a_1 + 9d = 62 \end{pmatrix}$$

to yield $a_1 = -1$. Thus, the first term is -1. $\qquad \blacksquare$

REMARK Perhaps you can think of another way to solve the problem in Example 5 without using a system of equations. [*Hint*: How many "differences" are there between the 3rd and 10th terms of an arithmetic sequence?]

Phrases such as "the set of odd whole numbers," "the set of even whole numbers," and "the set of positive multiples of five" are commonly used in mathematical literature to refer to various subsets of the whole numbers. Though no specific ordering of the numbers is implied by these phrases, most of us would probably react with a natural ordering. For example, if we were asked to list the set of *odd whole numbers*, our answer probably would be 1, 3, 5, 7, ... Using such an ordering, we can think of the set of odd whole numbers as an arithmetic sequence. Therefore, we can formulate a general representation for the set of odd whole numbers by using the general term formula. Thus,

$$a_n = a_1 + (n-1)d$$
$$= 1 + (n-1)2$$
$$= 1 + 2n - 2$$
$$= 2n - 1.$$

The final example of this section illustrates the use of an arithmetic sequence to solve a problem that deals with the *constant growth* of a man's yearly salary.

EXAMPLE 6   A man started to work in 1970 at an annual salary of $9500. He received a $700 raise each year. How much was his annual salary in 1991?

Solution   The following arithmetic sequence represents the annual salary beginning in 1970.

$$9500, 10,200, 10,900, 11,600, \ldots.$$

The general term of this sequence is

$$a_n = a_1 + (n - 1)d$$

$$= 9500 + (n - 1)700$$

$$= 9500 + 700n - 700$$

$$= 700n + 8800.$$

The man's 1991 salary is the 22nd term of this sequence. Thus,

$$a_{22} = 700(22) + 8800 = 24,200.$$

His salary in 1991 was $24,200.   ∎

Problem Set 15.1

For Problems 1–14, write the first five terms of each sequence that has the indicated general term.

1. $a_n = 3n - 4$ **2.** $a_n = 2n + 3$

3. $a_n = -2n + 5$ **4.** $a_n = -3n - 2$

5. $a_n = n^2 - 2$ **6.** $a_n = n^2 + 3$

7. $a_n = -n^2 + 1$ **8.** $a_n = -n^2 - 2$

9. $a_n = 2n^2 - 3$ **10.** $a_n = 3n^2 + 2$

11. $a_n = 2^{n-2}$ **12.** $a_n = 3^{n+1}$

13. $a_n = -2(3)^{n-2}$ **14.** $a_n = -3(2)^{n-3}$

15. Find the 8th and 12th terms of the sequence where $a_n = n^2 - n - 2$.

16. Find the 10th and 15th terms of the sequence where $a_n = -n^2 - 2n + 3$.

17. Find the 7th and 8th terms of the sequence where $a_n = (-2)^{n-2}$.

18. Find the 6th and 7th terms of the sequence where $a_n = -(3)^{n-1}$.

For Problems 19–28, find the general term (nth term) for each of the arithmetic sequences.

19. $1, 3, 5, 7, 9, \ldots$ **20.** $2, 4, 6, 8, 10, \ldots$

21. $-2, 2, 6, 10, 14, \ldots$ **22.** $-3, 2, 7, 12, 17, \ldots$

23. $5, 3, 1, -1, -3, \ldots$ **24.** $2, -1, -4, -7, -10, \ldots$

25. $-7, -10, -13, -16, -19, \ldots$

26. $-3, -5, -7, -9, -11, \ldots$

27. $1, \dfrac{3}{2}, 2, \dfrac{5}{2}, 3, \ldots$ **28.** $\dfrac{3}{2}, 3, \dfrac{9}{2}, 6, \dfrac{15}{2}, \ldots$

For Problems 29–34, find the indicated term of the arithmetic sequence.

29. The 10th term of 7, 10, 13, 16,...

30. The 12th term of 9, 11, 13, 15,...

31. The 20th term of 2, 6, 10, 14,...

32. The 50th term of -1, 4, 9, 14,...

33. The 75th term of -7, -9, -11, -13,...

34. The 100th term of -7, -10, -13, -16,...

For Problems 35–40, find the number of terms of each of the finite arithmetic sequences.

35. 1, 3, 5, 7,..., 211

36. 2, 4, 6, 8, ..., 312

37. 10, 13, 16, 19,..., 157

38. 9, 13, 17, 21,..., 849

39. -7, -9, -11, -13, ..., -345

40. -4, -7, -10, -13,..., -331

41. If the 6th term of an arithmetic sequence is 24 and the 10th term is 44, find the first term.

42. If the 5th term of an arithmetic sequence is 26 and the 12th term is 75, find the first term.

43. If the 4th term of an arithmetic sequence is -9 and the 9th term is -29, find the 5th term.

44. If the 6th term of an arithmetic sequence is -4 and the 14th term is -20, find the 10th term.

45. In the arithmetic sequence .97, 1.00, 1.03, 1.06,..., which term is 5.02?

46. In the arithmetic sequence 1, 1.2, 1.4, 1.6,..., which term is 35.4?

For Problems 47–50, set up an arithmetic sequence and use $a_n = a_1 + (n-1)d$ to solve each problem.

47. A woman started to work in 1975 at an annual salary of $12,500. She received a $900 raise each year. How much was her annual salary in 1992?

48. Math University had an enrollment of 8500 students in 1976. Each year the enrollment has increased by 350 students. What was the enrollment in 1990?

49. Suppose you are offered a job starting at $900 a month with a guaranteed increase of $30 a month every 6 months for the next 5 years. What will your monthly salary be for the last six months of the 5th year of your employment?

50. Between 1976 and 1990 a person invested $500 at 14% simple interest at the beginning of each year. By the end of 1990, how much interest had been earned by the $500 that was invested at the beginning of 1982?

15.2

Arithmetic Series

Let's solve a problem to begin this section. Study the solution very carefully.

PROBLEM 1   Find the sum of the first one hundred positive integers.

Solution

We are asked to find the sum of $1 + 2 + 3 + 4 + \cdots + 100$. Rather than use a calculator, let's find the sum in the following way.

$$
\begin{array}{r}
1 + 2 + 3 + 4 + \cdots + 100 \\
100 + 99 + 98 + 97 + \cdots + 1 \\
\hline
101 + 101 + 101 + 101 + \cdots + 101
\end{array}
$$

$$
\frac{\overset{50}{\cancel{100}}(101)}{\cancel{2}} = 5050.
$$

Note that we simply wrote the indicated sum *forward and backward* and then added the results. In so doing, 100 sums of 101 are produced, but half of them are "repeats." For example, $(100 + 1)$ and $(1 + 100)$ are both counted in this process. Thus, we divided the product $(100)(101)$ by 2, which yielded the final result of 5050.

■

Now let's introduce some terminology that applies to problems like the one above. The indicated sum of a sequence is called a **series**. Associated with the finite sequence

$$a_1, a_2, a_3, \ldots, a_n$$

is a **finite series**

$$a_1 + a_2 + a_3 + \cdots + a_n.$$

Likewise, from the infinite sequence

$$a_1, a_2, a_3, \ldots$$

we can form the **infinite series**

$$a_1 + a_2 + a_3 + \cdots.$$

In this section we will direct our attention to working with **arithmetic series**; that is, the indicated sums of arithmetic sequences.

Problem 1 above could have been stated as "find the sum of the first one hundred terms of the arithmetic series that has a general term (nth term) of $a_n = n$." In fact, the *forward-backward* approach used to solve that problem can be applied to the general arithmetic series

$$a_1 + a_2 + a_3 + \cdots + a_n$$

to produce a formula for finding the sum of the first n terms of any arithmetic series. Use S_n to represent the sum of the first n terms and proceed as follows.

$$S_n = a_1 + (a_1 + d) + (a_1 + 2d) + \cdots + (a_n - 2d) + (a_n - d) + a_n.$$

Now write this sum in reverse as

$$S_n = a_n + (a_n - d) + (a_n - 2d) + \cdots + (a_1 + 2d) + (a_1 + d) + a_1.$$

Add the two equations to produce

$$2S_n = (a_1 + a_n) + (a_1 + a_n) + (a_1 + a_n) + \cdots + (a_1 + a_n)$$
$$+ (a_1 + a_n) + (a_1 + a_n).$$

That is, we have n sums of $(a_1 + a_n)$, so

$$2S_n = n(a_1 + a_n)$$

from which we obtain

$$S_n = \frac{n(a_1 + a_n)}{2}.$$

Using the nth term formula $a_n = a_1 + (n - 1)d$ and the sum formula $\frac{n(a_1 + a_n)}{2}$, we can solve a variety of problems that involve arithmetic series.

EXAMPLE 1   Find the sum of the first 50 terms of the series $2 + 5 + 8 + 11 + \cdots$.

Solution   Using $a_n = a_1 + (n - 1)d$, we find the 50th term to be

$$a_{50} = 2 + 49(3) = 149.$$

Then using $S_n = \frac{n(a_1 + a_n)}{2}$, we obtain

$$S_{50} = \frac{50(2 + 149)}{2} = 3775. \qquad \blacksquare$$

EXAMPLE 2   Find the sum of all odd numbers between 7 and 433, inclusive.

Solution   We need to find the sum $7 + 9 + 11 + \cdots + 433$.

To use $S_n = \frac{n(a_1 + a_n)}{2}$, the number of terms, n, is needed. Perhaps you could figure this out without a formula (try it), but suppose we use the nth term formula.

$$a_n = a_1 + (n - 1)d$$
$$433 = 7 + (n - 1)2$$
$$433 = 7 + 2n - 2$$
$$433 = 2n + 5$$
$$428 = 2n$$
$$214 = n.$$

Then use $n = 214$ in the sum formula to yield

$$S_{214} = \frac{214(7 + 433)}{2} = 47,080.$$ ∎

EXAMPLE 3 Find the sum of the first 75 terms of the series that has a general term of $a_n = -5n + 9$.

Solution Using $a_n = 5n + 9$ we can generate the series as follows.

$$a_1 = -5(1) + 9 = 4,$$
$$a_2 = -5(2) + 9 = -1,$$
$$a_3 = -5(3) + 9 = -6,$$
$$\vdots$$
$$a_{75} = -5(75) + 9 = -366$$

Thus, we have the series

$$4 + (-1) + (-6) + \cdots + (-366).$$

Using the sum formula we obtain

$$S_{75} = \frac{75(4 + (-366))}{2} = -13,575.$$ ∎

EXAMPLE 4 Sue is saving quarters. She saves 1 quarter the first day, 2 quarters the second day, 3 quarters the third day, and so on. How much money will she have saved in 30 days?

Solution The total number of quarters will be the sum of the series

$$1 + 2 + 3 + \cdots + 30.$$

Using the sum formula yields

$$S_{30} = \frac{30(1 + 30)}{2} = 465.$$

So Sue will have saved $(465)(0.25) = \$116.25$ at the end of 30 days. ∎

REMARK The sum formula, $S_n = \dfrac{n(a_1 + a_n)}{2}$, was developed by using the **forward-backward** technique that we previously used on a specific problem. Now that we have the sum formula, we have two choices as we meet problems where the formula applies. We can either memorize the formula and use it as it applies, or disregard the formula and use the forward-backward approach. However, even if you choose to use the formula and some day your memory fails you and you forget the formula, use the forward-backward technique. In other words, once you understand the development of a formula you can do some problems even though the formula itself is forgotten.

Problem Set 15.2

For Problems 1–10, find the sum of the indicated number of terms of the given series.

1. First 50 terms of $2 + 4 + 6 + 8 + \cdots$

2. First 45 terms of $1 + 3 + 5 + 7 + \cdots$

3. First 60 terms of $3 + 8 + 13 + 18 + \cdots$

4. First 80 terms of $2 + 6 + 10 + 14 + \cdots$

5. First 65 terms of $(-1) + (-3) + (-5) + (-7) + \cdots$

6. First 100 terms of $(-1) + (-4) + (-7) + (-10) + \cdots$

7. First 40 terms of $\dfrac{1}{2} + 1 + \dfrac{3}{2} + 2 + \cdots$

8. First 50 terms of $1 + \dfrac{5}{2} + 4 + \dfrac{11}{2} + \cdots$

9. First 75 terms of $7 + 10 + 13 + 16 + \cdots$

10. First 90 terms of $(-8) + (-1) + 6 + 13 + \cdots$

For Problems 11–16, find the sum of each finite arithmetic series.

11. $4 + 8 + 12 + 16 + \cdots + 212$

12. $7 + 9 + 11 + 13 + \cdots + 179$

13. $(-4) + (-1) + 2 + 5 + \cdots + 173$

14. $5 + 10 + 15 + 20 + \cdots + 495$

15. $2.5 + 3.0 + 3.5 + 4.0 + \cdots + 18.5$

16. $1 + (-6) + (-13) + (-20) + \cdots + (-202)$

For Problems 17–22, find the sum of the indicated number of terms of the series with the given nth term.

17. First 50 terms of series with $a_n = 3n - 1$

18. First 150 terms of series with $a_n = 2n - 7$

19. First 125 terms of series with $a_n = 5n + 1$

20. First 75 terms of series with $a_n = 4n + 3$

21. First 65 terms of series with $a_n = -4n - 1$

22. First 90 terms of series with $a_n = -3n - 2$

For Problems 23–34, use arithmetic sequences and series to help solve the problem.

23. Find the sum of the first 350 positive even whole numbers.

24. Find the sum of the first 200 odd whole numbers.

25. Find the sum of all odd whole numbers between 15 and 397, inclusive.

26. Find the sum of all even whole numbers between 14 and 286, inclusive.

27. An auditorium has 20 seats in the front row, 24 seats in the second row, 28 seats in the third row, and so on, for 15 rows. How many seats are there in the last row? How many seats are there in the auditorium?

28. A pile of wood has 15 logs in the bottom row, 14 logs in the next to the bottom row, and so on, with one less log in each row until the top row, which consists of 1 log. How many logs are there in the pile?

29. A raffle is organized so that the amount paid for each ticket is determined by a number on the ticket. The tickets are numbered with the consecutive odd whole numbers, 1, 3, 5, 7, ... Each participant pays as many cents as the number on the ticket drawn. How much money will the raffle take in if 1000 tickets are sold?

30. A woman invests $700 at 13% simple interest at the beginning of each year for a period of 15 years. Find the total accumulated value of all of the investments at the end of the 15-year period.

31. A man started to work in 1970 at an annual salary of $18,500. He received a $1500 raise each year through 1982. What were his total earnings for the 13-year period?

32. A well-driller charges $9.00 per foot for the first 10 feet, $9.10 per foot for the next 10 feet, $9.20 per foot for the next 10 feet, and so on; he continues to increase the price by $0.10 per foot for succeeding intervals of 10 feet. How much would it cost to drill a well with a depth of 150 feet?

33. A display in a grocery store has cans stacked with 25 cans in the bottom row, 23 cans in the second row from the bottom, 21 cans in the third row from the bottom, and so on until there is only 1 can in the top row. How many cans are there in the display?

34. Suppose that a person starts on the first day of August and saves a dime the first day, $0.20 the second day, $0.30 the third day, and continues to save $0.10 more per day than the previous day. How much could be saved in the 31 days of August?

Miscellaneous Problems

35. We can express a series where the general term is known in a convenient and compact form using the symbol $\sum$ along with the general term expression. For example, consider the finite arithmetic series

$$1 + 3 + 5 + 7 + 9 + 11$$

where the general term is $a_n = 2n - 1$. This series can be expressed in *summation notation* as

$$\sum_{i=1}^{6} (2i - 1)$$

where the letter i is used as the **index of summation**. The individual terms of the series can be generated by successively replacing i in the expression $(2i - 1)$ with the numbers 1, 2, 3, 4, 5, and 6. Thus, the first term is $2(1) - 1 = 1$, the second term $2(2) - 1 = 3$, the third term is $2(3) - 1 = 5$, and so on. Write out the terms and find the sum of each of the following series.

(a) $\displaystyle\sum_{i=1}^{3} (5i + 2)$

(b) $\displaystyle\sum_{i=1}^{4} (6i - 7)$

(c) $\displaystyle\sum_{i=1}^{6} (-2i - 1)$

(d) $\displaystyle\sum_{i=1}^{5} (-3i + 4)$

(e) $\displaystyle\sum_{i=1}^{5} 3i$

(f) $\displaystyle\sum_{i=1}^{6} -4i$

36. Write each of the following in summation notation. For example, since $3 + 8 + 13 + 18 + 23 + 28$ is an arithmetic series, the general term formula $a_n = a_1 + (n-1)d$ yields

$$a_n = 3 + (n-1)5$$
$$= 3 + 5n - 5$$
$$= 5n - 2.$$

Now using i as an index of summation, we can write

$$\sum_{i=1}^{6} (5i - 2).$$

(a) $2 + 5 + 8 + 11 + 14$

(b) $8 + 15 + 22 + 29 + 36 + 43$

(c) $1 + (-1) + (-3) + (-5) + (-7)$

(d) $(-5) + (-9) + (-13) + (-17) + (-21) + (-25) + (-29)$

15.3

Geometric Sequences and Series

A **geometric sequence** or **geometric progression** is a sequence in which each term after the first is obtained by multiplying the preceding term by a common multiplier. The common multiplier is called the **common ratio** of the sequence. The following geometric sequences have common ratios of $2, 3, \frac{1}{2}$, and -4, respectively.

$$1, 2, 4, 8, 16, \ldots, \qquad 3, 9, 27, 81, 243, \ldots,$$

$$8, 4, 2, 1, \frac{1}{2}, \ldots, \qquad 1, -4, 16, -64, 256, \ldots$$

We find the common ratio of a geometric sequence by dividing a term (other than the first term) by the preceding term.

In a more general setting we say that the sequence

$$a_1, a_2, a_3, a_4, \ldots, a_n, \ldots$$

is a geometric sequence if and only if there is a nonzero real number r, such that

$$a_{k+1} = ra_k \tag{1}$$

for every positive integer k. The nonzero real number r is called the common ratio.

Equation (1) can be used to generate a general geometric sequence that has a_1 as a first term and r as a common ratio. We can proceed as follows.

first term	a_1
second term	$a_1 r$
third term	$a_1 r^2$ $(a_1 r)(r) = a_1 r^2$
fourth term	$a_1 r^3$ $(a_1 r^2)(r) = a_1 r^3$
$\vdots$	
nth term	$a_1 r^{n-1}$

Thus, the *general term of a geometric sequence* is given by

$$a_n = a_1 r^{n-1}$$

where a_1 is the first term and r is the common ratio.

EXAMPLE 1   Find the general term for the geometric sequence $2, 4, 8, 16, \ldots$.

Solution   Using $a_n = a_1 r^{n-1}$, we obtain

$$a_n = 2(2)^{n-1} \qquad r = 4/2 = 8/4 = 16/8 = 2$$

$$= 2^n. \qquad 2^1(2)^{n-1} = 2^{1+n-1} = 2^n$$ ∎

EXAMPLE 2   Find the 10th term of the geometric sequence $9, 3, 1, \ldots$.

Solution   Using $a_n = a_1 r^{n-1}$, we can find the 10th term as follows.

$$a_{10} = 9\left(\frac{1}{3}\right)^{10-1}$$

$$= 9\left(\frac{1}{3}\right)^9$$

$$= 9\left(\frac{1}{19,683}\right)$$

$$= \frac{1}{2187}$$ ∎

A **geometric series** is the indicated sum of a geometric sequence. The following are examples of geometric series.

$$1 + 2 + 4 + 8 + \cdots,$$
$$3 + 9 + 27 + 81 + \cdots,$$
$$8 + 4 + 2 + 1 + \cdots,$$
$$1 + (-4) + 16 + (-64) + \cdots$$

Before we develop a general formula for finding the sum of a geometric series, let's consider a specific example.

EXAMPLE 3   Find the sum of $1 + 2 + 4 + 8 + \cdots + 512$.

Solution   Let S represent the sum and we can proceed as follows.

$$S = 1 + 2 + 4 + 8 + \cdots + 512, \tag{2}$$
$$2S = \quad\; 2 + 4 + 8 + \cdots + 512 + 1024 \tag{3}$$

Equation (3) is the result of multiplying both sides of equation (2) by 2. Subtract equation (2) from equation (3) to yield

$$S = 1024 - 1 = 1023. \qquad\blacksquare$$

Now let's consider the general geometric series

$$a_1 + a_1 r + a_1 r^2 + \cdots + a_1 r^{n-1}.$$

By applying a procedure similar to the one used in Example 3 we can develop a formula for finding the sum of the first n terms of any geometric series.

Let S_n represent the sum of the first n terms. Thus,

$$S_n = a_1 + a_1 r + a_1 r^2 + \cdots + a_1 r^{n-1}. \tag{4}$$

Multiply both sides of equation (4) by the common ratio r to produce

$$rS_n = a_1 r + a_1 r^2 + a_1 r^3 + \cdots + a_1 r^{n-1} + a_1 r^n. \tag{5}$$

Subtract equation (4) from equation (5) to yield

$$rS_n - S_n = a_1 r^n - a_1.$$

Apply the distributive property on the left side and then solve for S_n to obtain

$$S_n(r - 1) = a_1 r^n - a_1$$
$$S_n = \frac{a_1 r^n - a_1}{r - 1}, \qquad r \neq 1.$$

Therefore, the sum of the first n terms of a geometric series that has a first term of a_1 and a common ratio of r is given by

$$S_n = \frac{a_1 r^n - a_1}{r - 1}, \qquad r \neq 1.$$

EXAMPLE 4   Find the sum of the first 7 terms of the geometric series $2 + 6 + 18 + \cdots$.

Solution   Use the sum formula to obtain

$$S_7 = \frac{2(3)^7 - 2}{3 - 1}$$

$$= \frac{2(3^7 - 1)}{2}$$

$$= 3^7 - 1$$

$$= 2187 - 1$$

$$= 2186.$$ ∎

If the common ratio of a geometric series is less than 1, it may be more convenient to change the form of the sum formula. That is, we can change the fraction $\frac{a_1 r^n - a_1}{r - 1}$ to $\frac{a_1 - a_1 r^n}{1 - r}$ by multiplying both the numerator and denominator by -1. Thus, using $S_n = \frac{a_1 - a_1 r^n}{1 - r}$ when $r < 1$ can sometimes avoid unnecessary work with negative numbers, as the next example demonstrates. ∎

EXAMPLE 5   Find the sum of the geometric series $1 + \frac{1}{2} + \frac{1}{4} + \cdots + \frac{1}{256}$.

Solution A   To use the sum formula we need to know the number of terms, which can be found by simply counting them or applying the nth term formula as follows.

$$u_n = a_1 r^{n-1}$$

$$\frac{1}{256} = 1\left(\frac{1}{2}\right)^{n-1}$$

$$\left(\frac{1}{2}\right)^8 = \left(\frac{1}{2}\right)^{n-1}$$

$$8 = n - 1 \qquad \text{Remember that "if } b^n = b^m,$$
$$\text{then } n = m.\text{"}$$
$$9 = n$$

Using $n = 9$, $a_1 = 1$, and $r = \dfrac{1}{2}$ in the form of the sum formula

$$S_n = \frac{a_1 - a_1 r^n}{1 - r},$$

we obtain

$$S_9 = \frac{1 - 1\left(\dfrac{1}{2}\right)^9}{1 - \dfrac{1}{2}}$$

$$= \frac{1 - \dfrac{1}{512}}{\dfrac{1}{2}} = \frac{\dfrac{511}{512}}{\dfrac{1}{2}}$$

$$= \left(\frac{511}{512}\right)\left(\frac{2}{1}\right) = \frac{511}{256} \qquad \text{or} \qquad 1\frac{255}{256}.$$

You should realize that a problem such as Example 5 can be done *without* using the sum formula; you can apply the general technique used to develop the formula. Solution B illustrates this approach.

Solution B   Let S represent the desired sum. Thus,

$$S = 1 + \frac{1}{2} + \frac{1}{4} + \cdots + \frac{1}{256}.$$

Multiply both sides by $\dfrac{1}{2}$ (the common ratio).

$$\frac{1}{2}S = \frac{1}{2} + \frac{1}{4} + \cdots + \frac{1}{256} + \frac{1}{512}$$

Subtract the second equation from the first equation to produce

$$\frac{1}{2}S = 1 - \frac{1}{512}$$

$$\frac{1}{2}S = \frac{511}{512}$$

$$S = \frac{511}{256} = 1\frac{255}{256}. \qquad \blacksquare$$

EXAMPLE 6   Suppose your employer agrees to pay you a penny for your first day's wages and then agrees to double your pay on each succeeding day. How much will you earn on the 15th day? What will be your total earnings for the first 15 days?

Solution   The terms of the geometric series $1 + 2 + 4 + 8 + \cdots$ depict your daily wages and the sum of the first 15 terms is your total earnings for the 15 days. The formula $a_n = a_1 r^{n-1}$ can be used to find the 15th day's wages.

$$a_{15} = (1)(2)^{14} = 16{,}384.$$

Since the terms of the series are expressed in cents, your wages for the 15th day would be $163.84. Now using the sum formula we can find your total earnings as follows.

$$S_n = \frac{a_1 r^n - a_1}{r - 1}$$

$$S_{15} = \frac{1(2)^{15} - 1}{1} = 32{,}768 - 1 = 32{,}767.$$

Thus, for the 15 days you would earn a total of $327.67.   ■

Problem Set 15.3

For Problems 1–12, find the general term (nth term) of each geometric sequence.

1. $1, 3, 9, 27, \ldots$

2. $1, 2, 4, 8, \ldots$

3. $2, 8, 32, 128, \ldots$

4. $3, 9, 27, 81, \ldots$

5. $1, \dfrac{1}{3}, \dfrac{1}{9}, \dfrac{1}{27}, \ldots$

6. $\dfrac{1}{2}, \dfrac{1}{4}, \dfrac{1}{8}, \dfrac{1}{16}, \ldots$

7. $.2, .04, .008, .0016, \ldots$

8. $1, .3, .09, .027, \ldots$

9. $9, 6, 4, \dfrac{8}{3}, \ldots$

10. $6, 2, \dfrac{2}{3}, \dfrac{2}{9}, \ldots$

11. $1, -4, 16, -64, \ldots$

12. $1, -2, 4, -8, \ldots$

For Problems 13–18, find the indicated term of the geometric sequence.

13. 12th term of $\dfrac{1}{9}, \dfrac{1}{3}, 1, 3, \ldots$

14. 9th term of $2, 4, 8, 16, \ldots$

15. 10th term of $1, -2, 4, -8, \ldots$

16. 8th term of $\dfrac{1}{2}, \dfrac{1}{8}, \dfrac{1}{32}, \dfrac{1}{128}, \cdots$

17. 9th term of $-1, -\dfrac{3}{2}, -\dfrac{9}{4}, -\dfrac{27}{8}, \cdots$

18. 11th term of $1, \dfrac{2}{3}, \dfrac{4}{9}, \dfrac{8}{27}, \cdots$

For Problems 19–24, find the sum of the indicated number of terms of each geometric series.

19. First 10 terms of $\dfrac{1}{2} + \dfrac{3}{2} + \dfrac{9}{2} + \dfrac{27}{2} + \cdots$

20. First 9 terms of $1 + 2 + 4 + 8 + \cdots$

21. First 9 terms of $-2 + 6 + (-18) + 54 + \cdots$

22. First 10 terms of $-4 + 8 + (-16) + 32 + \cdots$

23. First 7 terms of $1 + 3 + 9 + 27 + \cdots$

24. First 8 terms of $4 + 2 + 1 + \dfrac{1}{2} + \cdots$

For Problems 25–30, find the sum of the indicated number of terms of the geometric series with the given nth term.

25. First 9 terms of series where $a_n = 2^{n-1}$

26. First 8 terms of series where $a_n = 3^n$

27. First 8 terms of series where $a_n = 2(3)^n$

28. First 10 terms of series where $a_n = \dfrac{1}{2^{n-4}}$

29. First 12 terms of series where $a_n = (-2)^n$

30. First 9 terms of series where $a_n = (-3)^{n-1}$

For Problems 31–36, find the sum of each finite geometric series.

31. $1 + 3 + 9 + \cdots + 729$ **32.** $2 + 8 + 32 + \cdots + 2048$

33. $1 + \dfrac{1}{2} + \dfrac{1}{4} + \cdots + \dfrac{1}{1024}$ **34.** $1 + (-2) + 4 + \cdots + (-128)$

35. $8 + 4 + 2 + \cdots + \dfrac{1}{32}$ **36.** $2 + 6 + 18 + \cdots + 4374$

For Problems 37–48, use geometric sequences and series to help solve the problem.

37. Find the common ratio of a geometric sequence if the 2nd term is $\dfrac{1}{6}$ and the 5th term is $\dfrac{1}{48}$.

38. Find the first term of a geometric sequence if the 5th term is $\dfrac{32}{3}$ and the common ratio is 2.

39. Find the sum of the first 16 terms of the geometric series where $a_n = (-1)^n$. Also find the sum of the first 19 terms.

40. A fungus culture growing under controlled conditions doubles in size each day. How many units will the culture contain after 7 days if it originally contained 5 units?

41. A tank contains 16,000 liters of water. Each day one-half of the water in the tank is removed and not replaced. How much water remains in the tank at the end of the 7th day?

42. Suppose that you save 25 cents the first day of a week, 50 cents the second day, $1 the third day, and continue to double your savings each day. How much will you save on the 7th day? What will be your total savings for the week?

43. Suppose you save a nickel the first day of a month, a dime the second day, 20 cents the third day, and continue to double your savings each day. How much will you save on the 12th day of the month? What will be your total savings for the first 12 days?

44. Suppose an element has a half-life of 3 hours. This means that if n grams of it exist at a specific time, then only $\frac{1}{2}n$ grams remain 3 hours later. If at a particular moment we have 40 grams of the element, how much of it remains 24 hours later?

45. A rubber ball is dropped from a height of 486 meters and each time it rebounds one-third of the height from which it last fell. How far has the ball traveled by the time it strikes the ground for the 7th time?

46. A pump is attached to a container for the purpose of creating a vacuum. For each stroke of the pump, one-fourth of the air remaining in the container is removed.

(**a**) Form a geometric sequence where each term represents the fractional part of the air *that still remains* in the container after each stroke. Then use this sequence to find out how much of the air remains after 6 strokes.

(**b**) Form a geometric sequence where each term represents the fractional part of the air *being removed* from the container on each stroke of the pump. Then use this sequence (or the associated *series*) to find out how much of the air remains after 6 strokes.

47. If you pay $9500 for a car and its value depreciates 10% per year, how much will it be worth in 5 years?

48. Suppose that you could get a job that pays only a penny for the first day of employment, but doubles your wages each succeeding day. How much would you be earning on the 31st day of your employment?

Miscellaneous Problems

For Problems 49–54, use your calculator to help find the indicated term of each geometric sequence.

49. 20th term of $2, 4, 8, 16, \ldots$

50. 15th term of $3, 9, 27, 81, \ldots$

51. 12th term of $\frac{2}{3}, \frac{4}{9}, \frac{8}{27}, \frac{16}{81}, \ldots$

52. 10th term of $-\frac{3}{4}, \frac{9}{16}, -\frac{27}{64}, \frac{81}{256}, \ldots$

53. 11th term of $-\dfrac{3}{2}, \dfrac{9}{4}, -\dfrac{27}{8}, \dfrac{81}{16}, \ldots$

54. 6th term of the sequence where $a_n = (0.1)^n$

55. In Problem 35 of Problem Set 15.2 we introduced the summation notation, which can also be used with geometric series. For example, the series

$$2 + 4 + 8 + 16 + 32$$

can be expressed as

$$\sum_{i=1}^{5} 2^i.$$

Write out the terms and find the sum of each of the following.

(a) $\displaystyle\sum_{i=1}^{6} 2^i$ (b) $\displaystyle\sum_{i=1}^{5} 3^i$ (c) $\displaystyle\sum_{i=1}^{5} 2^{i-1}$

(d) $\displaystyle\sum_{i=1}^{6} \left(\frac{1}{2}\right)^{i+1}$ (e) $\displaystyle\sum_{i=1}^{4} \left(\frac{2}{3}\right)^i$ (f) $\displaystyle\sum_{i=1}^{5} \left(-\frac{3}{4}\right)^i$

15.4

Infinite Geometric Series

In the previous section we used the formula

$$S_n = \frac{a_1 - a_1 r^n}{1 - r}, \qquad r \neq 1 \tag{1}$$

to find the sum of the first n terms of a geometric series. By using the property, $\dfrac{a - b}{c} = \dfrac{a}{c} - \dfrac{b}{c}$, we can express the right side of equation (1) in terms of two fractions as follows.

$$S_n = \frac{a_1 - a_1 r^n}{1 - r} = \frac{a_1}{1 - r} - \frac{a_1 r^n}{1 - r}, \qquad r \neq 1 \tag{2}$$

Now let's examine the behavior of r^n for $|r| < 1$, that is, for $-1 < r < 1$. For example, suppose that $r = \dfrac{1}{3}$; then

$$r^2 = \left(\frac{1}{3}\right)^2 = \frac{1}{9}, \qquad r^3 = \left(\frac{1}{3}\right)^3 = \frac{1}{27},$$

$$r^4 = \left(\frac{1}{3}\right)^4 = \frac{1}{81}, \qquad r^5 = \left(\frac{1}{3}\right)^5 = \frac{1}{243},$$

and so on. We can make $\left(\dfrac{1}{3}\right)^n$ as close to 0 as we please by taking sufficiently large values for n. In general, for values of r such that $|r| < 1$, the expression r^n will approach 0 as n increases. Therefore, in equation (2) the fraction $\dfrac{a_1 r^n}{1 - r}$ will approach 0 as n increases and we say that the *sum of an infinite geometric series* is given by

$$S_\infty = \frac{a_1}{1 - r}, \qquad |r| < 1.$$

EXAMPLE 1   Find the sum of the infinite geometric series

$$1 + \frac{2}{3} + \frac{4}{9} + \frac{8}{27} + \cdots.$$

Solution   Since $a_1 = 1$ and $r = \dfrac{2}{3}$, we obtain

$$S_\infty = \frac{1}{1 - \dfrac{2}{3}}$$

$$= \frac{1}{\dfrac{1}{3}} = 3.$$

In Example 1, by stating $S_\infty = 3$ we mean that as we add more and more terms, the sum approaches 3 as follows.

first term: 1

sum of first 2 terms: $1 + \dfrac{2}{3} = 1\dfrac{2}{3}$

sum of first 3 terms: $1 + \dfrac{2}{3} + \dfrac{4}{9} = 2\dfrac{1}{9}$

sum of first 4 terms: $1 + \dfrac{2}{3} + \dfrac{4}{9} + \dfrac{8}{27} = 2\dfrac{11}{27}$

sum of first 5 terms: $1 + \dfrac{2}{3} + \dfrac{4}{9} + \dfrac{8}{27} + \dfrac{16}{81} = 2\dfrac{49}{81}$, etc.

EXAMPLE 2   Find the sum of the infinite geometric series

$$\frac{1}{2} - \frac{1}{4} + \frac{1}{8} - \frac{1}{16} + \cdots.$$

Solution   Since $a_1 = \frac{1}{2}$ and $r = -\frac{1}{2}$, we obtain

$$S_\infty = \frac{\frac{1}{2}}{1 - \left(-\frac{1}{2}\right)} = \frac{\frac{1}{2}}{\frac{3}{2}} = \frac{1}{3}.$$ ∎

If $|r| > 1$, the absolute value of r^n increases without bound as n increases. Consider the following two examples and notice the unbounded growth of the absolute value of r^n.

Let $r = 2$	*Let $r = -3$*	
$r^2 = 2^2 = 4$,	$r^2 = (-3)^2 = 9$,	
$r^3 = 2^3 = 8$,	$r^3 = (-3)^3 = -27$,	$|{-27}| = 27$,
$r^4 = 2^4 = 16$,	$r^4 = (-3)^4 = 81$,	
$r^5 = 2^5 = 32$,	$r^5 = (-3)^5 = -243$,	$|{-243}| = 243$,
etc.	etc.	

If $r = 1$, then $S_n = na_1$, and as n increases without bound, $|S_n|$ also increases without bound. If $r = -1$, then S_n will either be a_1 or 0. Therefore, we say that the sum of any infinite geometric sequence where $|r| \geq 1$ *does not exist*.

Repeating Decimals as Infinite Geometric Series

In Section 1.1 we learned that a rational number is a number that has either a terminating or repeating decimal representation. For example,

$$0.23, \quad 0.147, \quad 0.\overline{3}, \quad 0.\overline{14}, \quad \text{and} \quad 0.5\overline{81}$$

are examples of rational numbers. (Remember that $0.\overline{3}$ means $0.333\ldots$.) Our knowledge of place value provides the basis for changing terminating decimals such as 0.23 and 0.147 to $\frac{a}{b}$ form, where a and b are integers, $b \neq 0$.

$$0.23 = \frac{23}{100},$$

$$0.147 = \frac{147}{1000}$$

However, changing repeating decimals to $\frac{a}{b}$ form requires a different technique and our work with infinite geometric series provides the basis for one such approach. Consider the following examples.

EXAMPLE 3

Change $0.\overline{3}$ to $\frac{a}{b}$ form, where a and b are integers, $b \neq 0$.

Solution

We can write the repeating decimal $0.\overline{3}$ as the infinite geometric series

$$0.3 + 0.03 + 0.003 + 0.0003 + \cdots$$

with $a_1 = 0.3$ and $r = 0.1$. Therefore, we can use the sum formula and obtain

$$S_\infty = \frac{a_1}{1-r} = \frac{0.3}{1-0.1} = \frac{0.3}{0.9} = \frac{3}{9} = \frac{1}{3}.$$

So, $0.\overline{3} = \frac{1}{3}$. ∎

EXAMPLE 4

Change $0.\overline{14}$ to $\frac{a}{b}$ form, where a and b are integers, $b \neq 0$.

Solution

We can write the repeating decimal $0.\overline{14}$ as the infinite geometric series

$$0.14 + 0.0014 + 0.000014 + \cdots$$

with $a_1 = 0.14$ and $r = 0.01$. The sum formula produces

$$S_\infty = \frac{0.14}{1-0.01} = \frac{0.14}{0.99} = \frac{14}{99}.$$

Thus, $0.\overline{14} = \frac{14}{99}$. ∎

If the repeating block of digits does not begin immediately after the decimal point we can make a slight adjustment, as the final example illustrates.

EXAMPLE 5

Change $0.5\overline{81}$ to $\frac{a}{b}$ form, where a and b are integers, $b \neq 0$.

Solution

We can write the repeating decimal $0.5\overline{81}$ as

$$[0.5] + [0.081 + 0.00081 + 0.0000081 + \cdots]$$

where

$$0.081 + 0.00081 + 0.0000081 + \cdots$$

is an infinite geometric series with $a_1 = 0.081$ and $r = 0.01$. Thus,

$$S_\infty = \frac{0.081}{1 - 0.01} = \frac{0.081}{0.99} = \frac{81}{990} = \frac{9}{110}.$$

Therefore,

$$0.5\overline{81} = 0.5 + \frac{9}{110}$$

$$= \frac{5}{10} + \frac{9}{110}$$

$$= \frac{55}{110} + \frac{9}{110}$$

$$= \frac{64}{110}$$

$$= \frac{32}{55}.$$

■

Problem Set 15.4

For Problems 1–20, find the sum of the infinite geometric series. If the series has no sum, so state.

1. $1 + \dfrac{3}{4} + \dfrac{9}{16} + \dfrac{27}{64} + \cdots$

2. $\dfrac{2}{3} + \dfrac{2}{9} + \dfrac{2}{27} + \dfrac{2}{81} + \cdots$

3. $\dfrac{1}{2} + \dfrac{1}{4} + \dfrac{1}{8} + \dfrac{1}{16} + \cdots$

4. $1 + \dfrac{1}{2} + \dfrac{1}{4} + \dfrac{1}{8} + \cdots$

5. $\dfrac{2}{3} + \dfrac{4}{9} + \dfrac{8}{27} + \dfrac{16}{81} + \cdots$

6. $\dfrac{1}{3} + \dfrac{1}{9} + \dfrac{1}{27} + \dfrac{1}{81} + \cdots$

7. $1 - \dfrac{1}{2} + \dfrac{1}{4} - \dfrac{1}{8} + \cdots$

8. $1 + 2 + 4 + 8 + \cdots$

9. $6 + 2 + \dfrac{2}{3} + \dfrac{2}{9} + \cdots$

10. $4 + (-2) + 1 + \left(-\dfrac{1}{2}\right) + \cdots$

11. $2 + (-6) + 18 + (-54) + \cdots$

12. $4 + 2 + 1 + \dfrac{1}{2} + \cdots$

13. $1 + \left(-\dfrac{3}{4}\right) + \dfrac{9}{16} + \left(-\dfrac{27}{64}\right) + \cdots$

14. $9 - 3 + 1 - \dfrac{1}{3} + \cdots$

15. $8 - 4 + 2 - 1 + \cdots$

16. $5 + 3 + \dfrac{9}{5} + \dfrac{27}{25} + \cdots$

17. $1 + \dfrac{3}{2} + \dfrac{9}{4} + \dfrac{27}{8} + \cdots$

18. $1 - \dfrac{4}{3} + \dfrac{16}{9} - \dfrac{64}{27} + \cdots$

19. $27 + 9 + 3 + 1 + \cdots$

20. $9 + 3 + 1 + \dfrac{1}{3} + \cdots$

For Problems 21–34, change each repeating decimal to $\frac{a}{b}$ form, where a and b are integers,

$b \neq 0$. Express $\frac{a}{b}$ in reduced form.

21. $.\overline{4}$	**22.** $.\overline{7}$	**23.** $.\overline{47}$	**24.** $.\overline{23}$	**25.** $.\overline{45}$
26. $.\overline{72}$	**27.** $.\overline{427}$	**28.** $.\overline{129}$	**29.** $.4\overline{6}$	**30.** $.8\overline{6}$
31. $2.\overline{18}$	**32.** $2.9\overline{6}$	**33.** $.4\overline{27}$	**34.** $.2\overline{36}$	

Thoughts into Words

35. Explain the difference between an arithmetic sequence and a geometric sequence.

36. What does it mean to say that the sum of the infinite geometric series
$1 + \dfrac{1}{2} + \dfrac{1}{4} + \dfrac{1}{8} + \cdots$ is 2?

37. What do we mean when we state that the infinite geometric series $1 + 2 + 4 + 8 + \cdots$ has no sum?

15.5

Binomial Expansions

In Chapter 6 we used the pattern $(x + y)^2 = x^2 + 2xy + y^2$ to square binomials and used the pattern $(x + y)^3 = x^3 + 3x^2y + 3xy^2 + y^3$ to cube binomials. At this time, we can extend those ideas to arrive at a pattern that will allow us to write the expansion of $(x + y)^n$, where n is *any* positive integer. Let's begin by looking at some specific expansions that we can verify by direct multiplication.

$$(x + y)^1 = x + y,$$
$$(x + y)^2 = x^2 + 2xy + y^2,$$
$$(x + y)^3 = x^3 + 3x^2y + 3xy^2 + y^3,$$
$$(x + y)^4 = x^4 + 4x^3y + 6x^2y^2 + 4xy^3 + y^4,$$
$$(x + y)^5 = x^5 + 5x^4y + 10x^3y^2 + 10x^2y^3 + 5xy^4 + y^5$$

First, note the patterns of the exponents for x and y on a term-by-term basis. The exponents of x begin with the exponent of the binomial and term-by-term decrease by 1 until the last term has x^0, which is 1. The exponents of y begin with $0(y^0 = 1)$ and term-by-term increase by 1 until the last term contains y to the power of the original binomial. In other words, the variables in the expansion of $(x + y)^n$ have the following pattern.

$$x^n, \quad x^{n-1}y, \quad x^{n-2}y^2, \quad x^{n-3}y^3, \quad \ldots, \quad xy^{n-1}, \quad y^n.$$

Notice that the sum of the exponents of x and y for each term is n.

Next, let's arrange the **coefficients** in the following triangular formation that yields an easy-to-remember pattern.

$$
\begin{array}{ccccccc}
 & & & 1 & & 1 & & \\
 & & 1 & & 2 & & 1 & \\
 & 1 & & 3 & & 3 & & 1 \\
1 & & 4 & & 6 & & 4 & & 1 \\
\end{array}
$$

1   5   10   10   5   1

The number of the row of the formation contains the coefficients of the expansion of $(x + y)$ to that power. For example, the 5th row contains 1   5   10   10   5   1, which are the coefficients of the terms of the expansion of $(x + y)^5$. Furthermore, each row can be formed from the previous row as follows.

1. Start and end each row with 1.

2. All other entries result from adding the two numbers in the row immediately above, one number to the left and one number to the right.

Thus, from row 5 we can form row 6 as follows.

row 5:   1   5   10   10   5   1
               add   add   add   add   add
row 6:   1   6   15   20   15   6   1

We can use the row 6 coefficients and our previous discussion relative to the exponents and write out the expansion for $(x + y)^6$.

$$(x + y)^6 = x^6 + 6x^5y + 15x^4y^2 + 20x^3y^3 + 15x^2y^4 + 6xy^5 + y^6$$

REMARK We often refer to the triangular formation of numbers that we have been discussing as Pascal's triangle. This is in honor of Blaise Pascal, a 17th century mathematician, to whom the discovery of this pattern is attributed.

Although Pascal's triangle will work for any positive integral power of a binomial, it does become somewhat impractical for large powers. So we need another technique for determining the coefficients. Let's look at the following notational agreements. $n!$ (read "n factorial") means $n(n - 1)(n - 2) \cdots 1$, where n is any positive integer. For example,

$3!$ means $3 \cdot 2 \cdot 1 = 6$,

$5!$ means $5 \cdot 4 \cdot 3 \cdot 2 \cdot 1 = 120$.

We also agree that $0! = 1$. (Note that both $0!$ and $1!$ equal 1.)

Let us now use the factorial notation and state the expansion of the general case $(x + y)^n$, where n is any positive integer.

$$(x + y)^n = x^n + nx^{n-1}y + \frac{n(n - 1)}{2!}x^{n-2}y^2 + \frac{n(n - 1)(n - 2)}{3!}x^{n-3}y^3 + \cdots + y^n$$

The binomial expansion for the general case may look a little confusing, but actually it is quite easy to apply once you try it a few times on some specific examples. Remember the decreasing pattern for the exponents of x and the increasing pattern for the exponents of y. Furthermore, note the following pattern of the coefficients.

$$1, \quad n, \quad \frac{n(n-1)}{2!}, \quad \frac{n(n-1)(n-2)}{3!}, \quad \text{etc.}$$

Keep these ideas in mind as you study the following examples.

EXAMPLE 1   Expand $(x + y)^7$.

Solution

$$(x + y)^7 = x^7 + 7x^6y + \frac{7 \cdot 6}{2!} x^5y^2 + \frac{7 \cdot 6 \cdot 5}{3!} x^4y^3$$

$$+ \frac{7 \cdot 6 \cdot 5 \cdot 4}{4!} x^3y^4 + \frac{7 \cdot 6 \cdot 5 \cdot 4 \cdot 3}{5!} x^2y^5$$

$$+ \frac{7 \cdot 6 \cdot 5 \cdot 4 \cdot 3 \cdot 2}{6!} xy^6 + y^7$$

$$= x^7 + 7x^6y + 21x^5y^2 + 35x^4y^3 + 35x^3y^4 + 21x^2y^5 + 7xy^6 + y^7$$

∎

EXAMPLE 2   Expand $(x - y)^5$.

Solution   We shall treat $(x - y)^5$ as $(x + (-y))^5$.

$$(x + (-y))^5 = x^5 + 5x^4(-y) + \frac{5 \cdot 4}{2!} x^3(-y)^2 + \frac{5 \cdot 4 \cdot 3}{3!} x^2(-y)^3$$

$$+ \frac{5 \cdot 4 \cdot 3 \cdot 2}{4!} x(-y)^4 + (-y)^5$$

$$= x^5 - 5x^4y + 10x^3y^2 - 10x^2y^3 + 5xy^4 - y^5$$

∎

EXAMPLE 3   Expand and simplify $(2a + 3b)^4$.

Solution   Let $x = 2a$ and $y = 3b$.

$$(2a + 3b)^4 = (2a)^4 + 4(2a)^3(3b) + \frac{4 \cdot 3}{2!} (2a)^2(3b)^2$$

$$+ \frac{4 \cdot 3 \cdot 2}{3!} (2a)(3b)^3 + (3b)^4$$

$$= 16a^4 + 96a^3b + 216a^2b^2 + 216ab^3 + 81b^4$$

∎

Finding Specific Terms

Sometimes it is convenient to find a specific term of a binomial expansion without writing out the entire expansion. For example, suppose that we need the 6th term of the expansion $(x + y)^{12}$. We could proceed as follows.

The 6th term will contain y^5. (Note in the general expansion that the **exponent of y is always one less than the number of the term**.) Since the sum of the exponents for x and y must be 12 (the exponent of the binomial), the 6th term will also contain x^7. Again looking back at the general binomial expansion, note that the **denominators of the coefficients** are of the form $r!$ where the value of r agrees with the exponent of y for each term. Thus, if we have y^5, the denominator of the coefficient is $5!$ In the general expansion each **numerator of a coefficient** contains r factors where the first factor is the exponent of the binomial and each succeeding factor is one less than the preceding one. Thus, the 6th term of $(x + y)^{12}$ is $\dfrac{12 \cdot 11 \cdot 10 \cdot 9 \cdot 8}{5!} x^7 y^5$, which simplifies to $792 x^7 y^5$.

EXAMPLE 4   Find the 4th term of $(3a + 2b)^7$.

Solution   The 4th term will contain $(2b)^3$ and therefore $(3a)^4$. The coefficient is $\dfrac{7 \cdot 6 \cdot 5}{3!}$. Therefore, the 4th term is

$$\frac{7 \cdot 6 \cdot 5}{3!}(3a)^4(2b)^3,$$

which simplifies to $22{,}680 a^4 b^3$.   ∎

Problem Set 15.5

For Problems 1–6, use Pascal's triangle to help expand each of the following.

1. $(x + y)^8$ 2. $(x + y)^7$
3. $(3x + y)^4$ 4. $(x + 2y)^4$
5. $(x - y)^5$ 6. $(x - y)^4$

For Problems 7–20, expand and simplify.

7. $(x + y)^{10}$ 8. $(x + y)^9$

9. $(2x + y)^6$

10. $(x + 3y)^5$

11. $(x - 3y)^5$

12. $(2x - y)^6$

13. $(3a - 2b)^5$

14. $(2a - 3b)^4$

15. $(x + y^3)^6$

16. $(x^2 + y)^5$

17. $(x + 2)^7$

18. $(x + 3)^6$

19. $(x - 3)^4$

20. $(x - 1)^9$

For Problems 21–24, write the first four terms of the expansion.

21. $(x + y)^{15}$

22. $(x + y)^{12}$

23. $(a - 2b)^{13}$

24. $(x - y)^{20}$

For Problems 25–30, find the indicated term of the expansion.

25. 7th term of $(x + y)^{11}$

26. 4th term of $(x + y)^8$

27. 4th term of $(x - 2y)^6$

28. 5th term of $(x - y)^9$

29. 3rd term of $(2x - 5y)^5$

30. 6th term of $(3a + b)^7$

Chapter 15 Summary

(15.1) An **infinite sequence** is a function whose domain is the set of positive integers. We frequently express a general infinite sequence as

$$a_1, a_2, a_3, \ldots, a_n, \ldots$$

where a_1 is the first term, a_2 the second term, and so on, and a_n represents the general or nth term.

An **arithmetic sequence** is a sequence where there is a **common difference** between successive terms.

The **general term of an arithmetic sequence** is given by

$$a_n = a_1 + (n - 1)d$$

where a_1 is the first term and d is the common difference.

(15.2) The indicated sum of a sequence is called a **series**. The sum of the first n terms of an arithmetic series is given by

$$S_n = \frac{n(a_1 + a_n)}{2}.$$

(15.3) A **geometric sequence** is a sequence in which each term after the first is obtained by multiplying the preceding term by a common multiplier. The common multiplier is called the **common ratio** of the sequence.

The **general term of a geometric sequence** is given by

$$a_n = a_1 r^{n-1}$$

where a_1 is the first term and r is the common ratio.

The sum of the first n terms of a geometric series is given by

$$S_n = \frac{a_1 r^n - a_1}{r - 1}, \qquad r \neq 1.$$

(15.4) The sum of an infinite geometric series is given by

$$S_\infty = \frac{a_1}{1 - r}, \qquad |r| < 1.$$

Any infinite geometric series where $|r| \geq 1$ has no sum.

This sum formula can be used to change repeating decimals to $\dfrac{a}{b}$ form.

(15.5) The expansion of $(x + y)^n$, where n is a positive integer is given by

$$(x + y)^n = x^n + nx^{n-1}y + \frac{n(n-1)}{2!} x^{n-2}y^2$$

$$+ \frac{n(n-1)(n-2)}{3!} x^{n-3}y^3 + \cdots + y^n.$$

To find a specific term of a binomial expansion review Example 4.

Chapter 15 Review Problem Set

1. Find the first five terms of the sequence where the general term is given by $a_n = 2^n - 3$.

2. Find the general term of each of the following sequences.

   (a) $2, 8, 14, 20, \ldots$ (b) $3, -2, -7, -12, \ldots$

   (c) $4, 12, 36, 108, \ldots$ (d) $5, \dfrac{5}{2}, \dfrac{5}{4}, \dfrac{5}{8}, \ldots$

3. Find the indicated term of each of the following sequences.

   (a) The 8th term of $\dfrac{1}{2}, \dfrac{1}{6}, \dfrac{1}{18}, \dfrac{1}{54}, \ldots$

   (b) The 50th term of $8, 11, 14, 17, \ldots$

4. Find the sum of each of the following series.

   (a) $2 + 5 + 8 + \cdots + 149$

   (b) $4 + 8 + 16 + \cdots + 2048$

   (c) $-5 + (-9) + (-13) + \cdots + (-101)$

5. Find the sum of the first 300 odd whole numbers.

6. Find the sum of the first 6 terms of the series where $a_n = \left(\dfrac{1}{4}\right)^n$.

7. Find the sum of the first 25 terms of the series $7 + 11 + 15 + 19 + \cdots$.

8. Find the sum of all even numbers between 18 and 286, inclusive.

9. A tank contains 60,750 liters of water. Each day one-third of the water in the tank is removed and not replaced. How much water is in the tank at the end of the 6th day?

10. An object falling from rest in a vacuum falls 16 feet the first second, 48 feet the second second, 80 feet the third second, 112 feet the fourth second, and so on. How far will the object fall in 15 seconds?

11. Suppose you save a penny the first day of a month, 2 cents the second day, 3 cents the third day, and continue to increase your savings per day by a penny. What will be your total savings for the first 20 days?

12. Suppose you save a nickel the first day of a month, a dime the second day, 20 cents the third day, and continue to double your daily savings each day. What will be your total savings at the end of 13 days?

13. Find the sum of the infinite geometric series $16 + 8 + 4 + 2 \cdots$.

14. Express $0.\overline{29}$ in $\dfrac{a}{b}$ form, where a and b are integers, $b \neq 0$.

15. Expand and simplify $(2x + y)^7$.

16. Find the 6th term of $(a + b)^{10}$.

Appendices

676

A
Synthetic Division and the Factor Theorem

In Section 6.5 we discussed the process of dividing polynomials. If the divisor is of the form $x - k$, then the typical long division algorithm can be conveniently simplified into a process called synthetic division.

First, let's consider an example and use the usual division process. Then, in a step-by-step fashion, we can observe some shortcuts that will lead us into the synthetic division procedure. Consider the division problem $(2x^4 + x^3 - 17x^2 + 13x + 2) \div (x - 2)$.

$$
\begin{array}{r}
2x^3 + 5x^2 - 7x - 1 \\
x - 2\overline{\smash{)}2x^4 + x^3 - 17x^2 + 13x + 2} \\
\underline{2x^4 - 4x^3} \\
5x^3 - 17x^2 \\
\underline{5x^3 - 10x^2} \\
-7x^2 + 13x \\
\underline{-7x^2 + 14x} \\
-x + 2 \\
\underline{-x + 2}
\end{array}
$$

Notice that since the dividend $(2x^4 + x^3 - 17x^2 + 13x + 2)$ is written in descending powers of x, the quotient $(2x^3 + 5x^2 - 7x - 1)$ is produced, also in descending powers of x. In other words, the numerical coefficients are the *key issues*. So let's rewrite the above problem in terms of its coefficients.

$$
\begin{array}{r}
2 \ +5 \ -7 \ -1 \\
1 - 2\overline{\smash{)}2 \ +1 \ -17 \ +13 \ +2} \\
\underline{2 \ -4} \\
5 \ \boxed{-17} \\
\underline{5 \ -10} \\
-7 \ +13 \\
\underline{\boxed{-7} \ +14} \\
-1 \ +2 \\
\underline{\boxed{-1} \ +2}
\end{array}
$$

Now observe that the numbers circled are simply repetitions of the numbers directly above them in the format. Therefore, we can write the process in a more compact form as

$$
\begin{array}{r}
\ 2\quad\ \ 5\ -\ 7\ -\ 1 \\
\hline
-\,2)2\quad\ \ 1\ -17\ -13\ \ 2 \\
\ -4\ -10\quad\ 14\ \ 2 \\
\hline
\ \ 5\ -\ 7\ -\ 1\quad 0
\end{array}
$$

(1)
(2)
(3)
(4)

where the repetitions are omitted and where 1, the coefficient of x in the divisor, is omitted.

Notice that line (4) reveals all of the coefficients of the quotient (line (1)), except for the first coefficient of 2. Thus, we can begin line (4) with the first coefficient and then use the following form.

$$
\begin{array}{r}
-\,2)2\quad\ \ 1\ -17\quad 13\ \ 2 \\
\ -4\ -10\quad\ 14\ \ 2 \\
\hline
2\quad\ \ 5\ -\ 7\ -1\quad 0
\end{array}
$$

(5)
(6)
(7)

Line (7) contains the coefficients of the quotient where the 0 indicates the remainder.

Finally, by changing the constant in the divisor to 2 (instead of -2) we can add the corresponding entries in lines (5) and (6) rather than subtract. Thus, the final synthetic division form for this problem is as follows.

$$
\begin{array}{r}
2)2\quad\ 1\ -17\quad\ 13\quad\ 2 \\
\ \ 4\quad\ 10\ -14\ -2 \\
\hline
2\quad 5\ -\ 7\ -\ 1\quad\ 0
\end{array}
$$

Now let's consider another problem that indicates a step-by-step procedure for carrying out the synthetic division process. Suppose that we want to divide $3x^3 - 2x^2 + 6x - 5$ by $x + 4$.

Step 1. Write the coefficients of the dividend as follows.

$$
)3\quad -2\quad 6\quad -5
$$

Step 2. In the divisor, $(x + 4)$, use -4 instead of 4 so that later we can add rather than subtract.

$$
-4)3\quad -2\quad 6\quad -5
$$

Step 3. Bring down the first coefficient of the dividend (3).

$$
\begin{array}{r}
-4)3\quad -2\quad 6\quad -5 \\
\hline
3
\end{array}
$$

Step 4. Multiply $(3)(-4)$, which yields -12; this result is to be added to the second coefficient of the dividend (-2).

$$
\begin{array}{r}
-4\overline{)3 \quad -2 \quad 6 \quad -5} \\
-12 \quad\quad\quad\quad \\
\hline
3 \quad -14 \quad\quad\quad\quad
\end{array}
$$

Step 5. Multiply $(-14)(-4)$, which yields 56; this result is to be added to the third coefficient of the dividend (6).

$$
\begin{array}{r}
-4\overline{)3 \quad -2 \quad 6 \quad -5} \\
-12 \quad 56 \quad\quad \\
\hline
3 \quad -14 \quad 62 \quad\quad
\end{array}
$$

Step 6. Multiply $(62)(-4)$, which yields -248; this result is added to the last term of the dividend (-5).

$$
\begin{array}{r}
-4\overline{)3 \quad -2 \quad 6 \quad -5} \\
-12 \quad 56 \quad -248 \\
\hline
3 \quad -14 \quad 62 \quad -253
\end{array}
$$

The last row indicates a quotient of $3x^2 - 14x + 62$ and a remainder of -253. Thus, we have

$$
\frac{3x^3 - 2x^2 + 6x - 5}{x + 4} = 3x^2 - 14x + 62 - \frac{253}{x + 4}.
$$

Let's consider one more example showing only the final compact form for synthetic division.

EXAMPLE 1   Find the quotient and remainder for $(4x^4 - 2x^3 + 6x - 1) \div (x - 1)$.

Solution

$$
\begin{array}{r}
1\overline{)4 \quad -2 \quad 0 \quad 6 \quad -1} \\
4 \quad 2 \quad 2 \quad 8 \\
\hline
4 \quad 2 \quad 2 \quad 8 \quad 7
\end{array}
$$

Notice that a zero has been inserted as the coefficient of the missing x^2 term.

Therefore,

$$
\frac{4x^4 - 2x^3 + 6x - 1}{x - 1} = 4x^3 + 2x^2 + 2x + 8 + \frac{7}{x - 1}. \qquad\blacksquare
$$

Let us very briefly look at how the concept of synthetic division is used in subsequent mathematics courses. In general, if a polynomial, $P(x)$, is divided by $x - k$, then we can write

$$
P(x) = (x - k)Q(x) + R
$$

where $Q(x)$ is the quotient polynomial and R is the real number remainder. If k is substituted for x in the above relationship, we obtain

$$P(k) = (k - k)Q(k) + R$$

$$= 0 + R$$

$$= R.$$

In other words, the functional value $P(k)$ is equal to the remainder when $P(x)$ is divided by $x - k$. This conclusion is commonly referred to as the **Remainder Theorem**.

Remainder Theorem

If $P(x)$ is divided by $(x - k)$, the remainder is $P(k)$.

In Example 1, we obtained a remainder of 7 when dividing $4x^4 - 2x^3 + 6x - 1$ by $x - 1$. Therefore, according to the Remainder Theorem we know that the functional value of $4x^4 - 2x^3 + 6x - 1$ at $x = 1$ is 7. (Perhaps you should check this result by actually substituting 1 for x in $4x^4 - 2x^3 + 6x - 1$.) A direct consequence of the Remainder Theorem is the Factor Theorem, which can be stated as follows.

Factor Theorem

If $P(k) = 0$, then $(x - k)$ is a factor of $P(x)$.

Let's consider one final example where we use synthetic division, the Factor Theorem, and some previous equation solving ideas to find a solution set.

EXAMPLE 2   Show that $(x - 1)$, $(x + 2)$, and $(x - 3)$ are factors of $x^3 - 2x^2 - 5x + 6$, and therefore 1, -2, and 3 are solutions of $x^3 - 2x^2 - 5x + 6 = 0$.

Solution   Using synthetic division we can show that $(x - 1), (x + 2)$, and $(x - 3)$ are factors of $x^3 - 2x^2 - 5x + 6$ as follows.

$$
\begin{array}{r|rrrr}
1 & 1 & -2 & -5 & 6 \\
  &   & 1 & -1 & -6 \\
\hline
  & 1 & -1 & -6 & 0 \quad\longleftarrow \text{ remainder} \\
\end{array}
$$

$$
\begin{array}{r|rrrr}
-2 & 1 & -2 & -5 & 6 \\
   &   & -2 & 8 & -6 \\
\hline
   & 1 & -4 & 3 & 0 \quad\longleftarrow \text{ remainder} \\
\end{array}
$$

$$3\overline{)\begin{array}{rrr} 1 & -2 & -5 & 6 \\ & 3 & 3 & -6 \end{array}}$$

$$\begin{array}{rrrr} 1 & 1 & -2 & 0 \end{array} \longleftarrow \text{remainder}$$

Once we know that $(x - 1)$, $(x + 2)$, and $(x - 3)$ are factors of $x^3 - 2x^2 - 5x + 6$, then the solution set for $x^3 - 2x^2 - 5x + 6 = 0$ is easily obtained as follows.

$$x^3 - 2x^2 - 5x + 6 = 0$$

$$(x - 1)(x + 2)(x - 3) = 0$$

$$x - 1 = 0 \quad \text{or} \quad x + 2 = 0 \quad \text{or} \quad x - 3 = 0$$

$$x = 1 \quad \text{or} \quad x = -2 \quad \text{or} \quad x = 3$$

The solution set is $\{1, -2, 3\}$.   ∎

Practice Exercises

Use synthetic division to determine the quotient and remainder for each of the following.

1. $(x^2 - 8x + 12) \div (x - 2)$ **2.** $(x^2 + 9x + 18) \div (x + 3)$

3. $(x^2 + 2x - 10) \div (x - 4)$ **4.** $(x^2 - 10x + 15) \div (x - 8)$

5. $(x^3 - 2x^2 - x + 2) \div (x - 2)$ **6.** $(x^3 - 5x^2 + 2x + 8) \div (x + 1)$

7. $(x^3 - 7x - 6) \div (x + 2)$ **8.** $(x^3 + 6x^2 - 5x - 1) \div (x - 1)$

9. $(2x^3 - 5x^2 - 4x + 6) \div (x - 2)$ **10.** $(3x^4 - x^3 + 2x^2 - 7x - 1) \div (x + 1)$

11. $(x^4 + 4x^3 - 7x - 1) \div (x - 3)$ **12.** $(2x^4 + 3x^2 + 3) \div (x + 2)$

13. Use synthetic division and the Factor Theorem to show that for each of the following the given binomials are factors of the polynomial.

(a) $x + 1$, $x + 2$, and $x + 3$ are factors of $x^3 + 6x^2 + 11x + 6$

(b) $x - 2$, $x + 3$, and $x - 5$ are factors of $x^3 - 4x^2 - 11x + 30$

B
Square Root Table

n	n^2	$\sqrt{n}$	n	n^2	$\sqrt{n}$
1	1	1.000	6	36	2.449
2	4	1.414	7	49	2.646
3	9	1.732	8	64	2.828
4	16	2.000	9	81	3.000
5	25	2.236	10	100	3.162

Square Root Table
(continued)

n	n^2	$\sqrt{n}$	n	n^2	$\sqrt{n}$
11	121	3.317	56	3136	7.483
12	144	3.464	57	3249	7.550
13	169	3.606	58	3364	7.616
14	196	3.742	59	3481	7.681
15	225	3.873	60	3600	7.746
16	256	4.000	61	3721	7.810
17	289	4.123	62	3844	7.874
18	324	4.243	63	3969	7.937
19	361	4.359	64	4096	8.000
20	400	4.472	65	4225	8.062
21	441	4.583	66	4356	8.124
22	484	4.690	67	4489	8.185
23	529	4.796	68	4624	8.246
24	576	4.899	69	4761	8.307
25	625	5.000	70	4900	8.367
26	676	5.099	71	5041	8.426
27	729	5.196	72	5184	8.485
28	784	5.292	73	5329	8.544
29	841	5.385	74	5476	8.602
30	900	5.477	75	5625	8.660
31	961	5.568	76	5776	8.718
32	1024	5.657	77	5929	8.775
33	1089	5.745	78	6084	8.832
34	1156	5.831	79	6241	8.888
35	1225	5.916	80	6400	8.944
36	1296	6.000	81	6561	9.000
37	1369	6.083	82	6724	9.055
38	1444	6.164	83	6889	9.110
39	1521	6.245	84	7056	9.165
40	1600	6.325	85	7225	9.220
41	1681	6.403	86	7396	9.274
42	1764	6.481	87	7569	9.327
43	1849	6.557	88	7744	9.381
44	1936	6.633	89	7921	9.434
45	2025	6.708	90	8100	9.487
46	2116	6.782	91	8281	9.539
47	2209	6.856	92	8464	9.592
48	2304	6.928	93	8649	9.644
49	2401	7.000	94	8836	9.695
50	2500	7.071	95	9025	9.747
51	2601	7.141	96	9216	9.798
52	2704	7.211	97	9409	9.849
53	2809	7.280	98	9604	9.899
54	2916	7.348	99	9801	9.950
55	3025	7.416	100	10000	10.000

From the table we can find rational approximations as follows.

$$\sqrt{31} = 5.568 \qquad \text{rounded to three decimal places}$$

Locate 31   Locate this value for
in the column $\sqrt{31}$ in the column
labeled n.   labeled $\sqrt{n}$.

Be sure that you agree with the following values taken from the table. Each of these is rounded to three decimal places.

$$\sqrt{14} = 3.742, \qquad \sqrt{46} = 6.782, \qquad \sqrt{65} = 8.062$$

The column labeled n^2 contains the squares of the whole numbers from 1 through 100, inclusive. For example, from the table we obtain $25^2 = 625$, $62^2 = 3844$, and $89^2 = 7921$. Since $25^2 = 625$, we can state that $\sqrt{625} = 25$. Thus, the column labeled n^2 also provides us with additional square root facts.

$$\sqrt{1936} = 44$$

Locate 1936   Locate this value for
in the column $\sqrt{1936}$ in the column
labeled n^2.   labeled n.

Now suppose that you want to find a value for $\sqrt{500}$. Scanning the column labeled n^2, we see that 500 does not appear; however, it is between two values given in the table. We can reason as follows.

$$22^2 = 484$$
$$23^2 = 529 \qquad \longleftarrow \qquad 500$$

Since 500 is closer to 484 than to 529, we approximate $\sqrt{500}$ to be closer to 22 than to 23. Thus, we write $\sqrt{500} = 22$, to the nearest whole number. This is merely a whole number approximation, but for some purposes it may be sufficiently precise.

C

Common Logarithms

Using a table to find a common logarithm is relatively easy but it does require a little more effort than pushing a button as you would with a calculator. Let's consider a small part of the table that appears in the back of the book. Each number in the column headed n represents the first two significant digits of a number between 1 and 10 and each of the column headings 0 through 9 represents the third significant digit.

Table of Common Logarithms

n	0	1	2	3	4	5	6	7	8	9
1.0	0.0000	0.0043	0.0086	0.0128	0.0170	0.0212	0.0253	0.0294	0.0334	0.0374
1.1	0.0414	0.0453	0.0492	0.0531	0.0569	0.0607	0.0645	0.0682	0.0719	0.0755
1.2	0.0792	0.0828	0.0864	0.0899	0.0934	0.0969	0.1004	0.1038	0.1072	0.1106
1.3	0.1139	0.1173	0.1206	0.1239	0.1271	0.1303	0.1335	0.1367	0.1399	0.1430
1.4	0.1461	0.1492	0.1523	0.1553	0.1584	0.1614	0.1644	0.1673	0.1703	0.1732
1.5	0.1761	0.1790	0.1818	0.1847	0.1875	0.1903	0.1931	0.1959	0.1987	0.2014
1.6	0.2041	0.2068	0.2095	0.2122	0.2148	0.2175	0.2201	0.2227	0.2253	0.2279
1.7	0.2304	0.2330	0.2355	0.2380	0.2405	0.2430	0.2455	0.2480	0.2504	0.2529
1.8	0.2553	0.2577	0.2601	0.2625	0.2648	0.2672	0.2695	0.2718	0.2742	0.2765
1.9	0.2788	0.2810	0.2833	0.2856	0.2878	0.2900	0.2923	0.2945	0.2967	0.2989
2.0	0.3010	0.3032	0.3054	0.3075	0.3096	0.3118	0.3139	0.3160	0.3181	0.3201
2.1	0.3222	0.3243	0.3263	0.3284	0.3304	0.3324	0.3345	0.3365	0.3385	0.3404
2.2	0.3424	0.3444	0.3464	0.3483	0.3502	0.3522	0.3541	0.3560	0.3579	0.3598
2.3	0.3617	0.3636	0.3655	0.3674	0.3692	0.3711	0.3929	0.3747	0.3766	0.3784
2.4	0.3802	0.3820	0.3838	0.3856	0.3874	0.3892	0.3909	0.3927	0.3945	0.3962

To find the logarithm of a number such as 1.75, we look at the intersection of the row that contains 1.7 and the column headed 5. Thus, we obtain

$$\log 1.75 = 0.2430.$$

Similarly, we can find that

$$\log 2.09 = 0.3201 \quad \text{and} \quad \log 2.40 = 0.3802;$$

keep in mind that these values are also rounded to four decimal places.

Now suppose that we want to use the table to find the logarithm of a positive number greater than 10 or less than 1. To accomplish this we represent the number in scientific notation and then apply the property $\log rs = \log r + \log s$. For example, to find $\log 134$ we can proceed as follows.

$$\log 134 = \log(1.34 \cdot 10^2)$$

$$= \log 1.34 + \log 10^2$$

$$= 0.1271 + 2 = 2.1271$$

By inspection we know that the common logarithm of 10^2 is 2 (the exponent), and the common logarithm of 1.34 can be found in the table.

The decimal part (0.1271) of the logarithm 2.1271 is called the **mantissa**, and the integral part, (2), is called the **characteristic**. Thus, we can find the characteristic of a common logarithm by inspection (since it is the exponent of 10 when the number is written in scientific notation) and the mantissa we can get from a table. Let's consider two more examples.

$$\log 23.8 = \log(2.38 \cdot 10^1)$$

$$= \log 2.38 + \log 10^1$$

$$= 0.3766 + 1$$

↑
from the   exponent of 10
table

$$= 1.3766$$

$$\log 0.192 = \log(1.92 \cdot 10^{-1})$$

$$= \log 1.92 + \log 10^{-1}$$

$$= 0.2833 + (-1)$$

↑   ↑
from the   exponent of 10
table

$$= 0.2833 + (-1)$$

Notice that in the last example we expressed the logarithm of 0.192 as $0.2833 + (-1)$; we did not add 0.2833 and -1. This is normal procedure when using a table of common logarithms because the mantissas given in the table are positive numbers. However, you should recognize that adding 0.2833 and -1 produces -0.7167, which agrees with the result obtained earlier with a calculator.

We can also use the table to find a number when given the common logarithm of the number. That is to say, given $\log x$ we can determine x from the table. Traditionally, x is referred to as the **antilogarithm** (abbreviated **antilog**) of $\log x$. Let's consider some examples.

EXAMPLE 1   Determine antilog 1.3365.

Solution   Finding an antilogarithm simply reverses the process used before for finding a logarithm. Thus, antilog 1.3365 means that 1 is the characteristic and 0.3365 the mantissa. We look for 0.3365 in the body of the common logarithm table and we find that it is located at the intersection of the 2.1-row and the 7-column. Therefore, the antilogarithm is

$$2.17 \cdot 10^1 = 21.7.$$ ■

EXAMPLE 2   Determine antilog $(0.1523 + (-2))$.

Solution   The mantissa, 0.1523, is located at the intersection of the 1.4-row and 2-column. The characteristic is -2 and therefore the antilogarithm is

$$1.42 \cdot 10^{-2} = 0.0142.$$ ■

EXAMPLE 3   Determine antilog -2.6038.

Solution   The mantissas given in a table are *positive* numbers. Thus, we need to express -2.6038 in terms of a positive mantissa and this can be done by adding and subtracting 3 as follows.

$$(-2.6038 + 3) - 3 = 0.3962 + (-3)$$

Now we can look for 0.3962 and find it at the intersection of the 2.4-row and 9-column. Therefore, the antilogarithm is

$$2.49 \cdot 10^{-3} = 0.00249. \qquad \blacksquare$$

Linear Interpolation

Now suppose that we want to determine log 2.774 from the table in the inside back cover. Because the table contains only logarithms of numbers with, at most, three significant digits, we have a problem. However, by a process called **linear interpolation** we can extend the capabilities of the table to include numbers with four significant digits.

First, let's consider a geometric basis of linear interpolation and then we will use a systematic procedure for carrying out the necessary calculations. A portion of the graph of $y = \log x$, with the curvature exaggerated to help illustrate the principle involved, is shown in Figure C.1. The line segment that joins points P and Q is used to approximate the curve from P to Q. The actual value of log 2.744 is the ordinate of the point C, that is, the length of $\overline{AC}$. This cannot be determined from the table. Instead we will use the ordinate of point B (the length of $\overline{AB}$) as an approximation for log 2.744.

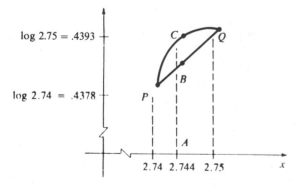

Figure C.1

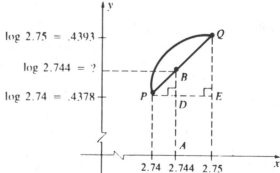

Figure C.2

Consider Figure C.2 where line segments $\overline{DB}$ and $\overline{EQ}$ are drawn perpendicular to $\overline{PE}$. The right triangles formed, $\triangle PDB$ and $\triangle PEQ$, are similar and therefore the lengths of their corresponding sides are proportional. Thus, we can write

$$\frac{PD}{PE} = \frac{DB}{EQ}. \tag{1}$$

From Figure C.2 we see that

$$PD = 2.744 - 2.74 = 0.004$$

$$PE = 2.75 - 2.74 = 0.01$$

$$EQ = 0.4393 - 0.4378 = 0.0015.$$

Therefore, the proportion (1) becomes

$$\frac{0.004}{0.01} = \frac{DB}{0.0015}.$$

Solving this proportion for DB yields

$$DB = 0.0006.$$

Since $AB = AD + DB$, we have

$$AB = 0.4378 + 0.0006 = 0.4384.$$

Thus, we obtain $\log 2.744 = 0.4384$.

Now let's suggest an abbreviated format for carrying out the calculations necessary to find $\log 2.744$.

$$
\begin{array}{c c}
\underline{\qquad x \qquad} & \underline{\qquad \log x \qquad} \\[4pt]
4\left\{\begin{array}{l} 2.740 \\ 2.744 \end{array}\right\} & k\left\{\begin{array}{l} 0.4378 \\ ? \end{array}\right\} \\[4pt]
\left.\right\}10 & \left.\right\}0.0015 \\[6pt]
2.750\bigg\} & 0.4393\bigg\}
\end{array}
$$

Notice that we have used 4 and 10 for the differences for values of x instead of 0.004 and 0.01 because the ratio $\dfrac{0.004}{0.01}$ equals $\dfrac{4}{10}$. Setting up a proportion and solving for k yields

$$\frac{4}{10} = \frac{k}{0.0015}$$

$$10k = 4(0.0015) = 0.0060$$

$$k = 0.0006.$$

Thus, $\log 2.744 = 0.4378 + 0.0006 = 0.4384$.

Let's do another example to make sure of the process.

EXAMPLE 4   Find $\log 617.6$.

Solution
$$\log 617.6 = \log(6.176 \cdot 10^2)$$
$$= \log 6.176 + \log 10^2$$

Thus, the characteristic is 2 and we can approximate the mantissa by using interpolation from the table as follows.

x	$\log x$

$$6\left\{\begin{matrix} 6.170 \\ \\ \\ 6.176 \\ 6.180 \end{matrix}\right\}10 \qquad k\left\{\begin{matrix} 0.7903 \\ \\ ? \\ 0.7910 \end{matrix}\right\}0.0007$$

$$\frac{6}{10} = \frac{k}{0.0007}$$

$$10k = 6(0.0007) = 0.0042$$

$$k = 0.00042 \approx 0.0004$$

Therefore, $\log 6.176 = 0.7903 + 0.0004 = 0.7907$ and we can complete the solution for $\log 617.6$ as follows.

$$\log 617.6 = \log(6.176 \cdot 10^2)$$
$$= \log 6.176 + \log 10^2$$
$$= 0.7907 + 2$$
$$= 2.7907 \qquad \blacksquare$$

The process of linear interpolation can also be used to approximate an antilogarithm when the mantissa is in between two values in the table. The following example illustrates this procedure.

EXAMPLE 5   Find antilog 1.6157.

Solution   From the table we see that the mantissa, 0.6157, is between 0.6149 and 0.6160. We can carry out the interpolation as follows.

$$h \left\{ \begin{array}{c} 4.120 \\ \\ \\ ? \\ 4.130 \end{array} \right\} 0.010 \qquad 8 \left\{ \begin{array}{c} 0.6149 \\ \\ \\ 0.6157 \\ 0.6160 \end{array} \right\} 11 \qquad \frac{0.0008}{0.0011} = \frac{8}{11}$$

$$\frac{h}{0.010} = \frac{8}{11}$$

$$h = 8(0.010) = 0.080$$

$$h = \frac{1}{11}(0.080) = 0.007 \qquad \text{nearest thousandth}$$

Thus, antilog $0.6157 = 4.120 + 0.007 = 4.127$.
Therefore,

$$\begin{aligned} \text{antilog } 1.6157 &= \text{antilog}(0.6157 + 1) \\ &= 4.127 \cdot 10^1 \\ &= 41.27. \end{aligned}$$

Computation with Common Logarithms

Let's first restate the basic properties of logarithms in terms of **common logarithms**. (Remember that we are writing $\log x$ instead of $\log_{10} x$.)

If x and y are positive real numbers, then

1. $\log xy = \log x + \log y$;

2. $\log\dfrac{x}{y} = \log x - \log y$;

3. $\log x^p = p \log x$. p is any real number

The following two properties of equality that pertain to logarithms will also be used.

4. If $x = y$ (x and y are positive), then $\log x = \log y$.

5. If $\log x = \log y$, then $x = y$.

EXAMPLE 6   Find the product $(49.1)(876)$.

Solution   Let $N = (49.1)(876)$. By Property 4.

$$\log N = \log(49.1)(876)$$

By Property 1.

$$\log N = \log 49.1 + \log 876$$

From the table at the back of the book we find that $\log 49.1 = 1.6911$ and that $\log 876 = 2.9425$. Thus,

$$\log N = 1.6911 + 2.9425$$
$$= 4.6336.$$

Therefore,

$$N = \text{antilog } 4.6336.$$

By using linear interpolation, we can determine antilog 0.6336 to four significant digits. Thus, we obtain

$$N = \text{antilog}(0.6336 + 4)$$
$$= 4.301 \cdot 10^4$$
$$= 43,010.$$

CHECK   By using a calculator we obtain

$$N = (49.1)(876) = 43011.6. \qquad \blacksquare$$

EXAMPLE 7   Find the quotient $\dfrac{942}{64.8}$.

Solution   Let $N = \dfrac{94.2}{64.8}$. Therefore,

$$\log N = \log \frac{942}{64.8}$$

$$= \log 942 - \log 64.8 \qquad \log \frac{x}{y} = \log x - \log y$$

$$= 2.9741 - 1.8116 \qquad \text{from the table}$$

$$= 1.1625.$$

Therefore,

$$N = \text{antilog } 1.1625$$
$$= \text{antilog}(0.01625 + 1)$$
$$= 1.454 \cdot 10^1$$
$$= 14.54.$$

CHECK   By using a calculator we obtain

$$N = \frac{942}{64.8} = 14.537037. \qquad \blacksquare$$

EXAMPLE 8   Evaluate $\dfrac{(571.4)(8.236)}{71.68}$.

Solution   Let $N = \dfrac{(571.4)(8.236)}{71.68}$. Therefore,

$$\log N = \log\frac{(571.4)(8.236)}{71.68}$$

$$= \log 571.4 + \log 8.236 - \log 71.68$$
$$= 2.7569 + 0.9157 - 1.8554$$
$$= 1.8172.$$

Therefore,

$$N = \text{antilog}\,1.8172$$
$$= \text{antilog}(0.8172 + 1)$$
$$= 6.564 \cdot 10^1$$
$$= 65.64.$$

CHECK By using a calculator we obtain

$$N = \frac{(571.4)(8.236)}{71.68} = 65.653605.$$

■

EXAMPLE 9   Evaluate $\sqrt[3]{3770}$.

Solution   Let $N = \sqrt[3]{3770} = (3770)^{\frac{1}{3}}$. Therefore,

$$\log N = \log(3770)^{\frac{1}{3}}$$

$$= \frac{1}{3}\log 3770 \qquad \log x^p = p \log x$$

$$= \frac{1}{3}(3.5763)$$

$$= 1.1921.$$

Therefore,

$$N = \text{antilog}\,1.1921$$
$$= \text{antilog}(0.1921 + 1)$$
$$= 1.556 \cdot 10^1$$
$$= 15.56.$$

CHECK By using a calculator we obtain

$$N = \sqrt[3]{3370} = 15.563733.$$ ■

When using tables of logarithms, it is sometimes necessary to change the form of writing a logarithm so that the decimal part (mantissa) is positive. The next example illustrates this idea.

EXAMPLE 10   Find the quotient $\dfrac{1.73}{5.08}$.

Solution   Let $N = \dfrac{1.73}{5.08}$. Therefore,

$$\log N = \log \frac{1.73}{5.08}$$

$$= \log 1.73 - \log 5.08$$
$$= 0.2380 - 0.7059 = -0.4679.$$

Now by adding 1 and subtracting 1, which changes the form but not the value, we obtain

$$\log N = -0.4679 + 1 - 1$$
$$= 0.5321 - 1$$
$$= 0.5321 + (-1).$$

Therefore,

$$N = \text{antilog}(0.5321 + (-1))$$
$$= 3.405 \cdot 10^{-1} = 0.3405.$$

CHECK By using a calculator we obtain

$$N = \frac{1.73}{5.08} = 0.34055118.$$ ■

Sometimes it is also necessary to change the form of a logarithm so that a subsequent calculation will produce an **integer for the characteristic part of the logarithm**. Let's consider an example to illustrate this idea.

EXAMPLE 11   Evaluate $\sqrt[4]{0.0767}$.

Solution   Let $N = \sqrt[4]{0.0767} = (0.0767)^{\frac{1}{4}}$. Therefore,

$$\log N = \log(0.0767)^{\frac{1}{4}}$$

$$= \frac{1}{4}\log 0.0767$$

$$= \frac{1}{4}(0.8848 + (-2))$$

$$= \frac{1}{4}(-2 + 0.8848).$$

At this stage we recognize that applying the distributive property will produce a nonintegral characteristic, namely, $-\frac{1}{2}$. Therefore, let's add 4 and subtract 4 inside the parentheses, which will change the form as follows.

$$\log N = \frac{1}{4}(-2 + 0.8848 + 4 - 4)$$

$$= \frac{1}{3}(4 - 2 + 0.8848 - 4)$$

$$= \frac{1}{4}(2.8848 - 4)$$

Now applying the distributive property we obtain

$$\log N = \frac{1}{4}(2.8848) - \frac{1}{4}(4)$$

$$= 0.7212 - 1 = 0.7212 + (-1).$$

Therefore,

$$N = \text{antilog}(0.7212 + (-1))$$
$$= 5.262 \cdot 10^{-1} = 0.5262.$$

CHECK By using a calculator we obtain

$$N = \sqrt[4]{0.0767} = 0.5262816.$$ ■

Practice Exercises

Use the table at the back of the book and linear interpolation to find each of the following common logarithms.

1. $\log 4.327$ **2.** $\log 27.43$ **3.** $\log 128.9$

4. $\log 3526$ **5.** $\log 0.8761$ **6.** $\log 0.07692$

7. $\log 0.005186$ **8.** $\log 0.0002558$

Use the table at the back of the book and linear interpolation to find each of the following antilogarithms to four significant digits.

9. antilog 0.4690

10. antilog 1.7971

11. antilog 2.1925

12. antilog 3.7225

13. antilog(0.5026 + (−1))

14. antilog(0.9397 + (−2))

Use common logarithms and linear interpolation to help evaluate each of the following. Express your answers with four significant digits. Check your answers by using a calculator.

15. (294)(71.2)

16. (192.6)(4.017)

17. $\dfrac{23.4}{4.07}$

18. $\dfrac{718.5}{8.248}$

19. $(17.3)^5$

20. $(48.02)^3$

21. $\dfrac{(108)(76.2)}{13.4}$

22. $\dfrac{(126.3)(24.32)}{8.019}$

23. $\sqrt[5]{0.821}$

24. $\sqrt[4]{645.3}$

25. $(79.3)^{\frac{3}{5}}$

26. $(176.8)^{\frac{3}{4}}$

27. $\sqrt{\dfrac{(7.05)(18.7)}{0.521}}$

28. $\sqrt[3]{\dfrac{(41.3)(0.271)}{8.05}}$

D

Natural Logarithms

The following table contains the natural logarithms for numbers between 0.1 and 10, inclusive, at intervals of 0.1. Be sure that you agree with the following values taken directly from the table.

$\ln 1.6 = 0.4700$

$\ln 0.5 = -0.6931$

$\ln 4.8 = 1.5686$

$\ln 9.2 = 2.2192$

Table of Natural Logarithms

n	$\ln n$	n	$\ln n$	n	$\ln n$	n	$\ln n$
0.1	−2.3026	2.6	0.9555	5.1	1.6292	7.6	2.0281
0.2	−1.6094	2.7	0.9933	5.2	1.6487	7.7	2.0412
0.3	−1.2040	2.8	1.0296	5.3	1.6677	7.8	2.0541
0.4	−0.9163	2.9	1.0647	5.4	1.6864	7.9	2.0669
0.5	−0.6931	3.0	1.0986	5.5	1.7047	8.0	2.0794
0.6	−0.5108	3.1	1.1314	5.6	1.7228	8.1	2.0919
0.7	−0.3567	3.2	1.1632	5.7	1.7405	8.2	2.1041

Table of Natural Logarithms (continued)

n	$\ln n$	n	$\ln n$	n	$\ln n$	n	$\ln n$
0.8	−0.2231	3.3	1.1939	5.8	1.7579	8.3	2.1163
0.9	−0.1054	3.4	1.2238	5.9	1.7750	8.4	2.1282
1.0	0.0000	3.5	1.2528	6.0	1.7918	8.5	2.1401
1.1	0.0953	3.6	1.2809	6.1	1.8083	8.6	2.1518
1.2	0.1823	3.7	1.3083	6.2	1.8245	8.7	2.1633
1.3	0.2624	3.8	1.3350	6.3	1.8405	8.8	2.1748
1.4	0.3365	3.9	1.3610	6.4	1.8563	8.9	2.1861
1.5	0.4055	4.0	1.3863	6.5	1.8718	9.0	2.1972
1.6	0.4700	4.1	1.4110	6.6	1.8871	9.1	2.2083
1.7	0.5306	4.2	1.4351	6.7	1.9021	9.2	2.2192
1.8	0.5878	4.3	1.4586	6.8	1.9169	9.3	2.2300
1.9	0.6419	4.4	1.4816	6.9	1.9315	9.4	2.2407
2.0	0.6931	4.5	1.5041	7.0	1.9459	9.5	2.2513
2.1	0.7419	4.6	0.5261	7.1	1.9601	9.6	2.2618
2.2	0.7885	4.7	1.5476	7.2	1.9741	9.7	2.2721
2.3	0.8329	4.8	1.5686	7.3	1.9879	9.8	2.2824
2.4	0.8755	4.9	1.5892	7.4	2.0015	9.9	2.2925
2.5	0.9163	5.0	1.6094	7.5	2.0149	10	2.3026

When using a table, the natural logarithm of a positive number less than 0.1 or greater than 10 can be approximated by using the property $\ln rs = \ln r + \ln s$ as follows.

$$\ln 190 = \ln(1.9 \cdot 10^2)$$
$$= \ln 1.9 + \ln 10^2$$
$$= \ln 1.9 + 2\ln 10$$
$$= 0.6419 + 2(2.3026)$$

    ↑   ↑
  from the from the
  table   table

$$= 5.2471.$$

$$\ln 0.0084 = \ln(8.4 \cdot 10^{-3})$$
$$= \ln 8.4 + \ln 10^{-3}$$
$$= \ln 8.4 + (-3)(\ln 10)$$
$$= 2.1282 - 3(2.3026)$$

    ↑   ↑
  from the from the
  table   table

$$= 2.1282 - 6.9078 = -4.7796.$$

Answers to Odd-Numbered Problems and All Chapter Review Problems

CHAPTER 1

Problem Set 1.1 (page 6)

1. 16   **3.** 35   **5.** 51   **7.** 72   **9.** 82   **11.** 55   **13.** 60   **15.** 66   **17.** 26
19. 2   **21.** 47   **23.** 21   **25.** 11   **27.** 15   **29.** 14   **31.** 79   **33.** 6   **35.** 74
37. 12   **39.** 187   **41.** 884   **43.** 9   **45.** 18   **47.** 55   **49.** 99   **51.** 72
53. 11   **55.** 48   **57.** 21   **59.** 40   **61.** 81   **63.** 100   **65.** 84   **67.** 170
69. 164   **71.** 153

Problem Set 1.2 (page 11)

1. True   **3.** False   **5.** True   **7.** True   **9.** True   **11.** False   **13.** True
15. False   **17.** True   **19.** False   **21.** Prime   **23.** Prime   **25.** Composite
27. Prime   **29.** Composite   **31.** $2 \cdot 13$ **33.** $2 \cdot 2 \cdot 3 \cdot 3$ **35.** $7 \cdot 7$ **37.** $2 \cdot 2 \cdot 2 \cdot 7$
39. $2 \cdot 2 \cdot 2 \cdot 3 \cdot 5$ **41.** $3 \cdot 3 \cdot 3 \cdot 5$ **43.** 4   **45.** 8   **47.** 9   **49.** 12   **51.** 18

53. 12   **55.** 24   **57.** 48   **59.** 140   **61.** 392   **63.** 168   **65.** 90
67. All other even numbers are divisible by 2.   **69.** 61   **71.** x **73.** xy **75.** $2 \cdot 2 \cdot 19$
77. $3 \cdot 41$ **79.** $5 \cdot 23$ **81.** $3 \cdot 3 \cdot 7 \cdot 7$ **83.** $3 \cdot 3 \cdot 17$

Problem Set 1.3 (page 19)

1. 2   **3.** -4 **5.** -7 **7.** 6   **9.** -6 **11.** 8   **13.** -11 **15.** -15 **17.** -7
19. -31 **21.** -19 **23.** 9   **25.** -61 **27.** -18 **29.** -92 **31.** -5
33. -13 **35.** 12   **37.** 6   **39.** -1 **41.** -45 **43.** -29 **45.** 27   **47.** -65
49. -29 **51.** -11 **53.** -1 **55.** -8 **57.** -13 **59.** -35 **61.** -15
63. -32 **65.** 2   **67.** -4 **69.** -31 **71.** -9 **73.** 18   **75.** 8   **77.** -29
79. -7 **81.** 15   **83.** 1   **85.** 36   **87.** -39 **89.** -24 **91.** 7   **93.** -1
95. 10   **97.** 9   **99.** -17 **101.** -3 **103.** -10 **105.** -3 **107.** 11   **109.** 5
111. -65 **113.** -100 **115.** -25 **117.** 130   **119.** 80   **121. (a)** 319   **(b)** 357

Problem Set 1.4 (page 25)

1. -30 **3.** -9 **5.** 7   **7.** -56 **9.** 60   **11.** -12 **13.** -126 **15.** 154
17. -9 **19.** 11   **21.** 225   **23.** -14 **25.** 0   **27.** 23   **29.** -19 **31.** 90
33. 14   **35.** undefined   **37.** -4 **39.** -972 **41.** -47 **43.** 18   **45.** 69   **47.** 4
49. 4   **51.** -6 **53.** 31   **55.** 4   **57.** 28   **59.** -7 **61.** 10   **63.** -59
65. 66   **67.** 7   **69.** 69   **71.** -7 **73.** 126   **75.** -70 **77.** 15   **79.** -10
81. -25 **83.** 77   **85.** 104   **87.** 14

Problem Set 1.5 (page 33)

1. Distributive property   **3.** Associative property for addition
5. Commutative property for multiplication   **7.** Additive inverse property
9. Identity property for addition   **11.** Associative property for multiplication   **13.** 56   **15.** 7
17. 1800   **19.** -14400 **21.** -3700 **23.** 5900   **25.** -338 **27.** -38 **29.** 7
31. $-5x$ **33.** $-3m$ **35.** $-11y$ **37.** $-3x - 2y$ **39.** $-16a - 4b$ **41.** $-7xy + 3x$
43. $10x + 5$ **45.** $6xy - 4$ **47.** $-6a - 5b$ **49.** $5ab - 11a$ **51.** $8x + 36$ **53.** $11x + 28$
55. $8x + 44$ **57.** $5a + 29$ **59.** $3m + 29$ **61.** $-8y + 6$ **63.** -5 **65.** -40
67. 72   **69.** -18 **71.** 37   **73.** -74 **75.** 180   **77.** 34   **79.** -65
85. (a) -340 **(c)** -1236 **(e)** 76   **(g)** 1684   **(i)** -2452

Chapter 1 Review Problem Set (page 36)

1. -3 **2.** -25 **3.** -5 **4.** -15 **5.** -1 **6.** 2   **7.** -156 **8.** 252   **9.** 6
10. -13 **11.** Prime   **12.** Composite   **13.** Composite   **14.** Composite   **15.** Composite
16. $2 \cdot 2 \cdot 2 \cdot 3$ **17.** $3 \cdot 3 \cdot 7$ **18.** $3 \cdot 19$ **19.** $2 \cdot 2 \cdot 2 \cdot 2 \cdot 2 \cdot 2$ **20.** $2 \cdot 2 \cdot 3 \cdot 7$ **21.** 18
22. 12   **23.** 180   **24.** 945   **25.** 66   **26.** -7 **27.** -2 **28.** 4   **29.** -18
30. 12   **31.** -34 **32.** -27 **33.** -38 **34.** -93 **35.** 2   **36.** 3   **37.** 35
38. 27   **39.** $8x$ **40.** $-5y - 9$ **41.** $-5x + 4y$ **42.** $13a - 6b$ **43.** $-ab - 2a$
44. $-3xy - y$ **45.** $10x + 74$ **46.** $2x + 7$ **47.** $-7x - 18$ **48.** $-3x + 12$
49. $-2a + 4$ **50.** $-2a - 4$ **51.** -59 **52.** -57 **53.** 2   **54.** 1   **55.** 12
56. 13   **57.** 22   **58.** 32   **59.** -9 **60.** 37   **61.** -39 **62.** -32 **63.** 9
64. -44

CHAPTER 2

Problem Set 2.1 (page 44)

1. $\frac{2}{3}$ **3.** $\frac{2}{3}$ **5.** $\frac{5}{3}$ **7.** $-\frac{1}{6}$ **9.** $-\frac{3}{4}$ **11.** $\frac{27}{28}$ **13.** $\frac{6}{11}$ **15.** $\frac{3x}{7y}$ **17.** $\frac{2x}{5}$

19. $-\frac{5a}{13c}$ **21.** $\frac{8z}{7x}$ **23.** $\frac{5b}{7}$ **25.** $\frac{15}{28}$ **27.** $\frac{10}{21}$ **29.** $\frac{3}{10}$ **31.** $-\frac{4}{3}$ **33.** $\frac{7}{5}$

35. $-\frac{3}{10}$ **37.** $\frac{1}{4}$ **39.** -27 **41.** $\frac{35}{27}$ **43.** $\frac{8}{21}$ **45.** $-\frac{5}{6y}$ **47.** $2a$ **49.** $\frac{2}{5}$

51. $\frac{y}{2x}$ **53.** $\frac{20}{13}$ **55.** $-\frac{7}{9}$ **57.** $\frac{2}{9}$ **59.** $\frac{2}{5}$ **61.** $\frac{13}{28}$ **63.** $\frac{8}{5}$ **65.** -4 **67.** $\frac{36}{49}$

69. 1 **71.** $\frac{2}{3}$ **73.** $\frac{20}{9}$ **75.** (a) 8 (c) 40 (e) 5 **77.** (a) $\frac{11}{13}$ (c) $-\frac{37}{41}$

(e) $\frac{6}{11}$ (g) $\frac{7}{11}$

Problem Set 2.2 (page 52)

1. $\frac{5}{7}$ **3.** $\frac{5}{9}$ **5.** 3 **7.** $\frac{2}{3}$ **9.** $-\frac{1}{2}$ **11.** $\frac{2}{3}$ **13.** $\frac{15}{x}$ **15.** $\frac{2}{y}$ **17.** $\frac{8}{15}$

19. $\frac{9}{16}$ **21.** $\frac{37}{30}$ **23.** $\frac{59}{96}$ **25.** $-\frac{19}{72}$ **27.** $-\frac{1}{24}$ **29.** $-\frac{1}{3}$ **31.** $-\frac{1}{6}$ **33.** $-\frac{31}{7}$

35. $-\frac{21}{4}$ **37.** $\frac{3y+4x}{xy}$ **39.** $\frac{7b-2a}{ab}$ **41.** $\frac{11}{2x}$ **43.** $\frac{4}{3x}$ **45.** $-\frac{2}{5x}$ **47.** $\frac{19}{6y}$

49. $\frac{1}{24y}$ **51.** $-\frac{17}{24n}$ **53.** $\frac{5y+7x}{3xy}$ **55.** $\frac{32y+15x}{20xy}$ **57.** $\frac{63y-20x}{36xy}$ **59.** $\frac{-6y-5x}{4xy}$

61. $\frac{3x+2}{x}$ **63.** $\frac{4x-3}{2x}$ **65.** $\frac{1}{4}$ **67.** $\frac{37}{30}$ **69.** $\frac{1}{3}$ **71.** $-\frac{12}{5}$ **73.** $-\frac{1}{30}$ **75.** 14

77. 68 **79.** $\frac{7}{26}$ **81.** $\frac{11}{15}x$ **83.** $\frac{5}{24}a$ **85.** $\frac{4}{3}x$ **87.** $\frac{13}{20}n$ **89.** $\frac{20}{9}n$ **91.** $-\frac{79}{36}n$

93. $\frac{13}{14}x+\frac{9}{8}y$ **95.** $-\frac{11}{45}x-\frac{9}{20}y$ **101.** (a) $\frac{41}{45}$ (c) $1\frac{5}{84}$ (e) $\frac{69}{143}$

(g) $-\frac{29}{48}$ (i) $\frac{113}{168}$

Problem Set 2.3 (page 63)

1. 0.62 **3.** 1.45 **5.** 3.8 **7.** -3.3 **9.** 7.5 **11.** 7.8 **13.** -0.9 **15.** -7.8
17. 1.16 **19.** -0.272 **21.** -24.3 **23.** 44.8 **25.** 0.0156 **27.** 1.2 **29.** -7.4
31. 0.38 **33.** 7.2 **35.** -0.42 **37.** 0.76 **39.** 4.7 **41.** 4.3 **43.** -14.8 **45.** 1.3
47. $-1.2x$ **49.** $3n$ **51.** $0.5t$ **53.** $-5.8x+2.8y$ **55.** $0.1x+1.2$ **57.** $-3x-2.3$

59. $4.6x-8$ **61.** $\frac{11}{12}$ **63.** $\frac{4}{3}$ **65.** 17.3 **67.** -97.8 **69.** 2.2 **71.** 13.75

73. 0.6 **77.** (a) The denominator has only factors of 2.   (c) $\frac{7}{8}, \frac{11}{16}, \frac{13}{32}, \frac{17}{40}, \frac{9}{20}$, and $\frac{3}{64}$

Problem Set 2.4 (page 69)

1. 64   **3.** 81   **5.** -8 **7.** -9 **9.** 16   **11.** $\dfrac{16}{81}$ **13.** $-\dfrac{1}{8}$ **15.** $\dfrac{9}{4}$ **17.** 0.027

19. -1.44 **21.** -47 **23.** -33 **25.** 11   **27.** -75 **29.** -60 **31.** 31

33. -13 **35.** $9x^2$ **37.** $12xy^2$ **39.** $-18x^4y$ **41.** $15xy$ **43.** $12x^4$ **45.** $8a^5$

47. $-8x^2$ **49.** $4y^3$ **51.** $-2x^2 + 6y^2$ **53.** $-\dfrac{11}{60}n^2$ **55.** $-2x^2 - 6x$

57. $7x^2 - 3x + 8$ **59.** $\dfrac{3y}{5}$ **61.** $\dfrac{11}{3y}$ **63.** $\dfrac{7b^2}{17a}$ **65.** $-\dfrac{3ac}{4}$ **67.** $\dfrac{x^2y^2}{4}$.   **69.** $\dfrac{4x}{9}$

71. $\dfrac{5}{12ab}$ **73.** $\dfrac{6y^2 + 5x}{xy^2}$ **75.** $\dfrac{5 - 7x^2}{x^4}$ **77.** $\dfrac{3 + 12x^2}{2x^3}$ **79.** $\dfrac{13}{12x^2}$ **81.** $\dfrac{11b^2 - 14a^2}{a^2b^2}$

83. $\dfrac{3 - 8x}{6x^3}$ **85.** $\dfrac{3y - 4x - 5}{xy}$ **87.** 79   **89.** $\dfrac{23}{36}$ **91.** $\dfrac{25}{4}$ **93.** -64 **95.** -25

97. -33 **99.** $.45$

Problem Set 2.5 (page 75)

Answers may vary somewhat for Problems 1–11.
1. The difference of a and b **3.** One-third of the product of B and h
5. Two times the quantity, l plus w **7.** The quotient of A divided by w
9. The quantity, a plus b, divided by 2   **11.** Two more than three times y **13.** $l + w$ **15.** ab

17. $\dfrac{d}{t}$ **19.** lwh **21.** $y - x$ **23.** $xy + 2$ **25.** $7 - y^2$ **27.** $\dfrac{x - y}{4}$ **29.** $10 - x$

31. $10(n + 2)$ **33.** $xy - 7$ **35.** $xy - 12$ **37.** $35 - n$ **39.** $n + 45$ **41.** $y + 10$

43. $2x - 3$ **45.** $10d + 25q$ **47.** $\dfrac{d}{t}$ **49.** $\dfrac{d}{p}$ **51.** $\dfrac{d}{12}$ **53.** $n + 1$ **55.** $n + 2$

57. $3y - 2$ **59.** $36y + 12f$ **61.** $\dfrac{f}{3}$ **63.** $8w$ **65.** $3l - 4$ **67.** $48f + 72$

Chapter 2 Review Problem Set (page 78)

1. 64   **2.** -27 **3.** -16 **4.** $\dfrac{9}{16}$ **5.** $\dfrac{49}{36}$ **6.** 0.216   **7.** 0.0144   **8.** 0.0036

9. $-\dfrac{8}{27}$ **10.** $\dfrac{1}{16}$ **11.** $\dfrac{19}{24}$ **12.** $\dfrac{39}{70}$ **13.** $\dfrac{1}{15}$ **14.** $\dfrac{14y + 9x}{2xy}$ **15.** $\dfrac{5x - 8y}{x^2y}$

16. $\dfrac{7y}{20}$ **17.** $\dfrac{4x^3}{5y^2}$ **18.** $\dfrac{2}{7}$ **19.** 1   **20.** $\dfrac{27n'}{28}$ **21.** $\dfrac{1}{24}$ **22.** $-\dfrac{13}{8}$ **23.** $\dfrac{7}{9}$

24. $\dfrac{29}{12}$ **25.** $\dfrac{1}{2}$ **26.** 0.67   **27.** 0.49   **28.** 2.4   **29.** -0.11 **30.** 1.76

31. $\dfrac{5}{56}x^2 + \dfrac{7}{20}y^2$ **32.** $-0.58ab + 0.36bc$ **33.** $\dfrac{11x}{24}$ **34.** $2.2a + 1.7b$ **35.** $-\dfrac{1}{10}n$

36. $\dfrac{41}{20}n$ **37.** $\dfrac{19}{42}$ **38.** $-\dfrac{1}{72}$ **39.** -0.75 **40.** -0.35 **41.** $\dfrac{1}{17}$ **42.** -8

43. $72 - n$ **44.** $p + 10d$ **45.** $\dfrac{x}{60}$ **46.** $2y - 3$ **47.** $5n + 3$ **48.** $36y + 12f$

49. $100m$ **50.** $5n + 10d + 25q$ **51.** $n - 5$ **52.** $5 - n$ **53.** $10(x - 2)$ **54.** $10x - 2$

55. $x - 3$ **56.** $\dfrac{d}{r}$ **57.** $x^2 + 9$ **58.** $(x + 9)^2$ **59.** $x^3 + y^3$ **60.** $xy - 4$

CHAPTER 3
Problem Set 3.1 (page 88)

1. $\{8\}$ **3.** $\{-6\}$ **5.** $\{-9\}$ **7.** $\{-6\}$ **9.** $\{13\}$ **11.** $\{48\}$ **13.** $\{23\}$

15. $\{-7\}$ **17.** $\left\{\dfrac{17}{21}\right\}$ **19.** $\left\{-\dfrac{4}{15}\right\}$ **21.** $\{.27\}$ **23.** $\{-3.5\}$ **25.** $\{-17\}$

27. $\{-35\}$ **29.** $\{-8\}$ **31.** $\{-17\}$ **33.** $\left\{\dfrac{37}{5}\right\}$ **35.** $\{-3\}$ **37.** $\left\{\dfrac{13}{2}\right\}$ **39.** $\{144\}$

41. $\{24\}$ **43.** $\{-15\}$ **45.** $\{24\}$ **47.** $\{-35\}$ **49.** $\left\{\dfrac{3}{10}\right\}$ **51.** $\left\{-\dfrac{9}{10}\right\}$ **53.** $\left\{\dfrac{1}{2}\right\}$

55. $\left\{-\dfrac{1}{3}\right\}$ **57.** $\left\{\dfrac{27}{32}\right\}$ **59.** $\left\{-\dfrac{5}{14}\right\}$ **61.** $\left\{-\dfrac{7}{5}\right\}$ **63.** $\left\{-\dfrac{1}{12}\right\}$ **65.** $\left\{-\dfrac{3}{20}\right\}$

Problem Set 3.2 (page 93)

1. $\{4\}$ **3.** $\{6\}$ **5.** $\{8\}$ **7.** $\{11\}$ **9.** $\left\{\dfrac{17}{6}\right\}$ **11.** $\left\{\dfrac{19}{2}\right\}$ **13.** $\{6\}$ **15.** $\{-1\}$

17. $\{-5\}$ **19.** $\{-6\}$ **21.** $\left\{\dfrac{11}{2}\right\}$ **23.** $\{-2\}$ **25.** $\left\{\dfrac{10}{7}\right\}$ **27.** $\{18\}$ **29.** $\left\{-\dfrac{25}{4}\right\}$

31. $\{-7\}$ **33.** $\left\{-\dfrac{24}{7}\right\}$ **35.** $\left\{\dfrac{5}{2}\right\}$ **37.** $\left\{\dfrac{4}{17}\right\}$ **39.** $\left\{-\dfrac{12}{5}\right\}$ **41.** 9   **43.** 22

45. $18   **47.** 35 years old   **49.** $6.50   **51.** 6   **53.** 5   **55.** 11   **57.** 8   **59.** 3

61. $300   **63.** 4 meters   **65.** 14 cars   **67.** 5 hours

Problem Set 3.3 (page 100)

1. $\{5\}$ **3.** $\{-8\}$ **5.** $\left\{\dfrac{8}{5}\right\}$ **7.** $\{-11\}$ **9.** $\left\{-\dfrac{5}{2}\right\}$ **11.** $\{-9\}$ **13.** $\{2\}$

15. $\{-3\}$ **17.** $\left\{\dfrac{13}{2}\right\}$ **19.** $\left\{\dfrac{5}{3}\right\}$ **21.** $\{17\}$ **23.** $\left\{-\dfrac{13}{2}\right\}$ **25.** $\left\{\dfrac{16}{3}\right\}$ **27.** $\{2\}$

29. $\left\{-\dfrac{1}{3}\right\}$ **31.** $\left\{-\dfrac{19}{10}\right\}$ **33.** 17   **35.** 35 and 37   **37.** 36, 38, and 40   **39.** $\dfrac{3}{2}$

41. -6 **43.** 32° and 58°   **45.** 50° and 130°   **47.** 65° and 75°   **49.** $26 per share

51. $9 per hour   **53.** 150 males and 450 females   **55.** 690 voters   **57.** 6 feet

59. 29 yards by 10 yards

Problem Set 3.4 (page 107)

1. $\{1\}$ **3.** $\{10\}$ **5.** $\{-9\}$ **7.** $\left\{\dfrac{29}{4}\right\}$ **9.** $\left\{-\dfrac{17}{3}\right\}$ **11.** $\{10\}$ **13.** $\{44\}$

15. $\{26\}$ **17.** $\{-38\}$ **19.** $\left\{\dfrac{11}{6}\right\}$ **21.** $\{3\}$ **23.** $\{-1\}$ **25.** $\{-2\}$ **27.** $\{16\}$

29. $\left\{\dfrac{22}{3}\right\}$ **31.** $\{-2\}$ **33.** $\left\{-\dfrac{1}{6}\right\}$ **35.** $\{-57\}$ **37.** $\left\{-\dfrac{7}{5}\right\}$ **39.** $\{2\}$ **41.** $\{-3\}$

43. $\left\{\dfrac{27}{10}\right\}$ **45.** $\left\{\dfrac{3}{28}\right\}$ **47.** $\left\{\dfrac{18}{5}\right\}$ **49.** $\left\{\dfrac{24}{7}\right\}$ **51.** $\{5\}$ **53.** $\{0\}$ **55.** $\left\{-\dfrac{51}{10}\right\}$

57. $\{-12\}$ **59.** $\{15\}$ **61.** 7 and 8 **63.** 14, 15, and 16 **65.** 6 and 11 **67.** 48

69. 12 and 18 **71.** 8 feet and 12 feet **73.** 12 nickels, 17 dimes, and 40 quarters

75. 40 nickels, 80 dimes, and 90 quarters **77.** 8 dimes and 10 quarters

79. 75 pennies, 150 nickels, and 225 dimes **81.** 30° **83.** 20°, 50°, and 110° **85.** 40°

87. **(a)** {all reals} **(c)** $\{0\}$ **(e)** $\varnothing$ **(g)** $\varnothing$ **(i)** $\{0\}$

Problem Set 3.5 (page 117)

1. $\{9\}$ **3.** $\{10\}$ **5.** $\left\{\dfrac{15}{2}\right\}$ **7.** $\{-22\}$ **9.** $\{-4\}$ **11.** $\{6\}$ **13.** $\{-28\}$

15. $\{34\}$ **17.** $\{6\}$ **19.** $\left\{-\dfrac{8}{5}\right\}$ **21.** $\{7\}$ **23.** $\left\{\dfrac{9}{2}\right\}$ **25.** $\left\{-\dfrac{53}{2}\right\}$ **27.** $\{50\}$

29. $\{120\}$ **31.** $\left\{\dfrac{9}{7}\right\}$ **33.** 55% **35.** 60% **37.** $16\dfrac{2}{3}\%$ **39.** $37\dfrac{1}{3}\%$ **41.** 150%

43. 240% **45.** 2.66 **47.** 42 **49.** 80% **51.** 60 **53.** 115% **55.** 90

57. 15 feet by $19\dfrac{1}{2}$ feet **59.** 330 miles **61.** 60 centimeters **63.** 7.5 pounds

65. $33\dfrac{1}{3}$ pounds **67.** 90,000 **69.** 15 inches by 10 inches **71.** \$300 **73.** \$150,000

75. $\varnothing$ **77.** $\varnothing$ **79.** **(a)** 62.5% **(c)** 42.9% **(e)** 30.8%

(g) 15.6% **(i)** 112.5% **(k)** 141.7%

Problem Set 3.6 (page 124)

1. $\{1.11\}$ **3.** $\{6.6\}$ **5.** $\{.48\}$ **7.** $\{80\}$ **9.** $\{3\}$ **11.** $\{50\}$ **13.** $\{70\}$ **15.** $\{200\}$

17. $\{450\}$ **19.** $\{150\}$ **21.** $\{2200\}$ **23.** \$50 **25.** \$36 **27.** \$20.80 **29.** 30%

31. \$8.50 **33.** \$4.65 **35.** \$1000 **37.** 40% **39.** \$400 at 9% and \$650 at 10%

41. \$500 at 9% and \$700 at 12% **43.** \$4000 **45.** \$800 at 10% and \$1500 at 12%

47. \$3000 at 8% and \$2400 at 10% **51.** Yes, if the profit is figured as a percent of the selling price.

55. $\{-2.3125\}$ **57.** $\{320\}$ **59.** $\{0.08\}$ **61.** $\{9.3\}$

Chapter 3 Review Problem Set (page 127)

1. $\{-3\}$ **2.** $\{1\}$ **3.** $\left\{-\dfrac{3}{4}\right\}$ **4.** $\{9\}$ **5.** $\{-4\}$ **6.** $\left\{\dfrac{40}{3}\right\}$ **7.** $\left\{\dfrac{9}{4}\right\}$ **8.** $\left\{-\dfrac{15}{8}\right\}$

9. $\{-7\}$ **10.** $\left\{\dfrac{2}{41}\right\}$ **11.** $\left\{\dfrac{19}{7}\right\}$ **12.** $\left\{\dfrac{1}{2}\right\}$ **13.** $\left\{\dfrac{17}{12}\right\}$ **14.** $\{5\}$ **15.** $\{-32\}$

16. $\{-12\}$ **17.** $\{21\}$ **18.** $\{-60\}$ **19.** $\{10\}$ **20.** $\left\{-\dfrac{11}{4}\right\}$ **21.** $\left\{-\dfrac{8}{5}\right\}$ **22.** $\{800\}$

23. $\{16\}$ **24.** $\{73\}$ **25.** $\left\{\dfrac{5}{21}\right\}$ **26.** 24 **27.** 7 **28.** 33 **29.** 8 **30.** 40

31. 16 and 24　　**32.** 60%　　**33.** 18　　**34.** 40 and 56　　**35.** 8 nickels and 22 dimes
36. 8 nickels, 25 dimes, and 50 quarters　　**37.** 52°　　**38.** 20°　　**39.** 30 gallons
40. $900 at 9% and $1200 at 11%　　**41.** $40　　**42.** 35%　　**43.** 700 miles

CHAPTER 4

Problem Set 4.1 (page 136)

1. 7　　**3.** 500　　**5.** 20　　**7.** 48　　**9.** 9　　**11.** 46 centimeters　　**13.** 15 inches
15. 504 square feet　　**17.** $2　　**19.** 7 inches　　**21.** 150π square centimeters

23. $\frac{1}{4}\pi$ square yards　　**25.** $S = 324\pi$ square inches and $V = 972\pi$ cubic inches

27. $V = 1152\pi$ cubic feet and $S = 416\pi$ square feet　　**29.** 12 inches　　**31.** 8 feet　　**33.** $h = \dfrac{V}{B}$

35. $B = \dfrac{3V}{h}$　　**37.** $w = \dfrac{P - 2l}{2}$　　**39.** $h = \dfrac{3V}{\pi r^2}$　　**41.** $C = \dfrac{5}{9}(F - 32)$　　**43.** $h = \dfrac{A - 2\pi r^2}{2\pi r}$

45. $x = \dfrac{9 - 7y}{3}$　　**47.** $y = \dfrac{9x - 13}{6}$　　**49.** $x = \dfrac{11y - 14}{2}$　　**51.** $x = \dfrac{-y - 4}{3}$　　**53.** $y = \dfrac{3}{2}x$

55. $y = \dfrac{ax - c}{b}$　　**57.** $x = \dfrac{2y - 22}{5}$　　**59.** $y = mx + b$　　**61.** 834.3 square meters

63. 78 square inches, 113 square inches and 154 square inches　　**65.** 201 square centimeters
67. 2562 cubic meters

Problem Set 4.2 (page 144)

1. $\left\{8\dfrac{1}{3}\right\}$　　**3.** {16}　　**5.** {25}　　**7.** {7}　　**9.** {24}　　**11.** {4}　　**13.** $12\dfrac{1}{2}$ years
15. 20 years　　**17.** The width is 14 inches and the length is 42 inches.
19. The width is 12 centimeters and the length is 34 centimeters.　　**21.** 80 square inches
23. 24 feet, 31 feet, and 45 feet　　**25.** 6 centimeters, 19 centimeters, and 21 centimeters

27. 12 centimeters　　**29.** 7 centimeters　　**31.** 9 hours　　**33.** $2\dfrac{1}{2}$ hours　　**35.** 55 miles per hour

37. 64 and 72 miles per hour　　**39.** 60 miles

Problem Set 4.3 (page 149)

1. {15}　　**3.** $\left\{\dfrac{20}{7}\right\}$　　**5.** $\left\{\dfrac{15}{4}\right\}$　　**7.** $\left\{\dfrac{5}{3}\right\}$　　**9.** {2}　　**11.** $\left\{\dfrac{33}{10}\right\}$　　**13.** 12.5 milliliters

15. 15 centiliters　　**17.** $7\dfrac{1}{2}$ quarts of the 30% solution and $2\dfrac{1}{2}$ quarts of the 50% solution

19. 5 gallons　　**21.** 3 quarts　　**23.** 12 gallons　　**25.** 16.25%　　**27.** The square is 6 inches by 6 inches and the rectangle is 9 inches long and 3 inches wide.　　**29.** 40 minutes
31. Pam is 9 and Bill is 18.　　**33.** Abby is 14 and her mother is 35.　　**35.** 1 year old and 10 years old
37. 56 miles

Problem Set 4.4 (page 156)

1. True **3.** False **5.** False **7.** True **9.** True
11. $\{x \mid x > -2\}$ **13.** $\{x \mid x \leq 3\}$

15. $\{x \mid x > 2\}$ **17.** $\{x \mid x \leq -2\}$

19. $\{x \mid x < -1\}$ **21.** $\{x \mid x < 2\}$

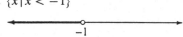

23. $\{x \mid x < -20\}$ **25.** $\{x \mid x \geq -9\}$ **27.** $\{x \mid x > 9\}$ **29.** $\left\{x \mid x < \dfrac{10}{3}\right\}$ **31.** $\{x \mid x < -8\}$

33. $\{n \mid n \geq 8\}$ **35.** $\left\{n \mid n > -\dfrac{24}{7}\right\}$ **37.** $\{n \mid n > 7\}$ **39.** $\{x \mid x > 5\}$ **41.** $\{x \mid x \leq 6\}$

43. $\{x \mid x \leq -21\}$ **45.** $\left\{x \mid x < \dfrac{8}{3}\right\}$ **47.** $\left\{x \mid x < \dfrac{5}{4}\right\}$ **49.** $\{x \mid x < 1\}$ **51.** $\{t \mid t \geq 4\}$

53. $\{x \mid x > 14\}$ **55.** $\left\{x \mid x > \dfrac{3}{2}\right\}$ **57.** $\left\{t \mid t \geq \dfrac{1}{4}\right\}$ **59.** $\left\{x \mid x < -\dfrac{9}{4}\right\}$ **61.** $\varnothing$

63. $\{x \mid x \text{ is any real number}\}$ **65.** $\varnothing$ **67.** $\{x \mid x \text{ is any real number}\}$

Problem Set 4.5 (page 162)

1. $\{x \mid x > 2\}$ **3.** $\{x \mid x < -1\}$ **5.** $\left\{x \mid x > -\dfrac{10}{3}\right\}$ **7.** $\{n \mid n \geq -11\}$ **9.** $\{t \mid t \leq 11\}$

11. $\left\{x \mid x > -\dfrac{11}{5}\right\}$ **13.** $\left\{x \mid x < \dfrac{5}{2}\right\}$ **15.** $\{x \mid x \leq 8\}$ **17.** $\left\{n \mid n > \dfrac{3}{2}\right\}$ **19.** $\{y \mid y > -3\}$

21. $\left\{x \mid x < \dfrac{5}{2}\right\}$ **23.** $\{x \mid x < 8\}$ **25.** $\{x \mid x < 21\}$ **27.** $\{x \mid x < 6\}$ **29.** $\left\{n \mid n > -\dfrac{17}{2}\right\}$

31. $\{n \mid n \leq 42\}$ **33.** $\left\{n \mid n > -\dfrac{9}{5}\right\}$ **35.** $\left\{x \mid x > \dfrac{4}{3}\right\}$ **37.** $\{n \mid n \geq 4\}$ **39.** $\{t \mid t > 300\}$

41. $\{x \mid x \leq 50\}$ **43.** $\{x \mid x > 0\}$ **45.** $\{x \mid x > 64\}$ **47.** $\left\{n \mid n > \dfrac{33}{5}\right\}$ **49.** $\left\{x \mid x \geq -\dfrac{16}{3}\right\}$

51.
$$\circ\!\!-\!\!-\!\!\circ \quad {-1 \quad 2}$$

53.
$$\circ\!\!-\!\!-\!\!\circ \quad {-2 \quad 1}$$

55.
$$\circ\!\!-\!\!-\!\!\bullet \quad {-2 \quad 2}$$

57.
$$\circ \quad 2$$

59.
$$\circ \quad {-4}$$

61. $\varnothing$

63.
$$\bullet\!\!-\!\!-\!\!\bullet \quad 0 \quad 2$$

65. all reals

67. All numbers greater than 7 **69.** 15 inches **71.** 158 or better **73.** Better than 90
75. More than 12% **77.** 77 or less

Problem Set 4.6 (page 168)

1. $\{-4, 4\}$

3. $\{x \mid x > -1 \text{ and } x < 1\}$

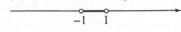

5. $\{x \mid x \le -2 \text{ or } x \ge 2\}$

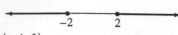

7. $\{-3, -1\}$

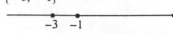

9. $\{-1, 3\}$

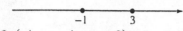

11. $\{x \mid x \ge 0 \text{ and } x \le 4\}$

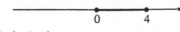

13. $\{x \mid x < -4 \text{ or } x > 2\}$

15. $\{-2, 1\}$

17. $\left\{-\dfrac{2}{5}, \dfrac{6}{5}\right\}$

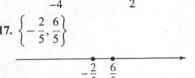

19. $\{x \mid x \le 1 \text{ or } x \ge 2\}$

21. $\left\{x \mid x > -\dfrac{5}{4} \text{ and } x < -\dfrac{1}{4}\right\}$

23. $\{-2\}$

25. $\left\{x \mid x \ne \dfrac{2}{3}\right\}$

27. $\left\{-\dfrac{16}{3}, 6\right\}$ **29.** $\{x \mid x < -5 \text{ or } x > 4\}$

31. $\left\{x \mid x > -\dfrac{14}{3} \text{ and } x < 8\right\}$ **33.** $\left\{-6, \dfrac{16}{3}\right\}$ **35.** $\left\{x \mid x \ge -6 \text{ and } x \le \dfrac{19}{2}\right\}$

37. $\left\{x \mid x \le -\dfrac{21}{5} \text{ or } x \ge 3\right\}$ **39.** $\{x \mid x > -6 \text{ and } x < 2\}$ **41.** $\left\{x \mid x < -\dfrac{5}{2} \text{ or } x > \dfrac{7}{2}\right\}$

43. $\left\{\dfrac{1}{12}, \dfrac{17}{12}\right\}$ **45.** $\{x \mid x > -5 \text{ and } x < 11\}$

47. $\left\{x \mid x \le -\dfrac{3}{2} \text{ or } x \ge \dfrac{1}{2}\right\}$ **49.** $\{x \mid x \ge -2 \text{ and } x \le 3\}$ **51.** $\{0\}$

53. $\{x \mid x \text{ is any real number}\}$ **55.** $\varnothing$ **57.** $\{-6\}$ **59.** $\{x \mid 4 < x < 14\}$

61. $\{x \mid -1 \le x \le 4\}$ **63.** $\{x \mid -11 < x < 7\}$ **65.** $\{x \mid -7 < x < -1\}$ **67.** $\{x \mid -1 < x < 6\}$

Chapter 4 Review Problem Set (page 171)

1. 6   **2.** 25   **3.** $t = \dfrac{A - P}{Pr}$ **4.** $x = \dfrac{13 + 3y}{2}$ **5.** 77 square inches   **6.** 6 centimeters

7. 15 feet   **8.** $\{x \mid x > 4\}$ **9.** $\{x \mid x > -4\}$ **10.** $\{x \mid x \ge 13\}$ **11.** $\left\{x \mid x \ge \dfrac{11}{2}\right\}$

12. $\{x \mid x > 35\}$ **13.** $\left\{x \mid x < \dfrac{26}{5}\right\}$ **14.** $\{n \mid n < 2\}$ **15.** $\left\{n \mid n > \dfrac{5}{11}\right\}$ **16.** $\{s \mid s \geq 6\}$

17. $\{t \mid t \leq 100\}$ **18.** $\{y \mid y < 24\}$ **19.** $\{x \mid x > 10\}$ **20.** $\left\{n \mid n < \dfrac{2}{11}\right\}$ **21.** $\{n \mid n > 33\}$

22. $\{n \mid n \leq 120\}$ **23.** $\left\{n \mid n \leq -\dfrac{180}{13}\right\}$ **24.** $\left\{x \mid x > \dfrac{9}{2}\right\}$ **25.** $\left\{x \mid x < -\dfrac{43}{3}\right\}$

26.
```
      ○━━━━━━━○
     -3       2
```

27.
```
  ←━━━━━━○       ○━━━━━→
        -1       4
```

28.   all reals
```
  ←━━━━━━━━━━━━━━━━━━→
```

29.
```
  ━━━━━━━━━○━━━━━━━━━→
           1
```

30.
```
  ━━━━━━━━●━━━━━●━━━━━→
         -2/3   4
```

31.
```
  ━━━━━━━━━○━━━○
           3   5
```

32.
```
  ←━━━━━━●━━━━━●━━━━━━
        -1     2
```

33.
```
  ━━━━━━●━━━━━●━━━━━━→
      -2/3    2
```

34.
```
  ━━●━━━━━━━━━━━●━━━━→
   -4           5
```

35.
```
  ←━━━━━━━━━●   ●━━━━━━
           -4/5  8/5
```

36. The width is 6 meters and the length is 17 meters. **37.** $1\dfrac{1}{2}$ hours **38.** 20 liters

39. 89 or better **40.** 34° and 99° **41.** 5 hours **42.** 18 gallons **43.** 88 or better
44. 26% **45.** 10 years old

CHAPTER 5
Problem Set 5.1 (page 181)

1. $y = \dfrac{13 - 3x}{7}$ **3.** $x = 3y + 9$ **5.** $y = \dfrac{x + 14}{5}$ **7.** $x = \dfrac{y - 7}{3}$ **9.** $y = \dfrac{2x - 5}{3}$

11.

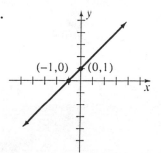

13.

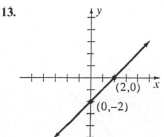

15.

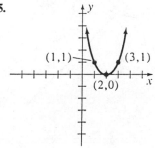

17.

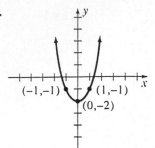

19.

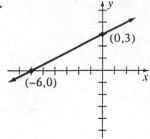

21.

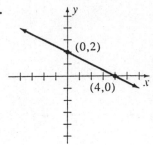

23.

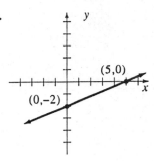

25.

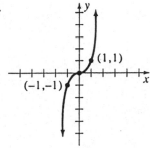

27.

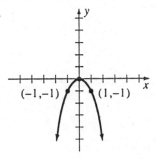

29.

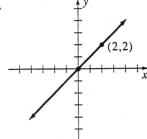

31.

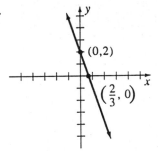

33.

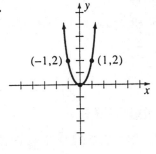

Problem Set 5.2 (page 186)

1.

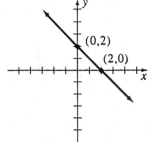

3.

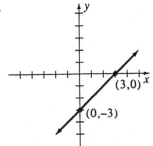

5.

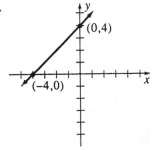

7.
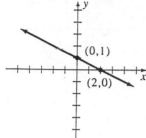
(0,1)
(2,0)

9.
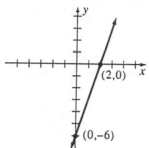
(2,0)
(0,−6)

11.
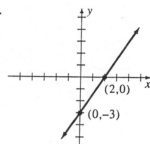
(2,0)
(0,−3)

13.
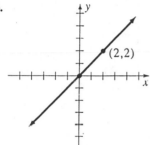
(2,2)

15.
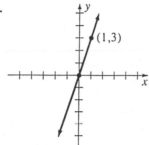
(1,3)

17.
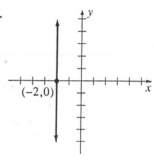
(−2,0)

19.
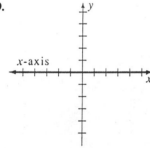
x-axis

21.
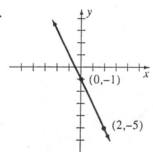
(0,−1)
(2,−5)

23.
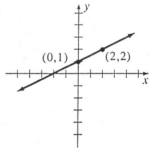
(0,1)   (2,2)

25.
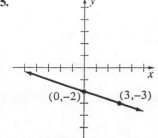
(0,−2)   (3,−3)

27.
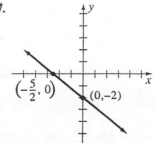
$\left(-\frac{5}{2}, 0\right)$ (0,−2)

29.
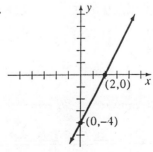
(2,0)
(0,−4)

31.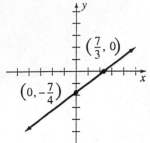

33.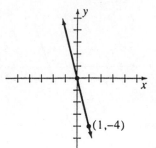

35.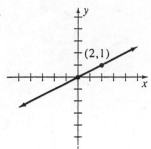

37.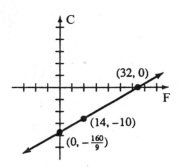

39.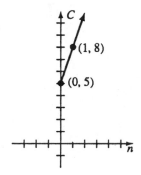

43.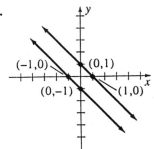

45.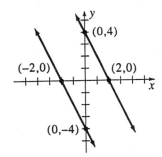

Problem Set 5.3 (page 192)

1. $\dfrac{3}{4}$ **3.** $\dfrac{7}{5}$ **5.** $-\dfrac{6}{5}$ **7.** $-\dfrac{10}{3}$ **9.** $\dfrac{3}{4}$ **11.** 0   **13.** $-\dfrac{3}{2}$ **15.** undefined

17. 1   **19.** $\dfrac{b-d}{a-c}$ or $\dfrac{d-b}{c-a}$ **21.** 4   **23.** -6 **25–31.** answers will vary   **33.** negative

35. positive   **37.** zero   **39.** negative   **41.** $-\dfrac{3}{2}$ **43.** $\dfrac{5}{4}$ **45.** $-\dfrac{1}{5}$ **47.** 2   **49.** 0

51. $\dfrac{2}{5}$ **53.** $\dfrac{6}{5}$ **55.** -3 **57.** 4   **59.** $\dfrac{2}{3}$ **61.** (a) 105.6 feet   **63.** 1.0 feet

Problem Set 5.4 (page 201)

1. $2x - 3y = -5$ **3.** $x - 2y = 7$ **5.** $x + 3y = 20$ **7.** $y = -7$ **9.** $4x + 9y = 0$
11. $3x - y = -16$ **13.** $7x - 5y = -1$ **15.** $3x + 2y = 5$ **17.** $x - y = 1$ **19.** $5x - 3y = 0$
21. $4x + 7y = 28$ **23.** $y = \frac{3}{5}x + 2$ **25.** $y = 2x - 1$ **27.** $y = -\frac{1}{6}x - 4$ **29.** $y = -x + \frac{5}{2}$
31. $y = -\frac{5}{9}x - \frac{1}{2}$ **33.** $m = -2, b = -5$ **35.** $m = \frac{3}{5}, b = -3$ **37.** $m = \frac{4}{9}, b = 2$
39. $m = \frac{3}{4}, b = -4$ **41.** $m = -\frac{2}{11}, b = -1$ **43.** $m = -\frac{9}{7}, b = 0$ **45.** perpendicular
47. intersecting lines that are not perpendicular **49.** parallel **51.** $2x - 3y = -1$
53. $x - 3y = 19$ **55.** $2x + 9y = 0$

Problem Set 5.5 (page 208)

1. No **3.** Yes **5.** Yes **7.** Yes **9.** No **11.** $\{(2, -1\}$ **13.** $\{(2, 1)\}$ **15.** $\varnothing$
17. $\{(0, 0)\}$ **19.** $\{(1, -1)\}$ **21.** Infinitely many **23.** $\{(1, 3)\}$ **25.** $\{(3, -2)\}$ **27.** $\{(2, 4)\}$
29. $\{(-2, -3)\}$

Problem Set 5.6 (page 215)

1. $\{(6, 8)\}$ **3.** $\{(-5, -4)\}$ **5.** $\{(-6, 12)\}$ **7.** $\{(5, -2)\}$ **9.** $\left\{\left(\frac{11}{4}, \frac{9}{8}\right)\right\}$ **11.** $\left\{\left(-\frac{2}{3}, \frac{2}{3}\right)\right\}$
13. $\{(-4, 5)\}$ **15.** $\{(4, 1)\}$ **17.** $\left\{\left(\frac{3}{2}, -3\right)\right\}$ **19.** $\left\{\left(-\frac{18}{71}, \frac{5}{71}\right)\right\}$ **21.** $\{(250, 500)\}$
23. $\{(100, 200)\}$ **25.** 9 and 21 **27.** 8 and 15 **29.** 6 and 12
31. \$.15 per lemon and \$.30 per apple **33.** 7 dimes and 3 quarters
35. 14 books at \$12 each and 21 books at \$14 each **37.** 4 gallons of 10% and 6 gallons of 15%
39. \$500 at 10% and \$800 at 12% **41.** $\{(12, 24)\}$ **43.** $\varnothing$

Problem Set 5.7 (page 226)

1. $\{(5, 9)\}$ **3.** $\{(-4, 10)\}$ **5.** $\left\{\left(\frac{3}{2}, 4\right)\right\}$ **7.** $\left\{\left(-\frac{1}{3}, \frac{4}{3}\right)\right\}$ **9.** $\{(18, 24)\}$
11. $\{(-10, -15)\}$ **13.** $\varnothing$ **15.** $\{(-2, 6)\}$ **17.** $\{(-6, -13)\}$ **19.** $\left\{\left(\frac{9}{8}, \frac{7}{12}\right)\right\}$
21. $\left\{\left(\frac{6}{31}, \frac{9}{31}\right)\right\}$ **23.** $\{(100, 400)\}$ **25.** $\{(3, 10)\}$ **27.** $\{(-2, -6)\}$ **29.** $\left\{\left(-\frac{6}{7}, \frac{6}{7}\right)\right\}$
31. $\{(10, 12)\}$ **33.** $\left\{\left(\frac{3}{5}, \frac{12}{5}\right)\right\}$ **35.** $\left\{\left(-\frac{5}{2}, 6\right)\right\}$ **37.** $\left\{\left(\frac{1}{2}, 4\right)\right\}$
39. infinitely many solutions **41.** $\left\{\left(5, \frac{5}{2}\right)\right\}$ **43.** $\{(12, 4)\}$ **45.** $\left\{\left(-\frac{1}{11}, -\frac{10}{11}\right)\right\}$
47. 12 and 34 **49.** 35 double rooms and 15 single rooms **51.** 45
53. 18 dimes and 41 quarters **55.** 93 **57.** \$150 at 8% and \$400 at 9%
59. 7.5 liters of 30% and 2.5 liters of 70% **63.** $\left\{\left(-1, -\frac{7}{6}\right)\right\}$ **65.** Infinitely many

Problem Set 5.8 (page 234)

1.
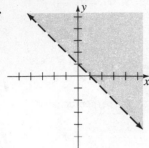

3.
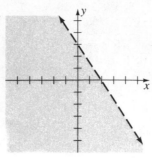

5.
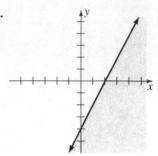

7.
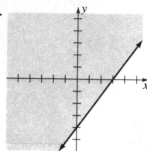

9.
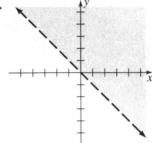

11.
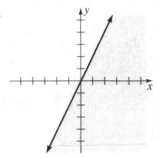

13.
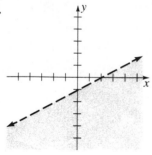

15.
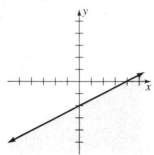

17.
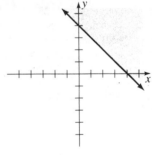

19.
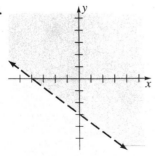

21.
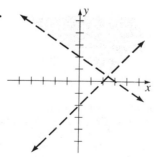

23.
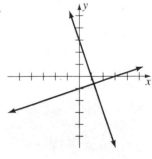

25.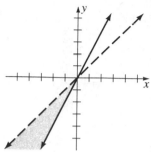

27.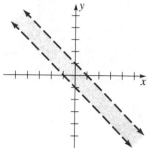

29.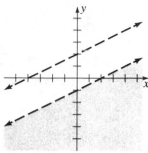

31.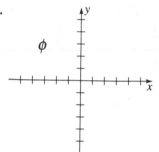

33.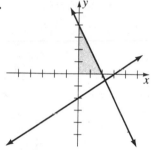

Chapter 5 Review Problem Set (page 236)

1.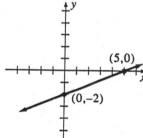

2.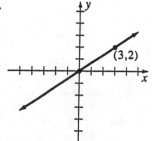

3.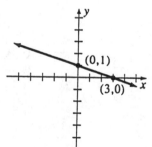

4.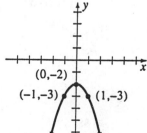

5.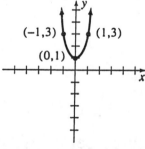

6.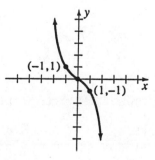

7. $-\dfrac{9}{5}$ **8.** $\dfrac{5}{6}$ **9.** $5x + 7y = -11$ **10.** $8x - 3y = 1$ **11.** $2x - 9y = 9$ **12.** $x = 2$

13. $\{(3, -2)\}$ **14.** $\{(7, 13)\}$ **15.** $\{(16, -5)\}$ **16.** $\left\{\left(\dfrac{41}{23}, \dfrac{19}{23}\right)\right\}$ **17.** $\{(10, 25)\}$

18. $\{(-6, -8)\}$ **19.** $\{(400, 600)\}$ **20.** $\varnothing$ **21.** $\left\{\left(\dfrac{5}{16}, -\dfrac{17}{16}\right)\right\}$ **22.** $t = 4$ and $u = 8$

23. $t = 8$ and $u = 4$ **24.** $t = 3$ and $u = 7$ **25.** $\{(-9, 6)\}$

26. 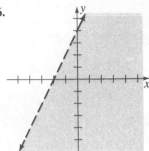   **27.** 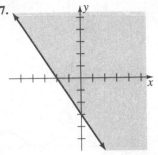   **28.**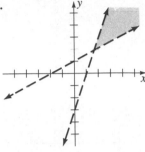

29. 38 and 75   **30.** $250 at 9% and $300 at 11%   **31.** 18 nickels and 25 dimes
32. length of 19 inches and width of 6 inches   **33.** 49   **34.** 36   **35.** 32° and 58°
36. 50° and 130°   **37.** $1.15 for a cheeseburger and $0.75 for a milkshake
38. $0.89 for prune juice and $0.59 for tomato juice

CHAPTER 6
Problem Set 6.1 (page 243)

1. 3   **3.** 2   **5.** 3   **7.** 2   **9.** $8x + 11$ **11.** $4y + 10$ **13.** $2x^2 - 2x - 23$
15. $17x - 19$ **17.** $6x^2 - 5x - 4$ **19.** $5n - 6$ **21.** $-7x^2 - 13x - 7$ **23.** $5x + 5$
25. $-2x - 5$ **27.** $-3x + 7$ **29.** $2x^2 + 15x - 6$ **31.** $5n^2 + 2n + 3$
33. $-3x^3 + 2x^2 - 11$ **35.** $9x - 2$ **37.** $2a + 15$ **39.** $-2x^2 + 5$ **41.** $6x^3 + 12x^2 - 5$
43. $4x^3 - 8x^2 + 13x + 12$ **45.** $x + 11$ **47.** $-3x - 14$ **49.** $-x^2 - 13x - 12$
51. $x^2 - 11x - 8$ **53.** $-10a - 3b$ **55.** $-n^2 + 2n - 17$ **57.** $8x + 1$ **59.** $-5n - 1$
61. $-2a + 6$ **63.** $11x + 6$ **65.** $6x + 7$ **67.** $-5n + 7$

Problem Set 6.2 (page 249)

1. $45x^2$ **3.** $21x^3$ **5.** $-6x^2y^2$ **7.** $14x^3y$ **9.** $-48a^3b^3$ **11.** $5x^4y$ **13.** $104a^3b^2c^2$

15. $30x^6$ **17.** $-56x^2y^3$ **19.** $-6a^2b^3$ **21.** $72c^3d^3$ **23.** $\dfrac{2}{5}x^3y^5$ **25.** $-\dfrac{2}{9}a^2b^5$

27. $0.28x^8$ **29.** $-6.4a^4b^2$ **31.** $4x^8$ **33.** $9a^4b^6$ **35.** $27x^6$ **37.** $-64x^{12}$
39. $81x^8y^{10}$ **41.** $16x^8y^4$ **43.** $81a^{12}b^8$ **45.** $x^{12}y^6$ **47.** $15x^2 + 10x$ **49.** $18x^3 - 6x^2$
51. $\{-28x^3 + 16x$ **53.** $2x^3 - 8x^2 + 12x$ **55.** $-18a^3 + 30a^2 + 42a$
57. $28x^3y - 7x^2y + 35xy$ **59.** $-9x^3y + 2x^2y + 6xy$ **61.** $13x + 22y$ **63.** $-2x - 9y$
65. $4x^3 - 3x^2 - 14x$ **67.** $-x + 14$ **69.** $-7x + 12$ **71.** $18x^5$ **73.** $-432x^5$
75. $25x^7y^8$ **77.** $-a^{12}b^5c^9$ **79.** $-16x^{11}y^{17}$ **81.** x^{4n} **83.** x^{5n+1} **85.** x^{4n-2}
87. $6x^{3n}$ **89.** $-30x^{5n}$

Problem Set 6.3 (page 254)

1. $xy + 3x + 2y + 6$ **3.** $xy + x - 4y - 4$ **5.** $xy - 6x - 5y + 30$
7. $xy + xz + x + 2y + 2z + 2$ **9.** $6xy + 2x + 9y + 3$ **11.** $x^2 + 10x + 21$ **13.** $x^2 + 5x - 24$
15. $x^2 - 6x - 7$ **17.** $n^2 - 10n + 24$ **19.** $3n^2 + 19n + 6$ **21.** $15x^2 + 29x - 14$
23. $x^3 + 7x^2 + 21x + 27$ **25.** $x^3 + 3x^2 - 10x - 24$ **27.** $2x^3 - 7x^2 - 22x + 35$
29. $8a^3 - 14a^2 + 23a - 9$ **31.** $3a^3 + 2a^2 - 8a - 5$ **33.** $x^4 + 7x^3 + 17x^2 + 23x + 12$
35. $x^4 - 3x^3 - 34x^2 + 33x + 63$ **37.** $x^2 + 11x + 18$ **39.** $x^2 + 4x - 12$ **41.** $x^2 - 8x - 33$
43. $n^2 - 7n + 12$ **45.** $n^2 + 18n + 72$ **47.** $y^2 - 4y - 21$ **49.** $y^2 - 19y + 84$
51. $x^2 + 2x - 35$ **53.** $x^2 - 6x - 112$ **55.** $a^2 + a - 90$ **57.** $2a^2 + 13a + 6$
59. $5x^2 + 33x - 14$ **61.** $6x^2 - 11x - 7$ **63.** $12a^2 - 7a - 12$ **65.** $12n^2 - 28n + 15$
67. $14x^2 + 13x - 12$ **69.** $45 - 19x + 2x^2$ **71.** $-8x^2 + 22x - 15$ **73.** $-9x^2 + 9x + 4$
75. $72n^2 - 5n - 12$ **77.** $27 - 21x + 2x^2$ **79.** $20x^2 - 7x - 6$ **81.** $x^2 + 14x + 49$
83. $x^2 - 10x + 25$ **85.** $9x^2 + 12x + 4$ **87.** $16x^2 - 24x + 9$ **89.** $1 + 2x + x^2$
91. $x^3 + 6x^2 + 12x + 8$ **93.** $x^3 - 9x^2 + 27x - 27$ **95.** $8x^3 + 36x^2 + 54x + 27$
99. $x^{2n} + 12x^n + 35$ **101.** $x^{2a} + 4x^a - 32$ **103.** $x^{4n} - x^{2n} - 12$ **105.** $6x^{2a} - 5x^a - 4$
107. $20x^{2n} - 39x^n + 18$

Problem Set 6.4 (page 260)

1. $x^2 + 18x + 81$ **3.** $x^2 - 14x + 49$ **5.** $25x^2 + 20x + 4$ **7.** $4x^2 - 28x + 49$
9. $x^3 - 6x^2 + x + 14$ **11.** $x^2 - 81$ **13.** $t^2 - 121$ **15.** $36t^2 - 1$
17. $x^3 + 18x^2 + 108x + 216$ **19.** $n^3 - 3n^2 + 3n - 1$ **21.** $27n^3 + 108n^2 + 144n + 64$
23. $6x^3 + x^2 - 14x - 8$ **25.** $4 - 28x + 49x^2$ **27.** $9 - 42x + 49x^2$ **29.** $1 - 49n^2$
31. $2x^3 + 9x^2 - 32x + 21$ **33.** $27y^3 - 135y^2 + 225y - 125$ **35.** $1 - 3n + 3n^2 - n^3$ **37.** x^8
39. $2x^2$ **41.** $-8n^4$ **43.** -8 **45.** $13xy^2$ **47.** $7ab^2$ **49.** $18xy^4$ **51.** $-32x^5y^2$
53. $-8x^5y^4$ **55.** -1 **57.** $4x^2 + 6x^3$ **59.** $3x^3 - 8x$ **61.** $-7n^3 + 9$
63. $5x^4 - 8x^3 - 12x$ **65.** $4n^5 - 8n^2 + 13$ **67.** $5a^6 + 8a^2$ **69.** $-3xy + 5y$
71. $-8ab - 10a^2b^3$ **73.** $-3bc + 13b^2c^4$ **75.** $-9xy^2 + 12x^2y^3$ **77.** $-3x^4 - 5x^2 + 7$
79. $-3a^2 + 7a + 13b$ **81.** $-1 + 5xy^2 - 7xy^5$
83. (a) 899   **(c)** 3591   **(e)** 6241   **(g)** 2704

Problem Set 6.5 (page 265)

1. $x + 12$ **3.** $x + 2$ **5.** $x + 8$ with a remainder of 4   **7.** $x + 4$ with a remainder of -7
9. $5n + 4$ **11.** $8y - 3$ with a remainder of 2   **13.** $4x - 7$ **15.** $3x + 2$ with a remainder of -6
17. $2x^2 + 3x + 4$ **19.** $5n^2 - 4n - 3$ **21.** $n^2 + 6n - 4$ **23.** $x^2 + 3x + 9$
25. $9x^2 + 12x + 16$ **27.** $3n - 8$ with a remainder of 17   **29.** $3t + 2$ with a remainder of 6
31. $3n^2 - n - 4$ **33.** $4x^2 - 5x + 5$ with a remainder of -3 **35.** $x + 4$ with a remainder of $5x - 1$
37. $2x - 12$ with a remainder of $49x - 5$ **39.** $x^3 - 2x^2 + 4x - 8$

Chapter 6 Review Problem Set (page 267)

1. $8x^2 - 13x + 2$ **2.** $3y^2 + 11y - 9$ **3.** $3x^2 + 2x - 9$ **4.** $-8x^2 + 18$ **5.** $11x + 8$
6. $-9x^2 + 8x - 20$ **7.** $2y^2 - 54y + 18$ **8.** $-13a - 30$ **9.** $-27a - 7$ **10.** $n - 2$
11. $-5n^2 - 2n$ **12.** $17n^2 - 14n - 16$ **13.** $35x^6$ **14.** $-54x^8$ **15.** $24x^3y^5$
16. $-6a^4b^9$ **17.** $8a^6b^9$ **18.** $9x^2y^4$ **19.** $35x^2 + 15x$ **20.** $-24x^3 + 3x^2$
21. $x^2 + 17x + 72$ **22.** $3x^2 + 10x + 7$ **23.** $x^2 - 3x - 10$ **24.** $y^2 - 13y + 36$

25. $14x^2 - x - 3$ **26.** $20a^2 - 3a - 56$ **27.** $9a^2 - 30a + 25$ **28.** $2x^3 + 17x^2 + 26x - 24$
29. $30n^2 + 19n - 5$ **30.** $12n^2 + 13n - 4$ **31.** $4n^2 - 1$ **32.** $16n^2 - 25$
33. $4a^2 + 28a + 49$ **34.** $9a^2 + 30a + 25$ **35.** $x^3 - 3x^2 + 8x - 12$ **36.** $2x^3 + 7x^2 + 10x - 7$
37. $a^3 + 15a^2 + 75a + 125$ **38.** $a^3 - 18a^2 + 108a - 216$ **39.** $x^4 + x^3 + 2x^2 - 7x - 5$
40. $n^4 - 5n^3 - 11n^2 - 30n - 4$ **41.** $-12x^3y^3$ **42.** $7a^3b^4$ **43.** $-3x^2y - 9x^4$
44. $10a^4b^9 - 13a^3b^7$ **45.** $14x^2 - 10x - 8$ **46.** $x + 4, R = -21$ **47.** $7x - 6$
48. $2x^2 + x + 4, R = 4$

CHAPTER 7

Problem Set 7.1 (page 275)

1. $6y$ **3.** $12xy$ **5.** $14ab^2$ **7.** $2x$ **9.** $8a^2b^2$ **11.** $4(2x + 3y)$ **13.** $7y(2x - 3)$
15. $9x(2x + 5)$ **17.** $6xy(2y - 5x)$ **19.** $12a^2b(3 - 5ab^3)$ **21.** $xy^2(16y + 25x)$
23. $8(8ab - 9cd)$ **25.** $9a^2b(b^3 - 3)$ **27.** $4x^4y(13y + 15x^2)$ **29.** $8x^2y(5y + 1)$
31. $3x(4 + 5y + 7x)$ **33.** $x(2x^2 - 3x + 4)$ **35.** $4y^2(11y^3 - 6y - 5)$ **37.** $7ab(2ab^2 + 5b - 7a^2)$
39. $(y + 1)(x + z)$ **41.** $(b - 4)(a - c)$ **43.** $(x + 3)(x + 6)$ **45.** $(x + 1)(2x - 3)$
47. $(x + y)(5 + b)$ **49.** $(x - y)(b - c)$ **51.** $(a + b)(c + 1)$ **53.** $(x + 5)(x + 12)$
55. $(x - 2)(x - 8)$ **57.** $(2x + 1)(x - 5)$ **59.** $(2n - 1)(3n - 4)$ **61.** $\{0, 8\}$ **63.** $\{-1, 0\}$
65. $\{0, 5\}$ **67.** $\left\{0, \dfrac{3}{2}\right\}$ **69.** $\left\{-\dfrac{3}{7}, 0\right\}$ **71.** $\{-5, 0\}$ **73.** $\left\{0, \dfrac{3}{2}\right\}$ **75.** $\{0, 7\}$
77. $\{0, 13\}$ **79.** $\left\{-\dfrac{5}{2}, 0\right\}$ **81.** 0 or 9   **83.** 20 units   **85.** $\dfrac{4}{\pi}$
87. The square is 3 inches by 3 inches and the rectangle is 3 inches by 6 inches.   **89.** (a) \$116
(c) \$750   **91.** $x = 0$ or $x = \dfrac{c}{b^2}$ **93.** $y = \dfrac{c}{1 + a - b}$

Problem Set 7.2 (page 282)

1. $(x - 1)(x + 1)$ **3.** $(x - 10)(x + 10)$ **5.** $(x - 2y)(x + 2y)$ **7.** $(3x - y)(3x + y)$
9. $(6a - 5b)(6a + 5b)$ **11.** $(1 - 2n)(1 + 2n)$ **13.** $5(x - 2)(x + 2)$ **15.** $8(x^2 + 4)$
17. $2(x - 3y)(x + 3y)$ **19.** $x(x - 5)(x + 5)$ **21.** Not factorable   **23.** $9x(5x - 4y)$
25. $4(3 - x)(3 + x)$ **27.** $4a^2(a^2 + 4)$ **29.** $(x - 3)(x + 3)(x^2 + 9)$ **31.** $x^2(x^2 + 1)$
33. $3x(x^2 + 16)$ **35.** $5x(1 - 2x)(1 + 2x)$ **37.** $4(x - 4)(x + 4)$ **39.** $3xy(5x - 2y)(5x + 2y)$
41. $\{-3, 3\}$ **43.** $\{-2, 2\}$ **45.** $\left\{-\dfrac{4}{3}, \dfrac{4}{3}\right\}$ **47.** $\{-11, 11\}$ **49.** $\left\{-\dfrac{2}{5}, \dfrac{2}{5}\right\}$ **51.** $\{-5, 5\}$
53. $\{-4, 0, 4\}$ **55.** $\{-4, 0, 4\}$ **57.** $\left\{-\dfrac{1}{3}, \dfrac{1}{3}\right\}$ **59.** $\{-10, 0, 10\}$ **61.** $\left\{-\dfrac{9}{8}, \dfrac{9}{8}\right\}$
63. $\left\{-\dfrac{1}{2}, 0, \dfrac{1}{2}\right\}$ **65.** -7 or 7 **67.** $-4, 0,$ or 4 **69.** 3 inches and 15 inches
71. The length is 20 centimeters and the width is 8 centimeters.   **73.** 4 meters and 8 meters
75. 5 centimeters   **79.** $(x + 1)(x^2 - x + 1)$ **81.** $(n - 3)(n^2 + 3n + 9)$
83. $(2x + 3y)(4x^2 - 6xy + 9y^2)$ **85.** $(1 - 2x)(1 + 2x + 4x^2)$ **87.** $(x + 2y)(x^2 - 2xy + 4y^2)$
89. $(ab - 1)(a^2b^2 + ab + 1)$ **91.** $(2 + n)(4 - 2n + n^2)$ **95.** $(3n - 5)(9n^2 + 15n + 25)$

Problem Set 7.3 (page 290)

1. $(x + 4)(x + 6)$ **3.** $(x + 5)(x + 8)$ **5.** $(x - 2)(x - 9)$ **7.** $(n - 7)(n - 4)$ **9.** $(n + 9)(n - 3)$
11. $(n - 10)(n + 4)$ **13.** Not factorable   **15.** $(x - 6)(x - 12)$ **17.** $(x + 11)(x - 6)$
19. $(y - 9)(y + 8)$ **21.** $(x + 5)(x + 16)$ **23.** $(x + 12)(x - 6)$ **25.** Not factorable
27. $(x - 2y)(x + 5y)$ **29.** $(a - 8b)(a + 4b)$ **31.** $\{-7, -3\}$ **33.** $\{3, 6\}$ **35.** $\{-2, 5\}$
37. $\{-9, 4\}$ **39.** $\{-4, 10\}$ **41.** $\{-8, 7\}$ **43.** $\{2, 14\}$ **45.** $\{-12, 1\}$ **47.** $\{2, 8\}$
49. $\{-6, 4\}$ **51.** 7 and 8 or -7 and -8 **53.** 12 and 14
55. $-4, -3, -2,$ and -1 or 7, 8, 9, and 10   **57.** 4 and 7 or 0 and 3
59. The length is 9 inches and the width is 6 inches.   **61.** 9 centimeters by 6 centimeters
63. 7 rows and 12 trees per row   **65.** 8 feet, 15 feet, and 17 feet   **67.** 6 inches and 8 inches
69. $(x^a + 6)(x^a + 4)$ **71.** $(x^a - 4)(x^a + 2)$ **73. (a)** $\{-18, -12\}$ **(c)** $\{16, 24\}$ **(e)** $\{-24, 18\}$

Problem Set 7.4 (page 297)

1. $(3x + 1)(x + 2)$ **3.** $(2x + 5)(3x + 2)$ **5.** $(4x - 1)(x - 6)$ **7.** $(4x - 5)(3x - 4)$
9. $(5y + 2)(y - 7)$ **11.** $(2n - 3)(n + 8)$ **13.** Not factorable   **15.** $(3x + 7)(6x + 1)$
17. $(7x - 2)(x - 4)$ **19.** $(4x + 7)(2x - 3)$ **21.** $(3t + 2)(3t - 7)$ **23.** $(12y - 5)(y + 7)$
25. Not factorable   **27.** $(7x + 3)(2x + 7)$ **29.** $(4x - 3)(5x - 4)$ **31.** $(4n + 3)(4n - 5)$
33. $(6x - 5)(4x - 5)$ **35.** $(2x + 9)(x + 8)$ **37.** $(3a + 1)(7a - 2)$ **39.** $(12a + 5)(a - 3)$
41. $(2x + 3)(2x + 3)$ **43.** $(2x - y)(3x - y)$ **45.** $(4x + 3y)(5x - 2y)$ **47.** $(5x - 2)(x - 6)$

49. $(8x + 1)(x - 7)$ **51.** $\left\{-6, -\dfrac{1}{2}\right\}$ **53.** $\left\{-\dfrac{2}{3}, -\dfrac{1}{4}\right\}$ **55.** $\left\{\dfrac{1}{3}, 8\right\}$ **57.** $\left\{\dfrac{2}{5}, \dfrac{7}{3}\right\}$

59. $\left\{-7, \dfrac{5}{6}\right\}$ **61.** $\left\{-\dfrac{3}{8}, \dfrac{3}{2}\right\}$ **63.** $\left\{-\dfrac{2}{3}, \dfrac{4}{3}\right\}$ **65.** $\left\{\dfrac{2}{5}, \dfrac{5}{2}\right\}$ **67.** $\left\{-\dfrac{5}{2}, -\dfrac{2}{3}\right\}$

69. $\left\{-\dfrac{5}{4}, \dfrac{1}{4}\right\}$ **71.** $\left\{-\dfrac{3}{7}, \dfrac{7}{5}\right\}$ **73.** $\left\{\dfrac{5}{4}, 10\right\}$ **75.** $\left\{-7, \dfrac{3}{7}\right\}$ **77.** $\left\{-\dfrac{5}{12}, 4\right\}$

79. $\left\{-\dfrac{7}{2}, \dfrac{4}{9}\right\}$

Problem Set 7.5 (page 306)

1. $(x + 2)^2$ **3.** $(x - 5)^2$ **5.** $(3n + 2)^2$ **7.** $(4a - 1)^2$ **9.** $(2 + 9x)^2$ **11.** $(4x - 3y)^2$
13. $(2x + 1)(x + 8)$ **15.** $2x(x - 6)(x + 6)$ **17.** $(n - 12)(n + 5)$ **19.** Not factorable
21. $8(x^2 + 9)$ **23.** $(3x + 5)^2$ **25.** $5(x + 2)(3x + 7)$ **27.** $(4x - 3)(6x + 5)$
29. $(x + 5)(y - 8)$ **31.** $(5x - y)(4x + 7y)$ **33.** $3(2x - 3)(4x + 9)$ **35.** $6(2x^2 + x + 5)$
37. $5(x - 2)(x + 2)(x^2 + 4)$ **39.** $(x + 6y)^2$ **41.** $\{0, 5\}$ **43.** $\{-3, 12\}$ **45.** $\{-2, 0, 2\}$
47. $\left\{-\dfrac{2}{3}, \dfrac{11}{2}\right\}$ **49.** $\left\{\dfrac{1}{3}, \dfrac{3}{4}\right\}$ **51.** $\{-3, -1\}$ **53.** $\{0, 6\}$ **55.** $\{-4, 0, 6\}$ **57.** $\left\{\dfrac{2}{5}, \dfrac{6}{5}\right\}$

59. $\{12, 16\}$ **61.** $\left\{-\dfrac{10}{3}, 1\right\}$ **63.** $\{0, 6\}$ **65.** $\left\{\dfrac{4}{3}\right\}$ **67.** $\{-5, 0\}$ **69.** $\left\{-\dfrac{4}{3}, \dfrac{5}{8}\right\}$

71. $\dfrac{5}{4}$ and 12 or -5 and -3 **73.** -1 and 1 or $-\dfrac{1}{2}$ and 2   **75.** 4 and 9 or $-\dfrac{24}{5}$ and $-\dfrac{43}{5}$

77. 6 rows and 9 chairs per row
79. One square is 6 feet by 6 feet and the other one is 18 feet by 18 feet.

81. 11 centimeters long and 5 centimeters wide

83. The side is 17 inches long and the altitude to that side is 6 inches long.

85. $1\frac{1}{2}$ inches   **87.** 6 inches and 12 inches

Chapter 7 Review Problem Set (page 309)

1. $(x - 2)(x - 7)$ **2.** $3x(x + 7)$ **3.** $(3x + 2)(3x - 2)$ **4.** $(2x - 1)(2x + 5)$ **5.** $(5x - 6)^2$

6. $n(n + 5)(n + 8)$ **7.** $(y + 12)(y - 1)$ **8.** $3xy(y + 2x)$ **9.** $(x + 1)(x - 1)(x^2 + 1)$

10. $(6n + 5)(3n - 1)$ **11.** Not factorable   **12.** $(4x - 7)(x + 1)$ **13.** $3(n + 6)(n - 5)$

14. $x(x + y)(x - y)$ **15.** $(2x - y)(x + 2y)$ **16.** $2(n - 4)(2n + 5)$ **17.** $(x + y)(5 + a)$

18. $(7t - 4)(3t + 1)$ **19.** $2x(x + 1)(x - 1)$ **20.** $3x(x + 6)(x - 6)$ **21.** $(4x + 5)^2$

22. $(y - 3)(x - 2)$ **23.** $(5x + y)(3x - 2y)$ **24.** $n^2(2n - 1)(3n - 1)$ **25.** $\{-6, 2\}$ **26.** $\{0, 11\}$

27. $\left\{-4, \frac{5}{2}\right\}$ **28.** $\left\{-\frac{8}{3}, \frac{1}{3}\right\}$ **29.** $\{-2, 2\}$ **30.** $\left\{-\frac{5}{4}\right\}$ **31.** $\{-1, 0, 1\}$ **32.** $\left\{-\frac{2}{7}, -\frac{9}{4}\right\}$

33. $\{-7, 4\}$ **34.** $\{-5, 5\}$ **35.** $\left\{-6, \frac{3}{5}\right\}$ **36.** $\left\{-\frac{7}{2}, 1\right\}$ **37.** $\{-2, 0, 2\}$ **38.** $\{8, 12\}$

39. $\left\{-5, \frac{3}{4}\right\}$ **40.** $\{-2, 3\}$ **41.** $\left\{-\frac{7}{3}, \frac{5}{2}\right\}$ **42.** $\{-9, 6\}$ **43.** $\left\{-5, \frac{3}{2}\right\}$ **44.** $\left\{\frac{4}{3}, \frac{5}{2}\right\}$

45. $-\frac{8}{3}$ and $-\frac{19}{3}$ or 4 and 7   **46.** The length is 8 centimeters and the width is 2 centimeters.

47. A 2-by-2-inch square and a 10-by-10 inch square   **48.** 8 by 15 by 17

49. $-\frac{13}{6}$ and -12 or 2 and 13   **50.** 7, 9, and 11   **51.** 4 shelves

52. A 5-by-5-yard square and a 5-by-40-yard rectangle   **53.** -18 and -17 or 17 and 18

54. 6 units   **55.** 2 meters and 7 meters   **56.** 9 and 11   **57.** 2 centimeters

58. 15 feet

CHAPTER 8
Problem Set 8.1 (page 317)

1. $\frac{1}{27}$ **3.** $-\frac{1}{100}$ **5.** 81   **7.** -27 **9.** -8 **11.** 1   **13.** $\frac{9}{49}$ **15.** 16

17. $\frac{1}{1000}$ **19.** $\frac{1}{1000}$ **21.** 27   **23.** $\frac{1}{125}$ **25.** $\frac{9}{8}$ **27.** $\frac{256}{25}$ **29.** $\frac{2}{25}$ **31.** $\frac{81}{4}$

33. 81   **35.** $\frac{1}{10000}$ **37.** $\frac{13}{36}$ **39.** $\frac{1}{2}$ **41.** $\frac{72}{17}$ **43.** $\frac{1}{x^6}$ **45.** $\frac{1}{a^3}$ **47.** $\frac{1}{a^8}$

49. $\frac{y^6}{x^2}$ **51.** $\frac{c^8}{a^4 b^{12}}$ **53.** $\frac{y^{12}}{8x^9}$ **55.** $\frac{x^3}{y^{12}}$ **57.** $\frac{4a^4}{9b^2}$ **59.** $\frac{1}{x^2}$ **61.** $a^5 b^2$ **63.** $\frac{6y^3}{x}$

65. $7b^2$ **67.** $\frac{7x}{y^2}$ **69.** $-\frac{12b^3}{a}$ **71.** $\frac{x^5 y^5}{5}$ **73.** $\frac{b^{20}}{81}$ **75.** $\frac{x + 1}{x^3}$ **77.** $\frac{y - x^3}{x^3 y}$

79. $\frac{3b + 4a^2}{a^2 b}$ **81.** $\frac{1 - x^2 y}{xy^2}$ **83.** $\frac{2x - 3}{x^2}$

Problem Set 8.2 (page 322)

1. 7   **3.** -8 **5.** 11   **7.** 60   **9.** -40 **11.** 80   **13.** 18   **15.** $\dfrac{5}{3}$ **17.** .4

19. 24   **21.** 48   **23.** 28   **25.** 65   **27.** 58   **29.** 4.36   **31.** 7.07   **33.** 8.66

35. 9.75   **37.** 66   **39.** 34   **41.** 97   **43.** 81   **45.** 58   **47.** $21\sqrt{2}$ **49.** $8\sqrt{7}$

51. $-9\sqrt{3}$ **53.** $6\sqrt{5}$ **55.** $-\sqrt{2}+2\sqrt{3}$ **57.** $-9\sqrt{7}-2\sqrt{10}$ **59.** 17.3   **61.** 13.4

63. -1.4 **65.** 26.5   **67.** 4.1   **69.** 2.1   **71.** -29.8

Problem Set 8.3 (page 329)

1. 4   **3.** -5 **5.** 2   **7.** $\dfrac{3}{4}$ **9.** 8   **11.** 2   **13.** $\dfrac{2}{3}$ **15.** -3

17. 16   **19.** $\dfrac{4}{3}$ **21.** $3\sqrt{3}$ **23.** $4\sqrt{2}$ **25.** $4\sqrt{5}$ **27.** $4\sqrt{10}$ **29.** $12\sqrt{2}$

31. $-12\sqrt{5}$ **33.** $2\sqrt{3}$ **35.** $3\sqrt{6}$ **37.** $-\dfrac{5}{3}\sqrt{7}$ **39.** $\dfrac{\sqrt{19}}{2}$ **41.** $\dfrac{3\sqrt{3}}{4}$ **43.** $\dfrac{5\sqrt{3}}{9}$

45. $\dfrac{\sqrt{14}}{7}$ **47.** $\dfrac{\sqrt{6}}{3}$ **49.** $\dfrac{\sqrt{15}}{6}$ **51.** $\dfrac{\sqrt{66}}{12}$ **53.** $\dfrac{\sqrt{6}}{3}$ **55.** $\sqrt{5}$ **57.** $\dfrac{2\sqrt{21}}{7}$

59. $-\dfrac{8\sqrt{15}}{5}$ **61.** $\dfrac{\sqrt{6}}{4}$ **63.** $-\dfrac{12}{25}$ **65.** $2\sqrt[3]{2}$ **67.** $6\sqrt[3]{3}$ **69.** $\dfrac{2\sqrt[3]{3}}{3}$ **71.** $\dfrac{3\sqrt[3]{2}}{2}$

73. $\dfrac{\sqrt[3]{12}}{2}$ **75.** (a) 1.414   (c) 12.490   (e) 57.000   (g) .374   (i) .930

Problem Set 8.4 (page 334)

1. $13\sqrt{2}$ **3.** $54\sqrt{3}$ **5.** $-30\sqrt{2}$ **7.** $-\sqrt{5}$ **9.** $-21\sqrt{6}$ **11.** $-\dfrac{7\sqrt{7}}{12}$ **13.** $\dfrac{37\sqrt{10}}{10}$

15. $\dfrac{41\sqrt{2}}{20}$ **17.** $-9\sqrt[3]{3}$ **19.** $10\sqrt[3]{2}$ **21.** $4\sqrt{2x}$ **23.** $5x\sqrt{3}$ **25.** $2x\sqrt{5y}$

27. $8xy^3\sqrt{xy}$ **29.** $3a^2b\sqrt{6b}$ **31.** $3x^3y^4\sqrt{7}$ **33.** $4a\sqrt{10a}$ **35.** $\dfrac{8y}{3}\sqrt{6xy}$ **37.** $\dfrac{\sqrt{10xy}}{5y}$

39. $\dfrac{\sqrt{15}}{6x^2}$ **41.** $\dfrac{5\sqrt{2y}}{6y}$ **43.** $\dfrac{\sqrt{14xy}}{4y^3}$ **45.** $\dfrac{3y\sqrt{2xy}}{4x}$ **47.** $\dfrac{2\sqrt{42ab}}{7b^2}$ **49.** $2\sqrt[3]{3y}$

51. $2x\sqrt[3]{2x}$ **53.** $2x^2y^2\sqrt[3]{7y^2}$ **55.** $\dfrac{\sqrt[3]{21x}}{3x}$ **57.** $\dfrac{\sqrt[3]{12x^2y}}{4x^2}$ **59.** $\dfrac{\sqrt[3]{4x^2y^2}}{xy^2}$ **61.** $2\sqrt{2x+3y}$

63. $4\sqrt{x+3y}$ **65.** $33\sqrt{x}$ **67.** $-30\sqrt{2x}$ **69.** $7\sqrt{3n}$ **71.** $-40\sqrt{ab}$ **73.** $-7x\sqrt{2x}$

75. 16.25   **77.** -2.78 **79.** 39.71

Problem Set 8.5 (page 340)

1. $6\sqrt{2}$ **3.** $18\sqrt{2}$ **5.** $-24\sqrt{10}$ **7.** $24\sqrt{6}$ **9.** 120   **11.** 24   **13.** $56\sqrt[3]{3}$

15. $\sqrt{6}+\sqrt{10}$ **17.** $6\sqrt{10}-3\sqrt{35}$ **19.** $24\sqrt{3}-60\sqrt{2}$ **21.** $-40-32\sqrt{15}$

23. $15\sqrt{2x} + 3\sqrt{xy}$ **25.** $5xy - 6x\sqrt{y}$ **27.** $2\sqrt{10xy} + 2y\sqrt{15y}$ **29.** $-25\sqrt{6}$
31. $-25 - 3\sqrt{3}$ **33.** $23 - 9\sqrt{5}$ **35.** $6\sqrt{35} + 3\sqrt{10} - 4\sqrt{21} - 2\sqrt{6}$
37. $8\sqrt{3} - 36\sqrt{2} + 6\sqrt{10} - 18\sqrt{15}$ **39.** $11 + 13\sqrt{30}$ **41.** $141 - 51\sqrt{6}$ **43.** -10

45. -8 **47.** $2x - 3y$ **49.** $10\sqrt[3]{12} + 2\sqrt[3]{18}$ **51.** $12 - 36\sqrt[3]{2}$ **53.** $\dfrac{\sqrt{7} - 1}{3}$

55. $\dfrac{-3\sqrt{2} - 15}{23}$ **57.** $\dfrac{\sqrt{7} - \sqrt{2}}{5}$ **59.** $\dfrac{2\sqrt{5} + \sqrt{6}}{7}$ **61.** $\dfrac{\sqrt{15} - 2\sqrt{3}}{2}$ **63.** $\dfrac{6\sqrt{7} + 4\sqrt{6}}{13}$

65. $\sqrt{3} - \sqrt{2}$ **67.** $\dfrac{2\sqrt{x} - 8}{x - 16}$ **69.** $\dfrac{x + 5\sqrt{x}}{x - 25}$ **71.** $\dfrac{x - 8\sqrt{x} + 12}{x - 36}$ **73.** $\dfrac{x - 2\sqrt{xy}}{x - 4y}$

75. $\dfrac{6\sqrt{xy} + 9y}{4x - 9y}$ **77.** 0.72 **79.** 0.17 **81.** -0.38

Problem Set 8.6 (page 345)

1. $\{20\}$ **3.** $\emptyset$ **5.** $\left\{\dfrac{25}{4}\right\}$ **7.** $\left\{\dfrac{4}{9}\right\}$ **9.** $\{5\}$ **11.** $\left\{\dfrac{39}{4}\right\}$ **13.** $\emptyset$ **15.** $\{1\}$

17. $\left\{\dfrac{3}{2}\right\}$ **19.** $\{3\}$ **21.** $\left\{\dfrac{61}{25}\right\}$ **23.** $\{-3, 3\}$ **25.** $\{-9, -4\}$ **27.** $\{0\}$ **29.** $\{3\}$

31. $\{4\}$ **33.** $\{-4, -3\}$ **35.** $\{12\}$ **37.** $\{25\}$ **39.** $\{29\}$ **41.** $\{-15\}$ **43.** $\left\{-\dfrac{1}{3}\right\}$

45. $\{-3\}$ **47.** $\{0\}$ **49.** $\{5\}$ **51.** $\{2, 6\}$

Problem Set 8.7 (page 350)

1. 9 **3.** 3 **5.** -2 **7.** -5 **9.** $\dfrac{1}{6}$ **11.** 3 **13.** 8 **15.** 81 **17.** -1

19. -32 **21.** $\dfrac{81}{16}$ **23.** 4 **25.** $\dfrac{1}{128}$ **27.** -125 **29.** 625 **31.** $\sqrt[3]{x^4}$ **33.** $3\sqrt{x}$

35. $\sqrt[3]{2y}$ **37.** $\sqrt{2x - 3y}$ **39.** $\sqrt[3]{(2a - 3b)^2}$ **41.** $\sqrt[3]{x^2 y}$ **43.** $-3\sqrt[5]{xy^2}$ **45.** $5^{\frac{1}{2}}y^{\frac{1}{2}}$

47. $3y^{\frac{1}{2}}$ **49.** $x^{\frac{1}{3}}y^{\frac{2}{3}}$ **51.** $a^{\frac{1}{2}}b^{\frac{3}{4}}$ **53.** $(2x - y)^{\frac{3}{5}}$ **55.** $5xy^{\frac{1}{2}}$ **57.** $-(x + y)^{\frac{1}{3}}$

59. $12x^{\frac{13}{20}}$ **61.** $y^{\frac{5}{12}}$ **63.** $\dfrac{4}{x^{\frac{1}{10}}}$ **65.** $16xy^2$ **67.** $2x^2 y$ **69.** $4x^{\frac{4}{15}}$ **71.** $\dfrac{4}{b^{\frac{5}{12}}}$

73. $\dfrac{36x^{\frac{4}{5}}}{49y^{\frac{4}{3}}}$ **75.** $\dfrac{y^{\frac{3}{2}}}{x}$ **77.** $4x^{\frac{1}{6}}$ **79.** $\dfrac{16}{a^{\frac{11}{10}}}$ **81.** $\sqrt[6]{243}$ **83.** $\sqrt[4]{216}$ **85.** $\sqrt[12]{3}$ **87.** $\sqrt{2}$

89. $\sqrt[4]{3}$ **91. (a)** 12 **(c)** 7 **(e)** 11 **93. (a)** 1024 **(c)** 512 **(e)** 49

Problem Set 8.8 (page 356)

1. $(8.9)(10)^1$ **3.** $(4.29)(10)^3$ **5.** $(6.12)(10)^6$ **7.** $(4)(10)^7$ **9.** $(3.764)(10)^2$
11. $(3.47)(10)^{-1}$ **13.** $(2.14)(10)^{-2}$ **15.** $(5)(10)^{-5}$ **17.** $(1.94)(10)^{-9}$ **19.** 23 **21.** 4190
23. $500,000,000$ **25.** $31,400,000,000$ **27.** $.43$ **29.** $.000914$ **31.** $.00000005123$
33. $.000000074$ **35.** $.77$ **37.** $300,000,000,000$ **39.** $.000000004$ **41.** 1000 **43.** 1000
45. 3000 **47.** 20 **49.** $27,000,000$ **51. (a)** 7000 **(c)** 120 **(e)** 30
53. (a) $(4.385)(10)^{14}$ **(c)** $(2.322)(10)^{17}$ **(e)** $(3.052)(10)^{12}$

Chapter 8 Review Problem Set (page 359)

1. $\dfrac{1}{64}$ **2.** $\dfrac{9}{4}$ **3.** 3   **4.** -2 **5.** $\dfrac{2}{3}$ **6.** 32   **7.** 1   **8.** $\dfrac{4}{9}$ **9.** -64 **10.** 32

11. 1   **12.** 27   **13.** $3\sqrt{6}$ **14.** $4x\sqrt{3xy}$ **15.** $2\sqrt{2}$ **16.** $\dfrac{\sqrt{15x}}{6x^2}$ **17.** $2\sqrt[3]{7}$

18. $\dfrac{\sqrt[3]{6}}{3}$ **19.** $\dfrac{3\sqrt{5}}{5}$ **20.** $\dfrac{x\sqrt{21x}}{7}$ **21.** $3xy^2\sqrt[3]{4xy^2}$ **22.** $\dfrac{15\sqrt{6}}{4}$ **23.** $2y\sqrt{5xy}$

24. $2\sqrt{x}$ **25.** $24\sqrt{10}$ **26.** 60   **27.** $24\sqrt{3} - 6\sqrt{14}$ **28.** $x - 2\sqrt{x} - 15$ **29.** 17

30. $12 - 8\sqrt{3}$ **31.** $6a - 5\sqrt{ab} - 4b$ **32.** 70   **33.** $\dfrac{2(\sqrt{7} + 1)}{3}$ **34.** $\dfrac{2\sqrt{6} - \sqrt{15}}{3}$

35. $\dfrac{3\sqrt{5} - 2\sqrt{3}}{11}$ **36.** $\dfrac{6\sqrt{3} + 3\sqrt{5}}{7}$ **37.** $\dfrac{x^6}{y^8}$ **38.** $\dfrac{27a^3b^{12}}{8}$ **39.** $20x^{\frac{7}{10}}$ **40.** $7a^{\frac{5}{12}}$

41. $\dfrac{y^{\frac{4}{3}}}{x}$ **42.** $\dfrac{x^{12}}{9}$ **43.** $\sqrt{5}$ **44.** $5\sqrt[3]{3}$ **45.** $\dfrac{29\sqrt{6}}{5}$ **46.** $-15\sqrt{3x}$ **47.** $\dfrac{y + x^2}{x^2 y}$

48. $\dfrac{b - 2a}{a^2 b}$ **49.** $\left\{\dfrac{19}{7}\right\}$ **50.** $\{4\}$ **51.** $\{8\}$ **52.** $\varnothing$ **53.** $\{14\}$ **54.** $\{-10, 1\}$

55. $\{2\}$ **56.** $\{8\}$ **57.** .000000006   **58.** 36,000,000,000   **59.** 6   **60.** .15

61. .000028   **62.** .002   **63.** .002   **64.** 8,000,000,000

CHAPTER 9

Problem Set 9.1 (page 366)

1. $\dfrac{3}{4}$ **3.** $\dfrac{5}{6}$ **5.** $-\dfrac{2}{5}$ **7.** $\dfrac{2}{7}$ **9.** $\dfrac{2x}{7}$ **11.** $\dfrac{2a}{5b}$ **13.** $-\dfrac{y}{4x}$ **15.** $-\dfrac{9c}{13d}$ **17.** $\dfrac{5x^2}{3y^3}$

19. $\dfrac{y}{x - 2}$ **21.** $\dfrac{x + 5}{2y}$ **23.** $\dfrac{xy}{2x - y}$ **25.** $2x + 3$ **27.** $\dfrac{x - 2}{x}$ **29.** $\dfrac{3x + 2}{2x - 1}$ **31.** $\dfrac{a + 5}{a - 9}$

33. $\dfrac{n - 3}{5n - 1}$ **35.** $\dfrac{5x^2 + 7}{10x}$ **37.** $\dfrac{3x + 5}{4x + 1}$ **39.** $\dfrac{3x}{x^2 + 4x + 16}$ **41.** $\dfrac{x + 6}{3x - 1}$ **43.** $\dfrac{x(2x + 7)}{y(x + 9)}$

45. $\dfrac{y + 4}{5y - 2}$ **47.** $\dfrac{3x(x - 1)}{x^2 + 1}$ **49.** $\dfrac{2(x + 3y)}{3x(3x + y)}$ **51.** $\dfrac{3n - 4}{7n + 2}$ **53.** $\dfrac{4 - x}{5 + 3x}$

55. $\dfrac{9x^2 + 3x + 1}{2(x + 2)}$ **57.** $\dfrac{-2(x - 1)}{x + 1}$ **59.** $\dfrac{y + b}{y + c}$ **61.** $\dfrac{x + 2y}{2x + y}$ **63.** $\dfrac{x + 1}{x} \quad \dfrac{}{6}$ **65.** $\dfrac{2s + 5}{3s + 1}$

67. -1 **69.** $-n - 7$ **71.** $-\dfrac{2}{x + 1}$ **73.** -2 **75.** $-\dfrac{n + 3}{n + 5}$

Problem Set 9.2 (page 372)

1. $\dfrac{1}{10}$ **3.** $-\dfrac{4}{15}$ **5.** $\dfrac{3}{16}$ **7.** $-\dfrac{5}{6}$ **9.** $-\dfrac{2}{3}$ **11.** $\dfrac{10}{11}$ **13.** $-\dfrac{5x^3}{12y^2}$ **15.** $\dfrac{2a^3}{3b}$

17. $\dfrac{3x^3}{4}$ **19.** $\dfrac{25x^3}{108y^2}$ **21.** $\dfrac{ac^2}{2b^2}$ **23.** $\dfrac{3x}{4y}$ **25.** $\dfrac{3(x^2 + 4)}{5y(x + 8)}$ **27.** $\dfrac{5(a + 3)}{a(a - 2)}$ **29.** $\dfrac{3}{2}$

31. $\dfrac{3xy}{4(x+6)}$ **33.** $\dfrac{5(x-2y)}{7y}$ **35.** $\dfrac{5+n}{3-n}$ **37.** $\dfrac{x^2+1}{x^2-10}$ **39.** $\dfrac{6x+5}{3x+4}$ **41.** $\dfrac{2t^2+5}{2(t^2+1)(t+1)}$

43. $\dfrac{t(t+6)}{4t+5}$ **45.** $\dfrac{n+3}{n(n-2)}$ **47.** $\dfrac{25x^3y^3}{4(x+1)}$ **49.** $\dfrac{2(a-2b)}{a(3a-2b)}$

Problem Set 9.3 (page 378)

1. $\dfrac{13}{12}$ **3.** $\dfrac{11}{40}$ **5.** $\dfrac{19}{20}$ **7.** $\dfrac{49}{75}$ **9.** $\dfrac{17}{30}$ **11.** $-\dfrac{11}{84}$ **13.** $\dfrac{2x+4}{x-1}$ **15.** 4

17. $\dfrac{7y-10}{7y}$ **19.** $\dfrac{5x+3}{6}$ **21.** $\dfrac{12a+1}{12}$ **23.** $\dfrac{n+14}{18}$ **25.** $-\dfrac{11}{15}$ **27.** $\dfrac{3x-25}{30}$

29. $\dfrac{43}{40x}$ **31.** $\dfrac{20y-77x}{28xy}$ **33.** $\dfrac{16y+15x-12xy}{12xy}$ **35.** $\dfrac{21+22x}{30x^2}$ **37.** $\dfrac{10n-21}{7n^2}$

39. $\dfrac{45-6n+20n^2}{15n^2}$ **41.** $\dfrac{11x-10}{6x^2}$ **43.** $\dfrac{42t+43}{35t^3}$ **45.** $\dfrac{20b^2-33a^3}{96a^2b}$

47. $\dfrac{14-24y^3+45xy}{18xy^3}$ **49.** $\dfrac{2x^2+3x-3}{x(x-1)}$ **51.** $\dfrac{a^2-a-8}{a(a+4)}$ **53.** $\dfrac{-41n-55}{(4n+5)(3n+5)}$

55. $\dfrac{-3x+17}{(x+4)(7x-1)}$ **57.** $\dfrac{-x+74}{(3x-5)(2x+7)}$ **59.** $\dfrac{38x+13}{(3x-2)(4x+5)}$ **61.** $\dfrac{5x+5}{2x+5}$ **63.** $\dfrac{x+15}{x-5}$

65. $\dfrac{-2x-4}{2x+1}$ **67. (a)** -1 **(c)** 0

Problem Set 9.4 (page 387)

1. $\dfrac{7x+20}{x(x+4)}$ **3.** $\dfrac{-x-3}{x(x+7)}$ **5.** $\dfrac{6x-5}{(x+1)(x-1)}$ **7.** $\dfrac{1}{a+1}$ **9.** $\dfrac{5n+15}{4(n+5)(n-5)}$

11. $\dfrac{x^2+60}{x(x+6)}$ **13.** $\dfrac{11x+13}{(x+2)(x+7)(2x+1)}$ **15.** $\dfrac{-3a+1}{(a-5)(a+2)(a+9)}$

17. $\dfrac{9a^2+17a+1}{(5a+1)(4a-3)(3a+4)}$ **19.** $\dfrac{3x^2+20x-111}{(x^2+3)(x+7)(x-3)}$ **21.** $\dfrac{-7y-14}{(y+8)(y-2)}$

23. $\dfrac{-2x^2-4x+3}{(x+2)(x-2)}$ **25.** $\dfrac{2x^2+14x-19}{(x+10)(x-2)}$ **27.** $\dfrac{2n+1}{n-6}$ **29.** $\dfrac{2x^2-32x+16}{(x+1)(2x-1)(3x-2)}$

31. $\dfrac{1}{(n^2+1)(n+1)}$ **33.** $\dfrac{-16x}{(5x-2)(x-1)}$ **35.** $\dfrac{t+1}{t-2}$ **37.** $\dfrac{2}{11}$ **39.** $-\dfrac{7}{27}$ **41.** $\dfrac{x}{4}$

43. $\dfrac{3y-2x}{4x-7}$ **45.** $\dfrac{6ab^2-5a^2}{12b^2+2a^2b}$ **47.** $\dfrac{2y-3xy}{3x+4xy}$ **49.** $\dfrac{3n+14}{5n+19}$ **51.** $\dfrac{5n-17}{4n-13}$

53. $\dfrac{-x+5y-10}{3y-10}$ **55.** $\dfrac{-x+15}{-2x-1}$ **57.** $\dfrac{3a^2-2a+1}{2a-1}$ **59.** $\dfrac{-x^2+6x-4}{3x-2}$

Problem Set 9.5 (page 395)

1. $\{2\}$ **3.** $\{-3\}$ **5.** $\{6\}$ **7.** $\left\{-\dfrac{85}{18}\right\}$ **9.** $\left\{\dfrac{7}{10}\right\}$ **11.** $\{5\}$ **13.** $\{58\}$

15. $\left\{\frac{1}{4}, 4\right\}$ **17.** $\left\{-\frac{2}{5}, 5\right\}$ **19.** $\{-16\}$ **21.** $\left\{-\frac{13}{3}\right\}$ **23.** $\{-3, 1\}$ **25.** $\left\{-\frac{5}{2}\right\}$

27. $\{-51\}$ **29.** $\left\{-\frac{5}{3}, 4\right\}$ **31.** $\varnothing$ **33.** $\left\{-\frac{11}{8}, 2\right\}$ **35.** $\{-29, 0\}$ **37.** $\{-9, 3\}$

39. $\left\{-2, \frac{23}{8}\right\}$ **41.** $\left\{\frac{11}{23}\right\}$ **43.** $\left\{3, \frac{7}{2}\right\}$ **45.** \$750 and \$1000 **47.** 48° and 72°

49. $\frac{2}{7}$ or $\frac{7}{2}$ **51.** \$1080 **53.** \$69 for Tammy and \$51.75 for Laura **55.** 8 and 82

57. 14 feet and 6 feet **59.** 690 females and 460 males

Problem Set 9.6 (page 404)

1. $\{-21\}$ **3.** $\{-1, 2\}$ **5.** $\{2\}$ **7.** $\left\{\frac{37}{15}\right\}$ **9.** $\{-1\}$ **11.** $\{-1\}$ **13.** $\left\{0, \frac{13}{2}\right\}$

15. $\left\{-2, \frac{19}{2}\right\}$ **17.** $\{-2\}$ **19.** $\left\{-\frac{1}{5}\right\}$ **21.** $\varnothing$ **23.** $\left\{\frac{7}{2}\right\}$ **25.** $\{-3\}$ **27.** $\left\{-\frac{7}{9}\right\}$

29. $\left\{-\frac{7}{6}\right\}$ **31.** $x = \frac{18y - 4}{15}$ **33.** $y = \frac{-5x + 22}{2}$ **35.** $M = \frac{IC}{100}$ **37.** $R = \frac{ST}{S + T}$

39. $y = \frac{bx - x - 3b + a}{a - 3}$ **41.** $y = \frac{ab - bx}{a}$ **43.** $y = \frac{-2x - 9}{3}$

45. 50 miles per hour for Dave and 54 miles per hour for Kent **47.** 60 minutes
49. 60 words per minute for Connie and 40 words per minute for Katie
51. Plane B could travel at 400 miles per hour for 5 hours and plane A at 350 miles per hour for 4 hours or plane B could travel at 250 miles per hour for 8 hours and plane A at 200 miles per hour for 7 hours.
53. 60 minutes for Nancy and 120 minutes for Amy **55.** 3 hours
57. 16 miles per hour on the way out and 12 miles per hour on the way back or 12 miles per hour out and 8 miles per hour back

Chapter 9 Review Problem Set (page 407)

1. $\frac{2y}{3x^2}$ **2.** $\frac{a - 3}{a}$ **3.** $\frac{n - 5}{n - 1}$ **4.** $\frac{x^2 + 1}{x}$ **5.** $\frac{2x + 1}{3}$ **6.** $\frac{x^2 - 10}{2x^2 + 1}$ **7.** $\frac{3}{22}$

8. $\frac{18y + 20x}{48y - 9x}$ **9.** $\frac{3x + 2}{3x - 2}$ **10.** $\frac{x - 1}{2x - 1}$ **11.** $\frac{2x}{7y^2}$ **12.** $3b$ **13.** $\frac{n(n + 5)}{n - 1}$

14. $\frac{x(x - 3y)}{x^2 + 9y^2}$ **15.** $\frac{23x - 6}{20}$ **16.** $\frac{57 - 2n}{18n}$ **17.** $\frac{3x^2 - 2x - 14}{x(x + 7)}$ **18.** $\frac{2}{x - 5}$

19. $\frac{5n - 21}{(n - 9)(n + 4)(n - 1)}$ **20.** $\frac{6y - 23}{(2y + 3)(y - 6)}$ **21.** $\left\{\frac{4}{13}\right\}$ **22.** $\left\{\frac{3}{16}\right\}$ **23.** $\varnothing$

24. $\{-17\}$ **25.** $\left\{\frac{2}{7}, \frac{7}{2}\right\}$ **26.** $\{22\}$ **27.** $\left\{-\frac{6}{7}, 3\right\}$ **28.** $\left\{\frac{3}{4}, \frac{5}{2}\right\}$

29. $\left\{\frac{9}{7}\right\}$ **30.** $\left\{-\frac{5}{4}\right\}$ **31.** $y = \frac{3x + 27}{4}$ **32.** $y = \frac{bx - ab}{a}$

33. $525 and $875   **34.** 20 minutes for Don and 30 minutes for Dan
35. 50 miles per hour and 55 miles per hour   **36.** 9 hours   **37.** 80 hours   **38.** 13 miles per hour

CHAPTER 10

Problem Set 10.1 (page 416)

1. False   **3.** True   **5.** True   **7.** True   **9.** $10 + 8i$ **11.** $-6 + 10i$ **13.** $-2 - 5i$

15. $-12 + 5i$ **17.** $-1 - 23i$ **19.** $-4 - 5i$ **21.** $1 + 3i$ **23.** $\dfrac{5}{3} - \dfrac{5}{12}i$ **25.** $-\dfrac{17}{9} + \dfrac{23}{30}i$

27. $9i$ **29.** $i\sqrt{14}$ **31.** $\dfrac{4}{5}i$ **33.** $3i\sqrt{2}$ **35.** $5i\sqrt{3}$ **37.** $6i\sqrt{7}$ **39.** $-8i\sqrt{5}$

41. $36i\sqrt{10}$ **43.** -8 **45.** $-\sqrt{15}$ **47.** $-3\sqrt{6}$ **49.** $-5\sqrt{3}$ **51.** $-3\sqrt{6}$

53. $4i\sqrt{3}$ **55.** $\dfrac{5}{2}$ **57.** $2\sqrt{2}$ **59.** $2i$ **61.** $-20 + 0i$ **63.** $42 + 0i$ **65.** $15 + 6i$

67. $-42 + 12i$ **69.** $7 + 22i$ **71.** $40 - 20i$ **73.** $-3 - 28i$ **75.** $-3 - 15i$

77. $-9 + 40i$ **79.** $-12 + 16i$ **81.** $85 + 0i$ **83.** $5 + 0i$ **85.** $\dfrac{3}{5} + \dfrac{3}{10}i$ **87.** $\dfrac{5}{17} - \dfrac{3}{17}i$

89. $2 + \dfrac{2}{3}i$ **91.** $0 - \dfrac{2}{7}i$ **93.** $\dfrac{22}{25} - \dfrac{4}{25}i$ **95.** $-\dfrac{18}{41} + \dfrac{39}{41}i$ **97.** $\dfrac{9}{2} - \dfrac{5}{2}i$ **99.** $\dfrac{4}{13} - \dfrac{1}{26}i$

Problem Set 10.2 (page 424)

1. $\{0, 9\}$ **3.** $\{-3, 0\}$ **5.** $\{-4, 0\}$ **7.** $\left\{0, \dfrac{9}{5}\right\}$ **9.** $\{-6, 5\}$ **11.** $\{7, 12\}$

13. $\left\{-8, -\dfrac{3}{2}\right\}$ **15.** $\left\{-\dfrac{7}{3}, \dfrac{2}{5}\right\}$ **17.** $\left\{\dfrac{3}{5}\right\}$ **19.** $\left\{-\dfrac{3}{2}, \dfrac{7}{3}\right\}$ **21.** $\{1, 4\}$ **23.** $\{8\}$

25. $\{12\}$ **27.** $\{0, 5k\}$ **29.** $\{0, 16k^2\}$ **31.** $\{5k, 7k\}$ **33.** $\left\{\dfrac{k}{2}, -3k\right\}$ **35.** $\{\pm 1\}$

37. $\{\pm 6i\}$ **39.** $\{\pm\sqrt{14}\}$ **41.** $\{\pm 2\sqrt{7}\}$ **43.** $\{\pm 3\sqrt{2}\}$ **45.** $\left\{\pm\dfrac{\sqrt{14}}{2}\right\}$ **47.** $\left\{\pm\dfrac{2\sqrt{3}}{3}\right\}$

49. $\left\{\pm\dfrac{2i\sqrt{30}}{5}\right\}$ **51.** $\left\{\pm\dfrac{\sqrt{6}}{2}\right\}$ **53.** $\{-1, 5\}$ **55.** $\{-8, 2\}$ **57.** $\{-6 \pm 2i\}$ **59.** $\{1, 2\}$

61. $\{4 \pm \sqrt{5}\}$ **63.** $\{-5 \pm 2\sqrt{3}\}$ **65.** $\left\{\dfrac{2 \pm 3i\sqrt{3}}{3}\right\}$ **67.** $\{-12, -2\}$ **69.** $\left\{\dfrac{2 \pm \sqrt{10}}{5}\right\}$

71. $2\sqrt{13}$ centimeters or approximately 7.2 centimeters
73. $4\sqrt{5}$ inches or approximately 8.9 inches
75. 8 yards   **77.** $6\sqrt{2}$ inches
79. $a = b = 4\sqrt{2}$ meters   **81.** $b = 3\sqrt{3}$ inches and $c = 6$ inches
83. $a = 7$ centimeters and $b = 7\sqrt{3}$ centimeters   **85.** $a = \dfrac{10\sqrt{3}}{3}$ feet and $c = \dfrac{20\sqrt{3}}{3}$ feet

Problem Set 10.3 (page 429)

1. $\{-6, 10\}$ **3.** $\{4, 10\}$ **5.** $\{-5, 10\}$ **7.** $\{-8, 1\}$ **9.** $\left\{-\dfrac{5}{2}, 3\right\}$ **11.** $\left\{-3, \dfrac{2}{3}\right\}$

13. $\{-16, 10\}$ **15.** $\{-2 \pm \sqrt{6}\}$ **17.** $\{-3 \pm 2\sqrt{3}\}$ **19.** $\{5 \pm \sqrt{26}\}$ **21.** $\{4 \pm i\}$

23. $\{-6 \pm 3\sqrt{3}\}$ **25.** $\{-1 \pm i\sqrt{5}\}$ **27.** $\left\{\dfrac{-3 \pm \sqrt{17}}{2}\right\}$ **29.** $\left\{\dfrac{-5 \pm \sqrt{21}}{2}\right\}$

31. $\left\{\dfrac{7 \pm \sqrt{37}}{2}\right\}$ **33.** $\left\{\dfrac{-2 \pm \sqrt{10}}{2}\right\}$ **35.** $\left\{\dfrac{3 \pm i\sqrt{6}}{3}\right\}$ **37.** $\left\{\dfrac{-5 \pm \sqrt{37}}{6}\right\}$ **39.** $\{-12, 4\}$

41. $\left\{\dfrac{4 \pm \sqrt{10}}{2}\right\}$ **43.** $\left\{-\dfrac{9}{2}, \dfrac{1}{3}\right\}$ **45.** $\{-3, 8\}$ **47.** $\{3 \pm 2\sqrt{3}\}$ **49.** $\left\{\dfrac{3 \pm i\sqrt{3}}{3}\right\}$

51. $\{-20, 12\}$ **53.** $\left\{-6, -\dfrac{11}{3}\right\}$ **55.** $\left\{-\dfrac{7}{3}, -\dfrac{3}{2}\right\}$ **57.** $\{-6 \pm 2\sqrt{10}\}$ **59.** $\left\{\dfrac{1}{4}, \dfrac{1}{3}\right\}$

61. $\left\{\dfrac{-b \pm \sqrt{b^2 - 4ac}}{2a}\right\}$ **63.** $x = \dfrac{a\sqrt{b^2 - y^2}}{b}$ **65.** $r = \dfrac{\sqrt{A\pi}}{\pi}$ **67.** $\{2a, 3a\}$ **69.** $\left\{\dfrac{a}{2}, -\dfrac{2a}{3}\right\}$

71. $\left\{\dfrac{2b}{3}\right\}$

Problem Set 10.4 (page 437)

1. Two real solutions; $\{-7, 3\}$ **3.** One real solution; $\left\{\dfrac{1}{3}\right\}$ **5.** Two complex solutions; $\left\{\dfrac{7 \pm i\sqrt{3}}{2}\right\}$

7. Two real solutions; $\left\{-\dfrac{4}{3}, \dfrac{1}{5}\right\}$ **9.** Two real solutions; $\left\{\dfrac{-2 \pm \sqrt{10}}{3}\right\}$ **11.** $\{-1 \pm \sqrt{2}\}$

13. $\left\{\dfrac{-5 \pm \sqrt{37}}{2}\right\}$ **15.** $\{4 \pm 2\sqrt{5}\}$ **17.** $\left\{\dfrac{-5 \pm i\sqrt{7}}{2}\right\}$ **19.** $\{8, 10\}$ **21.** $\left\{\dfrac{9 \pm \sqrt{61}}{2}\right\}$

23. $\left\{\dfrac{-1 \pm \sqrt{33}}{4}\right\}$ **25.** $\left\{\dfrac{-1 \pm i\sqrt{3}}{4}\right\}$ **27.** $\left\{\dfrac{4 \pm \sqrt{10}}{3}\right\}$ **29.** $\left\{-1, \dfrac{5}{2}\right\}$ **31.** $\left\{-5, -\dfrac{4}{3}\right\}$

33. $\left\{\dfrac{5}{6}\right\}$ **35.** $\left\{\dfrac{1 \pm \sqrt{13}}{4}\right\}$ **37.** $\left\{0, \dfrac{13}{5}\right\}$ **39.** $\left\{\pm \dfrac{\sqrt{15}}{3}\right\}$ **41.** $\left\{\dfrac{-1 \pm \sqrt{73}}{12}\right\}$

43. $\{-18, -14\}$ **45.** $\left\{\dfrac{11}{4}, \dfrac{10}{3}\right\}$ **47.** $\left\{\dfrac{2 \pm i\sqrt{2}}{2}\right\}$ **49.** $\left\{\dfrac{1 \pm \sqrt{7}}{6}\right\}$

55. $\{1.381, 17.381\}$ **57.** $\{-13.426, 3.426\}$ **59.** $\{-.347, -8.653\}$

61. $\{.119, 1.681\}$ **63.** $\{-.708, 4.708\}$ **65.** $k = 4$ or $k = -4$

Problem Set 10.5 (page 446)

1. $\{2 \pm \sqrt{10}\}$ **3.** $\left\{-9, \dfrac{4}{3}\right\}$ **5.** $\{9 \pm 3\sqrt{10}\}$ **7.** $\left\{\dfrac{3 \pm i\sqrt{23}}{4}\right\}$ **9.** $\{-15, -9\}$

11. $\{-8, 1\}$ **13.** $\left\{\dfrac{2 \pm i\sqrt{10}}{2}\right\}$ **15.** $\{9 \pm \sqrt{66}\}$ **17.** $\left\{-\dfrac{5}{4}, \dfrac{2}{5}\right\}$ **19.** $\left\{\dfrac{-1 \pm \sqrt{2}}{2}\right\}$

21. $\left\{\frac{3}{4}, 4\right\}$ **23.** $\left\{\frac{11 \pm \sqrt{109}}{2}\right\}$ **25.** $\left\{\frac{3}{7}, 4\right\}$ **27.** $\left\{\frac{7 \pm \sqrt{129}}{10}\right\}$ **29.** $\left\{-\frac{10}{7}, 3\right\}$

31. $\{1 \pm \sqrt{34}\}$ **33.** $\{\pm\sqrt{6}, \pm 2\sqrt{3}\}$ **35.** $\left\{\pm 3, \pm\frac{2\sqrt{6}}{3}\right\}$ **37.** $\left\{\pm\frac{i\sqrt{15}}{3}, \pm 2i\right\}$

39. $\left\{\pm\frac{\sqrt{14}}{2}, \pm\frac{2\sqrt{3}}{3}\right\}$ **41.** 8 and 9   **43.** 9 and 12   **45.** $5 + \sqrt{3}$ and $5 - \sqrt{3}$ **47.** 3 and 6

49. 9 inches and 12 inches   **51.** 1 meter   **53.** 8 inches by 14 inches

55. 20 miles per hour for Lorraine and 25 miles per hour for Charlotte or 45 miles per hour for Lorraine and 50 miles per hour for Charlotte

57. 55 miles per hour   **59.** 6 hours for Tom and 8 hours for Terry   **61.** 30 hours   **63.** 8 people

65. 40 shares at \$20 per share   **67.** 50 numbers   **69.** 9%   **71.** $\{9, 36\}$ **73.** $\{1\}$

75. $\left\{-\frac{8}{27}, \frac{27}{8}\right\}$ **77.** $\left\{-4, \frac{3}{5}\right\}$

Problem Set 10.6 (page 453)

1. $\{x \mid x < -2 \text{ or } x > 1\}$

3. $\{x \mid x > -4 \text{ and } x < -1\}$

5. $\left\{x \mid x \le -\frac{7}{3} \text{ or } x \ge \frac{1}{2}\right\}$

7. $\left\{x \mid x \ge -2 \text{ and } x \le \frac{3}{4}\right\}$

9. $\{x \mid x > -1 \text{ and } x < 1 \text{ or } x > 3\}$

11. $\{x \mid x \le -2 \text{ or } x \ge 0 \text{ and } x \le 4\}$

13. $\{x \mid x < -1 \text{ or } x > 2\}$

15. $\{x \mid x > -2 \text{ and } x < 3\}$

17. $\left\{x \mid x < 0 \text{ or } x \ge \frac{1}{2}\right\}$

19. $\{x \mid x < 1 \text{ or } x \ge 2\}$

21. $\{x \mid x > -7 \text{ and } x < 5\}$ **23.** $\{x \mid x < 4 \text{ or } x > 7\}$ **25.** $\left\{x \mid x \ge -5 \text{ and } x \le \frac{2}{3}\right\}$

27. $\left\{x \mid x \le -\frac{5}{2} \text{ or } x \ge -\frac{1}{4}\right\}$ **29.** $\left\{x \mid x < -\frac{4}{5} \text{ or } x > 8\right\}$ **31.** $\{x \mid x \text{ is any real number}\}$

33. $\left\{-\frac{5}{2}\right\}$ **35.** $\{x \mid x > -1 \text{ and } x < 3 \text{ or } x > 3\}$ **37.** $\{x \mid x > -6 \text{ and } x < -3\}$

39. $\{x \mid x < 5 \text{ or } x \geq 9\}$ **41.** $\left\{x \mid x < \dfrac{4}{3} \text{ or } x > 3\right\}$ **43.** $\{x \mid x > -4 \text{ and } x \leq 6\}$ **45.** $\{x \mid x < 2\}$

49. (a) $\{x \mid x < -7 \text{ or } x > 2\}$ (a) $\{x \mid x \geq -1 \text{ and } x \leq 6\}$ (e) $\{x \mid x < -4 \text{ or } x > 7\}$

Chapter 10 Review Problem Set (page 456)

1. $2 - 2i$ **2.** $-3 - i$ **3.** $30 + 15i$ **4.** $86 - 2i$ **5.** $-32 + 4i$ **6.** 25 **7.** $\dfrac{9}{20} + \dfrac{13}{20}i$

8. $-\dfrac{3}{29} + \dfrac{7}{29}i$ **9.** Two equal real solutions **10.** Two nonreal complex solutions

11. Two unequal real solutions **12.** Two unequal real solutions **13.** $\{0, 17\}$ **14.** $\{-4, 8\}$

15. $\left\{\dfrac{1 \pm 8i}{2}\right\}$ **16.** $\{-3, 7\}$ **17.** $\{-1 \pm \sqrt{10}\}$ **18.** $\{3 \pm 5i\}$ **19.** $\{25\}$ **20.** $\left\{-4, \dfrac{2}{3}\right\}$

21. $\{-10, 20\}$ **22.** $\left\{\dfrac{-1 \pm \sqrt{61}}{6}\right\}$ **23.** $\left\{\dfrac{1 \pm i\sqrt{11}}{2}\right\}$ **24.** $\left\{\dfrac{5 \pm i\sqrt{23}}{4}\right\}$ **25.** $\left\{\dfrac{-2 \pm \sqrt{14}}{2}\right\}$

26. $\{-9, 4\}$ **27.** $\{-2 \pm i\sqrt{5}\}$ **28.** $\{-6, 12\}$ **29.** $\{1 \pm \sqrt{10}\}$ **30.** $\left\{\pm\dfrac{\sqrt{14}}{2}, \pm 2\sqrt{2}\right\}$

31. $\left\{\dfrac{-3 \pm \sqrt{97}}{2}\right\}$ **32.** $\{x \mid x < -5 \text{ or } x > 2\}$ **33.** $\left\{x \mid x \geq -\dfrac{7}{2} \text{ and } x \leq 3\right\}$

34. $\{x \mid x < -6 \text{ or } x \geq 4\}$ **35.** $\left\{x \mid x > -\dfrac{5}{2} \text{ and } x < -1\right\}$ **36.** $3 + \sqrt{7}$ and $3 - \sqrt{7}$

37. 20 shares at $15 per share **38.** 45 miles per hour and 52 miles per hour **39.** 8 units
40. 8 and 10 **41.** 7 inches by 12 inches **42.** 4 hours for Janet and 6 hours for Billy
43. 10 meters

CHAPTER 11

Problem Set 11.1 (page 468)

1. (a) III (c) II **3.** (a) I and III (c) II and III **5.** $(-3, -1)$; $(3, 1)$; $(3, -1)$
7. $(7, 2)$; $(-7, -2)$; $(-7, 2)$ **9.** $(5, 0)$; $(-5, 0)$ **11.** x-axis **13.** y-axis
15. x-axis, y-axis, and origin **17.** x-axis **19.** none **21.** origin **23.** y-axis

25.

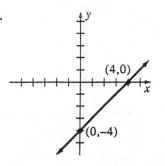

27.

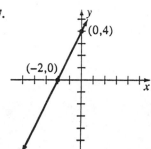

29.

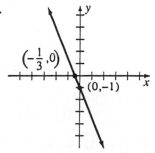

31.

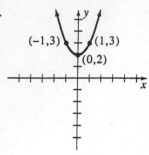

33.

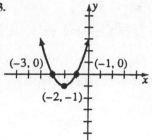

35.

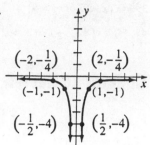

37.

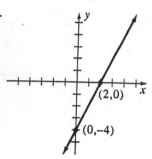

39.

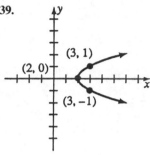

41.

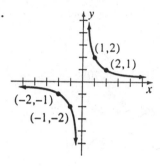

43.

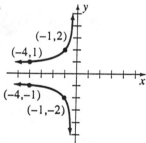

45.

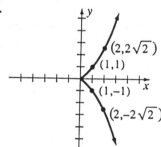

47.

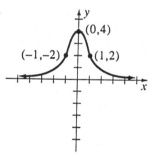

49.

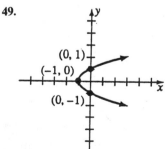

51.

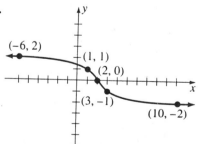

53. $41.76; $43.20; $44.40; $46.50; $46.92

Problem Set 11.2 (page 475)

1. 15   **3.** $\sqrt{13}$ **5.** $3\sqrt{2}$ **7.** $3\sqrt{5}$ **9.** 6   **11.** $3\sqrt{10}$

13. The lengths of the sides are 10, $5\sqrt{5}$, and 5. Since $10^2 + 5^2 = (5\sqrt{5})^2$, it is a right triangle.

15. The distances between (3, 6) and (7, 12), between (7, 12) and (11, 18), and between (11, 18) and (15, 24) are all $2\sqrt{13}$ units.

17. (1, 3), $r = 4$ **19.** $(-3, -5)$, $r = 4$ **21.** (0, 0), $r = \sqrt{10}$ **23.** $(8, -3)$, $r = \sqrt{2}$

25. $(-3, 4)$, $r = 5$ **27.** $\left(-\dfrac{1}{2}, 4\right)$, $r = 2\sqrt{2}$ **29.** $x^2 + y^2 - 6x - 10y + 9 = 0$

31. $x^2 + y^2 + 8x - 2y - 47 = 0$ **33.** $x^2 + y^2 + 4x + 12y + 22 = 0$ **35.** $x^2 + y^2 = 20$

37. $x^2 + y^2 - 10x + 16y - 7 = 0$ **39.** $x^2 + y^2 - 8y = 0$ **41.** $x^2 + y^2 + 8x - 6y = 0$

43. (a) $x + 3y = 10$ **(c)** Yes, because their slopes are negative reciprocals of each other.

45. (a) $(-2, 3)$, $r = \sqrt{17}$ **47. (a)** (1, 4), $r = 3$ **(c)** $(-6, -4)$, $r = 8$ **(e)** (0, 6), $r = 9$

Problem Set 11.3 (page 485)

1.

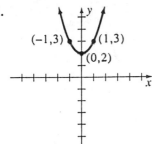

3.

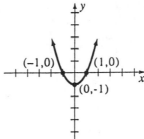

5.

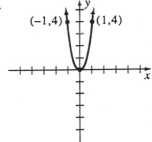

7.

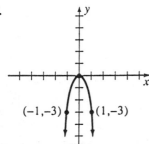

9.

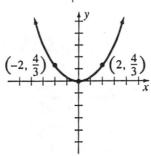

11.

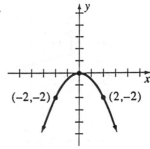

13.

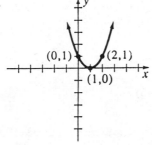

15.

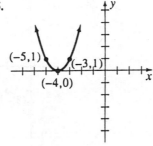

17.

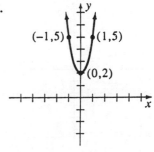

19.

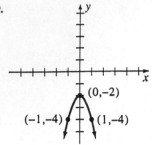

21.

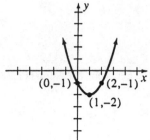

23.

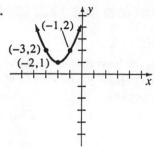

25.

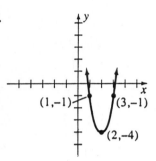

27.

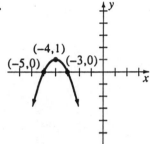

29.

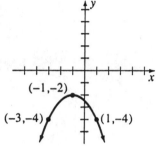

35. (a)

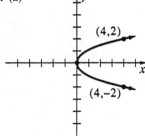

(c)

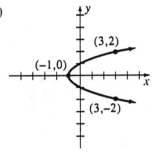

(e)

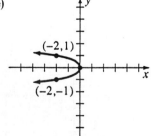

(g)

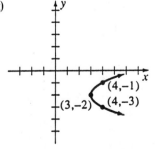

Problem Set 11.4 (page 491)

1.
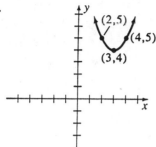

3.
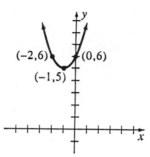

5.
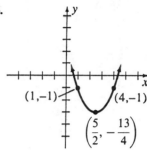

7.
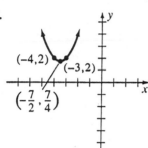

9.
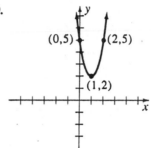

11.
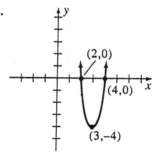

13.
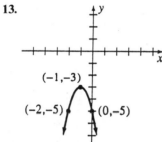

15.
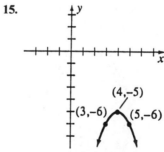

17.
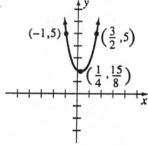

19.
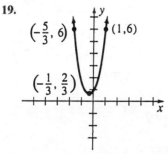

21.
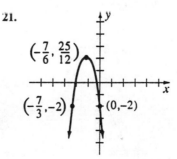

23.
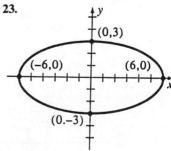

25.

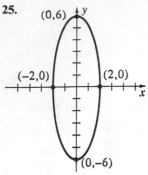

$(0,6)$ $(-2,0)$ $(2,0)$ $(0,-6)$

27.

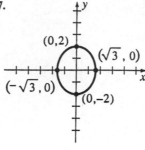

$(0,2)$ $(\sqrt{3}, 0)$ $(-\sqrt{3}, 0)$ $(0,-2)$

29.

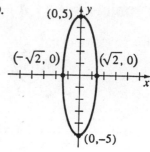

$(0,5)$ $(-\sqrt{2}, 0)$ $(\sqrt{2}, 0)$ $(0,-5)$

Problem Set 11.5 (page 497)

1.

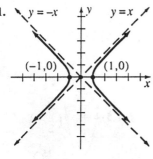

$y = -x$ $y = x$ $(-1,0)$ $(1,0)$

3.

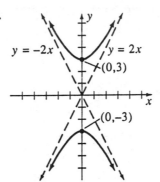

$y = -2x$ $y = 2x$ $(0,3)$ $(0,-3)$

5.

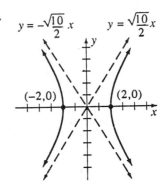

$y = -\dfrac{\sqrt{10}}{2} x$ $y = \dfrac{\sqrt{10}}{2} x$ $(-2,0)$ $(2,0)$

7.

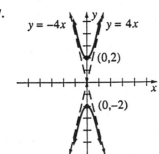

$y = -4x$ $y = 4x$ $(0,2)$ $(0,-2)$

9.

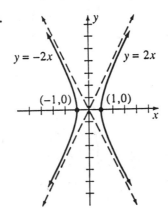

$y = -2x$ $y = 2x$ $(-1,0)$ $(1,0)$

11.

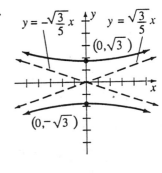

$y = -\sqrt{\dfrac{3}{5}} x$ $y = \sqrt{\dfrac{3}{5}} x$ $(0,\sqrt{3})$ $(0,-\sqrt{3})$

13.

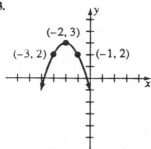

15.

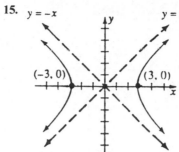

17.

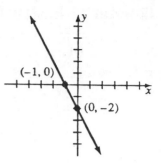

19.

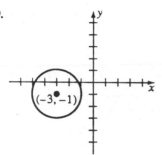

21.

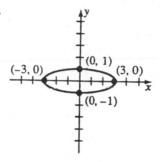

23. (a)

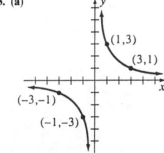

(c)

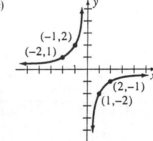

25. (a) Origin

(c)

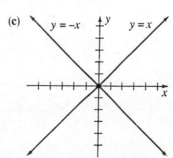

Chapter 11 Review Problem Set (page 499)

1. (a) $\dfrac{6}{5}$ **(b)** $-\dfrac{2}{3}$ **2. (a)** -4 **(b)** $\dfrac{2}{7}$ **3.** 5, 10, and $\sqrt{97}$ **4.** $7x + 4y = 1$

5. $3x + 7y = 28$ **6.** $2x - 3y = 16$ **7.** $x - 2y = -8$ **8.** $2x - 3y = 14$

9.

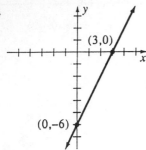

10.

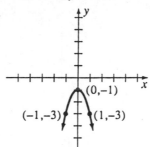

11.

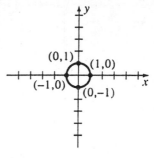

12.

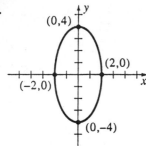

13.

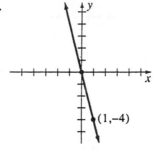

14.

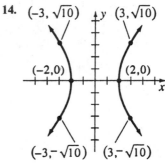

15.

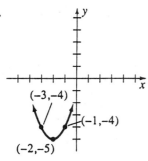

16.

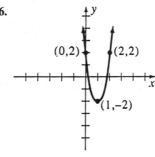

17.

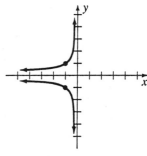

18.

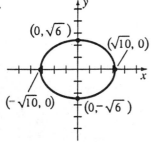

19.

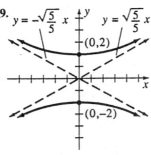

20.

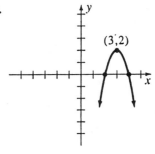

21. 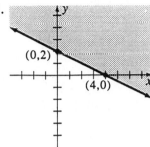   **22.** 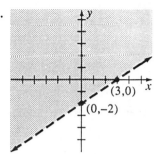   **23.** $(-3, 4)r = 3$

24. $(-4, -6)$ **25.** $y = \dfrac{3}{2}x$ and $y = -\dfrac{3}{2}x$ **26.** 6

CHAPTER 12

Problem Set 12.1 (page 507)

1. $D = \{1, 2, 3, 4\}$, $R = \{5, 8, 11, 14\}$ It is a function.

3. $D = \{0, 1\}$, $R = \{-2\sqrt{6}, -5, 5, 2\sqrt{6}\}$ It is not a function.

5. $D = \{1, 2, 3, 4, 5\}$, $R = \{2, 5, 10, 17, 26\}$ It is a function.

7. $D = \{\text{all reals}\}$, $R = \{\text{all reals}\}$ It is a function.   **9.** $D = \{\text{all reals}\}$, $R = \{y \mid y \geq 0\}$ It is a function.

11. $\{\text{all reals}\}$ **13.** $\{x \mid x \neq 1\}$ **15.** $\left\{x \mid x \neq \dfrac{3}{4}\right\}$ **17.** $\{x \mid x \neq -1 \text{ and } x \neq 4\}$

19. $\{x \mid x \neq -8 \text{ and } x \neq 5\}$ **21.** $\{x \mid x \neq -6 \text{ and } x \neq 0\}$ **23.** $\{\text{all reals}\}$

25. $\{t \mid t \neq -2 \text{ and } t \neq 2\}$ **27.** $\{x \mid x \geq -4\}$ **29.** $\left\{s \mid s \geq \dfrac{5}{4}\right\}$ **31.** $\{x \mid x \leq -4 \text{ or } x \geq 4\}$

33. $\{x \mid x \leq -3 \text{ or } x \geq 6\}$ **35.** $\{x \mid -1 \leq x \leq 1\}$

37. $f(0) = -2$, $f(2) = 8$, $f(-1) = -7$, $f(-4) = -22$

39. $f(-2) = -\dfrac{7}{4}$, $f(0) = -\dfrac{3}{4}$, $f\left(\dfrac{1}{2}\right) = -\dfrac{1}{2}$, $f\left(\dfrac{2}{3}\right) = -\dfrac{5}{12}$

41. $g(-1) = 0$; $g(2) = -9$; $g(-3) = 26$; $g(4) = 5$

43. $h(-2) = -2$; $h(-3) = -11$; $h(4) = -32$; $h(5) = -51$

45. $f(3) = \sqrt{7}$; $f(4) = 3$; $f(10) = \sqrt{21}$; $f(12) = 5$

47. $f(1) = -1$; $f(-1) = -2$; $f(3) = -\dfrac{2}{3}$; $f(-6) = \dfrac{4}{3}$

49. $f(-2) = 27$; $f(3) = 42$; $g(-4) = -37$; $g(6) = -17$

51. $f(-2) = 5$; $f(3) = 8$; $g(-4) = -3$; $g(5) = -4$ **53.** -3 **55.** $-2a - h$ **57.** $4a - 1 + 2h$

59. $-8a - 7 - 4h$ **61.** $h(1) = 48$; $h(2) = 64$; $h(3) = 48$; $h(4) = 0$

63. $C(75) = \$74$; $C(150) = \$98$; $C(225) = \$122$; $C(650) = \$258$

65. $I(.11) = 55$; $I(.12) = 60$; $I(.135) = 67.5$; $I(.15) = 75$

Problem Set 12.2 (page 517)

1.

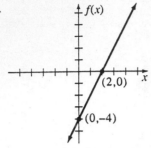

3.

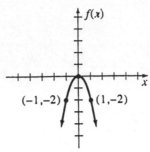

5.

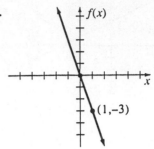

7.

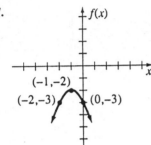

9.

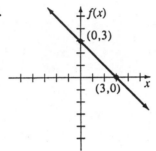

11.

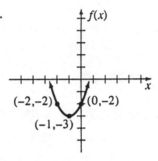

13.

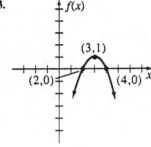

15.

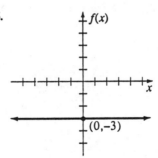

17.

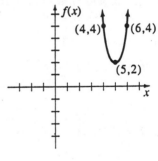

19.

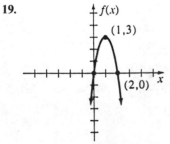

21.

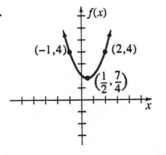

23.

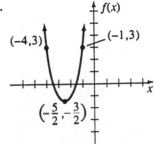

25.

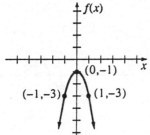

27.

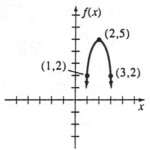

29.

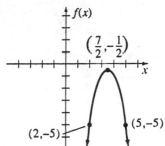

31. 80   **33.** 5 and 25   **35.** 60 meters by 60 meters
37. 1100 subscribers at \$13.75 per month
39. −4 and 26   **41.** −25; −5 and 5

Problem Set 12.3 (page 525)

1.

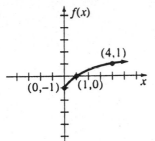

3.

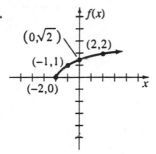

5.

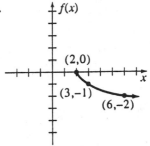

7.

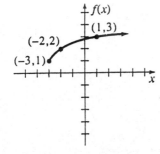

9.

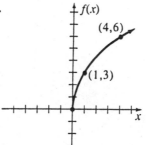

11.

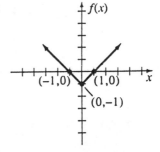

13.

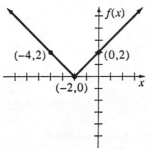

15.

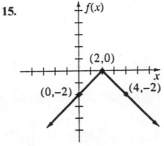

17.

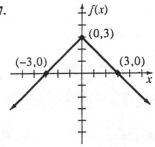

19.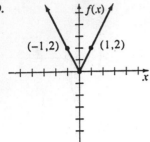

21.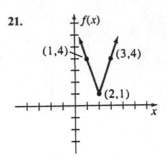

23.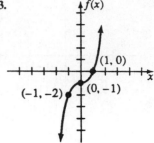

25.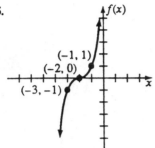

27.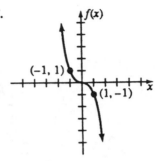

29.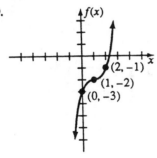

31. (a)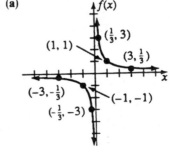

Problem Set 12.4 (page 531)

1. $8x - 2$; $-2x - 6$; $15x^2 - 14x - 8$; $\dfrac{3x - 4}{5x + 2}$

3. $x^2 - 7x + 3$; $x^2 - 5x + 5$; $-x^3 + 5x^2 + 2x - 4$; $\dfrac{x^2 - 6x + 4}{-x - 1}$

5. $2x^2 + 3x - 6$; $-5x + 4$; $x^4 + 3x^3 - 10x^2 + x + 5$; $\dfrac{x^2 - x - 1}{x^2 + 4x - 5}$

7. $\sqrt{x - 1} + \sqrt{x}$; $\sqrt{x - 1} - \sqrt{x}$; $\sqrt{x^2 - x}$; $\dfrac{\sqrt{x(x - 1)}}{x}$

9. 124 and -130 **11.** 323 and 257   **13.** $\dfrac{1}{2}$ and -1 **15.** undefined and undefined

17. $\sqrt{7}$ and 0   **19.** 37 and 27   **21.** $(f \circ g)(x) = 15x - 3, D = \{\text{all reals}\}$
$(g \circ f)(x) = 15x - 1, D = \{\text{all reals}\}$

23. $(f \circ g)(x) = -14x - 7, D = \{\text{all reals}\}$ **25.** $(f \circ g)(x) = 3x^2 + 11, D = \{\text{all reals}\}$
$(g \circ f)(x) = -14x + 11, D = \{\text{all reals}\}$ $(g \circ f)(x) = 9x^2 + 12x + 7, D = \{\text{all reals}\}$

27. $(f \circ g)(x) = 2x^2 - 11x + 17, D = \{\text{all reals}\}$ **29.** $(f \circ g)(x) = \dfrac{3}{4x - 9}, D = \left\{x \mid x \neq \dfrac{9}{4}\right\}$
$(g \circ f)(x) = -2x^2 + x + 1, D = \{\text{all reals}\}$
$(g \circ f)(x) = \dfrac{12 - 9x}{x}, D = \{x \mid x \neq 0\}$

31. $(f \circ g)(x) = \sqrt{5x + 4}, D = \left\{x \mid x \geq -\dfrac{4}{5}\right\}$ **33.** $(f \circ g)(x) = x - 4, D = \{x \mid x \neq 4\}$
$(g \circ f)(x) = 5\sqrt{x + 1} + 3, D = \{x \mid x \geq -1\}$ $(g \circ f)(x) = \dfrac{x}{1 - 4x}, D = \left\{x \mid x \neq 0 \text{ and } x \neq \dfrac{1}{4}\right\}$

35. $(f \circ g)(x) = \dfrac{2\sqrt{x}}{x}, D = \{x \mid x > 0\}$ **37.** $(f \circ g)(x) = \dfrac{3x + 3}{2}, D = \{x \mid x \neq -1\}$
$(g \circ f)(x) = \dfrac{4\sqrt{x}}{x}, D = \{x \mid x > 0\}$ $(g \circ f)(x) = \dfrac{2x}{2x + 3}, D = \left\{x \mid x \neq 0 \; x \neq -\dfrac{3}{2}\right\}$

Problem Set 12.5 (page 538)

1. Not a function   **3.** Function   **5.** Function   **7.** Function   **9.** One-to-one function
11. Not a one-to-one function   **13.** Not a one-to-one function   **15.** One-to-one function
17. Domain of f: $\{1, 2, 3, 4\}$ **19.** Domain of f: $\{-2, -1, 0, 5\}$
Range of f: $\{3, 6, 11, 18\}$ Range of f: $\{-1, 1, 5, 10\}$ **21.** $f^{-1}(x) = \dfrac{x + 4}{5}$
f^{-1}: $\{(3, 1), (6, 2), (11, 3), (18, 4)\}$ f^{-1}: $\{(-1, -2), (1, -1), (5, 0), (10, 5)\}$
Domain of f^{-1}: $\{3, 6, 11, 18\}$ Domain of f^{-1}: $\{-1, 1, 5, 10\}$ **23.** $f^{-1}(x) = \dfrac{1 - x}{2}$
Range of f^{-1}: $\{1, 2, 3, 4\}$ Range of f^{-1}: $\{-2, -1, 0, 5\}$

25. $f^{-1}(x) = \dfrac{5}{4}x$ **27.** $f^{-1}(x) = 2x - 8$ **29.** $f^{-1}(x) = \dfrac{15x + 6}{5}$ **31.** $f^{-1}(x) = \dfrac{x - 4}{9}$

33. $f^{-1}(x) = \dfrac{-x - 4}{5}$ **35.** $f^{-1}(x) = \dfrac{-3x + 21}{2}$ **37.** $f^{-1}(x) = \dfrac{3}{4}x + \dfrac{3}{16}$

39. $f^{-1}(x) = -\dfrac{7}{3}x - \dfrac{14}{9}$ **41.** $f^{-1}(x) = \dfrac{1}{4}x$ **43.** $f^{-1}(x) = -3x$ **45.** $f^{-1}(x) = \dfrac{x + 3}{3}$

47. $f^{-1}(x) = \dfrac{-x - 4}{2}$ **49.** $f^{-1}(x) = \sqrt{x}, x \geq 0$

51. Every nonconstant linear function is a one-to-one function.

Problem Set 12.6 (page 547)

1. $y = \dfrac{k}{x^2}$ **3.** $C = \dfrac{kg}{t^3}$ **5.** $V = kr^3$ **7.** $S = ke^2$ **9.** $V = khr^2$ **11.** $\dfrac{2}{3}$ **13.** -4

15. $\dfrac{1}{3}$ **17.** -2 **19.** 2   **21.** 5   **23.** 9   **25.** 9   **27.** $\dfrac{1}{6}$ **29.** 112

31. 12 cubic centimeters   **33.** 28   **35.** 2 seconds   **37.** 12 ohms   **39. (a)** $210
(c) $1050   **41.** 3560.76 cubic meters   **43.** .048

Chapter 12 Review Problem Set (page 552)

1. $D = \{1, 2, 4\}$ **2.** $D = \{x \mid x \neq 5\}$ **3.** $D = \{x \mid x \neq 0 \text{ and } x \neq -4\}$
4. $D = \{x \mid x \geq 5 \text{ or } x \leq -5\}$ **5.** $f(2) = -1$, $f(-3) = 14$; $f(a) = a^2 - 2a - 1$ **6.** $4a + 2h + 1$

7.
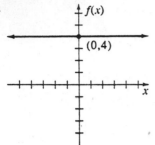

8.
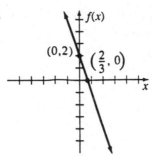

9.
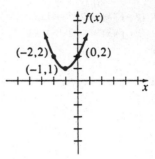

10.
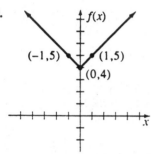

11.
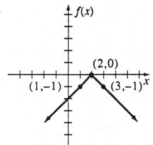

12.
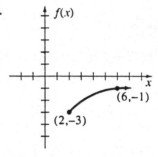

13.
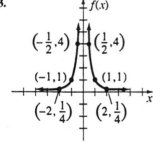

14.
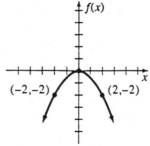

15.
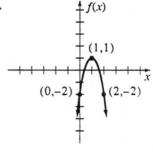

16.
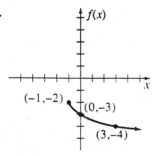

17. **(a)** $(-5, -28)$; $x = -5$ **(b)** $\left(-\dfrac{7}{6}, \dfrac{67}{2}\right)$; $x = -\dfrac{7}{2}$

18. $(f \circ g)(x) = 6x - 11$ and $(g \circ f)(x) = 6x - 13$

19. $(f \circ g)(x) = x^2 - 2x - 1$ and $(g \circ f)(x) = x^2 - 10x + 27$

20. $(f \circ g)(x) = 4x^2 - 20x + 20$ and $(g \circ f)(x) = -2x^2 + 15$ **21.** $f^{-1}(x) - \dfrac{x + 1}{6}$

22. $f^{-1}(x) = \dfrac{3x - 21}{2}$ **23.** $f^{-1}(x) = \dfrac{-35x - 10}{21}$ **24.** $k = 9$ **25.** $y = 120$ **26.** 128 pounds

27. 20 and 20   **28.** 3 and 47   **29.** 25 students   **30.** 600 square inches

CHAPTER 13

Problem Set 13.1 (page 564)

1. $\{3\}$ **3.** $\{2\}$ **5.** $\{4\}$ **7.** $\{1\}$ **9.** $\{5\}$ **11.** $\{1\}$ **13.** $\left\{\dfrac{3}{2}\right\}$ **15.** $\left\{\dfrac{5}{6}\right\}$

17. $\{-3\}$ **19.** $\{-3\}$ **21.** $\{0\}$ **23.** $\{1\}$ **25.** $\{-1\}$ **27.** $\left\{-\dfrac{2}{5}\right\}$ **29.** $\left\{\dfrac{5}{2}\right\}$

31. $\{3\}$ **33.** $\left\{\dfrac{1}{2}\right\}$

35.

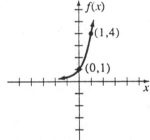

37.

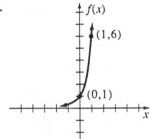

39.

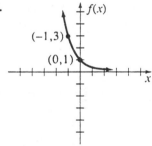

41.

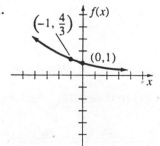

43.

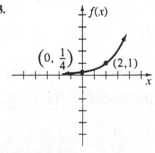

45.

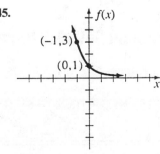

47.
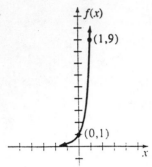

49.
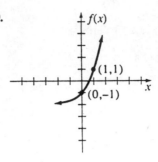

51.
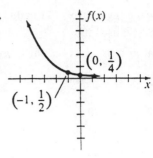

Problem Set 13.2 (page 572)

1. (a) $.62   **(c)** $2.06   **(e)** $10,950   **(g)** $658   **3.** $283.70   **5.** $865.84   **7.** $1782.25
9. $2725.05   **11.** $16,998.71   **13.** $22,553.65   **15.** $567.63   **17.** $1422.36
19. $8963.38   **21.** $17,547.35   **23.** $32,558.88

25.

	1 yr	5 yrs	10 yrs	20 yrs
compounded annually	$1120	1762	3106	9646
compounded semiannually	1124	1791	3207	10,286
compounded quarterly	1126	1806	3262	10,641
compounded monthly	1127	1817	3300	10,893
compounded continuously	1127	1822	3320	11,023

27. Nora will have $.24 more.   **29.** 9.54%   **31.** 50 grams; 37 grams

33.
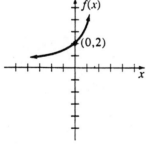

35.
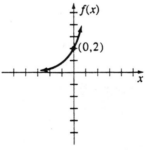

37.
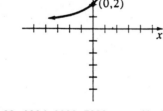

39. 2226; 3320; 7389   **41.** 2000   **43. (a)** 82,888   **(c)** 96,302
45. (a) 6.5 pounds per square inch   **(e)** 13.6 pounds per square inch

Problem Set 13.3 (page 582)

1. $\log_2 128 = 7$ **3.** $\log_5 125 = 3$ **5.** $\log_{10} 1000 = 3$ **7.** $\log_2\left(\dfrac{1}{4}\right) = -2$ **9.** $\log_{10} .1 = -1$

11. $3^4 = 81$ **13.** $4^3 = 64$ **15.** $10^4 = 10000$ **17.** $2^{-4} = \dfrac{1}{16}$ **19.** $10^{-3} = .001$ **21.** 4

23. 4   **25.** 3   **27.** $\dfrac{1}{2}$ **29.** 0   **31.** -1 **33.** 5   **35.** -5 **37.** 1   **39.** 0

41. $\{49\}$ **43.** $\{16\}$ **45.** $\{27\}$ **47.** $\left\{\dfrac{1}{8}\right\}$ **49.** $\{4\}$ **51.** 5.1293   **53.** 6.9657

55. 1.4037   **57.** 7.4512   **59.** 6.3219   **61.** $-.3791$ **63.** -5.766 **65.** 2.1531

67. .3949   **69.** $\log_b x + \log_b y + \log_b z$ **71.** $\log_b y - \log_b z$ **73.** $3\log_b y + 4\log_b z$

75. $\dfrac{1}{2}\log_b x + \dfrac{1}{3}\log_b y - 4\log_b z$ **77.** $\dfrac{2}{3}\log_b x + \dfrac{1}{3}\log_b z$ **79.** $\dfrac{3}{2}\log_b x - \dfrac{1}{2}\log_b y$ **81.** $\left\{\dfrac{9}{4}\right\}$

83. $\{25\}$ **85.** $\{4\}$ **87.** $\left\{\dfrac{19}{8}\right\}$ **89.** $\{9\}$ **91.** $\{1\}$

Problem Set 13.4 (page 589)

1. 0.8597   **3.** 1.7179   **5.** 3.5071   **7.** -0.1373 **9.** -3.4685 **11.** 411.43

13. 90095   **15.** 79.543   **17.** 0.048440   **19.** 0.0064150   **21.** 1.6094   **23.** 3.4843

25. 6.0638   **27.** -0.7765 **29.** -3.4609 **31.** 1.6034   **33.** 3.1346   **35.** 108.56

37. 0.48268   **39.** 0.035994

41.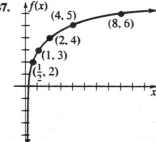

43.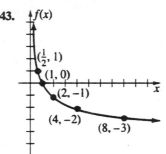

45.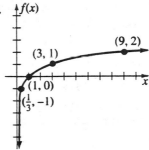

47.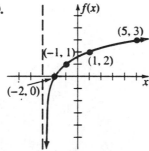

49.

51.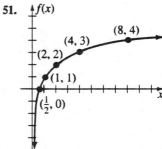

53.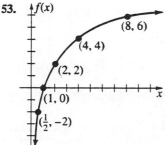

55. 0.36   **57.** 0.73   **59.** 23.10   **61.** 7.93

Problem Set 13.5 (page 597)

1. {3.15} **3.** {2.20} **5.** {4.18} **7.** {0.12} **9.** {1.69} **11.** {4.57} **13.** {2.46}

15. {4} **17.** $\left\{\dfrac{19}{47}\right\}$ **19.** {1} **21.** {8} **23.** 4.524 **25.** −.860 **27.** 3.105

29. −2.902 **31.** 5.989 **33.** 2.4 years **35.** 5.3 years **37.** 6.8 hours **39.** 1.5 hours
41. 34.7 years **43.** 6.7 **45.** Approximately 8 times

Chapter 13 Review Problem Set (page 600)

1. 7 **2.** 3 **3.** 4 **4.** −3 **5.** 2 **6.** 13 **7.** 0 **8.** −2 **9.** {−4}

10. {3} **11.** $\left\{-\dfrac{3}{4}\right\}$ **12.** {4} **13.** {9} **14.** $\left\{\dfrac{1}{2}\right\}$ **15.** {5} **16.** {8}

17. 1.8642 **18.** 4.7380 **19.** −4.2687 **20.** −2.6289 **21.** 1.1882 **22.** 2639.4
23. .013289 **24.** .077197 **25.** {3.40} **26.** {1.95} **27.** {5.30} **28.** {5.61}

29. 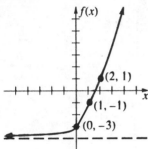 **30.** 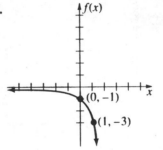 **31.**

32.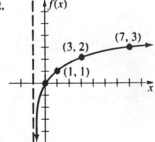

33. 2.842 **34.** $2913.99 **35.** $8178.72 **36.** $5656.26 **37.** 133 grams
38. Approximately 5.3 years **39.** Approximately 12.1 years **40.** 61,070; 67,493; 74,591
41. Approximately 4.8 hours **42.** 8.1

CHAPTER 14

Problem Set 14.1 (page 613)

1. {(4, −3)} **3.** {(−1, −3)} **5.** {(−8, 2)} **7.** {(−4, 0)} **9.** {(1, −1)} **11.** Inconsistent
13. {(−2, 5, 2)} **15.** {(4, −1, −2)} **17.** {(−1, 3, 5)} **19.** Infinitely many solutions **21.** ∅

23. $\left\{\left(-2, \dfrac{3}{2}, 1\right)\right\}$ **25.** $\left\{\left(\dfrac{1}{3}, -\dfrac{1}{2}, 1\right)\right\}$ **27.** $\left\{\left(\dfrac{2}{3}, -4, \dfrac{3}{4}\right)\right\}$

29. $\{(-2, 4, 0)\}$

31. $\left\{\left(\dfrac{1}{2}, \dfrac{1}{3}, \dfrac{1}{6}\right)\right\}$

33. 194   **35.** $.70 per bottle for catsup, $1 per jar of peanut butter, and $.80 per jar of pickles
37. -2, 6, and 16   **39.** $40°$, $60°$, and $80°$ **41.** $500 at 12%, $1000 at 13%, and $1500 at 14%

Problem Set 14.2 (page 622)

1. $\{(6, -4)\}$ **3.** $\{(-4, -4)\}$ **5.** $\left\{\left(-\dfrac{11}{7}, -\dfrac{13}{7}\right)\right\}$ **7.** $\{(1, -3)\}$ **9.** $\{(-1, -2, -3)\}$

11. $\{(3, 1, 4)\}$ **13.** $\{4, 3, -2)\}$ **15.** $\{(-5, 2, 0)\}$ **17.** $\{(-2, -1, 1)\}$ **19.** $\{(-1, 4, -1)\}$

21. $\{(2, 0, -3)\}$ **23.** $\varnothing$ **25.** $\{(1, -1, 2, -3)\}$

Problem Set 14.3 (page 627)

1. 10   **3.** -48 **5.** -29 **7.** 50   **9.** 8   **11.** 6   **13.** -32 **15.** -2 **17.** 5

19. $-\dfrac{7}{20}$ **21.** $\{3, 8)\}$ **23.** $\{(-2, 5)\}$ **25.** $\{(-2, 2)\}$ **27.** $\varnothing$ **29.** $\{(1, 6)\}$

31. $\left\{\left(-\dfrac{1}{4}, \dfrac{2}{3}\right)\right\}$ **33.** $\left\{\left(\dfrac{17}{62}, \dfrac{4}{31}\right)\right\}$ **35.** $\{-1, 3)\}$ **37.** $\{(9, -2)\}$ **39.** $\{(-4, -3)\}$

Problem Set 14.4 (page 635)

1. -57 **3.** 14   **5.** -41 **7.** -8 **9.** -96 **11.** $\{(-3, 1, -1)\}$ **13.** $\{(0, 2, -3)\}$

15. $\left\{\left(-2, \dfrac{1}{2}, -\dfrac{2}{3}\right)\right\}$ **17.** $\{(-1, -1, -1)\}$ **19.** $\varnothing$ **21.** $\{(-5, -2, 1)\}$ **23.** $\{(-2, 3, -4)\}$

25. $\left\{\left(-\dfrac{1}{2}, 0, \dfrac{2}{3}\right)\right\}$ **27.** $\{(-6, 7, 1)\}$ **29.** $\left\{\left(1, -6, -\dfrac{1}{2}\right)\right\}$ **31.** 0   **33.** (b) -20

Chapter 14 Review Problem Set (page 639)

1. $\{(2, 6)\}$ **2.** $\{(-3, 7)\}$ **3.** $\{(-9, 8)\}$ **4.** $\left\{\left(\dfrac{89}{23}, -\dfrac{12}{23}\right)\right\}$ **5.** $\{(4, -7)\}$ **6.** $\{(-2, 8)\}$

7. $\{(0, 3)\}$ **8.** $\{(6, 7)\}$ **9.** $\{(1, -6)\}$ **10.** $\{(-4, 0)\}$ **11.** $\{(-12, 18)\}$ **12.** $\{(24, 8)\}$

13. $\left\{\left(\dfrac{2}{3}, -\dfrac{3}{4}\right)\right\}$ **14.** $\left\{\left(-\dfrac{1}{2}, \dfrac{3}{5}\right)\right\}$ **15.** -18 **16.** 11   **17.** -29 **18.** 59

19. $\{(2, -4, -6)\}$ **20.** $\{(-1, 5, -7)\}$ **21.** $\{(2, -3, 1)\}$ **22.** $\{(2, -1, -2)\}$

23. $\left\{\left(-\dfrac{1}{3}, -1, 4\right)\right\}$ **24.** $\{(0, -2, -4)\}$

CHAPTER 15

Problem Set 15.1 (page 649)

1. $-1, 2, 5, 8, 11$ **3.** $3, 1, -1, -3, -5$ **5.** $-1, 2, 7, 14, 23$ **7.** $0, -3, -8, -15, -24$

9. $-1, 5, 15, 29, 47$ **11.** $\dfrac{1}{2}, 1, 2, 4, 8$ **13.** $-\dfrac{2}{3}, -2, -6, -18, -54$ **15.** $a_8 = 54$ and $a_{12} = 130$

17. $a_7 = -32$ and $a_8 = 64$ **19.** $a_n = 2n - 1$ **21.** $a_n = 4n - 6$ **23.** $a_n = -2n + 7$

25. $a_n = -3n - 4$ **27.** $a_n = \dfrac{1}{2}(n + 1)$ **29.** 34 **31.** 78 **33.** -155 **35.** 106

37. 50 **39.** 170 **41.** -1 **43.** -13 **45.** 136 **47.** $\$27{,}800$ **49.** $\$1170$

Problem Set 15.2 (page 654)

1. 2550 **3.** 9030 **5.** -4225 **7.** 410 **9.** 8850 **11.** 5724 **13.** 5070
15. 346.5 **17.** 3775 **19.** $39{,}500$ **21.** -8645 **23.** $122{,}850$ **25.** $39{,}552$
27. 76 seats, 720 seats   **29.** $\$10{,}000$ **31.** $\$357{,}500$ **33.** 169 **35.** **(a)** $7 + 12 + 17$; 36
(c) $(-3) + (-5) + (-7) + (-9) + (-11) + (-13)$; -48 **(e)** $3 + 6 + 9 + 12 + 15$; 45

Problem Set 15.3 (page 661)

1. $a_n = 3^{n-1}$ **3.** $a_n = 2(4)^{n-1}$ **5.** $a_n = \left(\dfrac{1}{3}\right)^{n-1}$ or $a_n = \dfrac{1}{3^{n-1}}$ **7.** $a_n = .2(.2)^{n-1}$ or $a_n = (.2)^n$

9. $a_n = 9\left(\dfrac{2}{3}\right)^{n-1}$ or $a_n = (3^{-n+3})(2)^{n-1}$ **11.** $a_n = (-4)^{n-1}$ **13.** $19{,}683$ **15.** -512

17. $-\dfrac{6561}{256}$ **19.** $14{,}762$ **21.** -9842 **23.** 1093 **25.** 511 **27.** $19{,}680$ **29.** 2730

31. 1093 **33.** $1\dfrac{1023}{1024}$ **35.** $15\dfrac{31}{32}$ **37.** $\dfrac{1}{2}$ **39.** $0; -1$ **41.** 125 liters

43. $\$102.40$; $\$204.75$ **45.** $971.\overline{3}$ meters   **47.** $\$5609.66$ **49.** $1{,}048{,}576$ **51.** $\dfrac{4096}{531{,}441}$

53. $-\dfrac{177{,}147}{2048}$ **55.** **(a)** 126 **(c)** 31 **(e)** $1\dfrac{49}{81}$

Problem Set 15.4 (page 668)

1. 4 **3.** 1 **5.** 2 **7.** $\dfrac{2}{3}$ **9.** 9 **11.** No sum   **13.** $\dfrac{4}{7}$ **15.** $\dfrac{16}{3}$ **17.** No sum

19. $\dfrac{81}{2}$ **21.** $\dfrac{4}{9}$ **23.** $\dfrac{47}{99}$ **25.** $\dfrac{5}{11}$ **27.** $\dfrac{427}{999}$ **29.** $\dfrac{7}{15}$ **31.** $\dfrac{72}{33}$ **33.** $\dfrac{47}{110}$

Problem Set 15.5 (page 672)

1. $x^8 + 8x^7y + 28x^6y^2 + 56x^5y^3 + 70x^4y^4 + 56x^3y^5 + 28x^2y^6 + 8xy^7 + y^8$
3. $81x^4 + 108x^3y + 54x^2y^2 + 12xy^3 + y^4$ **5.** $x^5 - 5x^4y + 10x^3y^2 - 10x^2y^3 + 5xy^4 - y^5$
7. $x^{10} + 10x^9y + 45x^8y^2 + 120x^7y^3 + 210x^6y^4 + 252x^5y^5 + 210x^4y^6 + 120x^3y^7 + 45x^2y^8 + 10xy^9 + y^{10}$
9. $64x^6 + 192x^5y + 240x^4y^2 + 160x^3y^3 + 60x^2y^4 + 12xy^5 + y^6$
11. $x^5 - 15x^4y + 90x^3y^2 - 270x^2y^3 + 405xy^4 - 243y^5$
13. $243a^5 - 810a^4b + 1080a^3b^2 - 720a^2b^3 + 240ab^4 - 32b^5$
15. $x^6 + 6x^5y^3 + 15x^4y^6 + 20x^3y^9 + 15x^2y^{12} + 6xy^{15} + y^{18}$
17. $x^7 + 14x^6 + 84x^5 + 280x^4 + 560x^3 + 672x^2 + 448x + 128$ **19.** $x^4 - 12x^3 + 54x^2 - 108x + 81$
21. $x^{15} + 15x^{14}y + 105x^{13}y^2 + 455x^{12}y^3$ **23.** $a^{13} - 26a^{12}b + 312a^{11}b^2 - 2288a^{10}b^3$
25. $462x^5y^6$ **27.** $-160x^3y^3$ **29.** $2000x^3y^2$

Chapter 15 Review Problem Set (page 674)

1. $-1, 1, 5, 13, 29$ **2. (a)** $a_n = 6n - 4$ **(b)** $a_n = -5n + 8$ **(c)** $a_n = 4(3)^{n-1}$
(d) $a_n - 5\left(\dfrac{1}{2}\right)^{n-1}$ **3. (a)** $\dfrac{1}{4374}$ **(b)** 155 **4. (a)** 3775 **(b)** 4092 **(c)** -1325

5. $90{,}000$ **6.** $\dfrac{1365}{4096}$ **7.** 1375 **8.** $20{,}520$ **9.** $5333.\overline{3}$ liters   **10.** 3600 feet   **11.** $\$2.10$

12. $\$409.55$ **13.** 32 **14.** $\dfrac{29}{99}$
15. $128x^7 + 448x^6y + 672x^5y^2 + 560x^4y^3 + 280x^3y^4 + 84x^2y^5 + 14xy^6 + y^7$ **16.** $252a^5b^5$

Practice Exercises for Appendix A (page 681)

1. $(x - 6)$ **2.** $x + 6$ **3.** $x + 6, R = 14$ **4.** $x - 2, R = -1$ **5.** $x^2 - 1$
6. $x^2 - 6x + 8$ **7.** $x^2 - 2x - 3$ **8.** $x^2 + 7x + 2, R = 1$
9. $2x^2 - x - 6, R = -6$ **10.** $3x^3 - 4x^2 + 6x - 13, R = 12$
11. $x^3 + 7x^2 + 21x + 56, R = 167$ **12.** $2x^3 - 4x^2 + 11x - 22, R = 47$
13. (a)

```
-1) 1    6    11    6
        -1   -5   -6        Therefore, (x + 1) is a factor.
    ─────────────────
     1    5    6    0


-2) 1    6    11    6
        -2   -8   -6        Therefore, (x + 2) is a factor.
    ─────────────────
     1    4    3    0


-3) 1    6    11    6
        -3   -9   -6        Therefore, (x + 3) is a factor.
    ─────────────────
     1    3    2    0
```

(b)

$$
\begin{array}{r|rrrr}
2) & 1 & -4 & -11 & 30 \\
  &   & 2 & -4 & -30 \\
\hline
  & 1 & -2 & -15 & 0 \\
\end{array}
$$

Therefore, $(x - 2)$ is a factor.

$$
\begin{array}{r|rrrr}
-3) & 1 & -4 & -11 & 30 \\
   &   & -3 & 21 & -30 \\
\hline
   & 1 & -7 & 10 & 0 \\
\end{array}
$$

Therefore, $(x + 3)$ is a factor.

$$
\begin{array}{r|rrrr}
5) & 1 & -4 & -11 & 30 \\
  &   & 5 & 5 & -30 \\
\hline
  & 1 & 1 & -6 & 0 \\
\end{array}
$$

Therefore, $(x - 5)$ is a factor.

Practice Exercises for Appendix C (page 693)

1. 0.6362   **2.** 1.4382   **3.** 2.1103   **4.** 3.5473   **5.** $-1 + 0.9426$ **6.** $-2 + 0.8861$
7. $-3 + 0.7148$ **8.** $-4 + 0.4079$ **9.** 2.945   **10.** 62.67   **11.** 155.8   **12.** 5279
13. 0.3181   **14.** 0.08704   **15.** 20,930   **16.** 773.8   **17.** 5.749   **18.** 87.12
19. 1,549,000   **20.** 110,700   **21.** 614.1   **22.** 383.1   **23.** 0.9614   **24.** 5.041
25. 13.79   **26.** 48.49   **27.** 15.91   **28.** 1.116

Index

Table of Common Logarithms

N	0	1	2	3	4	5	6	7	8	9
1.0	.0000	.0043	.0086	.0128	0.170	0.212	.0253	.0294	.0334	.0374
1.1	.0414	.0453	.0492	.0531	.0569	.0607	.0645	.0682	.0719	.0755
1.2	.0792	.0828	.0864	.0899	.0934	.0969	.1004	.1038	.1072	.1106
1.3	.1139	.1173	.1206	.1239	.1271	.1303	.1335	.1367	.1399	.1430
1.4	.1461	.1492	.1523	.1553	.1584	.1614	.1644	.1673	.1703	.1732
1.5	.1761	.1790	.1818	.1847	.1875	.1903	.1931	.1959	.1987	.2014
1.6	.2041	.2068	.2095	.2122	.2148	.2175	.2201	.2227	.2253	.2279
1.7	.2304	.2330	.2355	.2380	.2405	.2430	.2455	.2480	.2504	.2529
1.8	.2553	.2577	.2601	.2625	.2648	.2672	.2695	.2718	.2742	.2765
1.9	.2788	.2810	.2833	.2856	.2878	.2900	.2923	.2945	.2967	.2989
2.0	.3010	.3032	.3054	.3075	.3096	.3118	.3139	.3160	.3181	.3201
2.1	.3222	.3243	.3263	.3284	.3304	.3324	.3345	.3365	.3385	.3404
2.2	.3424	.3444	.3464	.3483	.3502	.3522	.3541	.3560	.3579	.3598
2.3	.3617	.3636	.3655	.3674	.3692	.3711	.3729	.3747	.3766	.3784
2.4	.3802	.3820	.3838	.3856	.3874	.3892	.3909	.3927	.3945	.3962
2.5	.3979	.3997	.4014	.4031	.4048	.4065	.4082	.4099	.4116	.4133
2.6	.4150	.4166	.4183	.4200	.4216	.4232	.4249	.4265	.4281	.4298
2.7	.4314	.4330	.4346	.4362	.4378	.4393	.4409	.4425	.4440	.4456
2.8	.4472	.4487	.4502	.4518	.4533	.4548	.4564	.4579	.4594	.4609
2.9	.4624	.4639	.4654	.4669	.4683	.4698	.4713	.4728	.4742	.4757
3.0	.4771	.4786	.4800	.4814	.4829	.4843	.4857	.4871	.4886	.4900
3.1	.4914	.4928	.4942	.4955	.4969	.4983	.4997	.5011	.5024	.5038
3.2	.5051	.5065	.5079	.5092	.5105	.5119	.5132	.5145	.5159	.5172
3.3	.5185	.5198	.5211	.5224	.5237	.5250	.5263	.5276	.5289	.5302
3.4	.5315	.5328	.5340	.5353	.5366	.5378	.5391	.5403	.5416	.5428
3.5	.5441	.5453	.5465	.5478	.5490	.5502	.5514	.5527	.5539	.5551
3.6	.5563	.5575	.5587	.5599	.5611	.5623	.5635	.5647	.5658	.5670
3.7	.5682	.5694	.5705	.5717	.5729	.5740	.5752	.5763	.5775	.5786
3.8	.5798	.5809	.5821	.5832	.5843	.5855	.5866	.5877	.5888	.5899
3.9	.5911	.5922	.5933	.5944	.5955	.5966	.5977	.5988	.5999	.6010
4.0	.6021	.6031	.6042	.6053	.6064	.6075	.6085	.6096	.6107	.6117
4.1	.6128	.6138	.6149	.6160	.6170	.6180	.6191	.6201	.6212	.6222
4.2	.6232	.6243	.6253	.6263	.6274	.6284	.6294	.6304	.6314	.6325
4.3	.6335	.6345	.6355	.6365	.6375	.6385	.6395	.6405	.6415	.6425
4.4	.6435	.6444	.6454	.6464	.6474	.6484	.6493	.6503	.6513	.6522
4.5	.6532	.6542	.6551	.6561	.6571	.6580	.6590	.6599	.6609	.6618
4.6	.6628	.6637	.6646	.6656	.6665	.6675	.6684	.6693	.6702	.6712
4.7	.6721	.6730	.6739	.6749	.6758	.6767	.6776	.6785	.6794	.6803
4.8	.6812	.6821	.6830	.6839	.6848	.6857	.6866	.6875	.6884	.6893
4.9	.6902	.6911	.6920	.6928	.6937	.6946	.6955	.6964	.6972	.6981
5.0	.6990	.6998	.7007	.7016	.7024	.7033	.7042	.7050	.7059	.7067
5.1	.7076	.7084	.7093	.7101	.7110	.7118	.7126	.7135	.7143	.7152
5.2	.7160	.7168	.7177	.7185	.7193	.7202	.7210	.7218	.7226	.7235
5.3	.7243	.7251	.7259	.7267	.7275	.7284	.7292	.7300	.7308	.7316
5.4	.7324	.7332	.7340	.7348	.7356	.7364	.7372	.7380	.7388	.7396